PROBABILITY AND STATISTICAL INFERENCE

Ninth Edition

Global Edition

ROBERT V. HOGG

ELLIOT A. TANIS

DALE L. ZIMMERMAN

PEARSON

Boston   Columbus   Indianapolis   New York   San Francisco
Upper Saddle River   Amsterdam   Cape Town   Dubai
London   Madrid   Milan   Munich   Paris   Montreal   Toronto
Delhi   Mexico City   Sao Paulo   Sydney   Hong Kong   Seoul
Singapore   Taipei   Tokyo

Editor in Chief: Deirdre Lynch
Acquisitions Editor: Christopher Cummings
Sponsoring Editor: Christina Lepre
Assistant Editor: Sonia Ashraf
Marketing Manager: Erin Lane
Marketing Assistant: Kathleen DeChavez
Senior Managing Editor: Karen Wernholm
Senior Production Editor: Beth Houston
Head of Learning Asset Acquisition, Global Editions:
    Laura Dent
Assistant Acquisitions Editor, Global Editions:
    Aditee Agarwal
Senior Project Editor, Global Editions: Shambhavi Thakur

Media Producer, Global Editions: M Vikram Kumar
Senior Manufacturing Controller, Production, Global
    Editions: Trudy Kimber
Procurement Manager: Vincent Scelta
Procurement Specialist: Carol Melville
Associate Director of Design, USHE EMSS/HSC/EDU:
    Andrea Nix
Art Director: Heather Scott
Interior Designer: Tamara Newnam
Cover Designer: Lumina Datamatics
Cover Image: © Sergey Mironov/Shutterstock
Full-Service Project Management: Integra Software Services
Composition: Integra Software Services

Pearson Education Limited
Edinburgh Gate
Harlow
Essex CM20 2JE
England

and Associated Companies throughout the world

Visit us on the World Wide Web at:
www.pearsonglobaleditions.com

© Pearson Education Limited 2015

The rights of Robert V. Hogg, Elliot A. Tanis, and Dale L. Zimmerman to be identified as the authors of this work have been asserted by them in accordance with the Copyright, Designs and Patents Act 1988.

Authorized adaptation from the United States edition entitled Probability and Statistical Inference, 9th edition, ISBN 978-0-321-92327-1, by Robert V. Hogg, Elliot A. Tanis, and Dale L. Zimmerman, published by Pearson Education © 2015.

British Library Cataloguing-in-Publication Data
A catalogue record for this book is available from the British Library

10 9 8 7 6 5 4 3 2 1
14

Typeset in 10/12 Times Ten by Integra Software Services

Printed and bound by Courier Kendallville in The United States of America

PEARSON

ISBN-10:   1-292-06235-5
ISBN-13: 978-1-292-06235-8

CONTENTS

PREFACE

In this Ninth Edition of *Probability and Statistical Inference*, Bob Hogg and Elliot Tanis are excited to add a third person to their writing team to contribute to the continued success of this text. Dale Zimmerman is the Robert V. Hogg Professor in the Department of Statistics and Actuarial Science at the University of Iowa. Dale has rewritten several parts of the text, making the terminology more consistent and contributing much to a substantial revision. The text is designed for a two-semester course, but it can be adapted for a one-semester course. A good calculus background is needed, but no previous study of probability or statistics is required.

CONTENT AND COURSE PLANNING

In this revision, the first five chapters on probability are much the same as in the eighth edition. They include the following topics: probability, conditional probability, independence, Bayes' theorem, discrete and continuous distributions, certain mathematical expectations, bivariate distributions along with marginal and conditional distributions, correlation, functions of random variables and their distributions, including the moment-generating function technique, and the central limit theorem. While this strong probability coverage of the course is important for all students, it has been particularly helpful to actuarial students who are studying for Exam P in the Society of Actuaries' series (or Exam 1 of the Casualty Actuarial Society).

The greatest change to this edition is in the statistical inference coverage, now Chapters 6–9. The first two of these chapters provide an excellent presentation of estimation. Chapter 6 covers point estimation, including descriptive and order statistics, maximum likelihood estimators and their distributions, sufficient statistics, and Bayesian estimation. Interval estimation is covered in Chapter 7, including the topics of confidence intervals for means and proportions, distribution-free confidence intervals for percentiles, confidence intervals for regression coefficients, and resampling methods (in particular, bootstrapping).

The last two chapters are about tests of statistical hypotheses. Chapter 8 considers terminology and standard tests on means and proportions, the Wilcoxon tests, the power of a test, best critical regions (Neyman/Pearson) and likelihood ratio tests. The topics in Chapter 9 are standard chi-square tests, analysis of variance including general factorial designs, and some procedures associated with regression, correlation, and statistical quality control.

The first semester of the course should contain most of the topics in Chapters 1–5. The second semester includes some topics omitted there and many of those in Chapters 6–9. A more basic course might omit some of the (optional) starred sections, but we believe that the order of topics will give the instructor the flexibility needed in his or her course. The usual nonparametric and Bayesian techniques are placed at appropriate places in the text rather than in separate chapters. We find that many persons like the applications associated with statistical quality control in the last section. Overall, one of the authors, Hogg, believes that the presentation (at a somewhat reduced mathematical level) is much like that given in the earlier editions of Hogg and Craig (see References).

The Prologue suggests many fields in which statistical methods can be used. In the Epilogue, the importance of understanding variation is stressed, particularly for its need in continuous quality improvement as described in the usual Six-Sigma programs. At the end of each chapter we give some interesting historical comments, which have proved to be very worthwhile in the past editions.

The answers given in this text for questions that involve the standard distributions were calculated using our probability tables which, of course, are rounded off for printing. If you use a statistical package, your answers may differ slightly from those given.

ANCILLARIES

Data sets from this textbook are available for download at http://www.pearson globaleditions.com/Hogg. This website also offers an **Instructor's Solutions Manual** containing worked-out solutions to the even-numbered exercises in the text. Some of the numerical exercises were solved with *Maple*. For additional exercises that involve simulations, a separate manual, *Probability & Statistics: Explorations with MAPLE*, second edition, by Zaven Karian and Elliot Tanis, can be downloaded from the website. Several exercises in that manual also make use of the power of *Maple* as a computer algebra system.

If you find any errors in this text, please send them to tanis@hope.edu so that they can be corrected in a future printing. These **errata** will also be posted on http://www.math.hope.edu/tanis/.

ACKNOWLEDGMENTS

We wish to thank our colleagues, students, and friends for many suggestions and for their generosity in supplying data for exercises and examples. In particular, we would like to thank the reviewers of the eighth edition who made suggestions for this edition. They are Steven T. Garren from James Madison University, Daniel C. Weiner from Boston University, and Kyle Siegrist from the University of Alabama in Huntsville. Mark Mills from Central College in Iowa also made some helpful comments. We also acknowledge the excellent suggestions from our copy editor, Kristen Cassereau Ng, and the fine work of our accuracy checkers, Kyle Siegrist and Steven Garren. We also thank the University of Iowa and Hope College for providing office space and encouragement. Finally, our families, through nine editions, have been most understanding during the preparation of all of this material. We would especially like to thank our wives, Ann, Elaine, and Bridget. We truly appreciate their patience and needed their love.

Robert V. Hogg

Elliot A. Tanis

tanis@hope.edu

Dale L. Zimmerman

dale-zimmerman@uiowa.edu

GLOBAL ACKNOWLEDGMENTS

Pearson gratefully acknowledges and thanks the following people for their work on the Global Edition:

Contributor
C. B. Gupta, *Birla institute of Technology and Science, Pilani*

Reviewers
Kalpana K. Mahajan, *Panjab University, Chandigarh*
Dinesh P. A., *M.S. Ramaiah Institute of Technology, Bangalore*
J. Subramani, *Pondicherry University, Puducherry*

PROLOGUE

The discipline of statistics deals with the collection and analysis of data. Advances in computing technology, particularly in relation to changes in science and business, have increased the need for more statistical scientists to examine the huge amount of data being collected. We know that data are not equivalent to information. Once data (hopefully of high quality) are collected, there is a strong need for statisticians to make sense of them. That is, data must be analyzed in order to provide information upon which decisions can be made. In light of this great demand, opportunities for the discipline of statistics have never been greater, and there is a special need for more bright young persons to go into statistical science.

If we think of fields in which data play a major part, the list is almost endless: accounting, actuarial science, atmospheric science, biological science, economics, educational measurement, environmental science, epidemiology, finance, genetics, manufacturing, marketing, medicine, pharmaceutical industries, psychology, sociology, sports, and on and on. Because statistics is useful in all of these areas, it really should be taught as an applied science. Nevertheless, to go very far in such an applied science, it is necessary to understand the importance of creating models for each situation under study. Now, no model is ever exactly right, but some are extremely useful as an approximation to the real situation. Most appropriate models in statistics require a certain mathematical background in probability. Accordingly, while alluding to applications in the examples and the exercises, this textbook is really about the mathematics needed for the appreciation of probabilistic models necessary for statistical inferences.

In a sense, statistical techniques are really the heart of the scientific method. Observations are made that suggest conjectures. These conjectures are tested, and data are collected and analyzed, providing information about the truth of the conjectures. Sometimes the conjectures are supported by the data, but often the conjectures need to be modified and more data must be collected to test the modifications, and so on. Clearly, in this iterative process, statistics plays a major role with its emphasis on the proper design and analysis of experiments and the resulting inferences upon which decisions can be made. Through statistics, information is provided that is relevant to taking certain actions, including improving manufactured products, providing better services, marketing new products or services, forecasting energy needs, classifying diseases better, and so on.

Statisticians recognize that there are often small errors in their inferences, and they attempt to quantify the probabilities of those mistakes and make them as small as possible. That these uncertainties even exist is due to the fact that there is variation in the data. Even though experiments are repeated under seemingly the same conditions, the results vary from trial to trial. We try to improve the quality of the data by making them as reliable as possible, but the data simply do not fall on given patterns. In light of this uncertainty, the statistician tries to determine the pattern in the best possible way, always explaining the error structures of the statistical estimates.

This is an important lesson to be learned: Variation is almost everywhere. It is the statistician's job to understand variation. Often, as in manufacturing, the desire is to reduce variation because the products will be more consistent. In other words, car doors will fit better in the manufacturing of automobiles if the variation is decreased by making each door closer to its target values.

Many statisticians in industry have stressed the need for "statistical thinking" in everyday operations. This need is based upon three points (two of which have been mentioned in the preceding paragraph): (1) Variation exists in all processes; (2) understanding and reducing undesirable variation is a key to success; and (3) all work occurs in a system of interconnected processes. W. Edwards Deming, an esteemed statistician and quality improvement "guru," stressed these three points, particularly the third one. He would carefully note that you could not maximize the total operation by maximizing the individual components unless they are independent of each other. However, in most instances, they are highly dependent, and persons in different departments must work together in creating the best products and services. If not, what one unit does to better itself could very well hurt others. He often cited an orchestra as an illustration of the need for the members to work together to create an outcome that is consistent and desirable.

Any student of statistics should understand the nature of variability and the necessity for creating probabilistic models of that variability. We cannot avoid making inferences and decisions in the face of this uncertainty; however, these inferences and decisions are greatly influenced by the probabilistic models selected. Some persons are better model builders than others and accordingly will make better inferences and decisions. The assumptions needed for each statistical model are carefully examined; it is hoped that thereby the reader will become a better model builder.

Finally, we must mention how modern statistical analyses have become dependent upon the computer. Statisticians and computer scientists really should work together in areas of exploratory data analysis and "data mining." Statistical software development is critical today, for the best of it is needed in complicated data analyses. In light of this growing relationship between these two fields, it is good advice for bright students to take substantial offerings in statistics and in computer science.

Students majoring in statistics, computer science, or a joint program are in great demand in the workplace and in graduate programs. Clearly, they can earn advanced degrees in statistics or computer science or both. But, more important, they are highly desirable candidates for graduate work in other areas: actuarial science, industrial engineering, finance, marketing, accounting, management science, psychology, economics, law, sociology, medicine, health sciences, etc. So many fields have been "mathematized" that their programs are begging for majors in statistics or computer science. Often, such students become "stars" in these other areas. We truly hope that we can interest students enough that they want to study more statistics. If they do, they will find that the opportunities for very successful careers are numerous.

PROBABILITY

1.1 PROPERTIES OF PROBABILITY

It is usually difficult to explain to the general public what statisticians do. Many think of us as "math nerds" who seem to enjoy dealing with numbers. And there is some truth to that concept. But if we consider the bigger picture, many recognize that statisticians can be extremely helpful in many investigations.

Consider the following:

1. There is some problem or situation that needs to be considered; so statisticians are often asked to work with investigators or research scientists.

2. Suppose that some measure (or measures) is needed to help us understand the situation better. The measurement problem is often extremely difficult, and creating good measures is a valuable skill. As an illustration, in higher education, how do we measure good teaching? This is a question to which we have not found a satisfactory answer, although several measures, such as student evaluations, have been used in the past.

3. After the measuring instrument has been developed, we must collect data through observation, possibly the results of a survey or an experiment.

4. Using these data, statisticians summarize the results, often with descriptive statistics and graphical methods.

5. These summaries are then used to analyze the situation. Here it is possible that statisticians make what are called statistical inferences.

6. Finally, a report is presented, along with some recommendations that are based upon the data and the analysis of them. Frequently such a recommendation might be to perform the survey or experiment again, possibly changing some of the questions or factors involved. This is how statistics is used in what is referred to as the scientific method, because often the analysis of the data suggests other experiments. Accordingly, the scientist must consider different possibilities in his or her search for an answer and thus performs similar experiments over and over again.

The discipline of statistics deals with the *collection* and *analysis of data*. When measurements are taken, even seemingly under the same conditions, the results usually vary. Despite this variability, a statistician tries to find a pattern; yet due to the "noise," not all of the data fit into the pattern. In the face of the variability, the statistician must still determine the best way to describe the pattern. Accordingly, statisticians know that mistakes will be made in data analysis, and they try to minimize those errors as much as possible and then give bounds on the possible errors. By considering these bounds, decision makers can decide how much confidence they want to place in the data and in their analysis of them. If the bounds are wide, perhaps more data should be collected. If, however, the bounds are narrow, the person involved in the study might want to make a decision and proceed accordingly.

Variability is a fact of life, and proper statistical methods can help us understand data collected under inherent variability. Because of this variability, many decisions have to be made that involve uncertainties. In medical research, interest may center on the effectiveness of a new vaccine for mumps; an agronomist must decide whether an increase in yield can be attributed to a new strain of wheat; a meteorologist is interested in predicting the probability of rain; the state legislature must decide whether decreasing speed limits will result in fewer accidents; the admissions officer of a college must predict the college performance of an incoming freshman; a biologist is interested in estimating the clutch size for a particular type of bird; an economist desires to estimate the unemployment rate; an environmentalist tests whether new controls have resulted in a reduction in pollution.

In reviewing the preceding (relatively short) list of possible areas of applications of statistics, the reader should recognize that good statistics is closely associated with careful thinking in many investigations. As an illustration, students should appreciate how statistics is used in the endless cycle of the scientific method. We observe nature and ask questions, we run experiments and collect data that shed light on these questions, we analyze the data and compare the results of the analysis with what we previously thought, we raise new questions, and on and on. Or if you like, statistics is clearly part of the important "plan–do–study–act" cycle: Questions are raised and investigations planned and carried out. The resulting data are studied and analyzed and then acted upon, often raising new questions.

There are many aspects of statistics. Some people get interested in the subject by collecting data and trying to make sense out of their observations. In some cases the answers are obvious and little training in statistical methods is necessary. But if a person goes very far in many investigations, he or she soon realizes that there is a need for some theory to help describe the error structure associated with the various estimates of the patterns. That is, at some point appropriate probability and mathematical models are required to make sense out of complicated data sets. Statistics and the probabilistic foundation on which statistical methods are based can provide the models to help people do this. So in this book, we are more concerned with the mathematical, rather than the applied, aspects of statistics. Still, we give enough real examples so that the reader can get a good sense of a number of important applications of statistical methods.

In the study of statistics, we consider experiments for which the outcome cannot be predicted with certainty. Such experiments are called **random experiments**. Although the specific outcome of a random experiment cannot be predicted with certainty before the experiment is performed, the collection of all possible outcomes *is* known and can be described and perhaps listed. The collection of all possible outcomes is denoted by S and is called the **outcome space**. Given an outcome space S, let A be a part of the collection of outcomes in S; that is, $A \subset S$. Then A is called an **event**. When the random experiment is performed and the outcome of the experiment is in A, we say that **event** A **has occurred**.

Since, in studying probability, the words *set* and *event* are interchangeable, the reader might want to review **algebra of sets**. Here we remind the reader of some terminology:

- $\emptyset$ denotes the **null** or **empty** set;
- $A \subset B$ means A is a **subset** of B;
- $A \cup B$ is the **union** of A and B;
- $A \cap B$ is the **intersection** of A and B;
- A' is the **complement** of A (i.e., all elements in S that are not in A).

Some of these sets are depicted by the shaded regions in Figure 1.1-1, in which S is the interior of the rectangles. Such figures are called **Venn diagrams**.

Special terminology associated with events that is often used by statisticians includes the following:

1. $A_1, A_2, \ldots, A_k$ are **mutually exclusive events** means that $A_i \cap A_j = \emptyset, i \neq j$; that is, $A_1, A_2, \ldots, A_k$ are disjoint sets;

2. $A_1, A_2, \ldots, A_k$ are **exhaustive events** means that $A_1 \cup A_2 \cup \cdots \cup A_k = S$.

So if $A_1, A_2, \ldots, A_k$ are **mutually exclusive and exhaustive** events, we know that $A_i \cap A_j = \emptyset, i \neq j$, and $A_1 \cup A_2 \cup \cdots \cup A_k = S$.

Set operations satisfy several properties. For example, if A, B, and C are subsets of S, we have the following:

Commutative Laws

$$A \cup B = B \cup A$$
$$A \cap B = B \cap A$$

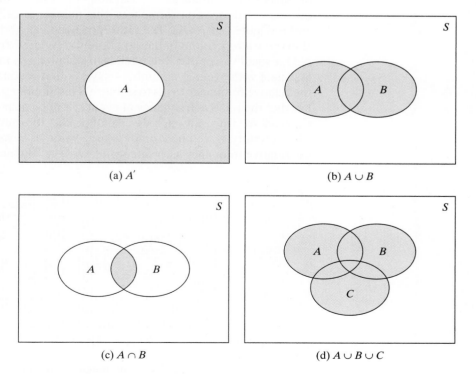

(a) A'

(b) $A \cup B$

(c) $A \cap B$

(d) $A \cup B \cup C$

Figure 1.1-1 Algebra of sets

Associative Laws

$$(A \cup B) \cup C = A \cup (B \cup C)$$
$$(A \cap B) \cap C = A \cap (B \cap C)$$

Distributive Laws

$$A \cap (B \cup C) = (A \cap B) \cup (A \cap C)$$
$$A \cup (B \cap C) = (A \cup B) \cap (A \cup C)$$

De Morgan's Laws

$$(A \cup B)' = A' \cap B'$$
$$(A \cap B)' = A' \cup B'$$

A Venn diagram will be used to justify the first of De Morgan's laws. In Figure 1.1-2(a), $A \cup B$ is represented by horizontal lines, and thus $(A \cup B)'$ is the region represented by vertical lines. In Figure 1.1-2(b), A' is indicated with horizontal lines, and B' is indicated with vertical lines. An element belongs to $A' \cap B'$ if it belongs to both A' and B'. Thus the crosshatched region represents $A' \cap B'$. Clearly, this crosshatched region is the same as that shaded with vertical lines in Figure 1.1-2(a).

We are interested in defining what is meant by the probability of event A, denoted by $P(A)$ and often called the chance of A occurring. To help us understand what is meant by the probability of A, consider repeating the experiment a number of times—say, n times. Count the number of times that event A actually occurred throughout these n performances; this number is called the frequency of event A and is denoted by $\mathcal{N}(A)$. The ratio $\mathcal{N}(A)/n$ is called the **relative frequency** of event A in these n repetitions of the experiment. A relative frequency is usually very unstable for small values of n, but it tends to stabilize as n increases. This suggests that we associate with event A a number—say, p—that is equal to the number about which the relative frequency tends to stabilize. This number p can then be taken as the number that the relative frequency of event A will be near in future performances of the experiment. Thus, although we cannot predict the outcome of a random experiment with certainty, if we know p, for a large value of n, we can predict fairly accurately the relative frequency associated with event A. The number p assigned to event A is

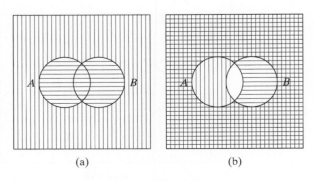

(a)　　　　　　　　(b)

Figure 1.1-2 Venn diagrams illustrating De Morgan's laws

called the **probability** of event A and is denoted by $P(A)$. That is, $P(A)$ represents the proportion of outcomes of a random experiment that terminate in the event A as the number of trials of that experiment increases without bound.

The next example will help to illustrate some of the ideas just presented.

Example 1.1-1

A fair six-sided die is rolled six times. If the face numbered k is the outcome on roll k for $k = 1, 2, \ldots, 6$, we say that a match has occurred. The experiment is called a success if at least one match occurs during the six trials. Otherwise, the experiment is called a failure. The sample space is $S = \{\text{success, failure}\}$. Let $A = \{\text{success}\}$. We would like to assign a value to $P(A)$. Accordingly, this experiment was simulated 500 times on a computer. Figure 1.1-3 depicts the results of this simulation, and the following table summarizes a few of the results:

n	$\mathcal{N}(A)$	$\mathcal{N}(A)/n$
50	37	0.740
100	69	0.690
250	172	0.688
500	330	0.660

The probability of event A is not intuitively obvious, but it will be shown in Example 1.4-6 that $P(A) = 1 - (1 - 1/6)^6 = 0.665$. This assignment is certainly supported by the simulation (although not proved by it). ■

Example 1.1-1 shows that at times intuition cannot be used to assign probabilities, although simulation can perhaps help to assign a probability empirically. The next example illustrates where intuition can help in assigning a probability to an event.

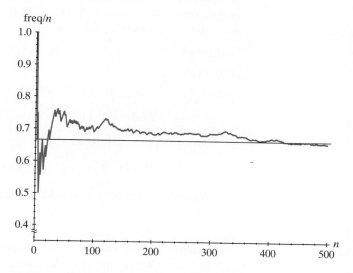

Figure 1.1-3 Fraction of experiments having at least one match

**Example
1.1-2**

A disk 2 inches in diameter is thrown at random on a tiled floor, where each tile is a square with sides 4 inches in length. Let C be the event that the disk will land entirely on one tile. In order to assign a value to $P(C)$, consider the center of the disk. In what region must the center lie to ensure that the disk lies entirely on one tile? If you draw a picture, it should be clear that the center must lie within a square having sides of length 2 and with its center coincident with the center of a tile. Since the area of this square is 4 and the area of a tile is 16, it makes sense to let $P(C) = 4/16$. ∎

Sometimes the nature of an experiment is such that the probability of A can be assigned easily. For example, when a state lottery randomly selects a three-digit integer, we would expect each of the 1000 possible three-digit numbers to have the same chance of being selected, namely, $1/1000$. If we let $A = \{233, 323, 332\}$, then it makes sense to let $P(A) = 3/1000$. Or if we let $B = \{234, 243, 324, 342, 423, 432\}$, then we would let $P(B) = 6/1000$, the probability of the event B. Probabilities of events associated with many random experiments are perhaps not quite as obvious and straightforward as was seen in Example 1.1-1.

So we wish to associate with A a number $P(A)$ about which the relative frequency $\mathcal{N}(A)/n$ of the event A tends to stabilize with large n. A function such as $P(A)$ that is evaluated for a set A is called a **set function**. In this section, we consider the probability set function $P(A)$ and discuss some of its properties. In succeeding sections, we shall describe how the probability set function is defined for particular experiments.

To help decide what properties the probability set function should satisfy, consider properties possessed by the relative frequency $\mathcal{N}(A)/n$. For example, $\mathcal{N}(A)/n$ is always nonnegative. If $A = S$, the sample space, then the outcome of the experiment will always belong to S, and thus $\mathcal{N}(S)/n = 1$. Also, if A and B are two mutually exclusive events, then $\mathcal{N}(A \cup B)/n = \mathcal{N}(A)/n + \mathcal{N}(B)/n$. Hopefully, these remarks will help to motivate the following definition.

Definition 1.1-1

Probability is a real-valued set function P that assigns, to each event A in the sample space S, a number $P(A)$, called the probability of the event A, such that the following properties are satisfied:

(a) $P(A) \geq 0$;

(b) $P(S) = 1$;

(c) if $A_1, A_2, A_3, \ldots$ are events and $A_i \cap A_j = \emptyset, i \neq j$, then

$$P(A_1 \cup A_2 \cup \cdots \cup A_k) = P(A_1) + P(A_2) + \cdots + P(A_k)$$

for each positive integer k, and

$$P(A_1 \cup A_2 \cup A_3 \cup \cdots) = P(A_1) + P(A_2) + P(A_3) + \cdots$$

for an infinite, but countable, number of events.

The theorems that follow give some other important properties of the probability set function. When one considers these theorems, it is important to understand the theoretical concepts and proofs. However, if the reader keeps the relative frequency concept in mind, the theorems should also have some intuitive appeal.

Theorem 1.1-1	For each event A,

$$P(A) = 1 - P(A').$$

Proof [See Figure 1.1-1(a).] We have

$$S = A \cup A' \qquad \text{and} \qquad A \cap A' = \emptyset.$$

Thus, from properties (b) and (c), it follows that

$$1 = P(A) + P(A').$$

Hence

$$P(A) = 1 - P(A').$$

Example 1.1-3	A fair coin is flipped successively until the same face is observed on successive flips. Let $A = \{x : x = 3, 4, 5, \ldots\}$; that is, A is the event that it will take three or more flips of the coin to observe the same face on two consecutive flips. To find $P(A)$, we first find the probability of $A' = \{x : x = 2\}$, the complement of A. In two flips of a coin, the possible outcomes are $\{HH, HT, TH, TT\}$, and we assume that each of these four points has the same chance of being observed. Thus,

$$P(A') = P(\{HH, TT\}) = \frac{2}{4}.$$

It follows from Theorem 1.1-1 that

$$P(A) = 1 - P(A') = 1 - \frac{2}{4} = \frac{2}{4}.$$

Theorem 1.1-2	$P(\emptyset) = 0.$

Proof In Theorem 1.1-1, take $A = \emptyset$ so that $A' = S$. Then

$$P(\emptyset) = 1 - P(S) = 1 - 1 = 0.$$

Theorem 1.1-3	If events A and B are such that $A \subset B$, then $P(A) \leq P(B)$.

Proof We have

$$B = A \cup (B \cap A') \qquad \text{and} \qquad A \cap (B \cap A') = \emptyset.$$

Hence, from property (c),

$$P(B) = P(A) + P(B \cap A') \geq P(A)$$

because, from property (a),

$$P(B \cap A') \geq 0.$$

Theorem 1.1-4

For each event A, $P(A) \leq 1$.

Proof Since $A \subset S$, we have, by Theorem 1.1-3 and property (b),

$$P(A) \leq P(S) = 1,$$

which gives the desired result. □

Property (a), along with Theorem 1.1-4, shows that, for each event A,

$$0 \leq P(A) \leq 1.$$

Theorem 1.1-5

If A and B are any two events, then

$$P(A \cup B) = P(A) + P(B) - P(A \cap B).$$

Proof [See Figure 1.1-1(b).] The event $A \cup B$ can be represented as a union of mutually exclusive events, namely,

$$A \cup B = A \cup (A' \cap B).$$

Hence, by property (c),

$$P(A \cup B) = P(A) + P(A' \cap B). \tag{1.1-1}$$

However,

$$B = (A \cap B) \cup (A' \cap B),$$

which is a union of mutually exclusive events. Thus,

$$P(B) = P(A \cap B) + P(A' \cap B)$$

and

$$P(A' \cap B) = P(B) - P(A \cap B).$$

If the right-hand side of this equation is substituted into Equation 1.1-1, we obtain

$$P(A \cup B) = P(A) + P(B) - P(A \cap B),$$

which is the desired result. □

Example 1.1-4

A faculty leader was meeting two students in Paris, one arriving by train from Amsterdam and the other arriving by train from Brussels at approximately the same time. Let A and B be the events that the respective trains are on time. Suppose we know from past experience that $P(A) = 0.93$, $P(B) = 0.89$, and $P(A \cap B) = 0.87$. Then

$$\begin{aligned} P(A \cup B) &= P(A) + P(B) - P(A \cap B) \\ &= 0.93 + 0.89 - 0.87 = 0.95 \end{aligned}$$

is the probability that at least one train is on time. ■

Theorem 1.1-6	If A, B, and C are any three events, then

$$P(A \cup B \cup C) = P(A) + P(B) + P(C) - P(A \cap B)$$
$$-P(A \cap C) - P(B \cap C) + P(A \cap B \cap C).$$

Proof [See Figure 1.1-1(d).] Write

$$A \cup B \cup C = A \cup (B \cup C)$$

and apply Theorem 1.1-5. The details are left as an exercise. □

Example 1.1-5 A survey was taken of a group's viewing habits of sporting events on TV during the last year. Let $A = \{$watched football$\}$, $B = \{$watched basketball$\}$, $C = \{$watched baseball$\}$. The results indicate that if a person is selected at random from the surveyed group, then $P(A) = 0.43$, $P(B) = 0.40$, $P(C) = 0.32$, $P(A \cap B) = 0.29$, $P(A \cap C) = 0.22$, $P(B \cap C) = 0.20$, and $P(A \cap B \cap C) = 0.15$. It then follows that

$$P(A \cup B \cup C) = P(A) + P(B) + P(C) - P(A \cap B) - P(A \cap C)$$
$$-P(B \cap C) + P(A \cap B \cap C)$$
$$= 0.43 + 0.40 + 0.32 - 0.29 - 0.22 - 0.20 + 0.15$$
$$= 0.59$$

is the probability that this person watched at least one of these sports. ■

Let a probability set function be defined on a sample space S. Let $S = \{e_1, e_2, \ldots, e_m\}$, where each e_i is a possible outcome of the experiment. The integer m is called the total number of ways in which the random experiment can terminate. If each of these outcomes has the same probability of occurring, we say that the m outcomes are **equally likely**. That is,

$$P(\{e_i\}) = \frac{1}{m}, \qquad i = 1, 2, \ldots, m.$$

If the number of outcomes in an event A is h, then the integer h is called the number of ways that are favorable to the event A. In this case, $P(A)$ is equal to the number of ways favorable to the event A divided by the total number of ways in which the experiment can terminate. That is, under this assumption of equally likely outcomes, we have

$$P(A) = \frac{h}{m} = \frac{N(A)}{N(S)},$$

where $h = N(A)$ is the number of ways A can occur and $m = N(S)$ is the number of ways S can occur. Exercise 1.1-15 considers this assignment of probability in a more theoretical manner.

It should be emphasized that in order to assign the probability h/m to the event A, we must assume that each of the outcomes $e_1, e_2, \ldots, e_m$ has the same probability $1/m$. This assumption is then an important part of our probability model; if it is not realistic in an application, then the probability of the event A cannot be computed in this way. Actually, we have used this result in the simple case given in Example 1.1-3 because it seemed realistic to assume that each of the possible outcomes in $S = \{HH, HT, TH, TT\}$ had the same chance of being observed.

Example
1.1-6
Let a card be drawn at random from an ordinary deck of 52 playing cards. Then the sample space S is the set of $m = 52$ different cards, and it is reasonable to assume that each of these cards has the same probability of selection, 1/52. Accordingly, if A is the set of outcomes that are kings, then $P(A) = 4/52 = 1/13$ because there are $h = 4$ kings in the deck. That is, 1/13 is the probability of drawing a card that is a king, provided that each of the 52 cards has the same probability of being drawn. ∎

In Example 1.1-6, the computations are very easy because there is no difficulty in the determination of the appropriate values of h and m. However, instead of drawing only one card, suppose that 13 are taken at random and without replacement. Then we can think of each possible 13-card hand as being an outcome in a sample space, and it is reasonable to assume that each of these outcomes has the same probability. For example, using the preceding method to assign the probability of a hand consisting of seven spades and six hearts, we must be able to count the number h of all such hands as well as the number m of possible 13-card hands. In these more complicated situations, we need better methods of determining h and m. We discuss some of these counting techniques in Section 1.2.

Exercises

1.1-1. Of a group of small-business owners, 30% consult both an accountant and a financial planner and 10% consult neither. Say that the probability of consulting an accountant exceeds the probability of consulting a financial planner by 20%. What is the probability of a randomly selected person from this group consulting an accountant?

1.1-2. An insurance company looks at its auto insurance customers and finds that (a) all insure at least one car, (b) 90% insure more than one car, (c) 25% insure a sports car, and (d) 15% insure more than one car, including a sports car. Find the probability that a customer selected at random insures exactly one car and it is not a sports car.

1.1-3. Draw one card at random from a standard deck of cards. The sample space S is the collection of the 52 cards. Assume that the probability set function assigns 1/52 to each of the 52 outcomes. Let

$A = \{x: x$ is a jack, queen, or king$\}$,

$B = \{x: x$ is a 9, 10, or jack and x is red$\}$,

$C = \{x: x$ is a club$\}$,

$D = \{x: x$ is a diamond, a heart, or a spade$\}$.

Find **(a)** $P(A)$, **(b)** $P(A \cap B)$, **(c)** $P(A \cup B)$, **(d)** $P(C \cup D)$, and **(e)** $P(C \cap D)$.

1.1-4. A fair coin is tossed five times, and the sequence of heads and tails is observed.

(a) List each of the 32 sequences in the sample space S.

(b) Let events A, B, C, and D be given by $A = \{$at least 4 heads$\}$, $B = \{$at most 3 heads$\}$, $C = \{$heads on the fourth toss$\}$, and $D = \{1$ head and 4 tails$\}$. If the

probability set function assigns 1/32 to each outcome in the sample space, find **(i)** $P(A)$, **(ii)** $P(A \cap B)$, **(iii)** $P(B)$, **(iv)** $P(A \cap C)$, **(v)** $P(D)$, **(vi)** $P(A \cup C)$, and **(vii)** $P(B \cap D)$.

1.1-5. Consider the trial on which a 4 is first observed in successive rolls of a six-sided die. Let A be the event that 4 is observed on the first trial. Let B be the event that at least two trials are required to observe a 4. Assuming that each side has probability 1/6, find **(a)** $P(A)$, **(b)** $P(B)$, and **(c)** $P(A \cup B)$.

1.1-6. If $P(A) = 0.4$, $P(B) = 0.5$, and $P(A \cap B) = 0.3$, find **(a)** $P(A \cup B)$, **(b)** $P(A \cap B')$, and **(c)** $P(A' \cup B')$.

1.1-7. Given that $P(A \cup B) = 0.76$ and $P(A \cup B') = 0.87$, find $P(A)$.

1.1-8. During a visit to a primary care physician's office, the probability of having neither lab work nor referral to a specialist is 0.21. Of those coming to that office, the probability of having lab work is 0.41 and the probability of having a referral is 0.53. What is the probability of having both lab work and a referral?

1.1-9. Roll a fair six-sided die three times. Let $A_1 = \{1$ or 2 on the first roll$\}$, $A_2 = \{3$ or 4 on the second roll$\}$, and $A_3 = \{5$ or 6 on the third roll$\}$. It is given that $P(A_i) = 1/3$, $i = 1, 2, 3$; $P(A_i \cap A_j) = (1/3)^2$, $i \neq j$; and $P(A_1 \cap A_2 \cap A_3) = (1/3)^3$.

(a) Use Theorem 1.1-6 to find $P(A_1 \cup A_2 \cup A_3)$.

(b) Show that $P(A_1 \cup A_2 \cup A_3) = 1 - (1 - 1/3)^3$.

1.1-10. Prove Theorem 1.1-6.

1.1-11. A typical roulette wheel used in a casino has 38 slots that are numbered $1, 2, 3, \ldots, 36, 0, 00$, respectively. The 0 and 00 slots are colored green. Half of the remaining slots are red and half are black. Also, half of the integers between 1 and 36 inclusive are odd, half are even, and 0 and 00 are defined to be neither odd nor even. A ball is rolled around the wheel and ends up in one of the slots; we assume that each slot has equal probability of 1/38, and we are interested in the number of the slot into which the ball falls.

(a) Define the sample space S.

(b) Let $A = \{0, 00\}$. Give the value of $P(A)$.

(c) Let $B = \{14, 15, 17, 18\}$. Give the value of $P(B)$.

(d) Let $D = \{x : x \text{ is odd}\}$. Give the value of $P(D)$.

1.1-12. Let x equal a number that is selected randomly from the closed interval from zero to three, $[0, 3]$. Use your intuition to assign values to

(a) $P(\{x : 0 \le x \le 5/4\})$.

(b) $P(\{x : 5/4 \le x \le 3\})$.

(c) $P(\{x : x = 5/4\})$.

(d) $P(\{x : 1/2 < x < 5\})$.

1.1-13. Divide a line segment into two parts by selecting a point at random. Use your intuition to assign a probability to the event that the longer segment is at least two times longer than the shorter segment.

1.1-14. Let the interval $[-r, r]$ be the base of a semicircle. If a point is selected at random from this interval, assign a probability to the event that the length of the perpendicular segment from the point to the semicircle is less than $r/2$.

1.1-15. Let $S = A_1 \cup A_2 \cup \cdots \cup A_m$, where events $A_1, A_2, \ldots, A_m$ are mutually exclusive and exhaustive.

(a) If $P(A_1) = P(A_2) = \cdots = P(A_m)$, show that $P(A_i) = 1/m$, $i = 1, 2, \ldots, m$.

(b) If $A = A_1 \cup A_2 \cup \cdots \cup A_h$, where $h < m$, and (a) holds, prove that $P(A) = h/m$.

1.1-16. Let p_n, $n = 0, 1, 2, \ldots$, be the probability that an automobile policyholder will file for n claims in a five-year period. The actuary involved makes the assumption that $p_{n+1} = (1/4)p_n$. What is the probability that the holder will file two or more claims during this period?

1.2 METHODS OF ENUMERATION

In this section, we develop counting techniques that are useful in determining the number of outcomes associated with the events of certain random experiments. We begin with a consideration of the multiplication principle.

Multiplication Principle: Suppose that an experiment (or procedure) E_1 has n_1 outcomes and, for each of these possible outcomes, an experiment (procedure) E_2 has n_2 possible outcomes. Then the composite experiment (procedure) $E_1 E_2$ that consists of performing first E_1 and then E_2 has $n_1 n_2$ possible outcomes.

Example 1.2-1

Let E_1 denote the selection of a rat from a cage containing one female (F) rat and one male (M) rat. Let E_2 denote the administering of either drug A (A), drug B (B), or a placebo (P) to the selected rat. Then the outcome for the composite experiment can be denoted by an ordered pair, such as (F, P). In fact, the set of all possible outcomes, namely, $(2)(3) = 6$ of them, can be denoted by the following rectangular array:

$$\text{(F, A)} \quad \text{(F, B)} \quad \text{(F, P)}$$
$$\text{(M, A)} \quad \text{(M, B)} \quad \text{(M, P)}$$

■

Another way of illustrating the multiplication principle is with a tree diagram like that in Figure 1.2-1. The diagram shows that there are $n_1 = 2$ possibilities (branches) for the sex of the rat and that, for each of these outcomes, there are $n_2 = 3$ possibilities (branches) for the drug.

Clearly, the multiplication principle can be extended to a sequence of more than two experiments or procedures. Suppose that the experiment E_i has

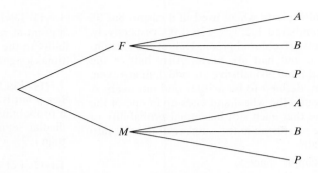

Figure 1.2-1 Tree diagram

n_i $(i = 1, 2, \ldots, m)$ possible outcomes after previous experiments have been performed. Then the composite experiment $E_1 E_2 \cdots E_m$ that consists of performing E_1, then $E_2, \ldots$, and finally E_m has $n_1 n_2 \cdots n_m$ possible outcomes.

Example 1.2-2

A certain food service gives the following choices for dinner: E_1, soup or tomato juice; E_2, steak or shrimp; E_3, French fried potatoes, mashed potatoes, or a baked potato; E_4, corn or peas; E_5, jello, tossed salad, cottage cheese, or coleslaw; E_6, cake, cookies, pudding, brownie, vanilla ice cream, chocolate ice cream, or orange sherbet; E_7, coffee, tea, milk, or punch. How many different dinner selections are possible if one of the listed choices is made for each of $E_1, E_2, \ldots$, and E_7? By the multiplication principle, there are

$$(2)(2)(3)(2)(4)(7)(4) = 2688$$

different combinations. ∎

Although the multiplication principle is fairly simple and easy to understand, it will be extremely useful as we now develop various counting techniques.

Suppose that n positions are to be filled with n different objects. There are n choices for filling the first position, $n - 1$ for the second, $\ldots$, and 1 choice for the last position. So, by the multiplication principle, there are

$$n(n - 1) \cdots (2)(1) = n!$$

possible arrangements. The symbol $n!$ is read "n factorial." We define $0! = 1$; that is, we say that zero positions can be filled with zero objects in one way.

> **Definition 1.2-1**
> Each of the $n!$ arrangements (in a row) of n different objects is called a **permutation** of the n objects.

Example 1.2-3

The number of permutations of the four letters a, b, c, and d is clearly $4! = 24$. However, the number of possible four-letter code words using the four letters a, b, c, and d if letters may be repeated is $4^4 = 256$, because in this case each selection can be performed in four ways. ∎

If only r positions are to be filled with objects selected from n different objects, $r \leq n$, then the number of possible ordered arrangements is

$$_nP_r = n(n-1)(n-2)\cdots(n-r+1).$$

That is, there are n ways to fill the first position, $(n-1)$ ways to fill the second, and so on, until there are $[n-(r-1)] = (n-r+1)$ ways to fill the rth position.

In terms of factorials, we have

$$_nP_r = \frac{n(n-1)\cdots(n-r+1)(n-r)\cdots(3)(2)(1)}{(n-r)\cdots(3)(2)(1)} = \frac{n!}{(n-r)!}.$$

Definition 1.2-2

Each of the $_nP_r$ arrangements is called a **permutation of n objects taken r at a time**.

Example 1.2-4

The number of possible four-letter code words, selecting from the 26 letters in the alphabet, in which all four letters are different is

$$_{26}P_4 = (26)(25)(24)(23) = \frac{26!}{22!} = 358{,}800.$$ ∎

Example 1.2-5

The number of ways of selecting a president, a vice president, a secretary, and a treasurer in a club consisting of 10 persons is

$$_{10}P_4 = 10 \cdot 9 \cdot 8 \cdot 7 = \frac{10!}{6!} = 5040.$$ ∎

Suppose that a set contains n objects. Consider the problem of drawing r objects from this set. The order in which the objects are drawn may or may not be important. In addition, it is possible that a drawn object is replaced before the next object is drawn. Accordingly, we give some definitions and show how the multiplication principle can be used to count the number of possibilities.

Definition 1.2-3

If r objects are selected from a set of n objects, and if the order of selection is noted, then the selected set of r objects is called an **ordered sample of size r**.

Definition 1.2-4

Sampling with replacement occurs when an object is selected and then replaced before the next object is selected.

By the multiplication principle, the number of possible ordered samples of size r taken from a set of n objects is n^r when sampling with replacement.

Example 1.2-6

A die is rolled seven times. The number of possible ordered samples is $6^7 = 279{,}936$. Note that rolling a die is equivalent to sampling with replacement from the set $\{1,2,3,4,5,6\}$. ∎

Example 1.2-7

An urn contains 10 balls numbered 0, 1, 2, ..., 9. If 4 balls are selected one at a time and with replacement, then the number of possible ordered samples is $10^4 = 10{,}000$. Note that this is the number of four-digit integers between 0000 and 9999, inclusive. ∎

> **Definition 1.2-5**
> **Sampling without replacement** occurs when an object is not replaced after it has been selected.

By the multiplication principle, the number of possible ordered samples of size r taken from a set of n objects without replacement is

$$n(n-1)\cdots(n-r+1) = \frac{n!}{(n-r)!},$$

which is equivalent to $_nP_r$, the number of permutations of n objects taken r at a time.

Example 1.2-8

The number of ordered samples of 5 cards that can be drawn without replacement from a standard deck of 52 playing cards is

$$(52)(51)(50)(49)(48) = \frac{52!}{47!} = 311{,}875{,}200.$$ ∎

REMARK Note that it must be true that $r \le n$ when sampling without replacement, but r can exceed n when sampling with replacement. ∎

Often the order of selection is not important and interest centers only on the selected set of r objects. That is, we are interested in the number of subsets of size r that can be selected from a set of n different objects. In order to find the number of (unordered) subsets of size r, we count, in two different ways, the number of ordered subsets of size r that can be taken from the n distinguishable objects. Then, equating the two answers, we are able to count the number of (unordered) subsets of size r.

Let C denote the number of (unordered) subsets of size r that can be selected from n different objects. We can obtain each of the $_nP_r$ ordered subsets by first selecting one of the C unordered subsets of r objects and then ordering these r objects. Since the latter ordering can be carried out in $r!$ ways, the multiplication principle yields $(C)(r!)$ ordered subsets; so $(C)(r!)$ must equal $_nP_r$. Thus, we have

$$(C)(r!) = \frac{n!}{(n-r)!},$$

or

$$C = \frac{n!}{r!\,(n-r)!}.$$

We denote this answer by either $_nC_r$ or $\binom{n}{r}$; that is,

$$_nC_r = \binom{n}{r} = \frac{n!}{r!\,(n-r)!}.$$

Accordingly, a set of n different objects possesses

$$\binom{n}{r} = \frac{n!}{r!\,(n-r)!}$$

unordered subsets of size $r \leq n$.

We could also say that the number of ways in which r objects can be selected without replacement from n objects when the order of selection is disregarded is $\binom{n}{r} = {}_nC_r$, and the latter expression can be read as "n choose r." This result motivates the next definition.

> **Definition 1.2-6**
> Each of the ${}_nC_r$ unordered subsets is called a **combination of n objects taken r at a time**, where
>
> $$_nC_r = \binom{n}{r} = \frac{n!}{r!\,(n-r)!}.$$

Example 1.2-9 The number of possible 5-card hands (in 5-card poker) drawn from a deck of 52 playing cards is

$$_{52}C_5 = \binom{52}{5} = \frac{52!}{5!\,47!} = 2{,}598{,}960.$$ ∎

Example 1.2-10 The number of possible 13-card hands (in bridge) that can be selected from a deck of 52 playing cards is

$$_{52}C_{13} = \binom{52}{13} = \frac{52!}{13!\,39!} = 635{,}013{,}559{,}600.$$ ∎

The numbers $\binom{n}{r}$ are frequently called **binomial coefficients**, since they arise in the expansion of a binomial. We illustrate this property by giving a justification of the binomial expansion

$$(a+b)^n = \sum_{r=0}^{n} \binom{n}{r} b^r a^{n-r}. \tag{1.2-1}$$

For each summand in the expansion of

$$(a+b)^n = (a+b)(a+b)\cdots(a+b),$$

either an a or a b is selected from each of the n factors. One possible product is then $b^r a^{n-r}$; this occurs when b is selected from each of r factors and a from each of the remaining $n-r$ factors. But the latter operation can be completed in $\binom{n}{r}$ ways, which then must be the coefficient of $b^r a^{n-r}$, as shown in Equation 1.2-1.

The binomial coefficients are given in Table I in Appendix B for selected values of n and r. Note that for some combinations of n and r, the table uses the fact that

$$\binom{n}{r} = \frac{n!}{r!\,(n-r)!} = \frac{n!}{(n-r)!\,r!} = \binom{n}{n-r}.$$

That is, the number of ways in which r objects can be selected out of n objects is equal to the number of ways in which $n - r$ objects can be selected out of n objects.

Example 1.2-11

Assume that each of the $\binom{52}{5} = 2{,}598{,}960$ five-card hands drawn from a deck of 52 playing cards has the same probability of being selected. Then the number of possible 5-card hands that are all spades (event A) is

$$N(A) = \binom{13}{5}\binom{39}{0},$$

because the 5 spades can be selected from the 13 spades in $\binom{13}{5}$ ways, after which zero nonspades can be selected in $\binom{39}{0} = 1$ way. We have

$$\binom{13}{5} = \frac{13!}{5!8!} = 1287$$

from Table I in Appendix B. Thus, the probability of an all-spade five-card hand is

$$P(A) = \frac{N(A)}{N(S)} = \frac{1287}{2{,}598{,}960} = 0.000495.$$

Suppose now that the event B is the set of outcomes in which exactly three cards are kings and exactly two cards are queens. We can select the three kings in any one of $\binom{4}{3}$ ways and the two queens in any one of $\binom{4}{2}$ ways. By the multiplication principle, the number of outcomes in B is

$$N(B) = \binom{4}{3}\binom{4}{2}\binom{44}{0},$$

where $\binom{44}{0}$ gives the number of ways in which 0 cards are selected out of the nonkings and nonqueens and of course is equal to 1. Thus,

$$P(B) = \frac{N(B)}{N(S)} = \frac{\binom{4}{3}\binom{4}{2}\binom{44}{0}}{\binom{52}{5}} = \frac{24}{2{,}598{,}960} = 0.0000092.$$

Finally, let C be the set of outcomes in which there are exactly two kings, two queens, and one jack. Then

$$P(C) = \frac{N(C)}{N(S)} = \frac{\binom{4}{2}\binom{4}{2}\binom{4}{1}\binom{40}{0}}{\binom{52}{5}} = \frac{144}{2{,}598{,}960} = 0.000055$$

because the numerator of this fraction is the number of outcomes in C. ∎

Now suppose that a set contains n objects of two types: r of one type and $n - r$ of the other type. The number of permutations of n different objects is $n!$. However, in this case, the objects are not all distinguishable. To count the number of distinguishable arrangements, first select r out of the n positions for the objects of the first type.

This can be done in $\binom{n}{r}$ ways. Then fill in the remaining positions with the objects of the second type. Thus, the number of distinguishable arrangements is

$$_nC_r = \binom{n}{r} = \frac{n!}{r!\,(n-r)!}.$$

> **Definition 1.2-7**
> Each of the $_nC_r$ permutations of n objects, r of one type and $n-r$ of another type, is called a **distinguishable permutation**.

Example 1.2-12

A coin is flipped 10 times and the sequence of heads and tails is observed. The number of possible 10-tuplets that result in four heads and six tails is

$$\binom{10}{4} = \frac{10!}{4!\,6!} = \frac{10!}{6!\,4!} = \binom{10}{6} = 210.$$ ∎

Example 1.2-13

In an orchid show, seven orchids are to be placed along one side of the greenhouse. There are four lavender orchids and three white orchids. Considering only the color of the orchids, we see that the number of lineups of the orchids is

$$\binom{7}{4} = \frac{7!}{4!\,3!} = 35.$$

If the colors of the seven orchids are white, lavender, yellow, mauve, crimson, orange, and pink, the number of different displays is $7! = 5040$. ∎

The foregoing results can be extended. Suppose that in a set of n objects, n_1 are similar, n_2 are similar, $\ldots$, n_s are similar, where $n_1 + n_2 + \cdots + n_s = n$. Then the number of distinguishable permutations of the n objects is (see Exercise 1.2-15)

$$\binom{n}{n_1, n_2, \ldots, n_s} = \frac{n!}{n_1!\,n_2!\,\cdots\,n_s!}. \qquad (1.2\text{-}2)$$

Example 1.2-14

Among nine orchids for a line of orchids along one wall, three are white, four lavender, and two yellow. The number of different color displays is then

$$\binom{9}{3, 4, 2} = \frac{9!}{3!\,4!\,2!} = 1260.$$ ∎

The argument used in determining the binomial coefficients in the expansion of $(a+b)^n$ can be extended to find the expansion of $(a_1 + a_2 + \cdots + a_s)^n$. The coefficient of $a_1^{n_1} a_2^{n_2} \cdots a_s^{n_s}$, where $n_1 + n_2 + \cdots + n_s = n$, is

$$\binom{n}{n_1, n_2, \ldots, n_s} = \frac{n!}{n_1!\,n_2!\,\cdots\,n_s!}.$$

This is sometimes called a **multinomial coefficient**.

When r objects are selected out of n objects, we are often interested in the number of possible outcomes. We have seen that for ordered samples, there are

n^r possible outcomes when sampling with replacement and $_nP_r$ outcomes when sampling without replacement. For unordered samples, there are $_nC_r$ outcomes when sampling without replacement. Each of the preceding outcomes is equally likely, provided that the experiment is performed in a fair manner.

REMARK   Although not needed as often in the study of probability, it is interesting to count the number of possible samples of size r that can be selected out of n objects when the order is irrelevant and when sampling with replacement. For example, if a six-sided die is rolled 10 times (or 10 six-sided dice are rolled once), how many possible unordered outcomes are there? To count the number of possible outcomes, think of listing r 0's for the r objects that are to be selected. Then insert $(n-1)$ |'s to partition the r objects into n sets, the first set giving objects of the first kind, and so on. So if $n = 6$ and $r = 10$ in the die illustration, a possible outcome is

$$0\,0\,|\,|\,0\,0\,0\,|\,0\,|\,0\,0\,0\,|\,0,$$

which says there are two 1's, zero 2's, three 3's, one 4, three 5's, and one 6. In general, each outcome is a permutation of r 0's and $(n-1)$ |'s. Each distinguishable permutation is equivalent to an unordered sample. The number of distinguishable permutations, and hence the number of unordered samples of size r that can be selected out of n objects when sampling with replacement, is

$$_{n-1+r}C_r = \frac{(n-1+r)!}{r!\,(n-1)!}.$$
∎

Exercises

1.2-1. A boy found a bicycle lock for which the combination was unknown. The correct combination is a four-digit number, $d_1 d_2 d_3 d_4$, where d_i, $i = 1, 2, 3, 4$, is selected from 1, 2, 3, 4, 5, 6, 7, and 8. How many different lock combinations are possible with such a lock?

1.2-2. In designing an experiment, the researcher can often choose many different levels of the various factors in order to try to find the best combination at which to operate. As an illustration, suppose the researcher is studying a certain chemical reaction and can choose five levels of temperature, eight different pressures, and four different catalysts.

(a) To consider all possible combinations, how many experiments would need to be conducted?

(b) Suppose that during an experiment, each factor is restricted to four levels. With the three factors noted, how many experiments would need to be run to cover all possible combinations with each of the three factors at four levels?

1.2-3. How many different license plates are possible if a state uses

(a) Two letters followed by a four-digit integer (leading zeros are permissible and the letters and digits can be repeated)?

(b) Three letters followed by a three-digit integer? (In practice, it is possible that certain "spellings" are ruled out.)

1.2-4. The "eating club" is hosting a make-your-own sundae at which the following are provided:

Ice Cream Flavors	Toppings
Chocolate	Caramel
Cookies 'n' cream	Hot fudge
Strawberry	Marshmallow
Vanilla	M&M's
	Nuts
	Strawberries

(a) How many sundaes are possible using one flavor of ice cream and three different toppings?

(b) How many sundaes are possible using one flavor of ice cream and from zero to six toppings?

(c) How many different combinations of flavors of three scoops of ice cream are possible if it is permissible to make all three scoops the same flavor?

1.2-5. How many four-letter code words are possible using the letters in IOWA if

(a) The letters may not be repeated?

(b) The letters may be repeated?

1.2-6. Suppose that Novak Djokovic and Roger Federer are playing a tennis match in which the first player to win three sets wins the match. Using **D** and **F** for the winning player of a set, in how many ways could this tennis match end?

1.2-7. In a state lottery, four digits are drawn at random one at a time with replacement from 0 to 9. Suppose that you win if any permutation of your selected integers is drawn. Give the probability of winning if you select

(a) 6, 7, 8, 9.

(b) 6, 7, 8, 8.

(c) 7, 7, 8, 8.

(d) 7, 8, 8, 8.

1.2-8. How many different varieties of pizza can be made if you have the following choice: extra-small, small, medium, or large size; thin 'n' crispy, hand-tossed, or pan crust; and 16 toppings (cheese is automatic), from which you may select from 0 to 16?

1.2-9. The World Series in baseball continues until either the American League team or the National League team wins four games. How many different orders are possible (e.g., *ANNAAA* means the American League team wins in six games) if the series goes

(a) Four games?

(b) Five games?

(c) Six games?

(d) Seven games?

1.2-10. Pascal's triangle gives a method for calculating the binomial coefficients; it begins as follows:

$$
\begin{array}{ccccccccccc}
 & & & & & 1 & & & & & \\
 & & & & 1 & & 1 & & & & \\
 & & & 1 & & 2 & & 1 & & & \\
 & & 1 & & 3 & & 3 & & 1 & & \\
 & 1 & & 4 & & 6 & & 4 & & 1 & \\
1 & & 5 & & 10 & & 10 & & 5 & & 1 \\
 & \vdots & & \vdots & & \vdots & & \vdots & & \vdots &
\end{array}
$$

The nth row of this triangle gives the coefficients for $(a + b)^{n-1}$. To find an entry in the table other than a 1 on the boundary, add the two nearest numbers in the row directly above. The equation

$$\binom{n}{r} = \binom{n-1}{r} + \binom{n-1}{r-1},$$

called **Pascal's equation**, explains why Pascal's triangle works. Prove that this equation is correct.

1.2-11. Three students (S) and six faculty members (F) are on a panel discussing a new college policy.

(a) In how many different ways can the nine participants be lined up at a table in the front of the auditorium?

(b) How many lineups are possible, considering only the labels S and F?

(c) For each of the nine participants, you are to decide whether the participant did a good job or a poor job stating his or her opinion of the new policy; that is, give each of the nine participants a grade of G or P. How many different "scorecards" are possible?

1.2-12. Prove

$$\sum_{r=0}^{n}(-1)^r\binom{n}{r} = 0 \qquad \text{and} \qquad \sum_{r=0}^{n}\binom{n}{r} = 2^n.$$

HINT: Consider $(1 - 1)^n$ and $(1 + 1)^n$, or use Pascal's equation and proof by induction.

1.2-13. A bridge hand is found by taking 13 cards at random and without replacement from a deck of 52 playing cards. Find the probability of drawing each of the following hands.

(a) One in which there are 5 spades, 4 hearts, 3 diamonds, and 1 club.

(b) One in which there are 5 spades, 4 hearts, 2 diamonds, and 2 clubs.

(c) One in which there are 5 spades, 4 hearts, 1 diamond, and 3 clubs.

(d) Suppose you are dealt 5 cards of one suit, 4 cards of another. Would the probability of having the other suits split 3 and 1 be greater than the probability of having them split 2 and 2?

1.2-14. A bag of 36 dum-dum pops (suckers) contains up to 10 flavors. That is, there are from 0 to 36 suckers of each of 10 flavors in the bag. How many different flavor combinations are possible?

1.2-15. Prove Equation 1.2-2. HINT: First select n_1 positions in $\binom{n}{n_1}$ ways. Then select n_2 from the remaining $n - n_1$ positions in $\binom{n-n_1}{n_2}$ ways, and so on. Finally, use the multiplication rule.

1.2-16. A box of candy hearts contains 52 hearts, of which 19 are white, 10 are tan, 7 are pink, 3 are purple, 5 are yellow, 2 are orange, and 6 are green. If you select nine pieces of candy randomly from the box, without replacement, give the probability that

(a) Three of the hearts are white.

(b) Three are white, two are tan, one is pink, one is yellow, and two are green.

1.2-17. A poker hand is defined as drawing 5 cards at random without replacement from a deck of 52 playing cards. Find the probability of each of the following poker hands:

(a) Four of a kind (four cards of equal face value and one card of a different value).

(b) Full house (one pair and one triple of cards with equal face value).

(c) Three of a kind (three equal face values plus two cards of different values).

(d) Two pairs (two pairs of equal face value plus one card of a different value).

(e) One pair (one pair of equal face value plus three cards of different values).

1.3 CONDITIONAL PROBABILITY

We introduce the idea of conditional probability by means of an example.

Example 1.3-1   Suppose that we are given 20 tulip bulbs that are similar in appearance and told that 8 will bloom early, 12 will bloom late, 13 will be red, and 7 will be yellow, in accordance with the various combinations listed in Table 1.3-1. If one bulb is selected at random, the probability that it will produce a red tulip (R) is given by $P(R) = 13/20$, under the assumption that each bulb is "equally likely." Suppose, however, that close examination of the bulb will reveal whether it will bloom early (E) or late (L). If we consider an outcome only if it results in a tulip bulb that will bloom early, only eight outcomes in the sample space are now of interest. Thus, under this limitation, it is natural to assign the probability 5/8 to R; that is, $P(R \mid E) = 5/8$, where $P(R \mid E)$ is read as the probability of R given that E has occurred. Note that

$$P(R \mid E) = \frac{5}{8} = \frac{N(R \cap E)}{N(E)} = \frac{N(R \cap E)/20}{N(E)/20} = \frac{P(R \cap E)}{P(E)},$$

where $N(R \cap E)$ and $N(E)$ are the numbers of outcomes in events $R \cap E$ and E, respectively.   ■

This example illustrates a number of common situations. That is, in some random experiments, we are interested only in those outcomes which are elements of a subset B of the sample space S. This means, for our purposes, that the sample space is effectively the subset B. We are now confronted with the problem of defining a probability set function with B as the "new" sample space. That is, for a given event

Table 1.3-1 Tulip combinations

	Early (E)	Late (L)	Totals
Red (R)	5	8	13
Yellow (Y)	3	4	7
Totals	8	12	20

A, we want to define $P(A \mid B)$, the probability of A, considering only those outcomes of the random experiment that are elements of B. The previous example gives us the clue to that definition. That is, for experiments in which each outcome is equally likely, it makes sense to define $P(A \mid B)$ by

$$P(A \mid B) = \frac{N(A \cap B)}{N(B)},$$

where $N(A \cap B)$ and $N(B)$ are the numbers of outcomes in $A \cap B$ and B, respectively. If we then divide the numerator and the denominator of this fraction by $N(S)$, the number of outcomes in the sample space, we have

$$P(A \mid B) = \frac{N(A \cap B)/N(S)}{N(B)/N(S)} = \frac{P(A \cap B)}{P(B)}.$$

We are thus led to the following definition.

Definition 1.3-1
The **conditional probability** of an event A, given that event B has occurred, is defined by

$$P(A \mid B) = \frac{P(A \cap B)}{P(B)},$$

provided that $P(B) > 0$.

A formal use of the definition is given in the next example.

Example 1.3-2
If $P(A) = 0.4$, $P(B) = 0.5$, and $P(A \cap B) = 0.3$, then $P(A \mid B) = 0.3/0.5 = 0.6$; $P(B \mid A) = P(A \cap B)/P(A) = 0.3/0.4 = 0.75$. ∎

We can think of "given B" as specifying the new sample space for which, to determine $P(A \mid B)$, we now want to calculate the probability of that part of A that is contained in B. The next two examples illustrate this idea.

Example 1.3-3
Suppose that $P(A) = 0.7$, $P(B) = 0.3$, and $P(A \cap B) = 0.2$. These probabilities are listed on the Venn diagram in Figure 1.3-1. Given that the outcome of the experiment belongs to B, what then is the probability of A? We are effectively restricting the sample space to B; of the probability $P(B) = 0.3$, 0.2 corresponds to $P(A \cap B)$ and hence to A. That is, $0.2/0.3 = 2/3$ of the probability of B corresponds to A. Of course, by the formal definition, we also obtain

$$P(A \mid B) = \frac{P(A \cap B)}{P(B)} = \frac{0.2}{0.3} = \frac{2}{3}.$$ ∎

Example 1.3-4
A pair of fair four-sided dice is rolled and the sum is determined. Let A be the event that a sum of 3 is rolled, and let B be the event that a sum of 3 or a sum of 5 is rolled. In a sequence of rolls, the probability that a sum of 3 is rolled before a sum of 5 is

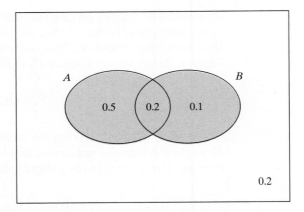

Figure 1.3-1 Conditional probability

rolled can be thought of as the conditional probability of a sum of 3 given that a sum of 3 or 5 has occurred; that is, the conditional probability of A given B is

$$P(A \mid B) = \frac{P(A \cap B)}{P(B)} = \frac{P(A)}{P(B)} = \frac{2/16}{6/16} = \frac{2}{6}.$$

Note that for this example, the only outcomes of interest are those having a sum of 3 or a sum of 5, and of these six equally likely outcomes, two have a sum of 3. (See Figure 1.3-2 and Exercise 1.3-13.) ■

It is interesting to note that conditional probability satisfies the axioms for a probability function, namely, with $P(B) > 0$,

(a) $P(A \mid B) \geq 0$;
(b) $P(B \mid B) = 1$;

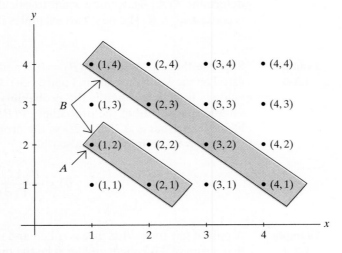

Figure 1.3-2 Dice example

(c) if $A_1, A_2, A_3, \ldots$ are mutually exclusive events, then

$$P(A_1 \cup A_2 \cup \cdots \cup A_k \mid B) = P(A_1 \mid B) + P(A_2 \mid B) + \cdots + P(A_k \mid B),$$

for each positive integer k, and

$$P(A_1 \cup A_2 \cup \cdots \mid B) = P(A_1 \mid B) + P(A_2 \mid B) + \cdots,$$

for an infinite, but countable, number of events.

Properties (a) and (b) are evident because

$$P(A \mid B) = \frac{P(A \cap B)}{P(B)} \geq 0,$$

since $P(A \cap B) \geq 0$, $P(B) > 0$, and

$$P(B \mid B) = \frac{P(B \cap B)}{P(B)} = \frac{P(B)}{P(B)} = 1.$$

Property (c) holds because, for the second part of (c),

$$P(A_1 \cup A_2 \cup \cdots \mid B) = \frac{P[(A_1 \cup A_2 \cup \cdots) \cap B]}{P(B)}$$

$$= \frac{P[(A_1 \cap B) \cup (A_2 \cap B) \cup \cdots]}{P(B)}.$$

But $(A_1 \cap B), (A_2 \cap B), \ldots$ are also mutually exclusive events; so

$$P(A_1 \cup A_2 \cup \cdots \mid B) = \frac{P(A_1 \cap B) + P(A_2 \cap B) + \cdots}{P(B)}$$

$$= \frac{P(A_1 \cap B)}{P(B)} + \frac{P(A_2 \cap B)}{P(B)} + \cdots$$

$$= P(A_1 \mid B) + P(A_2 \mid B) + \cdots.$$

The first part of property (c) is proved in a similar manner.

Note that, as a consequence, results analogous to those given by Theorems 1.1-1 through 1.1-6 hold for conditional probabilities. For example,

$$P(A' \mid B) = 1 - P(A \mid B)$$

is true.

Many times, the conditional probability of an event is clear because of the nature of an experiment. The next example illustrates this.

Example 1.3-5 At a county fair carnival game there are 25 balloons on a board, of which 10 balloons are yellow, 8 are red, and 7 are green. A player throws darts at the balloons to win a prize and randomly hits one of them. Given that the first balloon hit is yellow, what is the probability that the next balloon hit is also yellow? Of the 24 remaining balloons, 9 are yellow, so a natural value to assign to this conditional probability is 9/24. ∎

In Example 1.3-5, let A be the event that the first balloon hit is yellow, and let B be the event that the second balloon hit is yellow. Suppose that we are interested in the probability that both balloons hit are yellow. That is, we are interested in finding $P(A \cap B)$. We noted in that example that

$$P(B \mid A) = \frac{P(A \cap B)}{P(A)} = \frac{9}{24}.$$

Thus, multiplying through by $P(A)$, we have

$$P(A \cap B) = P(A)P(B \mid A) = P(A)\left(\frac{9}{24}\right), \qquad \text{(1.3-1)}$$

or

$$P(A \cap B) = \left(\frac{10}{25}\right)\left(\frac{9}{24}\right).$$

That is, Equation 1.3-1 gives us a general rule for the probability of the intersection of two events once we know the conditional probability $P(B \mid A)$.

> **Definition 1.3-2**
> The probability that two events, A and B, both occur is given by the **multiplication rule**,
>
> $$P(A \cap B) = P(A)P(B \mid A),$$
>
> provided $P(A) > 0$ or by
>
> $$P(A \cap B) = P(B)P(A \mid B)$$
>
> provided $P(B) > 0$.

Sometimes, after considering the nature of the random experiment, one can make reasonable assumptions so that it is easier to assign $P(B)$ and $P(A \mid B)$ rather than $P(A \cap B)$. Then $P(A \cap B)$ can be computed with these assignments. This approach will be illustrated in Examples 1.3-6 and 1.3-7.

Example 1.3-6 A bowl contains seven blue chips and three red chips. Two chips are to be drawn successively at random and without replacement. We want to compute the probability that the first draw results in a red chip (A) and the second draw results in a blue chip (B). It is reasonable to assign the following probabilities:

$$P(A) = \frac{3}{10} \text{ and } P(B \mid A) = \frac{7}{9}.$$

The probability of obtaining red on the first draw and blue on the second draw is

$$P(A \cap B) = \frac{3}{10} \cdot \frac{7}{9} = \frac{7}{30}. \qquad \blacksquare$$

Note that in many instances it is possible to compute a probability by two seemingly different methods. For instance, consider Example 1.3-6, but find the probability of drawing a red chip on each of the two draws. Following that example, it is

$$\frac{3}{10} \cdot \frac{2}{9} = \frac{1}{15}.$$

However, we can also find this probability by using combinations as follows:

$$\frac{\binom{3}{2}\binom{7}{0}}{\binom{10}{2}} = \frac{\frac{(3)(2)}{(1)(2)}}{\frac{(10)(9)}{(1)(2)}} = \frac{1}{15}.$$

Thus, we obtain the same answer, as we should, provided that our reasoning is consistent with the underlying assumptions.

Example 1.3-7

From an ordinary deck of playing cards, cards are to be drawn successively at random and without replacement. The probability that the third spade appears on the sixth draw is computed as follows: Let A be the event of two spades in the first five cards drawn, and let B be the event of a spade on the sixth draw. Thus, the probability that we wish to compute is $P(A \cap B)$. It is reasonable to take

$$P(A) = \frac{\binom{13}{2}\binom{39}{3}}{\binom{52}{5}} = 0.274 \qquad \text{and} \qquad P(B \mid A) = \frac{11}{47} = 0.234.$$

The desired probability, $P(A \cap B)$, is the product of those numbers:

$$P(A \cap B) = (0.274)(0.234) = 0.064.$$

Example 1.3-8

Continuing with Example 1.3-4, in which a pair of four-sided dice is rolled, the probability of rolling a sum of 3 on the first roll and then, continuing the sequence of rolls, rolling a sum of 3 before rolling a sum of 5 is

$$\frac{2}{16} \cdot \frac{2}{6} = \frac{4}{96} = \frac{1}{24}.$$

The multiplication rule can be extended to three or more events. In the case of three events, using the multiplication rule for two events, we have

$$P(A \cap B \cap C) = P[(A \cap B) \cap C]$$
$$= P(A \cap B)P(C \mid A \cap B).$$

But

$$P(A \cap B) = P(A)P(B \mid A).$$

Hence,

$$P(A \cap B \cap C) = P(A)P(B \mid A)P(C \mid A \cap B).$$

This type of argument can be used to extend the multiplication rule to more than three events, and the general formula for k events can be officially proved by mathematical induction.

Example 1.3-9

Four cards are to be dealt successively at random and without replacement from an ordinary deck of playing cards. The probability of receiving, in order, a spade, a heart, a diamond, and a club is

$$\frac{13}{52} \cdot \frac{13}{51} \cdot \frac{13}{50} \cdot \frac{13}{49},$$

a result that follows from the extension of the multiplication rule and reasonable assignments to the probabilities involved. ∎

Example 1.3-10 A grade school boy has five blue and four white marbles in his left pocket and four blue and five white marbles in his right pocket. If he transfers one marble at random from his left to his right pocket, what is the probability of his then drawing a blue marble from his right pocket? For notation, let BL, BR, and WL denote drawing blue from left pocket, blue from right pocket, and white from left pocket, respectively. Then

$$P(BR) = P(BL \cap BR) + P(WL \cap BR)$$
$$= P(BL)P(BR \mid BL) + P(WL)P(BR \mid WL)$$
$$= \frac{5}{9} \cdot \frac{5}{10} + \frac{4}{9} \cdot \frac{4}{10} = \frac{41}{90}$$

is the desired probability. ∎

Example 1.3-11 An insurance company sells several types of insurance policies, including auto policies and homeowner policies. Let A_1 be those people with an auto policy only, A_2 those people with a homeowner policy only, and A_3 those people with both an auto and homeowner policy (but no other policies). For a person randomly selected from the company's policy holders, suppose that $P(A_1) = 0.3$, $P(A_2) = 0.2$, and $P(A_3) = 0.2$. Further, let B be the event that the person will renew at least one of these policies. Say from past experience that we assign the conditional probabilities $P(B \mid A_1) = 0.6$, $P(B \mid A_2) = 0.7$, and $P(B \mid A_3) = 0.8$. Given that the person selected at random has an auto or homeowner policy, what is the conditional probability that the person will renew at least one of those policies? The desired probability is

$$P(B \mid A_1 \cup A_2 \cup A_3) = \frac{P(A_1 \cap B) + P(A_2 \cap B) + P(A_3 \cap B)}{P(A_1) + P(A_2) + P(A_3)}$$
$$= \frac{(0.3)(0.6) + (0.2)(0.7) + (0.2)(0.8)}{0.3 + 0.2 + 0.2}$$
$$= \frac{0.48}{0.70} = 0.686.$$ ∎

Example 1.3-12 An electronic device has two components, C_1 and C_2, but it will operate if at least one of the components is working properly. Each component has its own switch, and both switches must be turned on, one after the other, for the device to begin operating. Thus, the device can begin to operate by either switching C_1 on first and then C_2, or vice versa. If C_1 is switched on first, it fails immediately with probability 0.01, whereas if C_2 is switched on first, it fails immediately with probability 0.02. Furthermore, if C_1 is switched on first and fails, the probability that C_2 fails immediately when it is switched on is 0.025, due to added strain. Similarly, if C_2 is switched on first and fails, the probability that C_1 fails immediately when it is switched on is 0.015. Thus, the probability that the device fails to operate after switching on C_1 first and then C_2 is

$$P(C_1 \text{ fails})P(C_2 \text{ fails} \mid C_1 \text{ fails}) = (0.01)(0.025) = 0.00025,$$

while the probability that the device fails to operate after switching on C_2 first and then C_1 is

$$P(C_2 \text{ fails})P(C_1 \text{ fails} \mid C_2 \text{ fails}) = (0.02)(0.015) = 0.0003.$$

The device therefore is more likely to operate properly if C_1 is switched on first. ■

Exercises

1.3-1. A common test for AIDS is called the ELISA (enzyme-linked immunosorbent assay) test. Among 1 million people who are given the ELISA test, we can expect results similar to those given in the following table:

	B_1: Carry AIDS Virus	B_2: Do Not Carry Aids Virus	Totals
A_1: Test Positive	4,885	73,630	78,515
A_2: Test Negative	115	921,370	921,485
Totals	5,000	995,000	1,000,000

If one of these 1 million people is selected randomly, find the following probabilities: **(a)** $P(B_1)$, **(b)** $P(A_1)$, **(c)** $P(A_1 \mid B_2)$, **(d)** $P(B_1 \mid A_1)$. **(e)** In words, what do parts (c) and (d) say?

1.3-2. The following table classifies 1456 people by their gender and by whether or not they favor a gun law.

	Male (S_1)	Female (S_2)	Totals
Favor (A_1)	392	649	1041
Oppose (A_2)	241	174	415
Totals	633	823	1456

Compute the following probabilities if one of these 1456 persons is selected randomly: **(a)** $P(A_1)$, **(b)** $P(A_1 \mid S_1)$, **(c)** $P(A_1 \mid S_2)$. **(d)** Interpret your answers to parts (b) and (c).

1.3-3. Let A_1 and A_2 be the events that a person is left-eye dominant or right-eye dominant respectively. When a person folds his or her hands, let B_1 and B_2 be the events that the right thumb and the left thumb, respectively, are on top. A survey in one statistics class yielded the following table:

	B_1	B_2	Totals
A_1	10	15	25
A_2	25	10	35
Totals	35	25	60

If a student is selected randomly, find the following probabilities: **(a)** $P(A_2 \cap B_1)$, **(b)** $P(A_2 \cup B_1)$, **(c)** $P(A_2 \mid B_1)$, **(d)** $P(B_2 \mid A_1)$. **(e)** If the students had their hands folded and you hoped to select a left-eye-dominant student, would you select a "right thumb on top" or a "left thumb on top" student? Why?

1.3-4. Three cards are drawn successively and without replacement from an ordinary deck of playing cards. Compute the probability of drawing

(a) Three spades.

(b) A spade on the first draw, a heart on the second draw, and a diamond on the third draw.

(c) A spade on the first draw, a heart on the second draw, and an ace on the third draw.

HINT: In part (c), note that a spade or a heart can be drawn by getting the ace or one of the other 12 cards in that suit.

1.3-5. Suppose that the alleles for eye color for a certain male fruit fly are (R, W) and the alleles for eye color for the mating female fruit fly are (R, W), where R and W represent red and white, respectively. Their offspring receive one allele for eye color from each parent.

(a) Define the sample space of the alleles for eye color for the offspring.

(b) Assume that each of the four possible outcomes has equal probability. If an offspring ends up with either two white alleles or one red and one white allele for eye color, its eyes will look white. Given that an offspring's eyes look white, what is the conditional probability that it has two white alleles for eye color?

1.3-6. A researcher finds that, of 982 men who died in 2002, 221 died from some heart disease. Also, of the 982 men, 334 had at least one parent who had some heart disease. Of the latter 334 men, 111 died from some heart disease. A man is selected from the group of 982. Given that neither of his parents had some heart disease, find the conditional probability that this man died of some heart disease.

1.3-7. An urn contains 15 balls numbered 1 through 15. A ball is drawn from the urn and kept aside. Another ball is then drawn from the same urn. What is the probability that the first ball bears an even number and the second ball bears an odd number?

1.3-8. An urn contains 17 balls marked LOSE and 3 balls marked WIN. You and an opponent take turns selecting a single ball at random from the urn without replacement.

The person who selects the third WIN ball wins the game. It does not matter who selected the first two WIN balls.

(a) If you draw first, find the probability that you win the game on your second draw.

(b) If you draw first, find the probability that your opponent wins the game on his second draw.

(c) If you draw first, what is the probability that you win? HINT: You could win on your second, third, fourth, ..., or tenth draw, but not on your first.

(d) Would you prefer to draw first or second? Why?

1.3-9. An urn contains four balls numbered 1 through 4. The balls are selected one at a time without replacement. A match occurs if the ball numbered m is the mth ball selected. Let the event A_i denote a match on the ith draw, $i = 1, 2, 3, 4$.

(a) Show that $P(A_i) = \dfrac{3!}{4!}$ for each i.

(b) Show that $P(A_i \cap A_j) = \dfrac{2!}{4!}, i \neq j$.

(c) Show that $P(A_i \cap A_j \cap A_k) = \dfrac{1!}{4!}, i \neq j, i \neq k, j \neq k$.

(d) Show that the probability of at least one match is

$$P(A_1 \cup A_2 \cup A_3 \cup A_4) = 1 - \frac{1}{2!} + \frac{1}{3!} - \frac{1}{4!}.$$

(e) Extend this exercise so that there are n balls in the urn. Show that the probability of at least one match is

$$P(A_1 \cup A_2 \cup \cdots \cup A_n)$$

$$= 1 - \frac{1}{2!} + \frac{1}{3!} - \frac{1}{4!} + \cdots + \frac{(-1)^{n+1}}{n!}$$

$$= 1 - \left(1 - \frac{1}{1!} + \frac{1}{2!} - \frac{1}{3!} + \cdots + \frac{(-1)^n}{n!}\right).$$

(f) What is the limit of this probability as n increases without bound?

1.3-10. A single card is drawn at random from each of six well-shuffled decks of playing cards. Let A be the event that all six cards drawn are different.

(a) Find $P(A)$.

(b) Find the probability that at least two of the drawn cards match.

1.3-11. Consider the birthdays of the students in a class of size r. Assume that the year consists of 365 days.

(a) How many different ordered samples of birthdays are possible (r in sample) allowing repetitions (with replacement)?

(b) The same as part (a), except requiring that all the students have different birthdays (without replacement)?

(c) If we can assume that each ordered outcome in part (a) has the same probability, what is the probability that at least two students have the same birthday?

(d) For what value of r is the probability in part (c) about equal to 1/2? Is this number surprisingly small? HINT: Use a calculator or computer to find r.

1.3-12. You are a member of a class of 30 students. A bowl contains 30 chips: 2 blue and 28 red. Each student is to take 1 chip from the bowl without replacement. The student who draws the blue chip is guaranteed an A for the course.

(a) If you have a choice of drawing first, tenth, twentieth, or last, which position would you choose? Justify your choice on the basis of probability.

(b) Suppose the bowl contains 4 blue and 26 red chips. What position would you now choose?

1.3-13. In the gambling game "craps," a pair of dice is rolled and the outcome of the experiment is the sum of the points on the up sides of the six-sided dice. The bettor wins on the first roll if the sum is 7 or 11. The bettor loses on the first roll if the sum is 2, 3, or 12. If the sum is 4, 5, 6, 8, 9, or 10, that number is called the bettor's "point." Once the point is established, the rule is as follows: If the bettor rolls a 7 before the point, the bettor loses; but if the point is rolled before a 7, the bettor wins.

(a) List the 36 outcomes in the sample space for the roll of a pair of dice. Assume that each of them has a probability of 1/36.

(b) Find the probability that the bettor wins on the first roll. That is, find the probability of rolling a 7 or 11, $P(7 \text{ or } 11)$.

(c) Given that 8 is the outcome on the first roll, find the probability that the bettor now rolls the point 8 before rolling a 7 and thus wins. Note that at this stage in the game the only outcomes of interest are 7 and 8. Thus find $P(8 \mid 7 \text{ or } 8)$.

(d) The probability that a bettor rolls an 8 on the first roll and then wins is given by $P(8)P(8 \mid 7 \text{ or } 8)$. Show that this probability is $(5/36)(5/11)$.

(e) Show that the total probability that a bettor wins in the game of craps is 0.49293. HINT: Note that the bettor can win in one of several mutually exclusive ways: by rolling a 7 or an 11 on the first roll or by establishing one of the points 4, 5, 6, 8, 9, or 10 on the first roll and then obtaining that point on successive rolls before a 7 comes up.

1.3-14. Paper is often tested for "burst strength" and "tear strength." Say we classify these strengths as low, middle, and high. Then, after examining 100 pieces of paper, we find the following:

| | Burst Strength | | |
Tear Strength	A_1 (low)	A_2 (middle)	A_3 (high)
B_1 (low)	7	11	13
B_2 (middle)	11	21	9
B_3 (high)	12	9	7

If we select one of the pieces at random, what are the probabilities that it has the following characteristics:

(a) A_1?

(b) $A_3 \cap B_2$?

(c) $A_2 \cup B_3$?

(d) A_1, given that it is B_2?

(e) B_1, given that it is A_3?

1.3-15. An urn contains eight red and seven blue balls. A second urn contains an unknown number of red balls and nine blue balls. A ball is drawn from each urn at random, and the probability of getting two balls of the same color is 151/300. How many red balls are in the second urn?

1.3-16. College degrees are held by 60% of the employees of a company, and 10% of the degree holders work in sales. Among the employees who do not have college degrees, 80% work in sales. What is the probability that an employee selected at random is a member of a sales team?

1.4 INDEPENDENT EVENTS

For certain pairs of events, the occurrence of one of them may or may not change the probability of the occurrence of the other. In the latter case, they are said to be **independent events**. However, before giving the formal definition of independence, let us consider an example.

Example 1.4-1

Flip a fair coin twice and observe the sequence of heads and tails. The sample space is then

$$S = \{HH, HT, TH, TT\}.$$

It is reasonable to assign a probability of 1/4 to each of these four outcomes. Let

$$A = \{\text{heads on the first flip}\} = \{HH, HT\},$$
$$B = \{\text{tails on the second flip}\} = \{HT, TT\},$$
$$C = \{\text{tails on both flips}\} = \{TT\}.$$

Then $P(B) = 2/4 = 1/2$. Now, on the one hand, if we are given that C has occurred, then $P(B \mid C) = 1$, because $C \subset B$. That is, the knowledge of the occurrence of C has changed the probability of B. On the other hand, if we are given that A has occurred, then

$$P(B \mid A) = \frac{P(A \cap B)}{P(A)} = \frac{1/4}{2/4} = \frac{1}{2} = P(B).$$

So the occurrence of A has not changed the probability of B. Hence, the probability of B does not depend upon knowledge about event A, so we say that A and B are independent events. That is, events A and B are independent if the occurrence of one of them does not affect the probability of the occurrence of the other. A more mathematical way of saying this is

$$P(B \mid A) = P(B) \qquad \text{or} \qquad P(A \mid B) = P(A),$$

provided that $P(A) > 0$ or, in the latter case, $P(B) > 0$. With the first of these equalities and the multiplication rule (Definition 1.3-2), we have

$$P(A \cap B) = P(A)P(B \mid A) = P(A)P(B).$$

The second of these equalities, namely, $P(A \mid B) = P(A)$, gives us the same result:

$$P(A \cap B) = P(B)P(A \mid B) = P(B)P(A). \qquad \blacksquare$$

This example motivates the following definition of independent events.

Definition 1.4-1
Events A and B are **independent** if and only if $P(A \cap B) = P(A)P(B)$. Otherwise, A and B are called **dependent** events.

Events that are independent are sometimes called **statistically independent**, **stochastically independent**, or **independent in a probabilistic sense**, but in most instances we use *independent* without a modifier if there is no possibility of mis-understanding. It is interesting to note that the definition always holds if $P(A) = 0$ or $P(B) = 0$, because then $P(A \cap B) = 0$, since $(A \cap B) \subset A$ and $(A \cap B) \subset B$. Thus, the left-hand and right-hand members of $P(A \cap B) = P(A)P(B)$ are both equal to zero and thus are equal to each other.

Example 1.4-2 A red die and a white die are rolled. Let event $A = \{4$ on the red die$\}$ and event $B = \{$sum of dice is odd$\}$. Of the 36 equally likely outcomes, 6 are favorable to A, 18 are favorable to B, and 3 are favorable to $A \cap B$. Then, assuming the dice are fair,

$$P(A)P(B) = \frac{6}{36} \cdot \frac{18}{36} = \frac{3}{36} = P(A \cap B).$$

Hence, A and B are independent by Definition 1.4-1. $\qquad \blacksquare$

Example 1.4-3 A red die and a white die are rolled. Let event $C = \{5$ on red die$\}$ and event $D = \{$sum of dice is 11$\}$. Of the 36 equally likely outcomes, 6 are favorable to C, 2 are favorable to D, and 1 is favorable to $C \cap D$. Then, assuming the dice are fair,

$$P(C)P(D) = \frac{6}{36} \cdot \frac{2}{36} = \frac{1}{108} \neq \frac{1}{36} = P(C \cap D).$$

Hence, C and D are dependent events by Definition 1.4-1. $\qquad \blacksquare$

Theorem 1.4-1 If A and B are independent events, then the following pairs of events are also independent:

(a) A and B';

(b) A' and B;

(c) A' and B'.

Proof We know that conditional probability satisfies the axioms for a probability function. Hence, if $P(A) > 0$, then $P(B' \mid A) = 1 - P(B \mid A)$. Thus,

$$P(A \cap B') = P(A)P(B'\,|\,A) = P(A)[1 - P(B\,|\,A)]$$
$$= P(A)[1 - P(B)]$$
$$= P(A)P(B'),$$

since $P(B\,|\,A) = P(B)$ by hypothesis. Consequently, A and B' are independent events. The proofs of parts (b) and (c) are left as exercises.   □

Before extending the definition of independent events to more than two events, we present the following example.

Example 1.4-4

An urn contains four balls numbered 1, 2, 3, and 4. One ball is to be drawn at random from the urn. Let the events A, B, and C be defined by $A = \{1, 2\}$, $B = \{1, 3\}$, and $C = \{1, 4\}$. Then $P(A) = P(B) = P(C) = 1/2$. Furthermore,

$$P(A \cap B) = \frac{1}{4} = P(A)P(B),$$

$$P(A \cap C) = \frac{1}{4} = P(A)P(C),$$

$$P(B \cap C) = \frac{1}{4} = P(B)P(C),$$

which implies that A, B, and C are independent in pairs (called **pairwise independence**). However, since $A \cap B \cap C = \{1\}$, we have

$$P(A \cap B \cap C) = \frac{1}{4} \neq \frac{1}{8} = P(A)P(B)P(C).$$

That is, something seems to be lacking for the complete independence of A, B, and C.   ▪

This example illustrates the reason for the second condition in the next definition.

Definition 1.4-2
Events A, B, and C are **mutually independent** if and only if the following two conditions hold:

(a) A, B, and C are pairwise independent; that is,

$$P(A \cap B) = P(A)P(B), \qquad P(A \cap C) = P(A)P(C),$$

and

$$P(B \cap C) = P(B)P(C).$$

(b) $P(A \cap B \cap C) = P(A)P(B)P(C).$

Definition 1.4-2 can be extended to the mutual independence of four or more events. In such an extension, each pair, triple, quartet, and so on, must satisfy this type of multiplication rule. If there is no possibility of misunderstanding, *independent* is often used without the modifier *mutually* when several events are considered.

Example 1.4-5

A rocket has a built-in redundant system. In this system, if component K_1 fails, it is bypassed and component K_2 is used. If component K_2 fails, it is bypassed and component K_3 is used. (An example of a system with these kinds of components is three computer systems.) Suppose that the probability of failure of any one component is 0.15, and assume that the failures of these components are mutually independent events. Let A_i denote the event that component K_i fails for $i = 1, 2, 3$. Because the system fails if K_1 fails and K_2 fails and K_3 fails, the probability that the system does not fail is given by

$$
\begin{aligned}
P[(A_1 \cap A_2 \cap A_3)'] &= 1 - P(A_1 \cap A_2 \cap A_3) \\
&= 1 - P(A_1)P(A_2)P(A_3) \\
&= 1 - (0.15)^3 \\
&= 0.9966.
\end{aligned}
$$

One way to increase the reliability of such a system is to add more components (realizing that this also adds weight and takes up space). For example, if a fourth component K_4 were added to this system, the probability that the system does not fail is

$$
P[(A_1 \cap A_2 \cap A_3 \cap A_4)'] = 1 - (0.15)^4 = 0.9995.
$$

∎

If A, B, and C are mutually independent events, then the following events are also independent:

(a) A and $(B \cap C)$;

(b) A and $(B \cup C)$;

(c) A' and $(B \cap C')$.

In addition, A', B', and C' are mutually independent. (The proofs and illustrations of these results are left as exercises.)

Many experiments consist of a sequence of n trials that are mutually independent. If the outcomes of the trials, in fact, do not have anything to do with one another, then events, such that each is associated with a different trial, should be independent in the probability sense. That is, if the event A_i is associated with the ith trial, $i = 1, 2, \ldots, n$, then

$$
P(A_1 \cap A_2 \cap \cdots \cap A_n) = P(A_1)P(A_2) \cdots P(A_n).
$$

Example 1.4-6

A fair six-sided die is rolled six independent times. Let A_i be the event that side i is observed on the ith roll, called a match on the ith trial, $i = 1, 2, \ldots, 6$. Thus, $P(A_i) = 1/6$ and $P(A_i') = 1 - 1/6 = 5/6$. If we let B denote the event that at least one match occurs, then B' is the event that no matches occur. Hence,

$$
P(B) = 1 - P(B') = 1 - P(A_1' \cap A_2' \cap \cdots \cap A_6')
$$

$$
= 1 - \frac{5}{6} \cdot \frac{5}{6} \cdot \frac{5}{6} \cdot \frac{5}{6} \cdot \frac{5}{6} \cdot \frac{5}{6} = 1 - \left(\frac{5}{6}\right)^6.
$$

∎

Example 1.4-7

The probability that a company's workforce has at least one accident during a certain month is $(0.01)k$, where k is the number of days in that month (say, February has 28 days). Assume that the numbers of accidents is independent from month to month. If the company's year starts with January, the probability that the first accident is in April is

$$P(\text{none in Jan., none in Feb., none in March, at least one in April}) =$$

$$(1 - 0.31)(1 - 0.28)(1 - 0.31)(0.30) = (0.69)(0.72)(0.69)(0.30)$$
$$= 0.103.$$ ∎

Example 1.4-8

Three inspectors look at a critical component of a product. Their probabilities of detecting a defect are different, namely, 0.99, 0.98, and 0.96, respectively. If we assume independence, then the probability of at least one detecting the defect is

$$1 - (0.01)(0.02)(0.04) = 0.999992.$$

The probability of *only* one finding the defect is

$$(0.99)(0.02)(0.04) + (0.01)(0.98)(0.04) + (0.01)(0.02)(0.96) = 0.001376.$$

As an exercise, compute the following probabilities: **(a)** that exactly two find the defect, **(b)** that all three find the defect. ∎

Example 1.4-9

Suppose that on five consecutive days an "instant winner" lottery ticket is purchased and the probability of winning is 1/5 on each day. Assuming independent trials, we have

$$P(WWLLL) = \left(\frac{1}{5}\right)^2 \left(\frac{4}{5}\right)^3,$$

$$P(LWLWL) = \frac{4}{5} \cdot \frac{1}{5} \cdot \frac{4}{5} \cdot \frac{1}{5} \cdot \frac{4}{5} = \left(\frac{1}{5}\right)^2 \left(\frac{4}{5}\right)^3.$$

In general, the probability of purchasing two winning tickets and three losing tickets is

$$\binom{5}{2}\left(\frac{1}{5}\right)^2 \left(\frac{4}{5}\right)^3 = \frac{5!}{2!3!}\left(\frac{1}{5}\right)^2 \left(\frac{4}{5}\right)^3 = 0.2048,$$

because there are $\binom{5}{2}$ ways to select the positions (or the days) for the winning tickets and each of these $\binom{5}{2}$ ways has the probability $(1/5)^2(4/5)^3$. ∎

Exercises

1.4-1. Let A and B be independent events with $P(A) = 0.7$ and $P(B) = 0.2$. Compute **(a)** $P(A \cap B)$, **(b)** $P(A \cup B)$, and **(c)** $P(A' \cup B')$.

1.4-2. Let $P(A) = 0.3$ and $P(B) = 0.6$.

(a) Find $P(A \cup B)$ when A and B are independent.

(b) Find $P(A \mid B)$ when A and B are mutually exclusive.

1.4-3. Let A and B be independent events with $P(A) = 1/4$ and $P(B) = 2/3$. Compute **(a)** $P(A \cap B)$, **(b)** $P(A \cap B')$, **(c)** $P(A' \cap B')$, **(d)** $P[(A \cup B)']$, and **(e)** $P(A' \cap B)$.

1.4-4. Prove parts (b) and (c) of Theorem 1.4-1.

1.4-5. If $P(A) = 0.8$, $P(B) = 0.5$, and $P(A \cup B) = 0.9$, are A and B independent events? Why or why not?

1.4-6. Show that if A, B, and C are mutually independent, then the following pairs of events are independent: A and $(B \cap C)$, A and $(B \cup C)$, A' and $(B \cap C')$. Show also that A', B', and C' are mutually independent.

1.4-7. Three independent reviewers are reviewing a book. Let A_i denote the event that a favorable review is submitted by reviewer i, $i = 1, 2, 3$. Assume that A_1, A_2, and A_3 are mutually independent and that $P(A_1) = 0.6$, $P(A_2) = 0.57$, and $P(A_3) = 0.4$.

(a) Compute the probability that at least one of the reviewers submits a favorable review.

(b) Compute the probability that exactly two reviewers submit favorable reviews.

1.4-8. Die A has orange on two faces and blue on four faces, Die B has orange on three faces and blue on three faces, Die C has orange on four faces and blue on two faces. All are fair dice. If the three dice are rolled, find the probability that exactly two of the three dice come up orange.

1.4-9. Suppose that A, B, and C are mutually independent events and that $P(A) = 0.5$, $P(B) = 0.8$, and $P(C) = 0.9$. Find the probabilities that **(a)** all three events occur, **(b)** exactly two of the three events occur, and **(c)** none of the events occurs.

1.4-10. Let D_1, D_2, D_3 be three four-sided dice whose sides have been labeled as follows:

$$D_1 : 0\ 3\ 3\ 3 \qquad D_2 : 2\ 2\ 2\ 5 \qquad D_3 : 1\ 1\ 4\ 6$$

The three dice are rolled at random. Let A, B, and C be the events that the outcome on die D_1 is larger than the outcome on D_2, the outcome on D_2 is larger than the outcome on D_3, and the outcome on D_3 is larger than the outcome on D_1, respectively. Show that **(a)** $P(A) = 9/16$, **(b)** $P(B) = 9/16$, and **(c)** $P(C) = 10/16$. Do you find it interesting that each of the probabilities that D_1 "beats" D_2, D_2 "beats" D_3, and D_3 "beats" D_1 is greater than 1/2? Thus, it is difficult to determine the "best" die.

1.4-11. Let A and B be two events.

(a) If the events A and B are mutually exclusive, are A and B always independent? If the answer is no, can they ever be independent? Explain.

(b) If $A \subset B$, can A and B ever be independent events? Explain.

1.4-12. Flip an unbiased coin eight independent times. Compute the probability of

(a) HHHTHTTH.

(b) TTHHHHTT.

(c) HTHTHTHT.

(d) Four heads occurring in the eight trials.

1.4-13. An urn contains two red balls and four white balls. Sample successively five times at random and with replacement, so that the trials are independent. Compute

the probability of each of the two sequences $WWRWR$ and $RWWWR$.

1.4-14. In Example 1.4-5, suppose that the probability of failure of a component is $p = 0.4$. Find the probability that the system does not fail if the number of redundant components is

(a) 3.

(b) 8.

1.4-15. An urn contains 10 red and 10 white balls. The balls are drawn from the urn at random, one at a time. Find the probabilities that the fourth white ball is the fourth, fifth, sixth, or seventh ball drawn if the sampling is done

(a) With replacement.

(b) Without replacement.

(c) In the World Series, the American League (red) and National League (white) teams play until one team wins four games. Do you think that the urn model presented in this exercise could be used to describe the probabilities of a 4-, 5-, 6-, or 7-game series? (Note that either "red" or "white" could win.) If your answer is yes, would you choose sampling with or without replacement in your model? (For your information, the numbers of 4-, 5-, 6-, and 7-game series, up to and including 2012, were 21, 24, 23, 36. This ignores games that ended in a tie, which occurred in 1907, 1912, and 1922. Also, it does not include the 1903 and 1919–1921 series, in which the winner had to take five out of nine games. The World Series was canceled in 1994.)

1.4-16. An urn contains five balls, one marked WIN and four marked LOSE. You and another player take turns selecting a ball at random from the urn, one at a time. The first person to select the WIN ball is the winner. If you draw first, find the probability that you will win if the sampling is done

(a) With replacement.

(b) Without replacement.

1.4-17. Each of the 12 students in a class is given a fair 12-sided die. In addition, each student is numbered from 1 to 12.

(a) If the students roll their dice, what is the probability that there is at least one "match" (e.g., student 4 rolls a 4)?

(b) If you are a member of this class, what is the probability that at least one of the other 11 students rolls the same number as you do?

1.4-18. An eight-team single-elimination tournament is set up as follows:

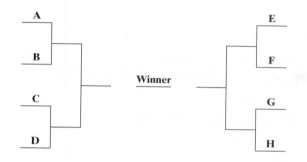

For example, eight students (called A–H) set up a tournament among themselves. The top-listed student in each bracket calls heads or tails when his or her opponent flips a coin. If the call is correct, the student moves on to the next bracket.

(a) How many coin flips are required to determine the tournament winner?

(b) What is the probability that you can predict all of the winners?

(c) In NCAA Division I basketball, after the "play-in" games, 64 teams participate in a single-elimination tournament to determine the national champion. Considering only the remaining 64 teams, how many games are required to determine the national champion?

(d) Assume that for any given game, either team has an equal chance of winning. (That is probably not true.) On page 43 of the March 22, 1999, issue, *Time* claimed that the "mathematical odds of predicting all 63 NCAA games correctly is 1 in 75 million." Do you agree with this statement? If not, why not?

1.4-19. Extend Example 1.4-6 to an n-sided die. That is, suppose that a fair n-sided die is rolled n independent times. A match occurs if side i is observed on the ith trial, $i = 1, 2, \ldots, n$.

(a) Show that the probability of at least one match is

$$1 - \left(\frac{n-1}{n}\right)^n = 1 - \left(1 - \frac{1}{n}\right)^n.$$

(b) Find the limit of this probability as n increases without bound.

1.4-20. Hunters A and B shoot at a target with probabilities of p_1 and p_2, respectively. Assuming independence, can p_1 and p_2 be selected so that $P(\text{zero hits}) = P(\text{one hit}) = P(\text{two hits})$?

1.5 BAYES' THEOREM

We begin this section by illustrating Bayes' theorem with an example.

Example 1.5-1

Bowl B_1 contains two red and four white chips, bowl B_2 contains one red and two white chips, and bowl B_3 contains five red and four white chips. Say that the probabilities for selecting the bowls are not the same but are given by $P(B_1) = 1/3$, $P(B_2) = 1/6$, and $P(B_3) = 1/2$, where B_1, B_2, and B_3 are the events that bowls B_1, B_2, and B_3 are respectively chosen. The experiment consists of selecting a bowl with these probabilities and then drawing a chip at random from that bowl. Let us compute the probability of event R, drawing a red chip—say, $P(R)$. Note that $P(R)$ is dependent first of all on which bowl is selected and then on the probability of drawing a red chip from the selected bowl. That is, the event R is the union of the mutually exclusive events $B_1 \cap R$, $B_2 \cap R$, and $B_3 \cap R$. Thus,

$$
\begin{aligned}
P(R) &= P(B_1 \cap R) + P(B_2 \cap R) + P(B_3 \cap R) \\
&= P(B_1)P(R \mid B_1) + P(B_2)P(R \mid B_2) + P(B_3)P(R \mid B_3) \\
&= \frac{1}{3} \cdot \frac{2}{6} + \frac{1}{6} \cdot \frac{1}{3} + \frac{1}{2} \cdot \frac{5}{9} = \frac{4}{9}.
\end{aligned}
$$

Suppose now that the outcome of the experiment is a red chip, but we do not know from which bowl it was drawn. Accordingly, we compute the conditional probability that the chip was drawn from bowl B_1, namely, $P(B_1 \mid R)$. From the definition of conditional probability and the preceding result, we have

$$P(B_1 \mid R) = \frac{P(B_1 \cap R)}{P(R)}$$

$$= \frac{P(B_1)P(R \mid B_1)}{P(B_1)P(R \mid B_1) + P(B_2)P(R \mid B_2) + P(B_3)P(R \mid B_3)}$$

$$= \frac{(1/3)(2/6)}{(1/3)(2/6) + (1/6)(1/3) + (1/2)(5/9)} = \frac{2}{8}.$$

Similarly,

$$P(B_2 \mid R) = \frac{P(B_2 \cap R)}{P(R)} = \frac{(1/6)(1/3)}{4/9} = \frac{1}{8}$$

and

$$P(B_3 \mid R) = \frac{P(B_3 \cap R)}{P(R)} = \frac{(1/2)(5/9)}{4/9} = \frac{5}{8}.$$

Note that the conditional probabilities $P(B_1 \mid R)$, $P(B_2 \mid R)$, and $P(B_3 \mid R)$ have changed from the original probabilities $P(B_1)$, $P(B_2)$, and $P(B_3)$ in a way that agrees with your intuition. Once the red chip has been observed, the probability concerning B_3 seems more favorable than originally because B_3 has a larger percentage of red chips than do B_1 and B_2. The conditional probabilities of B_1 and B_2 decrease from their original ones once the red chip is observed. Frequently, the original probabilities are called *prior probabilities* and the conditional probabilities are the *posterior probabilities*.   ■

We generalize the result of Example 1.5-1. Let B_1, B_2, ..., B_m constitute a *partition* of the sample space S. That is,

$$S = B_1 \cup B_2 \cup \cdots \cup B_m \text{ and } B_i \cap B_j = \emptyset, i \neq j.$$

Of course, the events B_1, B_2, ..., B_m are mutually exclusive and exhaustive (since the union of the disjoint sets equals the sample space S). Furthermore, suppose the **prior probability** of the event B_i is positive; that is, $P(B_i) > 0$, $i = 1, \ldots, m$. If A is an event, then A is the union of m mutually exclusive events, namely,

$$A = (B_1 \cap A) \cup (B_2 \cap A) \cup \cdots \cup (B_m \cap A).$$

Thus,

$$P(A) = \sum_{i=1}^{m} P(B_i \cap A)$$

$$= \sum_{i=1}^{m} P(B_i)P(A \mid B_i), \tag{1.5-1}$$

which is sometimes called the **law of total probability**. If $P(A) > 0$, then

$$P(B_k \mid A) = \frac{P(B_k \cap A)}{P(A)}, \qquad k = 1, 2, \ldots, m. \tag{1.5-2}$$

Using Equation 1.5-1 and replacing $P(A)$ in Equation 1.5-2, we have **Bayes' theorem**:

$$P(B_k \mid A) = \frac{P(B_k)P(A \mid B_k)}{\sum\limits_{i=1}^{m} P(B_i)P(A \mid B_i)}, \qquad k = 1, 2, \dots, m.$$

The conditional probability $P(B_k \mid A)$ is often called the **posterior probability** of B_k. The next example illustrates one application of Bayes' theorem.

Example 1.5-2
In a certain factory, machines I, II, and III are all producing springs of the same length. Of their production, machines I, II, and III respectively produce 2%, 1%, and 3% defective springs. Of the total production of springs in the factory, machine I produces 35%, machine II produces 25%, and machine III produces 40%. If one spring is selected at random from the total springs produced in a day, by the law of total probability, $P(D)$ equals, in an obvious notation,

$$P(D) = P(I)P(D \mid I) + P(II)P(D \mid II) + P(III)P(D \mid III)$$

$$= \left(\frac{35}{100}\right)\left(\frac{2}{100}\right) + \left(\frac{25}{100}\right)\left(\frac{1}{100}\right) + \left(\frac{40}{100}\right)\left(\frac{3}{100}\right) = \frac{215}{10,000}.$$

If the selected spring is defective, the conditional probability that it was produced by machine III is, by Bayes' formula,

$$P(III \mid D) = \frac{P(III)P(D \mid III)}{P(D)} = \frac{(40/100)(3/100)}{215/10,000} = \frac{120}{215}.$$

Note how the posterior probability of III increased from the prior probability of III after the defective spring was observed, because III produces a larger percentage of defectives than do I and II. ∎

Example 1.5-3
A Pap smear is a screening procedure used to detect cervical cancer. For women with this cancer, there are about 16% *false negatives*; that is,

$$P(T^- = \text{test negative} \mid C^+ = \text{cancer}) = 0.16.$$

Thus,

$$P(T^+ = \text{test positive} \mid C^+ = \text{cancer}) = 0.84.$$

For women without cancer, there are about 10% false positives; that is,

$$P(T^+ \mid C^- = \text{not cancer}) = 0.10.$$

Hence,

$$P(T^- \mid C^- = \text{not cancer}) = 0.90.$$

In the United States, there are about 8 women in 100,000 who have this cancer; that is,

$$P(C^+) = 0.00008; \text{ so } P(C^-) = 0.99992.$$

By Bayes' theorem and the law of total probability,

$$P(C^+ \mid T^+) = \frac{P(C^+ \text{ and } T^+)}{P(T^+)}$$

$$= \frac{(0.00008)(0.84)}{(0.00008)(0.84) + (0.99992)(0.10)}$$

$$= \frac{672}{672 + 999{,}920} = 0.000672.$$

What this means is that for every million positive Pap smears, only 672 represent true cases of cervical cancer. This low ratio makes one question the value of the procedure. The reason that it is ineffective is that the percentage of women having that cancer is so small and the error rates of the procedure—namely, 0.16 and 0.10—are so high. On the other hand, the test does give good information in a sense. The posterior probability of cancer, given a positive test, is about eight times the prior probability. ∎

Exercises

1.5-1. Bowl B_1 contains two white chips, bowl B_2 contains two red chips, bowl B_3 contains two white and two red chips, and bowl B_4 contains three white chips and one red chip. The probabilities of selecting bowl B_1, B_2, B_3, or B_4 are 1/2, 1/4, 1/8, and 1/8, respectively. A bowl is selected using these probabilities and a chip is then drawn at random. Find

(a) $P(W)$, the probability of drawing a white chip.

(b) $P(B_1 \mid W)$, the conditional probability that bowl B_1 had been selected, given that a white chip was drawn.

1.5-2. Bean seeds from supplier A have a 95% germination rate and those from supplier B have a 70% germination rate. A seed-packaging company purchases 30% of its bean seeds from supplier A and 70% from supplier B and mixes these seeds together.

(a) Find the probability P(G) that a seed selected at random from the mixed seeds will germinate.

(b) Given that a seed germinates, find the probability that the seed was purchased from supplier A.

1.5-3. A doctor is concerned about the relationship between blood pressure and irregular heartbeats. Among her patients, she classifies blood pressures as high, normal, or low and heartbeats as regular or irregular and finds that (a) 16% have high blood pressure; (b) 19% have low blood pressure; (c) 17% have an irregular heartbeat; (d) of those with an irregular heartbeat, 35% have high blood pressure; and (e) of those with normal blood pressure, 11% have an irregular heartbeat. What percentage of her patients have a regular heartbeat and low blood pressure?

1.5-4. Assume that an insurance company knows the following probabilities relating to automobile accidents (where the second column refers to the probability that the policyholder has at least one accident during the annual policy period):

Age of Driver	Probability of Accident	Fraction of Company's Insured Drivers
16–25	0.05	0.10
26–50	0.02	0.55
51–65	0.03	0.20
66–90	0.04	0.15

A randomly selected driver from the company's insured drivers has an accident. What is the conditional probability that the driver is in the 16–25 age group?

1.5-5. At a hospital's emergency room, patients are classified and 20% of them are critical, 30% are serious, and 50% are stable. Of the critical ones, 30% die; of the serious, 10% die; and of the stable, 1% die. Given that a patient dies, what is the conditional probability that the patient was classified as critical?

1.5-6. A life insurance company issues standard, preferred, and ultrapreferred policies. Of the company's policyholders of a certain age, 60% have standard policies and a probability of 0.01 of dying in the next year, 30% have preferred policies and a probability of 0.008 of dying in the next year, and 10% have ultrapreferred policies and a probability of 0.007 of dying in the next year. A policyholder of that age dies in the next year. What are the conditional probabilities of the deceased having had a standard, a preferred, and an ultrapreferred policy?

1.5-7. A chemist wishes to detect an impurity in a certain compound that she is making. There is a test that detects an impurity with probability 0.90; however, this test indicates that an impurity is there when it is not about 5% of the time. The chemist produces compounds with the impurity about 20% of the time; that is, 80% do not have the impurity. A compound is selected at random from the chemist's output. The test indicates that an impurity is present. What is the conditional probability that the compound actually has an impurity?

1.5-8. A store sells four brands of tablets. The least expensive brand, B_1, accounts for 40% of the sales. The other brands (in order of their price) have the following percentages of sales: $B_2, 30\%$; $B_3, 20\%$; and $B_4, 10\%$. The respective probabilities of needing repair during warranty are 0.10 for B_1, 0.05 for B_2, 0.03 for B_3, and 0.02 for B_4. A randomly selected purchaser has a tablet that needs repair under warranty. What are the four conditional probabilities of being brand B_i, $i = 1, 2, 3, 4$?

1.5-9. There is a new diagnostic test for a disease that occurs in about 0.05% of the population. The test is not perfect, but will detect a person with the disease 99% of the time. It will, however, say that a person without the disease has the disease about 3% of the time. A person is selected at random from the population, and the test indicates that this person has the disease. What are the conditional probabilities that

(a) the person has the disease?

(b) the person does not have the disease?

Discuss. HINT: Note that the fraction 0.0005 of diseased persons in the population is much smaller than the error probabilities of 0.01 and 0.03.

1.5-10. Suppose we want to investigate the percentage of abused children in a certain population. To do this, doctors examine some of these children taken at random from that population. However, doctors are not perfect: They sometimes classify an abused child (A^+) as one not abused (D^-) or they classify a nonabused child (A^-) as one that is abused (D^+). Suppose these error rates are $P(D^- | A^+) = 0.08$ and $P(D^+ | A^-) = 0.05$, respectively; thus, $P(D^+ | A^+) = 0.92$ and $P(D^- | A^-) = 0.95$ are the probabilities of the correct decisions. Let us pretend that only 2% of all children are abused; that is, $P(A^+) = 0.02$ and $P(A^-) = 0.98$.

(a) Select a child at random. What is the probability that the doctor classifies this child as abused? That is, compute

$$P(D^+) = P(A^+)P(D^+ | A^+) + P(A^-)P(D^+ | A^-).$$

(b) Compute $P(A^- | D^+)$ and $P(A^+ | D^+)$.

(c) Compute $P(A^- | D^-)$ and $P(A^+ | D^-)$.

(d) Are the probabilities in (b) and (c) alarming? This happens because the error rates of 0.08 and 0.05 are high relative to the fraction 0.02 of abused children in the population.

1.5-11. At the beginning of a certain study of a group of persons, 15% were classified as heavy smokers, 30% as light smokers, and 55% as nonsmokers. In the five-year study, it was determined that the death rates of the heavy and light smokers were five and three times that of the nonsmokers, respectively. A randomly selected participant died over the five-year period; calculate the probability that the participant was a nonsmoker.

1.5-12. A test indicates the presence of a particular disease 90% of the time when the disease is present and the presence of the disease 2% of the time when the disease is not present. If 0.5% of the population has the disease, calculate the conditional probability that a person selected at random has the disease if the test indicates the presence of the disease.

1.5-13. A salesperson, Jane, is assigned the task of selling two types of computers, C-I and C-II. During each sales transaction, the probability of the buyer purchasing a C-I type of computer is 40% and the probability of the buyer purchasing a C-II type of computer is 60%. Suppose Jane has been able to sell only one computer until now. What is the probability that this computer was of the type C-I?

1.5-14. Two processes of a company produce rolls of materials: The rolls of Process I are 4% defective and the rolls of Process II are 7% defective. Process I produces 55% of the company's output, Process II 45%. A roll is selected at random from the total output. Given that this roll is defective, what is the conditional probability that it is from Process I?

HISTORICAL COMMENTS Most probabilists would say that the mathematics of probability began when, in 1654, Chevalier de Méré, a French nobleman who liked to gamble, challenged Blaise Pascal to explain a puzzle and a problem created from his observations concerning rolls of dice. Of course, there was gambling well before this, and actually, almost 200 years before this challenge, a Franciscan monk, Luca Paccioli, proposed essentially the same puzzle. Here it is:

A and B are playing a fair game of balla. They agree to continue until one has six wins. However, the game actually stops when A has won five and B three. How should the stakes be divided?

And over 100 years before de Méré's challenge, a 16th-century doctor, Girolamo Cardano, who was also a gambler, had figured out the answers to many dice problems, but not the one that de Méré proposed. Chevalier de Méré had observed this: If a single fair die is tossed 4 times, the probability of obtaining at least one six was slightly greater than 1/2. However, keeping the same proportions, if a pair of dice is tossed 24 times, the probability of obtaining at least one double-six seemed to be slightly less than 1/2; at least de Méré was losing money betting on it. This is when he approached Blaise Pascal with the challenge. Not wanting to work on the problems alone, Pascal formed a partnership with Pierre de Fermat, a brilliant young mathematician. It was this 1654 correspondence between Pascal and Fermat that started the theory of probability.

Today an average student in probability could solve both problems easily. For the puzzle, note that B could win with six rounds only by winning the next three rounds, which has probability of $(1/2)^3 = 1/8$ because it was a fair game of balla. Thus, A's probability of winning six rounds is $1 - 1/8 = 7/8$, and stakes should be divided seven units to one. For the dice problem, the probability of at least one six in four rolls of a die is

$$1 - \left(\frac{5}{6}\right)^4 = 0.518,$$

while the probability of rolling at least one double-six in 24 rolls of a pair of dice is

$$1 - \left(\frac{35}{36}\right)^{24} = 0.491.$$

It seems amazing to us that de Méré could have observed enough trials of those events to detect the slight difference in those probabilities. However, he won betting on the first but lost by betting on the second.

Incidentally, the solution to the balla puzzle led to a generalization—namely, the binomial distribution—and to the famous Pascal triangle. Of course, Fermat was the great mathematician associated with "Fermat's last theorem."

The Reverend Thomas Bayes, who was born in 1701, was a Nonconformist (a Protestant who rejected most of the rituals of the Church of England). While he published nothing in mathematics when he was alive, two works were published after his death, one of which contained the essence of Bayes' theorem and a very original way of using data to modify prior probabilities to create posterior probabilities. It has had such an influence on modern statistics that many modern statisticians are associated with the neo-Bayesian movement and we devote Sections 6.8 and 6.9 to some of these methods.

Chapter 2

DISCRETE DISTRIBUTIONS

2.1 RANDOM VARIABLES OF THE DISCRETE TYPE

An outcome space S may be difficult to describe if the elements of S are not numbers. We shall now discuss how we can use a rule by which each outcome of a random experiment, an element s of S, may be associated with a real number x. We begin the discussion with an example.

Example 2.1-1 A rat is selected at random from a cage and its sex is determined. The set of possible outcomes is female and male. Thus, the outcome space is $S = \{$female, male$\} = \{$F, M$\}$. Let X be a function defined on S such that $X(\text{F}) = 0$ and $X(\text{M}) = 1$. X is then a real-valued function that has the outcome space S as its domain and the set of real numbers $\{x: x = 0, 1\}$ as its range. We call X a random variable, and in this example, the space associated with X is the set of numbers $\{x: x = 0, 1\}$. ■

We now formulate the definition of a random variable.

> **Definition 2.1-1**
> Given a random experiment with an outcome space S, a function X that assigns one and only one real number $X(s) = x$ to each element s in S is called a **random variable**. The **space** of X is the set of real numbers $\{x: X(s) = x, s \in S\}$, where $s \in S$ means that the element s belongs to the set S.

REMARK As we give examples of random variables and their probability distributions, the reader will soon recognize that, when observing a random experiment, the experimenter must take some type of measurement (or measurements). This measurement can be thought of as the outcome of a random variable. We would simply like to know the probability of a measurement resulting in A, a subset of the space of X. If this is known for all subsets A, then we know the probability distribution of the random variable. Obviously, in practice, we often do not know this distribution exactly. Hence, statisticians make conjectures about these distributions; that is, we construct probabilistic models for random variables. The ability of a statistician to model a real situation appropriately is a valuable trait. In this chapter we introduce some probability models in which the spaces of the random variables consist of sets of integers. ■

It may be that the set S has elements that are themselves real numbers. In such an instance, we could write $X(s) = s$, so that X is the identity function and the space of X is also S. This situation is illustrated in Example 2.1-2.

Example 2.1-2

Let the random experiment be the cast of a die. Then the outcome space associated with this experiment is $S = \{1, 2, 3, 4, 5, 6\}$, with the elements of S indicating the number of spots on the side facing up. For each $s \in S$, let $X(s) = s$. The space of the random variable X is then $\{1, 2, 3, 4, 5, 6\}$.

If we associate a probability of 1/6 with each outcome, then, for example, $P(X = 5) = 1/6$, $P(2 \leq X \leq 5) = 4/6$, and $P(X \leq 2) = 2/6$ seem to be reasonable assignments, where, in this example, $\{2 \leq X \leq 5\}$ means $\{X = 2, 3, 4, \text{ or } 5\}$ and $\{X \leq 2\}$ means $\{X = 1 \text{ or } 2\}$. ■

The student will no doubt recognize two major difficulties here:

1. In many practical situations, the probabilities assigned to the events are unknown.

2. Since there are many ways of defining a function X on S, which function do we want to use?

As a matter of fact, the solutions to these problems in particular cases are major concerns in applied statistics. In considering (2), statisticians try to determine what *measurement* (or measurements) should be taken on an outcome; that is, how best do we "mathematize" the outcome? These measurement problems are most difficult and can be answered only by getting involved in a practical project. For (1), we often need to estimate these probabilities or percentages through repeated observations (called sampling). For example, what percentage of newborn girls in the University of Iowa Hospital weigh less than 7 pounds? Here a newborn baby girl is the outcome, and we have measured her one way (by weight), but obviously there are many other ways of measuring her. If we let X be the weight in pounds, we are interested in the probability $P(X < 7)$, and we can estimate this probability only by repeated observations. One obvious way of estimating it is by the use of the relative frequency of $\{X < 7\}$ after a number of observations. If it is reasonable to make additional assumptions, we will study other ways of estimating that probability. It is this latter aspect with which the field of mathematical statistics is concerned. That is, if we assume certain models, we find that the theory of statistics can explain how best to draw conclusions or make predictions.

In many instances, it is clear exactly what function X the experimenter wants to define on the outcome space. For example, the caster in the dice game called craps is concerned about the sum of the spots (say X) that are facing upward on the pair of dice. Hence, we go directly to the space of X, which we shall denote by the same letter S. After all, in the dice game the caster is directly concerned only with the probabilities associated with X. Thus, for convenience, in many instances the reader can think of the space of X as being the outcome space.

Let X denote a random variable with space S. Suppose that we know how the probability is distributed over the various subsets A of S; that is, we can compute $P(X \in A)$. In this sense, we speak of the distribution of the random variable X, meaning, of course, the distribution of probability associated with the space S of X.

Let X denote a random variable with one-dimensional space S, a subset of the real numbers. Suppose that the space S contains a countable number of points; that is, either S contains a finite number of points, or the points of S can be put into a one-to-one correspondence with the positive integers. Such a set S is called a set of discrete points or simply a discrete outcome space. Furthermore, any random variable defined on such an S can assume at most a countable number of values, and is therefore called a random variable of the **discrete type**. The corresponding probability distribution likewise is said to be of the discrete type.

For a random variable X of the discrete type, the probability $P(X = x)$ is frequently denoted by $f(x)$, and this function $f(x)$ is called the **probability mass function**. Note that some authors refer to $f(x)$ as the probability function, the frequency function, or the probability density function. In the discrete case, we shall use "probability mass function," and it is hereafter abbreviated pmf.

Let $f(x)$ be the pmf of the random variable X of the discrete type, and let S be the space of X. Since $f(x) = P(X = x)$ for $x \in S$, $f(x)$ must be nonnegative for $x \in S$, and we want all these probabilities to add to 1 because each $P(X = x)$ represents the fraction of times x can be expected to occur. Moreover, to determine the probability associated with the event $A \in S$, we would sum the probabilities of the x values in A. This leads us to the following definition.

Definition 2.1-2

The pmf $f(x)$ of a discrete random variable X is a function that satisfies the following properties:

(a) $f(x) > 0, \qquad x \in S$;

(b) $\displaystyle\sum_{x \in S} f(x) = 1$;

(c) $P(X \in A) = \displaystyle\sum_{x \in A} f(x), \qquad$ where $A \subset S$.

Of course, we usually let $f(x) = 0$ when $x \notin S$; thus, the domain of $f(x)$ is the set of real numbers. When we define the pmf $f(x)$ and do not say "zero elsewhere," we tacitly mean that $f(x)$ has been defined at all x's in the space S and it is assumed that $f(x) = 0$ elsewhere; that is, $f(x) = 0$ when $x \notin S$. Since the probability $P(X = x) = f(x) > 0$ when $x \in S$, and since S contains all the outcomes with positive probabilities associated with X, we sometimes refer to S as the **support** of X as well as the space of X.

Cumulative probabilities are often of interest. We call the function defined by

$$F(x) = P(X \leq x), \qquad -\infty < x < \infty,$$

the **cumulative distribution function** and abbreviate it as cdf. The cdf is sometimes referred to as the **distribution function** of the random variable X. Values of the cdf of certain random variables are given in the appendix and will be pointed out as we use them (see Appendix B, Tables II, III, IV, Va, VI, VII, and IX).

When a pmf is constant on the space or support, we say that the distribution is **uniform** over that space. As an illustration, in Example 2.1-2 X has a discrete uniform distribution on $S = \{1, 2, 3, 4, 5, 6\}$ and its pmf is

$$f(x) = \frac{1}{6}, \qquad x = 1, 2, 3, 4, 5, 6.$$

We can generalize this result by letting X have a discrete uniform distribution over the first m positive integers, so that its pmf is

$$f(x) = \frac{1}{m}, \qquad x = 1, 2, 3, \ldots, m.$$

The cdf of X is defined as follows where $k = 1, 2, \ldots, m - 1$. We have

$$F(x) = P(X \le x) = \begin{cases} 0, & x < 1, \\ \dfrac{k}{m}, & k \le x < k+1, \\ 1, & m \le x. \end{cases}$$

Note that this is a step function with a jump of size $1/m$ for $x = 1, 2, \ldots, m$.

We now give an example in which X does not have a uniform distribution.

Example 2.1-3

Roll a fair four-sided die twice, and let X be the maximum of the two outcomes. The outcome space for this experiment is $S_0 = \{(d_1, d_2) : d_1 = 1, 2, 3, 4; d_2 = 1, 2, 3, 4\}$, where we assume that each of these 16 points has probability 1/16. Then $P(X = 1) = P[(1, 1)] = 1/16$, $P(X = 2) = P[\{(1, 2), (2, 1), (2, 2)\}] = 3/16$, and similarly $P(X = 3) = 5/16$ and $P(X = 4) = 7/16$. That is, the pmf of X can be written simply as

$$f(x) = P(X = x) = \frac{2x - 1}{16}, \qquad x = 1, 2, 3, 4. \tag{2.1-1}$$

We could add that $f(x) = 0$ elsewhere; but if we do not, the reader should take $f(x)$ to equal zero when $x \notin S = \{1, 2, 3, 4\}$. ∎

A better understanding of a particular probability distribution can often be obtained with a graph that depicts the pmf of X. Note that the graph of the pmf when $f(x) > 0$ would be simply the set of points $\{[x, f(x)] : x \in S\}$, where S is the space of X. Two types of graphs can be used to give a better visual appreciation of the pmf: a line graph and a probability histogram. A **line graph** of the pmf $f(x)$ of the random variable X is a graph having a vertical line segment drawn from $(x, 0)$ to $[x, f(x)]$ at each x in S, the space of X. If X can assume only integer values, a **probability histogram** of the pmf $f(x)$ is a graphical representation that has a rectangle of height $f(x)$ and a base of length 1, centered at x for each $x \in S$, the space of X. Thus, the area of each rectangle is equal to the respective probability $f(x)$, and the total area of a probability histogram is 1.

Figure 2.1-1 displays a line graph and a probability histogram for the pmf $f(x)$ defined in Equation 2.1-1.

Our next probability model uses the material in Section 1.2 on methods of enumeration. Consider a collection of $N = N_1 + N_2$ similar objects, N_1 of them belonging to one of two dichotomous classes (red chips, say) and N_2 of them belonging to the second class (blue chips, say). A collection of n objects is selected from these N objects at random and without replacement. Find the probability that exactly x (where the nonnegative integer x satisfies $x \le n$, $x \le N_1$, and $n - x \le N_2$) of these n objects belong to the first class and $n - x$ belong to the second. Of course, we can select x objects from the first class in any one of $\binom{N_1}{x}$ ways and $n - x$ objects from the second class in any one of $\binom{N_2}{n - x}$ ways. By the multiplication principle, the product $\binom{N_1}{x}\binom{N_2}{n - x}$ equals the number of ways the joint operation can be performed.

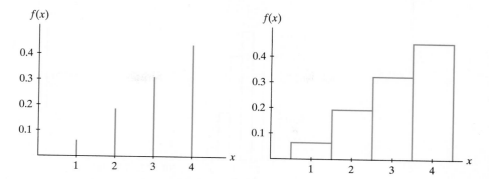

Figure 2.1-1 Line graph and probability histogram

If we assume that each of the $\binom{N}{n}$ ways of selecting n objects from $N = N_1 + N_2$ objects has the same probability, it follows that the desired probability is

$$f(x) = P(X = x) = \frac{\binom{N_1}{x}\binom{N_2}{n - x}}{\binom{N}{n}},$$

where the space S is the collection of nonnegative integers x that satisfies the inequalities $x \leq n$, $x \leq N_1$, and $n - x \leq N_2$. We say that the random variable X has a **hypergeometric distribution**.

Example 2.1-4 Some examples of hypergeometric probability histograms are given in Figure 2.1-2. The values of N_1, N_2, and n are given with each figure. ∎

Example 2.1-5 In a small pond there are 50 fish, 10 of which have been tagged. If a fisherman's catch consists of 7 fish selected at random and without replacement, and X denotes the number of tagged fish, the probability that exactly 2 tagged fish are caught is

$$P(X = 2) = \frac{\binom{10}{2}\binom{40}{5}}{\binom{50}{7}} = \frac{(45)(658{,}008)}{99{,}884{,}400} = \frac{246{,}753}{832{,}370} = 0.2964.$$ ∎

Example 2.1-6 A lot (collection) consisting of 100 fuses is inspected by the following procedure: Five fuses are chosen at random and tested; if all five blow at the correct amperage, the lot is accepted. Suppose that the lot contains 20 defective fuses. If X is a random variable equal to the number of defective fuses in the sample of 5, the probability of accepting the lot is

$$P(X = 0) = \frac{\binom{20}{0}\binom{80}{5}}{\binom{100}{5}} = \frac{19{,}513}{61{,}110} = 0.3193.$$

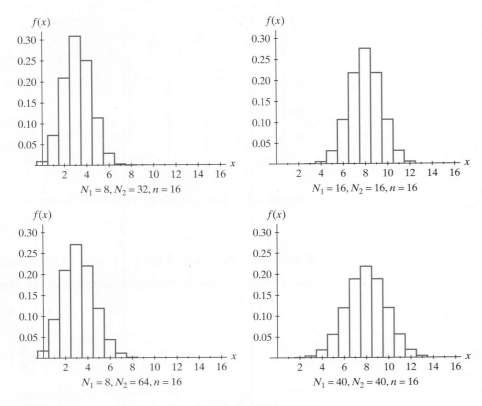

Figure 2.1-2 Hypergeometric probability histograms

More generally, the pmf of X is

$$f(x) = P(X = x) = \frac{\binom{20}{x}\binom{80}{5-x}}{\binom{100}{5}}, \qquad x = 0, 1, 2, 3, 4, 5.$$ ∎

In Section 1.1, we discussed the relationship between the probability $P(A)$ of an event A and the relative frequency $\mathcal{N}(A)/n$ of occurrences of event A in n repetitions of an experiment. We shall now extend those ideas.

Suppose that a random experiment is repeated n independent times. Let $A = \{X = x\}$, the event that x is the outcome of the experiment. Then we would expect the relative frequency $\mathcal{N}(A)/n$ to be close to $f(x)$. The next example illustrates this property.

Example 2.1-7 A fair four-sided die with outcomes 1, 2, 3, and 4 is rolled twice. Let X equal the sum of the two outcomes. Then the possible values of X are 2, 3, 4, 5, 6, 7, and 8. The following argument suggests that the pmf of X is given by $f(x) = (4 - |x - 5|)/16$, for $x = 2, 3, 4, 5, 6, 7, 8$ [i.e., $f(2) = 1/16$, $f(3) = 2/16$, $f(4) = 3/16$, $f(5) = 4/16$, $f(6) = 3/16$, $f(7) = 2/16$, and $f(8) = 1/16$]: Intuitively, these probabilities seem correct if we think of the 16 points (result on first roll, result on second roll) and

Table 2.1-1 Sum of two tetrahedral dice

x	Number of Observations of x	Relative Frequency of x	Probability of $\{X = x\}, f(x)$
2	71	0.071	0.0625
3	124	0.124	0.1250
4	194	0.194	0.1875
5	258	0.258	0.2500
6	177	0.177	0.1875
7	122	0.122	0.1250
8	54	0.054	0.0625

assume that each has probability 1/16. Then note that $X = 2$ only for the point $(1, 1)$, $X = 3$ for the two points $(2, 1)$ and $(1, 2)$, and so on. This experiment was simulated 1000 times on a computer. Table 2.1-1 lists the results and compares the relative frequencies with the corresponding probabilities.

A graph can be used to display the results shown in Table 2.1-1. The probability histogram of the pmf $f(x)$ of X is given by the dotted lines in Figure 2.1-3. It is superimposed over the shaded histogram that represents the observed relative frequencies of the corresponding x values. The shaded histogram is the **relative frequency histogram**. For random experiments of the discrete type, this relative frequency histogram of a set of data gives an estimate of the probability histogram of the associated random variable when the latter is unknown. (Estimation is considered in detail later in the book.) ∎

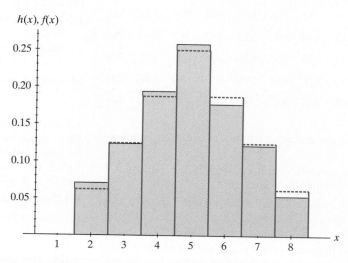

Figure 2.1-3 Sum of two tetrahedral dice

Exercises

2.1-1. Let the pmf of X be defined by $f(x) = x/9$, $x = 2, 3, 4$.

(a) Draw a line graph for this pmf.

(b) Draw a probability histogram for this pmf.

2.1-2. Let a chip be taken at random from a bowl that contains eight white chips, four red chips, and one blue chip. Let the random variable $X = 1$ if the outcome is a white chip, let $X = 4$ if the outcome is a red chip, and let $X = 8$ if the outcome is a blue chip.

(a) Find the pmf of X.

(b) Graph the pmf as a line graph.

2.1-3. For each of the following, determine the constant c so that $f(x)$ satisfies the conditions of being a pmf for a random variable X, and then depict each pmf as a line graph:

(a) $f(x) = x/c$, $x = 1, 2, 3, 4$.

(b) $f(x) = cx$, $x = 1, 2, 3, \ldots, 10$.

(c) $f(x) = c(1/4)^x$, $x = 1, 2, 3, \ldots$.

(d) $f(x) = c(x + 1)^2$, $x = 0, 1, 2, 3$.

(e) $f(x) = x/c$, $x = 1, 2, 3, \ldots, n$.

(f) $f(x) = \dfrac{c}{(x+1)(x+2)}$, $x = 0, 1, 2, 3, \ldots$.

HINT: In part (f), write $f(x) = 1/(x+1) - 1/(x+2)$.

2.1-4. The state of Michigan generates a three-digit number at random twice a day, seven days a week for its Daily 3 game. The numbers are generated one digit at a time. Consider the following set of 50 three-digit numbers as 150 one-digit integers that were generated at random:

169	938	506	757	594	656	444	809	321	545
732	146	713	448	861	612	881	782	209	752
571	701	852	924	766	633	696	023	601	789
137	098	534	826	642	750	827	689	979	000
933	451	945	464	876	866	236	617	418	988

Let X denote the outcome when a single digit is generated.

(a) With true random numbers, what is the pmf of X? Draw the probability histogram.

(b) For the 150 observations, determine the relative frequencies of 0, 1, 2, 3, 4, 5, 6, 7, 8, and 9, respectively.

(c) Draw the relative frequency histogram of the observations on the same graph paper as that of the probability histogram. Use a colored or dashed line for the relative frequency histogram.

2.1-5. The pmf of X is $f(x) = (5 - x)/10$, $x = 1, 2, 3, 4$.

(a) Graph the pmf as a line graph.

(b) Use the following independent observations of X, simulated on a computer, to construct a table like Table 2.1-1:

3	1	2	2	3	2	2	2	1	3	3	2	3	2	4	4	2	1	1	3
3	1	2	2	1	1	4	2	3	1	1	1	2	1	3	1	1	3	3	1
1	1	1	1	1	4	1	3	1	2	4	1	1	2	3	4	3	1	4	2
2	1	3	2	1	4	1	1	1	2	1	3	4	3	2	1	4	4	1	3
2	2	2	1	2	3	1	1	4	2	1	4	2	1	2	3	1	4	2	3

(c) Construct a probability histogram and a relative frequency histogram like Figure 2.1-3.

2.1-6. Let a random experiment be the casting of a pair of fair dice, each having six faces, and let the random variable X denote the sum of the dice.

(a) With reasonable assumptions, determine the pmf $f(x)$ of X. HINT: Picture the sample space consisting of the 36 points (result on first die, result on second die), and assume that each has probability 1/36. Find the probability of each possible outcome of X, namely, $x = 2, 3, 4, \ldots, 12$.

(b) Draw a probability histogram for $f(x)$.

2.1-7. Let a random experiment be the casting of a pair of fair six-sided dice and let X equal the minimum of the two outcomes.

(a) With reasonable assumptions, find the pmf of X.

(b) Draw a probability histogram of the pmf of X.

(c) Let Y equal the range of the two outcomes (i.e., the absolute value of the difference of the largest and the smallest outcomes). Determine the pmf $g(y)$ of Y for $y = 0, 1, 2, 3, 4, 5$.

(d) Draw a probability histogram for $g(y)$.

2.1-8. A fair six-sided die has three faces numbered 1 and three faces numbered 3. Another fair six-sided die has its faces numbered 0, 2, 4, 6, 8, and 10. The two dice are rolled. Let X and Y be the respective outcomes of the roll. Let $W = X + Y$.

(a) Determine the pmf of W.

(b) Draw a probability histogram of the pmf of W.

2.1-9. The pmf of X is $f(x) = (1 + |x - 3|)/11$, for $x = 1, 2, 3, 4, 5$. Graph this pmf as a line graph.

2.1-10. Suppose there are 4 defective items in a lot (collection) of 100 items. A sample of size 20 is taken at random and without replacement. Let X denote the number of defective items in the sample. Find the probability that the sample contains

(a) Exactly one defective item.

(b) At most one defective item.

2.1-11. In a lot (collection) of 50 light bulbs, there are 4 defective bulbs. An inspector inspects 8 bulbs selected at random. What is the probability that she will not find any defectives bulbs in this randomly selected sample?

2.1-12. Let X be the number of accidents per week in a factory. Let the pmf of X be

$$f(x) = \frac{1}{(x+1)(x+2)} = \frac{1}{x+1} - \frac{1}{x+2}, \qquad x = 0, 1, 2, \ldots.$$

Find the conditional probability of $X \geq 4$, given that $X \geq 1$.

2.1-13. A professor gives her students 15 essay questions, which include 4 that she intends to use in a test. A student has time to study for only five of these questions. What is the probability that, of the questions studied,

(a) Exactly three are selected for the test?

(b) Exactly two are selected?

(c) At least one is selected?

2.1-14. Often in buying a product at a supermarket, there is a concern about the item being underweight. Suppose there are 20 "one-pound" packages of frozen ground turkey on display and 3 of them are underweight. A consumer group buys 5 of the 20 packages at random. What is the probability of at least one of the five being underweight?

2.1-15. Five cards are selected at random without replacement from a standard, thoroughly shuffled 52-card deck of playing cards. Let X equal the number of face cards (kings, queens, jacks) in the hand. Forty observations of X yielded the following data:

2 1 2 1 0 0 1 0 1 1 0 2 0 2 3 0 1 1 0 3

1 2 0 2 0 2 0 1 0 1 1 2 1 0 1 1 2 1 1 0

(a) Argue that the pmf of X is

$$f(x) = \frac{\binom{12}{x}\binom{40}{5-x}}{\binom{52}{5}}, \qquad x = 0, 1, 2, 3, 4, 5,$$

and thus, that $f(0) = 2109/8330$, $f(1) = 703/1666$, $f(2) = 209/833$, $f(3) = 55/833$, $f(4) = 165/21{,}658$, and $f(5) = 33/108{,}290$.

(b) Draw a probability histogram for this distribution.

(c) Determine the relative frequencies of $0, 1, 2, 3$, and superimpose the relative frequency histogram on your probability histogram.

2.1-16. (Michigan Mathematics Prize Competition, 1992, Part II) From the set $\{1, 2, 3, \ldots, n\}$, k distinct integers are selected at random and arranged in numerical order (from lowest to highest). Let $P(i, r, k, n)$ denote the probability that integer i is in position r. For example, observe that $P(1, 2, k, n) = 0$, as it is impossible for the number 1 to be in the second position after ordering.

(a) Compute $P(2, 1, 6, 10)$.

(b) Find a general formula for $P(i, r, k, n)$.

2.1-17. A bag contains 144 ping-pong balls. More than half of the balls are painted orange and the rest are painted blue. Two balls are drawn at random without replacement. The probability of drawing two balls of the same color is the same as the probability of drawing two balls of different colors. How many orange balls are in the bag?

2.2 MATHEMATICAL EXPECTATION

An extremely important concept in summarizing important characteristics of distributions of probability is that of mathematical expectation, which we introduce with an example.

Example
2.2-1

An enterprising young man who needs a little extra money devises a game of chance in which some of his friends might wish to participate. The game that he proposes is to let the participant cast a fair die and then receive a payment according to the following schedule: If the event $A = \{1, 2, 3\}$ occurs, he receives one dollar; if $B = \{4, 5\}$ occurs, he receives two dollars; and if $C = \{6\}$ occurs, he receives three dollars. If X is a random variable that represents the payoff, then the pmf of X is given by

$$f(x) = (4 - x)/6, \qquad x = 1, 2, 3;$$

that is, $f(1) = 3/6, f(2) = 2/6, f(3) = 1/6$. If the game is repeated a large number of times, the payment of one dollar would occur about 3/6 of the times, two dollars about 2/6 of the times, and three dollars about 1/6 of the times. Thus, the average payment would be

$$(1)\left(\frac{3}{6}\right) + (2)\left(\frac{2}{6}\right) + (3)\left(\frac{1}{6}\right) = \frac{10}{6} = \frac{5}{3}.$$

That is, the young man expects to pay 5/3 of a dollar "on the average." This is called the mathematical expectation of the payment. If the young man could charge two dollars to play the game, he could make $2 - 5/3 = 1/3$ of a dollar on the average each play. Note that this mathematical expectation can be written

$$E(X) = \sum_{x=1}^{3} x f(x)$$

and is often denoted by the Greek letter μ, which is called the mean of X or of its distribution. ∎

Suppose that we are interested in another function of X, say $u(X)$. Let us call it $Y = u(X)$. Of course, Y is a random variable and has a pmf. For illustration, in Example 2.2-1, $Y = X^2$ has the pmf

$$g(y) = (4 - \sqrt{y})/6, \qquad y = 1, 4, 9;$$

that is, $g(1) = 3/6, g(4) = 2/6, g(9) = 1/6$. Moreover, where S_Y is the space of Y, the mean of Y is

$$\mu_Y = \sum_{y \in S_Y} y\, g(y) = (1)\left(\frac{3}{6}\right) + (4)\left(\frac{2}{6}\right) + (9)\left(\frac{1}{6}\right) = \frac{20}{6} = \frac{10}{3}.$$

Participants in the young man's game might be more willing to play this game for 4 dollars as they can win $9 - 4 = 5$ dollars and lose only $4 - 1 = 3$ dollars. Note that the young man can expect to win $4 - 10/3 = 2/3$ of a dollar on the average each play. A game based upon $Z = X^3$ might even be more attractive to participants if the young man charges 10 dollars to play this game. Then the participant could win $27 - 10 = 17$ dollars and lose only $10 - 1 = 9$ dollars. The details of this latter game are covered in Exercise 2.2-5.

In any case, it is important to note that

$$E(Y) = \sum_{y \in S_Y} y\, g(y) = \sum_{x \in S_X} x^2 f(x) = \frac{20}{6} = \frac{10}{3}.$$

That is, the same value is obtained by either formula. While we have not proved, for a general function $u(x)$, that if $Y = u(X)$, then

$$\sum_{y \in S_Y} y\, g(y) = \sum_{x \in S_X} u(x) f(x);$$

we have illustrated it in this simple case. This discussion suggests the more general definition of mathematical expectation of a function of X.

Definition 2.2-1
If $f(x)$ is the pmf of the random variable X of the discrete type with space S, and if the summation

$$\sum_{x \in S} u(x)f(x), \qquad \text{which is sometimes written} \qquad \sum_{S} u(x)f(x),$$

exists, then the sum is called the **mathematical expectation** or the **expected value** of $u(X)$, and it is denoted by $E[u(X)]$. That is,

$$E[u(X)] = \sum_{x \in S} u(x)f(x).$$

We can think of the expected value $E[u(X)]$ as a weighted mean of $u(x)$, $x \in S$, where the weights are the probabilities $f(x) = P(X = x)$, $x \in S$.

REMARK The usual definition of mathematical expectation of $u(X)$ requires that the sum converge absolutely—that is, that

$$\sum_{x \in S} |u(x)| f(x)$$

converge and be finite. The reason for the absolute convergence is that it allows one, in the advanced proof of

$$\sum_{x \in S_X} u(x)f(x) = \sum_{y \in S_Y} yg(y),$$

to rearrange the order of the terms in the x-summation. In this book, each $u(x)$ is such that the convergence is absolute. ∎

We provide another example.

Example 2.2-2 Let the random variable X have the pmf

$$f(x) = \frac{1}{3}, \qquad x \in S_X,$$

where $S_X = \{-1, 0, 1\}$. Let $u(X) = X^2$. Then

$$E(X^2) = \sum_{x \in S_X} x^2 f(x) = (-1)^2 \left(\frac{1}{3}\right) + (0)^2 \left(\frac{1}{3}\right) + (1)^2 \left(\frac{1}{3}\right) = \frac{2}{3}.$$

However, the support of the random variable $Y = X^2$ is $S_Y = \{0, 1\}$ and

$$P(Y = 0) = P(X = 0) = \frac{1}{3},$$

$$P(Y = 1) = P(X = -1) + P(X = 1) = \frac{1}{3} + \frac{1}{3} = \frac{2}{3}.$$

That is,

$$g(y) = \begin{cases} \dfrac{1}{3}, & y = 0, \\[2ex] \dfrac{2}{3}, & y = 1; \end{cases}$$

and $S_Y = \{0, 1\}$. Hence,

$$\mu_Y = E(Y) = \sum_{y \in S_Y} y\, g(y) = (0)\left(\frac{1}{3}\right) + (1)\left(\frac{2}{3}\right) = \frac{2}{3},$$

which again illustrates the preceding observation. ∎

Before presenting additional examples, we list some useful facts about mathematical expectation in the following theorem.

Theorem 2.2-1

When it exists, the mathematical expectation E satisfies the following properties:

(a) If c is a constant, then $E(c) = c$.

(b) If c is a constant and u is a function, then

$$E[c\, u(X)] = cE[u(X)].$$

(c) If c_1 and c_2 are constants and u_1 and u_2 are functions, then

$$E[c_1 u_1(X) + c_2 u_2(X)] = c_1 E[u_1(X)] + c_2 E[u_2(X)].$$

Proof First, for the proof of (a), we have

$$E(c) = \sum_{x \in S} cf(x) = c \sum_{x \in S} f(x) = c$$

because

$$\sum_{x \in S} f(x) = 1.$$

Next, to prove (b), we see that

$$E[c\, u(X)] = \sum_{x \in S} c\, u(x) f(x)$$

$$= c \sum_{x \in S} u(x) f(x)$$

$$= c\, E[u(X)].$$

Finally, the proof of (c) is given by

$$E[c_1 u_1(X) + c_2 u_2(X)] = \sum_{x \in S} [c_1 u_1(x) + c_2 u_2(x)] f(x)$$

$$= \sum_{x \in S} c_1 u_1(x) f(x) + \sum_{x \in S} c_2 u_2(x) f(x).$$

By applying (b), we obtain

$$E[c_1 u_1(X) + c_2 u_2(X)] = c_1 E[u_1(X)] + c_2 E[u_2(X)]. \qquad \square$$

Property (c) can be extended to more than two terms by mathematical induction; that is, we have

$$\textbf{(c')} \quad E\left[\sum_{i=1}^{k} c_i u_i(X)\right] = \sum_{i=1}^{k} c_i E[u_i(X)].$$

Because of property (c'), the mathematical expectation E is often called a **linear** or **distributive** operator.

Example 2.2-3

Let X have the pmf

$$f(x) = \frac{x}{10}, \qquad x = 1, 2, 3, 4.$$

Then the mean of X is

$$\mu = E(X) = \sum_{x=1}^{4} x\left(\frac{x}{10}\right)$$

$$= (1)\left(\frac{1}{10}\right) + (2)\left(\frac{2}{10}\right) + (3)\left(\frac{3}{10}\right) + (4)\left(\frac{4}{10}\right) = 3,$$

$$E(X^2) = \sum_{x=1}^{4} x^2\left(\frac{x}{10}\right)$$

$$= (1)^2\left(\frac{1}{10}\right) + (2)^2\left(\frac{2}{10}\right) + (3)^2\left(\frac{3}{10}\right) + (4)^2\left(\frac{4}{10}\right) = 10,$$

and

$$E[X(5 - X)] = 5E(X) - E(X^2) = (5)(3) - 10 = 5. \qquad \blacksquare$$

Example 2.2-4

Let $u(x) = (x - b)^2$, where b is not a function of X, and suppose $E[(X - b)^2]$ exists. To find that value of b for which $E[(X - b)^2]$ is a minimum, we write

$$g(b) = E[(X - b)^2] = E[X^2 - 2bX + b^2]$$
$$= E(X^2) - 2bE(X) + b^2$$

because $E(b^2) = b^2$. To find the minimum, we differentiate $g(b)$ with respect to b, set $g'(b) = 0$, and solve for b as follows:

$$g'(b) = -2E(X) + 2b = 0,$$
$$b = E(X).$$

Since $g''(b) = 2 > 0$, the mean of X, $\mu = E(X)$, is the value of b that minimizes $E[(X - b)^2]$. $\qquad \blacksquare$

Example 2.2-5

Let X have a hypergeometric distribution in which n objects are selected from $N = N_1 + N_2$ objects as described in Section 2.1. Then

$$\mu = E(X) = \sum_{x \in S} x \, \frac{\binom{N_1}{x}\binom{N_2}{n-x}}{\binom{N}{n}}.$$

Since the first term of this summation equals zero when $x = 0$, and since

$$\binom{N}{n} = \left(\frac{N}{n}\right)\binom{N-1}{n-1},$$

we can write

$$E(X) = \sum_{0 < x \in S} x \, \frac{N_1!}{x!(N_1 - x)!} \, \frac{\binom{N_2}{n-x}}{\left(\dfrac{N}{n}\right)\binom{N-1}{n-1}}.$$

Of course, $x/x! = 1/(x-1)!$ when $x \neq 0$; thus,

$$E(X) = \left(\frac{n}{N}\right) \sum_{0 < x \in S} \frac{(N_1)(N_1 - 1)!}{(x-1)!(N_1 - x)!} \, \frac{\binom{N_2}{n-x}}{\binom{N-1}{n-1}}$$

$$= n\left(\frac{N_1}{N}\right) \sum_{0 < x \in S} \frac{\binom{N_1 - 1}{x - 1}\binom{N_2}{n - 1 - (x-1)}}{\binom{N-1}{n-1}}.$$

However, when $x > 0$, the summand of this last expression represents the probability of obtaining, say, $x - 1$ red chips if $n - 1$ chips are selected from $N_1 - 1$ red chips and N_2 blue chips. Since the summation is over all possible values of $x - 1$, it must sum to 1, as it is the sum of all possible probabilities of $x - 1$. Thus,

$$\mu = E(X) = n\left(\frac{N_1}{N}\right),$$

which is a result that agrees with our intuition: We expect the number X of red chips to equal the product of the number n of selections and the fraction N_1/N of red chips in the original collection. ▪

Example 2.2-6 Say an experiment has probability of success p, where $0 < p < 1$, and probability of failure $1 - p = q$. This experiment is repeated independently until the first success occurs; say this happens on the X trial. Clearly the space of X is $S_X = \{1, 2, 3, 4, \ldots\}$. What is $P(X = x)$, where $x \in S_X$? We must observe $x - 1$ failures and then a success to have this happen. Thus, due to the independence, the probability is

$$f(x) = P(X = x) = \overbrace{q \cdot q \cdots q}^{x-1 \, q's} \cdot p = q^{x-1}p, \qquad x \in S_X.$$

Since p and q are positive, this is a pmf because

$$\sum_{x \in S_X} q^{x-1}p = p(1 + q + q^2 + q^3 + \cdots) = \frac{p}{1-q} = \frac{p}{p} = 1.$$

The mean of this **geometric distribution** is

$$\mu = \sum_{x=1}^{\infty} xf(x) = (1)p + (2)qp + (3)q^2p + \cdots$$

and

$$q\mu = (q)p + (2)q^2p + (3)q^3p + \cdots.$$

If we subtract the second of these two equations from the first, we have

$$(1-q)\mu = p + pq + pq^2 + pq^3 + \cdots$$
$$= (p)(1 + q + q^2 + q^3 + \cdots)$$
$$= (p)\left(\frac{1}{1-q}\right) = 1.$$

That is,

$$\mu = \frac{1}{1-q} = \frac{1}{p}.$$

For illustration, if $p = 1/10$, we would expect $\mu = 10$ trials are needed on the average to observe a success. This certainly agrees with our intuition. ∎

Exercises

2.2-1. Find $E(X)$ for each of the distributions given in Exercise 2.1-3.

2.2-2. Let the random variable X have the pmf

$$f(x) = \frac{(|x|+1)^2}{27}, \qquad x = -2, -1, 0, 1, 2.$$

Compute $E(X)$, $E(X^2)$, and $E(X^2 - 3X + 9)$.

2.2-3. Let the random variable X be the number of days that a certain patient needs to be in the hospital. Suppose X has the pmf

$$f(x) = \frac{5-x}{10}, \qquad x = 1, 2, 3, 4.$$

If the patient is to receive $200 from an insurance company for each of the first two days in the hospital and $100 for each day after the first two days, what is the expected payment for the hospitalization?

2.2-4. An insurance company sells an automobile policy with a deductible of one unit. Let X be the amount of the loss having pmf

$$f(x) = \begin{cases} 0.9, & x = 0, \\ \dfrac{c}{x}, & x = 1, 2, 3, 4, 5, 6, \end{cases}$$

where c is a constant. Determine c and the expected value of the amount the insurance company must pay.

2.2-5. In Example 2.2-1 let $Z = u(X) = X^3$.

(a) Find the pmf of Z, say $h(z)$.

(b) Find $E(Z)$.

(c) How much, on average, can the young man expect to win on each play if he charges $10 per play?

2.2-6. Let the pmf of X be defined by $f(x) = 6/(\pi^2 x^2)$, $x = 1, 2, 3, \ldots$. Show that $E(X) = +\infty$ and thus, does not exist.

2.2-7. An engineering firm needs to prepare a proposal for a research contract. The cost of preparing the proposal is $50,000. If this proposal is accepted, the probabilities of the firm earning potential gross profits of $500,000, $300,000, $100,000, or $0 from the contract are 0.20, 0.50, 0.20, and 0.10, respectively. Let 0.30 be the probability that the proposal is accepted. Calculate the net profit that the firm can earn.

2.2-8. Let X be a random variable with support $\{1, 2, 3, 5, 15, 25, 50\}$, each point of which has the same probability $1/7$. Argue that $c = 5$ is the value that minimizes $h(c) = E(|X - c|)$. Compare c with the value of b that minimizes $g(b) = E[(X - b)^2]$.

2.2-9. A roulette wheel used in a U.S. casino has 38 slots, of which 18 are red, 18 are black, and 2 are green. A roulette wheel used in a French casino has 37 slots, of which 18 are red, 18 are black, and 1 is green. A ball is rolled around the wheel and ends up in one of the slots with equal probability. Suppose that a player bets on red. If a $1 bet is placed, the player wins $1 if the ball ends up in a red slot. (The player's $1 bet is returned.) If the ball ends up in a black or green slot, the player loses $1. Find the expected value of this game to the player in

(a) The United States.

(b) France.

2.2-10. In the casino game called **high-low**, there are three possible bets. Assume that $15 is the size of the bet. A pair of fair six-sided dice is rolled and their sum is calculated. If you bet **low**, you win $15 if the sum of the dice is $\{2, 3, 4, 5, 6\}$. If you bet **high**, you win \$15 if the sum of the dice is $\{8, 9, 10, 11, 12\}$. If you bet on $\{7\}$, you win \$20 if a sum of 7 is rolled. Otherwise, you lose on each of the three bets. In all three cases, your original \$15 is returned if you win. Find the expected value of the game to the bettor for each of these three bets.

2.2-11. In the gambling game craps (see Exercise 1.3-13), the player wins \$1 with probability 0.49293 and loses \$1 with probability 0.50707 for each \$1 bet. What is the expected value of the game to the player?

2.2-12. Suppose that a school has 20 classes: 16 with 25 students in each, three with 100 students in each, and one with 300 students, for a total of 1000 students.

(a) What is the average class size?

(b) Select a student randomly out of the 1000 students. Let the random variable X equal the size of the class to which this student belongs, and define the pmf of X.

(c) Find $E(X)$, the expected value of X. Does this answer surprise you?

2.3 SPECIAL MATHEMATICAL EXPECTATIONS

Let us consider an example in which $x \in \{1, 2, 3\}$ and the pmf is given by $f(1) = 3/6, f(2) = 2/6, f(3) = 1/6$. That is, the probability that the random variable X equals 1, denoted by $P(X = 1)$, is $f(1) = 3/6$. Likewise, $P(X = 2) = f(2) = 2/6$ and $P(X = 3) = f(3) = 1/6$. Of course, $f(x) > 0$ when $x \in S$, and it must be the case that

$$\sum_{x \in S} f(x) = f(1) + f(2) + f(3) = 1.$$

We can think of the points 1, 2, 3 as having weights (probabilities) 3/6, 2/6, 1/6, and their weighted mean (weighted average) is

$$\mu = E(X) = 1 \cdot \frac{3}{6} + 2 \cdot \frac{2}{6} + 3 \cdot \frac{1}{6} = \frac{10}{6} = \frac{5}{3},$$

which, in this illustration, does not equal one of the x values in S. As a matter of fact, it is two thirds of the way between $x = 1$ and $x = 2$.

In Section 2.2 we called $\mu = E(X)$ the mean of the random variable X (or of its distribution). In general, suppose the random variable X has the space $S = \{u_1, u_2, \ldots, u_k\}$ and these points have respective probabilities $P(X = u_i) = f(u_i) > 0$, where $f(x)$ is the pmf. Of course,

$$\sum_{x \in S} f(x) = 1$$

and the **mean** of the random variable X (or of its distribution) is

$$\mu = \sum_{x \in S} x f(x) = u_1 f(u_1) + u_2 f(u_2) + \cdots + u_k f(u_k).$$

That is, in the notation of Section 2.2, $\mu = E(X)$.

Now, u_i is the distance of that ith point from the origin. In mechanics, the product of a distance and its weight is called a moment, so $u_i f(u_i)$ is a moment having a moment arm of length u_i. The sum of such products would be the moment of the system of distances and weights. Actually, it is called the first moment about the origin, since the distances are simply to the first power and the lengths of the arms (distances) are measured from the origin. However, if we compute the first moment about the mean μ, then, since here a moment arm equals $(x - \mu)$, we have

$$\sum_{x \in S} (x - \mu) f(x) = E[(X - \mu)] = E(X) - E(\mu)$$

$$= \mu - \mu = 0.$$

That is, that first moment about μ is equal to zero. In mechanics μ is called the centroid. The last equation implies that if a fulcrum is placed at the centroid μ, then the system of weights would balance, as the sum of the positive moments (when $x > \mu$) about μ equals the sum of the negative moments (when $x < \mu$). In our first illustration, $\mu = 10/6$ is the centroid, so the negative moment

$$\left(1 - \frac{10}{6}\right) \cdot \frac{3}{6} = -\frac{12}{36} = -\frac{1}{3}$$

equals the sum of the two positive moments

$$\left(2 - \frac{10}{6}\right) \cdot \frac{2}{6} + \left(3 - \frac{10}{6}\right) \cdot \frac{1}{6} = \frac{12}{36} = \frac{1}{3}.$$

Since $\mu = E(X)$, it follows from Example 2.2-4 that $b = \mu$ minimizes $E[(X-b)^2]$. Also, Example 2.2-5 shows that

$$\mu = n\left(\frac{N_1}{N}\right)$$

is the mean of the hypergeometric distribution. Moreover, $\mu = 1/p$ is the mean of the geometric distribution from Example 2.2-6.

Statisticians often find it valuable to compute the second moment about the mean μ. It is called the second moment because the distances are raised to the second power, and it is equal to $E[(X - \mu)^2]$; that is,

$$\sum_{x \in S} (x - \mu)^2 f(x) = (u_1 - \mu)^2 f(u_1) + (u_2 - \mu)^2 f(u_2) + \cdots + (u_k - \mu)^2 f(u_k).$$

This weighted mean of the squares of those distances is called the **variance** of the random variable X (or of its distribution). The positive square root of the variance is called the **standard deviation** of X and is denoted by the Greek letter σ (sigma). Thus, the variance is σ^2, sometimes denoted by Var(X). That is, $\sigma^2 = E[(X - \mu)^2] = $ Var(X). In our first illustration, since $\mu = 10/6$, the variance equals

$$\sigma^2 = \text{Var}(X) = \left(1 - \frac{10}{6}\right)^2 \cdot \frac{3}{6} + \left(2 - \frac{10}{6}\right)^2 \cdot \frac{2}{6} + \left(3 - \frac{10}{6}\right)^2 \cdot \frac{1}{6} = \frac{120}{216} = \frac{5}{9}.$$

Hence, the standard deviation is

$$\sigma = \sqrt{\sigma^2} = \sqrt{\frac{120}{216}} = 0.745.$$

It is worth noting that the variance can be computed in another way, because

$$\begin{aligned} \sigma^2 = E[(X - \mu)^2] &= E[X^2 - 2\mu X + \mu^2] \\ &= E(X^2) - 2\mu E(X) + \mu^2 \\ &= E(X^2) - \mu^2. \end{aligned}$$

That is, the variance σ^2 equals the difference of the second moment about the origin and the square of the mean. For our first illustration,

$$\sigma^2 = \sum_{x=1}^{3} x^2 f(x) - \mu^2$$

$$= 1^2\left(\frac{3}{6}\right) + 2^2\left(\frac{2}{6}\right) + 3^2\left(\frac{1}{6}\right) - \left(\frac{10}{6}\right)^2 = \frac{20}{6} - \frac{100}{36} = \frac{120}{216} = \frac{5}{9},$$

which agrees with our previous computation.

Example 2.3-1

Let X equal the number of spots on the side facing upward after a fair six-sided die is rolled. A reasonable probability model is given by the pmf

$$f(x) = P(X = x) = \frac{1}{6}, \qquad x = 1, 2, 3, 4, 5, 6.$$

The mean of X is

$$\mu = E(X) = \sum_{x=1}^{6} x\left(\frac{1}{6}\right) = \frac{1 + 2 + 3 + 4 + 5 + 6}{6} = \frac{7}{2}.$$

The second moment about the origin is

$$E(X^2) = \sum_{x=1}^{6} x^2\left(\frac{1}{6}\right) = \frac{1^2 + 2^2 + 3^2 + 4^2 + 5^2 + 6^2}{6} = \frac{91}{6}.$$

Thus, the variance equals

$$\sigma^2 = \frac{91}{6} - \left(\frac{7}{2}\right)^2 = \frac{182 - 147}{12} = \frac{35}{12}.$$

The standard deviation is $\sigma = \sqrt{35/12} = 1.708$. ◼

Although most students understand that $\mu = E(X)$ is, in some sense, a measure of the middle of the distribution of X, it is more difficult to get much of a feeling for the variance and the standard deviation. The next example illustrates that the standard deviation is a measure of the dispersion, or spread, of the points belonging to the space S.

Example 2.3-2

Let X have the pmf $f(x) = 1/3, x = -1, 0, 1$. Here the mean is

$$\mu = \sum_{x=-1}^{1} xf(x) = (-1)\left(\frac{1}{3}\right) + (0)\left(\frac{1}{3}\right) + (1)\left(\frac{1}{3}\right) = 0.$$

Accordingly, the variance, denoted by σ_X^2, is

$$\sigma_X^2 = E[(X - 0)^2]$$

$$= \sum_{x=-1}^{1} x^2 f(x)$$

$$= (-1)^2\left(\frac{1}{3}\right) + (0)^2\left(\frac{1}{3}\right) + (1)^2\left(\frac{1}{3}\right)$$

$$= \frac{2}{3},$$

so the standard deviation is $\sigma_X = \sqrt{2/3}$. Next, let another random variable Y have the pmf $g(y) = 1/3, y = -2, 0, 2$. Its mean is also zero, and it is easy to show that $\text{Var}(Y) = 8/3$, so the standard deviation of Y is $\sigma_Y = 2\sqrt{2/3}$. Here the standard deviation of Y is twice that of the standard deviation of X, reflecting the fact that the probability of Y is spread out twice as much as that of X. ◼

Example 2.3-3

Let X have a uniform distribution on the first m positive integers. The mean of X is

$$\mu = E(X) = \sum_{x=1}^{m} x\left(\frac{1}{m}\right) = \frac{1}{m}\sum_{x=1}^{m} x$$

$$= \left(\frac{1}{m}\right)\frac{m(m+1)}{2} = \frac{m+1}{2}.$$

To find the variance of X, we first find

$$E(X^2) = \sum_{x=1}^{m} x^2\left(\frac{1}{m}\right) = \frac{1}{m}\sum_{x=1}^{m} x^2$$

$$= \left(\frac{1}{m}\right)\frac{m(m+1)(2m+1)}{6} = \frac{(m+1)(2m+1)}{6}.$$

Thus, the variance of X is

$$\sigma^2 = \text{Var}(X) = E[(X-\mu)^2]$$

$$= E(X^2) - \mu^2 = \frac{(m+1)(2m+1)}{6} - \left(\frac{m+1}{2}\right)^2$$

$$= \frac{m^2-1}{12}.$$

For example, we find that if X equals the outcome when rolling a fair six-sided die, the pmf of X is

$$f(x) = \frac{1}{6}, \qquad x = 1, 2, 3, 4, 5, 6;$$

the respective mean and variance of X are

$$\mu = \frac{6+1}{2} = 3.5 \qquad \text{and} \qquad \sigma^2 = \frac{6^2-1}{12} = \frac{35}{12},$$

which agrees with calculations of Example 2.3-1. ∎

Now let X be a random variable with mean μ_X and variance σ_X^2. Of course, $Y = aX + b$, where a and b are constants, is a random variable, too. The mean of Y is

$$\mu_Y = E(Y) = E(aX + b) = aE(X) + b = a\mu_X + b.$$

Moreover, the variance of Y is

$$\sigma_Y^2 = E[(Y-\mu_Y)^2] = E[(aX + b - a\mu_X - b)^2] = E[a^2(X-\mu_X)^2] = a^2\sigma_X^2.$$

Thus, $\sigma_Y = |a|\sigma_X$. To illustrate, note in Example 2.3-2 that the relationship between the two distributions could be explained by defining $Y = 2X$, so that $\sigma_Y^2 = 4\sigma_X^2$ and consequently $\sigma_Y = 2\sigma_X$, which we had observed there. In addition, we see that adding or subtracting a constant from X does not change the variance. For illustration, $\text{Var}(X-1) = \text{Var}(X)$, because $a = 1$ and $b = -1$. Also note that $\text{Var}(-X) = \text{Var}(X)$ because here $a = -1$ and $b = 0$.

Let r be a positive integer. If

$$E(X^r) = \sum_{x\in S} x^r f(x)$$

is finite, it is called the rth **moment** of the distribution about the origin. In addition, the expectation

$$E[(X - b)^r] = \sum_{x \in S} (x - b)^r f(x)$$

is called the rth moment of the distribution about b.

For a given positive integer r,

$$E[(X)_r] = E[X(X - 1)(X - 2) \cdots (X - r + 1)]$$

is called the rth factorial moment. We note that the second factorial moment is equal to the difference of the second and first moments about 0:

$$E[X(X - 1)] = E(X^2) - E(X).$$

There is another formula that can be used to compute the variance. This formula uses the second factorial moment and sometimes simplifies the calculations. First find the values of $E(X)$ and $E[X(X - 1)]$. Then

$$\sigma^2 = E[X(X - 1)] + E(X) - [E(X)]^2,$$

since, by the distributive property of E, this becomes

$$\sigma^2 = E(X^2) - E(X) + E(X) - [E(X)]^2 = E(X^2) - \mu^2.$$

Example 2.3-4 In Example 2.2-5 concerning the hypergeometric distribution, we found that the mean of that distribution is

$$\mu = E(X) = n\left(\frac{N_1}{N}\right) = np,$$

where $p = N_1/N$, the fraction of red chips in the N chips. In Exercise 2.3-10, it is determined that

$$E[X(X - 1)] = \frac{(n)(n - 1)(N_1)(N_1 - 1)}{N(N - 1)}.$$

Thus, the variance of X is $E[X(X - 1)] + E(X) - [E(X)]^2$, namely,

$$\sigma^2 = \frac{n(n - 1)(N_1)(N_1 - 1)}{N(N - 1)} + \frac{nN_1}{N} - \left(\frac{nN_1}{N}\right)^2.$$

After some straightforward algebra, we find that

$$\sigma^2 = n\left(\frac{N_1}{N}\right)\left(\frac{N_2}{N}\right)\left(\frac{N - n}{N - 1}\right) = np(1 - p)\left(\frac{N - n}{N - 1}\right). \qquad \blacksquare$$

We now define a function that will help us generate the moments of a distribution. Thus, this function is called the moment-generating function. Although this generating characteristic is extremely important, there is a uniqueness property that is even more important. We first define the new function and then explain this uniqueness property before showing how it can be used to compute the moments of X.

Definition 2.3-1

Let X be a random variable of the discrete type with pmf $f(x)$ and space S. If there is a positive number h such that

$$E(e^{tX}) = \sum_{x \in S} e^{tx} f(x)$$

exists and is finite for $-h < t < h$, then the function defined by

$$M(t) = E(e^{tX})$$

is called the **moment-generating function of** X (or of the distribution of X). This function is often abbreviated as mgf.

First, it is evident that if we set $t = 0$, we have $M(0) = 1$. Moreover, if the space of S is $\{b_1, b_2, b_3, \ldots\}$, then the moment-generating function is given by the expansion

$$M(t) = e^{tb_1} f(b_1) + e^{tb_2} f(b_2) + e^{tb_3} f(b_3) + \cdots.$$

Thus, the coefficient of e^{tb_i} is the probability

$$f(b_i) = P(X = b_i).$$

Accordingly, if two random variables (or two distributions of probability) have the same moment-generating function, they must have the same distribution of probability. That is, if the two random variables had the two probability mass functions $f(x)$ and $g(y)$, as well as the same space $S = \{b_1, b_2, b_3, \ldots\}$, and if

$$e^{tb_1} f(b_1) + e^{tb_2} f(b_2) + \cdots = e^{tb_1} g(b_1) + e^{tb_2} g(b_2) + \cdots \qquad (2.3\text{-}1)$$

for all t, $-h < t < h$, then mathematical transform theory requires that

$$f(b_i) = g(b_i), \qquad i = 1, 2, 3, \ldots.$$

So we see that the moment-generating function of a discrete random variable uniquely determines the distribution of that random variable. In other words, if the mgf exists, there is one and only one distribution of probability associated with that mgf.

REMARK From elementary algebra, we can get some understanding of why Equation 2.3-1 requires that $f(b_i) = g(b_i)$. In that equation, let $e^t = w$ and say the points in the support, namely, $b_1, b_2, \ldots, b_k$, are positive integers, the largest of which is m. Then Equation 2.3-1 provides the equality of two mth-degree polynomials in w for an uncountable number of values of w. A fundamental theorem of algebra requires that the corresponding coefficients of the two polynomials be equal; that is, $f(b_i) = g(b_i), i = 1, 2, \ldots, k.$ ∎

Example 2.3-5

If X has the mgf

$$M(t) = e^t \left(\frac{3}{6} \right) + e^{2t} \left(\frac{2}{6} \right) + e^{3t} \left(\frac{1}{6} \right), \qquad -\infty < t < \infty,$$

then the support of X is $S = \{1, 2, 3\}$ and the associated probabilities are

$$P(X = 1) = \frac{3}{6}, \qquad P(X = 2) = \frac{2}{6}, \qquad P(X = 3) = \frac{1}{6}.$$

We could write this, if we choose to do so, by saying that X has the pmf

$$f(x) = \frac{4 - x}{6}, \qquad x = 1, 2, 3.$$ ∎

**Example
2.3-6** Suppose the mgf of X is

$$M(t) = \frac{e^t/2}{1 - e^t/2}, \qquad t < \ln 2.$$

Until we expand $M(t)$, we cannot detect the coefficients of $e^{b_i t}$. Recalling that

$$(1 - z)^{-1} = 1 + z + z^2 + z^3 + \cdots, \qquad -1 < z < 1,$$

we have

$$\frac{e^t}{2}\left(1 - \frac{e^t}{2}\right)^{-1} = \frac{e^t}{2}\left(1 + \frac{e^t}{2} + \frac{e^{2t}}{2^2} + \frac{e^{3t}}{2^3} + \cdots\right)$$

$$= \left(e^t\right)\left(\frac{1}{2}\right)^1 + \left(e^{2t}\right)\left(\frac{1}{2}\right)^2 + \left(e^{3t}\right)\left(\frac{1}{2}\right)^3 + \cdots$$

when $e^t/2 < 1$ and thus $t < \ln 2$. That is,

$$P(X = x) = \left(\frac{1}{2}\right)^x$$

when x is a positive integer, or, equivalently, the pmf of X is

$$f(x) = \left(\frac{1}{2}\right)^x, \qquad x = 1, 2, 3, \ldots.$$ ∎

From the theory of Laplace transforms, it can be shown that the existence of $M(t)$, for $-h < t < h$, implies that derivatives of $M(t)$ of all orders exist at $t = 0$; hence, $M(t)$ is continuous at $t = 0$. Moreover, it is permissible to interchange differentiation and summation as the series converges uniformly. Thus,

$$M'(t) = \sum_{x \in S} x e^{tx} f(x),$$

$$M''(t) = \sum_{x \in S} x^2 e^{tx} f(x),$$

and for each positive integer r,

$$M^{(r)}(t) = \sum_{x \in S} x^r e^{tx} f(x).$$

Setting $t = 0$, we see that

$$M'(0) = \sum_{x \in S} x f(x) = E(X),$$

$$M''(0) = \sum_{x \in S} x^2 f(x) = E(X^2),$$

and, in general,

$$M^{(r)}(0) = \sum_{x \in S} x^r f(x) = E(X^r).$$

In particular, if the moment-generating function exists, then

$$M'(0) = E(X) = \mu \qquad \text{and} \qquad M''(0) - [M'(0)]^2 = E(X^2) - [E(X)]^2 = \sigma^2.$$

The preceding argument shows that we can find the moments of X by differentiating $M(t)$. In using this technique, it must be emphasized that first we evaluate the summation representing $M(t)$ to obtain a closed-form solution and then we differentiate that solution to obtain the moments of X. The next example illustrates the use of the moment-generating function for finding the first and second moments and then the mean and variance of the geometric distribution.

Example 2.3-7

Suppose X has the geometric distribution of Example 2.2-6; that is, the pmf of X is

$$f(x) = q^{x-1}p, \qquad x = 1, 2, 3, \ldots.$$

Then the mgf of X is

$$M(t) = E(e^{tX}) = \sum_{x=1}^{\infty} e^{tx} q^{x-1} p = \left(\frac{p}{q}\right) \sum_{x=1}^{\infty} (qe^t)^x$$

$$= \left(\frac{p}{q}\right) [(qe^t) + (qe^t)^2 + (qe^t)^3 + \cdots]$$

$$= \left(\frac{p}{q}\right) \frac{qe^t}{1 - qe^t} = \frac{pe^t}{1 - qe^t}, \qquad \text{provided } qe^t < 1 \text{ or } t < -\ln q.$$

Note that $-\ln q = h$ is positive. To find the mean and the variance of X, we first differentiate $M(t)$ twice:

$$M'(t) = \frac{(1 - qe^t)(pe^t) - pe^t(-qe^t)}{(1 - qe^t)^2} = \frac{pe^t}{(1 - qe^t)^2}$$

and

$$M''(t) = \frac{(1 - qe^t)^2 pe^t - pe^t(2)(1 - qe^t)(-qe^t)}{(1 - qe^t)^4} = \frac{pe^t(1 + qe^t)}{(1 - qe^t)^3}.$$

Of course, $M(0) = 1$ and $M(t)$ is continuous at $t = 0$ as we were able to differentiate at $t = 0$. With $1 - q = p$,

$$M'(0) = \frac{p}{(1 - q)^2} = \frac{1}{p} = \mu$$

and

$$M''(0) = \frac{p(1 + q)}{(1 - q)^3} = \frac{1 + q}{p^2}.$$

Thus,

$$\sigma^2 = M''(0) - [M'(0)]^2 = \frac{1 + q}{p^2} - \frac{1}{p^2} = \frac{q}{p^2}.$$

∎

Exercises

2.3-1. Find the mean and variance for the following discrete distributions:

(a) $f(x) = \dfrac{1}{5}$, $x = 5, 10, 15, 20, 25$.

(b) $f(x) = 1$, $x = 5$.

(c) $f(x) = \dfrac{4 - x}{6}$, $x = 1, 2, 3$.

2.3-2. For each of the following distributions, find $\mu = E(X)$, $E[X(X-1)]$, and $\sigma^2 = E[X(X-1)] + E(X) - \mu^2$:

(a) $f(x) = \dfrac{3!}{x!(3-x)!}\left(\dfrac{1}{4}\right)^x \left(\dfrac{3}{4}\right)^{3-x}$, $x = 0, 1, 2, 3$.

(b) $f(x) = \dfrac{4!}{x!(4-x)!}\left(\dfrac{1}{2}\right)^4$, $x = 0, 1, 2, 3, 4$.

2.3-3. Given $E(X + 4) = 10$ and $E[(X + 4)^2] = 116$, determine **(a)** $\mathrm{Var}(X + 4)$, **(b)** $\mu = E(X)$, and **(c)** $\sigma^2 = \mathrm{Var}(X)$.

2.3-4. Let μ and σ^2 denote the mean and variance of the random variable X. Determine $E[(X - \mu)/\sigma]$ and $E\{[(X - \mu)/\sigma]^2\}$.

2.3-5. Consider an experiment that consists of selecting a card at random from an ordinary deck of cards. Let the random variable X equal the value of the selected card, where Ace $= 1$, Jack $= 11$, Queen $= 12$, and King $= 13$. Thus, the space of X is $S = \{1, 2, 3, \ldots, 13\}$. If the experiment is performed in an unbiased manner, assign probabilities to these 13 outcomes and compute the mean μ of this probability distribution.

2.3-6. Place eight chips in a bowl: Three have the number 1 on them, two have the number 2, and three have the number 3. Say each chip has a probability of 1/8 of being drawn at random. Let the random variable X equal the number on the chip that is selected, so that the space of X is $S = \{1, 2, 3\}$. Make reasonable probability assignments to each of these three outcomes, and compute the mean μ and the variance σ^2 of this probability distribution.

2.3-7. Let X equal an integer selected at random from the first m positive integers, $\{1, 2, \ldots, m\}$. Find the value of m for which $E(X) = \mathrm{Var}(X)$. (See Zerger in the references.)

2.3-8. Let X equal the larger outcome when a pair of fair four-sided dice is rolled. The pmf of X is

$$f(x) = \frac{2x - 1}{16}, \qquad x = 1, 2, 3, 4.$$

Find the mean, variance, and standard deviation of X.

2.3-9. A warranty is written on a product worth $10,000 so that the buyer is given $8000 if it fails in the first year, $6000 if it fails in the second, $4000 if it fails in the third, $2000 if it fails in the fourth, and zero after that. The probability that the product fails in the first year is 0.1, and the probability that it fails in any subsequent year, provided that it did not fail prior to that year, is 0.1. What is the expected value of the warranty?

2.3-10. To find the variance of a hypergeometric random variable in Example 2.3-4 we used the fact that

$$E[X(X-1)] = \frac{N_1(N_1 - 1)(n)(n - 1)}{N(N - 1)}.$$

Prove this result by making the change of variables $k = x - 2$ and noting that

$$\binom{N}{n} = \frac{N(N-1)}{n(n-1)}\binom{N-2}{n-2}.$$

2.3-11. If the moment-generating function of X is

$$M(t) = \frac{2}{5} e^t + \frac{1}{5} e^{2t} + \frac{2}{5} e^{3t},$$

find the mean, variance, and pmf of X.

2.3-12. Let X equal the number of times a computer can malfunction in a day. X can have the values 0, 1, 2, 3, 4, 5, and 6. The probabilities associated with each of these values are 0.15, 0.30, 0.25, 0.15, 0.05, 0.06, and 0.04, respectively.

(a) Compute the values of the mean, variance, and standard deviation of X.

(b) Find $P(X > 3)$ and $P(X < 3)$.

2.3-13. For each question on a multiple-choice test, there are five possible answers, of which exactly one is correct. If a student selects answers at random, give the probability that the first question answered correctly is question 4.

2.3-14. The probability that a machine produces a defective item is 0.01. Each item is checked as it is produced. Assume that these are independent trials, and compute the probability that at least 100 items must be checked to find one that is defective.

2.3-15. Apples are packaged automatically in 3-pound bags. Suppose that 4% of the time the bag of apples weighs less than 3 pounds. If you select bags randomly and weigh them in order to discover one underweight bag of apples, find the probability that the number of bags that must be selected is

(a) At least 20.

(b) At most 20.

(c) Exactly 20.

2.3-16. Let X equal the number of flips of a fair coin that are required to observe the same face on consecutive flips.

(a) Find the pmf of X. HINT: Draw a tree diagram.

(b) Find the moment-generating function of X.

(c) Use the mgf to find the values of **(i)** the mean and **(ii)** the variance of X.

(d) Find the values of **(i)** $P(X \leq 3)$, **(ii)** $P(X \geq 5)$, and **(iii)** $P(X = 3)$.

2.3-17. Let X equal the number of flips of a fair coin that are required to observe heads–tails on consecutive flips.

(a) Find the pmf of X. HINT: Draw a tree diagram.

(b) Show that the mgf of X is $M(t) = e^{2t}/(e^t - 2)^2$.

(c) Use the mgf to find the values of **(i)** the mean and **(ii)** the variance of X.

(d) Find the values of **(i)** $P(X \leq 3)$, **(ii)** $P(X \geq 5)$, and **(iii)** $P(X = 3)$.

2.3-18. Let X have a geometric distribution. Show that

$$P(X > k + j \mid X > k) = P(X > j),$$

where k and j are nonnegative integers. NOTE: We sometimes say that in this situation there has been loss of memory.

2.3-19. Given a random permutation of the integers in the set $\{1, 2, 3, 4, 5\}$, let X equal the number of integers that are in their natural position. The moment-generating function of X is

$$M(t) = \frac{44}{120} + \frac{45}{120}e^t + \frac{20}{120}e^{2t} + \frac{10}{120}e^{3t} + \frac{1}{120}e^{5t}.$$

(a) Find the mean and variance of X.

(b) Find the probability that at least one integer is in its natural position.

(c) Draw a graph of the probability histogram of the pmf of X.

2.4 THE BINOMIAL DISTRIBUTION

The probability models for random experiments that will be described in this section occur frequently in applications.

A **Bernoulli experiment** is a random experiment, the outcome of which can be classified in one of two mutually exclusive and exhaustive ways—say, success or failure (e.g., female or male, life or death, nondefective or defective). A sequence of **Bernoulli trials** occurs when a Bernoulli experiment is performed several *independent* times and the probability of success—say, p—remains the *same* from trial to trial. That is, in such a sequence we let p denote the probability of success on each trial. In addition, we shall frequently let $q = 1 - p$ denote the probability of failure; that is, we shall use q and $1 - p$ interchangeably.

Example 2.4-1 Suppose that the probability of germination of a beet seed is 0.8 and the germination of a seed is called a success. If we plant 10 seeds and can assume that the germination of one seed is independent of the germination of another seed, this would correspond to 10 Bernoulli trials with $p = 0.8$. ∎

Example 2.4-2 In the Michigan daily lottery the probability of winning when placing a six-way boxed bet is 0.006. A bet placed on each of 12 successive days would correspond to 12 Bernoulli trials with $p = 0.006$. ∎

Let X be a random variable associated with a Bernoulli trial by defining it as follows:

$$X(\text{success}) = 1 \quad \text{and} \quad X(\text{failure}) = 0.$$

That is, the two outcomes, success and failure, are denoted by one and zero, respectively. The pmf of X can be written as

$$f(x) = p^x(1 - p)^{1-x}, \qquad x = 0, 1,$$

and we say that X has a **Bernoulli distribution**. The expected value of X is

$$\mu = E(X) = \sum_{x=0}^{1} x p^x (1-p)^{1-x} = (0)(1-p) + (1)(p) = p,$$

and the variance of X is

$$\sigma^2 = \text{Var}(X) = \sum_{x=0}^{1} (x-p)^2 p^x (1-p)^{1-x}$$

$$= (0-p)^2(1-p) + (1-p)^2 p = p(1-p) = pq.$$

It follows that the standard deviation of X is

$$\sigma = \sqrt{p(1-p)} = \sqrt{pq}.$$

In a sequence of n Bernoulli trials, we shall let X_i denote the Bernoulli random variable associated with the ith trial. An observed sequence of n Bernoulli trials will then be an n-tuple of zeros and ones, and we often call this collection a **random sample** of size n from a Bernoulli distribution.

Example 2.4-3 Out of millions of instant lottery tickets, suppose that 20% are winners. If five such tickets are purchased, then $(0,0,0,1,0)$ is a possible observed sequence in which the fourth ticket is a winner and the other four are losers. Assuming independence among winning and losing tickets, we observe that the probability of this outcome is

$$(0.8)(0.8)(0.8)(0.2)(0.8) = (0.2)(0.8)^4. \qquad \blacksquare$$

Example 2.4-4 If five beet seeds are planted in a row, a possible observed sequence would be $(1, 0, 1, 0, 1)$ in which the first, third, and fifth seeds germinated and the other two did not. If the probability of germination is $p = 0.8$, the probability of this outcome is, assuming independence,

$$(0.8)(0.2)(0.8)(0.2)(0.8) = (0.8)^3(0.2)^2. \qquad \blacksquare$$

In a sequence of Bernoulli trials, we are often interested in the total number of successes but not the actual order of their occurrences. If we let the random variable X equal the number of observed successes in n Bernoulli trials, then the possible values of X are $0, 1, 2, \ldots, n$. If x successes occur, where $x = 0, 1, 2, \ldots, n$, then $n - x$ failures occur. The number of ways of selecting x positions for the x successes in the n trials is

$$\binom{n}{x} = \frac{n!}{x!(n-x)!}.$$

Since the trials are independent and since the probabilities of success and failure on each trial are, respectively, p and $q = 1 - p$, the probability of each of these ways is $p^x(1-p)^{n-x}$. Thus, $f(x)$, the pmf of X, is the sum of the probabilities of the $\binom{n}{x}$ mutually exclusive events; that is,

$$f(x) = \binom{n}{x} p^x (1-p)^{n-x}, \qquad x = 0, 1, 2, \ldots, n.$$

These probabilities are called binomial probabilities, and the random variable X is said to have a **binomial distribution**.

Summarizing, a binomial experiment satisfies the following properties:

1. A Bernoulli (success–failure) experiment is performed n times, where n is a (non-random) constant.
2. The trials are independent.
3. The probability of success on each trial is a constant p; the probability of failure is $q = 1 - p$.
4. The random variable X equals the number of successes in the n trials.

A binomial distribution will be denoted by the symbol $b(n, p)$, and we say that the distribution of X is $b(n, p)$. The constants n and p are called the **parameters** of the binomial distribution; they correspond to the number n of independent trials and the probability p of success on each trial. Thus, if we say that the distribution of X is $b(12, 1/4)$, we mean that X is the number of successes in a random sample of size $n = 12$ from a Bernoulli distribution with $p = 1/4$.

Example 2.4-5

In the instant lottery with 20% winning tickets, if X is equal to the number of winning tickets among $n = 8$ that are purchased, then the probability of purchasing two winning tickets is

$$f(2) = P(X = 2) = \binom{8}{2}(0.2)^2(0.8)^6 = 0.2936.$$

The distribution of the random variable X is $b(8, 0.2)$. ∎

Example 2.4-6

In order to obtain a better feeling for the effect of the parameters n and p on the distribution of probabilities, four probability histograms are displayed in Figure 2.4-1. ∎

Example 2.4-7

In Example 2.4-1, the number X of seeds that germinate in $n = 10$ independent trials is $b(10, 0.8)$; that is,

$$f(x) = \binom{10}{x}(0.8)^x(0.2)^{10-x}, \qquad x = 0, 1, 2, \ldots, 10.$$

In particular,

$$P(X \leq 8) = 1 - P(X = 9) - P(X = 10)$$
$$= 1 - 10(0.8)^9(0.2) - (0.8)^{10} = 0.6242.$$

Also, with a little more work, we could compute

$$P(X \leq 6) = \sum_{x=0}^{6} \binom{10}{x}(0.8)^x(0.2)^{10-x} = 0.1209.$$

∎

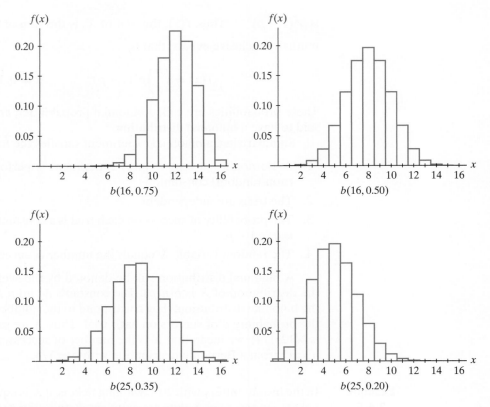

Figure 2.4-1 Binomial probability histograms

Recall that cumulative probabilities like those in the previous example are given by the cumulative distribution function (cdf) of X or sometimes called the distribution function (df) of X, defined by

$$F(x) = P(X \le x), \qquad -\infty < x < \infty.$$

We tend to use the cdf (rather than the pmf) to obtain probabilities of events involving a $b(n, p)$ random variable X. Tables of this cdf are given in Table II in Appendix B for selected values of n and p.

For the binomial distribution given in Example 2.4-7, namely, the $b(10, 0.8)$ distribution, the distribution function is defined by

$$F(x) = P(X \le x) = \sum_{y=0}^{\lfloor x \rfloor} \binom{10}{y} (0.8)^y (0.2)^{10-y},$$

where $\lfloor x \rfloor$ is the greatest integer in x. A graph of this cdf is shown in Figure 2.4-2. Note that the vertical jumps at the integers in this step function are equal to the probabilities associated with those respective integers.

Example 2.4-8 Leghorn chickens are raised for laying eggs. Let $p = 0.5$ be the probability that a newly hatched chick is a female. Assuming independence, let X equal the number of female chicks out of 10 newly hatched chicks selected at random. Then the distribution of X is $b(10, 0.5)$. From Table II in Appendix B, the probability of 5 or fewer female chicks is

$$P(X \le 5) = 0.6230.$$

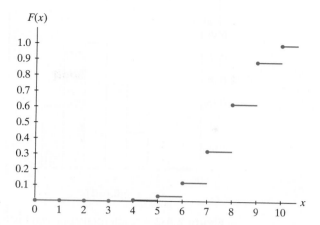

Figure 2.4-2 Distribution function for the $b(10, 0.8)$ distribution

The probability of exactly 6 female chicks is

$$P(X = 6) = \binom{10}{6}\left(\frac{1}{2}\right)^6\left(\frac{1}{2}\right)^4$$
$$= P(X \le 6) - P(X \le 5)$$
$$= 0.8281 - 0.6230 = 0.2051,$$

since $P(X \le 6) = 0.8281$. The probability of at least 6 female chicks is

$$P(X \ge 6) = 1 - P(X \le 5) = 1 - 0.6230 = 0.3770. \qquad \blacksquare$$

Although probabilities for the binomial distribution $b(n, p)$ are given in Table II in Appendix B for selected values of p that are less than or equal to 0.5, the next example demonstrates that this table can also be used for values of p that are greater than 0.5. In later sections we learn how to approximate certain binomial probabilities with those of other distributions. In addition, you may use your calculator and/or a statistical package such as Minitab to find binomial probabilities.

Example 2.4-9 Suppose that we are in one of those rare times when 65% of the American public approve of the way the president of the United States is handling the job. Take a random sample of $n = 8$ Americans and let Y equal the number who give approval. Then, to a very good approximation, the distribution of Y is $b(8, 0.65)$. (Y would have the stated distribution exactly if the sampling were done with replacement, but most public opinion polling uses sampling without replacement.) To find $P(Y \ge 6)$, note that

$$P(Y \ge 6) = P(8 - Y \le 8 - 6) = P(X \le 2),$$

where $X = 8 - Y$ counts the number who disapprove. Since $q = 1 - p = 0.35$ equals the probability of disapproval by each person selected, the distribution of X is $b(8, 0.35)$. (See Figure 2.4-3.) From Table II in Appendix B, since $P(X \le 2) = 0.4278$, it follows that $P(Y \ge 6) = 0.4278$.

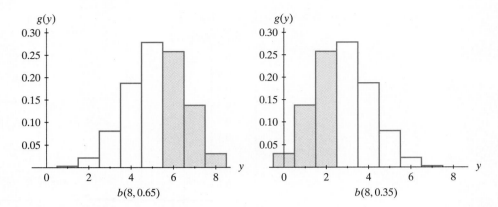

Figure 2.4-3 Presidential approval histogram

Similarly,

$$P(Y \leq 5) = P(8 - Y \geq 8 - 5)$$
$$= P(X \geq 3) = 1 - P(X \leq 2)$$
$$= 1 - 0.4278 = 0.5722$$

and

$$P(Y = 5) = P(8 - Y = 8 - 5)$$
$$= P(X = 3) = P(X \leq 3) - P(X \leq 2)$$
$$= 0.7064 - 0.4278 = 0.2786.$$ ∎

Recall that if n is a positive integer, then

$$(a + b)^n = \sum_{x=0}^{n} \binom{n}{x} b^x a^{n-x}.$$

Thus, if we use this binomial expansion with $b = p$ and $a = 1 - p$, then the sum of the binomial probabilities is

$$\sum_{x=0}^{n} \binom{n}{x} p^x (1 - p)^{n-x} = [(1 - p) + p]^n = 1,$$

a result that had to follow from the fact that $f(x)$ is a pmf.

We now use the binomial expansion to find the mgf for a binomial random variable and then the mean and variance.

The mgf is

$$M(t) = E(e^{tX}) = \sum_{x=0}^{n} e^{tx} \binom{n}{x} p^x (1 - p)^{n-x}$$

$$= \sum_{x=0}^{n} \binom{n}{x} (pe^t)^x (1 - p)^{n-x}$$

$$= [(1 - p) + pe^t]^n, \qquad -\infty < t < \infty,$$

from the expansion of $(a+b)^n$ with $a = 1 - p$ and $b = pe^t$. It is interesting to note that here and elsewhere the mgf is usually rather easy to compute if the pmf has a factor involving an exponential, like p^x in the binomial pmf.

The first two derivatives of $M(t)$ are

$$M'(t) = n[(1-p) + pe^t]^{n-1}(pe^t)$$

and

$$M''(t) = n(n-1)[(1-p) + pe^t]^{n-2}(pe^t)^2 + n[(1-p) + pe^t]^{n-1}(pe^t).$$

Thus,

$$\mu = E(X) = M'(0) = np$$

and

$$\sigma^2 = E(X^2) - [E(X)]^2 = M''(0) - [M'(0)]^2$$
$$= n(n-1)p^2 + np - (np)^2 = np(1-p).$$

Note that when p is the probability of success on each trial, the expected number of successes in n trials is np, a result that agrees with our intuition.

In the special case when $n = 1$, X has a Bernoulli distribution and

$$M(t) = (1-p) + pe^t$$

for all real values of t, $\mu = p$, and $\sigma^2 = p(1-p)$.

Example 2.4-10 Suppose that observation over a long period of time has disclosed that, on the average, 1 out of 10 items produced by a process is defective. Select five items independently from the production line and test them. Let X denote the number of defective items among the $n = 5$ items. Then X is $b(5, 0.1)$. Furthermore,

$$E(X) = 5(0.1) = 0.5, \qquad \mathrm{Var}(X) = 5(0.1)(0.9) = 0.45.$$

For example, the probability of observing at most one defective item is

$$P(X \leq 1) = \binom{5}{0}(0.1)^0(0.9)^5 + \binom{5}{1}(0.1)^1(0.9)^4 = 0.9185. \qquad \blacksquare$$

Suppose that an urn contains N_1 success balls and N_2 failure balls. Let $p = N_1/(N_1 + N_2)$, and let X equal the number of success balls in a random sample of size n that is taken from this urn. If the sampling is done one at a time with replacement, then the distribution of X is $b(n, p)$; if the sampling is done without replacement, then X has a hypergeometric distribution with pmf

$$f(x) = \frac{\binom{N_1}{x}\binom{N_2}{n-x}}{\binom{N_1 + N_2}{n}},$$

where x is a nonnegative integer such that $x \leq n$, $x \leq N_1$, and $n - x \leq N_2$. When $N_1 + N_2$ is large and n is relatively small, it makes little difference if the sampling is done with or without replacement. In Figure 2.4-4, the probability histograms are compared for different combinations of n, N_1, and N_2.

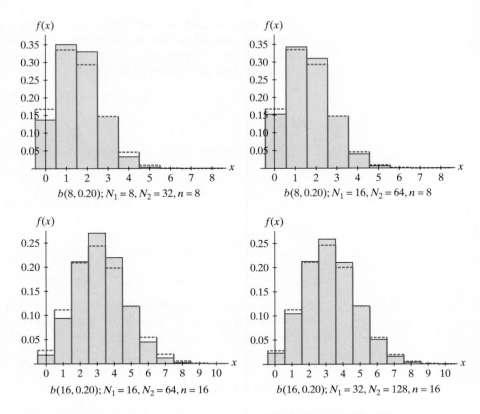

Figure 2.4-4 Binomial and hypergeometric (shaded) probability histograms

Exercises

2.4-1. An urn contains 7 red and 11 white balls. Draw one ball at random from the urn. Let $X = 1$ if a red ball is drawn, and let $X = 0$ if a white ball is drawn. Give the pmf, mean, and variance of X.

2.4-2. Suppose that in Exercise 2.4-1, $X = 5$ if a red ball is drawn and $X = -5$ if a white ball is drawn. Give the pmf, mean, and variance of X.

2.4-3. On a six-question multiple-choice test there are five possible answers for each question, of which one is correct (C) and four are incorrect (I). If a student guesses randomly and independently, find the probability of

(a) Being correct only on questions 1 and 4 (i.e., scoring C, I, I, C, I, I).

(b) Being correct on two questions.

2.4-4. It is claimed that 21% of the ducks in a particular region have patent schistosome infection. Suppose that nine ducks are selected at random. Let X equal the number of ducks that are infected.

(a) Assuming independence, how is X distributed?

(b) Find (i) $P(X \geq 3)$, (ii) $P(X = 1)$, and (iii) $P(X \leq 4)$.

2.4-5. In a lab experiment involving inorganic syntheses of molecular precursors to organometallic ceramics, the final step of a five-step reaction involves the formation of a metal–metal bond. The probability of such a bond forming is $p = 0.20$. Let X equal the number of successful reactions out of $n = 25$ such experiments.

(a) Find the probability that X is at most 4.

(b) Find the probability that X is at least 5.

(c) Find the probability that X is equal to 6.

(d) Give the mean, variance, and standard deviation of X.

2.4-6. It is believed that a health-care law has encouraged approximately 89% of the youth in a country to obtain insurance. Suppose this is true, and let X equal the number of youth with private health insurance in a random sample of $n = 50$.

(a) How is X distributed?

(b) Find the probability that X is at least 30.

(c) Find the probability that X is at most 30.

(d) Find the probability that X is equal to 30.

(e) Give the mean, variance, and standard deviation of X.

2.4-7. Suppose that 2000 points are selected independently and at random from the unit square $\{(x, y) : 0 \leq x < 1, 0 \leq y < 1\}$. Let W equal the number of points that fall into $A = \{(x, y) : x^2 + y^2 < 1\}$.

(a) How is W distributed?

(b) Give the mean, variance, and standard deviation of W.

(c) What is the expected value of $W/500$?

(d) Use the computer to select 2000 pairs of random numbers. Determine the value of W and use that value to find an estimate for π. (Of course, we know the real value of π, and more will be said about estimation later in this text.)

(e) How could you extend part (d) to estimate the volume $V = (4/3)\pi$ of a ball of radius 1 in 3-space?

(f) How could you extend these techniques to estimate the "volume" of a ball of radius 1 in n-space?

2.4-8. A boiler has five relief valves. The probability that each opens properly is 0.97.

(a) Find the probability that at least two open properly.

(b) Find the probability that all five open properly.

2.4-9. Suppose that the percentage of American drivers who are multitaskers (e.g., talk on cell phones, eat a snack, or text message at the same time they are driving) is approximately 80%. In a random sample of $n = 20$ drivers, let X equal the number of multitaskers.

(a) How is X distributed?

(b) Give the values of the mean, variance, and standard deviation of X.

(c) Find the following probabilities: **(i)** $P(X = 15)$, **(ii)** $P(X > 15)$, and **(iii)** $P(X \leq 15)$.

2.4-10. A certain type of mint has a label weight of 20.4 grams. Suppose that the probability is 0.90 that a mint weighs more than 20.7 grams. Let X equal the number of mints that weigh more than 20.7 grams in a sample of eight mints selected at random.

(a) How is X distributed if we assume independence?

(b) Find **(i)** $P(X = 8)$, **(ii)** $P(X \leq 6)$, and **(iii)** $P(X \geq 6)$.

2.4-11. A random variable X has a binomial distribution with mean 6 and variance 3.6. Find $P(X = 4)$.

2.4-12. In the casino game chuck-a-luck, three fair six-sided dice are rolled. One possible bet is $1 on fives, and the payoff is equal to $1 for each five on that roll. In addition, the dollar bet is returned if at least one five is rolled. The dollar that was bet is lost only if no fives are rolled. Let X denote the payoff for this game. Then X can equal $-1, 1, 2,$ or 3.

(a) Determine the pmf $f(x)$.

(b) Calculate μ, σ^2, and σ.

(c) Depict the pmf as a probability histogram.

2.4-13. It is claimed that for a particular lottery, 1/10 of the 50 million tickets will win a prize. What is the probability of winning at least one prize if you purchase **(a)** 10 tickets or **(b)** 15 tickets?

2.4-14. For the lottery described in Exercise 2.4-13, find the smallest number of tickets that must be purchased so that the probability of winning at least one prize is greater than **(a)** 0.50; **(b)** 0.95.

2.4-15. A hospital obtains 40% of its flu vaccine from Company A, 50% from Company B, and 10% from Company C. From past experience, it is known that 3% of the vials from A are ineffective, 2% from B are ineffective, and 5% from C are ineffective. The hospital tests five vials from each shipment. If at least one of the five is ineffective, find the conditional probability of that shipment's having come from C.

2.4-16. A company starts a fund of M dollars, from which it pays $2500 to each employee who achieves high performance during the year. The probability of each employee achieving this goal is 0.15 and is independent of the probabilities of the other employees doing so. If there are $n = 20$ employees, how much should M equal so that the fund has a probability of at least 98% of covering those payments?

2.4-17. Your stockbroker is free to take your calls about 60% of the time; otherwise, he is talking to another client or is out of the office. You call him at five random times during a given month. (Assume independence.)

(a) What is the probability that he will take every one of the five calls?

(b) What is the probability that he will accept exactly three of your five calls?

(c) What is the probability that he will accept at least one of the calls?

2.4-18. In group testing for a certain disease, a blood sample was taken from each of n individuals and part of each sample was placed in a common pool. The latter was then tested. If the result was negative, there was no more testing and all n individuals were declared negative with one test. If, however, the combined result was found positive, all individuals were tested, requiring $n+1$ tests. If $p = 0.05$ is the probability of a person's having the disease and $n = 5$, compute the expected number of tests needed, assuming independence.

2.4-19. Define the pmf and give the values of μ, σ^2, and σ when the moment-generating function of X is defined by

(a) $M(t) = 1/3 + (2/3)e^t$.

(b) $M(t) = (0.25 + 0.75e^t)^{12}$.

2.4-20. (i) Give the name of the distribution of X (if it has a name), **(ii)** find the values of μ and σ^2, and **(iii)** calculate $P(1 \leq X \leq 2)$ when the moment-generating function of X is given by

(a) $M(t) = (0.3 + 0.7e^t)^5$.

(b) $M(t) = \dfrac{0.3e^t}{1 - 0.7e^t}, \qquad t < -\ln(0.7)$.

(c) $M(t) = 0.45 + 0.55e^t$.

(d) $M(t) = 0.3e^t + 0.4e^{2t} + 0.2e^{3t} + 0.1e^{4t}$.

(e) $M(t) = \sum_{x=1}^{10} (0.1)e^{tx}$.

2.5 THE NEGATIVE BINOMIAL DISTRIBUTION

We turn now to the situation in which we observe a sequence of independent Bernoulli trials until exactly r successes occur, where r is a fixed positive integer. Let the random variable X denote the number of trials needed to observe the rth success. That is, X is the trial number on which the rth success is observed. By the multiplication rule of probabilities, the pmf of X—say, $g(x)$—equals the product of the probability

$$\binom{x-1}{r-1}p^{r-1}(1-p)^{x-r} = \binom{x-1}{r-1}p^{r-1}q^{x-r}$$

of obtaining exactly $r-1$ successes in the first $x-1$ trials and the probability p of a success on the rth trial. Thus, the pmf of X is

$$g(x) = \binom{x-1}{r-1}p^{r}(1-p)^{x-r} = \binom{x-1}{r-1}p^{r}q^{x-r}, \quad x = r, r+1, \ldots.$$

We say that X has a **negative binomial distribution**.

REMARK The reason for calling this distribution the negative binomial distribution is as follows: Consider $h(w) = (1-w)^{-r}$, the binomial $(1-w)$ with the negative exponent $-r$. Using Maclaurin's series expansion, we have

$$(1-w)^{-r} = \sum_{k=0}^{\infty} \frac{h^{(k)}(0)}{k!}w^{k} = \sum_{k=0}^{\infty}\binom{r+k-1}{r-1}w^{k}, \quad -1 < w < 1.$$

If we let $x = k + r$ in the summation, then $k = x - r$ and

$$(1-w)^{-r} = \sum_{x=r}^{\infty}\binom{r+x-r-1}{r-1}w^{x-r} = \sum_{x=r}^{\infty}\binom{x-1}{r-1}w^{x-r},$$

the summand of which is, except for the factor p^{r}, the negative binomial probability when $w = q$. In particular, the sum of the probabilities for the negative binomial distribution is 1 because

$$\sum_{x=r}^{\infty}g(x) = \sum_{x=r}^{\infty}\binom{x-1}{r-1}p^{r}q^{x-r} = p^{r}(1-q)^{-r} = 1. \qquad \blacksquare$$

If $r = 1$ in the negative binomial distribution, we note that X has a **geometric distribution**, since the pmf consists of terms of a geometric series, namely,

$$g(x) = p(1-p)^{x-1}, \quad x = 1, 2, 3, \ldots.$$

Recall that for a geometric series, the sum is given by

$$\sum_{k=0}^{\infty} ar^{k} = \sum_{k=1}^{\infty} ar^{k-1} = \frac{a}{1-r}$$

when $|r| < 1$. Thus, for the geometric distribution,

$$\sum_{x=1}^{\infty}g(x) = \sum_{x=1}^{\infty}(1-p)^{x-1}p = \frac{p}{1-(1-p)} = 1,$$

so that $g(x)$ does satisfy the properties of a pmf.

From the sum of a geometric series, we also note that when k is an integer,

$$P(X > k) = \sum_{x=k+1}^{\infty} (1-p)^{x-1}p = \frac{(1-p)^k p}{1-(1-p)} = (1-p)^k = q^k.$$

Thus, the value of the cdf at a positive integer k is

$$P(X \le k) = \sum_{x=1}^{k} (1-p)^{x-1}p = 1 - P(X > k) = 1 - (1-p)^k = 1 - q^k.$$

Example 2.5-1

Some biology students were checking eye color in a large number of fruit flies. For the individual fly, suppose that the probability of white eyes is 1/4 and the probability of red eyes is 3/4, and that we may treat these observations as independent Bernoulli trials. The probability that at least four flies have to be checked for eye color to observe a white-eyed fly is given by

$$P(X \ge 4) = P(X > 3) = q^3 = \left(\frac{3}{4}\right)^3 = \frac{27}{64} = 0.4219.$$

The probability that at most four flies have to be checked for eye color to observe a white-eyed fly is given by

$$P(X \le 4) = 1 - q^4 = 1 - \left(\frac{3}{4}\right)^4 = \frac{175}{256} = 0.6836.$$

The probability that the first fly with white eyes is the fourth fly considered is

$$P(X = 4) = q^{4-1}p = \left(\frac{3}{4}\right)^3 \left(\frac{1}{4}\right) = \frac{27}{256} = 0.1055.$$

It is also true that

$$P(X = 4) = P(X \le 4) - P(X \le 3)$$
$$= [1 - (3/4)^4] - [1 - (3/4)^3]$$
$$= \left(\frac{3}{4}\right)^3 \left(\frac{1}{4}\right). \qquad \blacksquare$$

We now show that the mean and the variance of a negative binomial random variable X are, respectively,

$$\mu = E(X) = \frac{r}{p} \quad \text{and} \quad \sigma^2 = \frac{rq}{p^2} = \frac{r(1-p)}{p^2}.$$

In particular, if $r = 1$, so that X has a geometric distribution, then

$$\mu = \frac{1}{p} \quad \text{and} \quad \sigma^2 = \frac{q}{p^2} = \frac{1-p}{p^2}.$$

The mean $\mu = 1/p$ agrees with our intuition. Let's check: If $p = 1/6$, then we would expect, on the average, $1/(1/6) = 6$ trials before the first success.

To find these moments, we determine the mgf of the negative binomial distribution. It is

$$M(t) = \sum_{x=r}^{\infty} e^{tx} \binom{x-1}{r-1} p^r (1-p)^{x-r}$$

$$= (pe^t)^r \sum_{x=r}^{\infty} \binom{x-1}{r-1} [(1-p)e^t]^{x-r}$$

$$= \frac{(pe^t)^r}{[1-(1-p)e^t]^r}, \qquad \text{where } (1-p)e^t < 1$$

(or, equivalently, when $t < -\ln(1-p)$). Thus,

$$M'(t) = (pe^t)^r(-r)[1-(1-p)e^t]^{-r-1}[-(1-p)e^t]$$
$$+ r(pe^t)^{r-1}(pe^t)[1-(1-p)e^t]^{-r}$$
$$= r(pe^t)^r[1-(1-p)e^t]^{-r-1}$$

and

$$M''(t) = r(pe^t)^r(-r-1)[1-(1-p)e^t]^{-r-2}[-(1-p)e^t]$$
$$+ r^2(pe^t)^{r-1}(pe^t)[1-(1-p)e^t]^{-r-1}.$$

Accordingly,

$$M'(0) = rp^r p^{-r-1} = rp^{-1}$$

and

$$M''(0) = r(r+1)p^r p^{-r-2}(1-p) + r^2 p^r p^{-r-1}$$
$$= rp^{-2}[(1-p)(r+1) + rp] = rp^{-2}(r+1-p).$$

Hence, we have

$$\mu = \frac{r}{p} \qquad \text{and} \qquad \sigma^2 = \frac{r(r+1-p)}{p^2} - \frac{r^2}{p^2} = \frac{r(1-p)}{p^2}.$$

Even these calculations are a little messy, so a somewhat easier way is given in Exercises 2.5-5 and 2.5-6.

Example 2.5-2 Suppose that during practice a basketball player can make a free throw 80% of the time. Furthermore, assume that a sequence of free-throw shooting can be thought of as independent Bernoulli trials. Let X equal the minimum number of free throws that this player must attempt to make a total of 10 shots. The pmf of X is

$$g(x) = \binom{x-1}{10-1}(0.80)^{10}(0.20)^{x-10}, \qquad x = 10, 11, 12, \ldots.$$

The mean, variance, and standard deviation of X are, respectively,

$$\mu = 10\left(\frac{1}{0.80}\right) = 12.5, \qquad \sigma^2 = \frac{10(0.20)}{0.80^2} = 3.125, \qquad \text{and} \qquad \sigma = 1.768.$$

And we have, for example,

$$P(X = 12) = g(12) = \binom{11}{9}(0.80)^{10}(0.20)^2 = 0.2362.$$ ∎

Example 2.5-3 To consider the effect of p and r on the negative binomial distribution, Figure 2.5-1 gives the probability histograms for four combinations of p and r. Note that since $r = 1$ in the first of these, it represents a geometric pmf. ∎

When the moment-generating function exists, derivatives of all orders exist at $t = 0$. Thus, it is possible to represent $M(t)$ as a Maclaurin series, namely,

$$M(t) = M(0) + M'(0)\left(\frac{t}{1!}\right) + M''(0)\left(\frac{t^2}{2!}\right) + M'''(0)\left(\frac{t^3}{3!}\right) + \cdots.$$

If the Maclaurin series expansion of $M(t)$ exists and the moments are given, we can sometimes sum the Maclaurin series to obtain the closed form of $M(t)$. This approach is illustrated in the next example.

Example 2.5-4 Let the moments of X be defined by

$$E(X^r) = 0.8, \qquad r = 1, 2, 3, \ldots.$$

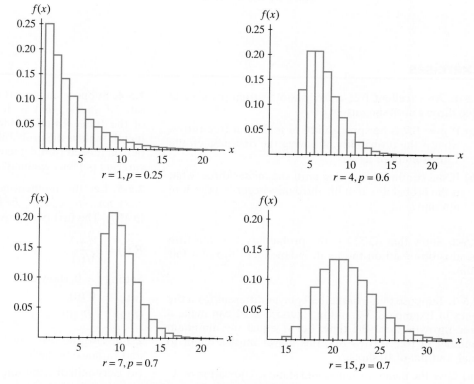

Figure 2.5-1 Negative binomial probability histograms

The moment-generating function of X is then

$$M(t) = M(0) + \sum_{r=1}^{\infty} 0.8\left(\frac{t^r}{r!}\right) = 1 + 0.8 \sum_{r=1}^{\infty} \frac{t^r}{r!}$$

$$= 0.2 + 0.8 \sum_{r=0}^{\infty} \frac{t^r}{r!} = 0.2e^{0t} + 0.8e^{1t}.$$

Thus,

$$P(X = 0) = 0.2 \quad \text{and} \quad P(X = 1) = 0.8.$$

This is an illustration of a Bernoulli distribution. ∎

The next example gives an application of the geometric distribution.

Example 2.5-5 A fair six-sided die is rolled until each face is observed at least once. On the average, how many rolls of the die are needed? It always takes one roll to observe the first outcome. To observe a different face from the first roll is like observing a geometric random variable with $p = 5/6$ and $q = 1/6$. So on the average it takes $1/(5/6) = 6/5$ rolls. After two different faces have been observed, the probability of observing a new face is $4/6$, so it will take, on the average, $1/(4/6) = 6/4$ rolls. Continuing in this manner, the answer is

$$1 + \frac{6}{5} + \frac{6}{4} + \frac{6}{3} + \frac{6}{2} + \frac{6}{1} = \frac{147}{10} = 14.7$$

rolls, on the average. ∎

Exercises

2.5-1. An excellent free-throw shooter attempts several free throws until she misses.

(a) If $p = 0.9$ is her probability of making a free throw, what is the probability of having the first miss on the 13th attempt or later?

(b) If she continues shooting until she misses three, what is the probability that the third miss occurs on the 30th attempt?

2.5-2. Show that $63/512$ is the probability that the fifth head is observed on the tenth independent flip of a fair coin.

2.5-3. Suppose that a basketball player different from the ones in Example 2.5-2 and in Exercise 2.5-1 can make a free throw 60% of the time. Let X equal the minimum number of free throws that this player must attempt to make a total of 10 shots.

(a) Give the mean, variance, and standard deviation of X.

(b) Find $P(X = 16)$.

2.5-4. Suppose an airport metal detector has a detection rate of 98.5%, that is, it misses detecting metal 1.5% of the time. Assume independence of people carrying metal. What is the probability that the first metal-carrying person missed (not detected) is among the first 20 metal-carrying persons scanned?

2.5-5. Let the moment-generating function $M(t)$ of X exist for $-h < t < h$. Consider the function $R(t) = \ln M(t)$. The first two derivatives of $R(t)$ are, respectively,

$$R'(t) = \frac{M'(t)}{M(t)} \quad \text{and} \quad R''(t) = \frac{M(t)M''(t) - [M'(t)]^2}{[M(t)]^2}.$$

Setting $t = 0$, show that

(a) $\mu = R'(0)$.

(b) $\sigma^2 = R''(0)$.

2.5-6. Use the result of Exercise 2.5-5 to find the mean and variance of the

(a) Bernoulli distribution.

(b) Binomial distribution.

(c) Geometric distribution.

(d) Negative binomial distribution.

2.5-7. If $E(X^r) = 5^r, r = 1, 2, 3, \ldots$, find the moment-generating function $M(t)$ of X and the pmf of X.

2.5-8. The probability that a company's work force has no accidents in a given month is 0.66. The numbers of accidents from month to month are independent. What is the probability that the fourth month in a year is the first month when at least one accident occurs?

2.5-9. One of four different prizes was randomly put into each box of a cereal. If a family decided to buy this cereal until it obtained at least one of each of the four different prizes, what is the expected number of boxes of cereal that must be purchased?

2.5-10. In 2012, Red Rose tea randomly began placing 1 of 12 English porcelain miniature figurines in a 100-bag box of the tea, selecting from 12 nautical figurines.

(a) On the average, how many boxes of tea must be purchased by a customer to obtain a complete collection consisting of the 12 nautical figurines?

(b) If the customer uses one tea bag per day, how long can a customer expect to take, on the average, to obtain a complete collection?

2.6 THE POISSON DISTRIBUTION

Some experiments result in counting the number of times particular events occur at given times or with given physical objects. For example, we could count the number of cell phone calls passing through a relay tower between 9 and 10 A.M., the number of flaws in 100 feet of wire, the number of customers that arrive at a ticket window between 12 noon and 2 P.M., or the number of defects in a 100-foot roll of aluminum screen that is 2 feet wide. Counting such events can be looked upon as observations of a random variable associated with an approximate Poisson process, provided that the conditions in the following definition are satisfied.

Definition 2.6-1

Let the number of occurrences of some event in a given continuous interval be counted. Then we have an **approximate Poisson process** with parameter $\lambda > 0$ if the following conditions are satisfied:

(a) The numbers of occurrences in nonoverlapping subintervals are independent.

(b) The probability of exactly one occurrence in a sufficiently short subinterval of length h is approximately λh.

(c) The probability of two or more occurrences in a sufficiently short subinterval is essentially zero.

REMARK We use *approximate* to modify the Poisson process since we use *approximately* in (b) and *essentially* in (c) to avoid the "little o" notation. Occasionally, we simply say "Poisson process" and drop *approximate*. ∎

Suppose that an experiment satisfies the preceding three conditions of an approximate Poisson process. Let X denote the number of occurrences in an interval of length 1 (where "length 1" represents one unit of the quantity under consideration). We would like to find an approximation for $P(X = x)$, where x is a nonnegative integer. To achieve this, we partition the unit interval into n subintervals of equal length $1/n$. If n is sufficiently large (i.e., much larger than x), we shall approximate the probability that there are x occurrences in this unit interval by finding the probability that exactly x of these n subintervals each has one occurrence. The probability of one occurrence in any one subinterval of length $1/n$ is approximately $\lambda(1/n)$, by condition (b). The probability of two or more occurrences in any one subinterval is essentially zero, by condition (c). So, for each subinterval, there is exactly one occurrence with a probability of approximately $\lambda(1/n)$. Consider the occurrence or nonoccurrence in each subinterval as a Bernoulli trial. By condition (a), we have a sequence of n Bernoulli trials with probability p approximately equal to $\lambda(1/n)$. Thus, an approximation for $P(X = x)$ is given by the binomial probability

$$\frac{n!}{x!\,(n-x)!}\left(\frac{\lambda}{n}\right)^x\left(1-\frac{\lambda}{n}\right)^{n-x}.$$

If n increases without bound, then

$$\lim_{n\to\infty}\frac{n!}{x!\,(n-x)!}\left(\frac{\lambda}{n}\right)^x\left(1-\frac{\lambda}{n}\right)^{n-x}$$

$$=\lim_{n\to\infty}\frac{n(n-1)\cdots(n-x+1)}{n^x}\frac{\lambda^x}{x!}\left(1-\frac{\lambda}{n}\right)^n\left(1-\frac{\lambda}{n}\right)^{-x}.$$

Now, for fixed x, we have

$$\lim_{n\to\infty}\frac{n(n-1)\cdots(n-x+1)}{n^x}=\lim_{n\to\infty}\left[(1)\left(1-\frac{1}{n}\right)\cdots\left(1-\frac{x-1}{n}\right)\right]=1,$$

$$\lim_{n\to\infty}\left(1-\frac{\lambda}{n}\right)^n=e^{-\lambda},$$

$$\lim_{n\to\infty}\left(1-\frac{\lambda}{n}\right)^{-x}=1.$$

Thus,

$$\lim_{n\to\infty}\frac{n!}{x!\,(n-x)!}\left(\frac{\lambda}{n}\right)^x\left(1-\frac{\lambda}{n}\right)^{n-x}=\frac{\lambda^x e^{-\lambda}}{x!}=P(X=x).$$

The distribution of probability associated with this process has a special name. We say that the random variable X has a **Poisson distribution** if its pmf is of the form

$$f(x)=\frac{\lambda^x e^{-\lambda}}{x!},\qquad x=0,1,2,\ldots,$$

where $\lambda > 0$.

It is easy to see that $f(x)$ has the properties of a pmf because, clearly, $f(x) \geq 0$ and, from the Maclaurin series expansion of e^λ, we have

$$\sum_{x=0}^{\infty}\frac{\lambda^x e^{-\lambda}}{x!}=e^{-\lambda}\sum_{x=0}^{\infty}\frac{\lambda^x}{x!}=e^{-\lambda}e^{\lambda}=1.$$

To discover the exact role of the parameter $\lambda > 0$, let us find some of the characteristics of the Poisson distribution. The mgf of X is

$$M(t)=E(e^{tX})=\sum_{x=0}^{\infty}e^{tx}\frac{\lambda^x e^{-\lambda}}{x!}=e^{-\lambda}\sum_{x=0}^{\infty}\frac{(\lambda e^t)^x}{x!}.$$

From the series representation of the exponential function, we have

$$M(t)=e^{-\lambda}e^{\lambda e^t}=e^{\lambda(e^t-1)}$$

for all real values of t. Now,

$$M'(t) = \lambda e^t e^{\lambda(e^t - 1)}$$

and

$$M''(t) = (\lambda e^t)^2 e^{\lambda(e^t - 1)} + \lambda e^t e^{\lambda(e^t - 1)}.$$

The values of the mean and variance of X are, respectively,

$$\mu = M'(0) = \lambda$$

and

$$\sigma^2 = M''(0) - [M'(0)]^2 = (\lambda^2 + \lambda) - \lambda^2 = \lambda.$$

That is, for the Poisson distribution, $\mu = \sigma^2 = \lambda$.

REMARK It is also possible to find the mean and the variance for the Poisson distribution directly, without using the mgf. The mean for the Poisson distribution is given by

$$E(X) = \sum_{x=0}^{\infty} x \frac{\lambda^x e^{-\lambda}}{x!} = e^{-\lambda} \sum_{x=1}^{\infty} \frac{\lambda^x}{(x - 1)!}$$

because $(0)f(0) = 0$ and $x/x! = 1/(x - 1)!$ when $x > 0$. If we let $k = x - 1$, then

$$E(X) = e^{-\lambda} \sum_{k=0}^{\infty} \frac{\lambda^{k+1}}{k!} = \lambda e^{-\lambda} \sum_{k=0}^{\infty} \frac{\lambda^k}{k!}$$

$$= \lambda e^{-\lambda} e^{\lambda} = \lambda.$$

To find the variance, we first determine the second factorial moment $E[X(X - 1)]$. We have

$$E[X(X - 1)] = \sum_{x=0}^{\infty} x(x - 1) \frac{\lambda^x e^{-\lambda}}{x!} = e^{-\lambda} \sum_{x=2}^{\infty} \frac{\lambda^x}{(x - 2)!}$$

because $(0)(0 - 1)f(0) = 0$, $(1)(1 - 1)f(1) = 0$, and $x(x - 1)/x! = 1/(x - 2)!$ when $x > 1$. If we let $k = x - 2$, then

$$E[X(X - 1)] = e^{-\lambda} \sum_{k=0}^{\infty} \frac{\lambda^{k+2}}{k!} = \lambda^2 e^{-\lambda} \sum_{k=0}^{\infty} \frac{\lambda^k}{k!}$$

$$= \lambda^2 e^{-\lambda} e^{\lambda} = \lambda^2.$$

Thus,

$$\text{Var}(X) = E(X^2) - [E(X)]^2 = E[X(X - 1)] + E(X) - [E(X)]^2$$
$$= \lambda^2 + \lambda - \lambda^2 = \lambda.$$

We again see that, for the Poisson distribution, $\mu = \sigma^2 = \lambda$. ∎

Table III in Appendix B gives values of the cdf of a Poisson random variable for selected values of λ. This table is illustrated in the next example.

Example 2.6-1

Let X have a Poisson distribution with a mean of $\lambda = 5$. Then, using Table III in Appendix B, we obtain

$$P(X \le 6) = \sum_{x=0}^{6} \frac{5^x e^{-5}}{x!} = 0.762,$$

$$P(X > 5) = 1 - P(X \le 5) = 1 - 0.616 = 0.384,$$

and

$$P(X = 6) = P(X \le 6) - P(X \le 5) = 0.762 - 0.616 = 0.146.$$ ■

Example 2.6-2

To see the effect of λ on the pmf $f(x)$ of X, Figure 2.6-1 shows the probability histograms of $f(x)$ for four different values of λ. ■

If events in an approximate Poisson process occur at a mean rate of λ per unit, then the expected number of occurrences in an interval of length t is λt. For example, let X equal the number of alpha particles emitted by barium-133 in one second and counted by a Geiger counter. If the mean number of emitted particles is 60 per second, then the expected number of emitted particles in 1/10 of a second is $60(1/10) = 6$. Moreover, the number of emitted particles, say X, in a time interval of length t has the Poisson pmf

$$f(x) = \frac{(\lambda t)^x e^{-\lambda t}}{x!}, \qquad x = 0, 1, 2, \ldots.$$

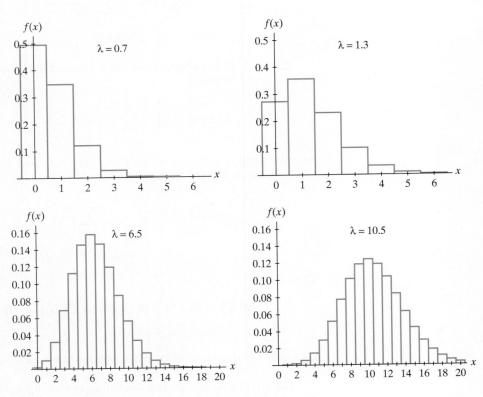

Figure 2.6-1 Poisson probability histograms

This equation follows if we treat the interval of length t as if it were the "unit interval" with mean λt instead of λ.

Example 2.6-3

A USB flash drive is sometimes used to back up computer files. However, in the past, a less reliable backup system that was used was a computer tape, and flaws occurred on these tapes. In a particular situation, flaws (bad records) on a used computer tape occurred on the average of one flaw per 1200 feet. If one assumes a Poisson distribution, what is the distribution of X, the number of flaws in a 4800-foot roll? The expected number of flaws in $4800 = 4(1200)$ feet is 4; that is, $E(X) = 4$. Thus, the pmf of X is

$$f(x) = \frac{4^x e^{-4}}{x!}, \qquad x = 0, 1, 2, \ldots,$$

and, in particular,

$$P(X = 0) = \frac{4^0 e^{-4}}{0!} = e^{-4} = 0.018,$$

$$P(X \le 4) = 0.629,$$

by Table III in Appendix B.

■

Example 2.6-4

In a large city, telephone calls to 911 come on the average of two every 3 minutes. If one assumes an approximate Poisson process, what is the probability of five or more calls arriving in a 9-minute period? Let X denote the number of calls in a 9-minute period. We see that $E(X) = 6$; that is, on the average, six calls will arrive during a 9-minute period. Thus,

$$P(X \ge 5) = 1 - P(X \le 4) = 1 - \sum_{x=0}^{4} \frac{6^x e^{-6}}{x!}$$

$$= 1 - 0.285 = 0.715,$$

by Table III in Appendix B.

■

Not only is the Poisson distribution important in its own right, but it can also be used to approximate probabilities for a binomial distribution. Earlier we saw that if X has a Poisson distribution with parameter λ, then with n large,

$$P(X = x) \approx \binom{n}{x} \left(\frac{\lambda}{n}\right)^x \left(1 - \frac{\lambda}{n}\right)^{n-x},$$

where $p = \lambda/n$, so that $\lambda = np$ in the above binomial probability. That is, if X has the binomial distribution $b(n, p)$ with large n and small p, then

$$\frac{(np)^x e^{-np}}{x!} \approx \binom{n}{x} p^x (1 - p)^{n-x}.$$

This approximation is reasonably good if n is large. But since λ was a fixed constant in that earlier argument, p should be small, because $np = \lambda$. In particular, the approximation is quite accurate if $n \ge 20$ and $p \le 0.05$ or if $n \ge 100$ and $p \le 0.10$, but it is not bad in other situations violating these bounds somewhat, such as $n = 50$ and $p = 0.12$.

Example 2.6-5

A manufacturer of Christmas tree light bulbs knows that 2% of its bulbs are defective. Assuming independence, the number of defective bulbs in a box of 100 bulbs has a binomial distribution with parameters $n = 100$ and $p = 0.02$. To approximate the probability that a box of 100 of these bulbs contains at most three defective bulbs, we use the Poisson distribution with $\lambda = 100(0.02) = 2$, which gives

$$\sum_{x=0}^{3} \frac{2^x e^{-2}}{x!} = 0.857,$$

from Table III in Appendix B. Using the binomial distribution, we obtain, after some tedious calculations,

$$\sum_{x=0}^{3} \binom{100}{x}(0.02)^x(0.98)^{100-x} = 0.859.$$

Hence, in this case, the Poisson approximation is extremely close to the true value, but much easier to find. ∎

REMARK With the availability of statistical computer packages and statistical calculators, it is often very easy to find binomial probabilities. So do not use the Poisson approximation if you are able to find the probability exactly. ∎

Example 2.6-6

In Figure 2.6-2, Poisson probability histograms have been superimposed on shaded binomial probability histograms so that we can see whether or not these are close to

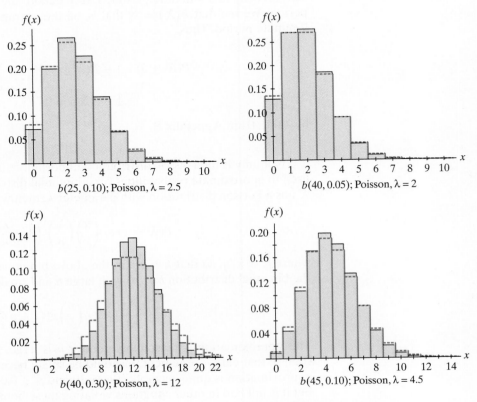

Figure 2.6-2 Binomial (shaded) and Poisson probability histograms

each other. If the distribution of X is $b(n,p)$, the approximating Poisson distribution has a mean of $\lambda = np$. Note that the approximation is not good when p is large (e.g., $p = 0.30$).

∎

Exercises

2.6-1. Let X have a Poisson distribution with a mean of 10. Find

(a) $P(4 \leq X \leq 9)$.

(b) $P(X \geq 4)$.

(c) $P(X \leq 4)$.

2.6-2. Let X have a Poisson distribution with a variance of 5. Find $P(X = 3)$.

2.6-3. Customers arrive at a travel agency at a mean rate of 11 per hour. Assuming that the number of arrivals per hour has a Poisson distribution, give the probability that more than 10 customers arrive in a given hour.

2.6-4. Let X have a Poisson distribution such that $P(X = 0) = P(X = 1)$. Find the mean of X.

2.6-5. Flaws in a certain type of drapery material appear on the average of one in 150 square feet. If we assume a Poisson distribution, find the probability of at most one flaw appearing in 225 square feet.

2.6-6. A certain type of aluminum screen that is 4 feet wide has, on the average, one flaw in an 85-foot roll. Find the probability that a 35-foot roll has no flaws.

2.6-7. With probability 0.001, a prize of $499 is won in the Michigan Daily Lottery when a $1 straight bet is placed. Let Y equal the number of $499 prizes won by a gambler after placing n straight bets. Note that Y is $b(n, 0.001)$. After placing $n = 2000$ $1 bets, the gambler is behind or even if $\{Y \leq 4\}$. Use the Poisson distribution to approximate $P(Y \leq 4)$ when $n = 2000$.

2.6-8. Suppose that the probability of suffering a side effect from a certain flu vaccine is 0.005. If 1000 persons are inoculated, find the approximate probability that

(a) At most 1 person suffers.

(b) 4, 5, or 6 persons suffer.

2.6-9. A store selling newspapers orders only $n = 4$ of a certain newspaper because the manager does not get many calls for that publication. If the number of requests per day follows a Poisson distribution with mean 3,

(a) What is the expected value of the number sold?

(b) What is the minimum number that the manager should order so that the chance of having more requests than available newspapers is less than 0.05?

2.6-10. The mean of a Poisson random variable X is $\mu = 9$. Compute

$$P(\mu - 2\sigma < X < \mu + 2\sigma).$$

2.6-11. An airline always overbooks if possible. A particular plane has 95 seats on a flight in which a ticket sells for $300. The airline sells 100 such tickets for this flight.

(a) If the probability of an individual not showing up is 0.05, assuming independence, what is the probability that the airline can accommodate all the passengers who do show up?

(b) If the airline must return the $300 price plus a penalty of $400 to each passenger that cannot get on the flight, what is the expected payout (penalty plus ticket refund) that the airline will pay?

2.6-12. A baseball team loses $88,000 for each consecutive day it rains. Say X, the number of consecutive days it rains at the beginning of the season, has a Poisson distribution with mean 0.32. What is the expected loss before the opening game?

2.6-13. Assume that a policyholder is four times more likely to file exactly two claims as to file exactly three claims. Assume also that the number X of claims of this policyholder is Poisson. Determine the expectation $E(X^2)$.

HISTORICAL COMMENTS   The next major items advanced in probability theory were by the Bernoullis, a remarkable Swiss family of mathematicians of the late 1600s to the late 1700s. There were eight mathematicians among them, but we shall mention just three of them: Jacob, Nicolaus II, and Daniel. While writing *Ars Conjectandi* (*The Art of Conjecture*), Jacob died in 1705, and a nephew, Nicolaus II, edited the work for publication. However, it was Jacob who discovered the important law of large numbers, which is included in our Section 5.8.

Another nephew of Jacob, Daniel, noted in his St. Petersburg paper that "expected values are computed by multiplying each possible gain by the number of ways in which it can occur and then dividing the sum of these products by the total number of cases." His cousin, Nicolaus II, then proposed the so-called St. Petersburg paradox: Peter continues to toss a coin until a head first appears—say, on the xth trial—and he then pays Paul 2^{x-1} units (originally ducats, but for convenience we use dollars). With each additional throw, the number of dollars has doubled. How much should another person pay Paul to take his place in this game? Clearly,

$$E(2^{X-1}) = \sum_{x=1}^{\infty} (2^{x-1})\left(\frac{1}{2^x}\right) = \sum_{x=1}^{\infty} \frac{1}{2} = \infty.$$

However, if we consider this as a practical problem, would someone be willing to give Paul $1000 to take his place even though there is this unlimited expected value? We doubt it and Daniel doubted it, and it made him think about the utility of money. For example, to most of us, $3 million is not worth three times $1 million. To convince you of that, suppose you had exactly $1 million and a very rich man offers to bet you $2 million against your $1 million on the flip of a coin. You will have zero or $3 million after the flip, so your expected value is

$$(\$0)\left(\frac{1}{2}\right) + (\$3,000,000)\left(\frac{1}{2}\right) = \$1,500,000,$$

much more than your $1 million. Seemingly, then, this is a great bet and one that Bill Gates might take. However, remember you have $1 million for certain and you could have zero with probability 1/2. None of us with limited resources should consider taking that bet, because the utility of that extra money to us is not worth the utility of the first $1 million. Now, each of us has our own utility function. Two dollars is worth twice as much as one dollar for practically all of us. But is $200,000 worth twice as much as $100,000? It depends upon your situation; so while the utility function is a straight line for the first several dollars, it still increases but begins to bend downward someplace as the amount of money increases. This occurs at different spots for all of us. Bob Hogg, one of the authors of this text, would bet $1000 against $2000 on a flip of the coin anytime, but probably not $100,000 against $200,000, so Hogg's utility function has started to bend downward someplace between $1000 and $100,000. Daniel Bernoulli made this observation, and it is extremely useful in all kinds of businesses.

As an illustration, in insurance, most of us know that the premium we pay for all types of insurance is greater than what the company expects to pay us; that is how they make money. Seemingly, insurance is a bad bet, but it really isn't always. It is true that we should self-insure less expensive items—those whose value is on that straight part of the utility function. We have even heard the "rule" that you not insure anything worth less than two months' salary; this is a fairly good guide, but each of us has our own utility function and must make that decision. Hogg can afford losses in the $5000 to $10,000 range (not that he likes them, of course), but he does not want to pay losses of $100,000 or more. So his utility function for negative values of the argument follows that straight line for relatively small negative amounts but again bends down for large negative amounts. If you insure expensive items, you will discover that the expected utility in absolute value will now exceed the premium. This is why most people insure their life, their home, and their car (particularly on the liability side). They should not, however, insure their golf clubs, eyeglasses, furs, or jewelry (unless the latter two items are extremely valuable).

CONTINUOUS DISTRIBUTIONS

3.1 RANDOM VARIABLES OF THE CONTINUOUS TYPE

Let the random variable X denote the outcome when a point is selected at random from an interval $[a, b]$, $-\infty < a < b < \infty$. If the experiment is performed in a fair manner, it is reasonable to assume that the probability that the point is selected from the interval $[a, x]$, $a \leq x \leq b$, is $(x-a)/(b-a)$. That is, the probability is proportional to the length of the interval, so the the cdf (cumulative distribution function) of X is

$$F(x) = \begin{cases} 0, & x < a, \\ \dfrac{x - a}{b - a}, & a \leq x < b, \\ 1, & b \leq x, \end{cases}$$

which can be written as

$$F(x) = \int_{-\infty}^{x} f(y)\, dy,$$

where

$$f(x) = \frac{1}{b - a}, \qquad a \leq x \leq b,$$

and equals zero elsewhere. That is, $F'(x) = f(x)$, and we call $f(x)$ the **probability density function** (pdf) of X.

Of course, students will note that $F'(x)$ does not exist at $x = a$ and at $x = b$. However, since $F(x)$ is continuous and cannot assign probability to individual points, we can define $f(x)$ at $x = a$ and at $x = b$ with any value. Most often, in this case, we take $f(a) = f(b) = 1/(b - a)$ or take $f(a) = f(b) = 0$. Because X is a continuous-type random variable, $F'(x)$ is equal to the pdf of X whenever $F'(x)$ exists; thus, when $a < x < b$, we have $f(x) = F'(x) = 1/(b - a)$.

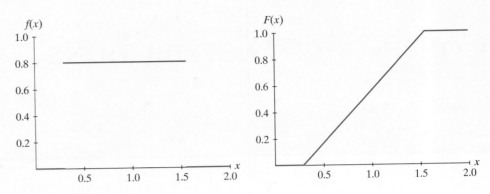

Figure 3.1-1 Uniform pdf and cdf

The random variable X has a **uniform distribution** if its pdf is equal to a constant on its support. In particular, if the support is the interval $[a, b]$, then

$$f(x) = \frac{1}{b - a}, \qquad a \leq x \leq b.$$

Moreover, we shall say that X is $U(a, b)$. This distribution is also referred to as **rectangular**, because the graph of $f(x)$ suggests that name. Figure 3.1-1 shows the graph of $f(x)$ and the cdf $F(x)$ when $a = 0.30$ and $b = 1.55$.

Now there are many probability density functions that could describe probabilities associated with a random variable X. Thus, we say that the **probability density function (pdf)** of a random variable X of the **continuous type**, with space S that is an interval or union of intervals, is an integrable function $f(x)$ satisfying the following conditions:

(a) $f(x) \geq 0, \qquad x \in S.$

(b) $\int_S f(x)\, dx = 1.$

(c) If $(a, b) \subseteq S$, then the probability of the event $\{a < X < b\}$ is

$$P(a < X < b) = \int_a^b f(x)\, dx.$$

The corresponding distribution of probability is said to be of the continuous type.

The **cumulative distribution function (cdf)** or **distribution function** of a random variable X of the continuous type, defined in terms of the pdf of X, is given by

$$F(x) = P(X \leq x) = \int_{-\infty}^x f(t)\, dt, \qquad -\infty < x < \infty.$$

Here, again, $F(x)$ accumulates (or, more simply, cumulates) all of the probability less than or equal to x. From the fundamental theorem of calculus, we have, for x values for which the derivative $F'(x)$ exists, $F'(x) = f(x)$.

Example 3.1-1 Let Y be a continuous random variable with pdf $g(y) = 2y, 0 < y < 1$. The cdf of Y is defined by

$$G(y) = \begin{cases} 0, & y < 0, \\ \displaystyle\int_0^y 2t\, dt = y^2, & 0 \leq y < 1, \\ 1, & 1 \leq y. \end{cases}$$

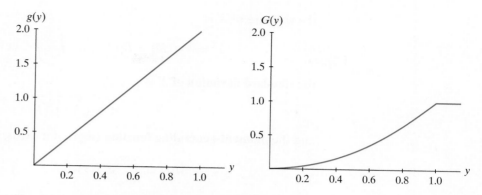

Figure 3.1-2 Continuous distribution pdf and cdf

Figure 3.1-2 gives the graph of the pdf $g(y)$ and the graph of the cdf $G(y)$. For examples of computations of probabilities, consider

$$P\left(\frac{1}{2} < Y \le \frac{3}{4}\right) = G\left(\frac{3}{4}\right) - G\left(\frac{1}{2}\right) = \left(\frac{3}{4}\right)^2 - \left(\frac{1}{2}\right)^2 = \frac{5}{16}$$

and

$$P\left(\frac{1}{4} \le Y < 2\right) = G(2) - G\left(\frac{1}{4}\right) = 1 - \left(\frac{1}{4}\right)^2 = \frac{15}{16}.$$ ■

Recall that the pmf $f(x)$ of a random variable of the discrete type is bounded by 1 because $f(x)$ gives a probability, namely,

$$f(x) = P(X = x).$$

For random variables of the continuous type, the pdf does not have to be bounded. [See Exercises 3.1-7(c) and 3.1-8(c).] The restriction is that *the area between the pdf and the x-axis must equal 1*. Furthermore, the pdf of a random variable X of the continuous type does not need to be a continuous function. For example, the function

$$f(x) = \begin{cases} \dfrac{1}{2}, & 0 < x < 1 \quad \text{or} \quad 2 < x < 3, \\[2mm] 0, & \text{elsewhere,} \end{cases}$$

enjoys the properties of a pdf of a distribution of the continuous type and yet has discontinuities at $x = 0, 1, 2$, and 3. However, the cdf associated with a distribution of the continuous type is always a continuous function.

For continuous-type random variables, the definitions associated with mathematical expectation are the same as those in the discrete case except that integrals replace summations. As an illustration, let X be a continuous random variable with a pdf $f(x)$. Then the **expected value of X**, or the **mean of X**, is

$$\mu = E(X) = \int_{-\infty}^{\infty} xf(x)\,dx,$$

the **variance of** X is

$$\sigma^2 = \text{Var}(X) = E[(X - \mu)^2] = \int_{-\infty}^{\infty} (x - \mu)^2 f(x)\, dx,$$

the **standard deviation of** X is

$$\sigma = \sqrt{\text{Var}(X)},$$

and the **moment-generating function (mgf)**, if it exists, is

$$M(t) = \int_{-\infty}^{\infty} e^{tx} f(x)\, dx, \qquad -h < t < h.$$

Moreover, results such as $\sigma^2 = E(X^2) - \mu^2$, $\mu = M'(0)$, and $\sigma^2 = M''(0) - [M'(0)]^2$ are still valid. Again, it is important to note that the mgf, if it is finite for $-h < t < h$ for some $h > 0$, completely determines the distribution.

REMARK In both the discrete and continuous cases, note that if the rth moment, $E(X^r)$, exists and is finite, then the same is true of all lower-order moments, $E(X^k), k = 1, 2, \ldots, r - 1$. However, the converse is not true; for example, the first moment can exist and be finite, but the second moment is not necessarily finite. (See Exercise 3.1-11.) Moreover, if $E(e^{tX})$ exists and is finite for $-h < t < h$, then all moments exist and are finite, but the converse is not necessarily true. ∎

The mean, variance, and moment-generating function of X, which is $U(0, 1)$, are not difficult to calculate. (See Exercise 3.1-1.) They are, respectively,

$$\mu = \frac{a + b}{2}, \qquad \sigma^2 = \frac{(b - a)^2}{12},$$

$$M(t) = \begin{cases} \dfrac{e^{tb} - e^{ta}}{t(b - a)}, & t \neq 0, \\ 1, & t = 0. \end{cases}$$

An important uniform distribution is that for which $a = 0$ and $b = 1$, namely, $U(0, 1)$. If X is $U(0, 1)$, approximate values of X can be simulated on most computers with the use of a random-number generator. In fact, it should be called a **pseudo-random-number generator** because the programs that produce the random numbers are usually such that if the starting number (the seed number) is known, all subsequent numbers in the sequence may be determined by simple arithmetical operations. Yet, despite their deterministic origin, these computer-produced numbers do behave as if they were truly randomly generated, and we shall not encumber our terminology by adding *pseudo*. (Examples of computer-produced random numbers are given in Appendix B in Table VIII.) Place a decimal point in front of each of the four-digit entries so that each is a number between 0 and 1.

Example 3.1-2 Let X have the pdf

$$f(x) = \frac{1}{100}, \qquad 0 < x < 100,$$

so that X is $U(0, 100)$. The mean and the variance are, respectively,

$$\mu = \frac{0 + 100}{2} = 50 \qquad \text{and} \qquad \sigma^2 = \frac{(100 - 0)^2}{12} = \frac{10,000}{12}.$$

The standard deviation is $\sigma = 100/\sqrt{12}$, which is 100 times that of the $U(0, 1)$ distribution. This agrees with our intuition, since the standard deviation is a measure of spread and $U(0, 100)$ is clearly spread out 100 times more than $U(0, 1)$. ∎

Example 3.1-3

For the random variable Y in Example 3.1-1,

$$\mu = E(Y) = \int_0^1 y\,(2y)\,dy = \left[\left(\frac{2}{3}\right)y^3\right]_0^1 = \frac{2}{3}$$

and

$$\sigma^2 = \text{Var}(Y) = E(Y^2) - \mu^2$$

$$= \int_0^1 y^2(2y)\,dy - \left(\frac{2}{3}\right)^2 = \left[\left(\frac{1}{2}\right)y^4\right]_0^1 - \frac{4}{9} = \frac{1}{18}$$

are the mean and variance, respectively, of Y. ∎

Example 3.1-4

Let X have the pdf

$$f(x) = \begin{cases} xe^{-x}, & 0 \le x < \infty, \\ 0, & \text{elsewhere.} \end{cases}$$

Then

$$M(t) = \int_0^\infty e^{tx}xe^{-x}\,dx = \lim_{b\to\infty}\int_0^b xe^{-(1-t)x}\,dx$$

$$= \lim_{b\to\infty}\left[-\frac{xe^{-(1-t)x}}{1-t} - \frac{e^{-(1-t)x}}{(1-t)^2}\right]_0^b$$

$$= \lim_{b\to\infty}\left[-\frac{be^{-(1-t)b}}{1-t} - \frac{e^{-(1-t)b}}{(1-t)^2}\right] + \frac{1}{(1-t)^2}$$

$$= \frac{1}{(1-t)^2},$$

provided that $t < 1$. Note that $M(0) = 1$, which is true for every mgf. Now,

$$M'(t) = \frac{2}{(1-t)^3} \qquad \text{and} \qquad M''(t) = \frac{6}{(1-t)^4}.$$

Thus,

$$\mu = M'(0) = 2$$

and

$$\sigma^2 = M''(0) - [M'(0)]^2 = 6 - 2^2 = 2.$$ ∎

The **(100p)th percentile** is a number π_p such that the area under $f(x)$ to the left of π_p is p. That is,

$$p = \int_{-\infty}^{\pi_p} f(x)\,dx = F(\pi_p).$$

The 50th percentile is called the **median**. We let $m = \pi_{0.50}$. The 25th and 75th percentiles are called the **first** and **third quartiles**, respectively, and are denoted by $q_1 = \pi_{0.25}$ and $q_3 = \pi_{0.75}$. Of course, the median $m = \pi_{0.50} = q_2$ is also called the **second quartile**.

Example 3.1-5

The time X in months until the failure of a certain product has the pdf

$$f(x) = \frac{3x^2}{4^3}\, e^{-(x/4)^3}, \qquad 0 < x < \infty.$$

Its cdf is

$$F(x) = \begin{cases} 0, & -\infty < x < 0, \\ 1 - e^{-(x/4)^3}, & 0 \le x < \infty. \end{cases}$$

For example, the 30th percentile, $\pi_{0.3}$, is given by

$$F(\pi_{0.3}) = 0.3$$

or, equivalently,

$$1 - e^{-(\pi_{0.3}/4)^3} = 0.3,$$

$$\ln(0.7) = -(\pi_{0.3}/4)^3,$$

$$\pi_{0.3} = -4\sqrt[3]{\ln(0.7)} = 2.84.$$

Likewise, $\pi_{0.9}$ is found by

$$F(\pi_{0.9}) = 0.9;$$

so

$$\pi_{0.9} = -4\sqrt[3]{\ln(0.1)} = 5.28.$$

Thus,

$$P(2.84 < X < 5.28) = 0.6.$$

The 30th and 90th percentiles are shown in Figure 3.1-3.

We conclude this section with an example that reviews its important ideas.

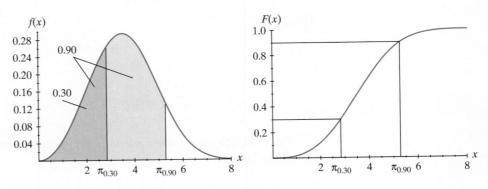

Figure 3.1-3 Illustration of percentiles $\pi_{0.30}$ and $\pi_{0.90}$

Example 3.1-6   Let X have the pdf

$$f(x) = e^{-x-1}, \qquad -1 < x < \infty.$$

Then

$$P(X \geq 1) = \int_1^\infty e^{-x-1} dx = e^{-1} \left[-e^{-x} \right]_1^\infty = e^{-2} = 0.135.$$

Also,

$$M(t) = E(e^{tX}) = \int_{-1}^\infty e^{tx} e^{-x-1} dx$$

$$= e^{-1} \left[\frac{-e^{-(1-t)x}}{1-t} \right]_{-1}^\infty$$

$$= e^{-t}(1-t)^{-1}, \qquad t < 1.$$

Thus, since

$$M'(t) = (e^{-t})(1-t)^{-2} - e^{-t}(1-t)^{-1}$$

and

$$M''(t) = (e^{-t})(2)(1-t)^{-3} - 2e^{-t}(1-t)^{-2} + e^{-t}(1-t)^{-1},$$

we have

$$\mu = M'(0) = 0 \qquad \text{and} \qquad \sigma^2 = M''(0) - [M'(0)]^2 = 1.$$

The cdf is

$$F(x) = \int_{-1}^x e^{-w-1} dw = e^{-1} \left[-e^{-w} \right]_{-1}^x$$

$$= e^{-1} \left[e^1 - e^{-x} \right] = 1 - e^{-x-1}, \qquad -1 < x < \infty,$$

and zero for $x \leq -1$.

As an illustration, the median, $\pi_{0.5}$, is found by solving

$$F(\pi_{0.5}) = 0.5,$$

which is equivalent to

$$-\pi_{0.5} - 1 = \ln(0.5);$$

so

$$\pi_{0.5} = \ln 2 - 1 = -0.307. \qquad \blacksquare$$

Exercises

3.1-1. Show that the mean, variance, and mgf of the uniform distribution are as given in this section.

3.1-2. Let $f(x) = 1/6, -3 \leq x \leq 3$, be the pdf of X. Graph the pdf and cdf, and record the mean and variance of X.

3.1-3. Customers arrive randomly at a bank teller's window. Given that one customer arrived during a particular 10-minute period, let X equal the time within the 10 minutes that the customer arrived. If X is $U(0,10)$, find

(a) The pdf of X.

(b) $P(X \geq 8)$.

(c) $P(2 \leq X < 8)$.

(d) $E(X)$.

(e) $\text{Var}(X)$.

3.1-4. If the mgf of X is

$$M(t) = \frac{e^{5t} - e^{4t}}{t}, \quad t \neq 0, \quad \text{and} \quad M(0) = 1,$$

find **(a)** $E(X)$, **(b)** $\text{Var}(X)$, and **(c)** $P(4.2 < X \leq 4.7)$.

3.1-5. Let Y have a uniform distribution $U(0, 1)$, and let

$$W = a + (b - a)Y, \quad a < b.$$

(a) Find the cdf of W.

HINT: Find $P[a + (b - a)Y \leq w]$.

(b) How is W distributed?

3.1-6. A grocery store has n watermelons to sell and makes \$1.00 on each sale. Say the number of consumers of these watermelons is a random variable with a distribution that can be approximated by

$$f(x) = \frac{1}{200}, \quad 0 < x < 200,$$

a pdf of the continuous type. If the grocer does not have enough watermelons to sell to all consumers, she figures that she loses \$5.00 in goodwill from each unhappy customer. But if she has surplus watermelons, she loses 50 cents on each extra watermelon. What should n be to maximize profit? HINT: If $X \leq n$, then her profit is $(1.00)X + (-0.50)(n - X)$; but if $X > n$, her profit is $(1.00)n + (-5.00)(X - n)$. Find the expected value of profit as a function of n, and then select n to maximize that function.

3.1-7. For each of the following functions, **(i)** find the constant c so that $f(x)$ is a pdf of a random variable X, **(ii)** find the cdf, $F(x) = P(X \leq x)$, **(iii)** sketch graphs of the pdf $f(x)$ and the cdf $F(x)$, and **(iv)** find μ and σ^2:

(a) $f(x) = 4x^c, \quad 0 \leq x \leq 1$.

(b) $f(x) = c\sqrt{x}, \quad 0 \leq x \leq 4$.

(c) $f(x) = c/x^{3/4}, \quad 0 < x < 1$.

3.1-8. For each of the following functions, **(i)** find the constant c so that $f(x)$ is a pdf of a random variable X, **(ii)** find the cdf, $F(x) = P(X \leq x)$, **(iii)** sketch graphs of the pdf $f(x)$ and the distribution function $F(x)$, and **(iv)** find μ and σ^2:

(a) $f(x) = x^3/4, \quad 0 < x < c$.

(b) $f(x) = (3/16)x^2, \quad -c < x < c$.

(c) $f(x) = c/\sqrt{x}, \quad 0 < x < 1$. Is this pdf bounded?

3.1-9. Let the random variable X have the pdf $f(x) = 2(1 - x), 0 \leq x \leq 1$, zero elsewhere.

(a) Sketch the graph of this pdf.

(b) Determine and sketch the graph of the cdf of X.

(c) Find **(i)** $P(0 \leq X \leq 1/2)$, **(ii)** $P(1/4 \leq X \leq 3/4)$, **(iii)** $P(X = 3/4)$, and **(iv)** $P(X \geq 3/4)$.

3.1-10. The pdf of X is $f(x) = c/x^2, 1 < x < \infty$.

(a) Calculate the value of c so that $f(x)$ is a pdf.

(b) Show that $E(X)$ is not finite.

3.1-11. The pdf of Y is $g(y) = dy^2, 0 \leq y \leq 1$.

(a) Calculate the value of d so that $g(y)$ is a pdf.

(b) Find $E(Y)$.

(c) Find $E(Y^2)$.

3.1-12. Sketch the graphs of the following pdfs and find and sketch the graphs of the cdfs associated with these distributions (note carefully the relationship between the shape of the graph of the pdf and the concavity of the graph of the cdf):

(a) $f(x) = \left(\dfrac{3}{2}\right)x^2, \quad -1 < x < 1$.

(b) $f(x) = \dfrac{1}{2}, \quad -1 < x < 1$.

(c) $f(x) = \begin{cases} x + 1, & -1 < x < 0, \\ 1 - x, & 0 \leq x < 1. \end{cases}$

3.1-13. The logistic distribution is associated with the cdf $F(x) = (1 + e^{-x})^{-1}, -\infty < x < \infty$. Find the pdf of the logistic distribution and show that its graph is symmetric about $x = 0$.

3.1-14. Let $f(x) = 1/2, 0 < x < 1$ or $2 < x < 3$, zero elsewhere, be the pdf of X.

(a) Sketch the graph of this pdf.

(b) Define the cdf of X and sketch its graph.

(c) Find $q_1 = \pi_{0.25}$.

(d) Find $m = \pi_{0.50}$. Is it unique?

(e) Find $q_3 = \pi_{0.75}$.

3.1-15. The life X (in years) of a voltage regulator of a car has the pdf

$$f(x) = \frac{3x^2}{7^3} e^{-(x/7)^3}, \quad 0 < x < \infty.$$

(a) What is the probability that this regulator will last at least 7 years?

(b) Given that it has lasted at least 7 years, what is the conditional probability that it will last at least another 3.5 years?

3.1-16. Let $f(x) = (x + 1)/2$, $-1 < x < 1$. Find **(a)** $\pi_{0.64}$, **(b)** $q_1 = \pi_{0.25}$, and **(c)** $\pi_{0.81}$.

3.1-17. An insurance agent receives a bonus if the loss ratio L on his business is less than 0.5, where L is the total losses (say, X) divided by the total premiums (say, T). The bonus equals $(0.5 - L)(T/30)$ if $L < 0.5$ and equals zero otherwise. If X (in \$100,000) has the pdf

$$f(x) = \frac{3}{x^4}, \qquad x > 1,$$

and if T (in \$100,000) equals 3, determine the expected value of the bonus.

3.1-18. The weekly demand X for propane gas (in thousands of gallons) has the pdf

$$f(x) = 4x^3 e^{-x^4}, \qquad 0 < x < \infty.$$

If the stockpile consists of two thousand gallons at the beginning of each week (and nothing extra is received during the week), what is the probability of not being able to meet the demand during a given week?

3.1-19. The total amount of medical claims (in \$100,000) of the employees of a company has the pdf that is given by $f(x) = 30x(1 - x)^4$, $0 < x < 1$. Find

(a) The mean and the standard deviation of the total in dollars.

(b) The probability that the total exceeds \$20,000.

3.1-20. Nicol (see References) lets the pdf of X be defined by

$$f(x) = \begin{cases} x, & 0 \le x \le 1, \\ c/x^3, & 1 \le x < \infty, \\ 0, & \text{elsewhere.} \end{cases}$$

Find

(a) The value of c so that $f(x)$ is a pdf.

(b) The mean of X (if it exists).

(c) The variance of X (if it exists).

(d) $P(1/2 \le X \le 2)$.

3.1-21. Let $X_1, X_2, \ldots, X_k$ be random variables of the continuous type, and let $f_1(x), f_2(x), \ldots, f_k(x)$ be their corresponding pdfs, each with sample space $S = (-\infty, \infty)$. Also, let $c_1, c_2, \ldots, c_k$ be nonnegative constants such that $\sum_{i=1}^{k} c_i = 1$.

(a) Show that $\sum_{i=1}^{k} c_i f_i(x)$ is a pdf of a continuous-type random variable on S.

(b) If X is a continuous-type random variable with pdf $\sum_{i=1}^{k} c_i f_i(x)$ on S, $E(X_i) = \mu_i$, and $\text{Var}(X_i) = \sigma_i^2$ for $i = 1, \ldots, k$, find the mean and the variance of X.

3.2 THE EXPONENTIAL, GAMMA, AND CHI-SQUARE DISTRIBUTIONS

We turn now to a continuous distribution that is related to the Poisson distribution. When previously observing a process of the (approximate) Poisson type, we counted the number of occurrences in a given interval. This number was a discrete-type random variable with a Poisson distribution. But not only is the number of occurrences a random variable; the waiting times between successive occurrences are also random variables. However, the latter are of the continuous type, since each of them can assume any positive value. In particular, let W denote the waiting time until the first occurrence during the observation of a Poisson process in which the mean number of occurrences in the unit interval is λ. Then W is a continuous-type random variable, and we proceed to find its cdf.

Because this waiting time is nonnegative, the cdf $F(w) = 0$, $w < 0$. For $w \ge 0$,

$$F(w) = P(W \le w) = 1 - P(W > w)$$

$$= 1 - P(\text{no occurrences in } [0, w])$$

$$= 1 - e^{-\lambda w},$$

since we previously discovered that $e^{-\lambda w}$ equals the probability of no occurrences in an interval of length w. That is, if the mean number of occurrences per unit interval is λ, then the mean number of occurrences in an interval of length w is proportional to w, and hence is given by λw. Thus, when $w > 0$, the pdf of W is

$$F'(w) = f(w) = \lambda e^{-\lambda w}.$$

We often let $\lambda = 1/\theta$ and say that the random variable X has an **exponential distribution** if its pdf is defined by

$$f(x) = \frac{1}{\theta} e^{-x/\theta}, \qquad 0 \leq x < \infty,$$

where the parameter $\theta > 0$. Accordingly, the waiting time W until the first occurrence in a Poisson process has an exponential distribution with $\theta = 1/\lambda$. To determine the exact meaning of the parameter θ, we first find the moment-generating function of X. It is

$$M(t) = \int_0^\infty e^{tx} \left(\frac{1}{\theta} \right) e^{-x/\theta} \, dx = \lim_{b \to \infty} \int_0^b \left(\frac{1}{\theta} \right) e^{-(1-\theta t)x/\theta} \, dx$$

$$= \lim_{b \to \infty} \left[-\frac{e^{-(1-\theta t)x/\theta}}{1 - \theta t} \right]_0^b = \frac{1}{1 - \theta t}, \qquad t < \frac{1}{\theta}.$$

Thus,

$$M'(t) = \frac{\theta}{(1 - \theta t)^2}$$

and

$$M''(t) = \frac{2\theta^2}{(1 - \theta t)^3}.$$

Hence, for an exponential distribution, we have

$$\mu = M'(0) = \theta \qquad \text{and} \qquad \sigma^2 = M''(0) - [M'(0)]^2 = \theta^2.$$

So if λ is the mean number of occurrences in the unit interval, then $\theta = 1/\lambda$ is the mean waiting time for the first occurrence. In particular, suppose that $\lambda = 7$ is the mean number of occurrences per minute; then the mean waiting time for the first occurrence is 1/7 of a minute, a result that agrees with our intuition.

Example 3.2-1

Let X have an exponential distribution with a mean of $\theta = 20$. Then the pdf of X is

$$f(x) = \frac{1}{20} e^{-x/20}, \qquad 0 \leq x < \infty.$$

The probability that X is less than 18 is

$$P(X < 18) = \int_0^{18} \frac{1}{20} e^{-x/20} \, dx = 1 - e^{-18/20} = 0.593. \qquad \blacksquare$$

Let X have an exponential distribution with mean $\mu = \theta$. Then the cdf of X is

$$F(x) = \begin{cases} 0, & -\infty < x < 0, \\ 1 - e^{-x/\theta}, & 0 \leq x < \infty. \end{cases}$$

The pdf and cdf are graphed in Figure 3.2-1 for $\theta = 5$. The median, m, is found by solving $F(m) = 0.5$. That is,

$$1 - e^{-m/\theta} = 0.5.$$

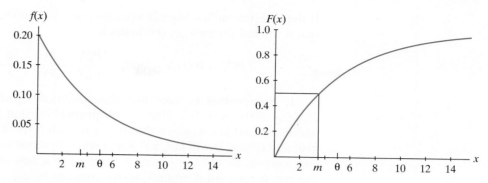

Figure 3.2-1 Exponential pdf, $f(x)$, and cdf, $F(x)$

Thus,

$$m = -\theta \ln(0.5) = \theta \ln(2).$$

So with $\theta = 5$, the median is $m = -5\ln(0.5) = 3.466$. Both the median and the mean $\theta = 5$ are indicated on the graphs.

It is useful to note that for an exponential random variable X with mean θ, we have

$$P(X > x) = 1 - F(x) = 1 - (1 - e^{-x/\theta})$$

$$= e^{-x/\theta} \qquad \text{when } x > 0.$$

Example 3.2-2   Customers arrive in a certain shop according to an approximate Poisson process at a mean rate of 20 per hour. What is the probability that the shopkeeper will have to wait more than 5 minutes for the arrival of the first customer? Let X denote the waiting time in minutes until the first customer arrives, and note that $\lambda = 1/3$ is the expected number of arrivals per minute. Thus,

$$\theta = \frac{1}{\lambda} = 3$$

and

$$f(x) = \frac{1}{3} e^{-(1/3)x}, \qquad 0 \le x < \infty.$$

Hence,

$$P(X > 5) = \int_5^\infty \frac{1}{3} e^{-(1/3)x}\, dx = e^{-5/3} = 0.1889.$$

The median time until the first arrival is

$$m = -3\ln(0.5) = 2.0794.$$ ■

Example 3.2-3   Suppose that a certain type of electronic component has an exponential distribution with a mean life of 500 hours. If X denotes the life of this component (or the time to failure of this component), then

$$P(X > x) = \int_x^\infty \frac{1}{500} e^{-t/500}\, dt = e^{-x/500}.$$

If the component has been in operation for 300 hours, the conditional probability that it will last for another 600 hours is

$$P(X > 900 \mid X > 300) = \frac{P(X > 900)}{P(X > 300)} = \frac{e^{-900/500}}{e^{-300/500}} = e^{-6/5}.$$

It is important to note that this conditional probability is exactly equal to $P(X > 600) = e^{-6/5}$. That is, the probability that the component will last an additional 600 hours, given that it has operated for 300 hours, is the same as the probability that it will last 600 hours when first put into operation. Thus, for such components, an old component is as good as a new one, and we say that the failure rate is constant. Certainly, with a constant failure rate, there is no advantage in replacing components that are operating satisfactorily. Obviously, however, this is not true in practice, because most components would have an increasing failure rate with time; hence, the exponential distribution is probably not the best model for the probability distribution of such a life. ∎

REMARK In Exercise 3.2-3, the result of Example 3.2-3 is generalized; that is, if the component has an exponential distribution, then the probability that it will last a time of at least $x + y$ units, given that it has lasted at least x units, is exactly the same as the probability that it will last at least y units when first put into operation. In effect, this statement says that the exponential distribution has a "forgetfulness" (or "no memory") property. It is also interesting to observe that, for continuous random variables whose support is $(0, \infty)$, the exponential distribution is the only continuous-type distribution with this forgetfulness property. Recall, however, that when we considered distributions of the discrete type, we noted that the geometric distribution has the property as well. (See Exercise 2.3-18.) ∎

In the (approximate) Poisson process with mean λ, we have seen that the waiting time until the first occurrence has an exponential distribution. We now let W denote the waiting time until the αth occurrence and find the distribution of W.

The cdf of W when $w \geq 0$ is given by

$$F(w) = P(W \leq w) = 1 - P(W > w)$$

$$= 1 - P(\text{fewer than } \alpha \text{ occurrences in } [0, w])$$

$$= 1 - \sum_{k=0}^{\alpha-1} \frac{(\lambda w)^k e^{-\lambda w}}{k!}, \tag{3.2-1}$$

since the number of occurrences in the interval $[0, w]$ has a Poisson distribution with mean λw. Because W is a continuous-type random variable, $F'(w)$, if it exists, is equal to the pdf of W. Also, provided that $w > 0$, we have

$$F'(w) = \lambda e^{-\lambda w} - e^{-\lambda w} \sum_{k=1}^{\alpha-1} \left[\frac{k(\lambda w)^{k-1} \lambda}{k!} - \frac{(\lambda w)^k \lambda}{k!} \right]$$

$$= \lambda e^{-\lambda w} - e^{-\lambda w} \left[\lambda - \frac{\lambda (\lambda w)^{\alpha-1}}{(\alpha - 1)!} \right]$$

$$= \frac{\lambda (\lambda w)^{\alpha-1}}{(\alpha - 1)!} e^{-\lambda w}.$$

If $w < 0$, then $F(w) = 0$ and $F'(w) = 0$. A pdf of this form is said to be one of the gamma type, and the random variable W is said to have a **gamma distribution**.

Before determining the characteristics of the gamma distribution, let us consider the gamma function for which the distribution is named. The **gamma function** is defined by

$$\Gamma(t) = \int_0^\infty y^{t-1} e^{-y} \, dy, \qquad 0 < t.$$

This integral is positive for $0 < t$ because the integrand is positive. Values of it are often given in a table of integrals. If $t > 1$, integration of the gamma function of t by parts yields

$$\Gamma(t) = \left[-y^{t-1} e^{-y} \right]_0^\infty + \int_0^\infty (t-1) y^{t-2} e^{-y} \, dy$$

$$= (t-1) \int_0^\infty y^{t-2} e^{-y} \, dy = (t-1)\Gamma(t-1).$$

For example, $\Gamma(6) = 5\Gamma(5)$ and $\Gamma(3) = 2\Gamma(2) = (2)(1)\Gamma(1)$. Whenever $t = n$, a positive integer, we have, by repeated application of $\Gamma(t) = (t-1)\Gamma(t-1)$,

$$\Gamma(n) = (n-1)\Gamma(n-1) = (n-1)(n-2)\cdots(2)(1)\Gamma(1).$$

However,

$$\Gamma(1) = \int_0^\infty e^{-y} \, dy = 1.$$

Thus, when n is a positive integer, we have

$$\Gamma(n) = (n-1)!.$$

For this reason, the gamma function is called the generalized factorial. [Incidentally, $\Gamma(1)$ corresponds to $0!$, and we have noted that $\Gamma(1) = 1$, which is consistent with earlier discussions.]

Let us now formally define the pdf of the gamma distribution and find its characteristics. The random variable X has a **gamma distribution** if its pdf is defined by

$$f(x) = \frac{1}{\Gamma(\alpha)\theta^\alpha} x^{\alpha-1} e^{-x/\theta}, \qquad 0 \le x < \infty.$$

Hence, W, the waiting time until the αth occurrence in an approximate Poisson process, has a gamma distribution with parameters α and $\theta = 1/\lambda$. To see that $f(x)$ actually has the properties of a pdf, note that $f(x) \ge 0$ and

$$\int_{-\infty}^\infty f(x) \, dx = \int_0^\infty \frac{x^{\alpha-1} e^{-x/\theta}}{\Gamma(\alpha)\theta^\alpha} \, dx,$$

which, by the change of variables $y = x/\theta$, equals

$$\int_0^\infty \frac{(\theta y)^{\alpha-1} e^{-y}}{\Gamma(\alpha)\theta^\alpha} \theta \, dy = \frac{1}{\Gamma(\alpha)} \int_0^\infty y^{\alpha-1} e^{-y} \, dy = \frac{\Gamma(\alpha)}{\Gamma(\alpha)} = 1.$$

The moment-generating function of X is (see Exercise 3.2-7)

$$M(t) = \frac{1}{(1 - \theta t)^\alpha}, \qquad t < 1/\theta.$$

The mean and variance are (see Exercise 3.2-10)

$$\mu = \alpha\theta \qquad \text{and} \qquad \sigma^2 = \alpha\theta^2.$$

Example 3.2-4

Suppose the number of customers per hour arriving at a shop follows a Poisson process with mean 30. That is, if a minute is our unit, then $\lambda = 1/2$. What is the probability that the shopkeeper will wait more than 5 minutes before both of the first two customers arrive? If X denotes the waiting time in minutes until the second customer arrives, then X has a gamma distribution with $\alpha = 2$, $\theta = 1/\lambda = 2$. Hence,

$$P(X > 5) = \int_5^\infty \frac{x^{2-1}e^{-x/2}}{\Gamma(2)2^2}\, dx = \int_5^\infty \frac{xe^{-x/2}}{4}\, dx$$

$$= \frac{1}{4}\left[(-2)xe^{-x/2} - 4e^{-x/2}\right]_5^\infty$$

$$= \frac{7}{2}e^{-5/2} = 0.287.$$

We could also have used Equation 3.2-1 with $\lambda = 1/\theta$ because α is an integer. From that equation, we have

$$P(X > x) = \sum_{k=0}^{\alpha-1} \frac{(x/\theta)^k e^{-x/\theta}}{k!}.$$

Thus, with $x = 5$, $\alpha = 2$, and $\theta = 2$, this is equal to

$$P(X > 5) = \sum_{k=0}^{2-1} \frac{(5/2)^k e^{-5/2}}{k!}$$

$$= e^{-5/2}\left(1 + \frac{5}{2}\right) = \left(\frac{7}{2}\right)e^{-5/2}. \qquad \blacksquare$$

Example 3.2-5

Telephone calls arrive at an office at a mean rate of $\lambda = 2$ per minute according to a Poisson process. Let X denote the waiting time in minutes until the fifth call arrives. The pdf of X, with $\alpha = 5$ and $\theta = 1/\lambda = 1/2$, is

$$f(x) = \frac{2^5 x^4}{4!}e^{-2x}, \qquad 0 \le x < \infty.$$

The mean and the variance of X are, respectively, $\mu = 5/2$ and $\sigma^2 = 5/4$. $\qquad \blacksquare$

In order to see the effect of the parameters on the shape of the gamma pdf, several combinations of α and θ have been used for graphs that are displayed in Figure 3.2-2. Note that for a fixed θ, as α increases, the probability moves to the right. The same is true for increasing θ with fixed α. Since $\theta = 1/\lambda$, as θ increases, λ decreases. That is, if $\theta_2 > \theta_1$, then $\lambda_2 = 1/\theta_2 < \lambda_1 = 1/\theta_1$. So if the mean number of

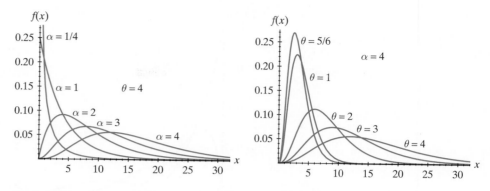

Figure 3.2-2 Gamma pdfs: $\theta = 4$, $\alpha = 1/4, 1, 2, 3, 4$; $\alpha = 4$, $\theta = 5/6, 1, 2, 3, 4$

changes per unit decreases, the waiting time to observe α changes can be expected to increase.

We now consider a special case of the gamma distribution that plays an important role in statistics. Let X have a gamma distribution with $\theta = 2$ and $\alpha = r/2$, where r is a positive integer. The pdf of X is

$$f(x) = \frac{1}{\Gamma(r/2)2^{r/2}} x^{r/2-1} e^{-x/2}, \qquad 0 < x < \infty.$$

We say that X has a **chi-square distribution with r degrees of freedom**, which we abbreviate by saying that X is $\chi^2(r)$. The mean and the variance of this chi-square distribution are, respectively,

$$\mu = \alpha\theta = \left(\frac{r}{2}\right)2 = r \qquad \text{and} \qquad \sigma^2 = \alpha\theta^2 = \left(\frac{r}{2}\right)2^2 = 2r.$$

That is, the mean equals the number of degrees of freedom, and the variance equals twice the number of degrees of freedom. An explanation of "number of degrees of freedom" is given later. From the results concerning the more general gamma distribution, we see that its moment-generating function is

$$M(t) = (1 - 2t)^{-r/2}, \qquad t < \frac{1}{2}.$$

In Figure 3.2-3, the graphs of chi-square pdfs for $r = 2, 3, 5$, and 8 are given. Note the relationship between the mean $\mu\ (= r)$ and the point at which the pdf attains its maximum. (See Exercise 3.2-15.)

Because the chi-square distribution is so important in applications, tables have been prepared, giving the values of the cdf,

$$F(x) = \int_0^x \frac{1}{\Gamma(r/2)2^{r/2}} w^{r/2-1} e^{-w/2}\, dw,$$

for selected values of r and x. (For an example, see Table IV in Appendix B.)

Example 3.2-6 Let X have a chi-square distribution with $r = 5$ degrees of freedom. Then, using Table IV in Appendix B, we obtain

$$P(1.145 \le X \le 12.83) = F(12.83) - F(1.145) = 0.975 - 0.050 = 0.925$$

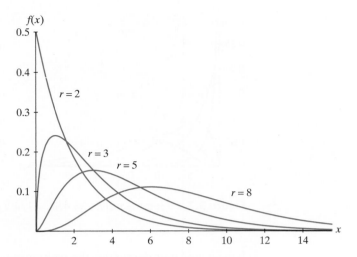

Figure 3.2-3 Chi-square pdfs with $r = 2, 3, 5, 8$

and

$$P(X > 15.09) = 1 - F(15.09) = 1 - 0.99 = 0.01.$$ ∎

Example 3.2-7

If X is $\chi^2(7)$, then two constants, a and b, such that

$$P(a < X < b) = 0.95$$

are $a = 1.690$ and $b = 16.01$. Other constants a and b can be found, and we are restricted in our choices only by the limited table. ∎

Probabilities like that of Example 3.2-7 are so important in statistical applications that we use special symbols for a and b. Let α be a positive probability (i.e., usually less than 0.5), and let X have a chi-square distribution with r degrees of freedom. Then $\chi_\alpha^2(r)$ is a number such that

$$P[X \geq \chi_\alpha^2(r)] = \alpha.$$

That is, $\chi_\alpha^2(r)$ is the $100(1 - \alpha)$th percentile (or upper 100αth percent point) of the chi-square distribution with r degrees of freedom. Then the 100α percentile is the number $\chi_{1-\alpha}^2(r)$ such that

$$P[X \leq \chi_{1-\alpha}^2(r)] = \alpha.$$

That is, the probability to the right of $\chi_{1-\alpha}^2(r)$ is $1 - \alpha$. (See Figure 3.2-4.)

Example 3.2-8

Let X have a chi-square distribution with five degrees of freedom. Then, using Table IV in Appendix B, we find that $\chi_{0.10}^2(5) = 9.236$ and $\chi_{0.90}^2(5) = 1.610$. These are the points, with $\alpha = 0.10$, that are indicated in Figure 3.2-4. ∎

Example 3.2-9

If customers arrive at a shop on the average of 30 per hour in accordance with a Poisson process, what is the probability that the shopkeeper will have to wait longer than 9.390 minutes for the first nine customers to arrive? Note that the mean rate of

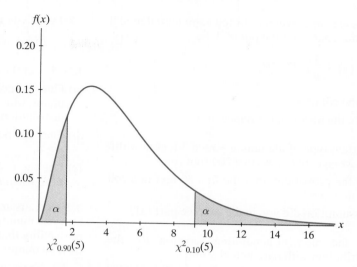

Figure 3.2-4 Chi-square tails, $r = 5, \alpha = 0.10$

arrivals per minute is $\lambda = 1/2$. Thus, $\theta = 2$ and $\alpha = r/2 = 9$. If X denotes the waiting time until the ninth arrival, then X is $\chi^2(18)$. Hence,

$$P(X > 9.390) = 1 - 0.05 = 0.95.$$ ∎

Example 3.2-10

If X has an exponential distribution with a mean of 2, then the pdf of X is

$$f(x) = \frac{1}{2} e^{-x/2} = \frac{x^{2/2-1}e^{-x/2}}{\Gamma(2/2)2^{2/2}}, \qquad 0 \le x < \infty.$$

That is, X is $\chi^2(2)$. As an illustration,

$$P(0.051 < X < 7.378) = 0.975 - 0.025 = 0.95.$$ ∎

Exercises

3.2-1. What are the pdf, the mean, and the variance of X if the moment-generating function of X is given by the following?

(a) $M(t) = \dfrac{1}{1 - 3t}, \qquad t < 1/3.$

(b) $M(t) = \dfrac{3}{3 - t}, \qquad t < 3.$

3.2-2. Telephone calls arrive at a doctor's office, according to a Poisson process, on the average of three every 5 minutes. Let X denote the waiting time until the first call that arrives after 9 A.M.

(a) What is the pdf of X?

(b) Find $P(X > 4)$.

3.2-3. Let X have an exponential distribution with mean $\theta > 0$. Show that

$$P(X > x + y \mid X > x) = P(X > y).$$

3.2-4. Let $F(x)$ be the cdf of the continuous-type random variable X, and assume that $F(x) = 0$ for $x \le 0$ and $0 < F(x) < 1$ for $0 < x$. Prove that if

$$P(X > x + y \mid X > x) = P(X > y),$$

then

$$F(x) = 1 - e^{-\lambda x}, \qquad 0 < x.$$

HINT: Show that $g(x) = 1 - F(x)$ satisfies the functional equation

$$g(x + y) = g(x)g(y),$$

which implies that $g(x) = a^{cx}$.

3.2-5. There are times when a shifted exponential model is appropriate. That is, let the pdf of X be

$$f(x) = \frac{1}{\theta} e^{-(x-\delta)/\theta}, \qquad \delta < x < \infty.$$

(a) Define the cdf of X.

(b) Calculate the mean and variance of X.

3.2-6. A certain type of aluminum screen 3 feet in width has, on the average, five flaws in a 200-foot roll.

(a) What is the probability that the first 50 feet in a roll contain no flaws?

(b) What assumption did you make to solve part (a)?

3.2-7. Find the moment-generating function for the gamma distribution with parameters α and θ.

HINT: In the integral representing $E(e^{tX})$, change variables by letting $y = (1 - \theta t)x/\theta$, where $1 - \theta t > 0$.

3.2-8. If X has a gamma distribution with $\theta = 5$ and $\alpha = 2$, find $P(X < 4)$.

3.2-9. If the moment-generating function of a random variable W is

$$M(t) = (1 - 7t)^{-20},$$

find the pdf, mean, and variance of W.

3.2-10. Use the moment-generating function of a gamma distribution to show that $E(X) = \alpha\theta$ and $\text{Var}(X) = \alpha\theta^2$.

3.2-11. If X is $\chi^2(17)$, find

(a) $P(X < 7.564)$.

(b) $P(X > 27.59)$.

(c) $P(6.408 < X < 27.59)$.

(d) $\chi^2_{0.95}(17)$.

(e) $\chi^2_{0.025}(17)$.

3.2-12. Let X equal the number of alpha particle emissions of carbon-14 that are counted by a Geiger counter each second. Assume that the distribution of X is Poisson with mean 16. Let W equal the time in seconds before the seventh count is made.

(a) Give the distribution of W.

(b) Find $P(W \leq 0.5)$. HINT: Use Equation 3.2-1 with $\lambda w = 8$.

3.2-13. If X is $\chi^2(23)$, find the following:

(a) $P(14.85 < X < 32.01)$.

(b) Constants a and b such that $P(a < X < b) = 0.95$ and $P(X < a) = 0.025$.

(c) The mean and variance of X.

(d) $\chi^2_{0.05}(23)$ and $\chi^2_{0.95}(23)$.

3.2-14. If X is $\chi^2(18)$, find constants a and b such that

$$P(a < X < b) = 0.82 \text{ and } P(X < a) = 0.04.$$

3.2-15. Let the distribution of X be $\chi^2(r)$.

(a) Find the point at which the pdf of X attains its maximum when $r \geq 2$. This is the mode of a $\chi^2(r)$ distribution.

(b) Find the points of inflection for the pdf of X when $r \geq 4$.

(c) Use the results of parts (a) and (b) to sketch the pdf of X when $r = 4$ and when $r = 10$.

3.2-16. Cars arrive at a tollbooth at a mean rate of 10 cars every 20 minutes according to a Poisson process. Find the probability that the toll collector will have to wait longer than 30 minutes before collecting the fifteenth toll.

3.2-17. If 15 observations are taken independently from a chi-square distribution with 4 degrees of freedom, find the probability that at most 3 of the 15 observations exceed 7.779.

3.2-18. Say the serum cholesterol level (X) of U.S. males ages 25–34 follows a translated gamma distribution with pdf

$$f(x) = \frac{x - 80}{50^2} e^{-(x - 80)/50}, \qquad 80 < x < \infty.$$

(a) What are the mean and the variance of this distribution?

(b) What is the mode?

(c) What percentage have a serum cholesterol level less than 200? HINT: Integrate by parts.

3.2-19. A bakery sells rolls in units of a dozen. The demand X (in 1000 units) for rolls has a gamma distribution with parameters $\alpha = 3, \theta = 0.5$, where θ is in units of days per 1000 units of rolls. It costs $2 to make a unit that sells for $5 on the first day when the rolls are fresh. Any leftover units are sold on the second day for $1. How many units should be made to maximize the expected value of the profit?

3.2-20. The initial value of an appliance is $700 and its dollar value in the future is given by

$$v(t) = 100 (2^{3-t} - 1), \qquad 0 \leq t \leq 3,$$

where t is time in years. Thus, after the first three years, the appliance is worth nothing as far as the warranty is concerned. If it fails in the first three years, the warranty pays $v(t)$. Compute the expected value of the payment on the warranty if T has an exponential distribution with mean 5.

3.2-21. A loss (in $100,000) due to fire in a building has a pdf $f(x) = (1/6)e^{-x/6}, 0 < x < \infty$. Given that the loss is greater than 5, find the probability that it is greater than 8.

3.2-22. Let X have a logistic distribution with pdf

$$f(x) = \frac{e^{-x}}{(1 + e^{-x})^2}, \qquad -\infty < x < \infty.$$

Show that

$$Y = \frac{1}{1 + e^{-X}}$$

has a $U(0, 1)$ distribution.

HINT: Find $G(y) = P(Y \le y) = P\left(\dfrac{1}{1 + e^{-X}} \le y\right)$ when $0 < y < 1$.

3.2-23. Some dental insurance policies cover the insurer only up to a certain amount, say, M. (This seems to us to be a dumb type of insurance policy because most people should want to protect themselves against large losses.) Say the dental expense X is a random variable with pdf $f(x) = (0.001)e^{-x/1000}$, $0 < x < \infty$. Find M so that $P(X < M) = 0.08$.

3.2-24. Let the random variable X be equal to the number of days that it takes a high-risk driver to have an accident. Assume that X has an exponential distribution. If $P(X < 50) = 0.25$, compute $P(X > 100 \,|\, X > 50)$.

3.3 THE NORMAL DISTRIBUTION

When observed over a large population, many variables have a "bell-shaped" relative frequency distribution, i.e., one that is approximately symmetric and relatively higher in the middle of the range of values than at the extremes. Examples include such variables as scholastic aptitude test scores, physical measurements (height, weight, length) of organisms, and repeated measurements of the same quantity on different occasions or by different observers. A very useful family of probability distributions for such variables are the normal distributions.

In this section, we give the definition of the pdf for the normal distribution, verify that it is a pdf, and then justify the use of μ and σ^2 in its formula. That is, we will show that μ and σ^2 are actually the mean and the variance of this distribution. Toward that end, the random variable X has a **normal distribution** if its pdf is defined by

$$f(x) = \frac{1}{\sigma \sqrt{2\pi}} \exp\left[-\frac{(x - \mu)^2}{2\sigma^2}\right], \qquad -\infty < x < \infty,$$

where μ and σ are parameters satisfying $-\infty < \mu < \infty$ and $0 < \sigma < \infty$, and also where $\exp[v]$ means e^v. Briefly, we say that X is $N(\mu, \sigma^2)$.

Clearly, $f(x) > 0$. We now evaluate the integral

$$I = \int_{-\infty}^{\infty} \frac{1}{\sigma \sqrt{2\pi}} \exp\left[-\frac{(x - \mu)^2}{2\sigma^2}\right] dx$$

and show that it is equal to 1. In I, change the variables of integration by letting $z = (x - \mu)/\sigma$. Then

$$I = \int_{-\infty}^{\infty} \frac{1}{\sqrt{2\pi}} e^{-z^2/2} \, dz.$$

Since $I > 0$, it follows that if $I^2 = 1$, then $I = 1$. Now,

$$I^2 = \frac{1}{2\pi} \left[\int_{-\infty}^{\infty} e^{-x^2/2} \, dx\right]\left[\int_{-\infty}^{\infty} e^{-y^2/2} \, dy\right],$$

or, equivalently,

$$I^2 = \frac{1}{2\pi} \int_{-\infty}^{\infty} \int_{-\infty}^{\infty} \exp\left(-\frac{x^2 + y^2}{2}\right) dx\,dy.$$

Letting $x = r\cos\theta$, $y = r\sin\theta$ (i.e., using polar coordinates), we have

$$I^2 = \frac{1}{2\pi} \int_0^{2\pi} \int_0^{\infty} e^{-r^2/2} r\,dr\,d\theta$$

$$= \frac{1}{2\pi} \int_0^{2\pi} d\theta = \frac{1}{2\pi} 2\pi = 1.$$

Thus, $I = 1$, and we have shown that $f(x)$ has the properties of a pdf. The moment-generating function of X is

$$M(t) = \int_{-\infty}^{\infty} \frac{e^{tx}}{\sigma\sqrt{2\pi}} \exp\left[-\frac{(x-\mu)^2}{2\sigma^2}\right] dx$$

$$= \int_{-\infty}^{\infty} \frac{1}{\sigma\sqrt{2\pi}} \exp\left\{-\frac{1}{2\sigma^2}[x^2 - 2(\mu + \sigma^2 t)x + \mu^2]\right\} dx.$$

To evaluate this integral, we complete the square in the exponent:

$$x^2 - 2(\mu + \sigma^2 t)x + \mu^2 = [x - (\mu + \sigma^2 t)]^2 - 2\mu\sigma^2 t - \sigma^4 t^2.$$

Hence,

$$M(t) = \exp\left(\frac{2\mu\sigma^2 t + \sigma^4 t^2}{2\sigma^2}\right) \int_{-\infty}^{\infty} \frac{1}{\sigma\sqrt{2\pi}} \exp\left\{-\frac{1}{2\sigma^2}[x - (\mu + \sigma^2 t)]^2\right\} dx.$$

Note that the integrand in the last integral is like the pdf of a normal distribution with μ replaced by $\mu + \sigma^2 t$. However, the normal pdf integrates to 1 for all real μ—in particular, when μ is replaced by $\mu + \sigma^2 t$. Thus,

$$M(t) = \exp\left(\frac{2\mu\sigma^2 t + \sigma^4 t^2}{2\sigma^2}\right) = \exp\left(\mu t + \frac{\sigma^2 t^2}{2}\right).$$

Now,

$$M'(t) = (\mu + \sigma^2 t) \exp\left(\mu t + \frac{\sigma^2 t^2}{2}\right)$$

and

$$M''(t) = [(\mu + \sigma^2 t)^2 + \sigma^2] \exp\left(\mu t + \frac{\sigma^2 t^2}{2}\right).$$

Consequently,

$$E(X) = M'(0) = \mu,$$

$$\mathrm{Var}(X) = M''(0) - [M'(0)]^2 = \mu^2 + \sigma^2 - \mu^2 = \sigma^2.$$

That is, the parameters μ and σ^2 in the pdf of X are the mean and the variance of X.

**Example
3.3-1**

If the pdf of X is

$$f(x) = \frac{1}{\sqrt{32\pi}} \exp\left[-\frac{(x+7)^2}{32}\right], \qquad -\infty < x < \infty,$$

then X is $N(-7, 16)$. That is, X has a normal distribution with a mean $\mu = -7$, a variance $\sigma^2 = 16$, and the moment-generating function

$$M(t) = \exp(-7t + 8t^2).$$

∎

**Example
3.3-2**

If the moment-generating function of X is

$$M(t) = \exp(5t + 12t^2),$$

then X is $N(5, 24)$, and its pdf is

$$f(x) = \frac{1}{\sqrt{48\pi}} \exp\left[-\frac{(x-5)^2}{48}\right], \qquad -\infty < x < \infty.$$

∎

If Z is $N(0, 1)$, we shall say that Z has a **standard normal distribution**. Moreover, the cdf of Z is

$$\Phi(z) = P(Z \leq z) = \int_{-\infty}^{z} \frac{1}{\sqrt{2\pi}} e^{-w^2/2} \, dw.$$

It is not possible to evaluate this integral by finding an antiderivative that can be expressed as an elementary function. However, numerical approximations for integrals of this type have been tabulated and are given in Tables Va and Vb in Appendix B. The bell-shaped curved in Figure 3.3-1 represents the graph of the pdf of Z, and the shaded area equals $\Phi(z_0)$.

Values of $\Phi(z)$ for $z \geq 0$ are given in Table Va in Appendix B. Because of the symmetry of the standard normal pdf, it is true that $\Phi(-z) = 1 - \Phi(z)$ for all real z. Thus, Table Va is enough. However, it is sometimes convenient to be able to read

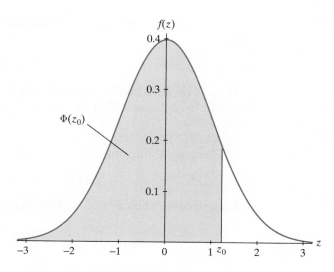

Figure 3.3-1 Standard normal pdf

$\Phi(-z)$, for $z > 0$, directly from a table. This can be done by using values in Table Vb in Appendix B, which lists right-tail probabilities. Again, because of the symmetry of the standard normal pdf, when $z > 0$, $\Phi(-z) = P(Z \le -z) = P(Z > z)$ can be read directly from Table Vb.

Example 3.3-3

If Z is $N(0, 1)$, then, using Table Va in Appendix B, we obtain

$$P(Z \le 1.24) = \Phi(1.24) = 0.8925,$$

$$P(1.24 \le Z \le 2.37) = \Phi(2.37) - \Phi(1.24) = 0.9911 - 0.8925 = 0.0986,$$

$$P(-2.37 \le Z \le -1.24) = P(1.24 \le Z \le 2.37) = 0.0986.$$

Now, using Table Vb, we find that

$$P(Z > 1.24) = 0.1075,$$
$$P(Z \le -2.14) = P(Z \ge 2.14) = 0.0162,$$

and using both tables, we obtain

$$P(-2.14 \le Z \le 0.77) = P(Z \le 0.77) - P(Z \le -2.14)$$

$$= 0.7794 - 0.0162 = 0.7632. \qquad \blacksquare$$

There are times when we want to read the normal probability table in the opposite way, essentially finding the inverse of the standard normal cdf. That is, given a probability p, we find a constant a so that $P(Z \le a) = p$. This situation is illustrated in the next example.

Example 3.3-4

If the distribution of Z is $N(0, 1)$, then to find constants a and b such that

$$P(Z \le a) = 0.9147 \qquad \text{and} \qquad P(Z \ge b) = 0.0526,$$

we find the respective probabilities in Tables Va and Vb in Appendix B and read off the corresponding values of z. From Table Va, we see that $a = 1.37$, and from Table Vb, we see that $b = 1.62$. $\qquad \blacksquare$

In statistical applications, we are often interested in finding a number z_α such that

$$P(Z \ge z_\alpha) = \alpha,$$

where Z is $N(0, 1)$ and α is usually less than 0.5. That is, z_α is the $100(1 - \alpha)$th percentile (sometimes called the upper 100α percent point) for the standard normal distribution. (See Figure 3.3-2.) The value of z_α is given in Table Va for selected values of α. For other values of α, z_α can be found in Table Vb.

Because of the symmetry of the normal pdf,

$$P(Z \le -z_\alpha) = P(Z \ge z_\alpha) = \alpha.$$

Also, since the subscript of z_α is the right-tail probability,

$$z_{1-\alpha} = -z_\alpha.$$

For example,

$$z_{0.95} = z_{1-0.05} = -z_{0.05}.$$

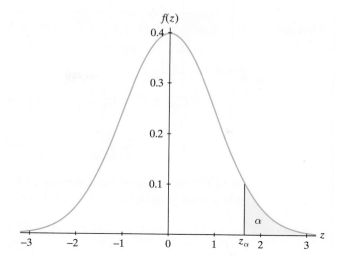

Figure 3.3-2 Upper 100α percent point, z_α

Example
3.3-5 To find $z_{0.0125}$, note that

$$P(Z \geq z_{0.0125}) = 0.0125.$$

Thus,

$$z_{0.0125} = 2.24,$$

from Table Vb in Appendix B. Also,

$$z_{0.05} = 1.645 \qquad \text{and} \qquad z_{0.025} = 1.960,$$

from the last rows of Table Va. ■

REMARK Recall that the $(100p)$th percentile, π_p, for a random variable X is a number such that $P(X \leq \pi_p) = p$. If Z is $N(0, 1)$, then since

$$P(Z \geq z_\alpha) = \alpha,$$

it follows that

$$P(Z < z_\alpha) = 1 - \alpha.$$

Thus, z_α is the $100(1 - \alpha)$th percentile for the standard normal distribution, $N(0, 1)$. For example, $z_{0.05} = 1.645$ is the $100(1 - 0.05) = 95$th percentile and $z_{0.95} = -1.645$ is the $100(1 - 0.95) = 5$th percentile. ■

The next theorem shows that if X is $N(\mu, \sigma^2)$, the random variable $(X - \mu)/\sigma$ is $N(0, 1)$. Hence, Tables Va and Vb in Appendix B can be used to find probabilities relating to X.

Theorem 3.3-1   If X is $N(\mu, \sigma^2)$, then $Z = (X - \mu)/\sigma$ is $N(0, 1)$.

Proof The cdf of Z is

$$P(Z \le z) = P\left(\frac{X - \mu}{\sigma} \le z\right) = P(X \le z\sigma + \mu)$$

$$= \int_{-\infty}^{z\sigma + \mu} \frac{1}{\sigma\sqrt{2\pi}} \exp\left[-\frac{(x - \mu)^2}{2\sigma^2}\right] dx.$$

Now, for the integral representing $P(Z \le z)$, we use the change of variable of integration given by $w = (x - \mu)/\sigma$ (i.e., $x = w\sigma + \mu$) to obtain

$$P(Z \le z) = \int_{-\infty}^{z} \frac{1}{\sqrt{2\pi}} e^{-w^2/2} dw.$$

But this is the expression for $\Phi(z)$, the cdf of a standardized normal random variable. Hence, Z is $N(0, 1)$. □

REMARK If X is any random variable for which $E(X) = \mu$ and $E[(X - \mu)^2] = \sigma^2$ exist and $Z = (X - \mu)/\sigma$, then

$$\mu_Z = E(Z) = E\left[\frac{X - \mu}{\sigma}\right] = \frac{E(X) - \mu}{\sigma} = 0$$

and

$$\sigma_Z^2 = E\left[\left(\frac{X - \mu}{\sigma}\right)^2\right] = \frac{E[(X - \mu)^2]}{\sigma^2} = \frac{\sigma^2}{\sigma^2} = 1.$$

That is, the mean and the variance of Z are 0 and 1, respectively, no matter what the distribution of X. The important aspect of the theorem is that if X is normally distributed, then Z is normally distributed—of course with zero mean and unit variance. Z is often called the **standard score** associated with X. ∎

Theorem 3.3-1 can be used to find probabilities relating to X, which is $N(\mu, \sigma^2)$, as follows:

$$P(a \le X \le b) = P\left(\frac{a - \mu}{\sigma} \le \frac{X - \mu}{\sigma} \le \frac{b - \mu}{\sigma}\right) = \Phi\left(\frac{b - \mu}{\sigma}\right) - \Phi\left(\frac{a - \mu}{\sigma}\right),$$

since $(X - \mu)/\sigma$ is $N(0, 1)$.

Example 3.3-6   If X is $N(3, 16)$, then

$$P(4 \le X \le 8) = P\left(\frac{4 - 3}{4} \le \frac{X - 3}{4} \le \frac{8 - 3}{4}\right)$$

$$= \Phi(1.25) - \Phi(0.25) = 0.8944 - 0.5987 = 0.2957,$$

$$P(0 \le X \le 5) = P\left(\frac{0 - 3}{4} \le Z \le \frac{5 - 3}{4}\right)$$

$$= \Phi(0.5) - \Phi(-0.75) = 0.6915 - 0.2266 = 0.4649,$$

and

$$P(-2 \leq X \leq 1) = P\left(\frac{-2-3}{4} \leq Z \leq \frac{1-3}{4}\right)$$

$$= \Phi(-0.5) - \Phi(-1.25) = 0.3085 - 0.1056 = 0.2029. \qquad \blacksquare$$

Example 3.3-7

If X is $N(25, 36)$, we find a constant c such that

$$P(|X - 25| \leq c) = 0.9544.$$

We want

$$P\left(\frac{-c}{6} \leq \frac{X - 25}{6} \leq \frac{c}{6}\right) = 0.9544.$$

Thus,

$$\Phi\left(\frac{c}{6}\right) - \left[1 - \Phi\left(\frac{c}{6}\right)\right] = 0.9544$$

and

$$\Phi\left(\frac{c}{6}\right) = 0.9772.$$

Hence, $c/6 = 2$ and $c = 12$. That is, the probability that X falls within two standard deviations of its mean is the same as the probability that the standard normal variable Z falls within two units (standard deviations) of zero. $\qquad \blacksquare$

In the next theorem, we give a relationship between the chi-square and normal distributions.

Theorem 3.3-2

If the random variable X is $N(\mu, \sigma^2)$, $\sigma^2 > 0$, then the random variable $V = (X - \mu)^2/\sigma^2 = Z^2$ is $\chi^2(1)$.

Proof Because $V = Z^2$, where $Z = (X - \mu)/\sigma$ is $N(0, 1)$, the cdf $G(v)$ of V is, for $v \geq 0$,

$$G(v) = P(Z^2 \leq v) = P(-\sqrt{v} \leq Z \leq \sqrt{v}).$$

That is, with $v \geq 0$,

$$G(v) = \int_{-\sqrt{v}}^{\sqrt{v}} \frac{1}{\sqrt{2\pi}} e^{-z^2/2}\, dz = 2\int_{0}^{\sqrt{v}} \frac{1}{\sqrt{2\pi}} e^{-z^2/2}\, dz.$$

If we change the variable of integration by writing $z = \sqrt{y}$, then, since

$$\frac{d}{dy}(\sqrt{y}) = \frac{1}{2\sqrt{y}},$$

we have

$$G(v) = \int_{0}^{v} \frac{1}{\sqrt{2\pi y}} e^{-y/2}\, dy, \qquad 0 \leq v.$$

Of course, $G(v) = 0$ when $v < 0$. Hence, the pdf $g(v) = G'(v)$ of the continuous-type random variable V is, by one form of the fundamental theorem of calculus,

$$g(v) = \frac{1}{\sqrt{\pi}\sqrt{2}} v^{1/2-1} e^{-v/2}, \qquad 0 < v < \infty.$$

Since $g(v)$ is a pdf, it must be true that

$$\int_0^\infty \frac{1}{\sqrt{\pi}\sqrt{2}} v^{1/2-1} e^{-v/2}\, dv = 1.$$

The change of variables $x = v/2$ yields

$$1 = \frac{1}{\sqrt{\pi}} \int_0^\infty x^{1/2-1} e^{-x}\, dx = \frac{1}{\sqrt{\pi}} \Gamma\left(\frac{1}{2}\right).$$

Hence, $\Gamma(1/2) = \sqrt{\pi}$, and it follows that V is $\chi^2(1)$. $\qquad\square$

Example 3.3-8

If Z is $N(0,1)$, then

$$P(|Z| < 1.96 = \sqrt{3.841}\,) = 0.95$$

and, of course,

$$P(Z^2 < 3.841) = 0.95$$

from the chi-square table with $r = 1$. ∎

Exercises

3.3-1. If Z is $N(0, 1)$, find

(a) $P(0 \leq Z \leq 1.40)$.

(b) $P(-2.00 \leq Z \leq 0)$.

(c) $P(-1.09 \leq Z \leq -0.06)$.

(d) $P(|Z| > 2.85)$.

(e) $P(|Z| < 1.96)$.

(f) $P(|Z| < 1)$.

(g) $P(|Z| < 2)$.

(h) $P(|Z| < 3)$.

3.3-2. If Z is $N(0, 1)$, find

(a) $P(0 \leq Z \leq 0.87)$.

(b) $P(-2.64 \leq Z \leq 0)$.

(c) $P(-2.13 \leq Z \leq -0.56)$.

(d) $P(|Z| > 1.39)$.

(e) $P(Z < -1.62)$.

(f) $P(|Z| > 1)$.

(g) $P(|Z| > 2)$.

(h) $P(|Z| > 3)$.

3.3-3. If Z is $N(0, 1)$, find values of c such that

(a) $P(Z \geq c) = 0.025$.

(b) $P(|Z| \leq c) = 0.95$.

(c) $P(Z > c) = 0.05$.

(d) $P(|Z| \leq c) = 0.90$.

3.3-4. Find the values of (a) $z_{0.05}$, (b) $z_{0.025}$, (c) $-z_{0.05}$, and (d) $-z_{0.025}$.

3.3-5. If X is normally distributed with a mean of 4 and a variance of 25, find

(a) $P(4 \leq X \leq 10)$.

(b) $P(-2 \leq X \leq 6)$.

(c) $P(-4 \leq X \leq -2)$.

(d) $P(X > 19)$.

(e) $P(|X - 4| < 5)$.

(f) $P(|X - 4| < 10)$.

(g) $P(|X - 4| < 15)$.

(h) $P(|X - 4| < 12.41)$.

3.3-6. If the moment-generating function of X is $M(t) = \exp(166t + 200t^2)$, find

(a) The mean of X.

(b) The variance of X.

(c) $P(170 < X < 200)$.

(d) $P(148 \leq X \leq 172)$.

3.3-7. If X is $N(650, 625)$, find

(a) $P(600 \leq X < 660)$.

(b) A constant $c > 0$ such that $P(|X - 650| \leq c) = 0.9544$.

3.3-8. Let the distribution of X be $N(\mu, \sigma^2)$. Show that the points of inflection of the graph of the pdf of X occur at $x = \mu \pm \sigma$.

3.3-9. Find the distribution of $W = X^2$ when

(a) X is $N(0, 4)$,

(b) X is $N(0, \sigma^2)$.

3.3-10. If X is $N(\mu, \sigma^2)$, show that the distribution of $Y = aX + b$ is $N(a\mu + b, a^2\sigma^2)$, $a \neq 0$. HINT: Find the cdf $P(Y \leq y)$ of Y, and in the resulting integral, let $w = ax+b$ or, equivalently, $x = (w - b)/a$.

3.3-11. A candy maker produces mints that have a label weight of 20.4 grams. Assume that the distribution of the weights of these mints is $N(21.37, 0.16)$.

(a) Let X denote the weight of a single mint selected at random from the production line. Find $P(X > 22.07)$.

(b) Suppose that 15 mints are selected independently and weighed. Let Y equal the number of these mints that weigh less than 20.857 grams. Find $P(Y \leq 2)$.

3.3-12. If the moment-generating function of X is given by $M(t) = e^{500t + 5000t^2}$, find $P[27, 060 \leq (X - 500)^2 \leq 50, 240]$.

3.3-13. The serum zinc level X in micrograms per deciliter for males between ages 15 and 17 has a distribution that is approximately normal with $\mu = 90$ and $\sigma = 15$. Compute the conditional probability $P(X > 120 \mid X > 105)$.

3.3-14. The strength X of a certain material is such that its distribution is found by $X = e^Y$, where Y is $N(10, 1)$. Find the cdf and pdf of X, and compute $P(10,000 < X < 20,000)$. NOTE: $F(x) = P(X \leq x) = P(e^Y \leq x) = P(Y \leq \ln x)$ so that the random variable X is said to have a **lognormal distribution**.

3.3-15. The "fill" problem is important in many industries, such as those making cereal, toothpaste, beer, and so on. If an industry claims that it is selling 12 ounces of its product in a container, it must have a mean greater than 12 ounces, or else the FDA will crack down, although the FDA will allow a very small percentage of the containers to have less than 12 ounces.

(a) If the content X of a container has a $N(12.1, \sigma^2)$ distribution, find σ so that $P(X < 12) = 0.01$.

(b) If $\sigma = 0.05$, find μ so that $P(X < 12) = 0.01$.

3.3-16. The graphs of the moment-generating functions of three normal distributions—$N(0, 1)$, $N(-1, 1)$, and $N(2, 1)$—are given in Figure 3.3-3(a). Identify them.

3.3-17. Figure 3.3-3(b) shows the graphs of the following three moment-generating functions near the origin:

$$g_1(t) = \frac{1}{1 - 4t}, \qquad t < 1/4,$$

$$g_2(t) = \frac{1}{(1 - 2t)^2}, \qquad t < 1/2,$$

$$g_3(t) = e^{4t + t^2/2}.$$

Why do these three graphs look so similar around $t = 0$?

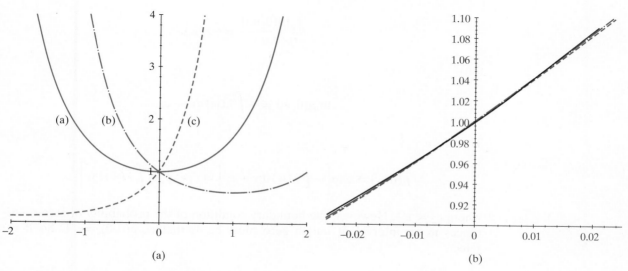

(a)   (b)

Figure 3.3-3 Moment-generating functions

3.4* ADDITIONAL MODELS

The binomial, Poisson, gamma, chi-square, and normal models are frequently used in statistics. However, many other interesting and very useful models can be found. We begin with a modification of one of the postulates of an approximate Poisson process as given in Section 2.6. In that definition, the numbers of occurrences in nonoverlapping intervals are independent, and the probability of at least two occurrences in a sufficiently small interval is essentially zero. We continue to use these postulates, but now we say that the probability of exactly one occurrence in a sufficiently short interval of length h is approximately λh, *where λ is a nonnegative function of the position of this interval.* To be explicit, say $p(x, w)$ is the probability of x occurrences in the interval $(0, w)$, $0 \leq w$. Then the last postulate, in more formal terms, becomes

$$p(x + 1, w + h) - p(x, w) \approx \lambda(w)h,$$

where $\lambda(w)$ is a nonnegative function of w. This means that if we want the approximate probability of zero occurrences in the interval $(0, w+h)$, we could take, from the independence of the occurrences, the probability of zero occurrences in the interval $(0, w)$ times that of zero occurrences in the interval $(w, w + h)$. That is,

$$p(0, w + h) \approx p(0, w)[1 - \lambda(w)h],$$

because the probability of one or more occurrences in $(w, w + h)$ is about equal to $\lambda(w)h$. Equivalently,

$$\frac{p(0, w + h) - p(0, w)}{h} \approx -\lambda(w)p(0, w).$$

Taking limits as $h \to 0$, we have

$$\frac{d}{dw}[p(0, w)] = -\lambda(w)p(0, w).$$

That is, the resulting equation is

$$\frac{\frac{d}{dw}[p(0, w)]}{p(0, w)} = -\lambda(w);$$

thus,

$$\ln p(0, w) = -\int \lambda(w) \, dw + c_1.$$

Therefore,

$$p(0, w) = \exp\left[-\int \lambda(w) \, dw + c_1\right] = c_2 \exp\left[-\int \lambda(w) \, dw\right],$$

where $c_2 = e^{c_1}$. However, the boundary condition of the probability of zero occurrences in an interval of length zero must be 1; that is, $p(0,0) = 1$. So if we select

$$H(w) = \int \lambda(w) \, dw$$

to be such that $H(0) = 0$, then $c_2 = 1$. That is,

$$p(0, w) = e^{-H(w)},$$

where $H'(w) = \lambda(w)$ and $H(0) = 0$. Hence,

$$H(w) = \int_0^w \lambda(t) \, dt.$$

Suppose that we now let the continuous-type random variable W be the interval necessary to produce the first occurrence. Then the cdf of W is

$$G(w) = P(W \leq w) = 1 - P(W > w), \qquad 0 \leq w.$$

Because zero occurrences in the interval $(0, w)$ are the same as $W > w$, then

$$G(w) = 1 - p(0, w) = 1 - e^{-H(w)}, \qquad 0 \leq w.$$

The pdf of W is

$$g(w) = G'(w) = H'(w)e^{-H(w)} = \lambda(w) \exp\left[-\int_0^w \lambda(t) \, dt\right], \qquad 0 \leq w.$$

From this formula, we see immediately that, in terms of $g(w)$ and $G(w)$,

$$\lambda(w) = \frac{g(w)}{1 - G(w)}.$$

In many applications of this result, W can be thought of as a random time interval. For example, if one occurrence means the "death" or "failure" of the item under consideration, then W is actually the length of life of the item. Usually, $\lambda(w)$, which is commonly called the **failure rate** or **force of mortality**, is an increasing function of w. That is, the larger w (the older the item), the better is the chance of failure within a short interval of length h, namely, $\lambda(w)h$. As we review the exponential distribution of Section 3.2, we note that there $\lambda(w)$ is a constant; that is, the failure rate or force of mortality does not increase as the item gets older. If this were true in human populations, it would mean that a person 80 years old would have as much chance of living another year as would a person 20 years old (sort of a mathematical "fountain of youth"). However, a constant failure rate (force of mortality) is not the case in most human populations or in most populations of manufactured items. That is, the failure rate $\lambda(w)$ is usually an increasing function of w. We give two important examples of useful probabilistic models.

Example 3.4-1 Let

$$H(w) = \left(\frac{w}{\beta}\right)^\alpha, \qquad 0 \leq w,$$

so that the failure rate is

$$\lambda(w) = H'(w) = \frac{\alpha w^{\alpha-1}}{\beta^\alpha},$$

where $\alpha > 0$, $\beta > 0$. Then the pdf of W is

$$g(w) = \frac{\alpha w^{\alpha-1}}{\beta^\alpha} \exp\left[-\left(\frac{w}{\beta}\right)^\alpha\right], \qquad 0 \leq w.$$

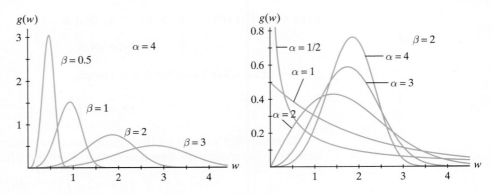

Figure 3.4-1 Weibull probability density functions

Frequently, in engineering, this distribution, with appropriate values of α and β, is excellent for describing the life of a manufactured item. Often α is greater than 1 but less than 5. This pdf is frequently called that of the **Weibull distribution** and, in model fitting, is a strong competitor of the gamma pdf.

The mean and variance of the Weibull distribution are

$$\mu = \beta \, \Gamma\left(1 + \frac{1}{\alpha}\right),$$

$$\sigma^2 = \beta^2 \left\{ \Gamma\left(1 + \frac{2}{\alpha}\right) - \left[\Gamma\left(1 + \frac{1}{\alpha}\right)\right]^2 \right\}.$$

Some graphs of Weibull pdfs are shown in Figure 3.4-1. ∎

Example 3.4-2 People are often shocked to learn that human mortality increases almost exponentially once a person reaches 25 years of age. Depending on which mortality table is used, one finds that the increase is about 10% each year, which means that the rate of mortality will double about every 7 years. (See the Rule of 72 as explained in the Historical Comments at the end of this chapter.) Although this fact can be shocking, we can be thankful that the force of mortality starts very low. The probability that a man in reasonably good health at age 63 dies within the next year is only about 1%. Now, assuming an exponential force of mortality, we have

$$\lambda(w) = H'(w) = ae^{bw}, \qquad a > 0, \quad b > 0.$$

Thus,

$$H(w) = \int_0^w ae^{bt}\, dt = \frac{a}{b}\, e^{bw} - \frac{a}{b}.$$

Hence,

$$G(w) = 1 - \exp\left[-\frac{a}{b}\, e^{bw} + \frac{a}{b}\right], \qquad 0 \le w,$$

and

$$g(w) = ae^{bw} \exp\left[-\frac{a}{b}\, e^{bw} + \frac{a}{b}\right], \qquad 0 \le w,$$

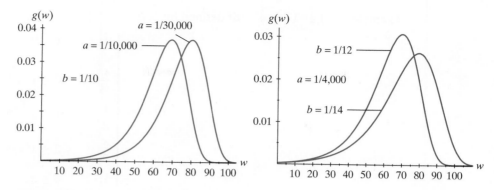

Figure 3.4-2 Gompertz law probability density functions

respectively, are, the cdf and the pdf associated with the famous **Gompertz law** found in actuarial science. Some graphs of pdfs associated with the Gompertz law are shown in Figure 3.4-2. Note that the mode of the Gompertz distribution is $\ln(b/a)/b$. ∎

Both the gamma and Weibull distributions are skewed. In many studies (life testing, response times, incomes, etc.), these are valuable distributions for selecting a model.

Thus far we have considered random variables that are either discrete or continuous. In most applications, these are the types that are encountered. However, on some occasions, combinations of the two types of random variables are found. That is, in some experiments, positive probability is assigned to each of certain points and also is spread over an interval of outcomes, each point of which has zero probability. An illustration will help clarify these remarks.

Example 3.4-3

A bulb for a projector is tested by turning it on, letting it burn for 1 hour, and then turning it off. Let X equal the length of time that the bulb performs satisfactorily during this test. There is a positive probability that the bulb will burn out when it is turned on; hence,

$$0 < P(X = 0) < 1,$$

It could also burn out during the 1-hour period during which it is lit; thus,

$$P(0 < X < 1) > 0,$$

with $P(X = x) = 0$ when $x \in (0, 1)$. In addition, $P(X = 1) > 0$. The act of turning the bulb off after 1 hour so that the actual failure time beyond 1 hour is not observed is called censoring, a phenomenon considered later in this section. ∎

The cdf for a distribution of the **mixed type** will be a combination of those for the discrete and continuous types. That is, at each point of positive probability the cdf will be discontinuous, so that the height of the step there equals the corresponding probability; at all other points the cdf will be continuous.

Example 3.4-4

Let X have a cdf defined by

$$F(x) = \begin{cases} 0, & x < 0, \\ \dfrac{x^2}{4}, & 0 \le x < 1, \\ \dfrac{1}{2}, & 1 \le x < 2, \\ \dfrac{x}{3}, & 2 \le x < 3, \\ 1, & 3 \le x. \end{cases}$$

This cdf is depicted in Figure 3.4-3 and can be used to compute probabilities. As an illustration, consider

$$P(0 < X < 1) = \frac{1}{4},$$

$$P(0 < X \le 1) = \frac{1}{2},$$

$$P(X = 1) = \frac{1}{4}.$$ ∎

Example 3.4-5

Consider the following game: A fair coin is tossed. If the outcome is heads, the player receives \$2. If the outcome is tails, the player spins a balanced spinner that has a scale from 0 to 1. The player then receives that fraction of a dollar associated with the point selected by the spinner. If X denotes the amount received, the space of X is $S = [0, 1) \cup \{2\}$. The cdf of X is defined by

$$F(x) = \begin{cases} 0, & x < 0, \\ \dfrac{x}{2}, & 0 \le x < 1, \\ \dfrac{1}{2}, & 1 \le x < 2, \\ 1, & 2 \le x. \end{cases}$$

The graph of the cdf $F(x)$ is given in Figure 3.4-3. ∎

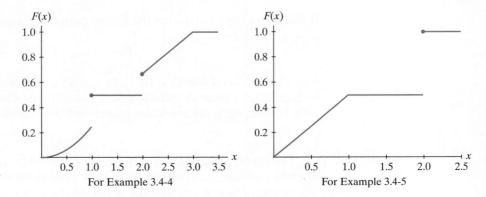

For Example 3.4-4   For Example 3.4-5

Figure 3.4-3 Mixed distribution functions

Suppose that the random variable X has a distribution of the mixed type. To find the expectation of the function $u(X)$ of X, a combination of a sum and a Riemann integral is used, as shown in Example 3.4-6.

Example 3.4-6

We shall find the mean and variance of the random variable given in Example 3.4-4. Note that there, $F'(x) = x/2$ when $0 < x < 1$ and $F'(x) = 1/3$ when $2 < x < 3$; also, $P(X = 1) = 1/4$ and $P(X = 2) = 1/6$. Accordingly, we have

$$\mu = E(X) = \int_0^1 x\left(\frac{x}{2}\right) dx + 1\left(\frac{1}{4}\right) + 2\left(\frac{1}{6}\right) + \int_2^3 x\left(\frac{1}{3}\right) dx$$

$$= \left[\frac{x^3}{6}\right]_0^1 + \frac{1}{4} + \frac{1}{3} + \left[\frac{x^2}{6}\right]_2^3$$

$$= \frac{19}{12}$$

and

$$\sigma^2 = E(X^2) - [E(X)]^2$$

$$= \int_0^1 x^2\left(\frac{x}{2}\right) dx + 1^2\left(\frac{1}{4}\right) + 2^2\left(\frac{1}{6}\right) + \int_2^3 x^2\left(\frac{1}{3}\right) dx - \left(\frac{19}{12}\right)^2$$

$$= \frac{31}{48}.$$

∎

Frequently, in life testing, we know that the length of life—say, X—exceeds the number b, but the exact value of X is unknown. This phenomenon is called **censoring**. It can happen, for instance, when a subject in a cancer study simply disappears; the investigator knows that the subject has lived a certain number of months, but the exact length of the subject's life is unknown. Or it might happen when an investigator does not have enough time to observe the moments of deaths of all the animals—say, rats—in some study. Censoring can also occur in the insurance industry, in the case of a loss with a limited-pay policy in which the top amount is exceeded but it is not known by how much.

Example 3.4-7

Reinsurance companies are concerned with large losses because they might agree, for example, to cover losses due to wind damages that are between $2 million and $10 million. Say that X equals the size of a wind loss in millions of dollars, and suppose that X has the cdf

$$F(x) = 1 - \left(\frac{10}{10 + x}\right)^3, \qquad 0 \leq x < \infty.$$

If losses beyond $10 million are reported only as 10, then $Y = X$, $X \leq 10$, and $Y = 10$, $X > 10$, and the cdf of this censored distribution is

$$G(y) = \begin{cases} 1 - \left(\dfrac{10}{10 + y}\right)^3, & 0 \leq y < 10, \\ 1, & 10 \leq y < \infty, \end{cases}$$

which has a jump of $[10/(10 + 10)]^3 = 1/8$ at $y = 10$.

∎

**Example
3.4-8**

A car worth 24 units (1 unit = $1000) is insured for a year with a one-unit deductible policy. The probability of no damage in a year is 0.95, and the probability of being totaled is 0.01. If the damage is partial, with probability 0.04, then this damage follows the pdf

$$f(x) = \frac{25}{24} \frac{1}{(x+1)^2}, \qquad 0 < x < 24.$$

In computing the expected payment, the insurance company recognizes that it will make zero payment if $X \leq 1, 24 - 1 = 23$ if the car is totaled, and $X - 1$ if $1 < X < 24$. Thus, the expected payment is

$$(0)(0.95) + (0)(0.04) \int_0^1 \frac{25}{24} \frac{1}{(x+1)^2} \, dx + (23)(0.01) + (0.04) \int_1^{24} (x-1) \frac{25}{24} \frac{1}{(x+1)^2} \, dx.$$

That is, the answer is

$$0.23 + (0.04)(1.67) = 0.297,$$

because the last integral is equal to

$$\int_1^{24} (x + 1 - 2) \frac{25}{24} \frac{1}{(x+1)^2} \, dx = (-2) \int_1^{24} \frac{25}{24} \frac{1}{(x+1)^2} \, dx + \int_1^{24} \frac{25}{24} \frac{1}{(x+1)} \, dx$$

$$= (-2) \left[\frac{25}{24} \frac{-1}{(x+1)} \right]_1^{24} + \left[\frac{25}{24} \ln(x+1) \right]_1^{24}$$

$$= 1.67. \qquad \blacksquare$$

Exercises

3.4-1. Let the life W (in years) of the usual family car have a Weibull distribution with $\alpha = 2$. Show that β must equal 10 for $P(W > 5) = e^{-1/4} \approx 0.7788$. HINT: $P(W > 5) = e^{-H(5)}$.

3.4-2. Suppose that the length W of a man's life does follow the Gompertz distribution with $\lambda(w) = a(1.1)^w = ae^{(\ln 1.1)w}$, $P(63 < W < 64) = 0.01$. Determine the constant a and $P(W \leq 71 \mid 70 < W)$.

3.4-3. Let Y_1 be the smallest observation of three independent random variables W_1, W_2, W_3, each with a Weibull distribution with parameters α and β. Show that Y_1 has a Weibull distribution. What are the parameters of this latter distribution? HINT:

$$G(y_1) = P(Y_1 \leq y_1) = 1 - P(y_1 < W_i, i = 1, 2, 3)$$

$$= 1 - [P(y_1 < W_1)]^3.$$

3.4-4. A frequent force of mortality used in actuarial science is $\lambda(w) = ae^{bw} + c$. Find the cdf and pdf associated with this **Makeham's law**.

3.4-5. From the graph of the first cdf of X in Figure 3.4-4, determine the indicated probabilities:

(a) $P(X < 0)$. **(b)** $P(X < -1)$. **(c)** $P(X \leq -1)$.

(d) $P(X < 1)$. **(e)** $P\left(-1 \leq X < \frac{1}{2}\right)$. **(f)** $P(-1 < X \leq 1)$.

3.4-6. Determine the indicated probabilities from the graph of the second cdf of X in Figure 3.4-4:

(a) $P\left(-\frac{1}{2} \leq X \leq \frac{1}{2}\right)$. **(b)** $P\left(\frac{1}{2} < X < 1\right)$. **(c)** $P\left(\frac{3}{4} < X < 2\right)$.

(d) $P(X > 1)$. **(e)** $P(2 < X < 3)$. **(f)** $P(2 < X \leq 3)$.

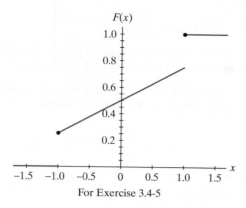
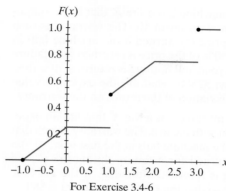

Figure 3.4-4 Mixed distribution functions

3.4-7. Let X be a random variable of the mixed type having the cdf

$$F(x) = \begin{cases} 0, & x < 0, \\ \dfrac{x^2}{4}, & 0 \le x < 1, \\ \dfrac{x+1}{4}, & 1 \le x < 2, \\ 1, & 2 \le x. \end{cases}$$

(a) Carefully sketch the graph of $F(x)$.

(b) Find the mean and the variance of X.

(c) Find $P(1/4 < X < 1)$, $P(X = 1)$, $P(X = 1/2)$, and $P(1/2 \le X < 2)$.

3.4-8. Find the mean and variance of X if the cdf of X is

$$F(x) = \begin{cases} 0, & x < 0, \\ 1 - \left(\dfrac{2}{3}\right)e^{-x}, & 0 \le x. \end{cases}$$

3.4-9. Consider the following game: A fair die is rolled. If the outcome is even, the player receives a number of dollars equal to the outcome on the die. If the outcome is odd, a number is selected at random from the interval $[0, 1)$ with a balanced spinner, and the player receives that fraction of a dollar associated with the point selected.

(a) Define and sketch the cdf of X, the amount received.

(b) Find the expected value of X.

3.4-10. The weekly gravel demand X (in tons) follows the pdf

$$f(x) = \left(\dfrac{1}{5}\right)e^{-x/5}, \qquad 0 < x < \infty.$$

However, the owner of the gravel pit can produce at most only 4 tons of gravel per week. Compute the expected value of the tons sold per week by the owner.

3.4-11. The lifetime X of a certain device has an exponential distribution with mean five years. However, the device is not observed on a continuous basis until after three years. Hence, we actually observe $Y = \max(X, 3)$. Compute $E(Y)$.

3.4-12. Let X have an exponential distribution with $\theta = 1$; that is, the pdf of X is $f(x) = e^{-x}$, $0 < x < \infty$. Let T be defined by $T = \ln X$, so that the cdf of T is

$$G(t) = P(\ln X \le t) = P(X \le e^t).$$

(a) Show that the pdf of T is

$$g(t) = e^t e^{-e^t}, \qquad -\infty < x < \infty,$$

which is the pdf of an extreme-value distribution.

(b) Let W be defined by $T = \alpha + \beta \ln W$, where $-\infty < \alpha < \infty$ and $\beta > 0$. Show that W has a Weibull distribution.

3.4-13. A loss X on a car has a mixed distribution with $p = 0.95$ on zero and $p = 0.05$ on an exponential distribution with a mean of $5000. If the loss X on a car is greater than the deductible of $500, the difference $X - 500$ is paid to the owner of the car. Considering zero (if $X \le 500$) as a possible payment, determine the mean and the standard deviation of the payment.

3.4-14. A customer buys a $1000 deductible policy on her $31,000 car. The probability of having an accident in which the loss is greater than $1000 is 0.03, and then that loss, as a fraction of the value of the car minus the deductible, has the pdf $f(x) = 6(1 - x)^5$, $0 < x < 1$.

(a) What is the probability that the insurance company must pay the customer more than $2000?

(b) What does the company expect to pay?

3.4-15. A certain machine has a life X that has an exponential distribution with mean 10. The warranty is such that 100% of the price is returned if the machine fails in the first year, and 50% of the price is returned for a failure during the second year, and nothing is returned after that. If the machine cost \$2500, what are the expected value and the standard deviation of the return on the warranty?

3.4-16. A certain machine has a life X that has an exponential distribution with mean 8. The warranty is such that \$$m$ is returned if the machine fails in the first year, $(0.5)m$ of the price is returned for a failure during the second year, and nothing is returned after that. If the machine cost \$1200, find m so that the expected payment is \$80.

3.4-17. Some banks now compound daily, but report only on a quarterly basis. It seems to us that it would be easier to compound every instant, for then a dollar invested at an annual rate of i for t years would be worth e^{ti}. [You might find it interesting to prove this statement by taking the limit of $(1 + i/n)^{nt}$ as $n \to \infty$.] If X is a random rate with pdf $f(x) = ce^{-x}$, $0.04 < x < 0.08$, find the pdf of the value of one dollar after three years invested at the rate of X.

3.4-18. The time X to failure of a machine has pdf $f(x) = (x/4)^3 e^{-(x/4)^4}$, $0 < x < \infty$. Compute $P(X > 5 \mid X > 4)$.

3.4-19. Suppose the birth weight (X) in grams of U.S. infants has an approximate Weibull model with pdf

$$f(x) = \frac{3x^2}{3500^3} e^{-(x/3500)^3}, \qquad 0 < x < \infty.$$

Given that a birth weight is greater than 3000, what is the conditional probability that it exceeds 4000?

3.4-20. Let X be the failure time (in months) of a certain insulating material. The distribution of X is modeled by the pdf

$$f(x) = \frac{2x}{50^2} e^{-(x/50)^2}, \qquad 0 < x < \infty.$$

Find

(a) $P(40 < X < 60)$,

(b) $P(X > 80)$.

3.4-21. In a medical experiment, a rat has been exposed to some radiation. The experimenters believe that the rat's survival time X (in weeks) has the pdf

$$f(x) = \frac{3x^2}{120^3} e^{-(x/120)^3}, \qquad 0 < x < \infty.$$

(a) What is the probability that the rat survives at least 100 weeks?

(b) Find the expected value of the survival time. HINT: In the integral representing $E(X)$, let $y = (x/120)^3$ and get the answer in terms of a gamma function.

HISTORICAL COMMENTS In this chapter, we studied several continuous distributions, including the very important normal distribution. Actually, the true importance of the normal distribution is given in Chapter 5, where we consider the central limit theorem and its generalizations. Together, that theorem and its generalizations imply that the sum of several random influences on some measurement suggests that the measurement has an approximate normal distribution. For example, in a study of the length of chicken eggs, different hens produce different eggs, the person measuring the eggs makes a difference, the way the egg is placed in a "holder" is a factor, the caliper used is important, and so on. Thus, the length of an egg might have an approximate normal distribution.

Sometimes instructors force grades to be normally distributed because they "grade on a (normal) curve." This is done too often, and it means that a certain percentage of the students should get A's, a certain percentage B's, etc. We believe that all students should be able to earn A's if they satisfy certain appropriate criteria. Thus, we think that it is wrong to restrict grades to a normal curve.

The normal distribution is symmetric, but many important distributions, like the gamma and Weibull, are skewed. We learned that the Weibull distribution has a failure rate equal to $\lambda(x) = \alpha x^{\alpha-1}/\beta^\alpha$, for $\alpha \geq 1$, and this distribution is appropriate for the length of the life of many manufactured products. It is interesting to note that if $\alpha = 1$, the failure rate is a constant, meaning that an old part is as good as a new one. If this were true for the lives of humans, an old man would have the same chance of living 50 more years as would a young man. That is, we would have a "mathematical fountain of youth." Unfortunately, as we learned in this chapter, the failure rate of humans is increasing with age and is close to being exponential [say, $\lambda(x) = ae^{bx}, a > 0, b > 0$], leading to the Gompertz distribution. As a matter of fact, most would find that the force of mortality is such that it increases about 10% each year; so by the Rule of 72, it would double about every $72/10 = 7.2$. Fortunately, it is very small for persons in their twenties.

The Rule of 72 comes from answering the following question: "How long does it take money to double in value if the interest rate is i?" Assuming the compounding is on an annual basis and that you begin with $1, after one year you have $(1 + i)$, and after two years the number of dollars you have is

$$(1 + i) + i(1 + i) = (1 + i)^2.$$

Continuing this process, the equation that we have to solve is

$$(1 + i)^n = 2,$$

the solution of which is

$$n = \frac{\ln 2}{\ln(1 + i)}.$$

To approximate the value of n, recall that $\ln 2 \approx 0.693$ and use the series expansion of $\ln(1 + i)$ to obtain

$$n \approx \frac{0.693}{i - \dfrac{i^2}{2} + \dfrac{i^3}{3} - \cdots}.$$

Due to the alternating series in the denominator, the denominator is a little less than i. Frequently, brokers increase the numerator a little (say to 0.72) and simply divide by i, obtaining the "well-known **Rule of 72**," namely,

$$n \approx \frac{72}{100i}.$$

For example, if $i = 0.08$, then $n \approx 72/8 = 9$ provides an excellent approximation (the answer is about 9.006). Many persons find that the Rule of 72 is extremely useful when dealing with money matters.

Chapter 4

BIVARIATE DISTRIBUTIONS

4.1 BIVARIATE DISTRIBUTIONS OF THE DISCRETE TYPE

So far, we have taken only one measurement on a single item under observation. However, it is clear in many practical cases that it is possible, and often very desirable, to take more than one measurement of a random observation. Suppose, for example, that we are observing female college students to obtain information about some of their physical characteristics, such as height, x, and weight, y, because we are trying to determine a relationship between those two characteristics. For instance, there may be some pattern between height and weight that can be described by an appropriate curve $y = u(x)$. Certainly, not all of the points observed will be on this curve, but we want to attempt to find the "best" curve to describe the relationship and then say something about the variation of the points around the curve.

Another example might concern high school rank—say, x—and the ACT (or SAT) score—say, y—of incoming college students. What is the relationship between these two characteristics? More importantly, how can we use those measurements to predict a third one, such as first-year college GPA—say, z—with a function $z = v(x, y)$? This is a very important problem for college admission offices, particularly when it comes to awarding an athletic scholarship, because the incoming student–athlete must satisfy certain conditions before receiving such an award.

Definition 4.1-1

Let X and Y be two random variables defined on a discrete space. Let S denote the corresponding two-dimensional space of X and Y, the two random variables of the discrete type. The probability that $X = x$ and $Y = y$ is denoted by $f(x, y) = P(X = x, Y = y)$. The function $f(x, y)$ is called the **joint probability mass function** (joint pmf) of X and Y and has the following properties:

(a) $0 \leq f(x,y) \leq 1$.

(b) $\displaystyle\sum\sum_{(x,y)\in S} f(x,y) = 1$.

(c) $P[(X,Y) \in A] = \displaystyle\sum\sum_{(x,y)\in A} f(x,y)$, where A is a subset of the space S.

The following example will make this definition more meaningful.

Example 4.1-1 Roll a pair of fair dice. For each of the 36 sample points with probability 1/36, let X denote the smaller and Y the larger outcome on the dice. For example, if the outcome is $(3,2)$, then the observed values are $X = 2$, $Y = 3$. The event $\{X = 2, Y = 3\}$ could occur in one of two ways—$(3,2)$ or $(2,3)$—so its probability is

$$\frac{1}{36} + \frac{1}{36} = \frac{2}{36}.$$

If the outcome is $(2,2)$, then the observed values are $X = 2$, $Y = 2$. Since the event $\{X = 2, Y = 2\}$ can occur in only one way, $P(X = 2, Y = 2) = 1/36$. The joint pmf of X and Y is given by the probabilities

$$f(x,y) = \begin{cases} \dfrac{1}{36}, & 1 \leq x = y \leq 6, \\[2mm] \dfrac{2}{36}, & 1 \leq x < y \leq 6, \end{cases}$$

when x and y are integers. Figure 4.1-1 depicts the probabilities of the various points of the space S. ∎

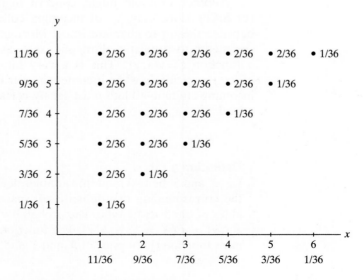

Figure 4.1-1 Discrete joint pmf

Notice that certain numbers have been recorded in the bottom and left-hand margins of Figure 4.1-1. These numbers are the respective column and row totals of the probabilities. The column totals are the respective probabilities that X will assume the values in the x space $S_X = \{1,2,3,4,5,6\}$, and the row totals are the respective probabilities that Y will assume the values in the y space $S_Y = \{1,2,3,4,5,6\}$. That is, the totals describe the probability mass functions of X and Y, respectively. Since each collection of these probabilities is frequently recorded in the margins and satisfies the properties of a pmf of one random variable, each is called a **marginal pmf.**

Definition 4.1-2

Let X and Y have the joint probability mass function $f(x,y)$ with space S. The probability mass function of X alone, which is called the **marginal probability mass function of** X, is defined by

$$f_X(x) = \sum_y f(x,y) = P(X = x), \qquad x \in S_X,$$

where the summation is taken over all possible y values for each given x in the x space S_X. That is, the summation is over all (x,y) in S with a given x value. Similarly, the **marginal probability mass function of** Y is defined by

$$f_Y(y) = \sum_x f(x,y) = P(Y = y), \qquad y \in S_Y,$$

where the summation is taken over all possible x values for each given y in the y space S_Y. The random variables X and Y are **independent** if and only if, for every $x \in S_X$ and every $y \in S_Y$,

$$P(X = x, Y = y) = P(X = x)P(Y = y)$$

or, equivalently,

$$f(x,y) = f_X(x)f_Y(y);$$

otherwise, X and Y are said to be **dependent**.

We note in Example 4.1-1 that X and Y are dependent because there are many x and y values for which $f(x,y) \neq f_X(x)f_Y(y)$. For instance,

$$f_X(1)f_Y(1) = \left(\frac{11}{36}\right)\left(\frac{1}{36}\right) \neq \frac{1}{36} = f(1,1).$$

Example 4.1-2

Let the joint pmf of X and Y be defined by

$$f(x,y) = \frac{x+y}{21}, \qquad x = 1,2,3, \qquad y = 1,2.$$

Then

$$f_X(x) = \sum_y f(x,y) = \sum_{y=1}^{2} \frac{x+y}{21}$$

$$= \frac{x+1}{21} + \frac{x+2}{21} = \frac{2x+3}{21}, \qquad x = 1,2,3,$$

and

$$f_Y(y) = \sum_x f(x,y) = \sum_{x=1}^{3} \frac{x+y}{21} = \frac{6+3y}{21} = \frac{2+y}{7}, \qquad y = 1, 2.$$

Note that both $f_X(x)$ and $f_Y(y)$ satisfy the properties of a probability mass function. Since $f(x,y) \neq f_X(x)f_Y(y)$, X and Y are dependent. ∎

Example 4.1-3 Let the joint pmf of X and Y be

$$f(x,y) = \frac{xy^2}{30}, \qquad x = 1, 2, 3, \qquad y = 1, 2.$$

The marginal probability mass functions are

$$f_X(x) = \sum_{y=1}^{2} \frac{xy^2}{30} = \frac{x}{6}, \qquad x = 1, 2, 3,$$

and

$$f_Y(y) = \sum_{x=1}^{3} \frac{xy^2}{30} = \frac{y^2}{5}, \qquad y = 1, 2.$$

Then $f(x,y) = f_X(x)f_Y(y)$ for $x = 1, 2, 3$ and $y = 1, 2$; thus, X and Y are independent. (See Figure 4.1-2.) ∎

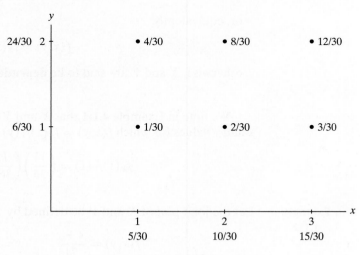

Figure 4.1-2 Joint pmf $f(x,y) = \dfrac{xy^2}{30}$, $x = 1, 2, 3$ and $y = 1, 2$

Example 4.1-4 Let the joint pmf of X and Y be

$$f(x,y) = \frac{xy^2}{13}, \qquad (x,y) = (1,1), (1,2), (2,2).$$

Then the pmf of X is

$$f_X(x) = \begin{cases} \dfrac{5}{13}, & x = 1, \\[2ex] \dfrac{8}{13}, & x = 2, \end{cases}$$

and that of Y is

$$f_Y(y) = \begin{cases} \dfrac{1}{13}, & y = 1, \\[2ex] \dfrac{12}{13}, & y = 2. \end{cases}$$

X and Y are dependent because $f_X(2)f_Y(1) = (8/13)(1/13) \neq 0 = f(2,1)$. ∎

Note that in Example 4.1-4 the support S of X and Y is "triangular." Whenever the support S is not "rectangular," the random variables must be dependent, because S cannot then equal the product set $\{(x, y) : x \in S_X, y \in S_Y\}$. That is, if we observe that the support S of X and Y is not a product set, then X and Y must be dependent. For example, in Example 4.1-4, X and Y are dependent because $S = \{(1,1),(1,2),(2,2)\}$ is not a product set. On the other hand, if S equals the product set $\{(x,y) : x \in S_X, y \in S_Y\}$ and if the formula for $f(x,y)$ is the product of an expression in x alone and an expression in y alone, then X and Y are independent, as shown in Example 4.1-3. Example 4.1-2 illustrates the fact that the support can be rectangular, but the formula for $f(x, y)$ is not such a product, and thus X and Y are dependent.

It is possible to define a probability histogram for a joint pmf just as we did for a pmf for a single random variable. Suppose that X and Y have a joint pmf $f(x,y)$ with space S, where S is a set of pairs of integers. At a point (x, y) in S, construct a "rectangular column" that is centered at (x, y) and has a one-unit-by-one-unit base and a height equal to $f(x,y)$. Note that $f(x, y)$ is equal to the "volume" of this rectangular column. Furthermore, the sum of the volumes of the rectangular columns in this probability histogram is equal to 1.

Example 4.1-5 Let the joint pmf of X and Y be that of Example 4.1-3, namely,

$$f(x, y) = \frac{xy^2}{30}, \qquad x = 1, 2, 3, \qquad y = 1, 2.$$

The probability histogram is shown in Figure 4.1-3. ∎

Sometimes it is convenient to replace the symbols X and Y representing random variables by X_1 and X_2. This is particularly true in situations in which we have more than two random variables; so we use X and Y sometimes and then X_1 and X_2 at other times. The reader will see the advantage of the use of subscripts as we go further in the text.

Let X_1 and X_2 be random variables of the discrete type with the joint pmf $f(x_1, x_2)$ on the space S. If $u(X_1, X_2)$ is a function of these two random variables, then

$$E[u(X_1, X_2)] = \sum_{(x_1, x_2) \in S} \sum u(x_1, x_2) f(x_1, x_2),$$

if it exists, is called the **mathematical expectation** (or **expected value**) of $u(X_1, X_2)$.

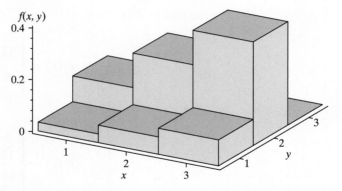

Figure 4.1-3 Joint pmf $f(x, y) = \dfrac{xy^2}{30}$, $x = 1, 2, 3$ and $y = 1, 2$

REMARK The same remarks can be made here that were made in the univariate case, namely, that

$$\sum_{(x_1, x_2) \in S} \sum |u(x_1, x_2)| f(x_1, x_2)$$

must converge and be finite in order for the expectation to exist. Also, $Y = u(X_1, X_2)$ is a random variable—say, with pmf $g(y)$ on space S_Y—and it is true that

$$\sum_{(x_1, x_2) \in S} \sum u(x_1, x_2) f(x_1, x_2) = \sum_{y \in S_Y} y \, g(y). \qquad \blacksquare$$

Example 4.1-6

There are eight similar chips in a bowl: three marked $(0, 0)$, two marked $(1, 0)$, two marked $(0, 1)$, and one marked $(1, 1)$. A player selects a chip at random and is given the sum of the two coordinates in dollars. If X_1 and X_2 represent those two coordinates, respectively, their joint pmf is

$$f(x_1, x_2) = \frac{3 - x_1 - x_2}{8}, \qquad x_1 = 0, 1 \qquad \text{and} \qquad x_2 = 0, 1.$$

Thus,

$$E(X_1 + X_2) = \sum_{x_2=0}^{1} \sum_{x_1=0}^{1} (x_1 + x_2) \frac{3 - x_1 - x_2}{8}$$

$$= (0)\left(\frac{3}{8}\right) + (1)\left(\frac{2}{8}\right) + (1)\left(\frac{2}{8}\right) + (2)\left(\frac{1}{8}\right) = \frac{3}{4}.$$

That is, the expected payoff is 75¢. $\qquad \blacksquare$

The following mathematical expectations, if they exist, have special names:

(a) If $u_i(X_1, X_2) = X_i$ for $i = 1, 2$, then

$$E[u_i(X_1, X_2)] = E(X_i) = \mu_i$$

is called the **mean** of X_i, for $i = 1, 2$.

(b) If $u_i(X_1, X_2) = (X_i - \mu_i)^2$ for $i = 1, 2$, then

$$E[u_i(X_1, X_2)] = E[(X_i - \mu_i)^2] = \sigma_i^2 = \text{Var}(X_i)$$

is called the **variance** of X_i, for $i = 1, 2$.

The mean μ_i and the variance σ_i^2 can be computed from the joint pmf $f(x_1, x_2)$ or the marginal pmf $f_i(x_i)$, $i = 1, 2$.

We give extensions of two important univariate distributions—the hypergeometric distribution and the binomial distribution—through examples.

Example 4.1-7

Consider a population of 200 students who have just finished a first course in calculus. Of these 200, 40 have earned A's, 60 B's, and 100 C's, D's, or F's. A sample of size 25 is taken at random and without replacement from this population in a way that each possible sample has probability

$$\frac{1}{\binom{200}{25}}$$

of being selected. Within the sample of 25, let X be the number of A students, Y the number of B students, and $25 - X - Y$ the number of other students. The space S of (X, Y) is defined by the collection of nonnegative integers (x, y) such that $x + y \leq 25$. The joint pmf of X, Y is

$$f(x, y) = \frac{\binom{40}{x}\binom{60}{y}\binom{100}{25 - x - y}}{\binom{200}{25}},$$

for $(x, y) \in S$, where it is understood that $\binom{k}{j} = 0$ if $j > k$. Without actually summing, we know that the marginal pmf of X is

$$f_X(x) = \frac{\binom{40}{x}\binom{160}{25 - x}}{\binom{200}{25}}, \qquad x = 0, 1, 2, \ldots, 25,$$

since X alone has a hypergeometric distribution. Of course, the function $f_Y(y)$ is also a hypergeometric pmf and

$$f(x, y) \neq f_X(x)f_Y(y),$$

so X and Y are dependent. Note that the space S is not "rectangular," which implies that the random variables are dependent. ∎

We now extend the binomial distribution to a trinomial distribution. Here we have three mutually exclusive and exhaustive ways for an experiment to terminate: perfect, "seconds," and defective. We repeat the experiment n independent times, and the probabilities $p_X, p_Y, p_Z = 1 - p_X - p_Y$ of perfect, seconds, and defective, respectively, remain the same from trial to trial. In the n trials, let $X =$ number of perfect items, $Y =$ number of seconds, and $Z = n - X - Y =$ number of defectives. If

x and y are nonnegative integers such that $x + y \leq n$, then the probability of having x perfects, y seconds, and $n - x - y$ defectives, in that order, is

$$p_X^x p_Y^y (1 - p_X - p_Y)^{n-x-y}.$$

However, if we want $P(X = x, Y = y)$, then we must recognize that $X = x, Y = y$ can be achieved in

$$\binom{n}{x, y, n-x-y} = \frac{n!}{x!y!(n-x-y)!}$$

different ways. Hence, the **trinomial** pmf is

$$f(x, y) = P(X = x, Y = y)$$
$$= \frac{n!}{x!y!(n-x-y)!} p_X^x p_Y^y (1 - p_X - p_Y)^{n-x-y},$$

where x and y are nonnegative integers such that $x + y \leq n$. Without summing, we know that X is $b(n, p_X)$ and Y is $b(n, p_Y)$; thus, X and Y are dependent, as the product of these marginal probability mass functions is not equal to $f(x, y)$.

Example 4.1-8

In manufacturing a certain item, it is found that about 95% of the items are good ones, 4% are "seconds," and 1% are defective. A company has a program of quality control by statistical methods, and each hour an online inspector observes 20 items selected at random, counting the number X of seconds and the number Y of defectives. Let us find the probability that, in this sample of size $n = 20$, at least two seconds or at least two defective items are discovered. If we let $A = \{(x, y) : x \geq 2 \text{ or } y \geq 2\}$, then

$$P(A) = 1 - P(A')$$
$$= 1 - P(X = 0 \text{ or } 1 \text{ and } Y = 0 \text{ or } 1)$$
$$= 1 - \frac{20!}{0!0!20!}(0.04)^0(0.01)^0(0.95)^{20} - \frac{20!}{1!0!19!}(0.04)^1(0.01)^0(0.95)^{19}$$
$$- \frac{20!}{0!1!19!}(0.04)^0(0.01)^1(0.95)^{19} - \frac{20!}{1!1!18!}(0.04)^1(0.01)^1(0.95)^{18}$$
$$= 0.204. \qquad \blacksquare$$

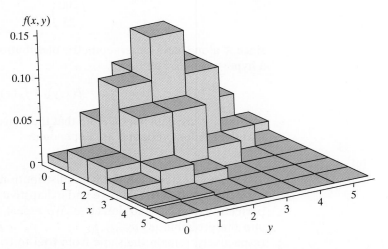

Figure 4.1-4 Trinomial distribution, $p_X = 1/5$, $p_Y = 2/5$, and $n = 5$

Example
4.1-9

Let X and Y have a trinomial distribution with parameters $p_X = 1/5$, $p_Y = 2/5$, and $n = 5$. The probability histogram for the joint pmf of X and Y is shown in Figure 4.1-4. ∎

Exercises

4.1-1. For each of the following functions, determine the constant c so that $f(x, y)$ satisfies the conditions of being a joint pmf for two discrete random variables X and Y:

(a) $f(x, y) = c(x + 2y)$, $x = 1, 2$, $y = 1, 2, 3$.

(b) $f(x, y) = c(x + y)$, $x = 1, 2, 3$, $y = 1, \ldots, x$.

(c) $f(x, y) = c$, x and y are integers such that $6 \le x + y$ ≤ 8, $0 \le y \le 5$.

(d) $f(x, y) = c \left(\dfrac{1}{4}\right)^x \left(\dfrac{1}{3}\right)^y$, $x = 1, 2, \ldots$, $y = 1, 2, \ldots$.

4.1-2. Roll a pair of four-sided dice, one red and one black, each of which has possible outcomes 1, 2, 3, 4 that have equal probabilities. Let X equal the outcome on the red die, and let Y equal the outcome on the black die.

(a) On graph paper, show the space of X and Y.

(b) Define the joint pmf on the space (similar to Figure 4.1-1).

(c) Give the marginal pmf of X in the margin.

(d) Give the marginal pmf of Y in the margin.

(e) Are X and Y dependent or independent? Why or why not?

4.1-3. Let the joint pmf of X and Y be defined by

$$f(x, y) = \frac{x + y}{32}, \quad x = 1, 2, \quad y = 1, 2, 3, 4.$$

(a) Find $f_X(x)$, the marginal pmf of X.

(b) Find $f_Y(y)$, the marginal pmf of Y.

(c) Find $P(X > Y)$.

(d) Find $P(Y = 2X)$.

(e) Find $P(X + Y = 3)$.

(f) Find $P(X \le 3 - Y)$.

(g) Are X and Y independent or dependent? Why or why not?

(h) Find the means and the variances of X and Y.

4.1-4. Select an (even) integer randomly from the set $\{12, 14, 16, 18, 20, 22\}$. Then select an integer randomly from the set $\{12, 13, 14, 15, 16, 17\}$. Let X equal the integer that is selected from the first set and let Y equal the sum of the two integers.

(a) Show the joint pmf of X and Y on the space of X and Y.

(b) Compute the marginal pmfs.

(c) Are X and Y independent? Why or why not?

4.1-5. Roll a pair of four-sided dice, one red and one black. Let X equal the outcome on the red die and let Y equal the sum of the two dice.

(a) On graph paper, describe the space of X and Y.

(b) Define the joint pmf on the space (similar to Figure 4.1-1).

(c) Give the marginal pmf of X in the margin.

(d) Give the marginal pmf of Y in the margin.

(e) Are X and Y dependent or independent? Why or why not?

4.1-6. The torque required to remove bolts in a steel plate is rated as very high, high, average, and low, and these occur about 25%, 35%, 20%, and 20% of the time, respectively. Suppose $n = 31$ bolts are rated; what is the probability of rating 9 very high, 10 high, 7 average, and 5 low? Assume independence of the 31 trials.

4.1-7. A particle starts at $(0, 0)$ and moves in one-unit independent steps with equal probabilities of $1/4$ in each of the four directions: north, south, east, and west. Let S equal the east–west position and T the north–south position after n steps.

(a) Define the joint pmf of S and T with $n = 3$. On a two–dimensional graph, give the probabilities of the joint pmf and the marginal pmfs (similar to Figure 4.1-1).

(b) What are the marginal distributions of X and Y?

4.1-8. In a smoking survey among men between the ages of 25 and 30, 63% prefer to date nonsmokers, 13% prefer to date smokers, and 24% don't care. Suppose nine such men are selected randomly. Let X equal the number who prefer to date nonsmokers and Y equal the number who prefer to date smokers.

(a) Determine the joint pmf of X and Y. Be sure to include the support of the pmf.

(b) Find the marginal pmf of X. Again include the support.

4.1-9. A manufactured item is classified as good, a "second," or defective with probabilities 6/10, 3/10, and 1/10, respectively. Fifteen such items are selected at random from the production line. Let X denote the number of good items, Y the number of seconds, and $15 - X - Y$ the number of defective items.

(a) Give the joint pmf of X and Y, $f(x, y)$.

(b) Sketch the set of integers (x, y) for which $f(x, y) > 0$. From the shape of this region, can X and Y be independent? Why or why not?

(c) Find $P(X = 10, Y = 4)$.

(d) Give the marginal pmf of X.

(e) Find $P(X \leq 11)$.

4.2 THE CORRELATION COEFFICIENT

In Section 4.1, we introduced the mathematical expectation of a function of two random variables—say, X, Y. We gave the respective special names of the mean and variance of X and Y to

$$\mu_X = E(X); \quad \mu_Y = E(Y) \qquad \text{and} \qquad \sigma_X^2 = E[(X - \mu_X)^2]; \quad \sigma_Y^2 = E[(Y - \mu_Y)^2].$$

Now we introduce two more special names:

(a) If $u(X, Y) = (X - \mu_X)(Y - \mu_Y)$, then

$$E[u(X, Y)] = E[(X - \mu_X)(Y - \mu_Y)] = \sigma_{XY} = \text{Cov}(X, Y)$$

is called the **covariance** of X and Y.

(b) If the standard deviations σ_X and σ_Y are positive, then

$$\rho = \frac{\text{Cov}(X, Y)}{\sigma_X \sigma_Y} = \frac{\sigma_{XY}}{\sigma_X \sigma_Y}$$

is called the **correlation coefficient** of X and Y.

It is convenient that the mean and the variance of X can be computed from either the joint pmf (or pdf) or the marginal pmf (or pdf) of X. For example, in the discrete case,

$$\mu_X = E(X) = \sum_x \sum_y x f(x, y)$$

$$= \sum_x x \left[\sum_y f(x, y) \right] = \sum_x x f_X(x).$$

However, to compute the covariance, we need the joint pmf (or pdf).

Before considering the significance of the covariance and the correlation coefficient, let us note a few simple facts. First,

$$E[(X - \mu_X)(Y - \mu_Y)] = E(XY - \mu_X Y - \mu_Y X + \mu_X \mu_Y)$$
$$= E(XY) - \mu_X E(Y) - \mu_Y E(X) + \mu_X \mu_Y,$$

because, even in the bivariate situation, E is still a linear or distributive operator. (See Exercise 4.4-12.) Thus,

$$\text{Cov}(X, Y) = E(XY) - \mu_X \mu_Y - \mu_Y \mu_X + \mu_X \mu_Y = E(XY) - \mu_X \mu_Y.$$

Since $\rho = \text{Cov}(X, Y)/\sigma_X \sigma_Y$, we also have

$$E(XY) = \mu_X \mu_Y + \rho \sigma_X \sigma_Y.$$

That is, the expected value of the product of two random variables is equal to the product $\mu_X \mu_Y$ of their expectations, plus their covariance $\rho \sigma_X \sigma_Y$.

A simple example at this point would be helpful.

Example 4.2-1

Let X and Y have the joint pmf

$$f(x,y) = \frac{x+2y}{18}, \qquad x = 1, 2, \qquad y = 1, 2.$$

The marginal probability mass functions are, respectively,

$$f_X(x) = \sum_{y=1}^{2} \frac{x+2y}{18} = \frac{2x+6}{18} = \frac{x+3}{9}, \qquad x = 1, 2,$$

and

$$f_Y(y) = \sum_{x=1}^{2} \frac{x+2y}{18} = \frac{3+4y}{18}, \qquad y = 1, 2.$$

Since $f(x,y) \neq f_X(x)f_Y(y)$, X and Y are dependent. The mean and the variance of X are, respectively,

$$\mu_X = \sum_{x=1}^{2} x \frac{x+3}{9} = (1)\left(\frac{4}{9}\right) + (2)\left(\frac{5}{9}\right) = \frac{14}{9}$$

and

$$\sigma_X^2 = \sum_{x=1}^{2} x^2 \frac{x+3}{9} - \left(\frac{14}{9}\right)^2 = \frac{24}{9} - \frac{196}{81} = \frac{20}{81}.$$

The mean and the variance of Y are, respectively,

$$\mu_Y = \sum_{y=1}^{2} y \frac{3+4y}{18} = (1)\left(\frac{7}{18}\right) + (2)\left(\frac{11}{18}\right) = \frac{29}{18}$$

and

$$\sigma_Y^2 = \sum_{y=1}^{2} y^2 \frac{3+4y}{18} - \left(\frac{29}{18}\right)^2 = \frac{51}{18} - \frac{841}{324} = \frac{77}{324}.$$

The covariance of X and Y is

$$\text{Cov}(X, Y) = \sum_{x=1}^{2} \sum_{y=1}^{2} xy \frac{x+2y}{18} - \left(\frac{14}{9}\right)\left(\frac{29}{18}\right)$$

$$= (1)(1)\left(\frac{3}{18}\right) + (2)(1)\left(\frac{4}{18}\right) + (1)(2)\left(\frac{5}{18}\right)$$

$$+ (2)(2)\left(\frac{6}{18}\right) - \left(\frac{14}{9}\right)\left(\frac{29}{18}\right)$$

$$= \frac{45}{18} - \frac{406}{162} = -\frac{1}{162}.$$

Hence, the correlation coefficient is

$$\rho = \frac{-1/162}{\sqrt{(20/81)(77/324)}} = \frac{-1}{\sqrt{1540}} = -0.025. \qquad \blacksquare$$

Insight into the correlation coefficient ρ of two discrete random variables X and Y may be gained by thoughtfully examining the definition of ρ, namely,

$$\rho = \frac{\sum_x \sum_y (x - \mu_X)(y - \mu_Y)f(x,y)}{\sigma_X \sigma_Y},$$

where μ_X, μ_Y, σ_X, and σ_Y denote the respective means and standard deviations. If positive probabilities are assigned to pairs (x, y) in which both x and y are either simultaneously above or simultaneously below their respective means, then the corresponding terms in the summation that defines ρ are positive because both factors $(x - \mu_X)$ and $(y - \mu_Y)$ will be positive or both will be negative. If, on the one hand, pairs (x, y), which yield large positive products $(x - \mu_X)(y - \mu_Y)$, contain most of the probability of the distribution, then the correlation coefficient will tend to be positive. If, on the other hand, the points (x, y), in which one component is below its mean and the other above its mean, have most of the probability, then the coefficient of correlation will tend to be negative because the products $(x - \mu_X)(y - \mu_Y)$ having higher probabilities are negative. (See Exercise 4.2-4.) This interpretation of the sign of the correlation coefficient will play an important role in subsequent work.

To gain additional insight into the meaning of the correlation coefficient ρ, consider the following problem: Think of the points (x, y) in the space S and their corresponding probabilities. Let us consider all possible lines in two-dimensional space, each with finite slope, that pass through the point associated with the means, namely, (μ_X, μ_Y). These lines are of the form $y - \mu_Y = b(x - \mu_X)$ or, equivalently, $y = \mu_Y + b(x - \mu_X)$. For each point in S—say, (x_0, y_0), so that $f(x_0, y_0) > 0$—consider the vertical distance from that point to one of the aforesaid lines. Since y_0 is the height of the point above the x-axis and $\mu_Y + b(x_0 - \mu_X)$ is the height of the point on the line that is directly above or below the point (x_0, y_0), the absolute value of the difference of these two heights is the vertical distance from the point (x_0, y_0) to the line $y = \mu_Y + b(x - \mu_X)$. That is, the required distance is $|y_0 - \mu_Y - b(x_0 - \mu_X)|$. Let us now square this distance and take the weighted average of all such squares; in other words, let us consider the mathematical expectation

$$E\{[(Y - \mu_Y) - b(X - \mu_X)]^2\} = K(b).$$

The problem is to find that line (or that b) which minimizes this expectation of the square $[Y - \mu_Y - b(X - \mu_X)]^2$. This is an application of the principle of least squares, and the line is sometimes called the least squares regression line.

The solution of the problem is very easy, since

$$K(b) = E[(Y - \mu_Y)^2 - 2b(X - \mu_X)(Y - \mu_Y) + b^2(X - \mu_X)^2]$$
$$= \sigma_Y^2 - 2b\rho\sigma_X\sigma_Y + b^2\sigma_X^2,$$

because E is a linear operator and $E[(X - \mu_X)(Y - \mu_Y)] = \rho\sigma_X\sigma_Y$. Accordingly, the derivative

$$K'(b) = -2\rho\sigma_X\sigma_Y + 2b\sigma_X^2$$

equals zero at $b = \rho\sigma_Y/\sigma_X$, and we see that $K(b)$ obtains its minimum for that b, since $K''(b) = 2\sigma_X^2 > 0$. Consequently, the **least squares regression line** (the line of the given form that is the best fit in the foregoing sense) is

$$y = \mu_Y + \rho \frac{\sigma_Y}{\sigma_X}(x - \mu_X).$$

Of course, if $\rho > 0$, the slope of the line is positive; but if $\rho < 0$, the slope is negative. It is also instructive to note the value of the minimum of

$$K(b) = E\left\{[(Y - \mu_Y) - b(X - \mu_X)]^2\right\} = \sigma_Y^2 - 2b\rho\sigma_X\sigma_Y + b^2\sigma_X^2.$$

This minimum is

$$K\left(\rho \frac{\sigma_Y}{\sigma_X}\right) = \sigma_Y^2 - 2\rho \frac{\sigma_Y}{\sigma_X}\rho\sigma_X\sigma_Y + \left(\rho \frac{\sigma_Y}{\sigma_X}\right)^2\sigma_X^2$$

$$= \sigma_Y^2 - 2\rho^2\sigma_Y^2 + \rho^2\sigma_Y^2 = \sigma_Y^2(1 - \rho^2).$$

Since $K(b)$ is the expected value of a square, it must be nonnegative for all b, and we see that $\sigma_Y^2(1 - \rho^2) \geq 0$; that is, $\rho^2 \leq 1$, and hence $-1 \leq \rho \leq 1$, which is an important property of the correlation coefficient ρ. On the one hand, if $\rho = 0$, then $K(\rho\sigma_Y/\sigma_X) = \sigma_Y^2$; on the other hand, if ρ is close to 1 or -1, then $K(\rho\sigma_Y/\sigma_X)$ is relatively small. That is, the vertical deviations of the points with positive probability from the line $y = \mu_Y + \rho(\sigma_Y/\sigma_X)(x - \mu_X)$ are small if ρ is close to 1 or -1 because $K(\rho\sigma_Y/\sigma_X)$ is the expectation of the square of those deviations. Thus, ρ measures, in this sense, the amount of *linearity* in the probability distribution. As a matter of fact, in the discrete case, all the points of positive probability lie on this straight line if and only if ρ is equal to 1 or -1.

REMARK More generally, we could have fitted the line $y = a + bx$ by the same application of the principle of least squares. We would then have proved that the "best" line actually passes through the point (μ_X, μ_Y). Recall that, in the preceding discussion, we assumed our line to be of that form. Students will find this derivation to be an interesting exercise using partial derivatives. (See Exercise 4.2-5.) ∎

The next example illustrates a joint discrete distribution for which ρ is negative. In Figure 4.2-1, the line of best fit, or the least squares regression line, is also drawn.

Example 4.2-2 Let X equal the number of ones and Y the number of twos and threes when a pair of fair four-sided dice is rolled. Then X and Y have a trinomial distribution with joint pmf

$$f(x, y) = \frac{2!}{x!y!(2 - x - y)!}\left(\frac{1}{4}\right)^x\left(\frac{2}{4}\right)^y\left(\frac{1}{4}\right)^{2-x-y}, \qquad 0 \leq x + y \leq 2,$$

where x and y are nonnegative integers. Since the marginal pmf of X is $b(2, 1/4)$ and the marginal pmf of Y is $b(2, 1/2)$, it follows that $\mu_X = 1/2$, $\text{Var}(X) = 6/16 = 3/8$, $\mu_Y = 1$, and $\text{Var}(Y) = 1/2$. Also, since $E(XY) = (1)(1)(4/16) = 1/4$, we have $\text{Cov}(X, Y) = 1/4 - (1/2)(1) = -1/4$; therefore, the correlation coefficient is $\rho = -1/\sqrt{3}$. Using these values for the parameters, we obtain the line of best fit, namely,

$$y = 1 + \left(-\frac{1}{\sqrt{3}}\right)\sqrt{\frac{1/2}{3/8}}\left(x - \frac{1}{2}\right) = -\frac{2}{3}x + \frac{4}{3}.$$

The joint pmf is displayed in Figure 4.2-1 along with the line of best fit. ∎

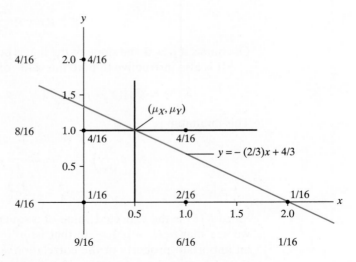

Figure 4.2-1 Trinomial distribution

Suppose that X and Y are independent, so that $f(x, y) \equiv f_X(x)f_Y(y)$. Suppose also that we want to find the expected value of the product $u(X)v(Y)$. Subject to the existence of the expectations, we know that

$$
\begin{aligned}
E[u(X)v(Y)] &= \sum_{S_X} \sum_{S_Y} u(x)v(y)f(x, y) \\
&= \sum_{S_X} \sum_{S_Y} u(x)v(y)f_X(x)f_Y(y) \\
&= \sum_{S_X} u(x)f_X(x) \sum_{S_Y} v(y)f_Y(y) \\
&= E[u(X)]E[v(Y)].
\end{aligned}
$$

This formula can be used to show that the correlation coefficient of two independent variables is zero. For, in standard notation, we have

$$
\begin{aligned}
\text{Cov}(X, Y) &= E[(X - \mu_X)(Y - \mu_Y)] \\
&= E(X - \mu_X)E(Y - \mu_Y) = 0.
\end{aligned}
$$

The converse of this equation is not necessarily true, however: Zero correlation does not, in general, imply independence. It is most important to keep the relationship straight: Independence implies zero correlation, but zero correlation does not necessarily imply independence. We now illustrate the latter proposition.

Example 4.2-3

Let X and Y have the joint pmf

$$
f(x, y) = \frac{1}{3}, \qquad (x, y) = (0, 1), (1, 0), (2, 1).
$$

Since the support is not "rectangular," X and Y must be dependent. The means of X and Y are $\mu_X = 1$ and $\mu_Y = 2/3$, respectively. Hence,

$$
\begin{aligned}
\text{Cov}(X, Y) &= E(XY) - \mu_X\mu_Y \\
&= (0)(1)\left(\frac{1}{3}\right) + (1)(0)\left(\frac{1}{3}\right) + (2)(1)\left(\frac{1}{3}\right) - (1)\left(\frac{2}{3}\right) = 0.
\end{aligned}
$$

That is, $\rho = 0$, but X and Y are dependent. ∎

Exercises

4.2-1. Let the random variables X and Y have the joint pmf

$$f(x,y) = \frac{x+y}{32}, \qquad x = 1,2, \quad y = 1,2,3,4.$$

Find the means μ_X and μ_Y, the variances σ_X^2 and σ_Y^2, and the correlation coefficient ρ.

4.2-2. Let X and Y have the joint pmf defined by $f(0,0) = f(1,2) = 0.2, f(0,1) = f(1,1) = 0.3$.

(a) Depict the points and corresponding probabilities on a graph.

(b) Give the marginal pmfs in the "margins."

(c) Compute $\mu_X, \mu_Y, \sigma_X^2, \sigma_Y^2, \text{Cov}(X,Y)$, and ρ.

(d) Find the equation of the least squares regression line and draw it on your graph. Does the line make sense to you intuitively?

4.2-3. Roll a fair four-sided die twice. Let X equal the outcome on the first roll, and let Y equal the sum of the two rolls.

(a) Determine $\mu_X, \mu_Y, \sigma_X^2, \sigma_Y^2, \text{Cov}(X,Y)$, and ρ.

(b) Find the equation of the least squares regression line and draw it on your graph. Does the line make sense to you intuitively?

4.2-4. Let X and Y have a trinomial distribution with parameters $n = 6, p_X = 1/7$, and $p_Y = 1/3$. Find

(a) $E(X)$.

(b) $E(Y)$.

(c) $\text{Var}(X)$.

(d) $\text{Var}(Y)$.

(e) $\text{Cov}(X,Y)$.

(f) ρ.

Note that $\rho = -\sqrt{p_X p_Y / [(1-p_X)(1-p_Y)]}$ in this case. (Indeed, the formula holds in general for the trinomial distribution; see Example 4.3-3.)

4.2-5. Let X and Y be random variables with respective means μ_X and μ_Y, respective variances σ_X^2 and σ_Y^2, and correlation coefficient ρ. Fit the line $y = a + bx$ by the method of least squares to the probability distribution by minimizing the expectation

$$K(a,b) = E[(Y - a - bX)^2]$$

with respect to a and b. HINT: Consider $\partial K / \partial a = 0$ and $\partial K / \partial b = 0$, and solve simultaneously.

4.2-6. The joint pmf of X and Y is $f(x,y) = 1/6, 0 \leq x + y \leq 2$, where x and y are nonnegative integers.

(a) Sketch the support of X and Y.

(b) Record the marginal pmfs $f_X(x)$ and $f_Y(y)$ in the "margins."

(c) Compute $\text{Cov}(X,Y)$.

(d) Determine ρ, the correlation coefficient.

(e) Find the best-fitting line and draw it on your figure.

4.2-7. Let the joint pmf of X and Y be

$$f(x,y) = 1/4, \quad (x,y) \in S = \{(0,0),(1,1),(1,-1),(2,0)\}.$$

(a) Are X and Y independent?

(b) Calculate $\text{Cov}(X,Y)$ and ρ.

This exercise also illustrates the fact that dependent random variables can have a correlation coefficient of zero.

4.2-8. A certain raw material is classified as to moisture content X (in percent) and impurity Y (in percent). Let X and Y have the joint pmf given by

y	x			
	1	2	3	4
2	0.10	0.20	0.30	0.05
1	0.05	0.05	0.15	0.10

(a) Find the marginal pmfs, the means, and the variances.

(b) Find the covariance and the correlation coefficient of X and Y.

(c) If additional heating is needed with high moisture content and additional filtering with high impurity such that the additional cost is given by the function $C = 2X + 10Y^2$ in dollars, find $E(C)$.

4.2-9. A car dealer sells X cars each day and always tries to sell an extended warranty on each of these cars. (In our opinion, most of these warranties are not good deals.) Let Y be the number of extended warranties sold; then $Y \leq X$. The joint pmf of X and Y is given by

$$f(x,y) = c(x+1)(4-x)(y+1)(3-y),$$
$$x = 0,1,2,3, \quad y = 0,1,2, \text{ with } y \leq x.$$

(a) Find the value of c.

(b) Sketch the support of X and Y.

(c) Record the marginal pmfs $f_X(x)$ and $f_Y(y)$ in the "margins."

(d) Are X and Y independent?

(e) Compute μ_X and σ_X^2.

(f) Compute μ_Y and σ_Y^2.

(g) Compute $\text{Cov}(X, Y)$.

(h) Determine ρ, the correlation coefficient.

(i) Find the best-fitting line and draw it on your figure.

4.2-10. If the correlation coefficient ρ exists, show that ρ satisfies the inequality $-1 \leq \rho \leq 1$. HINT: Consider the discriminant of the nonnegative quadratic function that is given by $h(v) = E\{[(X - \mu_X) + v(Y - \mu_Y)]^2\}$.

4.3 CONDITIONAL DISTRIBUTIONS

Let X and Y have a joint discrete distribution with pmf $f(x, y)$ on space S. Say the marginal probability mass functions are $f_X(x)$ and $f_Y(y)$ with spaces S_X and S_Y, respectively. Let event $A = \{X = x\}$ and event $B = \{Y = y\}$, $(x, y) \in S$. Thus, $A \cap B = \{X = x, Y = y\}$. Because

$$P(A \cap B) = P(X = x, Y = y) = f(x, y)$$

and

$$P(B) = P(Y = y) = f_Y(y) > 0 \qquad \text{(since } y \in S_Y\text{)},$$

the conditional probability of event A given event B is

$$P(A \mid B) = \frac{P(A \cap B)}{P(B)} = \frac{f(x, y)}{f_Y(y)}.$$

This formula leads to the following definition.

Definition 4.3-1

The **conditional probability mass function of** X, given that $Y = y$, is defined by

$$g(x \mid y) = \frac{f(x, y)}{f_Y(y)}, \qquad \text{provided that } f_Y(y) > 0.$$

Similarly, the **conditional probability mass function of** Y, given that $X = x$, is defined by

$$h(y \mid x) = \frac{f(x, y)}{f_X(x)}, \qquad \text{provided that } f_X(x) > 0.$$

Example 4.3-1

Let X and Y have the joint pmf

$$f(x, y) = \frac{x + y}{21}, \qquad x = 1, 2, 3, \qquad y = 1, 2.$$

In Example 4.1-2, we showed that

$$f_X(x) = \frac{2x + 3}{21}, \qquad x = 1, 2, 3,$$

and

$$f_Y(y) = \frac{y + 2}{7}, \qquad y = 1, 2.$$

Thus, the conditional pmf of X, given that $Y = y$, is equal to

$$g(x \mid y) = \frac{(x+y)/21}{(y+2)/7} = \frac{x+y}{3y+6}, \qquad x = 1,2,3, \text{ when } y = 1 \text{ or } 2.$$

For example,

$$P(X = 2 \mid Y = 2) = g(2 \mid 2) = \frac{4}{12} = \frac{1}{3}.$$

Similarly, the conditional pmf of Y, given that $X = x$, is equal to

$$h(y \mid x) = \frac{x+y}{2x+3}, \qquad y = 1,2, \text{ when } x = 1,2, \text{ or } 3.$$

The joint pmf $f(x,y)$ is depicted in Figure 4.3-1(a) along with the marginal pmfs. Now, if $y = 2$, we would expect the outcomes of x—namely, 1, 2, and 3—to occur in the ratio 3:4:5. This is precisely what $g(x \mid y)$ does:

$$g(1 \mid 2) = \frac{1+2}{12}, \qquad g(2 \mid 2) = \frac{2+2}{12}, \qquad g(3 \mid 2) = \frac{3+2}{12}.$$

Figure 4.3-1(b) displays $g(x \mid 1)$ and $g(x \mid 2)$, while Figure 4.3-1(c) gives $h(y \mid 1)$, $h(y \mid 2)$, and $h(y \mid 3)$. Compare the probabilities in Figure 4.3-1(c) with those in Figure 4.3-1(a). They should agree with your intuition as well as with the formula for $h(y \mid x)$. ∎

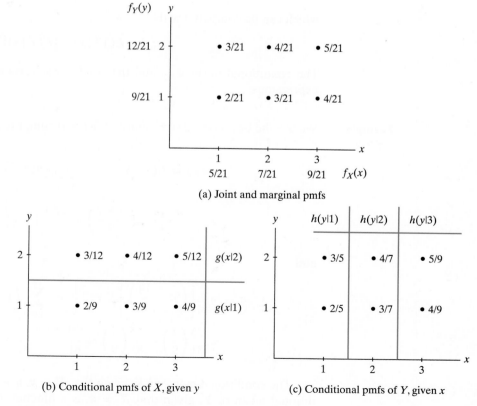

(a) Joint and marginal pmfs

(b) Conditional pmfs of X, given y

(c) Conditional pmfs of Y, given x

Figure 4.3-1 Joint, marginal, and conditional pmfs

Note that $0 \leq h(y \mid x)$. If we sum $h(y \mid x)$ over y for that fixed x, we obtain

$$\sum_y h(y \mid x) = \sum_y \frac{f(x,y)}{f_X(x)} = \frac{f_X(x)}{f_X(x)} = 1.$$

Thus, $h(y \mid x)$ satisfies the conditions of a probability mass function, and we can compute conditional probabilities such as

$$P(a < Y < b \mid X = x) = \sum_{\{y:a<y<b\}} h(y \mid x)$$

and conditional expectations such as

$$E[u(Y) \mid X = x] = \sum_y u(y)h(y \mid x)$$

in a manner similar to those associated with unconditional probabilities and expectations.

Two special conditional expectations are the **conditional mean** of Y, given that $X = x$, defined by

$$\mu_{Y|x} = E(Y \mid x) = \sum_y y\, h(y \mid x),$$

and the **conditional variance** of Y, given that $X = x$, defined by

$$\sigma^2_{Y|x} = E\{[Y - E(Y \mid x)]^2 \mid x\} = \sum_y [y - E(Y \mid x)]^2 h(y \mid x),$$

which can be computed with

$$\sigma^2_{Y|x} = E(Y^2 \mid x) - [E(Y \mid x)]^2.$$

The conditional mean $\mu_{X|y}$ and the conditional variance $\sigma^2_{X|y}$ are given by similar expressions.

Example 4.3-2

We use the background of Example 4.3-1 and compute $\mu_{Y|x}$ and $\sigma^2_{Y|x}$ when $x = 3$:

$$\mu_{Y|3} = E(Y \mid X = 3) = \sum_{y=1}^{2} y\, h(y \mid 3)$$

$$= \sum_{y=1}^{2} y\left(\frac{3+y}{9}\right) = 1\left(\frac{4}{9}\right) + 2\left(\frac{5}{9}\right) = \frac{14}{9},$$

and

$$\sigma^2_{Y|3} = E\left[\left(Y - \frac{14}{9}\right)^2 \middle| X = 3\right] = \sum_{y=1}^{2}\left(y - \frac{14}{9}\right)^2\left(\frac{3+y}{9}\right)$$

$$= \frac{25}{81}\left(\frac{4}{9}\right) + \frac{16}{81}\left(\frac{5}{9}\right) = \frac{20}{81}. \qquad \blacksquare$$

The conditional mean of X, given that $Y = y$, is a function of y alone; the conditional mean of Y, given that $X = x$, is a function of x alone. Suppose that the latter conditional mean is a linear function of x; that is, $E(Y \mid x) = a + bx$. Let us find the constants a and b in terms of characteristics μ_X, μ_Y, σ_X^2, σ_Y^2, and ρ. This development will shed additional light on the correlation coefficient ρ; accordingly, we assume that the respective standard deviations σ_X and σ_Y are both positive, so that the correlation coefficient will exist.

It is given that

$$\sum_y y\, h(y\,|\,x) = \sum_y y\, \frac{f(x,y)}{f_X(x)} = a + bx, \qquad \text{for } x \in S_X,$$

where S_X is the space of X and S_Y is the space of Y. Hence,

$$\sum_y y f(x,y) = (a + bx) f_X(x), \qquad \text{for } x \in S_X, \qquad (4.3\text{-}1)$$

and

$$\sum_{x \in S_X} \sum_y y f(x,y) = \sum_{x \in S_X} (a + bx) f_X(x).$$

That is, with μ_X and μ_Y representing the respective means, we have

$$\mu_Y = a + b\mu_X. \qquad (4.3\text{-}2)$$

In addition, if we multiply both members of Equation 4.3-1 by x and sum the resulting products, we obtain

$$\sum_{x \in S_X} \sum_y xy f(x,y) = \sum_{x \in S_X} (ax + bx^2) f_X(x).$$

That is,

$$E(XY) = aE(X) + bE(X^2)$$

or, equivalently,

$$\mu_X\mu_Y + \rho\sigma_X\sigma_Y = a\mu_X + b(\mu_X^2 + \sigma_X^2). \qquad (4.3\text{-}3)$$

The solution of Equations 4.3-2 and 4.3-3 is

$$a = \mu_Y - \rho\,\frac{\sigma_Y}{\sigma_X}\,\mu_X \qquad \text{and} \qquad b = \rho\,\frac{\sigma_Y}{\sigma_X},$$

which implies that if $E(Y\,|\,x)$ is linear, it is given by

$$E(Y\,|\,x) = \mu_Y + \rho\,\frac{\sigma_Y}{\sigma_X}(x - \mu_X).$$

So if the conditional mean of Y, given that $X = x$, is linear, it is exactly the same as the best-fitting line (least squares regression line) considered in Section 4.2.

By symmetry, if the conditional mean of X, given that $Y = y$, is linear, then

$$E(X\,|\,y) = \mu_X + \rho\,\frac{\sigma_X}{\sigma_Y}(y - \mu_Y).$$

We see that the point $[x = \mu_X, E(Y\,|\,X = \mu_X) = \mu_Y]$ satisfies the expression for $E(Y\,|\,x)$ and $[E(X\,|\,Y = \mu_Y) = \mu_X, y = \mu_Y]$ satisfies the expression for $E(X\,|\,y)$. That is, the point (μ_X, μ_Y) is on each of the two lines. In addition, we note that the product of the coefficient of x in $E(Y\,|\,x)$ and the coefficient of y in $E(X\,|\,y)$ equals ρ^2 and the ratio of these two coefficients equals σ_Y^2/σ_X^2. These observations sometimes prove useful in particular problems.

Example 4.3-3

Let X and Y have the trinomial pmf with parameters n, p_X, p_Y, and $1 - p_X - p_Y = p_Z$. That is,

$$f(x, y) = \frac{n!}{x!\, y!\, (n - x - y)!} p_X^x p_Y^y p_Z^{n-x-y},$$

where x and y are nonnegative integers such that $x + y \leq n$. From the development of the trinomial distribution, we note that X and Y have marginal binomial distributions $b(n, p_X)$ and $b(n, p_Y)$, respectively. Thus,

$$h(y \mid x) = \frac{f(x, y)}{f_X(x)} = \frac{(n - x)!}{y!\, (n - x - y)!} \left(\frac{p_Y}{1 - p_X} \right)^y \left(\frac{p_Z}{1 - p_X} \right)^{n-x-y},$$

$$y = 0, 1, 2, \ldots, n - x.$$

That is, the conditional pmf of Y, given that $X = x$, is binomial, or

$$b\left[n - x, \frac{p_Y}{1 - p_X} \right],$$

and thus has conditional mean

$$E(Y \mid x) = (n - x) \frac{p_Y}{1 - p_X}.$$

In a similar manner, we obtain

$$E(X \mid y) = (n - y) \frac{p_X}{1 - p_Y}.$$

Since each of the conditional means is linear, the product of the respective coefficients of x and y is

$$\rho^2 = \left(\frac{-p_Y}{1 - p_X} \right) \left(\frac{-p_X}{1 - p_Y} \right) = \frac{p_X p_Y}{(1 - p_X)(1 - p_Y)}.$$

However, ρ must be negative because the coefficients of x and y are negative; thus,

$$\rho = -\sqrt{\frac{p_X p_Y}{(1 - p_X)(1 - p_Y)}}. \qquad \blacksquare$$

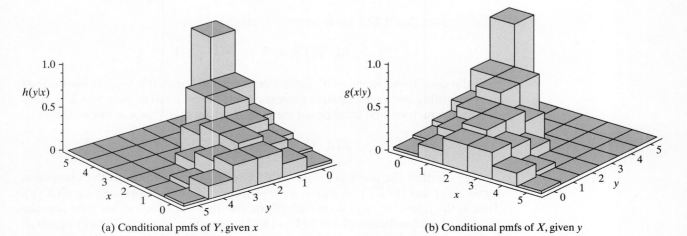

(a) Conditional pmfs of Y, given x (b) Conditional pmfs of X, given y

Figure 4.3-2 Conditional pmfs for the trinomial distribution

In the next example we again look at conditional pmfs when X and Y have a trinomial distribution.

Example 4.3-4 Let X and Y have a trinomial distribution with $p_X = 1/3$, $p_Y = 1/3$, and $n = 5$. Using a result from the last example, we find that the conditional distribution of Y, given that $X = x$, is $b(5 - x, (1/3)/(1 - 1/3))$, or $b(5 - x, 1/2)$. The pmfs $h(y \mid x)$ for $x = 0, 1, \ldots, 5$ are plotted in Figure 4.3-2(a). Note that the orientation of the axes was selected so that the shapes of these pmfs can be seen. Similarly, the conditional distribution of X, given that $Y = y$, is $b(n - y, 1/2)$. The pmfs $g(x \mid y)$ for $y = 0, 1, \ldots, 5$ are shown in Figure 4.3-2(b). ∎

Exercises

4.3-1. Let X and Y have the joint pmf

$$f(x, y) = \frac{x + y}{32}, \qquad x = 1, 2, \qquad y = 1, 2, 3, 4.$$

(a) Display the joint pmf and the marginal pmfs on a graph like Figure 4.3-1(a).

(b) Find $g(x \mid y)$ and draw a figure like Figure 4.3-1(b), depicting the conditional pmfs for $y = 1, 2, 3$, and 4.

(c) Find $h(y \mid x)$ and draw a figure like Figure 4.3-1(c), depicting the conditional pmfs for $x = 1$ and 2.

(d) Find $P(1 \leq Y \leq 3 \mid X = 1)$, $P(Y \leq 2 \mid X = 2)$, and $P(X = 2 \mid Y = 3)$.

(e) Find $E(Y \mid X = 1)$ and $\mathrm{Var}(Y \mid X = 1)$.

4.3-2. Let the joint pmf $f(x, y)$ of X and Y be given by the following:

(x, y)	$f(x, y)$
$(1, 1)$	3/8
$(2, 1)$	1/8
$(1, 2)$	1/8
$(2, 2)$	3/8

Find the two conditional probability mass functions and the corresponding means and variances.

4.3-3. Let W equal the weight of laundry soap in a 1-kilogram box that is distributed in Southeast Asia. Suppose that $P(W < 1) = 0.02$ and $P(W > 1.072) = 0.08$. Call a box of soap light, good, or heavy depending on whether $\{W < 1\}$, $\{1 \leq W \leq 1.072\}$, or $\{W > 1.072\}$, respectively. In $n = 50$ independent observations of these

boxes, let X equal the number of light boxes and Y the number of good boxes.

(a) What is the joint pmf of X and Y?

(b) Give the name of the distribution of Y along with the values of the parameters of this distribution.

(c) Given that $X = 3$, how is Y distributed conditionally?

(d) Determine $E(Y \mid X = 3)$.

(e) Find ρ, the correlation coefficient of X and Y.

4.3-4. The alleles for eye color in a certain male fruit fly are (R, W). The alleles for eye color in the mating female fruit fly are (R, W). Their offspring receive one allele for eye color from each parent. If an offspring ends up with either (W, W), (R, W), or (W, R), its eyes will look white. Let X equal the number of offspring having white eyes. Let Y equal the number of white-eyed offspring having (R, W) or (W, R) alleles.

(a) If the total number of offspring is $n = 400$, how is X distributed?

(b) Give the values of $E(X)$ and $\mathrm{Var}(X)$.

(c) Given that $X = 300$, how is Y distributed?

(d) Give the value of $E(Y \mid X = 300)$ and the value of $\mathrm{Var}(Y \mid X = 300)$.

4.3-5. Let X and Y have a trinomial distribution with $n = 2$, $p_X = 1/4$, and $p_Y = 1/2$.

(a) Give $E(Y \mid x)$.

(b) Compare your answer in part (a) with the equation of the line of best fit in Example 4.2-2. Are they the same? Why or why not?

4.3-6. An insurance company sells both homeowners' insurance and automobile deductible insurance. Let X be the deductible on the homeowners' insurance and Y the deductible on automobile insurance. Among those who take both types of insurance with this company, we find the following probabilities:

		x	
y	100	500	1000
1000	0.05	0.10	0.15
500	0.10	0.20	0.05
100	0.20	0.10	0.05

(a) Compute the following probabilities:
$P(X = 500)$, $P(Y = 500)$, $P(Y = 500 | X = 500)$, $P(Y = 100 | X = 500)$.

(b) Compute the means μ_X, μ_Y, and the variances σ_X^2, σ_Y^2.

(c) Compute the conditional means $E(X | Y = 100)$, $E(Y | X = 500)$.

(d) Compute $\text{Cov}(X, Y)$.

(e) Find the correlation coefficient, ρ.

4.3-7. Using the joint pmf from Exercise 4.2-3, find the value of $E(Y | x)$ for $x = 1, 2, 3, 4$. Do the points $[x, E(Y | x)]$ lie on the best-fitting line?

4.3-8. A fair six-sided die is rolled 42 independent times. Let X be the number of threes and Y the number of fives.

(a) What is the joint pmf of X and Y?

(b) Find the conditional pmf of X, given $Y = y$.

(c) Compute $E(X^2 - 2XY + 3Y^2)$.

4.3-9. Let X and Y have a uniform distribution on the set of points with integer coordinates in $S = \{(x, y): 0 \leq x \leq 7, x \leq y \leq x + 2\}$. That is, $f(x, y) = 1/24$, $(x, y) \in S$, and both x and y are integers. Find

(a) $f_X(x)$.

(b) $h(y | x)$.

(c) $E(Y | x)$.

(d) $\sigma_{Y|x}^2$.

(e) $f_Y(y)$.

4.3-10. Let $f_X(x) = 1/10$, $x = 0, 1, 2, \ldots, 9$, and $h(y | x) = 1/(10 - x)$, $y = x, x + 1, \ldots, 9$. Find

(a) $f(x, y)$.

(b) $f_Y(y)$.

(c) $E(Y | x)$.

4.3-11. Choose a random integer X from the interval $[0, 4]$. Then choose a random integer Y from the interval $[0, x]$, where x is the observed value of X. Make assumptions about the marginal pmf $f_X(x)$ and the conditional pmf $h(y | x)$ and compute $P(X + Y > 4)$.

4.4 BIVARIATE DISTRIBUTIONS OF THE CONTINUOUS TYPE

The idea of joint distributions of two random variables of the discrete type can be extended to that of two random variables of the continuous type. The definitions are really the same except that integrals replace summations. The **joint probability density function** (joint pdf) of two continuous-type random variables is an integrable function $f(x, y)$ with the following properties:

(a) $f(x, y) \geq 0$, where $f(x, y) = 0$ when (x, y) is not in the support (space) S of X and Y.

(b) $\int_{-\infty}^{\infty} \int_{-\infty}^{\infty} f(x, y) \, dx \, dy = 1$.

(c) $P[(X, Y) \in A] = \iint_A f(x, y) \, dx \, dy$, where $\{(X, Y) \in A\}$ is an event defined in the plane.

Property (c) implies that $P[(X, Y) \in A]$ is the volume of the solid over the region A in the xy-plane and bounded by the surface $z = f(x, y)$.

The mathematical expectations are the same as the discrete case, with integrals replacing summations. The following, if they exist, have special names.

The respective **marginal pdfs** of continuous-type random variables X and Y are given by

$$f_X(x) = \int_{-\infty}^{\infty} f(x, y) \, dy, \qquad x \in S_X,$$

and

$$f_Y(y) = \int_{-\infty}^{\infty} f(x, y)\, dx, \qquad y \in S_Y,$$

where S_X and S_Y are the respective spaces of X and Y. The definitions associated with mathematical expectations in the continuous case are the same as those associated with the discrete case after replacing the summations with integrations.

Example 4.4-1

Let X and Y have the joint pdf

$$f(x, y) = \left(\frac{4}{3}\right)(1 - xy), \qquad 0 \le x \le 1, \qquad 0 \le y \le 1.$$

The marginal pdfs are

$$f_X(x) = \int_0^1 \left(\frac{4}{3}\right)(1 - xy)\, dy = \left(\frac{4}{3}\right)\left(1 - \frac{x}{2}\right), \qquad 0 \le x \le 1,$$

and

$$f_Y(y) = \int_0^1 \left(\frac{4}{3}\right)(1 - xy)\, dx = \left(\frac{4}{3}\right)\left(1 - \frac{y}{2}\right), \qquad 0 \le y \le 1.$$

The following probability is computed by a double integral:

$$P(Y \le X/2) = \int_0^1 \int_0^{x/2} \left(\frac{4}{3}\right)(1 - xy)\, dy\, dx$$

$$= \int_0^1 \left(\frac{4}{3}\right)\left(\frac{x}{2} - \frac{x^3}{8}\right) dx$$

$$= \left(\frac{4}{3}\right)\left(\frac{1}{4} - \frac{1}{32}\right) = \frac{7}{24}.$$

The mean of X is

$$\mu_X = E(X) = \int_0^1 \int_0^1 x\left(\frac{4}{3}\right)(1 - xy)\, dy = \int_0^1 x\left(\frac{4}{3}\right)\left(1 - \frac{x}{2}\right) dx$$

$$= \left(\frac{4}{3}\right)\left(\frac{1}{2} - \frac{1}{6}\right) = \frac{4}{9}.$$

Likewise, the mean of Y is

$$\mu_Y = E(Y) = \frac{4}{9}.$$

The variance of X is

$$\mathrm{Var}(X) = \sigma_X^2 = E(X^2) - [E(X)]^2 = \int_0^1 \int_0^1 x^2\left(\frac{4}{3}\right)(1 - xy)\, dy - \left(\frac{4}{9}\right)^2$$

$$= \int_0^1 x^2\left(\frac{4}{3}\right)\left(1 - \frac{x}{2}\right) dx - \frac{16}{81}$$

$$= \left(\frac{4}{3}\right)\left(\frac{1}{3} - \frac{1}{8}\right) - \frac{16}{81} = \frac{13}{162}.$$

Likewise, the variance of Y is

$$\text{Var}(Y) = \sigma_Y^2 = \frac{13}{162}.$$

From these calculations, we see that the means and variances could be calculated using the marginal pdfs instead of the joint pdf. ∎

Example 4.4-2

Let X and Y have the joint pdf

$$f(x,y) = \frac{3}{2}x^2(1 - |y|), \qquad -1 < x < 1, \qquad -1 < y < 1.$$

The graph of $z = f(x,y)$ is given in Figure 4.4-1. Let $A = \{(x,y) : 0 < x < 1, 0 < y < x\}$. Then the probability that (X,Y) falls into A is given by

$$P[(X,Y) \in A] = \int_0^1 \int_0^x \frac{3}{2}x^2(1 - y)\,dy\,dx = \int_0^1 \frac{3}{2}x^2\left[y - \frac{y^2}{2}\right]_0^x dx$$

$$= \int_0^1 \frac{3}{2}\left(x^3 - \frac{x^4}{2}\right)dx = \frac{3}{2}\left[\frac{x^4}{4} - \frac{x^5}{10}\right]_0^1 = \frac{9}{40}.$$

The means are

$$\mu_X = E(X) = \int_{-1}^1 x \cdot \frac{3}{2}x^2\,dx = \left[\frac{3}{8}x^4\right]_{-1}^1 = 0$$

and

$$\mu_Y = E(Y) = \int_{-1}^1 y(1 - |y|)\,dy = \int_{-1}^0 y(1 + y)\,dy + \int_0^1 y(1 - y)\,dy$$

$$= -\frac{1}{2} + \frac{1}{3} + \frac{1}{2} - \frac{1}{3} = 0.$$

We leave the computation of the variances as Exercise 4.4-6. ∎

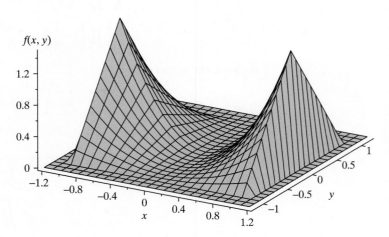

Figure 4.4-1 Joint pdf $f(x,y) = \frac{3}{2}x^2(1 - |y|), -1 < x < 1, -1 < y < 1$

**Example
4.4-3**

Let X and Y have the joint pdf

$$f(x, y) = 2, \qquad 0 \le x \le y \le 1.$$

Then $S = \{(x, y): 0 \le x \le y \le 1\}$ is the support; for example,

$$P\left(0 \le X \le \frac{1}{2}, 0 \le Y \le \frac{1}{2}\right) = P\left(0 \le X \le Y, 0 \le Y \le \frac{1}{2}\right)$$

$$= \int_0^{1/2} \int_0^y 2\, dx\, dy = \int_0^{1/2} 2y\, dy = \frac{1}{4}.$$

You should draw a figure to illustrate the set of points for which $f(x, y) > 0$ and then shade the region over which the integral is taken. The given probability is the volume above this shaded region under the surface $z = 2$. The marginal pdfs are given by

$$f_X(x) = \int_x^1 2\, dy = 2(1 - x), \qquad 0 \le x \le 1,$$

and

$$f_Y(y) = \int_0^y 2\, dx = 2y, \qquad 0 \le y \le 1.$$

Three illustrations of expected values are

$$E(X) = \int_0^1 \int_x^1 2x\, dy\, dx = \int_0^1 2x(1 - x)\, dx = \frac{1}{3},$$

$$E(Y) = \int_0^1 \int_0^y 2y\, dx\, dy = \int_0^1 2y^2\, dy = \frac{2}{3},$$

and

$$E(Y^2) = \int_0^1 \int_0^y 2y^2\, dx\, dy = \int_0^1 2y^3\, dy = \frac{1}{2}.$$

From these calculations we see that $E(X)$, $E(Y)$, and $E(Y^2)$ could be calculated from the marginal pdfs instead of the joint one. ∎

The definition of independent random variables of the continuous type carries over naturally from the discrete case. That is, X and Y are **independent** if and only if the joint pdf factors into the product of their marginal pdfs; namely,

$$f(x, y) \equiv f_X(x)f_Y(y), \qquad x \in S_X,\ y \in S_Y.$$

**Example
4.4-4**

(In this example, as in most, it is helpful to draw the support.) Let X and Y have the joint pdf

$$f(x, y) = cx^2y, \qquad -y \le x \le 1, \qquad 0 \le y \le 1.$$

To determine the value of the constant c, we evaluate

$$\int_0^1 \int_{-y}^1 cx^2 y \, dx \, dy = \int_0^1 \frac{c}{3}(y + y^4) \, dy$$

$$= \frac{c}{3}\left(\frac{1}{2} + \frac{1}{5}\right) = \frac{7c}{30};$$

so

$$\frac{7c}{30} = 1 \qquad \text{and thus} \qquad c = \frac{30}{7}.$$

The marginal pdfs are

$$f_X(x) = \begin{cases} \int_{-x}^1 \frac{30}{7} x^2 y \, dy = \frac{15}{7} x^2(1 - x^2), & -1 \leq x \leq 0, \\[2mm] \int_0^1 \frac{30}{7} x^2 y \, dy = \frac{15}{7} x^2, & 0 < x < 1, \end{cases}$$

and

$$f_Y(y) = \int_{-y}^1 \frac{30}{7} x^2 y \, dx = \frac{10}{7}(y + y^4), \qquad 0 \leq y \leq 1.$$

For illustration, we calculate two probabilities, in the first of which we use the marginal pdf $f_X(x)$. We have

$$P(X \leq 0) = \int_{-1}^0 \frac{15}{7} x^2(1 - x^2) \, dx = \frac{15}{7}\left(\frac{1}{3} - \frac{1}{5}\right) = \frac{2}{7}$$

and

$$P(0 \leq Y \leq X \leq 1) = \int_0^1 \int_0^x \frac{30}{7} x^2 y \, dy \, dx = \int_0^1 \frac{15}{7} x^4 \, dx = \frac{3}{7}.$$

In Exercise 4.4-8, the reader is asked to calculate the means and the variances of X and Y. ∎

In Section 4.2, we used discrete random variables to define the correlation coefficient and related concepts. These ideas carry over to the continuous case with the usual modifications—in particular, with integrals replacing summations. We illustrate the continuous relationships in the following example and also in Exercises 4.4-13 and 4.4-14.

Example 4.4-5   Let X and Y have the joint pdf

$$f(x, y) = 1, \qquad x \leq y \leq x + 1, \qquad 0 \leq x \leq 1.$$

Then

$$f_X(x) = \int_x^{x+1} 1 \, dy = 1, \qquad 0 \leq x \leq 1,$$

and

$$f_Y(y) = \begin{cases} \displaystyle\int_0^y 1\,dx = y, & 0 \le y \le 1, \\[2ex] \displaystyle\int_{y-1}^1 1\,dx = 2 - y, & 1 \le y \le 2. \end{cases}$$

Also,

$$\mu_X = \int_0^1 x \cdot 1\,dx = \frac{1}{2},$$

$$\mu_Y = \int_0^1 y \cdot y\,dy + \int_1^2 y(2 - y)\,dy = \frac{1}{3} + \frac{2}{3} = 1,$$

$$E(X^2) = \int_0^1 x^2 \cdot 1\,dx = \frac{1}{3},$$

$$E(Y^2) = \int_0^1 y^2 \cdot y\,dy + \int_1^2 y^2(2 - y)\,dy = \frac{1}{4} + \left(\frac{14}{3} - \frac{15}{4}\right) = \frac{7}{6},$$

$$E(XY) = \int_0^1 \int_x^{x+1} xy \cdot 1\,dy\,dx = \int_0^1 \frac{1}{2} x\,(2x + 1)\,dx = \frac{7}{12}.$$

Thus,

$$\sigma_X^2 = \frac{1}{3} - \left(\frac{1}{2}\right)^2 = \frac{1}{12}, \qquad \sigma_Y^2 = \frac{7}{6} - 1^2 = \frac{1}{6},$$

and

$$\text{Cov}(X, Y) = \frac{7}{12} - \left(\frac{1}{2}\right)(1) = \frac{1}{12}.$$

Therefore,

$$\rho = \frac{1/12}{\sqrt{(1/12)(1/6)}} = \frac{1}{\sqrt{2}} = \frac{\sqrt{2}}{2},$$

and the least squares regression line is

$$y = 1 + \frac{\sqrt{2}}{2}\frac{\sqrt{1/6}}{\sqrt{1/12}}\left(x - \frac{1}{2}\right) = x + \frac{1}{2}.$$

The latter expression agrees with our intuition since the joint pdf is constant on its support. ∎

In Section 4.3, we used random variables of the discrete type to introduce the new definitions. These definitions also hold for random variables of the continuous type. Let X and Y have a distribution of the continuous type with joint pdf $f(x, y)$

and marginal pdfs $f_X(x)$ and $f_Y(y)$, respectively. Then the conditional pdf, mean, and variance of Y, given that $X = x$, are, respectively,

$$h(y \mid x) = \frac{f(x, y)}{f_X(x)}, \qquad \text{provided that } f_X(x) > 0;$$

$$E(Y \mid x) = \int_{-\infty}^{\infty} y \, h(y \mid x) \, dy;$$

and

$$
\begin{aligned}
\text{Var}(Y \mid x) &= E\{[Y - E(Y \mid x)]^2 \mid x\} \\
&= \int_{-\infty}^{\infty} [y - E(Y \mid x)]^2 \, h(y \mid x) \, dy \\
&= E[Y^2 \mid x] - [E(Y \mid x)]^2.
\end{aligned}
$$

Similar expressions are associated with the conditional distribution of X, given that $Y = y$.

Example 4.4-6   Let X and Y be the random variables of Example 4.4-3. Thus,

$$
\begin{aligned}
f(x, y) &= 2, & 0 &\le x \le y \le 1, \\
f_X(x) &= 2(1 - x), & 0 &\le x \le 1, \\
f_Y(y) &= 2y, & 0 &\le y \le 1.
\end{aligned}
$$

Before we actually find the conditional pdf of Y, given that $X = x$, we shall give an intuitive argument. The joint pdf is constant over the triangular region bounded by $y = x$, $y = 1$, and $x = 0$. If the value of X is known (say, $X = x$), then the possible values of Y are between x and 1. Furthermore, we would expect Y to be uniformly distributed on the interval $[x, 1]$. That is, we would anticipate that $h(y \mid x) = 1/(1-x)$, $x \le y \le 1$.

More formally now, by definition, we have

$$h(y \mid x) = \frac{f(x, y)}{f_X(x)} = \frac{2}{2(1 - x)} = \frac{1}{1 - x}, \qquad x \le y \le 1, \qquad 0 \le x \le 1.$$

The conditional mean of Y, given that $X = x$, is

$$E(Y \mid x) = \int_x^1 y \, \frac{1}{1 - x} \, dy = \left[\frac{y^2}{2(1 - x)} \right]_{y=x}^{y=1} = \frac{1 + x}{2}, \qquad 0 \le x \le 1.$$

Similarly, it can be shown that

$$E(X \mid y) = \frac{y}{2}, \qquad 0 \le y \le 1.$$

The conditional variance of Y, given that $X = x$, is

$$E\{[Y - E(Y\,|\,x)]^2\,|\,x\} = \int_x^1 \left(y - \frac{1+x}{2}\right)^2 \frac{1}{1-x}\,dy$$

$$= \left[\frac{1}{3(1-x)}\left(y - \frac{1+x}{2}\right)^3\right]_{y=x}^{y=1}$$

$$= \frac{(1-x)^2}{12}.$$

Recall that if a random variable W is $U(a,b)$, then $E(W) = (a+b)/2$ and $\mathrm{Var}(W) = (b-a)^2/12$. Since the conditional distribution of Y, given that $X = x$, is $U(x,1)$, we could have inferred immediately that $E(Y\,|\,x) = (x+1)/2$ and $\mathrm{Var}(Y\,|\,x) = (1-x)^2/12$.

An illustration of a computation of a conditional probability is

$$P\left(\frac{3}{4} \le Y \le \frac{7}{8}\,\Big|\,X = \frac{1}{4}\right) = \int_{3/4}^{7/8} h\left(y\,\Big|\,\frac{1}{4}\right)dy$$

$$= \int_{3/4}^{7/8} \frac{1}{3/4}\,dy = \frac{1}{6}.$$ ∎

In general, if $E(Y\,|\,x)$ is linear, then

$$E(Y\,|\,x) = \mu_Y + \rho\left(\frac{\sigma_Y}{\sigma_X}\right)(x - \mu_X).$$

If $E(X\,|\,y)$ is linear, then

$$E(X\,|\,y) = \mu_X + \rho\left(\frac{\sigma_X}{\sigma_Y}\right)(y - \mu_Y).$$

Thus, in Example 4.4-6, the product of the coefficients of x in $E(Y\,|\,x)$ and y in $E(X\,|\,y)$ is $\rho^2 = 1/4$. It follows that $\rho = 1/2$, since each coefficient is positive. Because the ratio of those coefficients is equal to $\sigma_Y^2/\sigma_X^2 = 1$, we have $\sigma_X^2 = \sigma_Y^2$.

Exercises

4.4-1. Let $f(x,y) = (3/16)xy^2$, $0 \le x \le 2$, $0 \le y \le 2$, be the joint pdf of X and Y.

(a) Find $f_X(x)$ and $f_Y(y)$, the marginal probability density functions.

(b) Are the two random variables independent? Why or why not?

(c) Compute the means and variances of X and Y.

(d) Find $P(X \le Y)$.

4.4-2. Let X and Y have the joint pdf $f(x,y) = x + y$, $0 \le x \le 1$, $0 \le y \le 1$.

(a) Find the marginal pdfs $f_X(x)$ and $f_Y(y)$ and show that $f(x,y) \ne f_X(x)f_Y(y)$. Thus, X and Y are dependent.

(b) Compute **(i)** μ_X, **(ii)** μ_Y, **(iii)** σ_X^2, and **(iv)** σ_Y^2.

4.4-3. Let $f(x,y) = 2e^{-x-y}$, $0 \le x \le y < \infty$, be the joint pdf of X and Y. Find $f_X(x)$ and $f_Y(y)$, the marginal pdfs of X and Y, respectively. Are X and Y independent?

4.4-4. Let $f(x,y) = 1/240$, $8.5 \le x \le 10.5$, $120 \le y \le 240$, be the joint pdf of X and Y.

(a) Find $P(100 \le X \le 150)$.

(b) Find $P(150 \le Y \le 200)$.

(c) Find $P(X \ge 150, Y \ge 150)$.

(d) Are X and Y independent? Why or why not?

4.4-5. For each of the following functions, determine the value of c for which the function is a joint pdf of two continuous random variables X and Y.

(a) $f(x,y) = c(6-x-y),$ $\qquad 0 \le x \le 2,$ $\quad 2 \le y \le 4.$

(b) $f(x,y) = c,$ $\qquad\qquad 0 < x < 1,$ $\quad 0 < y < x.$

(c) $f(x,y) = cx^2 y,$ $\qquad\quad 0 < x < 1,$ $\quad 0 < y < 1.$

(d) $f(x,y) = c(x+y^2),$ $\qquad 0 < x < 1,$ $\quad 0 < y < 1.$

4.4-6. Using Example 4.4-2,

(a) Determine the variances of X and Y.

(b) Find $P(-X \le Y)$.

4.4-7. Let $f(x,y) = 4/3,$ $0 < x < 1,$ $x^3 < y < 1$, zero elsewhere.

(a) Sketch the region where $f(x,y) > 0$.

(b) Find $P(X > Y)$.

4.4-8. Using the background of Example 4.4-4, calculate the means and variances of X and Y.

4.4-9. Two construction companies make bids of X and Y (in \$100,000's) on a remodeling project. The joint pdf of X and Y is uniform on the space $2 < x < 2.5, 2 < y < 2.3$. If X and Y are within 0.1 of each other, the companies will be asked to rebid; otherwise, the low bidder will be awarded the contract. What is the probability that they will be asked to rebid?

4.4-10. Let T_1 and T_2 be random times for a company to complete two steps in a certain process. Say T_1 and T_2 are measured in days and they have the joint pdf that is uniform over the space $1 < t_1 < 15, 3 < t_2 < 7, t_1 + 3t_2 < 17$. What is $P(T_1 + T_2 > 15)$?

4.4-11. Let X and Y have the joint pdf $f(x,y) = cx(1-y),$ $0 < y < 1$, and $0 < x < 1 - y$.

(a) Determine c.

(b) Compute $P(Y < X \mid X \le 1/4)$.

4.4-12. Show that in the bivariate situation, E is a linear or distributive operator. That is, for constants a_1 and a_2, show that

$$E[a_1 u_1(X,Y) + a_2 u_2(X,Y)] = a_1 E[u_1(X,Y)] + a_2 E[u_2(X,Y)].$$

4.4-13. Let X and Y be random variables of the continuous type having the joint pdf

$$f(x,y) = 2, \qquad 0 \le y \le x \le 1.$$

Draw a graph that illustrates the domain of this pdf.

(a) Find the marginal pdfs of X and Y.

(b) Compute $\mu_X, \mu_Y, \sigma_X^2, \sigma_Y^2,$ Cov(X,Y), and ρ.

(c) Determine the equation of the least squares regression line and draw it on your graph. Does the line make sense to you intuitively?

4.4-14. Let X and Y be random variables of the continuous type having the joint pdf

$$f(x,y) = 8xy, \qquad 0 \le x \le y \le 1.$$

Draw a graph that illustrates the domain of this pdf.

(a) Find the marginal pdfs of X and Y.

(b) Compute $\mu_X, \mu_Y, \sigma_X^2, \sigma_Y^2,$ Cov(X,Y), and ρ.

(c) Determine the equation of the least squares regression line and draw it on your graph. Does the line make sense to you intuitively?

4.4-15. An automobile repair shop makes an initial estimate X (in thousands of dollars) of the amount of money needed to fix a car after an accident. Say X has the pdf

$$f(x) = 2e^{-2(x-0.2)}, \qquad 0.2 < x < \infty.$$

Given that $X = x$, the final payment Y has a uniform distribution between $x - 0.1$ and $x + 0.1$. What is the expected value of Y?

4.4-16. For the random variables defined in Example 4.4-3, calculate the correlation coefficient directly from the definition

$$\rho = \frac{\text{Cov}(X,Y)}{\sigma_X \sigma_Y}.$$

4.4-17. Let $f(x,y) = 1/40, 0 \le x \le 10, 10 - x \le y \le 14 - x$, be the joint pdf of X and Y.

(a) Sketch the region for which $f(x,y) > 0$.

(b) Find $f_X(x)$, the marginal pdf of X.

(c) Determine $h(y \mid x)$, the conditional pdf of Y, given that $X = x$.

(d) Calculate $E(Y \mid x)$, the conditional mean of Y, given that $X = x$.

4.4-18. Let $f(x,y) = 1/8, 0 \le y \le 4, y \le x \le y+2$, be the joint pdf of X and Y.

(a) Sketch the region for which $f(x,y) > 0$.

(b) Find $f_X(x)$, the marginal pdf of X.

(c) Find $f_Y(y)$, the marginal pdf of Y.

(d) Determine $h(y \mid x)$, the conditional pdf of Y, given that $X = x$.

(e) Determine $g(x \mid y)$, the conditional pdf of X, given that $Y = y$.

(f) Compute $E(Y \mid x)$, the conditional mean of Y, given that $X = x$.

(g) Compute $E(X \mid y)$, the conditional mean of X, given that $Y = y$.

(h) Graph $y = E(Y \mid x)$ on your sketch in part (a). Is $y = E(Y \mid x)$ linear?

(i) Graph $x = E(X \mid y)$ on your sketch in part (a). Is $x = E(X \mid y)$ linear?

4.4-19. Let X have a uniform distribution $U(0,2)$, and let the conditional distribution of Y, given that $X = x$, be $U(0, x^2)$.

(a) Determine $f(x,y)$, the joint pdf of X and Y.

(b) Calculate $f_Y(y)$, the marginal pdf of Y.

(c) Compute $E(X \mid y)$, the conditional mean of X, given that $Y = y$.

(d) Find $E(Y \mid x)$, the conditional mean of Y, given that $X = x$.

4.4-20. Let X have a uniform distribution on the interval $(0, 1)$. Given that $X = x$, let Y have a uniform distribution on the interval $(0, x + 1)$.

(a) Find the joint pdf of X and Y. Sketch the region where $f(x, y) > 0$.

(b) Find $E(Y \mid x)$, the conditional mean of Y, given that $X = x$. Draw this line on the region sketched in part (a).

(c) Find $f_Y(y)$, the marginal pdf of Y. Be sure to include the domain.

4.5 THE BIVARIATE NORMAL DISTRIBUTION

Let X and Y be random variables with joint pdf $f(x, y)$ of the continuous type. Many applications are concerned with the conditional distribution of one of the random variables—say, Y, given that $X = x$. For example, X and Y might be a student's grade point averages from high school and from the first year in college, respectively. Persons in the field of educational testing and measurement are extremely interested in the conditional distribution of Y, given that $X = x$, in such situations.

Suppose that we have an application in which we can make the following three assumptions about the conditional distribution of Y, given $X = x$:

(a) It is normal for each real x.

(b) Its mean, $E(Y \mid x)$, is a linear function of x.

(c) Its variance is constant; that is, it does not depend upon the given value of x.

Of course, assumption (b), along with a result given in Section 4.3, implies that

$$E(Y \mid x) = \mu_Y + \rho \frac{\sigma_Y}{\sigma_X}(x - \mu_X).$$

Let us consider the implication of assumption (c). The conditional variance is given by

$$\sigma_{Y|x}^2 = \int_{-\infty}^{\infty} \left[y - \mu_Y - \rho \frac{\sigma_Y}{\sigma_X}(x - \mu_X) \right]^2 h(y \mid x)\, dy,$$

where $h(y \mid x)$ is the conditional pdf of Y given that $X = x$. Multiply each member of this equation by $f_X(x)$ and integrate on x. Since $\sigma_{Y|x}^2$ is a constant, the left-hand member is equal to $\sigma_{Y|x}^2$. Thus, we have

$$\sigma_{Y|x}^2 = \int_{-\infty}^{\infty} \int_{-\infty}^{\infty} \left[y - \mu_Y - \rho \frac{\sigma_Y}{\sigma_X}(x - \mu_X) \right]^2 h(y \mid x) f_X(x)\, dy\, dx. \tag{4.5-1}$$

However, $h(y \mid x) f_X(x) = f(x, y)$; hence, the right-hand member is just an expectation and Equation 4.5-1 can be written as

$$\sigma_{Y|x}^2 = E \left[(Y - \mu_Y)^2 - 2\rho \frac{\sigma_Y}{\sigma_X}(X - \mu_X)(Y - \mu_Y) + \rho^2 \frac{\sigma_Y^2}{\sigma_X^2}(X - \mu_X)^2 \right].$$

But using the fact that the expectation E is a linear operator and recalling that $E[(X - \mu_X)(Y - \mu_Y)] = \rho\sigma_X\sigma_Y$, we have

$$\sigma_{Y|x}^2 = \sigma_Y^2 - 2\rho \frac{\sigma_Y}{\sigma_X} \rho\sigma_X\sigma_Y + \rho^2 \frac{\sigma_Y^2}{\sigma_X^2} \sigma_X^2$$

$$= \sigma_Y^2 - 2\rho^2\sigma_Y^2 + \rho^2\sigma_Y^2 = \sigma_Y^2(1 - \rho^2).$$

That is, the conditional variance of Y, for each given x, is $\sigma_Y^2(1 - \rho^2)$. These facts about the conditional mean and variance, along with assumption (a), require that the conditional pdf of Y, given that $X = x$, be

$$h(y \mid x) = \frac{1}{\sigma_Y\sqrt{2\pi}\sqrt{1 - \rho^2}} \exp\left[-\frac{[y - \mu_Y - \rho(\sigma_Y/\sigma_X)(x - \mu_X)]^2}{2\sigma_Y^2(1 - \rho^2)}\right],$$

$-\infty < y < \infty$, for every real x.

Before we make any assumptions about the distribution of X, we give an example and a figure to illustrate the implications of our current assumptions.

Example 4.5-1 Let $\mu_X = 10$, $\sigma_X^2 = 9$, $\mu_Y = 12$, $\sigma_Y^2 = 16$, and $\rho = 0.6$. We have seen that assumptions (a), (b), and (c) imply that the conditional distribution of Y, given that $X = x$, is

$$N\left[12 + (0.6)\left(\frac{4}{3}\right)(x - 10), 16(1 - 0.6^2)\right].$$

In Figure 4.5-1, the conditional mean line

$$E(Y \mid x) = 12 + (0.6)\left(\frac{4}{3}\right)(x - 10) = 0.8x + 4$$

has been graphed. For each of $x = 5, 10$, and 15, the conditional pdf of Y, given that $X = x$, is displayed. ∎

Up to this point, nothing has been said about the distribution of X other than that it has mean μ_X and positive variance σ_X^2. Suppose, in addition, we assume that this distribution is also normal; that is, the marginal pdf of X is

$$f_X(x) = \frac{1}{\sigma_X\sqrt{2\pi}} \exp\left[-\frac{(x - \mu_X)^2}{2\sigma_X^2}\right], \qquad -\infty < x < \infty.$$

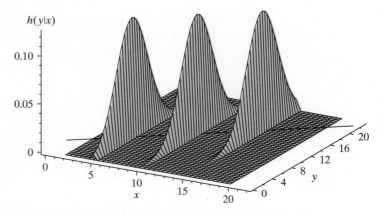

Figure 4.5-1 Conditional pdf of Y, given that $x = 5, 10, 15$

Hence, the joint pdf of X and Y is given by the product

$$f(x, y) = h(y \mid x) f_X(x) = \frac{1}{2\pi \sigma_X \sigma_Y \sqrt{1 - \rho^2}} \exp\left[-\frac{q(x, y)}{2}\right], \qquad (4.5\text{-}2)$$

where it can be shown (see Exercise 4.5-2) that

$$q(x, y) = \frac{1}{1 - \rho^2}\left[\left(\frac{x - \mu_X}{\sigma_X}\right)^2 - 2\rho\left(\frac{x - \mu_X}{\sigma_X}\right)\left(\frac{y - \mu_Y}{\sigma_Y}\right) + \left(\frac{y - \mu_Y}{\sigma_Y}\right)^2\right].$$

A joint pdf of this form is called a **bivariate normal pdf.**

Example 4.5-2
Let us assume that in a certain population of college students, the respective grade point averages—say, X and Y—in high school and the first year in college have an approximate bivariate normal distribution with parameters $\mu_X = 2.9$, $\mu_Y = 2.4$, $\sigma_X = 0.4$, $\sigma_Y = 0.5$, and $\rho = 0.8$.

Then, for example,

$$P(2.1 < Y < 3.3) = P\left(\frac{2.1 - 2.4}{0.5} < \frac{Y - 2.4}{0.5} < \frac{3.3 - 2.4}{0.5}\right)$$

$$= \Phi(1.8) - \Phi(-0.6) = 0.6898.$$

Since the conditional pdf of Y, given that $X = 3.2$, is normal with mean

$$2.4 + (0.8)\left(\frac{0.5}{0.4}\right)(3.2 - 2.9) = 2.7$$

and standard deviation $(0.5)\sqrt{1 - 0.64} = 0.3$, we have

$$P(2.1 < Y < 3.3 \mid X = 3.2)$$

$$= P\left(\frac{2.1 - 2.7}{0.3} < \frac{Y - 2.7}{0.3} < \frac{3.3 - 2.7}{0.3} \,\middle|\, X = 3.2\right)$$

$$= \Phi(2) - \Phi(-2) = 0.9544.$$

From a practical point of view, however, the reader should be warned that the correlation coefficient of these grade point averages is, in many instances, much smaller than 0.8.　■

Since x and y enter the bivariate normal pdf in a similar manner, the roles of X and Y could have been interchanged. That is, Y could have been assigned the marginal normal pdf $N(\mu_Y, \sigma_Y^2)$, and the conditional pdf of X, given that $Y = y$, would have then been normal, with mean $\mu_X + \rho(\sigma_X/\sigma_Y)(y - \mu_Y)$ and variance $\sigma_X^2(1 - \rho^2)$. Although this property is fairly obvious, we do want to make special note of it.

In order to have a better understanding of the geometry of the bivariate normal distribution, consider the graph of $z = f(x, y)$, where $f(x, y)$ is given by Equation 4.5-2. If we intersect this surface with planes parallel to the yz-plane (i.e., with $x = x_0$), we have

$$f(x_0, y) = f_X(x_0) h(y \mid x_0).$$

In this equation, $f_X(x_0)$ is a constant and $h(y \mid x_0)$ is a normal pdf. Thus, $z = f(x_0, y)$ is bell-shaped; that is, has the shape of a normal pdf. However, note that it is not necessarily a pdf, because of the factor $f_X(x_0)$. Similarly, intersections of the surface $z = f(x, y)$ with planes $y = y_0$ parallel to the xz-plane will be bell-shaped.

If

$$0 < z_0 < \frac{1}{2\pi \sigma_X \sigma_Y \sqrt{1 - \rho^2}},$$

then

$$0 < z_0 2\pi \sigma_X \sigma_Y \sqrt{1 - \rho^2} < 1.$$

If we intersect $z = f(x, y)$ with the plane $z = z_0$, which is parallel to the xy-plane, then we have

$$z_0 2\pi \sigma_X \sigma_Y \sqrt{1 - \rho^2} = \exp\left[\frac{-q(x, y)}{2}\right].$$

We show that these intersections are ellipses by taking the natural logarithm of each side to obtain

$$\left(\frac{x - \mu_X}{\sigma_X}\right)^2 - 2\rho \left(\frac{x - \mu_X}{\sigma_X}\right)\left(\frac{y - \mu_Y}{\sigma_Y}\right) + \left(\frac{y - \mu_Y}{\sigma_Y}\right)^2$$

$$= -2(1 - \rho^2)\ln(z_0 2\pi \sigma_X \sigma_Y \sqrt{1 - \rho^2}).$$

Example 4.5-3

With $\mu_X = 10$, $\sigma_X^2 = 9$, $\mu_Y = 12$, $\sigma_Y^2 = 16$, and $\rho = 0.6$, the bivariate normal pdf has been graphed in Figure 4.5-2(a). For $\rho = 0.6$, level curves, or contours, are given in Figure 4.5-2(b). The conditional mean line,

$$E(Y \mid x) = 12 + (0.6)\left(\frac{4}{3}\right)(x - 10) = 0.8x + 4,$$

is also drawn on Figure 4.5-2(b). Note that this line intersects the level curves at points through which vertical tangents can be drawn to the ellipses. ∎

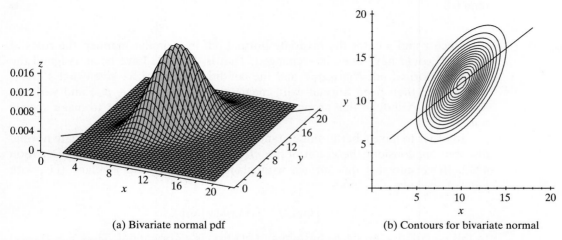

(a) Bivariate normal pdf   (b) Contours for bivariate normal

Figure 4.5-2 Bivariate normal, $\mu_X = 10$, $\sigma_X^2 = 9$, $\mu_Y = 12$, $\sigma_Y^2 = 16$, $\rho = 0.6$

We close this section by observing another important property of the correlation coefficient ρ if X and Y have a bivariate normal distribution. In Equation 4.5-2 of the product $h(y \mid x)f_X(x)$, let us consider the factor $h(y \mid x)$ if $\rho = 0$. We see that this product, which is the joint pdf of X and Y, equals $f_X(x)f_Y(y)$ because when $\rho = 0$, $h(y \mid x)$ is a normal pdf with mean μ_Y and variance σ_Y^2. That is, if $\rho = 0$, then the joint pdf factors into the product of the two marginal probability density functions, and hence X and Y are independent random variables. Of course, if X and Y are *any* independent random variables (not necessarily normal), then ρ, if it exists, is always equal to zero. Hence, we have proved the following theorem.

Theorem 4.5-1	If X and Y have a bivariate normal distribution with correlation coefficient ρ, then X and Y are independent if and only if $\rho = 0$.

Thus, in the bivariate normal case, $\rho = 0$ does imply independence of X and Y. Note that these characteristics of the bivariate normal distribution can be extended to the trivariate normal distribution or, more generally, the multivariate normal distribution. This is done in more advanced texts that assume some knowledge of matrices [e.g., Hogg, McKean, and Craig (2013)].

Exercises

4.5-1. Let X and Y have a bivariate normal distribution with parameters $\mu_X = -3$, $\mu_Y = 10$, $\sigma_X^2 = 25$, $\sigma_Y^2 = 9$, and $\rho = 3/5$. Compute

(a) $P(-5 < X < 5)$.

(b) $P(-5 < X < 5 \mid Y = 13)$.

(c) $P(7 < Y < 16)$.

(d) $P(7 < Y < 16 \mid X = 2)$.

4.5-2. Show that the expression in the exponent of Equation 4.5-2 is equal to the function $q(x, y)$ given in the text.

4.5-3. Let X and Y have a bivariate normal distribution with parameters $\mu_X = 2.00$, $\mu_Y = 120$, $\sigma_X^2 = 0.25$, $\sigma_Y^2 = 256$, and $\rho = 0.8$. Compute

(a) $P(1.9 < X < 2.2)$.

(b) $P(1.9 < X < 2.2 \mid Y = 109)$.

4.5-4. Let X and Y have a bivariate normal distribution with $\mu_X = 75$, $\mu_Y = 5$, $\sigma_X^2 = 144$, $\sigma_Y^2 = 225$, and $\rho = 0.6$. Find

(a) $E(Y \mid X = 76)$.

(b) $\mathrm{Var}(Y \mid X = 76)$.

(c) $P(Y \le 90, \mid X = 76)$.

4.5-5. Let X denote the height in centimeters and Y the weight in kilograms of male college students. Assume that X and Y have a bivariate normal distribution with parameters $\mu_X = 185$, $\sigma_X^2 = 100$, $\mu_Y = 84$, $\sigma_Y^2 = 64$, and $\rho = 3/5$.

(a) Determine the conditional distribution of Y, given that $X = 190$.

(b) Find $P(86.4 < Y < 95.36 \mid X = 190)$.

4.5-6. For a freshman taking introductory statistics and majoring in psychology, let X equal the student's ACT mathematics score and Y the student's ACT verbal score. Assume that X and Y have a bivariate normal distribution with $\mu_X = 22.7$, $\sigma_X^2 = 17.64$, $\mu_Y = 22.7$, $\sigma_Y^2 = 12.25$, and $\rho = 0.78$.

(a) Find $P(18.5 < Y < 25.5)$.

(b) Find $E(Y \mid x)$.

(c) Find $\mathrm{Var}(Y \mid x)$.

(d) Find $P(18.5 < Y < 25.5 \mid X = 23)$.

(e) Find $P(18.5 < Y < 25.5 \mid X = 25)$.

(f) For $x = 21, 23$, and 25, draw a graph of $z = h(y \mid x)$ similar to Figure 4.5-1.

4.5-7. For a pair of gallinules, let X equal the weight in grams of the male and Y the weight in grams of the female. Assume that X and Y have a bivariate normal distribution with $\mu_X = 415$, $\sigma_X^2 = 611$, $\mu_Y = 347$, $\sigma_Y^2 = 689$, and $\rho = -0.25$. Find

(a) $P(309.2 < Y < 380.6)$.

(b) $E(Y \mid x)$.

(c) $\mathrm{Var}(Y \mid x)$.

(d) $P(309.2 < Y < 380.6 \mid X = 385.1)$.

4.5-8. Let X and Y have a bivariate normal distribution with parameters $\mu_X = 12$, $\mu_Y = 15$, $\sigma_X^2 = 16$, $\sigma_Y^2 = 25$, and $\rho = 0$. Find

(a) $P(14 < Y < 16)$.

(b) $E(Y \mid x)$.

(c) $\text{Var}(Y \mid x)$.

(d) $P(14 < Y < 16 \mid X = 12)$.

4.5-9. Let X and Y have a bivariate normal distribution. Find two different lines, $a(x)$ and $b(x)$, parallel to and equidistant from $E(Y \mid x)$, such that

$$P[a(x) < Y < b(x) \mid X = x] = 0.9544$$

for all real x. Plot $a(x)$, $b(x)$, and $E(Y \mid x)$ when $\mu_X = 2$, $\mu_Y = -1$, $\sigma_X = 3$, $\sigma_Y = 5$, and $\rho = 3/5$.

4.5-10. In a college health fitness program, let X denote the weight in kilograms of a male freshman at the beginning of the program and Y denote his weight change during a semester. Assume that X and Y have a bivariate normal distribution with $\mu_X = 75.00$, $\mu_Y = 5.00$, $\sigma_X^2 = 100$, $\sigma_Y^2 = 2.56$, and $\rho = -0.6$. (The lighter students tend to gain weight, while the heavier students tend to lose weight.) Find

(a) $P(1.8 \le Y \le 8.2)$.

(b) $P(1.8 \le Y \le 8.2 \mid X = 80)$.

4.5-11. For a female freshman in a health fitness program, let X equal her percentage of body fat at the beginning of the program and Y equal the change in her percentage of body fat measured at the end of the program. Assume that X and Y have a bivariate normal distribution with $\mu_X = 3.9$, $\mu_Y = 3.4$, $\sigma_X^2 = 0.5$, $\sigma_Y^2 = 0.6$, and $\rho = 0.8$. Find

(a) $P(3.1 \le Y \le 3.7)$.

(b) $\mu_{Y \mid x}$, the conditional mean of Y, given that $X = x$.

(c) $\sigma_{Y \mid x}^2$, the conditional variance of Y, given that $X = x$.

(d) $P(3.1 \le Y \le 3.7 \mid X = 4)$.

4.5-12. Let

$$f(x,y) = \left(\frac{1}{2\pi}\right) e^{-(x^2 + y^2)/2} \left[1 + xy e^{-(x^2 + y^2 - 2)/2}\right],$$
$$-\infty < x < \infty, -\infty < y < \infty.$$

Show that $f(x,y)$ is a joint pdf and the two marginal pdfs are each normal. Note that X and Y can each be normal, but their joint pdf is not bivariate normal.

4.5-13. An obstetrician does ultrasound examinations on her patients between their 16th and 25th weeks of pregnancy to check the growth of each fetus. Let X equal the widest diameter of the fetal head, and let Y equal the length of the femur, both measurements in mm. Assume that X and Y have a bivariate normal distribution with $\mu_X = 60.6$, $\sigma_X = 11.2$, $\mu_Y = 46.8$, $\sigma_Y = 8.4$, and $\rho = 0.94$.

(a) Find $P(40.5 < Y < 48.9)$.

(b) Find $P(40.5 < Y < 48.9 \mid X = 68.6)$.

HISTORICAL COMMENTS Now that we have studied conditional distributions, it might be appropriate to point out that there is a group of statisticians called Bayesians who believe in the following approach (it is considered again in Chapter 6): They treat the parameter θ (such as μ, σ^2, α, and β in the various distributions) as a random variable with a pmf (or pdf), say, $g(\theta)$. Suppose another random variable X, given θ, has the pmf (or pdf) $f(x \mid \theta)$. Say the prior probabilities can be described by $g(\theta)$ so that the marginal pmf (or pdf) of X is given by the sum (or integral)

$$h(x) = \sum_{\theta} g(\theta) f(x \mid \theta).$$

Thus, the conditional pmf (or pdf) of θ, given that $X = x$, is

$$k(\theta \mid x) = \frac{g(\theta) f(x \mid \theta)}{h(x)} = \frac{g(\theta) f(x \mid \theta)}{\sum_{\tau} g(\tau) f(x \mid \tau)}.$$

With a little thought you can recognize this formula as Bayes' theorem. Here the posterior probabilities, $k(\theta \mid x)$, of θ change from the prior probabilities given by $g(\theta)$, after X is observed to be x. Repeating the experiment n independent times (see Chapter 5), we obtain n values of x—say, $x_1, x_2, \ldots, x_n$. The Bayesians use the posterior distribution $k(\theta \mid x_1, x_2, \ldots, x_n)$ to make their inferences about the parameter θ because they then know the conditional probabilities of θ, given $x_1, x_2, \ldots, x_n$.

It is interesting to note that the Reverend Thomas Bayes, a minister who started this method of thinking, never published a mathematical article in his lifetime, and his famous paper was published about two years after his death. Clearly, he was not working for tenure at some university! More will be noted about Bayesian methods in Chapter 6.

DISTRIBUTIONS OF FUNCTIONS OF RANDOM VARIABLES

5.1 FUNCTIONS OF ONE RANDOM VARIABLE

Let X be a random variable of the continuous type. If we consider a function of X— say, $Y = u(X)$—then Y must also be a random variable that has its own distribution. If we can find its cdf, say,

$$G(y) = P(Y \leq y) = P[u(X) \leq y],$$

then its pdf is given by $g(y) = G'(y)$. We now illustrate the **distribution function technique** by two examples.

Example 5.1-1   Let X have a gamma distribution with pdf

$$f(x) = \frac{1}{\Gamma(\alpha)\theta^\alpha} x^{\alpha-1} e^{-x/\theta}, \qquad 0 < x < \infty,$$

where $\alpha > 0$, $\theta > 0$. Let $Y = e^X$, so that the support of Y is $1 < y < \infty$. For each y in the support, the cdf of Y is

$$G(y) = P(Y \leq y) = P(e^X \leq y) = P(X \leq \ln y).$$

That is,

$$G(y) = \int_0^{\ln y} \frac{1}{\Gamma(\alpha)\theta^\alpha} x^{\alpha-1} e^{-x/\theta} dx,$$

and thus the pdf $g(y) = G'(y)$ of Y is

$$g(y) = \frac{1}{\Gamma(\alpha)\theta^\alpha} (\ln y)^{\alpha-1} e^{-(\ln y)/\theta} \left(\frac{1}{y}\right), \qquad 1 < y < \infty.$$

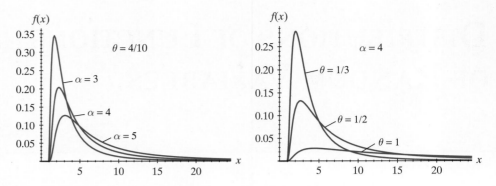

Figure 5.1-1 Loggamma probability density functions

Equivalently, we have

$$g(y) = \frac{1}{\Gamma(\alpha)\theta^\alpha} \frac{(\ln y)^{\alpha-1}}{y^{1+1/\theta}}, \qquad 1 < y < \infty,$$

which is called a **loggamma** pdf. (See Figure 5.1-1 for some graphs.) Note that $\alpha\theta$ and $\alpha\theta^2$ are the mean and the variance, not of Y, but of the original random variable $X = \ln Y$. For the loggamma distribution,

$$\mu = \frac{1}{(1-\theta)^\alpha}, \qquad \theta < 1,$$

$$\sigma^2 = \frac{1}{(1-2\theta)^\alpha} - \frac{1}{(1-\theta)^{2\alpha}}, \qquad \theta < \frac{1}{2}. \qquad \blacksquare$$

There is another interesting distribution, this one involving a transformation of a uniform random variable.

Example 5.1-2

A spinner is mounted at the point $(0, 1)$. Let w be the smallest angle between the y-axis and the spinner. (See Figure 5.1-2.) Assume that w is the value of a random variable W that has a uniform distribution on the interval $(-\pi/2, \pi/2)$. That is, W is $U(-\pi/2, \pi/2)$, and the cdf of W is

$$P(W \le w) = F(w) = \begin{cases} 0, & -\infty < w < -\dfrac{\pi}{2}, \\ \left(w + \dfrac{\pi}{2}\right)\left(\dfrac{1}{\pi}\right), & -\dfrac{\pi}{2} \le w < \dfrac{\pi}{2}, \\ 1, & \dfrac{\pi}{2} \le w < \infty. \end{cases}$$

The relationship between x and w is given by $x = \tan w$; that is, x is the point on the x-axis which is the intersection of that axis and the linear extension of the spinner. To find the distribution of the random variable $X = \tan W$, we note that the cdf of X is given by

$$G(x) = P(X \le x) = P(\tan W \le x) = P(W \le \arctan x)$$

$$= F(\arctan x) = \left(\arctan x + \frac{\pi}{2}\right)\left(\frac{1}{\pi}\right), \qquad -\infty < x < \infty.$$

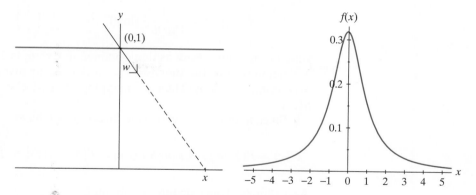

Figure 5.1-2 Spinner and Cauchy pdf

The last equality follows because $-\pi/2 < w = \arctan x < \pi/2$. The pdf of X is given by

$$g(x) = G'(x) = \frac{1}{\pi(1+x^2)}, \qquad -\infty < x < \infty.$$

Figure 5.1-2 shows the graph of this **Cauchy pdf**. In Exercise 5.1-12, you will be asked to show that $E(X)$ does not exist, because the tails of the Cauchy pdf contain too much probability for this pdf to "balance" at $x = 0$. ∎

Thus far, the examples have illustrated the use of the cdf technique. By making a simple observation, we can sometimes shortcut the cdf technique by using what is frequently called the **change-of-variable technique.** Let X be a continuous-type random variable with pdf $f(x)$ with support $c_1 < x < c_2$. We begin this discussion by taking $Y = u(X)$ as a continuous increasing function of X with inverse function $X = v(Y)$. Say the support of X, namely, $c_1 < x < c_2$, maps onto $d_1 = u(c_1) < y < d_2 = u(c_2)$, the support of Y. Then the cdf of Y is

$$G(y) = P(Y \leq y) = P[u(X) \leq y] = P[X \leq v(y)], \qquad d_1 < y < d_2,$$

since u and v are continuous increasing functions. Of course, $G(y) = 0$, $y \leq d_1$, and $G(y) = 1$, $y \geq d_2$. Thus,

$$G(y) = \int_{c_1}^{v(y)} f(x)\, dx, \qquad d_1 < y < d_2.$$

The derivative, $G'(y) = g(y)$, of such an expression is given by

$$G'(y) = g(y) = f[v(y)][v'(y)], \qquad d_1 < y < d_2.$$

Of course, $G'(y) = g(y) = 0$ if $y < d_1$ or $y > d_2$. We may let $g(d_1) = g(d_2) = 0$.

To illustrate, let us consider again Example 5.1-1 with $Y = e^X$, where X has pdf

$$f(x) = \frac{1}{\Gamma(\alpha)\theta^\alpha} x^{\alpha-1} e^{-x/\theta}, \qquad 0 < x < \infty.$$

Here $c_1 = 0$ and $c_2 = \infty$; thus, $d_1 = 1$ and $d_2 = \infty$. Also, $X = \ln Y = v(Y)$. Since $v'(y) = 1/y$, the pdf of Y is

$$g(y) = \frac{1}{\Gamma(\alpha)\theta^\alpha} (\ln y)^{\alpha-1} e^{-(\ln y)/\theta} \left(\frac{1}{y}\right), \qquad 1 < y < \infty,$$

which is the same result as that obtained in Example 5.1-1.

Suppose now the function $Y = u(X)$ and its inverse $X = v(Y)$ are continuous *decreasing* functions. Then the mapping of $c_1 < x < c_2$ is $d_1 = u(c_1) > y > d_2 = u(c_2)$.

Since u and v are decreasing functions, we have

$$G(y) = P(Y \le y) = P[u(X) \le y] = P[X \ge v(y)] = \int_{v(y)}^{c_2} f(x)\, dx, \qquad d_2 < y < d_1.$$

Accordingly, from calculus, we obtain

$$G'(y) = g(y) = f[v(y)][-v'(y)], \qquad d_2 < y < d_1,$$

and $G'(y) = g(y) = 0$ elsewhere. Note that in both the increasing and decreasing cases, we could write

$$g(y) = f[v(y)]\, |v'(y)|, \qquad y \in S_Y, \tag{5.1-1}$$

where S_Y is the support of Y found by mapping the support of X (say, S_X) onto S_Y. The absolute value $|v'(y)|$ assures that $g(y)$ is nonnegative.

The change-of-variable technique thus consists of using Equation 5.1-1 to obtain the pdf of Y directly, bypassing the determination of an expression for the cdf of Y.

Example 5.1-3   Let X have the pdf

$$f(x) = 3(1 - x)^2, \qquad 0 < x < 1.$$

Say $Y = (1 - X)^3 = u(X)$, a decreasing function of X. Thus, $X = 1 - Y^{1/3} = v(Y)$ and $0 < x < 1$ is mapped onto $0 < y < 1$. Since

$$v'(y) = -\frac{1}{3y^{2/3}},$$

we have

$$g(y) = 3[1 - (1 - y^{1/3})]^2 \left| \frac{-1}{3y^{2/3}} \right|, \qquad 0 < y < 1.$$

That is,

$$g(y) = 3y^{2/3} \left(\frac{1}{3y^{2/3}} \right) = 1, \qquad 0 < y < 1;$$

so $Y = (1 - X)^3$ has the uniform distribution $U(0, 1)$. ∎

As we have seen, it is sometimes easier to use the change-of-variable technique than the cdf technique. However, there are many occasions in which the latter is more convenient to use. As a matter of fact, we had to use the cdf in finding the gamma distribution from the Poisson process. (See Section 3.2.) We again use the cdf technique to prove two theorems involving the uniform distribution.

Theorem 5.1-1

Let Y have a distribution that is $U(0, 1)$. Let $F(x)$ have the properties of a cdf of the continuous type with $F(a) = 0$, $F(b) = 1$, and suppose that $F(x)$ is strictly increasing on the support $a < x < b$, where a and b could be $-\infty$ and ∞, respectively. Then the random variable X defined by $X = F^{-1}(Y)$ is a continuous-type random variable with cdf $F(x)$.

Proof The cdf of X is

$$P(X \leq x) = P[F^{-1}(Y) \leq x], \qquad a < x < b.$$

Since $F(x)$ is strictly increasing, $\{F^{-1}(Y) \leq x\}$ is equivalent to $\{Y \leq F(x)\}$. It follows that

$$P(X \leq x) = P[Y \leq F(x)], \qquad a < x < b.$$

But Y is $U(0, 1)$; so $P(Y \leq y) = y$ for $0 < y < 1$, and accordingly,

$$P(X \leq x) = P[Y \leq F(x)] = F(x), \qquad 0 < F(x) < 1.$$

That is, the cdf of X is $F(x)$. □

The next example illustrates how Theorem 5.1-1 can be used to simulate observations from a given distribution.

Example 5.1-4

To help appreciate the large probability in the tails of the Cauchy distribution, it is useful to simulate some observations of a Cauchy random variable. We can begin with a random number, Y, that is an observation from the $U(0, 1)$ distribution. From the distribution function of X, namely, $G(x)$, which is that of the Cauchy distribution given in Example 5.1-2, we have

$$y = G(x) = \left(\arctan x + \frac{\pi}{2} \right)\left(\frac{1}{\pi} \right), \qquad -\infty < x < \infty,$$

or, equivalently,

$$x = \tan\left(\pi y - \frac{\pi}{2} \right). \tag{5.1-2}$$

The latter expression provides observations of X.

In Table 5.1-1, the values of y are the first 10 random numbers in the last column of Table IX in Appendix B. The corresponding values of x are given by Equation 5.1-2. Although most of these observations from the Cauchy distribution are relatively small in magnitude, we see that a very large value (in magnitude) occurs occasionally. Another way of looking at this situation is by considering

Table 5.1-1 Cauchy observations

y	x	y	x
0.1514	−1.9415	0.2354	−1.0962
0.6697	0.5901	0.9662	9.3820
0.0527	−5.9847	0.0043	−74.0211
0.4749	−0.0790	0.1003	−3.0678
0.2900	−0.7757	0.9192	3.8545

sightings (or firing of a gun) from an observation tower, here with coordinates $(0, 1)$, at independent random angles, each with the uniform distribution $U(-\pi/2, \pi/2)$; the target points would then be at Cauchy observations. ∎

The following probability integral transformation theorem is the converse of Theorem 5.1-1.

Theorem 5.1-2	Let X have the cdf $F(x)$ of the continuous type that is strictly increasing on the support $a < x < b$. Then the random variable Y, defined by $Y = F(X)$, has a distribution that is $U(0, 1)$.

Proof Since $F(a) = 0$ and $F(b) = 1$, the cdf of Y is

$$P(Y \le y) = P[F(X) \le y], \qquad 0 < y < 1.$$

However, $\{F(X) \le y\}$ is equivalent to $\{X \le F^{-1}(y)\}$; thus,

$$P(Y \le y) = P[X \le F^{-1}(y)], \qquad 0 < y < 1.$$

Since $P(X \le x) = F(x)$, we have

$$P(Y \le y) = P[X \le F^{-1}(y)] = F[F^{-1}(y)] = y, \qquad 0 < y < 1,$$

which is the cdf of a $U(0, 1)$ random variable. □

REMARK Although in our statements and proofs of Theorems 5.1-1 and 5.1-2, we required $F(x)$ to be strictly increasing, this restriction can be dropped and both theorems are still true. In our exposition, we did not want to bother students with certain difficulties that are experienced if $F(x)$ is not strictly increasing. ∎

Another observation concerns the situation in which the transformation $Y = u(X)$ is not one-to-one, as it has been up to this point in this section. For example, let $Y = X^2$, where X is Cauchy. Here $-\infty < x < \infty$ maps onto $0 \le y < \infty$, so

$$G(y) = P(X^2 \le y) = P(-\sqrt{y} \le X \le \sqrt{y})$$

$$= \int_{-\sqrt{y}}^{\sqrt{y}} f(x)\, dx, \qquad 0 \le y < \infty,$$

where

$$f(x) = \frac{1}{\pi(1 + x^2)}, \qquad -\infty < x < \infty.$$

Thus,

$$G'(y) = g(y) = f(\sqrt{y}) \left| \frac{1}{2\sqrt{y}} \right| + f(-\sqrt{y}) \left| \frac{-1}{2\sqrt{y}} \right|$$

$$= \frac{1}{\pi(1 + y)\sqrt{y}}, \qquad 0 \le y < \infty.$$

That is, in this case of a two-to-one transformation, there is a need to sum two terms, each of which is similar to a counterpart term in the one-to-one case; but here $x_1 = \sqrt{y}$ and $x_2 = -\sqrt{y}$, $0 < y < \infty$, give the two inverse functions, respectively.

With careful thought, we can handle many situations that generalize this particular example.

Example 5.1-5

Let X have the pdf

$$f(x) = \frac{x^2}{3}, \qquad -1 < x < 2.$$

Then the random variable $Y = X^2$ will have the support $0 \leq y < 4$. Now, on the one hand, for $0 < y < 1$, we obtain the two-to-one transformation represented by $x_1 = -\sqrt{y}$ for $-1 < x_1 < 0$ and by $x_2 = \sqrt{y}$ for $0 < x_2 < 1$. On the other hand, if $1 < y < 4$, the one-to-one transformation is represented by $x_2 = \sqrt{y}, 1 < x_2 < 2$. Since

$$\frac{dx_1}{dy} = \frac{-1}{2\sqrt{y}} \qquad \text{and} \qquad \frac{dx_2}{dy} = \frac{1}{2\sqrt{y}},$$

it follows that the pdf of $Y = X^2$ is

$$g(y) = \begin{cases} \dfrac{(-\sqrt{y})^2}{3}\left|\dfrac{-1}{2\sqrt{y}}\right| + \dfrac{(\sqrt{y})^2}{3}\left|\dfrac{1}{2\sqrt{y}}\right| = \dfrac{\sqrt{y}}{3}, & 0 < y < 1, \\[4mm] \dfrac{(\sqrt{y})^2}{3}\left|\dfrac{1}{2\sqrt{y}}\right| = \dfrac{\sqrt{y}}{6}, & 1 < y < 4. \end{cases}$$

Note that if $0 < y < 1$, the pdf is the sum of two terms, but if $1 < y < 4$, then there is only one term. These different expressions in $g(y)$ correspond to the two types of transformations, namely, the two-to-one and the one-to-one transformations, respectively. ∎

The change-of-variable technique can be used for a variable X of the discrete type, but there is one major difference: The pmf $f(x) = P(X = x)$, $x \in S_X$, represents probability. Note that the support S_X consists of a countable number of points, say, $c_1, c_2, c_3, \ldots$. Let $Y = u(X)$ be a one-to-one transformation with inverse $X = v(Y)$. The function $y = u(x)$ maps S_X onto $d_1 = u(c_1), d_2 = u(c_2), d_3 = u(c_3), \ldots$, which we denote by S_Y. Hence, the pmf of Y is

$$g(y) = P(Y = y) = P[u(X) = y] = P[X = v(y)], \qquad y \in S_Y.$$

Since $P(X = x) = f(x)$, we have $g(y) = f[v(y)]$, $y \in S_Y$. Note that, in this discrete case, the value of the derivative, namely, $|v'(y)|$, is not needed.

Example 5.1-6

Let X have a Poisson distribution with $\lambda = 4$; thus, the pmf is

$$f(x) = \frac{4^x e^{-4}}{x!}, \qquad x = 0, 1, 2, \ldots.$$

If $Y = \sqrt{X}$, then, since $X = Y^2$, we have

$$g(y) = \frac{4^{y^2} e^{-4}}{(y^2)!}, \qquad y = 0, 1, \sqrt{2}, \sqrt{3}, \ldots. \qquad ∎$$

Example 5.1-7 Let the distribution of X be binomial with parameters n and p. Since X has a discrete distribution, $Y = u(X)$ will also have a discrete distribution, with the same probabilities as those in the support of X. For example, with $n = 3$, $p = 1/4$, and $Y = X^2$, we have

$$g(y) = \binom{3}{\sqrt{y}} \left(\frac{1}{4}\right)^{\sqrt{y}} \left(\frac{3}{4}\right)^{3-\sqrt{y}}, \qquad y = 0, 1, 4, 9. \qquad \blacksquare$$

Exercises

5.1-1. Let X have the pdf $f(x) = 2x$, $0 < x < 1$. Find the pdf of $Y = 3X + 1$.

5.1-2. Let X have the pdf $f(x) = xe^{-x^2/2}$, $0 < x < \infty$. Find the pdf of $Y = X^2$.

5.1-3. Let X have a gamma distribution with $\alpha = 3$ and $\theta = 2$. Determine the pdf of $Y = \sqrt{X}$.

5.1-4. The pdf of X is $f(x) = x^2 - 3$, $0 < x < 1$.
(a) Find the cdf of X.
(b) Describe how an observation of X can be simulated.
(c) Simulate 10 observations of X.

5.1-5. The pdf of X is $f(x) = \theta x^{\theta-1}$, $0 < x < 1$, $0 < \theta < \infty$. Let $Y = -2\theta \ln X$. How is Y distributed?

5.1-6. Let X have a **logistic distribution** with pdf

$$f(x) = \frac{e^{-x}}{(1 + e^{-x})^2}, \qquad -\infty < x < \infty.$$

Show that

$$Y = \frac{1}{1 + e^{-X}}$$

has a $U(0, 1)$ distribution.

5.1-7. A sum of \$50,000 is invested at a rate R, selected from a uniform distribution on the interval $(0.03, 0.07)$. Once R is selected, the sum is compounded instantaneously for a year, so that $X = 50000\, e^R$ dollars is the amount at the end of that year.
(a) Find the cdf and pdf of X.
(b) Verify that $X = 50000\, e^R$ is defined correctly if the compounding is done instantaneously. HINT: Divide the year into n equal parts, calculate the value of the amount at the end of each part, and then take the limit as $n \to \infty$.

5.1-8. The lifetime (in years) of a manufactured product is $Y = 5X^{0.7}$, where X has an exponential distribution with mean 1. Find the cdf and pdf of Y.

5.1-9. Statisticians frequently use the **extreme value distribution** given by the cdf

$$F(x) = 1 - \exp\left[-e^{(x - \theta_1)/\theta_2}\right], \qquad -\infty < x < \infty.$$

A simple case is when $\theta_1 = 0$ and $\theta_2 = 1$, giving

$$F(x) = 1 - \exp\left[-e^x\right], \qquad -\infty < x < \infty.$$

Let $Y = e^X$ or $X = \ln Y$; then the support of Y is $0 < y < \infty$.
(a) Show that the distribution of Y is exponential when $\theta_1 = 0$ and $\theta_2 = 1$.
(b) Find the cdf and the pdf of Y when $\theta_1 \neq 0$ and $\theta_2 > 0$.
(c) Let $\theta_1 = \ln \beta$ and $\theta_2 = 1/\alpha$ in the cdf and pdf of Y. What is this distribution?
(d) As suggested by its name, the extreme value distribution can be used to model the longest home run, the deepest mine, the greatest flood, and so on. Suppose the length X (in feet) of the maximum of someone's home runs was modeled by an extreme value distribution with $\theta_1 = 550$ and $\theta_2 = 25$. What is the probability that X exceeds 500 feet?

5.1-10. Let X have the uniform distribution $U(-1, 3)$. Find the pdf of $Y = X^2$.

5.1-11. Let X have a Cauchy distribution. Find
(a) $P(X > 1)$.
(b) $P(X > 5)$.
(c) $P(X > 10)$.

5.1-12. Let $f(x) = 1/[\pi(1 + x^2)]$, $-\infty < x < \infty$, be the pdf of the Cauchy random variable X. Show that $E(X)$ does not exist.

5.1-13. If the distribution of X is $N(\mu, \sigma^2)$, then $M(t) = E(e^{tX}) = \exp(\mu t + \sigma^2 t^2/2)$. We then say that $Y = e^X$ has a **lognormal distribution** because $X = \ln Y$.
(a) Show that the pdf of Y is

$$g(y) = \frac{1}{y\sqrt{2\pi\sigma^2}} \exp[-(\ln y - \mu)^2/2\sigma^2], \qquad 0 < y < \infty.$$

(b) Using $M(t)$, find **(i)** $E(Y) = E(e^X) = M(1)$, **(ii)** $E(Y^2) = E(e^{2X}) = M(2)$, and **(iii)** $\text{Var}(Y)$.

5.1-14. Let X be $N(0,1)$. Find the pdf of $Y = |X|$, a distribution that is often called the **half-normal**. HINT: Here $y \in S_y = \{y: 0 < y < \infty\}$. Consider the two transformations $x_1 = -y, -\infty < x_1 < 0$, and $x_2 = y, 0 < y < \infty$.

5.1-15. Let $Y = X^2$.

(a) Find the pdf of Y when the distribution of X is $N(0,1)$.

(b) Find the pdf of Y when the pdf of X is $f(x) = (3/2)x^2$, $-1 < x < 1$.

5.2 TRANSFORMATIONS OF TWO RANDOM VARIABLES

In Section 5.1, we considered the transformation of one random variable X with pdf $f(x)$. In particular, in the continuous case, if $Y = u(X)$ was an increasing or decreasing function of X, with inverse $X = v(Y)$, then the pdf of Y was

$$g(y) = |v'(y)| f[v(y)], \qquad c < y < d,$$

where the support $c < y < d$ corresponds to the support of X, say, $a < x < b$, through the transformation $x = v(y)$.

There is one note of warning here: If the function $Y = u(X)$ does not have a single-valued inverse, the determination of the distribution of Y will not be as simple. As a matter of fact, we did consider two examples in Section 5.1 in which there were two inverse functions, and we exercised special care in those examples. Here, we will not consider problems with many inverses; however, such a warning is nonetheless appropriate.

When two random variables are involved, many interesting problems can result. In the case of a single-valued inverse, the rule is about the same as that in the one-variable case, with the derivative being replaced by the Jacobian. That is, if X_1 and X_2 are two continuous-type random variables with joint pdf $f(x_1, x_2)$, and if $Y_1 = u_1(X_1, X_2)$, $Y_2 = u_2(X_1, X_2)$ has the single-valued inverse $X_1 = v_1(Y_1, Y_2)$, $X_2 = v_2(Y_1, Y_2)$, then the joint pdf of Y_1 and Y_2 is

$$g(y_1, y_2) = |J| f[v_1(y_1, y_2), v_2(y_1, y_2)], \qquad (y_1, y_2) \in S_Y,$$

where the Jacobian J is the determinant

$$J = \begin{vmatrix} \dfrac{\partial x_1}{\partial y_1} & \dfrac{\partial x_1}{\partial y_2} \\[2mm] \dfrac{\partial x_2}{\partial y_1} & \dfrac{\partial x_2}{\partial y_2} \end{vmatrix}.$$

Of course, we find the support S_Y of Y_1, Y_2 by considering the mapping of the support S_X of X_1, X_2 under the transformation $y_1 = u_1(x_1, x_2)$, $y_2 = u_2(x_1, x_2)$. This method of finding the distribution of Y_1 and Y_2 is called the **change-of-variables technique**.

It is often the mapping of the support S_X of X_1, X_2 into that (say, S_Y) of Y_1, Y_2 which causes the biggest challenge. That is, in most cases, it is easy to solve for x_1 and x_2 in terms of y_1 and y_2, say,

$$x_1 = v_1(y_1, y_2), \qquad x_2 = v_2(y_1, y_2),$$

and then to compute the Jacobian

$$J = \begin{vmatrix} \dfrac{\partial v_1(y_1, y_2)}{\partial y_1} & \dfrac{\partial v_1(y_1, y_2)}{\partial y_2} \\[2ex] \dfrac{\partial v_2(y_1, y_2)}{\partial y_1} & \dfrac{\partial v_2(y_1, y_2)}{\partial y_2} \end{vmatrix}.$$

However, the mapping of $(x_1, x_2) \in S_X$ into $(y_1, y_2) \in S_Y$ can be more difficult. Let us consider two simple examples.

Example 5.2-1

Let X_1, X_2 have the joint pdf

$$f(x_1, x_2) = 2, \qquad 0 < x_1 < x_2 < 1.$$

Consider the transformation

$$Y_1 = \frac{X_1}{X_2}, \qquad Y_2 = X_2.$$

It is certainly easy enough to solve for x_1 and x_2, namely,

$$x_1 = y_1 y_2, \qquad x_2 = y_2,$$

and compute

$$J = \begin{vmatrix} y_2 & y_1 \\ 0 & 1 \end{vmatrix} = y_2.$$

Let us now consider S_X, which is depicted in Figure 5.2-1(a). The boundaries of S_X are not part of the support, but let us see how they map. The points for which $x_1 = 0$, $0 < x_2 < 1$, map into $y_1 = 0$, $0 < y_2 < 1$; the points for which $x_2 = 1$, $0 \le x_1 < 1$, map into $y_2 = 1$, $0 \le y_1 < 1$; and $0 < x_1 = x_2 \le 1$ maps into $y_1 = 1$, $0 < y_2 \le 1$. We depict these line segments in Figure 5.2-1(b) and mark them with the symbols corresponding to the line segments in Figure 5.2-1(a).

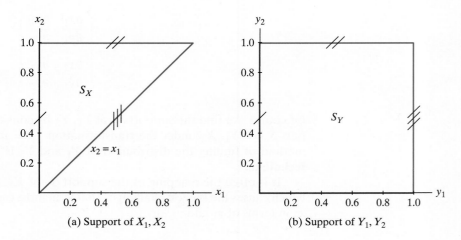

(a) Support of X_1, X_2 (b) Support of Y_1, Y_2

Figure 5.2-1 Mapping from x_1, x_2 to y_1, y_2

We note that the points for which $y_2 = 0$, $0 \le y_1 < 1$, all map into the single point $x_1 = 0$, $x_2 = 0$. That is, this is a many-to-one mapping, and yet we are restricting ourselves to one-to-one mappings. However, the boundaries are not part of our support! Thus, S_Y is as depicted in Figure 5.2-1(b), and, according to the rule, the joint pdf of Y_1, Y_2 is

$$g(y_1, y_2) = |y_2| \cdot 2 = 2y_2, \qquad 0 < y_1 < 1, \ 0 < y_2 < 1.$$

It is interesting to note that the marginal probability density functions are

$$g_1(y_1) = \int_0^1 2y_2 \, dy_2 = 1, \qquad 0 < y_1 < 1,$$

and

$$g_2(y_2) = \int_0^1 2y_2 \, dy_1 = 2y_2, \qquad 0 < y_2 < 1.$$

Hence, $Y_1 = X_1/X_2$ and $Y_2 = X_2$ are independent. Even though the computation of Y_1 depends very much on the value of Y_2, still Y_1 and Y_2 are independent in the probability sense. ∎

Example 5.2-2

Let X_1 and X_2 be independent random variables, each with pdf

$$f(x) = e^{-x}, \qquad 0 < x < \infty.$$

Hence, their joint pdf is

$$f(x_1)f(x_2) = e^{-x_1 - x_2}, \qquad 0 < x_1 < \infty, \ 0 < x_2 < \infty.$$

Let us consider

$$Y_1 = X_1 - X_2, \qquad Y_2 = X_1 + X_2.$$

Thus,

$$x_1 = \frac{y_1 + y_2}{2}, \qquad x_2 = \frac{y_2 - y_1}{2},$$

with

$$J = \begin{vmatrix} \dfrac{1}{2} & \dfrac{1}{2} \\[2mm] -\dfrac{1}{2} & \dfrac{1}{2} \end{vmatrix} = \frac{1}{2}.$$

The region S_X is depicted in Figure 5.2-2(a). The line segments on the boundary, namely, $x_1 = 0$, $0 < x_2 < \infty$, and $x_2 = 0$, $0 < x_1 < \infty$, map into the line segments $y_1 + y_2 = 0$, $y_2 > y_1$ and $y_1 = y_2$, $y_2 > -y_1$, respectively. These are shown in Figure 5.2-2(b), and the support of S_Y is depicted there. Since the region S_Y is not bounded by horizontal and vertical line segments, Y_1 and Y_2 are dependent.

The joint pdf of Y_1 and Y_2 is

$$g(y_1, y_2) = \frac{1}{2} e^{-y_2}, \qquad -y_2 < y_1 < y_2, \ 0 < y_2 < \infty.$$

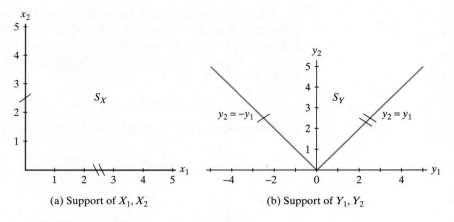

(a) Support of X_1, X_2 (b) Support of Y_1, Y_2

Figure 5.2-2 Mapping from x_1, x_2 to y_1, y_2

The probability $P(Y_1 \geq 0, Y_2 \leq 4)$ is given by

$$\int_0^4 \int_{y_1}^4 \frac{1}{2} e^{-y_2}\, dy_2\, dy_1 \quad \text{or} \quad \int_0^4 \int_0^{y_2} \frac{1}{2} e^{-y_2}\, dy_1\, dy_2.$$

While neither of these integrals is difficult to evaluate, we choose the latter one to obtain

$$\int_0^4 \frac{1}{2} y_2 e^{-y_2}\, dy_2 = \left[\frac{1}{2}(-y_2) e^{-y_2} - \frac{1}{2} e^{-y_2} \right]_0^4$$

$$= \frac{1}{2} - 2e^{-4} - \frac{1}{2} e^{-4} = \frac{1}{2}\left[1 - 5e^{-4} \right].$$

The marginal pdf of Y_2 is

$$g_2(y_2) = \int_{-y_2}^{y_2} \frac{1}{2} e^{-y_2}\, dy_1 = y_2 e^{-y_2}, \qquad 0 < y_2 < \infty.$$

This is a gamma pdf with shape parameter 2 and scale parameter 1. The pdf of Y_1 is

$$g_1(y_1) = \begin{cases} \displaystyle\int_{-y_1}^{\infty} \frac{1}{2} e^{-y_2}\, dy_2 = \frac{1}{2} e^{y_1}, & -\infty < y_1 \leq 0, \\[2ex] \displaystyle\int_{y_1}^{\infty} \frac{1}{2} e^{-y_2}\, dy_2 = \frac{1}{2} e^{-y_1}, & 0 < y_1 < \infty. \end{cases}$$

That is, the expression for $g_1(y_1)$ depends on the location of y_1, although this could be written as

$$g_1(y_1) = \frac{1}{2} e^{-|y_1|}, \qquad -\infty < y_1 < \infty,$$

which is called a **double exponential** pdf, or sometimes the **Laplace** pdf.

We now consider two examples that yield two important distributions. The second of these uses the cdf technique rather than the change-of-variable method.

Example 5.2-3

Let X_1 and X_2 have independent gamma distributions with parameters α, θ and β, θ, respectively. That is, the joint pdf of X_1 and X_2 is

$$f(x_1, x_2) = \frac{1}{\Gamma(\alpha)\Gamma(\beta)\theta^{\alpha+\beta}} x_1^{\alpha-1} x_2^{\beta-1} \exp\left(-\frac{x_1 + x_2}{\theta}\right), \quad 0 < x_1 < \infty, \ 0 < x_2 < \infty.$$

Consider

$$Y_1 = \frac{X_1}{X_1 + X_2}, \qquad Y_2 = X_1 + X_2,$$

or, equivalently,

$$X_1 = Y_1 Y_2, \qquad X_2 = Y_2 - Y_1 Y_2.$$

The Jacobian is

$$J = \begin{vmatrix} y_2 & y_1 \\ -y_2 & 1 - y_1 \end{vmatrix} = y_2(1 - y_1) + y_1 y_2 = y_2.$$

Thus, the joint pdf $g(y_1, y_2)$ of Y_1 and Y_2 is

$$g(y_1, y_2) = |y_2| \frac{1}{\Gamma(\alpha)\Gamma(\beta)\theta^{\alpha+\beta}} (y_1 y_2)^{\alpha-1} (y_2 - y_1 y_2)^{\beta-1} e^{-y_2/\theta},$$

where the support is $0 < y_1 < 1$, $0 < y_2 < \infty$, which is the mapping of $0 < x_i < \infty$, $i = 1, 2$. To see the shape of this joint pdf, $z = g(y_1, y_2)$ is graphed in Figure 5.2-3(a) with $\alpha = 4$, $\beta = 7$, and $\theta = 1$ and in Figure 5.2-3(b) with $\alpha = 8$, $\beta = 3$, and $\theta = 1$. To find the marginal pdf of Y_1, we integrate this joint pdf on y_2. We see that the marginal pdf of Y_1 is

$$g_1(y_1) = \frac{y_1^{\alpha-1}(1 - y_1)^{\beta-1}}{\Gamma(\alpha)\Gamma(\beta)} \int_0^\infty \frac{y_2^{\alpha+\beta-1}}{\theta^{\alpha+\beta}} e^{-y_2/\theta} \, dy_2.$$

But the integral in this expression is that of a gamma pdf with parameters $\alpha + \beta$ and θ, except for $\Gamma(\alpha + \beta)$ in the denominator; hence, the integral equals $\Gamma(\alpha + \beta)$, and we have

$$g_1(y_1) = \frac{\Gamma(\alpha + \beta)}{\Gamma(\alpha)\Gamma(\beta)} y_1^{\alpha-1}(1 - y_1)^{\beta-1}, \qquad 0 < y_1 < 1.$$

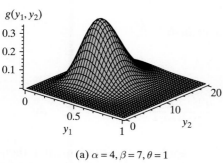

(a) $\alpha = 4, \beta = 7, \theta = 1$

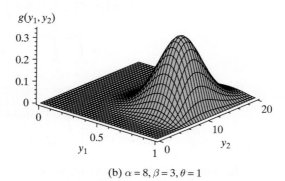

(b) $\alpha = 8, \beta = 3, \theta = 1$

Figure 5.2-3 Joint pdf of $z = g(y_1, y_2)$

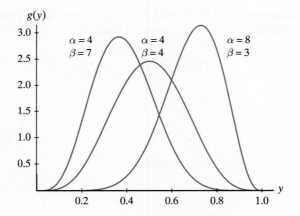

Figure 5.2-4 Beta distribution pdfs

We say that Y_1 has a **beta pdf** with parameters α and β. (See Figure 5.2-4.) Note the relationship between Figure 5.2-3 and Figure 5.2-4. ∎

The next example illustrates the distribution function technique. You will calculate the same results in Exercise 5.2-2, but using the change-of-variable technique.

Example 5.2-4

We let

$$F = \frac{U/r_1}{V/r_2},$$

where U and V are independent chi-square variables with r_1 and r_2 degrees of freedom, respectively. Thus, the joint pdf of U and V is

$$g(u,v) = \frac{u^{r_1/2-1}e^{-u/2}}{\Gamma(r_1/2)2^{r_1/2}}\frac{v^{r_2/2-1}e^{-v/2}}{\Gamma(r_2/2)2^{r_2/2}}, \qquad 0 < u < \infty, \; 0 < v < \infty.$$

In this derivation, we let $W = F$ to avoid using f as a symbol for a variable. The cdf $F(w) = P(W \le w)$ of W is

$$F(w) = P\left(\frac{U/r_1}{V/r_2} \le w\right) = P\left(U \le \frac{r_1}{r_2}wV\right)$$

$$= \int_0^\infty \int_0^{(r_1/r_2)wv} g(u,v)\,du\,dv.$$

That is,

$$F(w) = \frac{1}{\Gamma(r_1/2)\Gamma(r_2/2)}\int_0^\infty\left[\int_0^{(r_1/r_2)wv}\frac{u^{r_1/2-1}e^{-u/2}}{2^{(r_1+r_2)/2}}\,du\right]v^{r_2/2-1}e^{-v/2}\,dv.$$

The pdf of W is the derivative of the cdf; so, applying the fundamental theorem of calculus to the inner integral, exchanging the operations of integration and differentiation (which is permissible in this case), we have

$$f(w) = F'(w)$$

$$= \frac{1}{\Gamma(r_1/2)\Gamma(r_2/2)} \int_0^\infty \frac{[(r_1/r_2)vw]^{r_1/2-1}}{2^{(r_1+r_2)/2}} e^{-(r_1/2r_2)(vw)} \left(\frac{r_1}{r_2} v\right) v^{r_2/2-1} e^{-v/2} \, dv$$

$$= \frac{(r_1/r_2)^{r_1/2} w^{r_1/2-1}}{\Gamma(r_1/2)\Gamma(r_2/2)} \int_0^\infty \frac{v^{(r_1+r_2)/2-1}}{2^{(r_1+r_2)/2}} e^{-(v/2)[1+(r_1/r_2)w]} \, dv.$$

In the integral, we make the change of variable

$$y = \left(1 + \frac{r_1}{r_2}w\right)v, \qquad \text{so that} \qquad \frac{dv}{dy} = \frac{1}{1+(r_1/r_2)w}.$$

Thus, we have

$$f(w) = \frac{(r_1/r_2)^{r_1/2}\Gamma[(r_1+r_2)/2]w^{r_1/2-1}}{\Gamma(r_1/2)\Gamma(r_2/2)[1+(r_1w/r_2)]^{(r_1+r_2)/2}} \int_0^\infty \frac{y^{(r_1+r_2)/2-1}e^{-y/2}}{\Gamma[(r_1+r_2)/2]2^{(r_1+r_2)/2}} \, dy$$

$$= \frac{(r_1/r_2)^{r_1/2}\Gamma[(r_1+r_2)/2]w^{r_1/2-1}}{\Gamma(r_1/2)\Gamma(r_2/2)[1+(r_1w/r_2)]^{(r_1+r_2)/2}},$$

the pdf of the $W = F$ **distribution** with r_1 and r_2 degrees of freedom. Note that the integral in this last expression for $f(w)$ is equal to 1 because the integrand is like a pdf of a chi-square distribution with $r_1 + r_2$ degrees of freedom. Graphs of pdfs for the F distribution are given in Figure 5.2-5. ■

To find probabilities for an F random variable with r_1 (numerator) and r_2 (denominator) degrees of freedom, use your calculator, or a computer program, or Table VII in Appendix B. Table VII is limited but is adequate for most of the applications in this text. For notation, if W has an F distribution with r_1 and r_2 degrees of freedom, we say that the distribution of W is $F(r_1, r_2)$. For a right-tail probability of α, we write

$$P[W \geq F_\alpha(r_1, r_2)] = \alpha.$$

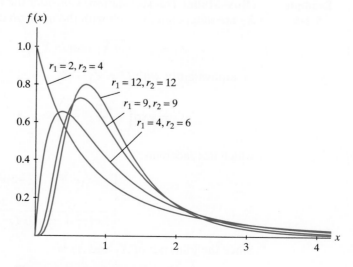

Figure 5.2-5 Graphs of F pdfs

For a left-tail probability of α, where α is generally small, we note that if the distribution of W is $F(r_1, r_2)$, then the distribution of $1/W$ is $F(r_2, r_1)$. Since

$$\alpha = P[W \leq F_{1-\alpha}(r_1, r_2)] = P[1/W \geq 1/F_{1-\alpha}(r_1, r_2)]$$

and

$$P[1/W \geq F_\alpha(r_2, r_1)] = \alpha,$$

it follows that

$$\frac{1}{F_{1-\alpha}(r_1, r_2)} = F_\alpha(r_2, r_1) \quad \text{or} \quad F_{1-\alpha}(r_1, r_2) = \frac{1}{F_\alpha(r_2, r_1)}.$$

Example 5.2-5

Let the distribution of W be $F(4, 6)$. From Table VII, we see that

$$F_{0.05}(4, 6) = 4.53;$$
$$P(W \leq 9.15) = 0.99.$$

That is, $F_{0.01}(4, 6) = 9.15$. We also note that

$$F_{0.95}(4, 6) = \frac{1}{F_{0.05}(6, 4)} = \frac{1}{6.16} = 0.1623;$$

$$F_{0.99}(4, 6) = \frac{1}{F_{0.01}(6, 4)} = \frac{1}{15.21}.$$

It follows that

$$P(1/15.21 \leq W \leq 9.15) = 0.98;$$
$$P(1/6.16 \leq W \leq 4.53) = 0.90.$$ ∎

Example 5.2-6

(**Box–Muller Transformation**) Consider the following transformation, where X_1 and X_2 are independent, each with the uniform distribution $U(0, 1)$: Let

$$Z_1 = \sqrt{-2 \ln X_1} \, \cos(2\pi X_2), \qquad Z_2 = \sqrt{-2 \ln X_1} \, \sin(2\pi X_2)$$

or, equivalently, with $Q = Z_1^2 + Z_2^2$,

$$X_1 = \exp\left(-\frac{Z_1^2 + Z_2^2}{2}\right) = e^{-Q/2}, \qquad X_2 = \frac{1}{2\pi} \arctan\left(\frac{Z_2}{Z_1}\right),$$

which has Jacobian

$$J = \begin{vmatrix} -z_1 e^{-q/2} & -z_2 e^{-q/2} \\ \dfrac{-z_2}{2\pi(z_1^2 + z_2^2)} & \dfrac{z_1}{2\pi(z_1^2 + z_2^2)} \end{vmatrix} = \frac{-1}{2\pi} e^{-q/2}.$$

Since the joint pdf of X_1 and X_2 is

$$f(x_1, x_2) = 1, \qquad 0 < x_1 < 1, \ 0 < x_2 < 1,$$

it follows that the joint pdf of Z_1 and Z_2 is

$$g(z_1, z_2) = \left| \frac{-1}{2\pi} e^{-q/2} \right| \quad (1)$$

$$= \frac{1}{2\pi} \exp\left(-\frac{z_1^2 + z_2^2}{2} \right), \qquad -\infty < z_1 < \infty, \quad -\infty < z_2 < \infty.$$

Note that there is some difficulty with the definition of this transformation, particularly when $z_1 = 0$. However, these difficulties occur at events with probability zero and hence cause no problems. (See Exercise 5.2-15.) Summarizing, from two independent $U(0,1)$ random variables we have generated two independent $N(0,1)$ random variables through this **Box–Muller transformation.** ∎

Exercises

5.2-1. Let X_1, X_2 denote two independent random variables, each with a $\chi^2(2)$ distribution. Find the joint pdf of $Y_1 = X_1$ and $Y_2 = X_2 + X_1$. Note that the support of Y_1, Y_2 is $0 < y_1 < y_2 < \infty$. Also, find the marginal pdf of each of Y_1 and Y_2. Are Y_1 and Y_2 independent?

5.2-2. Let X_1 and X_2 be independent chi-square random variables with r_1 and r_2 degrees of freedom, respectively. Let $Y_1 = (X_1/r_1)/(X_2/r_2)$ and $Y_2 = X_2$.

(a) Find the joint pdf of Y_1 and Y_2.

(b) Determine the marginal pdf of Y_1 and show that Y_1 has an F distribution. (This is another, but equivalent, way of finding the pdf of F.)

5.2-3. Find the mean and the variance of an F random variable with r_1 and r_2 degrees of freedom by first finding $E(U)$, $E(1/V)$, $E(U^2)$, and $E(1/V^2)$.

5.2-4. Let the distribution of W be $F(9, 20)$. Find the following:

(a) $F_{0.01}(9, 20)$.

(b) $F_{0.99}(9, 20)$.

(c) $P(0.34 \le W \le 2.30)$.

5.2-5. Let the distribution of W be $F(12, 6)$. Find the following:

(a) $F_{0.05}(12, 6)$.

(b) $F_{0.95}(12, 6)$.

(c) $P(0.33 \le W \le 4.00)$.

5.2-6. Let X_1 and X_2 have independent gamma distributions with parameters α, θ and β, θ, respectively. Let $W = X_1/(X_1 + X_2)$. Use a method similar to that given in the derivation of the F distribution (Example 5.2-4) to show that the pdf of W is

$$g(w) = \frac{\Gamma(\alpha + \beta)}{\Gamma(\alpha)\Gamma(\beta)} w^{\alpha-1}(1-w)^{\beta-1}, \qquad 0 < w < 1.$$

We say that W has a beta distribution with parameters α and β. (See Example 5.2-3.)

5.2-7. Let X_1 and X_2 be independent chi-square random variables with r_1 and r_2 degrees of freedom, respectively. Show that

(a) $U = X_1/(X_1 + X_2)$ has a beta distribution with $\alpha = r_1/2$ and $\beta = r_2/2$.

(b) $V = X_2/(X_1 + X_2)$ has a beta distribution with $\alpha = r_2/2$ and $\beta = r_1/2$.

5.2-8. Let X have a beta distribution with parameters α and β. (See Example 5.2-3.)

(a) Show that the mean and variance of X are, respectively,

$$\mu = \frac{\alpha}{\alpha + \beta} \quad \text{and} \quad \sigma^2 = \frac{\alpha\beta}{(\alpha + \beta + 1)(\alpha + \beta)^2}.$$

(b) Show that when $\alpha > 1$ and $\beta > 1$, the mode is at $x = (\alpha - 1)/(\alpha + \beta - 2)$.

5.2-9. Determine the constant c such that $f(x) = cx^3(1 - x)^6, 0 < x < 1$, is a pdf.

5.2-10. When α and β are integers and $0 < p < 1$, we have

$$\int_0^p \frac{\Gamma(\alpha + \beta)}{\Gamma(\alpha)\Gamma(\beta)} y^{\alpha-1}(1 - y)^{\beta-1} \, dy = \sum_{y=\alpha}^{n} \binom{n}{y} p^y (1 - p)^{n-y},$$

where $n = \alpha + \beta - 1$. Verify this formula when $\alpha = 4$ and $\beta = 3$. HINT: Integrate the left member by parts several times.

5.2-11. Evaluate

$$\int_0^{0.4} \frac{\Gamma(7)}{\Gamma(4)\Gamma(3)} y^3 (1 - y)^2 \, dy$$

(a) Using integration.

(b) Using the result of Exercise 5.2-10.

5.2-12. Let W_1, W_2 be independent, each with a Cauchy distribution. In this exercise we find the pdf of the sample mean, $(W_1 + W_2)/2$.

(a) Show that the pdf of $X_1 = (1/2)W_1$ is

$$f(x_1) = \frac{2}{\pi\left(1 + 4x_1^2\right)}, \qquad -\infty < x_1 < \infty.$$

(b) Let $Y_1 = X_1 + X_2 = \overline{W}$ and $Y_2 = X_1$, where $X_2 = (1/2)W_2$. Show that the joint pdf of Y_1 and Y_2 is

$$g(y_1, y_2) = f(y_1 - y_2)f(y_2), \qquad -\infty < y_1 < \infty,$$

$$-\infty < y_2 < \infty.$$

(c) Show that the pdf of $Y_1 = \overline{W}$ is given by the **convolution formula**,

$$g_1(y_1) = \int_{-\infty}^{\infty} f(y_1 - y_2)f(y_2)\, dy_2.$$

(d) Show that

$$g_1(y_1) = \frac{1}{\pi\left(1 + y_1^2\right)}, \qquad -\infty < y_1 < \infty.$$

That is, the pdf of $\overline{W}$ is the same as that of an individual W.

5.2-13. Let X_1, X_2 be independent random variables representing lifetimes (in hours) of two key components of a device that fails when and only when both components fail. Say each X_i has an exponential distribution with mean 1000. Let $Y_1 = \min(X_1, X_2)$ and $Y_2 = \max(X_1, X_2)$, so that the space of Y_1, Y_2 is $0 < y_1 < y_2 < \infty$.

(a) Find $G(y_1, y_2) = P(Y_1 \leq y_1, Y_2 \leq y_2)$.

(b) Compute the probability that the device fails after 1200 hours; that is, compute $P(Y_2 > 1200)$.

5.2-14. A company provides earthquake insurance. The premium X is modeled by the pdf

$$f(x) = \frac{x}{5^2}\, e^{-x/5}, \qquad 0 < x < \infty,$$

while the claims Y have the pdf

$$g(y) = \frac{1}{5}\, e^{-y/5}, \qquad 0 < y < \infty.$$

If X and Y are independent, find the pdf of $Z = X/Y$.

5.2-15. In Example 5.2-6, verify that the given transformation maps $\{(x_1, x_2) : 0 < x_1 < 1, 0 < x_2 < 1\}$ onto $\{(z_1, z_2) : -\infty < z_1 < \infty, -\infty < z_2 < \infty\}$, except for a set of points that has probability 0. HINT: What is the image of vertical line segments? What is the image of horizontal line segments?

5.2-16. Let W have an F distribution with parameters r_1 and r_2. Show that $Z = 1/[1 + (r_1/r_2)W]$ has a beta distribution.

5.3 SEVERAL RANDOM VARIABLES

In Section 5.2, we introduced several distributions concerning two random variables. Each of these random variables could be thought of as a measurement in some random experiment. In this section, we consider the possibility of performing several random experiments or one random experiment several times in which each trial results in one measurement that can be considered a random variable. That is, we obtain one random variable from each experiment, and thus we obtain a collection of several random variables from the several experiments. Further, suppose that these experiments are performed in such a way that the events associated with any one of them are independent of the events associated with others, and hence the corresponding random variables are, in the probabilistic sense, independent.

Recall from Section 5.2 that if X_1 and X_2 are random variables of the discrete type with probability mass functions $f_1(x_1)$ and $f_2(x_2)$, respectively, and if

$$P(X_1 = x_1 \text{ and } X_2 = x_2) = P(X_1 = x_1)\, P(X_2 = x_2)$$

$$= f_1(x_1)f_2(x_2), \qquad x_1 \in S_1, \ x_2 \in S_2,$$

then X_1 and X_2 are said to be independent, and the joint pmf is $f_1(x_1)f_2(x_2)$.

Sometimes the two random experiments are exactly the same. For example, we could roll a fair die twice, resulting first in X_1 and then in X_2. It is reasonable to say that the pmf of X_1 is $f(x_1) = 1/6$, $x_1 = 1, 2, \ldots, 6$, and the pmf of X_2 is $f(x_2) = 1/6$, $x_2 = 1, 2, \ldots, 6$. Assuming independence, which would be a fair way to perform the experiments, the joint pmf is then

$$f(x_1)f(x_2) = \left(\frac{1}{6}\right)\left(\frac{1}{6}\right) = \frac{1}{36}, \qquad x_1 = 1,2,\ldots,6; \ \ x_2 = 1,2,\ldots,6.$$

In general, if the pmf $f(x)$ of the independent random variables X_1 and X_2 is the same, then the joint pmf is $f(x_1)f(x_2)$. Moreover, in this case, the collection of the two random variables X_1, X_2 is called a **random sample of size** $n = 2$ from a distribution with pmf $f(x)$. Hence, in the two rolls of the fair die, we say that we have a random sample of size $n = 2$ from the uniform distribution on the space $\{1,2,3,4,5,6\}$.

Example 5.3-1

Let X_1 and X_2 be two independent random variables resulting from two rolls of a fair die. That is, X_1, X_2 is a random sample of size $n = 2$ from a distribution with pmf $f(x) = 1/6$, $x = 1,2,\ldots,6$. We have

$$E(X_1) = E(X_2) = \sum_{x=1}^{6} xf(x) = 3.5.$$

Moreover,

$$\text{Var}(X_1) = \text{Var}(X_2) = \sum_{x=1}^{6} (x - 3.5)^2 f(x) = \frac{35}{12}.$$

In addition, from independence,

$$E(X_1 X_2) = E(X_1)E(X_2) = (3.5)(3.5) = 12.25$$

and

$$E[(X_1 - 3.5)(X_2 - 3.5)] = E(X_1 - 3.5)E(X_2 - 3.5) = 0.$$

If $Y = X_1 + X_2$, then

$$E(Y) = E(X_1) + E(X_2) = 3.5 + 3.5 = 7$$

and

$$\begin{aligned}
\text{Var}(Y) = E\left[(X_1 + X_2 - 7)^2\right] &= E\left\{[(X_1 - 3.5) + (X_2 - 3.5)]^2\right\} \\
&= E\left[(X_1 - 3.5)^2\right] + E[2(X_1 - 3.5)(X_2 - 3.5)] + E\left[(X_2 - 3.5)^2\right] \\
&= \text{Var}(X_1) + (2)(0) + \text{Var}(X_2) \\
&= (2)\left(\frac{35}{12}\right) = \frac{35}{6}.
\end{aligned}$$

■

In Example 5.3-1, we can find the pmf $g(y)$ of $Y = X_1 + X_2$. Since the space of Y is $\{2,3,4,\ldots,12\}$, we have, by a rather straightforward calculation,

$$g(2) = P(X_1 = 1, X_2 = 1) = f(1)f(1) = \left(\frac{1}{6}\right)\left(\frac{1}{6}\right) = \frac{1}{36},$$

$$g(3) = P(X_1 = 1, X_2 = 2 \text{ or } X_1 = 2, X_2 = 1) = \left(\frac{1}{6}\right)\left(\frac{1}{6}\right) + \left(\frac{1}{6}\right)\left(\frac{1}{6}\right) = \frac{2}{36},$$

$$g(4) = P(X_1 = 1, X_2 = 3 \text{ or } X_1 = 2, X_2 = 2 \text{ or } X_1 = 3, X_2 = 1) = \frac{3}{36},$$

and so on. This results in the pmf given by

y	2	3	4	5	6	7	8	9	10	11	12
$g(y)$	$\frac{1}{36}$	$\frac{2}{36}$	$\frac{3}{36}$	$\frac{4}{36}$	$\frac{5}{36}$	$\frac{6}{36}$	$\frac{5}{36}$	$\frac{4}{36}$	$\frac{3}{36}$	$\frac{2}{36}$	$\frac{1}{36}$

With this pmf, it is simple to calculate

$$E(Y) = \sum_{y=2}^{12} y\, g(y) = 7$$

and

$$\text{Var}(Y) = \sum_{y=2}^{12} (y-7)^2\, g(y) = \frac{35}{6},$$

which agrees with the results of Example 5.3-1.

All of the definitions and results concerning two random variables of the discrete type can be carried over to two random variables of the continuous type. Moreover, the notions about two independent random variables can be extended to n independent random variables, which can be thought of as measurements on the outcomes of n random experiments. That is, if $X_1, X_2, \ldots, X_n$ are **independent**, then the joint pmf or pdf is the product of the respective pmfs or pdfs, namely, $f_1(x_1)f_2(x_2)\cdots f_n(x_n)$.

If all n of the distributions are the same, then the collection of n independent and identically distributed random variables, $X_1, X_2, \ldots, X_n$, is said to be a **random sample of size n from that common distribution**. If $f(x)$ is the common pmf or pdf of these n random variables, then the joint pmf or pdf is $f(x_1)f(x_2)\cdots f(x_n)$.

Example 5.3-2

Let X_1, X_2, X_3 be a random sample from a distribution with pdf

$$f(x) = e^{-x}, \qquad 0 < x < \infty.$$

The joint pdf of these three random variables is

$$f(x_1, x_2, x_3) = (e^{-x_1})(e^{-x_2})(e^{-x_3}) = e^{-x_1-x_2-x_3}, \qquad 0 < x_i < \infty, \ i = 1, 2, 3.$$

The probability
$$P(0 < X_1 < 1, 2 < X_2 < 4, 3 < X_3 < 7)$$

$$= \left(\int_0^1 e^{-x_1}\, dx_1 \right)\left(\int_2^4 e^{-x_2}\, dx_2 \right)\left(\int_3^7 e^{-x_3}\, dx_3 \right)$$

$$= (1 - e^{-1})(e^{-2} - e^{-4})(e^{-3} - e^{-7}),$$

because of the independence of X_1, X_2, X_3. ∎

Example 5.3-3

An electronic device runs until one of its three components fails. The lifetimes (in weeks), X_1, X_2, X_3, of these components are independent, and each has the Weibull pdf

$$f(x) = \frac{2x}{25} e^{-(x/5)^2}, \qquad 0 < x < \infty.$$

The probability that the device stops running in the first three weeks is equal to

$$1 - P(X_1 > 3, X_2 > 3, X_3 > 3) = 1 - P(X_1 > 3)P(X_2 > 3)P(X_3 > 3)$$

$$= 1 - \left(\int_3^\infty f(x)\,dx \right)^3$$

$$= 1 - \left(\left[-e^{-(x/5)^2} \right]_3^\infty \right)^3$$

$$= 1 - \left[e^{-(3/5)^2} \right]^3 = 0.660.$$

∎

Of course, when we are dealing with n random variables that are not independent, the joint pmf (or pdf) could be represented as

$$f(x_1, x_2, \ldots, x_n), \qquad (x_1, x_2, \ldots, x_n) \in S.$$

The **mathematical expectation** (or **expected value**) of $u(X_1, X_2, \ldots, X_n)$ is given by

$$E[u(X_1, X_2, \ldots, X_n)] = \sum \sum \cdots \sum_S u(x_1, x_2, \ldots, x_n) f(x_1, x_2, \ldots, x_n)$$

in the discrete case. (Replace summations with integrals in the continuous case.) Of course, $Y = u(X_1, X_2, \ldots, X_n)$ has a distribution with a pmf (or pdf), say $g(y)$. It is true, but we do not prove it, that

$$E(Y) = \sum_y y\, g(y) = \sum \sum \cdots \sum_S u(x_1, x_2, \ldots, x_n) f(x_1, x_2, \ldots, x_n).$$

In case $X_1, X_2, \ldots, X_n$ are independent random variables with pmfs (or pdfs) $f_1(x_1)$, $f_2(x_2), \ldots, f_n(x_n)$, respectively, then

$$f(x_1, x_2, \ldots, x_n) = f_1(x_1),\ f_2(x_2), \ldots, f_n(x_n).$$

The next theorem proves that the expected value of the product of functions of n independent random variables is the product of their expected values.

Theorem 5.3-1	Say $X_1, X_2, \ldots, X_n$ are independent random variables and $Y = u_1(X_1) u_2(X_2) \cdots u_n(X_n)$. If $E[u_i(X_i)]$, $i = 1, 2, \ldots, n$, exist, then

$$E(Y) = E[u_1(X_1)\, u_2(X_2) \cdots u_n(X_n)] = E[u_1(X_1)]E[u_2(X_2)] \cdots E[u_n(X_n)].$$

Proof In the discrete case, we have

$$E[u_1(X_1)u_2(X_2) \cdots u_n(X_n)]$$

$$= \sum_{x_1} \sum_{x_2} \cdots \sum_{x_n} u_1(x_1)\, u_2(x_2) \cdots u_n(x_n) f_1(x_1) f_2(x_2) \cdots f_n(x_n)$$

$$= \sum_{x_1} u_1(x_1) f_1(x_1) \sum_{x_2} u_2(x_2) f_2(x_2) \cdots \sum_{x_n} u_n(x_n) f_n(x_n)$$

$$= E[u_1(X_1)]E[u_2(x_2)] \cdots E[u_n(X_n)].$$

In the proof of the continuous case, obvious changes are made; in particular, integrals replace summations. □

REMARK Sometimes students recognize that $X^2 = (X)(X)$ and thus believe that $E(X^2)$ is equal to $[E(X)][E(X)] = [E(X)]^2$ because Theorem 5.3-1 states that the expected value of the product is the product of the expected values. However, note the hypothesis of independence in the theorem, and certainly X is not independent of itself. Incidentally, if $E(X^2)$ did equal $[E(X)]^2$, then the variance of X, or

$$\sigma^2 = E(X^2) - [E(X)]^2,$$

would always equal zero. This happens only in the case of degenerate (one-point) distributions. ∎

We now prove an important theorem about the mean and the variance of a linear combination of random variables.

Theorem 5.3-2

If $X_1, X_2, \ldots, X_n$ are n independent random variables with respective means $\mu_1, \mu_2, \ldots, \mu_n$ and variances $\sigma_1^2, \sigma_2^2, \ldots, \sigma_n^2$, then the mean and the variance of $Y = \sum_{i=1}^{n} a_i X_i$, where $a_1, a_2, \ldots, a_n$ are real constants, are, respectively,

$$\mu_Y = \sum_{i=1}^{n} a_i \mu_i \qquad \text{and} \qquad \sigma_Y^2 = \sum_{i=1}^{n} a_i^2 \sigma_i^2.$$

Proof We have

$$\mu_Y = E(Y) = E\left(\sum_{i=1}^{n} a_i X_i\right) = \sum_{i=1}^{n} a_i E(X_i) = \sum_{i=1}^{n} a_i \mu_i,$$

because the expected value of the sum is the sum of the expected values (i.e., E is a linear operator). Also,

$$\sigma_Y^2 = E[(Y - \mu_Y)^2] = E\left[\left(\sum_{i=1}^{n} a_i X_i - \sum_{i=1}^{n} a_i \mu_i\right)^2\right]$$

$$= E\left\{\left[\sum_{i=1}^{n} a_i (X_i - \mu_i)\right]^2\right\} = E\left[\sum_{i=1}^{n} \sum_{j=1}^{n} a_i a_j (X_i - \mu_i)(X_j - \mu_j)\right].$$

Again using the fact that E is a linear operator, we obtain

$$\sigma_Y^2 = \sum_{i=1}^{n} \sum_{j=1}^{n} a_i a_j E\left[(X_i - \mu_i)(X_j - \mu_j)\right].$$

However, if $i \neq j$, then from the independence of X_i and X_j, we have

$$E\left[(X_i - \mu_i)(X_j - \mu_j)\right] = E(X_i - \mu_i)E(X_j - \mu_j) = (\mu_i - \mu_i)(\mu_j - \mu_j) = 0.$$

Thus, the variance can be written as

$$\sigma_Y^2 = \sum_{i=1}^{n} a_i^2 E[(X_i - \mu_i)^2] = \sum_{i=1}^{n} a_i^2 \sigma_i^2. \qquad \square$$

REMARK Although Theorem 5.3-2 gives the mean and the variance of a linear function of independent random variables, the proof can easily be modified to the case in which X_i and X_j are correlated. Then

$$E\left[(X_i - \mu_i)(X_j - \mu_j)\right] = \rho_{ij}\sigma_i\sigma_j,$$

instead of zero, where ρ_{ij} is the correlation coefficient of X_i and X_j. Thus,

$$\sigma_Y^2 = \sum_{i=1}^{n} a_i^2 \sigma_i^2 + 2\sum\sum_{i<j} a_i a_j \rho_{ij}\sigma_i\sigma_j,$$

where the factor 2 appears because the sum is over $i < j$ and

$$a_i a_j \rho_{ij}\sigma_i\sigma_j = a_j a_i \rho_{ji}\sigma_j\sigma_i.$$

The mean of Y is still the same in both cases, namely,

$$\mu_Y = \sum_{i=1}^{n} a_i \mu_i. \qquad \blacksquare$$

We give two illustrations of the theorem.

Example 5.3-4

Let the independent random variables X_1 and X_2 have respective means $\mu_1 = -4$ and $\mu_2 = 3$ and variances $\sigma_1^2 = 4$ and $\sigma_2^2 = 9$. Then the mean and the variance of $Y = 3X_1 - 2X_2$ are, respectively,

$$\mu_Y = (3)(-4) + (-2)(3) = -18$$

and

$$\sigma_Y^2 = (3)^2(4) + (-2)^2(9) = 72. \qquad \blacksquare$$

Example 5.3-5

Let X_1, X_2 be a random sample from a distribution with mean μ and variance σ^2. Let $Y = X_1 - X_2$; then

$$\mu_Y = \mu - \mu = 0$$

and

$$\sigma_Y^2 = (1)^2\sigma^2 + (-1)^2\sigma^2 = 2\sigma^2. \qquad \blacksquare$$

Now consider the **mean of a random sample**, $X_1, X_2, \ldots, X_n$, from a distribution with mean μ and variance σ^2, namely,

$$\overline{X} = \frac{X_1 + X_2 + \cdots + X_n}{n},$$

which is a linear function with each $a_i = 1/n$. Then

$$\mu_{\overline{X}} = \sum_{i=1}^{n} \left(\frac{1}{n}\right)\mu = \mu \qquad \text{and} \qquad \sigma_{\overline{X}}^2 = \sum_{i=1}^{n} \left(\frac{1}{n}\right)^2 \sigma^2 = \frac{\sigma^2}{n}.$$

That is, the mean of $\overline{X}$ is that of the distribution from which the sample arose, but the variance of $\overline{X}$ is that of the underlying distribution divided by n. Any function of the sample observations, $X_1, X_2, \ldots, X_n$, that does not have any unknown parameters is called a **statistic**, so here $\overline{X}$ is a statistic and also an **estimator** of the distribution mean μ. Another important statistic is the **sample variance**

$$S^2 = \frac{1}{n-1} \sum_{i=1}^{n} (X_i - \overline{X})^2,$$

and later we find that S^2 is an estimator of σ^2.

Exercises

5.3-1. Let X_1 and X_2 be independent Poisson random variables with respective means $\lambda_1 = 2$ and $\lambda_2 = 3$. Find

(a) $P(X_1 = 3, X_2 = 5)$.

(b) $P(X_1 + X_2 = 1)$.
   HINT. Note that this event can occur if and only if $\{X_1 = 1, X_2 = 0\}$ or $\{X_1 = 0, X_2 = 1\}$.

5.3-2. Let X_1 and X_2 be independent random variables with respective binomial distributions $b(3, 1/2)$ and $b(5, 1/2)$. Determine

(a) $P(X_1 = 2, X_2 = 4)$.

(b) $P(X_1 + X_2 = 7)$.

5.3-3. Let X_1 and X_2 be independent random variables with probability density functions $f_1(x_1) = 2x_1$, $0 < x_1 < 1$, and $f_2(x_2) = 4x_2^3$, $0 < x_2 < 1$, respectively. Compute

(a) $P(0.5 < X_1 < 1 \text{ and } 0.4 < X_2 < 0.8)$.

(b) $E(X_1^2 X_2^3)$.

5.3-4. Let X_1 and X_2 be a random sample of size $n = 2$ from the exponential distribution with pdf $f(x) = 2e^{-2x}$, $0 < x < \infty$. Find

(a) $P(0.5 < X_1 < 1.0, 0.7 < X_2 < 1.2)$.

(b) $E[X_1(X_2 - 0.5)^2]$.

5.3-5. Let X_1 and X_2 be observations of a random sample of size $n = 2$ from a distribution with pmf $f(x) = x/6$, $x = 1, 2, 3$. Then find the pmf of $Y = X_1 + X_2$. Determine the mean and the variance of the sum in two ways.

5.3-6. Let X_1 and X_2 be random samples of size $n = 9$ each from a distribution with pdf $f(x) = 3x(2 - x)$, $0 < x < 1$. Find the mean and the variance of $Y = X_1 + X_2$.

5.3-7. The distributions of incomes in two cities follow the two Pareto-type pdfs

$$f(x) = \frac{2}{x^3}, \quad 1 < x < \infty, \quad \text{and} \quad g(y) = \frac{3}{y^4}, \quad 1 < y < \infty,$$

respectively. Here one unit represents \$20,000. One person with income is selected at random from each city. Let X and Y be their respective incomes. Compute $P(X < Y)$.

5.3-8. Suppose two independent claims are made on two insured homes, where each claim has pdf

$$f(x) = \frac{4}{x^5}, \quad 1 < x < \infty,$$

in which the unit is \$1000. Find the expected value of the larger claim. HINT: If X_1 and X_2 are the two independent claims and $Y = \max(X_1, X_2)$, then

$$G(y) = P(Y \le y) = P(X_1 \le y)P(X_2 \le y) = [P(X \le y)]^2.$$

Find $g(y) = G'(y)$ and $E(Y)$.

5.3-9. Let X_1, X_2 be a random sample of size $n = 2$ from a distribution with pdf $f(x) = 3x^2$, $0 < x < 1$. Determine

(a) $P(\max X_i < 3/4) = P(X_1 < 3/4, X_2 < 3/4)$.

(b) The mean and the variance of $Y = X_1 + X_2$.

5.3-10. Let X_1, X_2, X_3 denote a random sample of size $n = 3$ from a distribution with the geometric pmf

$$f(x) = \left(\frac{3}{4}\right)\left(\frac{1}{4}\right)^{x-1}, \quad x = 1, 2, 3, \ldots.$$

(a) Compute $P(X_1 = 1, X_2 = 3, X_3 = 1)$.

(b) Determine $P(X_1 + X_2 + X_3 = 5)$.

(c) If Y equals the maximum of X_1, X_2, X_3, find

$$P(Y \le 2) = P(X_1 \le 2)P(X_2 \le 2)P(X_3 \le 2).$$

5.3-11. Let X_1, X_2, X_3 be three independent random variables with binomial distributions $b(4, 1/2)$, $b(6, 1/3)$, and $b(12, 1/6)$, respectively. Find

(a) $P(X_1 = 2, X_2 = 2, X_3 = 5)$.

(b) $E(X_1 X_2 X_3)$.

(c) The mean and the variance of $Y = X_1 + X_2 + X_3$.

5.3-12. Let X_1, X_2, X_3 be a random sample of size $n = 3$ from the exponential distribution with pdf $f(x) = e^{-x}$, $0 < x < \infty$. Find

$$P(1 < \min X_i) = P(1 < X_1, 1 < X_2, 1 < X_3).$$

5.3-13. A device contains three components, each of which has a lifetime in hours with the pdf

$$f(x) = \frac{2x}{10^2} e^{-(x/10)^2}, \qquad 0 < x < \infty.$$

The device fails with the failure of one of the components. Assuming independent lifetimes, what is the probability that the device fails in the first hour of its operation? HINT: $G(y) = P(Y \le y) = 1 - P(Y > y) = 1 - P$ (all three $> y$).

5.3-14. Let X_1, X_2, X_3 be independent random variables that represent lifetimes (in hours) of three key components of a device. Say their respective distributions are exponential with means 3250, 3750, and 4250. Let Y be the minimum of X_1, X_2, X_3 and compute $P(Y > 3250)$.

5.3-15. Three drugs are being tested for use as the treatment of a certain disease. Let $p_1, p_2,$ and p_3 represent the probabilities of success for the respective drugs. As three patients come in, each is given one of the drugs in a random order. After $n = 10$ "triples" and assuming independence, compute the probability that the maximum number of successes with one of the drugs exceeds eight if, in fact, $p_1 = p_2 = p_3 = 0.7$.

5.3-16. Each of eight bearings in a bearing assembly has a diameter (in millimeters) that has the pdf

$$f(x) = 10x^9, \qquad 0 < x < 1.$$

Assuming independence, find the cdf and the pdf of the maximum diameter (say, Y) of the eight bearings and compute $P(0.9999 < Y < 1)$.

5.3-17. In considering medical insurance for a certain operation, let X equal the amount (in dollars) paid for the doctor and let Y equal the amount paid to the hospital. In the past, the variances have been $\text{Var}(X) = 8100$, $\text{Var}(Y) = 10,000$, and $\text{Var}(X + Y) = 20,000$. Due to increased expenses, it was decided to increase the doctor's fee by \$500 and increase the hospital charge Y by

8%. Calculate the variance of $X + 500 + (1.08)Y$, the new total claim.

5.3-18. The lifetime in months of a certain part has a gamma distribution with $\alpha = \theta = 2$. A company buys three such parts and uses one until it fails, replacing it with a second part. When the latter fails, it is replaced by the third part. What are the mean and the variance of the total lifetime (the sum of the lifetimes of the three parts) associated with this situation?

5.3-19. Two components operate in parallel in a device, so the device fails when and only when both components fail. The lifetimes, X_1 and X_2, of the respective components are independent and identically distributed with an exponential distribution with $\theta = 2$. The cost of operating the device is $Z = 2Y_1 + Y_2$, where $Y_1 = \min(X_1, X_2)$ and $Y_2 = \max(X_1, X_2)$. Compute $E(Z)$.

5.3-20. Let X and Y be independent random variables with nonzero variances. Find the correlation coefficient of $W = XY$ and $V = X$ in terms of the means and variances of X and Y.

5.3-21. Flip $n = 8$ fair coins and remove all that came up heads. Flip the remaining coins (that came up tails) and remove the heads again. Continue flipping the remaining coins until each has come up heads. We shall find the pmf of Y, the number of trials needed. Let X_i equal the number of flips required to observe heads on coin i, $i = 1, 2, \ldots, 8$. Then $Y = \max(X_1, X_2, \ldots, X_8)$.

(a) Show that $P(Y \le y) = [1 - (1/2)^y]^8$.

(b) Show that the pmf of Y is defined by $P(Y = y) = [1 - (1/2)^y]^8 - [1 - (1/2)^{y-1}]^8$, $y = 1, 2, \ldots$.

(c) Use a computer algebra system such as *Maple* or *Mathematica* to show that the mean of Y is $E(Y) = 13,315,424/3,011,805 = 4.421$.

(d) What happens to the expected value of Y as the number of coins is doubled?

5.4 THE MOMENT-GENERATING FUNCTION TECHNIQUE

The first three sections of this chapter presented several techniques for determining the distribution of a function of random variables with known distributions. Another technique for this purpose is the moment-generating function technique. If $Y = u(X_1, X_2, \ldots, X_n)$, we have noted that we can find $E(Y)$ by evaluating $E[u(X_1, X_2, \ldots, X_n)]$. It is also true that we can find $E[e^{tY}]$ by evaluating $E[e^{tu(X_1, X_2, \ldots, X_n)}]$. We begin with a simple example.

Example 5.4-1   Let X_1 and X_2 be independent random variables with uniform distributions on $\{1, 2, 3, 4\}$. Let $Y = X_1 + X_2$. For example, Y could equal the sum when two fair four-sided dice are rolled. The mgf of Y is

$$M_Y(t) = E\left(e^{tY}\right) = E\left[e^{t(X_1+X_2)}\right] = E\left(e^{tX_1}e^{tX_2}\right).$$

The independence of X_1 and X_2 implies that

$$M_Y(t) = E\left(e^{tX_1}\right) E\left(e^{tX_2}\right).$$

In this example, X_1 and X_2 have the same pmf, namely,

$$f(x) = \frac{1}{4}, \qquad x = 1, 2, 3, 4,$$

and thus the same mgf,

$$M_X(t) = \frac{1}{4}e^t + \frac{1}{4}e^{2t} + \frac{1}{4}e^{3t} + \frac{1}{4}e^{4t}.$$

It then follows that $M_Y(t) = [M_X(t)]^2$ equals

$$\frac{1}{16}e^{2t} + \frac{2}{16}e^{3t} + \frac{3}{16}e^{4t} + \frac{4}{16}e^{5t} + \frac{3}{16}e^{6t} + \frac{2}{16}e^{7t} + \frac{1}{16}e^{8t}.$$

Note that the coefficient of e^{bt} is equal to the probability $P(Y = b)$; for example, $4/16 = P(Y = 5)$. Thus, we can find the distribution of Y by determining its mgf. ∎

In some applications, it is sufficient to know the mean and variance of a linear combination of random variables, say, Y. However, it is often helpful to know exactly how Y is distributed. The next theorem can frequently be used to find the distribution of a linear combination of independent random variables.

Theorem 5.4-1 If $X_1, X_2, \ldots, X_n$ are independent random variables with respective moment-generating functions $M_{X_i}(t)$, $i = 1, 2, 3, \ldots, n$, where $-h_i < t < h_i$, $i = 1, 2, \ldots, n$, for positive numbers h_i, $i = 1, 2, \ldots, n$, then the moment-generating function of $Y = \sum_{i=1}^{n} a_i X_i$ is

$$M_Y(t) = \prod_{i=1}^{n} M_{X_i}(a_i t), \quad \text{where} \quad -h_i < a_i t < h_i, \ i = 1, 2, \ldots, n.$$

Proof From Theorem 5.3-1, the mgf of Y is given by

$$M_Y(t) = E\left[e^{tY}\right] = E\left[e^{t(a_1 X_1 + a_2 X_2 + \cdots + a_n X_n)}\right]$$

$$= E\left[e^{a_1 t X_1} e^{a_2 t X_2} \cdots e^{a_n t X_n}\right]$$

$$= E\left[e^{a_1 t X_1}\right] E\left[e^{a_2 t X_2}\right] \cdots E\left[e^{a_n t X_n}\right].$$

However, since

$$E\left(e^{tX_i}\right) = M_{X_i}(t),$$

it follows that

$$E\left(e^{a_i t X_i}\right) = M_{X_i}(a_i t).$$

Thus, we have

$$M_Y(t) = M_{X_1}(a_1 t) M_{X_2}(a_2 t) \cdots M_{X_n}(a_n t) = \prod_{i=1}^{n} M_{X_i}(a_i t). \qquad \square$$

A corollary follows immediately, and it will be used in some important examples.

> **Corollary 5.4-1**
>
> If $X_1, X_2, \ldots, X_n$ are observations of a random sample from a distribution with moment-generating function $M(t)$, where $-h < t < h$, then
>
> (a) the moment-generating function of $Y = \sum_{i=1}^n X_i$ is
>
> $$M_Y(t) = \prod_{i=1}^n M(t) = [M(t)]^n, \quad -h < t < h;$$
>
> (b) the moment-generating function of $\overline{X} = \sum_{i=1}^n (1/n) X_i$ is
>
> $$M_{\overline{X}}(t) = \prod_{i=1}^n M\left(\frac{t}{n}\right) = \left[M\left(\frac{t}{n}\right)\right]^n, \quad -h < \frac{t}{n} < h.$$
>
> **Proof** For (a), let $a_i = 1, i = 1, 2, \ldots, n$, in Theorem 5.4-1. For (b), take $a_i = 1/n$, $i = 1, 2, \ldots, n$. ◄

The next two examples and the exercises give some important applications of Theorem 5.4-1 and its corollary. Recall that the mgf, once found, uniquely determines the distribution of the random variable under consideration.

Example 5.4-2

Let $X_1, X_2, \ldots, X_n$ denote the outcomes of n Bernoulli trials, each with probability of success p. The mgf of $X_i, i = 1, 2, \ldots, n$, is

$$M(t) = q + pe^t, \quad -\infty < t < \infty.$$

If

$$Y = \sum_{i=1}^n X_i,$$

then

$$M_Y(t) = \prod_{i=1}^n (q + pe^t) = (q + pe^t)^n, \quad -\infty < t < \infty.$$

Thus, we again see that Y is $b(n, p)$. ∎

Example 5.4-3

Let X_1, X_2, X_3 be the observations of a random sample of size $n = 3$ from the exponential distribution having mean θ and, of course, mgf $M(t) = 1/(1 - \theta t), t < 1/\theta$. The mgf of $Y = X_1 + X_2 + X_3$ is

$$M_Y(t) = \left[(1 - \theta t)^{-1}\right]^3 = (1 - \theta t)^{-3}, \quad t < 1/\theta,$$

which is that of a gamma distribution with parameters $\alpha = 3$ and θ. Thus, Y has this distribution. On the other hand, the mgf of $\overline{X}$ is

$$M_{\overline{X}}(t) = \left[\left(1 - \frac{\theta t}{3}\right)^{-1}\right]^3 = \left(1 - \frac{\theta t}{3}\right)^{-3}, \quad t < 3/\theta.$$

Hence, the distribution of $\overline{X}$ is gamma with the parameters $\alpha = 3$ and $\theta/3$, respectively. ∎

Theorem 5.4-2

Let $X_1, X_2, \ldots, X_n$ be independent chi-square random variables with $r_1, r_2, \ldots, r_n$ degrees of freedom, respectively. Then $Y = X_1 + X_2 + \cdots + X_n$ is $\chi^2(r_1 + r_2 + \cdots + r_n)$.

Proof By Theorem 5.4-1 with each $a = 1$, the mgf of Y is

$$M_Y(t) = \prod_{i=1}^{n} M_{X_i}(t) = (1 - 2t)^{-r_1/2}(1 - 2t)^{-r_2/2} \cdots (1 - 2t)^{-r_n/2}$$

$$= (1 - 2t)^{-\Sigma r_i/2}, \qquad \text{with } t < 1/2,$$

which is the mgf of a $\chi^2(r_1 + r_2 + \cdots + r_n)$. Thus, Y is $\chi^2(r_1 + r_2 + \cdots + r_n)$. □

The next two corollaries combine and extend the results of Theorems 3.3-2 and 5.4-2 and give one interpretation of degrees of freedom.

Corollary 5.4-2

Let $Z_1, Z_2, \ldots, Z_n$ have standard normal distributions, $N(0, 1)$. If these random variables are independent, then $W = Z_1^2 + Z_2^2 + \cdots + Z_n^2$ has a distribution that is $\chi^2(n)$.

Proof By Theorem 3.3-2, Z_i^2 is $\chi^2(1)$ for $i = 1, 2, \ldots, n$. From Theorem 5.4-2, with $Y = W$ and $r_i = 1$, it follows that W is $\chi^2(n)$. ◄

Corollary 5.4-3

If $X_1, X_2, \ldots, X_n$ are independent and have normal distributions $N(\mu_i, \sigma_i^2)$, $i = 1, 2, \ldots, n$, respectively, then the distribution of

$$W = \sum_{i=1}^{n} \frac{(X_i - \mu_i)^2}{\sigma_i^2}$$

is $\chi^2(n)$.

Proof This follows from Corollary 5.4-2, since $Z_i = (X_i - \mu_i)/\sigma_i$ is $N(0, 1)$, and thus

$$Z_i^2 = \frac{(X_i - \mu_i)^2}{\sigma_i^2}$$

is $\chi^2(1)$, $i = 1, 2, \ldots, n$. ◄

Exercises

5.4-1. Let X_1, X_2, X_3 be a random sample of size 3 from the distribution with pmf $f(x) = 1/4$, $x = 1, 2, 3, 4$. For example, observe three independent rolls of a fair four-sided die.

(a) Find the pmf of $Y = X_1 + X_2 + X_3$.

(b) Sketch a bar graph of the pmf of Y.

5.4-2. Let X_1 and X_2 have independent distributions $b(n_1, p)$ and $b(n_2, p)$. Find the mgf of $Y = X_1 + X_2$. How is Y distributed?

5.4-3. Let X_1, X_2, X_3 be mutually independent random variables with Poisson distributions having means 2, 1, and 4, respectively.

(a) Find the mgf of the sum $Y = X_1 + X_2 + X_3$.

(b) How is Y distributed?

(c) Compute $P(3 \leq Y \leq 9)$.

5.4-4. Generalize Exercise 5.4-3 by showing that the sum of n independent Poisson random variables with respective means $\mu_1, \mu_2, \ldots, \mu_n$ is Poisson with mean

$$\mu_1 + \mu_2 + \cdots + \mu_n.$$

5.4-5. Let $Z_1, Z_2, \ldots, Z_7$ be a random sample from the standard normal distribution $N(0,1)$. Let $W = Z_1^2 + Z_2^2 + \cdots + Z_7^2$. Find $P(1.69 < W < 14.07)$.

5.4-6. Let X_1, X_2, X_3, X_4, X_5 be a random sample of size 5 from a geometric distribution with $p = 1/3$.

(a) Find the mgf of $Y = X_1 + X_2 + X_3 + X_4 + X_5$.

(b) How is Y distributed?

5.4-7. Let X_1, X_2, X_3 denote a random sample of size 3 from a gamma distribution with $\alpha = 7$ and $\theta = 5$.

(a) Find the mgf of $Y = X_1 + X_2 + X_3$.

(b) How is Y distributed?

5.4-8. Let $W = X_1 + X_2 + \cdots + X_h$, a sum of h mutually independent and identically distributed exponential random variables with mean θ. Show that W has a gamma distribution with parameters $\alpha = h$ and θ, respectively.

5.4-9. Let X and Y, with respective pmfs $f(x)$ and $g(y)$, be independent discrete random variables, each of whose support is a subset of the nonnegative integers $0, 1, 2, \ldots$. Show that the pmf of $W = X + Y$ is given by the **convolution formula**

$$h(w) = \sum_{x=0}^{w} f(x)g(w-x), \qquad w = 0, 1, 2, \ldots.$$

HINT: Argue that $h(w) = P(W = w)$ is the probability of the $w + 1$ mutually exclusive events $(x, y = w - x)$, $x = 0, 1, \ldots, w$.

5.4-10. Let X equal the outcome when a fair four-sided die that has its faces numbered 0, 1, 2, and 3 is rolled. Let Y equal the outcome when a fair four-sided die that has its faces numbered 0, 4, 8, and 12 is rolled.

(a) Define the mgf of X.

(b) Define the mgf of Y.

(c) Let $W = X + Y$, the sum when the pair of dice is rolled. Find the mgf of W.

(d) Give the pmf of W; that is, determine $P(W = w)$, $w = 0, 1, \ldots, 15$, from the mgf of W.

5.4-11. Let X and Y equal the outcomes when two fair six-sided dice are rolled. Let $W = X + Y$. Assuming independence, find the pmf of W when

(a) The first die has three faces numbered 0 and three faces numbered 2, and the second die has its faces numbered 0, 1, 4, 5, 8, and 9.

(b) The faces on the first die are numbered 0, 1, 2, 3, 4, and 5, and the faces on the second die are numbered 0, 6, 12, 18, 24, and 30.

5.4-12. Let X and Y be the outcomes when a pair of fair six-sided dice is rolled. Let $W = X + Y$. How should the faces of the dice be numbered so that W has a uniform distribution on $0, 1, 2, \ldots, 12$?

5.4-13. Let $X_1, X_2, \ldots, X_8$ be a random sample from a distribution having pmf $f(x) = (x + 1)/6$, $x = 0, 1, 2$.

(a) Use Exercise 5.4-9 to find the pmf of $W_1 = X_1 + X_2$.

(b) What is the pmf of $W_2 = X_3 + X_4$?

(c) Now find the pmf of $W = W_1 + W_2 = X_1 + X_2 + X_3 + X_4$.

(d) Find the pmf of $Y = X_1 + X_2 + \cdots + X_8$.

(e) Construct probability histograms for X_1, W_1, W, and Y. Are these histograms skewed or symmetric?

5.4-14. The number of accidents in a period of one week follows a Poisson distribution with mean 3. The numbers of accidents from week to week are independent. What is the probability of exactly eight accidents in a given four weeks? HINT: See Exercise 5.4-4.

5.4-15. Given a fair four-sided die, let Y equal the number of rolls needed to observe each face at least once.

(a) Argue that $Y = X_1 + X_2 + X_3 + X_4$, where X_i has a geometric distribution with $p_i = (5-i)/4$, $i = 1, 2, 3, 4$, and X_1, X_2, X_3, X_4 are independent.

(b) Find the mean and variance of Y.

(c) Find $P(Y = y)$, $y = 4, 5, 6, 7$.

5.4-16. The number X of sick days taken during a year by an employee follows a Poisson distribution with mean 2. Let us observe four such employees. Assuming independence, compute the probability that their total number of sick days exceeds 10.

5.4-17. In a study concerning a new treatment of a certain disease, two groups of 25 participants in each were followed for five years. Those in one group took the old treatment and those in the other took the new treatment. The theoretical dropout rate for an individual was 50% in both groups over that 5-year period. Let X be the number that dropped out in the first group and Y the number in the second group. Assuming independence where needed, give the sum that equals the probability that $Y \geq X + 2$. HINT: What is the distribution of $Y - X + 25$?

5.4-18. The number of cracks on a highway averages 0.5 per mile and follows a Poisson distribution.

Assuming independence (which may not be a good assumption; why?), what is the probability that, in a 40-mile stretch of that highway, there are fewer than 15 cracks?

5.4-19. A doorman at a hotel is trying to get three taxicabs for three different couples. The arrival of empty cabs has an exponential distribution with mean 2 minutes. Assuming independence, what is the probability that the doorman will get all three couples taken care of within 6 minutes?

5.4-20. The time X in minutes of a visit to a cardiovascular disease specialist by a patient is modeled by a gamma pdf with $\alpha = 1.5$ and $\theta = 10$. Suppose that you are such a patient and have four patients ahead of you. Assuming

independence, what integral gives the probability that you will wait more than 90 minutes?

5.4-21. Let X and Y be independent with distributions $N(5, 16)$ and $N(6, 9)$, respectively. Evaluate $P(X > Y) = P(X - Y > 0)$.

5.4-22. Let X_1 and X_2 be two independent random variables. Let X_1 and $Y = X_1 + X_2$ be $\chi^2(r_1)$ and $\chi^2(r)$, respectively, where $r_1 < r$.

(a) Find the mgf of X_2.

(b) What is its distribution?

5.4-23. Let X be $N(0, 1)$. Use the mgf technique to show that $Y = X^2$ is $\chi^2(1)$. HINT: Evaluate the integral representing $E(e^{tX^2})$ by writing $w = x\sqrt{1 - 2t}$.

5.5 RANDOM FUNCTIONS ASSOCIATED WITH NORMAL DISTRIBUTIONS

In statistical applications, it is often assumed that the population from which a sample is taken is normally distributed, $N(\mu, \sigma^2)$. There is then interest in estimating the parameters μ and σ^2 or in testing conjectures about these parameters. The usual statistics that are used in these activities are the sample mean $\overline{X}$ and the sample variance S^2; thus, we need to know something about the distribution of these statistics or functions of these statistics.

We now use the mgf technique of Section 5.4 to prove a theorem that deals with linear functions of independent normally distributed random variables.

| Theorem 5.5-1 | If $X_1, X_2, \ldots, X_n$ are n mutually independent normal variables with means $\mu_1, \mu_2, \ldots, \mu_n$ and variances $\sigma_1^2, \sigma_2^2, \ldots, \sigma_n^2$, respectively, then the linear function |

$$Y = \sum_{i=1}^{n} c_i X_i$$

has the normal distribution

$$N\left(\sum_{i=1}^{n} c_i \mu_i, \sum_{i=1}^{n} c_i^2 \sigma_i^2\right).$$

Proof By Theorem 5.4-1, we have, with $-\infty < c_i t < \infty$, or $-\infty < t < \infty$,

$$M_Y(t) = \prod_{i=1}^{n} M_{X_i}(c_i t) = \prod_{i=1}^{n} \exp\left(\mu_i c_i t + \sigma_i^2 c_i^2 t^2 / 2\right)$$

because $M_{X_i}(t) = \exp(\mu_i t + \sigma_i^2 t^2 / 2)$, $i = 1, 2, \ldots, n$. Thus,

$$M_Y(t) = \exp\left[\left(\sum_{i=1}^{n} c_i \mu_i\right) t + \left(\sum_{i=1}^{n} c_i^2 \sigma_i^2\right)\left(\frac{t^2}{2}\right)\right].$$

This is the mgf of a distribution that is

$$N\left(\sum_{i=1}^{n} c_i\mu_i, \sum_{i=1}^{n} c_i^2\sigma_i^2\right).$$

Thus, Y has this normal distribution. □

From Theorem 5.5-1, we observe that the difference of two independent normally distributed random variables, say, $Y = X_1 - X_2$, has the normal distribution $N(\mu_1 - \mu_2, \sigma_1^2 + \sigma_2^2)$.

Example 5.5-1

Let X_1 and X_2 equal the number of pounds of butterfat produced by two Holstein cows (one selected at random from those on the Koopman farm and one selected at random from those on the Vliestra farm, respectively) during the 305-day lactation period following the births of calves. Assume that the distribution of X_1 is $N(693.2, 22820)$ and the distribution of X_2 is $N(631.7, 19205)$. Moreover, let X_1 and X_2 be independent. We shall find $P(X_1 > X_2)$. That is, we shall find the probability that the butterfat produced by the Koopman farm cow exceeds that produced by the Vliestra farm cow. (Sketch pdfs on the same graph for these two normal distributions.) If we let $Y = X_1 - X_2$, then the distribution of Y is $N(693.2 - 631.7, 22820 + 19205)$. Thus,

$$P(X_1 > X_2) = P(Y > 0) = P\left(\frac{Y - 61.5}{\sqrt{42025}} > \frac{0 - 61.5}{205}\right)$$

$$= P(Z > -0.30) = 0.6179.$$ ■

Corollary 5.5-1

If $X_1, X_2, \ldots, X_n$ are observations of a random sample of size n from the normal distribution $N(\mu, \sigma^2)$, then the distribution of the sample mean $\overline{X} = (1/n)\sum_{i=1}^{n} X_i$ is $N(\mu, \sigma^2/n)$.

Proof Let $c_i = 1/n$, $\mu_i = \mu$, and $\sigma_i^2 = \sigma^2$ in Theorem 5.5-1. ◄

Corollary 5.5-1 shows that if $X_1, X_2, \ldots, X_n$ is a random sample from the normal distribution, $N(\mu, \sigma^2)$, then the probability distribution of $\overline{X}$ is also normal with the same mean μ but a variance σ^2/n. This means that $\overline{X}$ has a greater probability of falling into an interval containing μ than does a single observation, say, X_1. For example, if $\mu = 50$, $\sigma^2 = 16$, and $n = 64$, then $P(49 < \overline{X} < 51) = 0.9544$, whereas $P(49 < X_1 < 51) = 0.1974$. This property is illustrated again in the next example.

Example 5.5-2

Let $X_1, X_2, \ldots, X_n$ be a random sample from the $N(50, 16)$ distribution. We know that the distribution of $\overline{X}$ is $N(50, 16/n)$. To illustrate the effect of n, the graph of the pdf of $\overline{X}$ is given in Figure 5.5-1 for $n = 1, 4, 16$, and 64. When $n = 64$, compare the areas that represent $P(49 < \overline{X} < 51)$ and $P(49 < X_1 < 51)$. ■

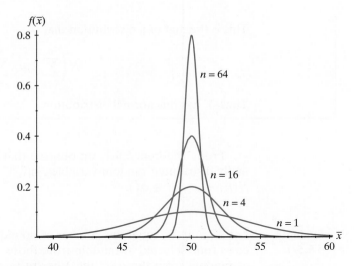

Figure 5.5-1 pdfs of means of samples from $N(50, 16)$

The next theorem gives an important result that will be used in statistical applications. In connection with those applications, we will use the sample variance S^2 to estimate the variance, σ^2, when sampling from the normal distribution, $N(\mu, \sigma^2)$. (More will be said about S^2 at the time of its use.)

Theorem 5.5-2

Let $X_1, X_2, \ldots, X_n$ be observations of a random sample of size n from the normal distribution $N(\mu, \sigma^2)$. Then the sample mean,

$$\overline{X} = \frac{1}{n} \sum_{i=1}^{n} X_i,$$

and the sample variance,

$$S^2 = \frac{1}{n-1} \sum_{i=1}^{n} (X_i - \overline{X})^2,$$

are independent and

$$\frac{(n-1)S^2}{\sigma^2} = \frac{\sum_{i=1}^{n} (X_i - \overline{X})^2}{\sigma^2} \text{ is } \chi^2(n-1).$$

Proof We are not prepared to prove the independence of $\overline{X}$ and S^2 at this time (see Section 6.7 for a proof), so we accept it without proof here. To prove the second part, note that

$$W = \sum_{i=1}^{n} \left(\frac{X_i - \mu}{\sigma} \right)^2 = \sum_{i=1}^{n} \left[\frac{(X_i - \overline{X}) + (\overline{X} - \mu)}{\sigma} \right]^2$$

$$= \sum_{i=1}^{n} \left(\frac{X_i - \overline{X}}{\sigma} \right)^2 + \frac{n(\overline{X} - \mu)^2}{\sigma^2} \tag{5.5-1}$$

because the cross-product term is equal to

$$2 \sum_{i=1}^{n} \frac{\left(\overline{X} - \mu\right)\left(X_i - \overline{X}\right)}{\sigma^2} = \frac{2\left(\overline{X} - \mu\right)}{\sigma^2} \sum_{i=1}^{n} \left(X_i - \overline{X}\right) = 0.$$

But $Y_i = (X_i - \mu)/\sigma$, $i = 1, 2, \ldots n$, are standardized normal variables that are independent. Hence, $W = \sum_{i=1}^{n} Y_i^2$ is $\chi^2(n)$ by Corollary 5.4-3. Moreover, since $\overline{X}$ is $N(\mu, \sigma^2/n)$, it follows that

$$Z^2 = \left(\frac{\overline{X} - \mu}{\sigma/\sqrt{n}}\right)^2 = \frac{n(\overline{X} - \mu)^2}{\sigma^2}$$

is $\chi^2(1)$ by Theorem 3.3-2. In this notation, Equation 5.5-1 becomes

$$W = \frac{(n-1)S^2}{\sigma^2} + Z^2.$$

However, from the fact that $\overline{X}$ and S^2 are independent, it follows that Z^2 and S^2 are also independent. In the mgf of W, this independence permits us to write

$$E\left[e^{tW}\right] = E\left[e^{t\left\{(n-1)S^2/\sigma^2 + Z^2\right\}}\right] = E\left[e^{t(n-1)S^2/\sigma^2} e^{tZ^2}\right]$$

$$= E\left[e^{t(n-1)S^2/\sigma^2}\right] E\left[e^{tZ^2}\right].$$

Since W and Z^2 have chi-square distributions, we can substitute their mgfs to obtain

$$(1 - 2t)^{-n/2} = E\left[e^{t(n-1)S^2/\sigma^2}\right] (1 - 2t)^{-1/2}.$$

Equivalently, we have

$$E\left[e^{t(n-1)S^2/\sigma^2}\right] = (1 - 2t)^{-(n-1)/2}, \qquad t < \frac{1}{2}.$$

This, of course, is the mgf of a $\chi^2(n-1)$-variable; accordingly, $(n-1)S^2/\sigma^2$ has that distribution. $\qquad \square$

Combining the results of Corollary 5.4-3 and Theorem 5.5-2, we see that when sampling is from a normal distribution,

$$U = \sum_{i=1}^{n} \frac{(X_i - \mu)^2}{\sigma^2}$$

is $\chi^2(n)$ and

$$W = \sum_{i=1}^{n} \frac{(X_i - \overline{X})^2}{\sigma^2}$$

is $\chi^2(n-1)$. That is, when the population mean, μ, in $\sum_{i=1}^{n}(X_i - \mu)^2$ is replaced by the sample mean, $\overline{X}$, one degree of freedom is lost. There are more general situations in which a degree of freedom is lost for each parameter estimated in certain chi-square random variables, some of which are found in Chapters 7, 8, and 9.

Example 5.5-3

Let X_1, X_2, X_3, X_4 be a random sample of size 4 from the normal distribution, $N(76.4, 383)$. Then

$$U = \sum_{i=1}^{4} \frac{(X_i - 76.4)^2}{383} \qquad \text{is} \qquad \chi^2(4),$$

$$W = \sum_{i=1}^{4} \frac{(X_i - \overline{X})^2}{383} \qquad \text{is} \qquad \chi^2(3),$$

and, for examples,

$$P(0.711 \leq U \leq 7.779) = 0.90 - 0.05 = 0.85,$$

$$P(0.352 \leq W \leq 6.251) = 0.90 - 0.05 = 0.85. \qquad \blacksquare$$

In later sections, we shall illustrate the importance of the chi-square distribution in applications.

We now prove a theorem that is the basis for some of the most important inferences in statistics.

Theorem 5.5-3

(Student's t distribution) Let

$$T = \frac{Z}{\sqrt{U/r}},$$

where Z is a random variable that is $N(0,1)$, U is a random variable that is $\chi^2(r)$, and Z and U are independent. Then T has a t distribution with pdf

$$f(t) = \frac{\Gamma((r+1)/2)}{\sqrt{\pi r}\,\Gamma(r/2)} \frac{1}{(1 + t^2/r)^{(r+1)/2}}, \qquad -\infty < t < \infty.$$

Proof The joint pdf of Z and U is

$$g(z, u) = \frac{1}{\sqrt{2\pi}} e^{-z^2/2} \frac{1}{\Gamma(r/2)2^{r/2}} u^{r/2-1} e^{-u/2}, \qquad -\infty < z < \infty, \ 0 < u < \infty.$$

The cdf $F(t) = P(T \leq t)$ of T is given by

$$F(t) = P\left(Z/\sqrt{U/r} \leq t\right)$$

$$= P\left(Z \leq \sqrt{U/r}\, t\right)$$

$$= \int_0^\infty \int_{-\infty}^{\sqrt{(u/r)}\,t} g(z, u)\, dz\, du.$$

That is,

$$F(t) = \frac{1}{\sqrt{\pi}\,\Gamma(r/2)} \int_0^\infty \left[\int_{-\infty}^{\sqrt{(u/r)}\,t} \frac{e^{-z^2/2}}{2^{(r+1)/2}}\, dz\right] u^{r/2-1} e^{-u/2}\, du.$$

The pdf of T is the derivative of the cdf; so, applying the fundamental theorem of calculus to the inner integral (interchanging the derivative and integral operators is permitted here), we find that

$$f(t) = F'(t) = \frac{1}{\sqrt{\pi}\,\Gamma(r/2)} \int_0^\infty \frac{e^{-(u/2)(t^2/r)}}{2^{(r+1)/2}} \sqrt{\frac{u}{r}}\, u^{r/2-1} e^{-u/2}\, du$$

$$= \frac{1}{\sqrt{\pi r}\,\Gamma(r/2)} \int_0^\infty \frac{u^{(r+1)/2-1}}{2^{(r+1)/2}} e^{-(u/2)(1+t^2/r)}\, du.$$

In the integral, make the change of variables

$$y = (1 + t^2/r)u, \qquad \text{so that} \qquad \frac{du}{dy} = \frac{1}{1 + t^2/r}.$$

Thus,

$$f(t) = \frac{\Gamma[(r+1)/2]}{\sqrt{\pi r}\,\Gamma(r/2)} \left[\frac{1}{(1+t^2/r)^{(r+1)/2}} \right] \int_0^\infty \frac{y^{(r+1)/2-1}}{\Gamma[(r+1)/2]\,2^{(r+1)/2}} e^{-y/2}\, dy.$$

The integral in this last expression for $f(t)$ is equal to 1 because the integrand is like the pdf of a chi-square distribution with $r + 1$ degrees of freedom. Hence, the pdf is

$$f(t) = \frac{\Gamma[(r+1)/2]}{\sqrt{\pi r}\,\Gamma(r/2)} \frac{1}{(1+t^2/r)^{(r+1)/2}}, \qquad -\infty < t < \infty. \qquad \square$$

Graphs of the pdf of T when $r = 1, 3,$ and 7, along with the $N(0,1)$ pdf, are given in Figure 5.5-2(a). In this figure, we see that the tails of the t distribution are heavier than those of a normal one; that is, there is more extreme probability in the t distribution than in the standardized normal one.

To find probabilities for a t random variable with r degrees of freedom, use your calculator, a computer program, or Table VI in Appendix B. If T has a t distribution with r degrees of freedom, we say that the distribution of T is $t(r)$. Furthermore, right-tail probabilities of size α are denoted by $t_\alpha(r)$. [See Figure 5.5-2(b).]

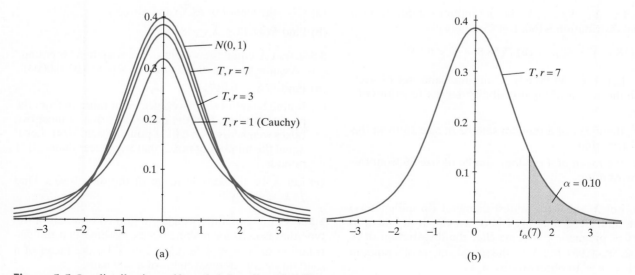

Figure 5.5-2 t distribution pdfs and right-tail probability

Example 5.5-4

Let the distribution of T be $t(11)$. Then

$$t_{0.05}(11) = 1.796 \qquad \text{and} \qquad -t_{0.05}(11) = -1.796.$$

Thus,

$$P(-1.796 \le T \le 1.796) = 0.90.$$

We can also find values of the cdf such as

$$P(T \le 2.201) = 0.975 \qquad \text{and} \qquad P(T \le -1.363) = 0.10. \qquad \blacksquare$$

We can use the results of Corollary 5.5-1 and Theorems 5.5-2 and 5.5-3 to construct an important T random variable. Given a random sample $X_1, X_2, \ldots, X_n$ from a normal distribution, $N(\mu, \sigma^2)$, let

$$Z = \frac{\overline{X} - \mu}{\sigma/\sqrt{n}} \qquad \text{and} \qquad U = \frac{(n-1)S^2}{\sigma^2}.$$

Then the distribution of Z is $N(0, 1)$ by Corollary 5.5-1. Theorem 5.5-2 tells us that the distribution of U is $\chi^2(n-1)$ and that Z and U are independent. Thus,

$$T = \frac{\dfrac{\overline{X} - \mu}{\sigma/\sqrt{n}}}{\sqrt{\dfrac{(n-1)S^2}{\sigma^2} \Big/ (n-1)}} = \frac{\overline{X} - \mu}{S/\sqrt{n}} \tag{5.5-2}$$

has a Student's t distribution (see Historical Comments) with $r = n - 1$ degrees of freedom by Theorem 5.5-3. We use this T in Section 7.1 to construct confidence intervals for an unknown mean μ of a normal distribution. (See also Exercise 5.5-16.)

Exercises

5.5-1. Let $X_1, X_2, \ldots, X_9$ be a random sample from a normal distribution $N(60, 16)$. Compute

(a) $P(58 < \overline{X} < 62)$.   **(b)** $P(58.8 < \overline{X} < 62.4)$.

5.5-2. Let X be $N(75, 64)$. Using the same set of axes, sketch the graphs of the probability density functions of

(a) X.

(b) $\overline{X}$, the mean of a random sample of size 16 from this distribution.

(c) $\overline{X}$, the mean of a random sample of size 36 from this distribution.

5.5-3. Let X equal the widest diameter (in millimeters) of the fetal head measured between the 16th and 25th weeks of pregnancy. Assume that the distribution of X is $N(46.58, 40.96)$. Let $\overline{X}$ be the sample mean of a random sample of $n = 16$ observations of X.

(a) Give the values of $E(\overline{X})$ and $\text{Var}(\overline{X})$.

(b) Find $P(44.42 \le \overline{X} \le 48.98)$.

5.5-4. Let X equal the weight of the soap in a "6-pound" box. Assume that the distribution of X is $N(6.05, 0.0004)$.

(a) Find $P(X < 6.0171)$.

(b) If nine boxes of soap are selected at random from the production line, find the probability that at most two boxes weigh less than 6.0171 pounds each. HINT: Let Y equal the number of boxes that weigh less than 6.0171 pounds.

(c) Let $\overline{X}$ be the sample mean of the nine boxes. Find $P(\overline{X} \le 6.035)$.

5.5-5. Let X equal the weight (in grams) of a nail of the type that is used for making decks. Assume that the distribution of X is $N(8.78, 0.16)$. Let $\overline{X}$ be the mean of a random sample of the weights of $n = 9$ nails.

(a) Sketch, on the same set of axes, the graphs of the pdfs of X and of $\overline{X}$.

(b) Let S^2 be the sample variance of the nine weights. Find constants a and b so that $P(a \leq S^2 \leq b) = 0.90$.

HINT: Because $8S^2/0.16$ is $\chi^2(8)$ and $P(a \leq S^2 \leq b)$ is equivalent to $P(8a/0.16 \leq 8S^2/0.16 \leq 8b/0.16)$, you can find $8a/0.16$ and $8b/0.16$ in Table IV in Appendix B.

5.5-6. At a heat-treating company, iron castings and steel forgings are heat-treated to achieve desired mechanical properties and machinability. One steel forging is annealed to soften the part for each machining. Two lots of this part, made of 1020 steel, are heat-treated in two different furnaces. The specification for this part is 36-66 on the Rockwell G scale. Let X_1 and X_2 equal the respective hardness measurements for parts selected randomly from furnaces 1 and 2. Assume that the distributions of X_1 and X_2 are $N(47.88, 2.19)$ and $N(43.04, 14.89)$, respectively.

(a) Sketch the pdfs of X_1 and X_2 on the same graph.

(b) Compute $P(X_1 > X_2)$, assuming independence of X_1 and X_2.

5.5-7. Suppose that the distribution of the weight of a prepackaged "1-pound bag" of carrots is $N(1.18, 0.07^2)$ and the distribution of the weight of a prepackaged "3-pound bag" of carrots is $N(3.22, 0.09^2)$. Selecting bags at random, find the probability that the sum of three 1-pound bags exceeds the weight of one 3-pound bag. HINT: First determine the distribution of Y, the sum of the three, and then compute $P(Y > W)$, where W is the weight of the 3-pound bag.

5.5-8. Let X denote the wing length in millimeters of a male and Y the wing length in millimeters of a female of a new species of birds. Assume that X is $N(162.05, 35.24)$ and Y is $N(145.93, 49.39)$ and that X and Y are independent. If a male and a female of this species are captured, what is the probability that X is greater than Y?

5.5-9. Suppose that the length of life in hours (say, X) of a light bulb manufactured by company A is $N(800, 14400)$ and the length of life in hours (say, Y) of a light bulb manufactured by company B is $N(850, 2500)$. One bulb is randomly selected from each company and is burned until "death."

(a) Find the probability that the length of life of the bulb from company A exceeds the length of life of the bulb from company B by at least 15 hours.

(b) Find the probability that at least one of the bulbs "lives" for at least 920 hours.

5.5-10. A consumer buys n light bulbs, each of which has a lifetime that has a mean of 800 hours, a standard deviation of 100 hours, and a normal distribution. A light bulb is replaced by another as soon as it burns out. Assuming

independence of the lifetimes, find the smallest n so that the succession of light bulbs produces light for at least 10,000 hours with a probability of 0.90.

5.5-11. A marketing research firm suggests to a company that two possible competing products can generate incomes X and Y (in millions) that are $N(3, 1)$ and $N(3.5, 4)$, respectively. Clearly, $P(X < Y) > 1/2$. However, the company would prefer the one with the smaller variance if, in fact, $P(X > 2) > P(Y > 2)$. Which product does the company select?

5.5-12. Let the independent random variables X_1 and X_2 be $N(0, 1)$ and $\chi^2(r)$, respectively. Let $Y_1 = X_1/\sqrt{X_2/r}$ and $Y_2 = X_2$.

(a) Find the joint pdf of Y_1 and Y_2.

(b) Determine the marginal pdf of Y_1 and show that Y_1 has a t distribution. (This is another, equivalent, way of finding the pdf of T.)

5.5-13. Let Z_1, Z_2, and Z_3 have independent standard normal distributions, $N(0, 1)$.

(a) Find the distribution of

$$W = \frac{Z_1}{\sqrt{(Z_2^2 + Z_3^2)/2}}.$$

(b) Show that

$$V = \frac{Z_1}{\sqrt{(Z_1^2 + Z_2^2)/2}}$$

has pdf $f(v) = 1/\left(\pi\sqrt{2 - v^2}\right)$, $-\sqrt{2} < v < \sqrt{2}$.

(c) Find the mean of V.

(d) Find the standard deviation of V.

(e) Why are the distributions of W and V so different?

5.5-14. Let T have a t distribution with r degrees of freedom. Show that $E(T) = 0$ provided that $r \geq 2$, and $\text{Var}(T) = r/(r - 2)$ provided that $r \geq 3$, by first finding $E(Z)$, $E(1/\sqrt{U})$, $E(Z^2)$, and $E(1/U)$.

5.5-15. Let the distribution of T be $t(17)$. Find

(a) $t_{0.01}(17)$.

(b) $t_{0.95}(17)$.

(c) $P(-1.740 \leq T \leq 1.740)$.

5.5-16. Let $n = 9$ in the T statistic defined in Equation 5.5-2.

(a) Find $t_{0.025}$ so that $P(-t_{0.025} \leq T \leq t_{0.025}) = 0.95$.

(b) Solve the inequality $[-t_{0.025} \leq T \leq t_{0.025}]$ so that μ is in the middle.

5.6 THE CENTRAL LIMIT THEOREM

In Section 5.4, we found that the mean $\overline{X}$ of a random sample of size n from a distribution with mean μ and variance $\sigma^2 > 0$ is a random variable with the properties that

$$E(\overline{X}) = \mu \qquad \text{and} \qquad \text{Var}(\overline{X}) = \frac{\sigma^2}{n}.$$

As n increases, the variance of $\overline{X}$ decreases. Consequently, the distribution of $\overline{X}$ clearly depends on n, and we see that we are dealing with sequences of distributions.

In Theorem 5.5-1, we considered the pdf of $\overline{X}$ when sampling is from the normal distribution $N(\mu, \sigma^2)$. We showed that the distribution of $\overline{X}$ is $N(\mu, \sigma^2/n)$, and in Figure 5.5-1, by graphing the pdfs for several values of n, we illustrated the property that as n increases, the probability becomes concentrated in a small interval centered at μ. That is, as n increases, $\overline{X}$ tends to converge to μ, or $(\overline{X} - \mu)$ tends to converge to 0 in a probability sense. (See Section 5.8.)

In general, if we let

$$W = \frac{\sqrt{n}}{\sigma}(\overline{X} - \mu) = \frac{\overline{X} - \mu}{\sigma/\sqrt{n}} = \frac{Y - n\mu}{\sqrt{n}\,\sigma},$$

where Y is the sum of a random sample of size n from some distribution with mean μ and variance σ^2, then, for each positive integer n,

$$E(W) = E\left[\frac{\overline{X} - \mu}{\sigma/\sqrt{n}}\right] = \frac{E(\overline{X}) - \mu}{\sigma/\sqrt{n}} = \frac{\mu - \mu}{\sigma/\sqrt{n}} = 0$$

and

$$\text{Var}(W) = E(W^2) = E\left[\frac{(\overline{X} - \mu)^2}{\sigma^2/n}\right] = \frac{E\left[(\overline{X} - \mu)^2\right]}{\sigma^2/n} = \frac{\sigma^2/n}{\sigma^2/n} = 1.$$

Thus, while $\overline{X} - \mu$ tends to "degenerate" to zero, the factor $\sqrt{n}/\sigma$ in $\sqrt{n}(\overline{X}-\mu)/\sigma$ "spreads out" the probability enough to prevent this degeneration. What, then, is the distribution of W as n increases? One observation that might shed some light on the answer to this question can be made immediately. If the sample arises from a normal distribution, then, from Theorem 5.5-1, we know that $\overline{X}$ is $N(\mu, \sigma^2/n)$, and hence W is $N(0, 1)$ for each positive n. Thus, in the limit, the distribution of W must be $N(0, 1)$. So if the solution of the question does not depend on the underlying distribution (i.e., it is unique), the answer must be $N(0, 1)$. As we will see, that is exactly the case, and this result is so important that it is called the central limit theorem, the proof of which is given in Section 5.9.

Theorem 5.6-1	**(Central Limit Theorem)** If $\overline{X}$ is the mean of a random sample $X_1, X_2, \ldots, X_n$ of size n from a distribution with a finite mean μ and a finite positive variance σ^2, then the distribution of $$W = \frac{\overline{X} - \mu}{\sigma/\sqrt{n}} = \frac{\sum_{i=1}^{n} X_i - n\mu}{\sqrt{n}\,\sigma}$$ is $N(0, 1)$ in the limit as $n \to \infty$.

When n is "sufficiently large," a practical use of the central limit theorem is approximating the cdf of W, namely,

$$P(W \leq w) \approx \int_{-\infty}^{w} \frac{1}{\sqrt{2\pi}} e^{-z^2/2} \, dz = \Phi(w).$$

We present some illustrations of this application, discuss the notion of "sufficiently large," and try to give an intuitive feeling for the central limit theorem.

Example 5.6-1

Let $\overline{X}$ be the mean of a random sample of $n = 25$ currents (in milliamperes) in a strip of wire in which each measurement has a mean of 15 and a variance of 4. Then $\overline{X}$ has an approximate $N(15, 4/25)$ distribution. As an illustration,

$$P(14.4 < \overline{X} < 15.6) = P\left(\frac{14.4 - 15}{0.4} < \frac{\overline{X} - 15}{0.4} < \frac{15.6 - 15}{0.4}\right)$$

$$\approx \Phi(1.5) - \Phi(-1.5) = 0.9332 - 0.0668 = 0.8664. \quad \blacksquare$$

Example 5.6-2

Let $X_1, X_2, \ldots, X_{20}$ denote a random sample of size 20 from the uniform distribution $U(0, 1)$. Here $E(X_i) = 1/2$ and $\text{Var}(X_i) = 1/12$, for $i = 1, 2, \ldots, 20$. If $Y = X_1 + X_2 + \cdots + X_{20}$, then

$$P(Y \leq 9.1) = P\left(\frac{Y - 20(1/2)}{\sqrt{20/12}} \leq \frac{9.1 - 10}{\sqrt{20/12}}\right) = P(W \leq -0.697)$$

$$\approx \Phi(-0.697)$$

$$= 0.2429.$$

Also,

$$P(8.5 \leq Y \leq 11.7) = P\left(\frac{8.5 - 10}{\sqrt{5/3}} \leq \frac{Y - 10}{\sqrt{5/3}} \leq \frac{11.7 - 10}{\sqrt{5/3}}\right)$$

$$= P(-1.162 \leq W \leq 1.317)$$

$$\approx \Phi(1.317) - \Phi(-1.162)$$

$$= 0.9061 - 0.1226 = 0.7835. \quad \blacksquare$$

Example 5.6-3

Let $\overline{X}$ denote the mean of a random sample of size 25 from the distribution whose pdf is $f(x) = x^3/4$, $0 < x < 2$. It is easy to show that $\mu = 8/5 = 1.6$ and $\sigma^2 = 8/75$. Thus,

$$P(1.5 \leq \overline{X} \leq 1.65) = P\left(\frac{1.5 - 1.6}{\sqrt{8/75}/\sqrt{25}} \leq \frac{\overline{X} - 1.6}{\sqrt{8/75}/\sqrt{25}} \leq \frac{1.65 - 1.6}{\sqrt{8/75}/\sqrt{25}}\right)$$

$$= P(-1.531 \leq W \leq 0.765)$$

$$\approx \Phi(0.765) - \Phi(-1.531)$$

$$= 0.7779 - 0.0629 = 0.7150. \quad \blacksquare$$

These examples show how the central limit theorem can be used for approximating certain probabilities concerning the mean $\overline{X}$ or the sum $Y = \sum_{i=1}^{n} X_i$ of a random sample. That is, $\overline{X}$ is approximately $N(\mu, \sigma^2/n)$, and Y is approximately $N(n\mu, n\sigma^2)$, when n is "sufficiently large," where μ and σ^2 are, respectively, the mean and the variance of the underlying distribution from which the sample arose. Generally, if n is greater than 25 or 30, these approximations will be good. However, if the underlying distribution is symmetric, unimodal, and of the continuous type, a value of n as small as 4 or 5 can yield an adequate approximation. Moreover, if the original distribution is approximately normal, $\overline{X}$ would have a distribution very close to normal when n equals 2 or 3. In fact, we know that if the sample is taken from $N(\mu, \sigma^2)$, $\overline{X}$ is exactly $N(\mu, \sigma^2/n)$ for every $n = 1, 2, 3, \ldots$.

The examples that follow will help to illustrate the previous remarks and will give the reader a better intuitive feeling about the central limit theorem. In particular, we shall see how the size of n affects the distribution of $\overline{X}$ and $Y = \sum_{i=1}^{n} X_i$ for samples from several underlying distributions.

Example 5.6-4

Let X_1, X_2, X_3, X_4 be a random sample of size 4 from the uniform distribution $U(0, 1)$ with pdf $f(x) = 1, 0 < x < 1$. Then $\mu = 1/2$ and $\sigma^2 = 1/12$. We shall compare the graph of the pdf of

$$Y = \sum_{i=1}^{n} X_i$$

with the graph of the $N[n(1/2), n(1/12)]$ pdf for $n = 2$ and 4, respectively.

By methods given in Section 5.2, we can determine that the pdf of $Y = X_1 + X_2$ is

$$g(y) = \begin{cases} y, & 0 < y \leq 1, \\ 2 - y, & 1 < y < 2. \end{cases}$$

This is the triangular pdf that is graphed in Figure 5.6-1(a). In this figure, the $N[2(1/2), 2(1/12)]$ pdf is also graphed.

Moreover, the pdf of $Y = X_1 + X_2 + X_3 + X_4$ is

$$g(y) = \begin{cases} \dfrac{y^3}{6}, & 0 \leq y < 1, \\[2mm] \dfrac{-3y^3 + 12y^2 - 12y + 4}{6}, & 1 \leq y < 2, \\[2mm] \dfrac{3y^3 - 24y^2 + 60y - 44}{6}, & 2 \leq y < 3, \\[2mm] \dfrac{-y^3 + 12y^2 - 48y + 64}{6}, & 3 \leq y \leq 4. \end{cases}$$

This pdf is graphed in Figure 5.6-1(b) along with the $N[4(1/2), 4(1/12)]$ pdf. If we are interested in finding $P(1.7 \leq Y \leq 3.2)$, we can do so by evaluating

$$\int_{1.7}^{3.2} g(y)\, dy,$$

which is tedious. (See Exercise 5.6-9.) It is much easier to use a normal approximation, which results in a number close to the exact value. ∎

In Example 5.6-4 and Exercise 5.6-9, we show that even for a small value of n, such as $n = 4$, the sum of the sample items has an approximate normal distribution.

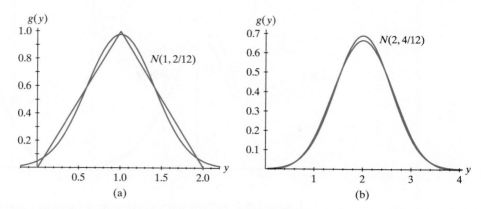

Figure 5.6-1 pdfs of sums of uniform random variables

The next example illustrates that, for some underlying distributions (particularly skewed ones), n must be quite large in order for us to obtain a satisfactory approximation. To keep the scale on the horizontal axis the same for each value of n, we will use the following result: Let $f(x)$ and $F(x)$, respectively, be the pdf and cdf of a random variable, X, of the continuous type having mean μ and variance σ^2. Let $W = (X - \mu)/\sigma$. The cdf of W is given by

$$G(w) = P(W \le w) = P\left(\frac{X - \mu}{\sigma} \le w\right)$$

$$= P(X \le \sigma w + \mu) = F(\sigma w + \mu).$$

Thus, the pdf of W is given by

$$g(w) = F'(\sigma w + \mu) = \sigma f(\sigma w + \mu).$$

Example 5.6-5   Let $X_1, X_2, \ldots, X_n$ be a random sample of size n from a chi-square distribution with one degree of freedom. If

$$Y = \sum_{i=1}^{n} X_i,$$

then Y is $\chi^2(n)$, and it follows that $E(Y) = n$ and $\text{Var}(Y) = 2n$. Let

$$W = \frac{Y - n}{\sqrt{2n}}.$$

The pdf of W is given by

$$g(w) = \sqrt{2n}\, \frac{(\sqrt{2n}\, w + n)^{n/2 - 1}}{\Gamma\left(\frac{n}{2}\right) 2^{n/2}}\, e^{-(\sqrt{2n}\, w + n)/2}, \qquad -n/\sqrt{2n} < w < \infty.$$

Note that $w > -n/\sqrt{2n}$ corresponds to $y > 0$. In Figure 5.6-2, the graphs of W are given along with the $N(0, 1)$ pdf for $n = 20$ and 100, respectively.   ∎

In order to gain an intuitive idea about how the sample size n affects the distribution of $W = (\overline{X} - \mu)/(\sigma/\sqrt{n})$, it is helpful to simulate values of W on a computer

Figure 5.6-2 pdfs of sums of chi-square random variables

using different values of n and different underlying distributions. The next example illustrates this simulation.

REMARK Recall that we simulate observations from a distribution of X having a continuous-type cdf $F(x)$ as follows: Suppose $F(a) = 0$, $F(b) = 1$, and $F(x)$ is strictly increasing for $a < x < b$. Let $Y = F(x)$ and let the distribution of Y be $U(0,1)$. If y is an observed value of Y, then $x = F^{-1}(y)$ is an observed value of X. (See Section 5.1.) Thus, if y is the value of a computer-generated random number, then $x = F^{-1}(y)$ is the simulated value of X. ∎

Example 5.6-6

It is often difficult to find the exact distribution of the random variable $W = (\overline{X} - \mu)/(\sigma/\sqrt{n})$, unless you use a computer algebra system such as *Maple*. In this example, we give some empirical evidence about the distribution of W by simulating random samples on the computer. We also superimpose the theoretical pdf of W, which we found by using *Maple*. Let $X_1, X_2, \ldots, X_n$ denote a random sample of size n from the distribution with pdf $f(x)$, cdf $F(x)$, mean μ, and variance σ^2. We simulated 1000 random samples of size $n = 2$ and $n = 7$ from each of two distributions. We then computed the value of W for each sample, thus obtaining 1000 observed values of W. Next, we constructed a histogram of these 1000 values by using 21 intervals of equal length. A relative frequency histogram of the observations of W, the pdf for the standard normal distribution, and the theoretical pdf of W are given in Figures 5.6-3 and 5.6-4.

(a) In Figure 5.6-3, $f(x) = (x+1)/2$ and $F(x) = (x+1)^2/4$ for $-1 < x < 1$; $\mu = 1/3$, $\sigma^2 = 2/9$; and $n = 2$ and 7. This underlying distribution is skewed to the left.

(b) In Figure 5.6-4, $f(x) = (3/2)x^2$ and $F(x) = (x^3 + 1)/2$ for $-1 < x < 1$; $\mu = 0$, $\sigma^2 = 3/5$; and $n = 2$ and 7. [Sketch the graph of $y = f(x)$. Give an argument as to why the histogram for $n = 2$ looks the way it does.] This underlying distribution is U-shaped; thus, W does not follow a normal distribution with small n. ∎

Note that these examples have not *proved* anything. They are presented to give *evidence* of the truth of the central limit theorem, and they do give a nice feeling for what is happening.

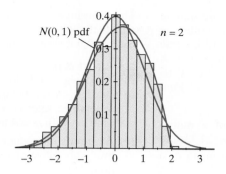 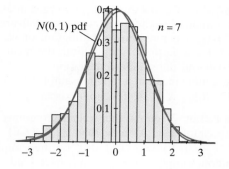

Figure 5.6-3 pdfs of $(\overline{X} - \mu)/(\sigma/\sqrt{n})$, underlying distribution triangular

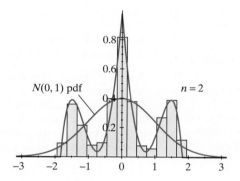 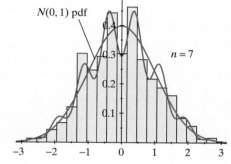

Figure 5.6-4 pdfs of $(\overline{X} - \mu)/(\sigma/\sqrt{n})$, underlying distribution U-shaped

So far, all the illustrations have concerned distributions of the continuous type. However, the hypotheses for the central limit theorem do not require the distribution to be continuous. We shall consider applications of the central limit theorem for discrete-type distributions in the next section.

Exercises

5.6-1. Let $\overline{X}$ be the mean of a random sample of size 12 from the uniform distribution on the interval $(0,1)$. Approximate $P(1/2 \leq \overline{X} \leq 2/3)$.

5.6-2. Let $Y = X_1 + X_2 + \cdots + X_{15}$ be the sum of a random sample of size 15 from the distribution whose pdf is $f(x) = (3/2)x^2$, $-1 < x < 1$. Using the pdf of Y, we find that $P(-0.3 \leq Y \leq 1.5) = 0.22788$. Use the central limit theorem to approximate this probability.

5.6-3. Let $\overline{X}$ be the mean of a random sample of size 36 from an exponential distribution with mean 3. Approximate $P(2.5 \leq \overline{X} \leq 4)$.

5.6-4. Approximate $P(39.75 \leq \overline{X} \leq 41.25)$, where $\overline{X}$ is the mean of a random sample of size 32 from a distribution with mean $\mu = 40$ and variance $\sigma^2 = 8$.

5.6-5. Let $X_1, X_2, \ldots, X_{18}$ be a random sample of size 18 from a chi-square distribution with $r = 1$. Recall that $\mu = 1$ and $\sigma^2 = 2$.

(a) How is $Y = \sum_{i=1}^{18} X_i$ distributed?

(b) Using the result of part (a), we see from Table IV in Appendix B that

$$P(Y \leq 9.390) = 0.05 \quad \text{and} \quad P(Y \leq 34.80) = 0.99.$$

Compare these two probabilities with the approximations found with the use of the central limit theorem.

5.6-6. A random sample of size $n = 25$ is taken from the distribution with pdf $f(x) = 2e^{-2x}$, $x > 0$.

(a) Find μ and σ^2. **(b)** Find $P(0.20 \leq \overline{X} \leq 0.25)$, approximately.

5.6-7. Let X equal the maximal oxygen intake of a human on a treadmill, where the measurements are in milliliters of oxygen per minute per kilogram of weight. Assume that, for a particular population, the mean of X is $\mu = 54.030$ and the standard deviation is $\sigma = 5.8$. Let $\overline{X}$ be the sample mean of a random sample of size $n = 47$. Find $P(52.761 \le \overline{X} \le 54.453)$, approximately.

5.6-8. Let X equal the weight in grams of a miniature candy bar. Assume that $\mu = E(X) = 24.43$ and $\sigma^2 = \text{Var}(X) = 2.20$. Let $\overline{X}$ be the sample mean of a random sample of $n = 30$ candy bars. Find

(a) $E(\overline{X})$. **(b)** $\text{Var}(\overline{X})$. **(c)** $P(24.17 \le \overline{X} \le 24.82)$, approximately.

5.6-9. In Example 5.6-4, compute $P(1.7 \le Y \le 3.2)$ with $n = 4$ and compare your answer with the normal approximation of this probability.

5.6-10. Let X and Y equal the respective numbers of hours a randomly selected child watches movies or cartoons on TV during a certain month. From experience, it is known that $E(X) = 30$, $E(Y) = 50$, $\text{Var}(X) = 52$, $\text{Var}(Y) = 64$, and $\text{Cov}(X, Y) = 14$. Twenty-five children are selected at random. Let Z equal the total number of hours these 25 children watch TV movies or cartoons in the next month. Approximate $P(1970 < Z < 2090)$. HINT: Use the remark after Theorem 5.3-2.

5.6-11. A company has a one-year group life policy that divides its employees into two classes as follows:

Class	Probability of Death	Benefit	Number in Class
A	0.01	$20,000	1000
B	0.03	$10,000	500

The insurance company wants to collect a premium that equals the 90th percentile of the distribution of the total claims. What should that premium be?

5.6-12. At certain times during the year, a bus company runs a special van holding 25 passengers from Iowa City to Chicago. After ticket sales open, the time (in minutes) between sales of tickets for the trip has a gamma distribution with $\alpha = 4$ and $\theta = 3$.

(a) Assuming independence, record an integral that gives the probability of tickets being sold out within one hour.

(b) Approximate the answer in part (a).

5.6-13. The tensile strength X of paper, in pounds per square inch, has $\mu = 30$ and $\sigma = 3$. A random sample of size $n = 100$ is taken from the distribution of tensile strengths. Compute the probability that the sample mean $\overline{X}$ is greater than 29.5 pounds per square inch.

5.6-14. Suppose that the sick leave taken by the typical worker per year has $\mu = 10$, $\sigma = 2$, measured in days. A firm has $n = 20$ employees. Assuming independence, how many sick days should the firm budget if the financial officer wants the probability of exceeding the number of days budgeted to be less than 20%?

5.6-15. Let X_1, X_2, X_3, X_4 represent the random times in days needed to complete four steps of a project. These times are independent and have gamma distributions with common $\theta = 2$ and $\alpha_1 = 3, \alpha_2 = 2, \alpha_3 = 5, \alpha_4 = 3$, respectively. One step must be completed before the next can be started. Let Y equal the total time needed to complete the project.

(a) Find an integral that represents $P(Y \le 25)$.

(b) Using a normal distribution, approximate the answer to part (a). Is this approach justified?

5.7 APPROXIMATIONS FOR DISCRETE DISTRIBUTIONS

In this section, we illustrate how the normal distribution can be used to approximate probabilities for certain discrete-type distributions. One of the more important discrete distributions is the binomial distribution. To see how the central limit theorem can be applied, recall that a binomial random variable can be described as the sum of Bernoulli random variables. That is, let $X_1, X_2, \ldots, X_n$ be a random sample from a Bernoulli distribution with mean $\mu = p$ and variance $\sigma^2 = p(1-p)$, where $0 < p < 1$. Then $Y = \sum_{i=1}^{n} X_i$ is $b(n, p)$. The central limit theorem states that the distribution of

$$W = \frac{Y - np}{\sqrt{np(1-p)}} = \frac{\overline{X} - p}{\sqrt{p(1-p)/n}}$$

is $N(0, 1)$ in the limit as $n \to \infty$. Thus, if n is "sufficiently large," the distribution of Y is approximately $N[np, np(1-p)]$, and probabilities for the binomial distribution

$b(n,p)$ can be approximated with this normal distribution. A rule often stated is that n is sufficiently large if $np \geq 5$ and $n(1-p) \geq 5$.

Note that we shall be approximating probabilities for a discrete distribution with probabilities for a continuous distribution. Let us discuss a reasonable procedure in this situation. If V is $N(\mu, \sigma^2)$, then $P(a < V < b)$ is equivalent to the area bounded by the pdf of V, the v-axis, $v = a$, and $v = b$. Now recall that for a Y that is $b(n,p)$, the probability histogram for Y was defined as follows: For each y such that $k - 1/2 < y = k < k + 1/2$, let

$$f(k) = \frac{n!}{k!(n-k)!} p^k (1-p)^{n-k}, \qquad k = 0, 1, 2, \ldots, n.$$

Then $P(Y = k)$ can be represented by the area of the rectangle with a height of $P(Y = k)$ and a base of length 1 centered at k. Figure 5.7-1 shows the graph of the probability histogram for the binomial distribution $b(4, 1/4)$. In using the normal distribution to approximate probabilities for the binomial distribution, areas under the pdf for the normal distribution will be used to approximate areas of rectangles in the probability histogram for the binomial distribution. Since these rectangles have unit base centered at the integers, this is called a **half-unit correction for continuity**. Note that, for an integer k,

$$P(Y = k) = P(k - 1/2 < Y < k + 1/2).$$

Example 5.7-1

Let the distribution of Y be $b(10, 1/2)$. Then, by the central limit theorem, $P(a < Y < b)$ can be approximated with the use of the normal distribution with mean $10(1/2) = 5$ and variance $10(1/2)(1/2) = 5/2$. Figure 5.7-2(a) shows the graph of the probability histogram for $b(10, 1/2)$ and the graph of the pdf of the normal distribution $N(5, 5/2)$. Note that the area of the rectangle whose base is

$$\left(k - \frac{1}{2}, k + \frac{1}{2} \right)$$

and the area under the normal curve between $k - 1/2$ and $k + 1/2$ are approximately equal for each integer k in Figure 5.7-2(a), illustrating the half-unit correction for continuity. ∎

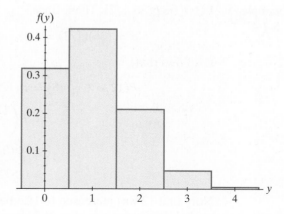

Figure 5.7-1 Probability histogram for $b(4, 1/4)$

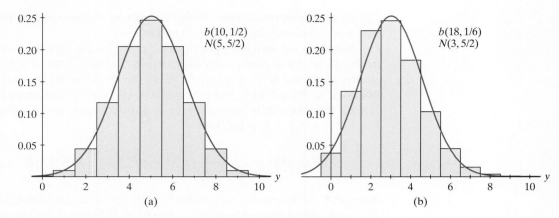

Figure 5.7-2 Normal approximation for the binomial distribution

Example 5.7-2   Let Y be $b(18, 1/6)$. Because $np = 18(1/6) = 3 < 5$, the normal approximation is not as good here. Figure 5.7-2(b) illustrates this by depicting the skewed probability histogram for $b(18, 1/6)$ and the symmetric pdf of the normal distribution $N(3, 5/2)$. ∎

Example 5.7-3   Let Y have the binomial distribution of Example 5.7-1 and Figure 5.7-2(a), namely, $b(10, 1/2)$. Then

$$P(3 \leq Y < 6) = P(2.5 \leq Y \leq 5.5),$$

because $P(Y = 6)$ is not in the desired answer. But the latter equals

$$P\left(\frac{2.5 - 5}{\sqrt{10/4}} \leq \frac{Y - 5}{\sqrt{10/4}} \leq \frac{5.5 - 5}{\sqrt{10/4}}\right) \approx \Phi(0.316) - \Phi(-1.581)$$

$$= 0.6240 - 0.0570 = 0.5670.$$

Using Table II in Appendix B, we find that $P(3 \leq Y < 6) = 0.5683$. ∎

Example 5.7-4   Let Y be $b(36, 1/2)$. Then, since

$$\mu = (36)(1/2) = 18 \quad \text{and} \quad \sigma^2 = (36)(1/2)(1/2) = 9,$$

it follows that

$$P(12 < Y \leq 18) = P(12.5 \leq Y \leq 18.5)$$

$$= P\left(\frac{12.5 - 18}{\sqrt{9}} \leq \frac{Y - 18}{\sqrt{9}} \leq \frac{18.5 - 18}{\sqrt{9}}\right)$$

$$\approx \Phi(0.167) - \Phi(-1.833)$$

$$= 0.5329.$$

Note that 12 was increased to 12.5 because $P(Y = 12)$ is not included in the desired probability. Using the binomial formula, we find that

$$P(12 < Y \leq 18) = P(13 \leq Y \leq 18) = 0.5334.$$

(You may verify this answer using your calculator or Minitab.) Also,

$$P(Y = 20) = P(19.5 \leq Y \leq 20.5)$$

$$= P\left(\frac{19.5 - 18}{\sqrt{9}} \leq \frac{Y - 18}{\sqrt{9}} \leq \frac{20.5 - 18}{\sqrt{9}}\right)$$

$$\approx \Phi(0.833) - \Phi(0.5)$$

$$= 0.1060.$$

Using the binomial formula, we have $P(Y = 20) = 0.1063$. So, in this situation, the approximations are extremely good. ∎

Note that, in general, if Y is $b(n,p)$, n is large, and $k = 0, 1, \ldots, n$, then

$$P(Y \leq k) \approx \Phi\left(\frac{k + 1/2 - np}{\sqrt{npq}}\right)$$

and

$$P(Y < k) \approx \Phi\left(\frac{k - 1/2 - np}{\sqrt{npq}}\right),$$

because in the first case k is included and in the second it is not.

We now show how the Poisson distribution with large enough mean can be approximated with the use of a normal distribution.

Example 5.7-5

A random variable having a Poisson distribution with mean 20 can be thought of as the sum Y of the observations of a random sample of size 20 from a Poisson distribution with mean 1. Thus,

$$W = \frac{Y - 20}{\sqrt{20}}$$

has a distribution that is approximately $N(0, 1)$, and the distribution of Y is approximately $N(20, 20)$. (See Figure 5.7-3.) For example, using a half-unit correction for continuity,

$$P(16 < Y \leq 21) = P(16.5 \leq Y \leq 21.5)$$

$$= P\left(\frac{16.5 - 20}{\sqrt{20}} \leq \frac{Y - 20}{\sqrt{20}} \leq \frac{21.5 - 20}{\sqrt{20}}\right)$$

$$\approx \Phi(0.335) - \Phi(-0.783)$$

$$= 0.4142.$$

Note that 16 is increased to 16.5 because $Y = 16$ is not included in the event $\{16 < Y \leq 21\}$. The answer obtained with the Poisson formula is 0.4226. ∎

In general, if Y has a Poisson distribution with mean λ, then the distribution of

$$W = \frac{Y - \lambda}{\sqrt{\lambda}}$$

is approximately $N(0, 1)$ when λ is sufficiently large.

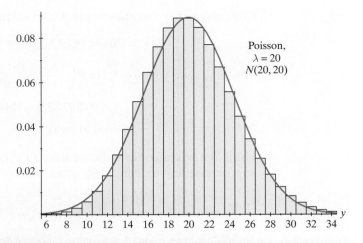

Figure 5.7-3 Normal approximation of Poisson, $\lambda = 20$

REMARK If you have a statistical calculator or a statistics computer package, use it to compute discrete probabilities. However, it is important to learn how to apply the central limit theorem. ■

Exercises

5.7-1. Let the distribution of Y be $b(25, 1/2)$. Find the given probabilities in two ways: exactly, using Table II in Appendix B; and approximately, using the central limit theorem. Compare the two results in each of the three cases.

(a) $P(10 < Y \le 12)$. **(b)** $P(12 \le Y < 15)$. **(c)** $P(Y = 12)$.

5.7-2. Suppose that among gifted seventh-graders who score very high on a mathematics exam, approximately 20% are left-handed or ambidextrous. Let X equal the number of left-handed or ambidextrous students among a random sample of $n = 25$ gifted seventh-graders. Find $P(2 < X < 9)$

(a) Using Table II in Appendix B.

(b) Approximately, using the central limit theorem.

REMARK Since X has a skewed distribution, the approximation is not as good as that for the symmetrical distribution where $p = 0.50$, even though $np = 5$. ■

5.7-3. A public opinion poll in a country was conducted to determine whether the citizens are prepared for the "big earthquake" that experts predict will devastate the country sometime in the next 25 years. It was learned that "50% have not secured objects in their homes that might fall and cause injury and damage during a temblor." In a random sample of $n = 900$ citizens of the country, let X

equal the number who "have not secured objects in their homes." Find $P(420 \le X \le 495)$ approximately.

5.7-4. Let X equal the number out of $n = 48$ mature aster seeds that will germinate when $p = 0.75$ is the probability that a particular seed germinates. Approximate $P(35 \le X \le 40)$.

5.7-5. Let $X_1, X_2, \ldots, X_{48}$ be a random sample of size 48 from the distribution with pdf $f(x) = 1/x^2$, $1 < x < \infty$. Approximate the probability that at most 10 of these random variables have values greater than 4. HINT: Let the ith trial be a success if $X_i > 4$, $i = 1, 2, \ldots, 48$, and let Y equal the number of successes.

5.7-6. In adults, the pneumococcus bacterium causes 70% of pneumonia cases. In a random sample of $n = 84$ adults who have pneumonia, let X equal the number whose pneumonia was caused by the pneumococcus bacterium. Use the normal distribution to find $P(X \le 52)$, approximately.

5.7-7. Let X equal the number of alpha particles emitted by barium-133 per second and counted by a Geiger counter. Assume that X has a Poisson distribution with $\lambda = 49$. Approximate $P(45 < X < 60)$.

5.7-8. A candy maker produces mints that have a label weight of 25 grams. Assume that the distribution of the weights of these mints is $N(26, 0.25)$.

(a) Let X denote the weight of a single mint selected at random from the production line. Find $P(X < 25.25)$.

(b) During a particular shift, 235 mints are selected at random and weighed. Let Y equal the number of these mints that weigh less than 25.25 grams. Approximate $P(Y \leq 10)$.

(c) Let $\overline{X}$ equal the sample mean of the 125 mints selected and weighed during another shift. Find $P(26 \leq \overline{X} \leq 28)$.

5.7-9. Let $X_1, X_2, \ldots, X_{30}$ be a random sample of size 30 from a Poisson distribution with a mean of 2/3. Approximate

(a) $P\left(15 < \sum_{i=1}^{30} X_i \leq 22\right).$ **(b)** $P\left(21 \leq \sum_{i=1}^{30} X_i < 27\right).$

5.7-10. In the casino game roulette, the probability of winning with a bet on red is $p = 0.50$. Let Y equal the number of winning bets out of 900 independent bets that are placed. Find $P(Y > 440)$, approximately.

5.7-11. About 60% of all Americans have a sedentary lifestyle. Select $n = 96$ Americans at random. (Assume independence.) What is the probability that between 50 and 60, inclusive, do not exercise regularly?

5.7-12. If X is $b(100, 0.1)$, find the approximate value of $P(12 \leq X \leq 14)$, using

(a) The normal approximation.

(b) The Poisson approximation.

(c) The binomial.

5.7-13. Let $X_1, X_2, \ldots, X_{36}$ be a random sample of size 36 from the geometric distribution with pmf $f(x) = (1/4)^{x-1}(3/4)$, $x = 1, 2, 3, \ldots$. Approximate

(a) $P\left(46 \leq \sum_{i=1}^{36} X_i \leq 49\right).$ **(b)** $P(1.25 \leq \overline{X} \leq 1.50)$.

HINT: Observe that the distribution of the sum is of the discrete type.

5.7-14. A die is rolled 36 independent times. Let Y be the sum of the 36 resulting values. Recalling that Y is a random variable of the discrete type, approximate

(a) $P(Y \geq 129)$. **(b)** $P(Y < 129)$. **(c)** $P(105 < Y \leq 129)$.

5.7-15. In the United States, the probability that a child dies in his or her first year of life is about $p = 0.01$. (It is actually slightly less than this.) Consider a group of 5000 such infants. What is the probability that between 45 and 53, inclusive, die in the first year of life?

5.7-16. Let Y equal the sum of $n = 100$ Bernoulli trials. That is, Y is $b(100, p)$. For each of **(i)** $p = 0.1$, **(ii)** $p = 0.5$, and **(iii)** $p = 0.8$,

(a) Draw the approximating normal pdfs, all on the same graph.

(b) Find $P(|Y/100 - p| \leq 0.015)$, approximately.

5.7-17. The number of trees in one acre has a Poisson distribution with mean 60. Assuming independence, compute $P(5950 \leq X \leq 6100)$, approximately, where X is the number of trees in 100 acres.

5.7-18. Assume that the background noise X of a digital signal has a normal distribution with $\mu = 0$ volts and $\sigma = 0.5$ volt. If we observe $n = 100$ independent measurements of this noise, what is the probability that at least 7 of them exceed 0.98 in absolute value?

(a) Use the Poisson distribution to approximate this probability.

(b) Use the normal distribution to approximate this probability.

(c) Use the binomial distribution to approximate this probability.

PROBABILISTIC COMMENTS (**Simulation: Central Limit Theorem**) As we think about what de Moivre, Laplace, and Gauss discovered in their times, we see that it was truly amazing. Of course, de Moivre could compute the probabilities associated with various binomial distributions and see how they "piled up" in that bell shape, and he came up with the normal formula. Now, Laplace and Gauss had an even tougher task, as they could not easily find the probabilities associated with the sample mean $\overline{X}$, even with simple underlying distributions. As an illustration, suppose the random sample $X_1, X_2, \ldots, X_n$ arises from a uniform distribution on the space $[0, 1)$. It is extremely difficult to compute probabilities about $\overline{X}$ unless n is very small, such as $n = 2$ or $n = 3$. Today, of course, we can use a computer algebra system (CAS) to simulate the distribution of $\overline{X}$ for any sample size and get fairly accurate estimates of the probabilities associated with $\overline{X}$. We did this 10,000 times for $n = 6$, which resulted in Figure 5.7-4. In this chapter, we learned that $\overline{X}$ has an

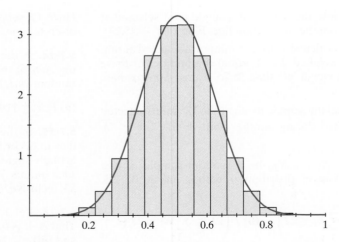

Figure 5.7-4 Simulation of 10,000 $\bar{x}$s for samples of size 6 from $U(0,1)$

approximate normal distribution with mean 1/2 and variance 1/72. We could then superimpose this normal pdf on the graph of the histogram for comparison. (Rather than superimposing the normal pdf, we used the actual pdf of $\overline{X}$ that is shown in the display that follows.) From the histogram, we see the bell-shaped curve resulting from this simulation. None of these three outstanding mathematicians had the advantage of the computer. Today, a researcher with a new idea about a probability distribution can check easily with a simulation to see if it is worth devoting more time to the idea.

The simulated $\bar{x}$s were grouped into 18 classes with equal lengths of 1/18. The frequencies of the 18 classes are

0, 1, 9, 72, 223, 524, 957, 1467, 1744, 1751, 1464, 954, 528, 220, 69, 15, 2, 0.

Using these data, we can estimate certain probabilities, for example,

$$P\left(\frac{1}{6} \leq \overline{X} \leq \frac{2}{6}\right) \approx \frac{72 + 223 + 524}{10{,}000} = 0.0819$$

and

$$P\left(\frac{11}{18} \leq \overline{X} \leq 1\right) \approx \frac{954 + 528 + 220 + 69 + 15 + 2 + 0}{10{,}000} = 0.1788.$$

With a CAS, it is sometimes possible to find pdfs that involve rather complex calculations. For example, *Maple* was used to find the actual pdf of $\overline{X}$ for this example. It is this pdf that is superimposed on the histogram. Letting $u = \bar{x}$, we find that the pdf is given by

$$g(u) = \begin{cases} 6\left(\dfrac{324u^5}{5}\right), & 0 < u < 1/6, \\[2ex] 6\left(\dfrac{1}{20} - 324u^5 + 324u^4 - 108u^3 + 18u^2 - \dfrac{3u}{2}\right), & 1/6 \le u < 2/6, \\[2ex] 6\left(-\dfrac{79}{20} + \dfrac{117u}{2} + 648u^5 - 1296u^4 + 972u^3 - 342u^2\right), & 2/6 \le u < 3/6, \\[2ex] 6\left(\dfrac{731}{20} - \dfrac{693u}{2} - 648u^5 + 1944u^4 - 2268u^3\right), & 3/6 \le u < 4/6, \\[2ex] 6\left(-\dfrac{1829}{20} + \dfrac{1227u}{2} - 1602u^2 + 2052u^3 + 324u^5 - 1296u^4\right), & 4/6 \le u < 5/6, \\[2ex] 6\left(\dfrac{324}{5} - 324u + 648u^2 - 648u^3 + 324u^4 - \dfrac{324u^5}{5}\right), & 5/6 \le u < 1. \end{cases}$$

We can also calculate

$$\int_{1/6}^{2/6} g(u)\,du = \frac{19}{240} = 0.0792$$

and

$$\int_{11/18}^{1} g(u)\,du = \frac{5,818}{32,805} = 0.17735.$$

Although these integrations are not difficult, they are tedious to do by hand. ∎

5.8 CHEBYSHEV'S INEQUALITY AND CONVERGENCE IN PROBABILITY

In this section, we use Chebyshev's inequality to show, in another sense, that the sample mean, $\overline{X}$, is a good statistic to use to estimate a population mean μ; the relative frequency of success in n independent Bernoulli trials, Y/n, is a good statistic for estimating p. We examine the effect of the sample size n on these estimates.

We begin by showing that Chebyshev's inequality gives added significance to the standard deviation in terms of bounding certain probabilities. The inequality is valid for *all* distributions for which the standard deviation exists. The proof is given for the discrete case, but it holds for the continuous case, with integrals replacing summations.

Theorem 5.8-1	**(Chebyshev's Inequality)** If the random variable X has a mean μ and variance σ^2, then, for every $k \ge 1$, $$P(|X - \mu| \ge k\sigma) \le \frac{1}{k^2}.$$

Proof Let $f(x)$ denote the pmf of X. Then

$$\sigma^2 = E[(X - \mu)^2] = \sum_{x \in S}(x - \mu)^2 f(x)$$

$$= \sum_{x \in A}(x - \mu)^2 f(x) + \sum_{x \in A'}(x - \mu)^2 f(x), \qquad (5.8\text{-}1)$$

where

$$A = \{x : |x - \mu| \geq k\sigma\}.$$

The second term in the right-hand member of Equation 5.8-1 is the sum of non-negative numbers and thus is greater than or equal to zero. Hence,

$$\sigma^2 \geq \sum_{x \in A}(x - \mu)^2 f(x).$$

However, in A, $|x - \mu| \geq k\sigma$; so

$$\sigma^2 \geq \sum_{x \in A}(k\sigma)^2 f(x) = k^2\sigma^2 \sum_{x \in A} f(x).$$

But the latter summation equals $P(X \in A)$; thus,

$$\sigma^2 \geq k^2\sigma^2 P(X \in A) = k^2\sigma^2 P(|X - \mu| \geq k\sigma).$$

That is,

$$P(|X - \mu| \geq k\sigma) \leq \frac{1}{k^2}.$$

$\square$

Corollary 5.8-1 If $\varepsilon = k\sigma$, then

$$P(|X - \mu| \geq \varepsilon) \leq \frac{\sigma^2}{\varepsilon^2}.$$

◄

In words, Chebyshev's inequality states that the probability that X differs from its mean by at least k standard deviations is less than or equal to $1/k^2$. It follows that the probability that X differs from its mean by less than k standard deviations is at least $1 - 1/k^2$. That is,

$$P(|X - \mu| < k\sigma) \geq 1 - \frac{1}{k^2}.$$

From the corollary, it also follows that

$$P(|X - \mu| < \varepsilon) \geq 1 - \frac{\sigma^2}{\varepsilon^2}.$$

Thus, Chebyshev's inequality can be used as a bound for certain probabilities. However, in many instances, the bound is not very close to the true probability.

Example 5.8-1 If it is known that X has a mean of 25 and a variance of 16, then, since $\sigma = 4$, a lower bound for $P(17 < X < 33)$ is given by

$$P(17 < X < 33) = P(|X - 25| < 8)$$
$$= P(|X - 25| < 2\sigma) \geq 1 - \frac{1}{4} = 0.75$$

and an upper bound for $P(|X - 25| \geq 12)$ is found to be

$$P(|X - 25| \geq 12) = P(|X - \mu| \geq 3\sigma) \leq \frac{1}{9}. \qquad \blacksquare$$

Note that the results of the last example hold for any distribution with mean 25 and standard deviation 4. But, even stronger, the probability that any random variable X differs from its mean by 3 or more standard deviations is at most 1/9, which may be seen by letting $k = 3$ in the theorem. Also, the probability that any random variable X differs from its mean by less than two standard deviations is at least 3/4, which may be seen by letting $k = 2$.

The following consideration partially indicates the value of Chebyshev's inequality in theoretical discussions: If Y is the number of successes in n independent Bernoulli trials with probability p of success on each trial, then Y is $b(n, p)$. Furthermore, Y/n gives the relative frequency of success, and when p is unknown, Y/n can be used as an estimate of its mean p. To gain some insight into the closeness of Y/n to p, we shall use Chebyshev's inequality. With $\varepsilon > 0$, we note from Corollary 5.8-1 that, since $\text{Var}(Y/n) = pq/n$, it follows that

$$P\left(\left|\frac{Y}{n} - p\right| \geq \varepsilon\right) \leq \frac{pq/n}{\varepsilon^2}$$

or, equivalently,

$$P\left(\left|\frac{Y}{n} - p\right| < \varepsilon\right) \geq 1 - \frac{pq}{n\varepsilon^2}. \qquad (5.8\text{-}2)$$

On the one hand, when p is completely unknown, we can use the fact that $pq = p(1 - p)$ is a maximum when $p = 1/2$ in order to find a lower bound for the probability in Equation 5.8-2. That is,

$$1 - \frac{pq}{n\varepsilon^2} \geq 1 - \frac{(1/2)(1/2)}{n\varepsilon^2}.$$

For example, if $\varepsilon = 0.05$ and $n = 400$, then

$$P\left(\left|\frac{Y}{400} - p\right| < 0.05\right) \geq 1 - \frac{(1/2)(1/2)}{400(0.0025)} = 0.75.$$

On the other hand, if it is known that p is equal to 1/10, we would have

$$P\left(\left|\frac{Y}{400} - p\right| < 0.05\right) \geq 1 - \frac{(0.1)(0.9)}{400(0.0025)} = 0.91.$$

Note that Chebyshev's inequality is applicable to all distributions with a finite variance, and thus the bound is not always a tight one; that is, the bound is not necessarily close to the true probability.

In general, however, it should be noted that, with fixed $\varepsilon > 0$ and $0 < p < 1$, we have

$$\lim_{n \to \infty} P\left(\left|\frac{Y}{400} - p\right| < \varepsilon\right) \geq \lim_{n \to \infty} \left(1 - \frac{pq}{n\varepsilon^2}\right) = 1.$$

But since the probability of every event is less than or equal to 1, it must be that

$$\lim_{n\to\infty} P\left(\left|\frac{Y}{400} - p\right| < \varepsilon\right) = 1.$$

That is, the probability that the relative frequency Y/n is within ε of p is arbitrarily close to 1 when n is large enough. This is one form of the **law of large numbers**, and we say that Y/n **converges in probability** to p.

A more general form of the law of large numbers is found by considering the mean $\overline{X}$ of a random sample from a distribution with mean μ and variance σ^2. This form of the law is more general because the relative frequency Y/n can be thought of as $\overline{X}$ when the sample arises from a Bernoulli distribution. To derive it, we note that

$$E(\overline{X}) = \mu \qquad \text{and} \qquad \text{Var}(\overline{X}) = \frac{\sigma^2}{n}.$$

Thus, from Corollary 5.8-1, for every $\varepsilon > 0$, we have

$$P[\,|\overline{X} - \mu| \geq \varepsilon\,] \leq \frac{\sigma^2/n}{\varepsilon^2} = \frac{\sigma^2}{n\varepsilon^2}.$$

Since probability is nonnegative, it follows that

$$\lim_{n\to\infty} P(|\overline{X} - \mu| \geq \varepsilon) \leq \lim_{n\to\infty} \frac{\sigma^2}{\varepsilon^2 n} = 0,$$

which implies that

$$\lim_{n\to\infty} P(|\overline{X} - \mu| \geq \varepsilon) = 0,$$

or, equivalently,

$$\lim_{n\to\infty} P(|\overline{X} - \mu| < \varepsilon) = 1.$$

The preceding discussion shows that the probability associated with the distribution of $\overline{X}$ becomes concentrated in an arbitrarily small interval centered at μ as n increases. This is a more general form of the law of large numbers, and we say that $\overline{X}$ *converges in probability* to μ.

Exercises

5.8-1. If X is a random variable with mean 33 and variance 16, use Chebyshev's inequality to find

(a) A lower bound for $P(23 < X < 43)$.

(b) An upper bound for $P(|X - 33| \geq 14)$.

5.8-2. If $E(X) = 18$ and $E(X^2) = 333$, use Chebyshev's inequality to determine

(a) A lower bound for $P(9 \leq X \leq 27)$.

(b) An upper bound for $P(|X - 18| \geq 25)$.

5.8-3. Let X denote the outcome when a fair die is rolled. Then $\mu = 7/2$ and $\sigma^2 = 35/12$. Note that the maximum deviation of X from μ equals 5/2. Express this deviation in terms of the number of standard deviations; that is,

find k, where $k\sigma = 5/2$. Determine a lower bound for $P(|X - 3.5| < 2.5)$.

5.8-4. If the distribution of Y is $b(n, 0.5)$, give a lower bound for $P(|Y/n - 0.5| < 0.08)$ when

(a) $n = 100$.

(b) $n = 500$.

(c) $n = 1000$.

5.8-5. If the distribution of Y is $b(n, 0.25)$, give a lower bound for $P(|Y/n - 0.25| < 0.05)$ when

(a) $n = 100$.

(b) $n = 500$.

(c) $n = 1000$.

5.8-6. Let $\overline{X}$ be the mean of a random sample of size $n = 19$ from a distribution with mean $\mu = 98$ and variance $\sigma^2 = 72$. Use Chebyshev's inequality to find a lower bound for $P(93 < \overline{X} < 103)$.

5.8-7. Suppose that W is a continuous random variable with mean 0 and a symmetric pdf $f(w)$ and cdf $F(w)$, but for which the variance is not specified (and may not exist). Suppose further that W is such that

$$P(|W - 0| < k) = 1 - \frac{1}{k^2}$$

for $k \geq 1$. (Note that this equality would be equivalent to the equality in Chebyshev's inequality if the variance of W were equal to 1.) Then the cdf satisfies

$$F(w) - F(-w) = 1 - \frac{1}{w^2}, \qquad w \geq 1.$$

Also, the symmetry assumption implies that

$$F(-w) = 1 - F(w).$$

(a) Show that the pdf of W is

$$f(w) = \begin{cases} \dfrac{1}{|w|^3}, & |w| > 1, \\[2mm] 0, & |w| \leq 1. \end{cases}$$

(b) Find the mean and the variance of W and interpret your results.

(c) Graph the cdf of W.

5.9 LIMITING MOMENT-GENERATING FUNCTIONS

We begin this section by showing that the binomial distribution can be approximated by the Poisson distribution when n is sufficiently large and p is fairly small. Of course, we proved this in Section 2.6 by showing that, under these conditions, the binomial pmf is close to that of the Poisson. Here, however, we show that the binomial mgf is close to that of the Poisson distribution. We do so by taking the limit of a mgf.

Consider the mgf of Y, which is $b(n, p)$. We shall take the limit of this function as $n \to \infty$ such that $np = \lambda$ is a constant; thus, $p \to 0$. The mgf of Y is

$$M(t) = (1 - p + pe^t)^n.$$

Because $p = \lambda/n$, we have

$$M(t) = \left[1 - \frac{\lambda}{n} + \frac{\lambda}{n}e^t\right]^n$$

$$= \left[1 + \frac{\lambda(e^t - 1)}{n}\right]^n.$$

Since

$$\lim_{n \to \infty} \left(1 + \frac{b}{n}\right)^n = e^b,$$

we have

$$\lim_{n \to \infty} M(t) = e^{\lambda(e^t - 1)},$$

which exists for all real t. But this is the mgf of a Poisson random variable with mean λ. Hence, this Poisson distribution seems like a reasonable approximation to the binomial distribution when n is large and p is small. That approximation is usually found to be fairly successful if $n \geq 20$ and $p \leq 0.05$ and is found to be very successful if $n \geq 100$ and $p \leq 0.10$, but it is not bad if these bounds are violated somewhat. That is, the approximation could be used in other situations, too; we only want to stress that it becomes better with larger n and smaller p.

The preceding result illustrates the theorem we now state without proof.

> **Theorem 5.9-1**   If a sequence of mgfs approaches a certain mgf, say, $M(t)$, for t in an open interval around 0, then the limit of the corresponding distributions must be the distribution corresponding to $M(t)$.

REMARK This theorem certainly appeals to one's intuition! In a more advanced course, the theorem is proven, and there the existence of the mgf is not even needed, for we would use the characteristic function $\phi(t) = E(e^{itX})$ instead. ■

The next example illustrates graphically the convergence of the binomial mgfs to that of a Poisson distribution.

Example 5.9-1   Consider the mgf for the Poisson distribution with $\lambda = 5$ and those for three binomial distributions for which $np = 5$, namely, $b(50, 1/10)$, $b(100, 1/20)$, and $b(200, 1/40)$. These four mgfs are, respectively,

$$M(t) = e^{5(e^t - 1)}, \qquad -\infty < t < \infty,$$
$$M(t) = (0.9 + 0.1e^t)^{50}, \qquad -\infty < t < \infty,$$
$$M(t) = (0.75 + 0.25e^t)^{20}, \qquad -\infty < t < \infty,$$
$$M(t) = (0.5 + 0.5e^t)^{10}, \qquad -\infty < t < \infty.$$

The graphs of these mgfs are shown in Figure 5.9-1. Although the proof and the figure show the convergence of the binomial mgfs to that of the Poisson distribution,

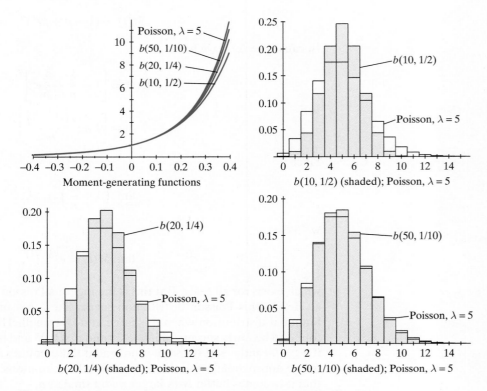

Figure 5.9-1 Poisson approximation to the binomial distribution

the last three graphs in Figure 5.9-1 show more clearly how the Poisson distribution can be used to approximate binomial probabilities with large n and small p. ■

The next example gives a numerical approximation.

Example 5.9-2

Let Y be $b(50, 1/25)$. Then

$$P(Y \leq 1) = \left(\frac{24}{25}\right)^{50} + 50\left(\frac{1}{25}\right)\left(\frac{24}{25}\right)^{49} = 0.400.$$

Since $\lambda = np = 2$, the Poisson approximation is

$$P(Y \leq 1) \approx 0.406,$$

from Table III in Appendix B, or note that $3e^{-2} = 0.406$. ■

Theorem 5.9-1 is used to prove the central limit theorem. To help in understanding this proof, let us first consider a different problem: that of the limiting distribution of the mean $\overline{X}$ of a random sample $X_1, X_2, \ldots, X_n$ from a distribution with mean μ. If the distribution has mgf $M(t)$, then the mgf of $\overline{X}$ is $[M(t/n)]^n$. But, by Taylor's expansion, there exists a number t_1 between 0 and t/n such that

$$M\left(\frac{t}{n}\right) = M(0) + M'(t_1)\frac{t}{n}$$

$$= 1 + \frac{\mu t}{n} + \frac{[M'(t_1) - M'(0)]t}{n},$$

because $M(0) = 1$ and $M'(0) = \mu$. Since $M'(t)$ is continuous at $t = 0$ and since $t_1 \to 0$ as $n \to \infty$, it follows that

$$\lim_{n \to \infty} [M'(t_1) - M'(0)] = 0.$$

Thus, using a result from advanced calculus, we obtain

$$\lim_{n \to \infty}\left[M\left(\frac{t}{n}\right)\right]^n = \lim_{n \to \infty}\left\{1 + \frac{\mu t}{n} + \frac{[M'(t_1) - M'(0)]t}{n}\right\}^n = e^{\mu t}$$

for all real t. But this limit is the mgf of a degenerate distribution with all of the probability on μ. Accordingly, $\overline{X}$ has this limiting distribution, indicating that $\overline{X}$ converges to μ in a certain sense. This is one form of the law of large numbers.

We have seen that, in some probability sense, $\overline{X}$ converges to μ in the limit, or, equivalently, $\overline{X} - \mu$ converges to zero. Let us multiply the difference $\overline{X} - \mu$ by some function of n so that the result will not converge to zero. In our search for such a function, it is natural to consider

$$W = \frac{\overline{X} - \mu}{\sigma/\sqrt{n}} = \frac{\sqrt{n}(\overline{X} - \mu)}{\sigma} = \frac{Y - n\mu}{\sqrt{n}\,\sigma},$$

where Y is the sum of the observations of the random sample. The reason for this is that, by the remark after the proof of Theorem 3.3-1, W is a standardized random variable. That is, W has mean 0 and variance 1 for each positive integer n. We are now ready to prove the central limit theorem, which is stated in Section 5.6.

Proof (of the Central Limit Theorem):
We first consider

$$E[\exp(tW)] = E\left\{\exp\left[\left(\frac{t}{\sqrt{n}\sigma}\right)\left(\sum_{i=1}^{n} X_i - n\mu\right)\right]\right\}$$

$$= E\left\{\exp\left[\left(\frac{t}{\sqrt{n}}\right)\left(\frac{X_1 - \mu}{\sigma}\right)\right]\cdots\exp\left[\left(\frac{t}{\sqrt{n}}\right)\left(\frac{X_n - \mu}{\sigma}\right)\right]\right\}$$

$$= E\left\{\exp\left[\left(\frac{t}{\sqrt{n}}\right)\left(\frac{X_1 - \mu}{\sigma}\right)\right]\right\}\cdots E\left\{\exp\left[\left(\frac{t}{\sqrt{n}}\right)\left(\frac{X_n - \mu}{\sigma}\right)\right]\right\},$$

which follows from the independence of $X_1, X_2, \ldots, X_n$. Then

$$E[\exp(tW)] = \left[m\left(\frac{t}{\sqrt{n}}\right)\right]^n, \qquad -h < \frac{t}{\sqrt{n}} < h,$$

where

$$m(t) = E\left\{\exp\left[t\left(\frac{X_i - \mu}{\sigma}\right)\right]\right\}, \qquad -h < t < h,$$

is the common mgf of each

$$Y_i = \frac{X_i - \mu}{\sigma}, \qquad i = 1, 2, \ldots, n.$$

Since $E(Y_i) = 0$ and $E(Y_i^2) = 1$, it must be that

$$m(0) = 1, \qquad m'(0) = E\left(\frac{X_i - \mu}{\sigma}\right) = 0, \qquad m''(0) = E\left[\left(\frac{X_i - \mu}{\sigma}\right)^2\right] = 1.$$

Hence, using Taylor's formula with a remainder, we know that there exists a number t_1 between 0 and t such that

$$m(t) = m(0) + m'(0)t + \frac{m''(t_1)t^2}{2} = 1 + \frac{m''(t_1)t^2}{2}.$$

Adding and subtracting $t^2/2$, we have

$$m(t) = 1 + \frac{t^2}{2} + \frac{[m''(t_1) - 1]t^2}{2}.$$

Using this expression of $m(t)$ in $E[\exp(tW)]$, we can represent the mgf of W by

$$E[\exp(tW)] = \left\{1 + \frac{1}{2}\left(\frac{t}{\sqrt{n}}\right)^2 + \frac{1}{2}[m''(t_1) - 1]\left(\frac{t}{\sqrt{n}}\right)^2\right\}^n$$

$$= \left\{1 + \frac{t^2}{2n} + \frac{[m''(t_1) - 1]t^2}{2n}\right\}^n, \qquad -\sqrt{n}\,h < t < \sqrt{n}\,h,$$

where now t_1 is between 0 and $t/\sqrt{n}$. Since $m''(t)$ is continuous at $t = 0$ and $t_1 \to 0$ as $n \to \infty$, we have

$$\lim_{n \to \infty}[m''(t_1) - 1] = 1 - 1 = 0.$$

Thus, using a result from advanced calculus, we obtain

$$\lim_{n \to \infty} E[\exp(tW)] = \lim_{n \to \infty} \left\{ 1 + \frac{t^2}{2n} + \frac{[m''(t_1) - 1]t^2}{2n} \right\}^n$$

$$= \lim_{n \to \infty} \left\{ 1 + \frac{t^2}{2n} \right\}^n = e^{t^2/2}$$

for all real t. We know that $e^{t^2/2}$ is the mgf of the standard normal distribution, $N(0, 1)$. It then follows that the limiting distribution of

$$W = \frac{\overline{X} - \mu}{\sigma/\sqrt{n}} = \frac{\sum_{i=1}^{n} X_i - n\mu}{\sqrt{n}\,\sigma}$$

is $N(0, 1)$. This completes the proof of the central limit theorem. □

Examples of the use of the central limit theorem as an approximating distribution were given in Sections 5.6 and 5.7.

To help appreciate the proof of the central limit theorem, the next example graphically illustrates the convergence of the mgfs for two distributions.

Example 5.9-3 Let $X_1, X_2, \ldots, X_n$ be a random sample of size n from an exponential distribution with $\theta = 2$. The mgf of $(\overline{X} - \theta)/(\theta/\sqrt{n})$ is

$$M_n(t) = \frac{e^{-t\sqrt{n}}}{(1 - t/\sqrt{n})^n}, \qquad t < \sqrt{n}.$$

The central limit theorem says that, as n increases, this mgf approaches that of the standard normal distribution, namely,

$$M(t) = e^{t^2/2}.$$

The mgfs for $M(t)$ and $M_n(t), n = 5, 15, 50$, are shown in Figure 5.9-2(a). [See also Figure 5.6-2, in which samples were taken from a $\chi^2(1)$ distribution, and recall that the exponential distribution with $\theta = 2$ is $\chi^2(2)$.]

In Example 5.6-6, a U-shaped distribution was considered for which the pdf is $f(x) = (3/2)x^2, -1 < x < 1$. For this distribution, $\mu = 0$ and $\sigma^2 = 3/5$. Its mgf, for $t \neq 0$, is

$$M(t) = \left(\frac{3}{2} \right) \frac{e^t t^2 - 2e^t t + 2e^t - e^{-t} t^2 - 2e^{-t} t - 2e^{-t}}{t^3}.$$

Of course, $M(0) = 1$. The mgf of

$$W_n = \frac{\overline{X} - 0}{\sqrt{3/(5n)}}$$

is

$$E[e^{tW_n}] = \left\{ E\left[\exp\left(\sqrt{\frac{5}{3n}}\,t \right) \right] \right\}^n = \left[M\left(\sqrt{\frac{5}{3n}}\,t \right) \right]^n.$$

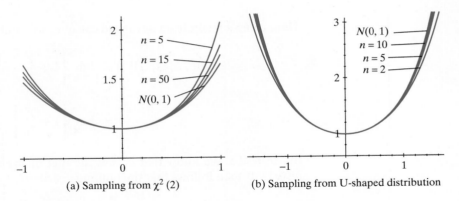

(a) Sampling from χ^2 (2)   (b) Sampling from U-shaped distribution

Figure 5.9-2 Convergence of moment-generating functions

The graphs of these mgfs when $n = 2, 5, 10$ and the graph of the mgf for the standard normal distribution are shown in Figure 5.9-2(b). Note how much more quickly the mgfs converge compared with those for the exponential distribution. ∎

Exercises

5.9-1. Let Y be the number of defectives in a box of 50 articles taken from the output of a machine. Each article is defective with probability 0.01. Find the probability that $Y = 0, 1, 2,$ or 3

(a) By using the binomial distribution.

(b) By using the Poisson approximation.

5.9-2. The probability that a certain type of inoculation takes effect is 0.825. Use the Poisson distribution to approximate the probability that at most 3 out of 350 people given the inoculation find that it has not taken effect. HINT: Let $p = 1 - 0.825 = 0.175$.

5.9-3. Let S^2 be the sample variance of a random sample of size n from $N(\mu, \sigma^2)$. Show that the limit, as $n \to \infty$, of the mgf of S^2 is $e^{\sigma^2 t}$. Thus, in the limit, the distribution of S^2 is degenerate with probability 1 at σ^2.

5.9-4. Let Y be $\chi^2(n)$. Use the central limit theorem to demonstrate that $W = (Y - n)/\sqrt{2n}$ has a limiting cdf that is $N(0, 1)$. HINT: Think of Y as being the sum of a random sample from a certain distribution.

5.9-5. Let Y have a Poisson distribution with mean $3n$. Use the central limit theorem to show that the limiting distribution of $W = (Y - 3n)/\sqrt{3n}$ is $N(0, 1)$.

HISTORICAL COMMENTS   In this chapter, we have discussed the t and F distributions, among many other important ones. However, we should make a few comments about both of them. W. S. Gosset published his work on the t distribution under the pseudonym "A Student" because Guinness did not want other breweries to know that they were using statistical methods. We have also heard the story that Gosset did not want Guinness to know that he was spending all his extra time on statistics; so he used "A Student," and it has become Student's t ever since. Whichever account is true, Gosset, in a sense, was lucky to discover the t distribution because he made two educated guesses, one of which involved the Pearson family of distributions. Incidentally, Karl Pearson, another famous statistician, had proposed this family a few years earlier. A few years later, the great statistician Sir Ronald A. Fisher (possibly the greatest statistician) actually proved that T had the t distribution that Gosset had discovered. Concerning the F distribution, Fisher had worked with a function of what is now called F. It was George Snedecor of Iowa State University who put it in its present form and called it F (probably to honor Fisher), and Snedecor's was a much more useful form, as we will see later. We should also note that Fisher had been knighted, but so were three other statisticians: Sir Maurice G. Kendall, Sir David R. Cox, and Sir Adrian Smith. (The latter two are still alive.) Their knighthood at least proves that the monarch of England appreciated some statistical efforts.

Another important person in the history of probability is Abraham de Moivre, who was born in France in 1667, but, as a Protestant, he did not fare very well in that Catholic country. As a matter of fact, he was imprisoned for about two years for his beliefs. After his release, he went to England, but led a gloomy life there, as he could not find an academic position. So de Moivre supported himself by tutoring or consulting with gamblers or insurance brokers. After publishing *The Doctrine of Chance*, he turned to a project that Nicolaus Bernoulli had suggested to him. Using the fact that Y is $b(n, p)$, he discovered that the relative frequency of successes, namely, Y/n, which Jacob Bernoulli had proved converged in probability to p, had an interesting approximating distribution itself. De Moivre had discovered the well-known bell-shaped curve called the normal distribution, and a special case of the central limit theorem. Although de Moivre did not have computers in his day, we show that if X is $b(100, 1/2)$, then $X/100$ has the pmf displayed in Figure 5.9-3.

This distribution allowed de Moivre to determine a measure of spread, which we now call a standard deviation. Also, he could determine the approximate probability of Y/n falling into given intervals containing p. De Moivre was truly impressed with the orderliness of these random relative frequencies, which he attributed to the plan of the Almighty. Despite his great works, Abraham de Moivre died a bitter and antisocial man, blind and in poverty, at the age of 87.

Two additional persons whom we would like to mention in the history of probability are Carl Friedrich Gauss and Marquis Pierre Simon de Laplace. Gauss was 29 years junior to Laplace, and Gauss was so secretive about his work that it is difficult to tell who discovered the central limit theorem first. The theorem was a generalization of de Moivre's result. In de Moivre's case, he was sampling from a Bernoulli distribution where $X_i = 1$ or 0 on the ith trial. Then

$$\frac{Y}{n} = \sum_{i=1}^{n} \frac{X_i}{n} = \overline{X},$$

the relative frequency of success, has that approximate normal distribution of de Moivre's. Laplace and Gauss were sampling from any distribution, provided that the second moment existed, and they found that the sample mean $\overline{X}$ had an approximate normal distribution. Seemingly, the central limit theorem was published in 1809 by

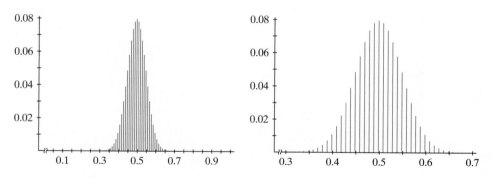

Figure 5.9-3 Line graphs for the distribution of $X/100$

Laplace, just before Gauss's *Theoria Motus* in 1810. For some reason, the normal distribution is often referred to as the Gaussian distribution; people seem to forget about Laplace's contribution and (worse than that) de Moivre's original work 83 years earlier. Since then, there have been many more generalizations of the central limit theorem; in particular, most estimators of parameters in the regular case have approximate normal distributions.

There are many good histories of probability and statistics. However, the two that we find particularly interesting are Peter L. Bernstein's *Against the Gods: The Remarkable Story of Risk* (New York: John Wiley & Sons, Inc., 1996) and Stephen M. Stigler's *The History of Statistics: The Measurement of Uncertainty Before 1900* (Cambridge, MA: Harvard University Press, 1986).

The reason we try to model any random phenomenon with a probability distribution is that if our model is reasonably good, we know the approximate percentages of being above or below certain marks. Having such information helps us make certain decisions—sometimes very important ones.

With these models, we learned how to simulate random variables having certain distributions. In many situations in practice, we cannot calculate exact solutions of equations that have numerous random variables. Thus, we simulate the random variables in question many times, leading to an approximate distribution of the random solution. "Monte Carlo" is a term often attached to such a simulation, and we believe that it was first used in a computer simulation of nuclear fission associated with the atom bomb in World War II. Of course, the name "Monte Carlo" was taken from that city, which is famous for gambling in its casinos.

POINT ESTIMATION

6.1 DESCRIPTIVE STATISTICS

In Chapter 2, we considered probability distributions of random variables whose space S contains a countable number of outcomes: either a finite number of outcomes or outcomes that can be put into a one-to-one correspondence with the positive integers. Such a random variable is said to be of the **discrete type**, and its distribution of probabilities is of the discrete type.

Of course, many experiments or observations of random phenomena do not have integers or other discrete numbers as outcomes, but instead are measurements selected from an interval of numbers. For example, you could find the length of time that it takes when waiting in line to buy frozen yogurt. Or the weight of a "1-pound" package of hot dogs could be any number between 0.94 pounds and 1.25 pounds. The weight of a miniature Baby Ruth candy bar could be any number between 20 and 27 grams. Even though such times and weights could be selected from an interval of values, times and weights are generally rounded off so that the data often look like discrete data. If, conceptually, the measurements could come from an interval of possible outcomes, we call them data from a distribution of the continuous type or, more simply, **continuous-type data**.

Given a set of continuous-type data, we shall group the data into classes and then construct a histogram of the grouped data. This will help us better visualize the data. The following guidelines and terminology will be used to group continuous-type data into classes of equal length (these guidelines can also be used for sets of discrete data that have a large range).

1. Determine the largest (maximum) and smallest (minimum) observations. The **range** is the difference, $R =$ maximum − minimum.

2. In general, select from $k = 5$ to $k = 20$ classes, which are nonoverlapping intervals, usually of equal length. These classes should cover the interval from the minimum to the maximum.

3. Each interval begins and ends halfway between two possible values of the measurements, which have been rounded off to a given number of decimal places.

4. The first interval should begin about as much below the smallest value as the last interval ends above the largest.

5. The intervals are called **class intervals** and the boundaries are called **class boundaries**. We shall denote these k class intervals by

$$(c_0, c_1], (c_1, c_2], \ldots, (c_{k-1}, c_k].$$

6. The **class limits** are the smallest and the largest possible observed (recorded) values in a class.

7. The **class mark** is the midpoint of a class.

A frequency table is constructed that lists the class intervals, the class limits, a tabulation of the measurements in the various classes, the frequency f_i of each class, and the class marks. A column is sometimes used to construct a relative frequency (density) histogram. With class intervals of equal length, a frequency histogram is constructed by drawing, for each class, a rectangle having as its base the class interval and a height equal to the frequency of the class. For the relative frequency histogram, each rectangle has an **area** equal to the relative frequency f_i/n of the observations for the class. That is, the function defined by

$$h(x) = \frac{f_i}{(n)(c_i - c_{i-1})}, \quad \text{for } c_{i-1} < x \le c_i, \quad i = 1, 2, \ldots, k,$$

is called a **relative frequency histogram** or **density histogram**, where f_i is the frequency of the ith class and n is the total number of observations. Clearly, if the class intervals are of equal length, the relative frequency histogram, $h(x)$, is proportional to the **frequency histogram** f_i, for $c_{i-1} < x \le c_i, i = 1, 2, \ldots, k$. The frequency histogram should be used only in those situations in which the class intervals are of equal length. A relative frequency histogram can be treated as an estimate of the underlying pdf.

Example 6.1-1

The weights in grams of 40 miniature Baby Ruth candy bars, with the weights ordered, are given in Table 6.1-1.

We shall group these data and then construct a histogram to visualize the distribution of weights. The range of the data is $R = 26.7 - 20.5 = 6.2$. The interval $(20.5, 26.7)$ could be covered with $k = 8$ classes of width 0.8 or with $k = 9$ classes of width 0.7. (There are other possibilities.) We shall use $k = 7$ classes of width 0.9. The first class interval will be $(20.45, 21.35)$ and the last class interval will be $(25.85, 26.75)$. The data are grouped in Table 6.1-2.

A relative frequency histogram of these data is given in Figure 6.1-1. Note that the total area of this histogram is equal to 1. We could also construct a frequency histogram in which the heights of the rectangles would be equal to the frequencies of the classes. The shape of the two histograms is the same. Later we will see

Table 6.1-1 Candy bar weights									
20.5	20.7	20.8	21.0	21.0	21.4	21.5	22.0	22.1	22.5
22.6	22.6	22.7	22.7	22.9	22.9	23.1	23.3	23.4	23.5
23.6	23.6	23.6	23.9	24.1	24.3	24.5	24.5	24.8	24.8
24.9	24.9	25.1	25.1	25.2	25.6	25.8	25.9	26.1	26.7

Table 6.1-2 Frequency table of candy bar weights

Class Interval	Class Limits	Tabulation	Frequency (f_i)	$h(x)$	Class Marks
(20.45, 21.35)	20.5–21.3	⦀⦀	5	5/36	20.9
(21.35, 22.25)	21.4–22.2	‖‖	4	4/36	21.8
(22.25, 23.15)	22.3–23.1	⦀⦀ ‖‖	8	8/36	22.7
(23.15, 24.05)	23.2–24.0	⦀⦀ ‖	7	7/36	23.6
(24.05, 24.95)	24.1–24.9	⦀⦀ ‖‖	8	8/36	24.5
(24.95, 25.85)	25.0–25.8	⦀⦀	5	5/36	25.4
(25.85, 26.75)	25.9–26.7	‖‖	3	3/36	26.3

the reason for preferring the relative frequency histogram. In particular, we will be superimposing on the relative frequency histogram the graph of a pdf. ∎

Suppose that we now consider the situation in which we actually perform a certain random experiment n times, obtaining n observed values of the random variable—say, $x_1, x_2, \ldots, x_n$. Often the collection is referred to as a **sample**. It is possible that some of these values might be the same, but we do not worry about this at this time. We artificially create a probability distribution by placing the weight $1/n$ on each of these x-values. Note that these weights are positive and sum to 1, so we have a distribution we call the **empirical distribution**, since it is determined by the data $x_1, x_2, \ldots, x_n$. The mean of the empirical distribution is

$$\sum_{i=1}^{n} x_i \left(\frac{1}{n} \right) = \frac{1}{n} \sum_{i=1}^{n} x_i,$$

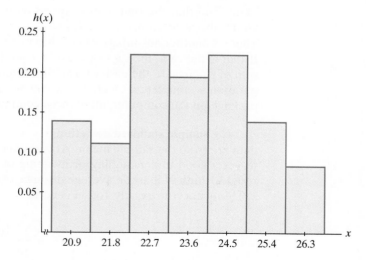

Figure 6.1-1 Relative frequency histogram of weights of candy bars

which is the arithmetic mean of the observations $x_1, x_2, \ldots, x_n$. We denote this mean by $\overline{x}$ and call it the **sample mean** (or mean of the sample $x_1, x_2, \ldots, x_n$). That is, the sample mean is

$$\overline{x} = \frac{1}{n} \sum_{i=1}^{n} x_i,$$

which is, in some sense, an estimate of μ if the latter is unknown.

Likewise, the **variance of the empirical distribution** is

$$v = \sum_{i=1}^{n} (x_i - \overline{x})^2 \left(\frac{1}{n}\right) = \frac{1}{n} \sum_{i=1}^{n} (x_i - \overline{x})^2,$$

which can be written as

$$v = \sum_{i=1}^{n} x_i^2 \left(\frac{1}{n}\right) - \overline{x}^2 = \frac{1}{n} \sum_{i=1}^{n} x_i^2 - \overline{x}^2,$$

that is, the second moment about the origin minus the square of the mean. However, v is not called the sample variance, but

$$s^2 = \left[\frac{n}{n-1}\right] v = \frac{1}{n-1} \sum_{i=1}^{n} (x_i - \overline{x})^2$$

is, because we will see later that, in some sense, s^2 is a better estimate of an unknown σ^2 than is v. Thus, the **sample variance** is

$$s^2 = \frac{1}{n-1} \sum_{i=1}^{n} (x_i - \overline{x})^2.$$

REMARK It is easy to expand the sum of squares; we have

$$\sum_{i=1}^{n} (x_i - \overline{x})^2 = \sum_{i=1}^{n} x_i^2 - \frac{\left(\sum_{i=1}^{n} x_i\right)^2}{n}.$$

Many find that the right-hand expression makes the computation easier than first taking the n differences, $x_i - \overline{x}$, $i = 1, 2, \ldots, n$; squaring them; and then summing. There is another advantage when $\overline{x}$ has many digits to the right of the decimal point. If that is the case, then $x_i - \overline{x}$ must be rounded off, and that creates an error in the sum of squares. In the easier form, that rounding off is not necessary until the computation is completed. Of course, if you are using a statistical calculator or statistics package on the computer, all of these computations are done for you. ■

The **sample standard deviation**, $s = \sqrt{s^2} \geq 0$, is a measure of how dispersed the data are from the sample mean. At this stage of your study of statistics, it is difficult to get a good understanding or meaning of the standard deviation s, but you can roughly think of it as the average distance of the values $x_1, x_2, \ldots, x_n$ from the mean $\overline{x}$. This is not true exactly, for, in general,

$$s \geq \frac{1}{n} \sum_{i=1}^{n} |x_i - \overline{x}|,$$

but it is fair to say that s is somewhat larger, yet of the same magnitude, as the average of the distances of $x_1, x_2, \ldots, x_n$ from $\overline{x}$.

Example 6.1-2

Rolling a fair six-sided die five times could result in the following sample of $n = 5$ observations:

$$x_1 = 3, \quad x_2 = 1, \quad x_3 = 2, \quad x_4 = 6, \quad x_5 = 3.$$

In this case,

$$\bar{x} = \frac{3 + 1 + 2 + 6 + 3}{5} = 3$$

and

$$s^2 = \frac{(3-3)^2 + (1-3)^2 + (2-3)^2 + (6-3)^2 + (3-3)^2}{4} = \frac{14}{4} = 3.5.$$

It follows that $s = \sqrt{14/4} = 1.87$. We had noted that s can roughly be thought of as the average distance that the x-values are away from the sample mean $\bar{x}$. In this example, the distances from the sample mean, $\bar{x} = 3$, are 0, 2, 1, 3, 0, with an average of 1.2, which is less than $s = 1.87$. In general, s will be somewhat larger than this average distance. ∎

There is an alternative way of computing s^2, because $s^2 = [n/(n-1)]v$ and

$$v = \frac{1}{n} \sum_{i=1}^{n} (x_i - \bar{x})^2 = \frac{1}{n} \sum_{i=1}^{n} x_i^2 - \bar{x}^2.$$

It follows that

$$s^2 = \frac{\sum_{i=1}^{n} x_i^2 - n\bar{x}^2}{n-1} = \frac{\sum_{i=1}^{n} x_i^2 - \frac{1}{n}\left(\sum_{i=1}^{n} x_i\right)^2}{n-1}.$$

Given a set of measurements, the sample mean is the center of the data such that the deviations from that center sum to zero; that is, $\sum_{i=1}^{n}(x_i - \bar{x}) = 0$, where $x_1, x_2, \ldots, x_n$ and $\bar{x}$ are a given set of observations of $X_1, X_2, \ldots, X_n$ and $\bar{X}$. The sample standard deviation s, an observed value of S, gives a measure of how spread out the data are from the sample mean. If the histogram is "mound-shaped" or "bell-shaped," the following empirical rule gives rough approximations to the percentages of the data that fall between certain points. These percentages clearly are associated with the normal distribution.

Empirical Rule: Let $x_1, x_2, \ldots, x_n$ have a sample mean $\bar{x}$ and sample standard deviation s. If the histogram of these data is "bell-shaped," then, for large samples,

- approximately 68% of the data are in the interval $(\bar{x} - s, \bar{x} + s)$,
- approximately 95% of the data are in the interval $(\bar{x} - 2s, \bar{x} + 2s)$,
- approximately 99.7% of the data are in the interval $(\bar{x} - 3s, \bar{x} + 3s)$.

For the data in Example 6.1-1, the sample mean is $\bar{x} = 23.505$ and the standard deviation is $s = 1.641$. The number of weights that fall within one standard deviation of the mean, $(23.505 - 1.641, 23.505 + 1.641)$, is 27, or 67.5%. For these particular weights, 100% fall within two standard deviations of $\bar{x}$. Thus, the histogram is missing part of the "bell" in the tails in order for the empirical rule to hold.

When you draw a histogram, it is useful to indicate the location of $\bar{x}$, as well as that of the points $\bar{x} \pm s$ and $\bar{x} \pm 2s$.

There is a refinement of the relative frequency histogram that can be made when the class intervals are of equal length. The **relative frequency polygon** smooths out the corresponding histogram somewhat. To form such a polygon, mark the midpoints at the top of each "bar" of the histogram. Connect adjacent midpoints with straight-line segments. On each of the two end bars, draw a line segment from the top middle mark through the middle point of the outer vertical line of the bar. Of course, if the area underneath the tops of the relative frequency histogram is equal to 1, which it should be, then the area underneath the relative frequency polygon is also equal to 1, because the areas lost and gained cancel out by a consideration of congruent triangles. This idea is made clear in the next example.

Example 6.1-3

A manufacturer of fluoride toothpaste regularly measures the concentration of fluoride in the toothpaste to make sure that it is within the specification of 0.85 to 1.10 mg/g. Table 6.1-3 lists 100 such measurements.

The minimum of these measurements is 0.85 and the maximum is 1.06. The range is $1.06 - 0.85 = 0.21$. We shall use $k = 8$ classes of length 0.03. Note that $8(0.03) = 0.24 > 0.21$. We start at 0.835 and end at 1.075. These boundaries are the same distance below the minimum and above the maximum. In Table 6.1-4, we also give the values of the heights of each rectangle in the relative frequency histogram, so that the total area of the histogram is 1. These heights are given by the formula

$$h(x) = \frac{f_i}{(0.03)(100)} = \frac{f_i}{3}.$$

The plots of the relative frequency histogram and polygon are given in Figure 6.1-2.

If you are using a computer program to analyze a set of data, it is very easy to find the sample mean, the sample variance, and the sample standard deviation. However, if you have only grouped data or if you are not using a computer, you can obtain close approximations of these values by computing the mean $\bar{u}$ and

Table 6.1-3 Concentrations of fluoride in mg/g in toothpaste

0.98	0.92	0.89	0.90	0.94	0.99	0.86	0.85	1.06	1.01
1.03	0.85	0.95	0.90	1.03	0.87	1.02	0.88	0.92	0.88
0.88	0.90	0.98	0.96	0.98	0.93	0.98	0.92	1.00	0.95
0.88	0.90	1.01	0.98	0.85	0.91	0.95	1.01	0.88	0.89
0.99	0.95	0.90	0.88	0.92	0.89	0.90	0.95	0.93	0.96
0.93	0.91	0.92	0.86	0.87	0.91	0.89	0.93	0.93	0.95
0.92	0.88	0.87	0.98	0.98	0.91	0.93	1.00	0.90	0.93
0.89	0.97	0.98	0.91	0.88	0.89	1.00	0.93	0.92	0.97
0.97	0.91	0.85	0.92	0.87	0.86	0.91	0.92	0.95	0.97
0.88	1.05	0.91	0.89	0.92	0.94	0.90	1.00	0.90	0.93

Table 6.1-4 Frequency table of fluoride concentrations

Class Interval	Class Mark (u_i)	Tabulation	Frequency (f_i)	$h(x) = f_i/3$
$(0.835, 0.865)$	0.85	卌 ‖	7	7/3
$(0.865, 0.895)$	0.88	卌 卌 卌 卌	20	20/3
$(0.895, 0.925)$	0.91	卌 卌 卌 卌 卌 ‖	27	27/3
$(0.925, 0.955)$	0.94	卌 卌 卌 ‖‖	18	18/3
$(0.955, 0.985)$	0.97	卌 卌 ‖‖‖	14	14/3
$(0.985, 1.015)$	1.00	卌 ‖‖‖	9	9/3
$(1.015, 1.045)$	1.03	‖‖	3	3/3
$(1.045, 1.075)$	1.06	‖	2	2/3

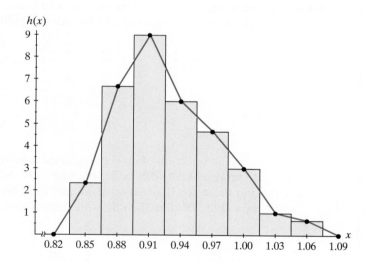

Figure 6.1-2 Concentrations of fluoride in toothpaste

variance s_u^2 of the grouped data, using the class marks weighted with their respective frequencies. We have

$$\overline{u} = \frac{1}{n} \sum_{i=1}^{k} f_i u_i$$

$$= \frac{1}{100} \sum_{i=1}^{8} f_i u_i = \frac{92.83}{100} = 0.9283,$$

$$s_u^2 = \frac{1}{n-1} \sum_{i=1}^{k} f_i(u_i - \bar{u})^2 = \frac{\sum_{i=1}^{k} f_i u_i^2 - \frac{1}{n}\left(\sum_{i=1}^{k} f_i u_i\right)^2}{n-1}$$

$$= \frac{0.237411}{99} = 0.002398.$$

Thus,

$$s_u = \sqrt{0.002398} = 0.04897.$$

These results compare rather favorably with $\bar{x} = 0.9293$ and $s_x = 0.04895$ of the original data. ∎

In some situations, it is not necessarily desirable to use class intervals of equal widths in the construction of the frequency distribution and histogram. This is particularly true if the data are skewed with a very long tail. We now present an illustration in which it seems desirable to use class intervals of unequal widths; thus, we cannot use the relative frequency polygon.

Example 6.1-4

The following 40 losses, due to wind-related catastrophes, were recorded to the nearest \$1 million (these data include only losses of \$2 million or more; for convenience, they have been ordered and recorded in millions):

2	2	2	2	2	2	2	2	2	2
2	2	3	3	3	3	4	4	4	5
5	5	5	6	6	6	6	8	8	9
15	17	22	23	24	24	25	27	32	43

The selection of class boundaries is more subjective in this case. It makes sense to let $c_0 = 1.5$ and $c_1 = 2.5$ because only values of \$2 million or more are recorded and there are 12 observations equal to 2. We could then let $c_2 = 6.5$, $c_3 = 29.5$, and $c_4 = 49.5$, yielding the following relative frequency histogram:

$$h(x) = \begin{cases} \dfrac{12}{40}, & 1.5 < x \le 2.5, \\[2mm] \dfrac{15}{(40)(4)}, & 2.5 < x \le 6.5, \\[2mm] \dfrac{11}{(40)(23)}, & 6.5 < x \le 29.5, \\[2mm] \dfrac{2}{(40)(20)}, & 29.5 < x \le 49.5. \end{cases}$$

This histogram is displayed in Figure 6.1-3. It takes some experience before a person can display a relative frequency histogram that is most meaningful.

The areas of the four rectangles—0.300, 0.375, 0.275, and 0.050—are the respective relative frequencies. It is important to note in the case of unequal widths among class intervals that the *areas*, not the heights, of the rectangles are proportional to the frequencies. In particular, the first and second classes have frequencies $f_1 = 12$

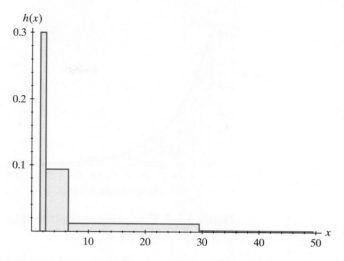

Figure 6.1-3 Relative frequency histogram of losses

and $f_2 = 15$, yet the height of the first is greater than the height of the second, while here $f_1 < f_2$. If we have equal widths among the class intervals, then the heights are proportional to the frequencies. ■

For continuous-type data, the interval with the largest class height is called the **modal class** and the respective class mark is called the **mode**. Hence, in the last example, $x = 2$ is the mode and $(1.5, 2.5)$ the modal class.

Example 6.1-5

The following table lists 105 observations of X, the times in minutes between calls to 911:

30	17	65	8	38	35	4	19	7	14	12	4	5	4	2
7	5	12	50	33	10	15	2	10	1	5	30	41	21	31
1	18	12	5	24	7	6	31	1	3	2	22	1	30	2
1	3	12	12	9	28	6	50	63	5	17	11	23	2	46
90	13	21	55	43	5	19	47	24	4	6	27	4	6	37
16	41	68	9	5	28	42	3	42	8	52	2	11	41	4
35	21	3	17	10	16	1	68	105	45	23	5	10	12	17

To help determine visually whether the exponential model in Example 3.2-1 is perhaps appropriate for this situation, we shall look at two graphs. First, we have constructed a relative frequency histogram, $h(x)$, of these data in Figure 6.1-4(a), with $f(x) = (1/20)e^{-x/20}$ superimposed. Second, we have also constructed the empirical cdf of these data in Figure 6.1-4(b), with the theoretical cdf superimposed. Note

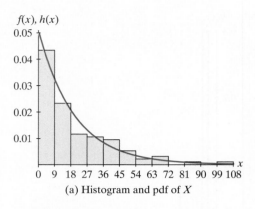
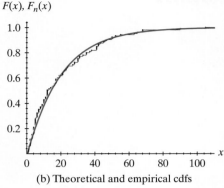

(a) Histogram and pdf of X (b) Theoretical and empirical cdfs

Figure 6.1-4 Times between calls to 911

that $F_n(x)$, the **empirical cumulative distribution function**, is a step function with a vertical step of size $1/n$ at each observation of X. If k observations are equal, the step at that value is k/n. ■

STATISTICAL COMMENTS **(Simpson's Paradox)** While most of the first five chapters were about probability and probability distributions, we now mention some statistical concepts. The relative frequency, f/n, is called a **statistic** and is used to **estimate** a probability, p, which is usually unknown. For example, if a major league batter gets $f = 152$ hits in $n = 500$ official at bats during the season, then the relative frequency $f/n = 0.304$ is an estimate of his probability of getting a hit and is called his batting average for that season.

Once while speaking to a group of coaches, one of us (Hogg) made the comment that it would be possible for batter A to have a higher average than batter B for each season during their careers and yet B could have a better overall average at the end of their careers. While no coach spoke up, you could tell that they were thinking, "And that guy is supposed to know something about math."

Of course, the following simple example convinced them that the statement was true: Suppose A and B played only two seasons, with these results:

Season	Player A			Player B		
	AB	Hits	Average	AB	Hits	Average
1	500	126	0.252	300	75	0.250
2	300	90	0.300	500	145	0.290
Totals	800	216	0.270	800	220	0.275

Clearly, A beats B in the two individual seasons, but B has a better overall average. Note that during their better season (the second), B had more at bats than did A. This kind of result is often called **Simpson's paradox** and it can happen in real life. (See Exercises 6.1-10 and 6.1-11.) ■

Exercises

6.1-1. One characteristic of a car's storage console that is checked by the manufacturer is the time in seconds that it takes for the lower storage compartment door to open completely. A random sample of size $n = 5$ yielded the following times:

$$1.1 \quad 0.9 \quad 1.4 \quad 1.1 \quad 1.0$$

(a) Find the sample mean, $\bar{x}$.

(b) Find the sample variance, s^2.

(c) Find the sample standard deviation, s.

6.1-2. A leakage test was conducted to determine the effectiveness of a seal designed to keep the inside of a plug airtight. An air needle was inserted into the plug, which was then placed underwater. Next, the pressure was increased until leakage was observed. The magnitude of this pressure in psi was recorded for 10 trials:

$$4.2 \quad 3.1 \quad 2.4 \quad 3.6 \quad 4.3 \quad 3.2 \quad 3.5 \quad 2.9 \quad 3.4 \quad 3.1$$

Find the sample mean and sample standard deviation for these 10 measurements.

6.1-3. During the course of an internship at a company that manufactures diesel engine fuel injector pumps, a student had to measure the category "plungers that force the fuel out of the pumps." This category is based on a relative scale, measuring the difference in diameter (in microns or micrometers) of a plunger from that of an absolute minimum acceptable diameter. For 96 plungers randomly taken from the production line, the data are as follows:

17.1 19.3 18.0 19.4 16.5 14.4 15.8 16.6 18.5 14.9
14.8 16.3 20.8 17.8 14.8 15.6 16.7 16.1 17.1 16.5
18.8 19.3 18.1 16.1 18.0 17.2 16.8 17.3 14.4 14.1
16.9 17.6 15.5 17.8 17.2 17.4 18.1 18.4 17.8 16.7
17.2 13.7 18.0 15.6 17.8 17.0 17.7 11.9 15.9 17.8
15.5 14.6 15.6 15.1 15.4 16.1 16.6 17.1 19.1 15.0
17.6 19.7 17.1 13.6 15.6 16.3 14.8 17.4 14.8 14.9
14.1 17.8 19.8 18.9 15.6 16.1 15.9 15.7 22.1 16.1
18.9 21.5 17.4 12.3 20.2 14.9 17.1 15.0 14.4 14.7
15.9 19.0 16.6 15.3 17.7 15.8

(a) Calculate the sample mean and the sample standard deviation of these measurements.

(b) Use the class boundaries $10.95, 11.95, \ldots, 22.95$ to construct a histogram of the data.

6.1-4. Ledolter and Hogg (see References) report that a manufacturer of metal alloys is concerned about customer complaints regarding the lack of uniformity in the melting points of one of the firm's alloy filaments. Fifty filaments are selected and their melting points determined. The following results were obtained:

320 326 325 318 322 320 329 317 316 331
320 320 317 329 316 308 321 319 322 335
318 313 327 314 329 323 327 323 324 314
308 305 328 330 322 310 324 314 312 318
313 320 324 311 317 325 328 319 310 324

(a) Construct a frequency distribution and display the histogram of the data.

(b) Calculate the sample mean and sample standard deviation.

(c) Locate $\bar{x}$ and $\bar{x} \pm s$, and $\bar{x} \pm 2s$ on your histogram. How many observations lie within one standard deviation of the mean? How many lie within two standard deviations of the mean?

6.1-5. In the casino game roulette, if a player bets $1 on red, the probability of winning $1 is 18/38 and the probability of losing $1 is 20/38. Let X equal the number of successive $1 bets that a player makes before losing $5. One hundred observations of X were simulated on a computer, yielding the following data:

23 127 877 65 101 45 61 95 21 43
53 49 89 9 75 93 71 39 25 91
15 131 63 63 41 7 37 13 19 413
65 43 35 23 135 703 83 7 17 65
49 177 61 21 9 27 507 7 5 87
13 213 85 83 75 95 247 1815 7 13
71 67 19 615 11 15 7 131 47 25
25 5 471 11 5 13 75 19 307 33
57 65 9 57 35 19 9 33 11 51
27 9 19 63 109 515 443 11 63 9

(a) Find the sample mean and sample standard deviation of these data.

(b) Construct a relative frequency histogram of the data, using about 10 classes. The classes do not need to be of the same length.

(c) Locate $\bar{x}$, $\bar{x} \pm s$, $\bar{x} \pm 2s$, and $\bar{x} \pm 3s$ on your histogram.

(d) In your opinion, does the median or sample mean give a better measure of the center of these data?

6.1-6. An insurance company experienced the following mobile home losses in 10,000's of dollars for 50 catastrophic events:

1	2	2	3	3	4	4	5	5	5
5	6	7	7	9	9	9	10	11	12
22	24	28	29	31	33	36	38	38	38
39	41	48	49	53	55	74	82	117	134
192	207	224	225	236	280	301	308	351	527

(a) Using class boundaries 0.5, 5.5, 17.5, 38.5, 163.5, and 549.5, group these data into five classes.

(b) Construct a relative frequency histogram of the data.

(c) Describe the distribution of losses.

6.1-7. Ledolter and Hogg (see References) report 64 observations that are a sample of daily weekday afternoon (3 to 7 P.M.) lead concentrations (in micrograms per cubic meter, $\mu g/m^3$). The following data were recorded at an air-monitoring station near the San Diego Freeway in Los Angeles during the fall of 1976:

6.7	5.4	5.2	6.0	8.7	6.0	6.4	8.3	5.3	5.9	7.6
5.0	6.9	6.8	4.9	6.3	5.0	6.0	7.2	8.0	8.1	7.2
10.9	9.2	8.6	6.2	6.1	6.5	7.8	6.2	8.5	6.4	8.1
2.1	6.1	6.5	7.9	14.1	9.5	10.6	8.4	8.3	5.9	6.0
6.4	3.9	9.9	7.6	6.8	8.6	8.5	11.2	7.0	7.1	6.0
9.0	10.1	8.0	6.8	7.3	9.7	9.3	3.2	6.4		

(a) Construct a frequency distribution of the data and display the results in the form of a histogram. Is this distribution symmetric?

(b) Calculate the sample mean and sample standard deviation.

(c) Locate $\bar{x}$ and $\bar{x} \pm s$ on your histogram. How many observations lie within one standard deviation of the mean? How many lie within two standard deviations of the mean?

6.1-8. A small part for an automobile rearview mirror was produced on two different punch presses. In order to describe the distribution of the weights of those parts, a random sample was selected, and each piece was weighed in grams, resulting in the following data set:

3.968	3.534	4.032	3.912	3.572	4.014	3.682	3.608
3.669	3.705	4.023	3.588	3.945	3.871	3.744	3.711
3.645	3.977	3.888	3.948	3.551	3.796	3.657	3.667
3.799	4.010	3.704	3.642	3.681	3.554	4.025	4.079
3.621	3.575	3.714	4.017	4.082	3.660	3.692	3.905
3.977	3.961	3.948	3.994	3.958	3.860	3.965	3.592
3.681	3.861	3.662	3.995	4.010	3.999	3.993	4.004
3.700	4.008	3.627	3.970	3.647	3.847	3.628	3.646
3.674	3.601	4.029	3.603	3.619	4.009	4.015	3.615
3.672	3.898	3.959	3.607	3.707	3.978	3.656	4.027
3.645	3.643	3.898	3.635	3.865	3.631	3.929	3.635
3.511	3.539	3.830	3.925	3.971	3.646	3.669	3.931
4.028	3.665	3.681	3.984	3.664	3.893	3.606	3.699
3.997	3.936	3.976	3.627	3.536	3.695	3.981	3.587
3.680	3.888	3.921	3.953	3.847	3.645	4.042	3.692
3.910	3.672	3.957	3.961	3.950	3.904	3.928	3.984
3.721	3.927	3.621	4.038	4.047	3.627	3.774	3.983
3.658	4.034	3.778					

(a) Using about 10 (say, 8 to 12) classes, construct a frequency distribution.

(b) Draw a histogram of the data.

(c) Describe the shape of the distribution represented by the histogram.

6.1-9. Old Faithful is a geyser in Yellowstone National Park. Tourists always want to know when the next eruption will occur, so data have been collected to help make those predictions. In the following data set, observations were made on several consecutive days, and the data recorded give the starting time of the eruption (STE); the duration of the eruption, in seconds (DIS); the predicted time until the next eruption, in minutes (PTM); the actual time until the next eruption, in minutes (ATM); and the duration of the eruption, in minutes (DIM).

STE	DIS	PTM	ATM	DIM	STE	DIS	PTM	ATM	DIM
706	150	65	72	2.500	1411	110	55	65	1.833
818	268	89	88	4.467	616	289	89	97	4.817
946	140	65	62	2.333	753	114	58	52	1.900
1048	300	95	87	5.000	845	271	89	94	4.517
1215	101	55	57	1.683	1019	120	58	60	2.000
1312	270	89	94	4.500	1119	279	89	84	4.650
651	270	89	91	4.500	1253	109	55	63	1.817
822	125	59	51	2.083	1356	295	95	91	4.917
913	262	89	98	4.367	608	240	85	83	4.000
1051	95	55	59	1.583	731	259	86	84	4.317
1150	270	89	93	4.500	855	128	60	71	2.133
637	273	89	86	4.550	1006	287	92	83	4.783
803	104	55	70	1.733	1129	253	65	70	4.217
913	129	62	63	2.150	1239	284	89	81	4.733
1016	264	89	91	4.400	608	120	58	60	2.000
1147	239	82	82	3.983	708	283	92	91	4.717
1309	106	55	58	1.767	839	115	58	51	1.917
716	259	85	97	4.317	930	254	85	85	4.233
853	115	55	59	1.917	1055	94	55	55	1.567
952	275	89	90	4.583	1150	274	89	98	4.567
1122	110	55	58	1.833	1328	128	64	49	2.133
1220	286	92	98	4.767	557	270	93	85	4.500
735	115	55	55	1.917	722	103	58	65	1.717
830	266	89	107	4.433	827	287	89	102	4.783
1017	105	55	61	1.750	1009	111	55	56	1.850
1118	275	89	82	4.583	1105	275	89	86	4.583
1240	226	79	91	3.767	1231	104	55	62	1.733

(a) Construct a histogram of the durations of the eruptions, in seconds. Use 10 to 12 classes.

(b) Calculate the sample mean and locate it on your histogram. Does it give a good measure of the average length of an eruption? Why or why not?

(c) Construct a histogram of the lengths of the times between eruptions. Use 10 to 12 classes.

(d) Calculate the sample mean and locate it on your histogram. Does it give a good measure of the average length of the times between eruptions?

6.1-10. In 1985, Kent Hrbek of the Minnesota Twins and Dion James of the Milwaukee Brewers had the following numbers of hits (H) and official at bats (AB) on grass and artificial turf:

	Hrbek			James		
Playing Surface	AB	H	BA	AB	H	BA
Grass	204	50		329	93	
Artificial Turf	355	124		58	21	
Total	559	174		387	114	

(a) Find the batting average BA (namely, H/AB) of each player on grass.

(b) Find the BA of each player on artificial turf.

(c) Find the season batting averages for the two players.

(d) Interpret your results.

6.1-11. In 1985, Al Bumbry of the Baltimore Orioles and Darrell Brown of the Minnesota Twins had the following numbers of hits (H) and official at bats (AB) on grass and artificial turf:

	Bumbry			Brown		
Playing Surface	AB	H	BA	AB	H	BA
Grass	295	77		92	18	
Artificial Turf	49	16		168	53	
Total	344	93		260	71	

(a) Find the batting average BA (namely, H/AB) of each player on grass.

(b) Find the BA of each player on artificial turf.

(c) Find the season batting averages for the two players.

(d) Interpret your results.

6.2 EXPLORATORY DATA ANALYSIS

To explore the other characteristics of an unknown distribution, we need to take a sample of n observations, $x_1, x_2, \ldots, x_n$, from that distribution and often need to order them from the smallest to the largest. One convenient way of doing this is to use a stem-and-leaf display, a method that was started by John W. Tukey. [For more details, see the books by Tukey (1977) and Velleman and Hoaglin (1981).]

Possibly the easiest way to begin is with an example to which all of us can relate. Say we have the following 50 test scores on a statistics examination:

93	77	67	72	52	83	66	84	59	63
75	97	84	73	81	42	61	51	91	87
34	54	71	47	79	70	65	57	90	83
58	69	82	76	71	60	38	81	74	69
68	76	85	58	45	73	75	42	93	65

We can do much the same thing as a frequency table and histogram can, but keep the original values, through a **stem-and-leaf display**. For this particular data set, we could use the following procedure: The first number in the set, 93, is recorded by treating the 9 (in the tens place) as the stem and the 3 (in the units place) as the corresponding leaf. Note that this leaf of 3 is the first digit after the stem of 9 in Table 6.2-1. The second number, 77, is that given by the leaf of 7 after the stem of 7; the third number, 67, by the leaf of 7 after the stem of 6; the fourth number, 72, as the leaf of 2 after the stem of 7 (note that this is the second leaf on the 7 stem); and so on. Table 6.2-1 is an example of a stem-and-leaf display. If the leaves are carefully aligned vertically, this table has the same effect as a histogram, but the original numbers are not lost.

It is useful to modify the stem-and-leaf display by ordering the leaves in each row from smallest to largest. The resulting stem-and-leaf diagram is called an **ordered stem-and-leaf display**. Table 6.2-2 uses the data from Table 6.2-1 to produce an ordered stem-and-leaf display.

There is another modification that can also be helpful. Suppose that we want two rows of leaves with each original stem. We can do this by recording leaves 0, 1, 2, 3, and 4 with a stem adjoined with an asterisk ($*$) and leaves 5, 6, 7, 8, and 9 with

Table 6.2-1 Stem-and-leaf display of scores from 50 statistics examinations

Stems	Leaves	Frequency
3	4 8	2
4	2 7 5 2	4
5	2 9 1 4 7 8 8	7
6	7 6 3 1 5 9 0 9 8 5	10
7	7 2 5 3 1 9 0 6 1 4 6 3 5	13
8	3 4 4 1 7 3 2 1 5	9
9	3 7 1 0 3	5

Table 6.2-2 Ordered stem-and-leaf display of statistics examinations

Stems	Leaves	Frequency
3	4 8	2
4	2 2 5 7	4
5	1 2 4 7 8 8 9	7
6	0 1 3 5 5 6 7 8 9 9	10
7	0 1 1 2 3 3 4 5 5 6 6 7 9	13
8	1 1 2 3 3 4 4 5 7	9
9	0 1 3 3 7	5

a stem adjoined with a dot (•). Of course, in our example, by going from 7 original classes to 14 classes, we lose a certain amount of smoothness with this particular data set, as illustrated in Table 6.2-3, which is also ordered.

Tukey suggested another modification, which is used in the next example.

Table 6.2-3 Ordered stem-and-leaf display of statistics examinations

Stems	Leaves	Frequency
3*	4	1
3•	8	1
4*	2 2	2
4•	5 7	2
5*	1 2 4	3
5•	7 8 8 9	4
6*	0 1 3	3
6•	5 5 6 7 8 9 9	7
7*	0 1 1 2 3 3 4	7
7•	5 5 6 6 7 9	6
8*	1 1 2 3 3 4 4	7
8•	5 7	2
9*	0 1 3 3	4
9•	7	1

Example 6.2-1

The following numbers represent ACT composite scores for 60 entering freshmen at a certain college:

26	19	22	28	31	29	25	23	20	33	23	26
30	27	26	29	20	23	18	24	29	27	32	24
25	26	22	29	21	24	20	28	23	26	30	19
27	21	32	28	29	23	25	21	28	22	25	24
19	24	35	26	25	20	31	27	23	26	30	29

An ordered stem-and-leaf display of these scores is given in Table 6.2-4, where leaves are recorded as zeros and ones with a stem adjoined with an asterisk (∗), twos and threes with a stem adjoined with t, fours and fives with a stem adjoined with f, sixes and sevens with a stem adjoined with s, and eights and nines with a stem adjoined with a dot (•). ∎

There is a reason for constructing ordered stem-and-leaf diagrams. For a sample of n observations, $x_1, x_2, \ldots, x_n$, when the observations are ordered from smallest to largest, the resulting ordered data are called the **order statistics** of the sample. Statisticians have found that order statistics and certain of their functions are extremely valuable; we will provide some theory concerning them in Section 6.3. It is very easy to determine the values of the sample in order from an ordered stem-and-leaf display. As an illustration, consider the values in Table 6.2-2 or Table 6.2-3. The order statistics of the 50 test scores are given in Table 6.2-5.

Sometimes we give ranks to these order statistics and use the rank as the subscript on y. The first order statistic $y_1 = 34$ has rank 1; the second order statistic $y_2 = 38$ has rank 2; the third order statistic $y_3 = 42$ has rank 3; the fourth order statistic $y_4 = 42$ has rank 4, ...; and the 50th order statistic $y_{50} = 97$ has rank 50. It is also about as easy to determine these values from the ordered stem-and-leaf display. We see that $y_1 \leq y_2 \leq \cdots \leq y_{50}$.

Table 6.2-4 Ordered stem-and-leaf display of 60 ACT scores

Stems	Leaves	Frequency
1•	8 9 9 9	4
2∗	0 0 0 0 1 1 1	7
2t	2 2 2 3 3 3 3 3 3	9
2f	4 4 4 4 4 5 5 5 5 5	10
2s	6 6 6 6 6 6 7 7 7 7	11
2•	8 8 8 8 9 9 9 9 9 9	10
3∗	0 0 0 1 1	5
3t	2 2 3	3
3f	5	1

Table 6.2-5 Order statistics of 50 exam scores

34	38	42	42	45	47	51	52	54	57
58	58	59	60	61	63	65	65	66	67
68	69	69	70	71	71	72	73	73	74
75	75	76	76	77	79	81	81	82	83
83	84	84	85	87	90	91	93	93	97

From either these order statistics or the corresponding ordered stem-and-leaf display, it is rather easy to find the **sample percentiles**. If $0 < p < 1$, then the $(100p)$th sample percentile has *approximately* np sample observations less than it and also $n(1 - p)$ sample observations greater than it. One way of achieving this is to take the $(100p)$th sample percentile as the $(n + 1)p$th order statistic, provided that $(n + 1)p$ is an integer. If $(n + 1)p$ is not an integer but is equal to r plus some proper fraction— say, a/b—use a weighted average of the rth and the $(r + 1)$st order statistics. That is, define the $(100p)$th sample percentile as

$$\widetilde{\pi}_p = y_r + (a/b)(y_{r+1} - y_r) = (1 - a/b)y_r + (a/b)y_{r+1}.$$

Note that this formula is simply a linear interpolation between y_r and y_{r+1}. [If $p < 1/(n + 1)$ or $p > n/(n + 1)$, that sample percentile is not defined.]

As an illustration, consider the 50 ordered test scores. With $p = 1/2$, we find the 50th percentile by averaging the 25th and 26th order statistics, since $(n + 1)p = (51)(1/2) = 25.5$. Thus, the 50th percentile is

$$\widetilde{\pi}_{0.50} = (1/2)y_{25} + (1/2)y_{26} = (71 + 71)/2 = 71.$$

With $p = 1/4$, we have $(n + 1)p = (51)(1/4) = 12.75$, and the 25th sample percentile is then

$$\widetilde{\pi}_{0.25} = (1 - 0.75)y_{12} + (0.75)y_{13} = (0.25)(58) + (0.75)(59) = 58.75.$$

With $p = 3/4$, so that $(n + 1)p = (51)(3/4) = 38.25$, the 75th sample percentile is

$$\widetilde{\pi}_{0.75} = (1 - 0.25)y_{38} + (0.25)y_{39} = (0.75)(81) + (0.25)(82) = 81.25.$$

Note that *approximately* 50%, 25%, and 75% of the sample observations are less than 71, 58.75, and 81.25, respectively.

Special names are given to certain percentiles. The 50th percentile is the **median** of the sample. The 25th, 50th, and 75th percentiles are, respectively, the **first, second**, and **third quartiles** of the sample. For notation, we let $\widetilde{q}_1 = \widetilde{\pi}_{0.25}$, $\widetilde{q}_2 = \widetilde{m} = \widetilde{\pi}_{0.50}$, and $\widetilde{q}_3 = \widetilde{\pi}_{0.75}$. The 10th, 20th, ..., and 90th percentiles are the **deciles** of the sample, so note that the 50th percentile is also the median, the second quartile, and the fifth decile. With the set of 50 test scores, since $(51)(2/10) = 10.2$ and $(51)(9/10) = 45.9$, the second and ninth deciles are, respectively,

$$\widetilde{\pi}_{0.20} = (0.8)y_{10} + (0.2)y_{11} = (0.8)(57) + (0.2)(58) = 57.2$$

and

$$\widetilde{\pi}_{0.90} = (0.1)y_{45} + (0.9)y_{46} = (0.1)(87) + (0.9)(90) = 89.7.$$

The second decile is commonly called the 20th percentile, and the ninth decile is the 90th percentile.

Example 6.2-2

We illustrate the preceding ideas with the fluoride data given in Table 6.1-3. For convenience, we use 0.02 as the length of a class interval. The ordered stem-and-leaf display is given in Table 6.2-6.

This ordered stem-and-leaf diagram is useful for finding sample percentiles of the data. ∎

We now find some of the sample percentiles associated with the fluoride data. Since $n = 100$, $(n+1)(0.25) = 25.25$, $(n+1)(0.50) = 50.5$, and $(n+1)(0.75) = 75.75$, so that the 25th, 50th, and 75th percentiles are, respectively,

$$\tilde{\pi}_{0.25} = (0.75)y_{25} + (0.25)y_{26} = (0.75)(0.89) + (0.25)(0.89) = 0.89,$$
$$\tilde{\pi}_{0.50} = (0.50)y_{50} + (0.50)y_{51} = (0.50)(0.92) + (0.50)(0.92) = 0.92,$$
$$\tilde{\pi}_{0.75} = (0.25)y_{75} + (0.75)y_{76} = (0.25)(0.97) + (0.75)(0.97) = 0.97.$$

These three percentiles are often called the **first quartile**, the **median** or **second quartile**, and the **third quartile**, respectively. Along with the smallest (the **minimum**) and largest (the **maximum**) values, they give the **five-number summary** of a set of data. Furthermore, the difference between the third and first quartiles is called the **interquartile range**, **IQR**. Here, it is equal to

$$\tilde{q}_3 - \tilde{q}_1 = \tilde{\pi}_{0.75} - \tilde{\pi}_{0.25} = 0.97 - 0.89 = 0.08.$$

Table 6.2-6 Ordered stem-and-leaf diagram of fluoride concentrations

Stems	Leaves	Frequency
0.8*f*	5 5 5 5	4
0.8*s*	6 6 6 7 7 7 7	7
0.8•	8 8 8 8 8 8 8 8 9 9 9 9 9 9 9 9	16
0.9*∗	0 0 0 0 0 0 0 0 0 1 1 1 1 1 1 1 1	17
0.9*t*	2 2 2 2 2 2 2 2 2 2 3 3 3 3 3 3 3 3 3	19
0.9*f*	4 4 5 5 5 5 5 5 5	9
0.9*s*	6 6 7 7 7 7	6
0.9•	8 8 8 8 8 8 8 8 9 9	10
1.0*∗	0 0 0 0 1 1 1	7
1.0*t*	2 3 3	3
1.0*f*	5	1
1.0*s*	6	1

One graphical means for displaying the five-number summary of a set of data is called a **box-and-whisker diagram**. To construct a horizontal box-and-whisker diagram, or, more simply, a **box plot**, draw a horizontal axis that is scaled to the data. Above the axis, draw a rectangular box with the left and right sides drawn at $\tilde{q}_1$ and $\tilde{q}_3$ and with a vertical line segment drawn at the median, $\tilde{q}_2 = \tilde{m}$. A left whisker is drawn as a horizontal line segment from the minimum to the midpoint of the left side of the box, and a right whisker is drawn as a horizontal line segment from the midpoint of the right side of the box to the maximum. Note that the length of the box is equal to the IQR. The left and right whiskers represent the first and fourth quarters of the data, while the two middle quarters of the data are represented, respectively, by the two sections of the box, one to the left and one to the right of the median line.

Example 6.2-3

Using the fluoride data shown in Table 6.2-6, we found that the five-number summary is given by

$$y_1 = 0.85, \tilde{q}_1 = 0.89, \tilde{q}_2 = \tilde{m} = 0.92, \tilde{q}_3 = 0.97, y_{100} = 1.06.$$

The box plot of these data is given in Figure 6.2-1. The fact that the long whisker is to the right and the right half of the box is larger than the left half of the box leads us to say that these data are slightly *skewed to the right*. Note that this skewness can also be seen in the histogram and in the stem-and-leaf diagram. ■

The next example illustrates how the box plot depicts data that are *skewed to the left*.

Example 6.2-4

The following data give the ordered weights (in grams) of 39 gold coins that were produced during the reign of Verica, a pre-Roman British king:

4.90	5.06	5.07	5.08	5.15	5.17	5.18	5.19	5.24	5.25
5.25	5.25	5.25	5.27	5.27	5.27	5.27	5.28	5.28	5.28
5.29	5.30	5.30	5.30	5.30	5.31	5.31	5.31	5.31	5.31
5.32	5.32	5.33	5.34	5.35	5.35	5.35	5.36	5.37	

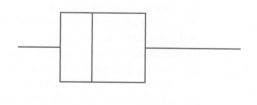

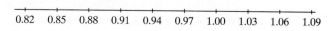

0.82   0.85   0.88   0.91   0.94   0.97   1.00   1.03   1.06   1.09

Figure 6.2-1 Box plot of fluoride concentrations

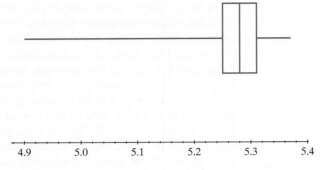

Figure 6.2-2 Box plot for weights of 39 gold coins

For these data, the minimum is 4.90 and the maximum is 5.37. Since

$$(39 + 1)(1/4) = 10, \qquad (39 + 1)(2/4) = 20, \qquad (39 + 1)(3/4) = 30,$$

we have

$$\tilde{q}_1 = y_{10} = 5.25,$$
$$\tilde{m} = y_{20} = 5.28,$$
$$\tilde{q}_3 = y_{30} = 5.31.$$

Thus, the five-number summary is given by

$$y_1 = 4.90, \tilde{q}_1 = 5.25, \tilde{q}_2 = \tilde{m} = 5.28, \tilde{q}_3 = 5.31, y_{39} = 5.37.$$

The box plot associated with the given data is shown in Figure 6.2-2. Note that the box plot indicates that the data are skewed to the left. ∎

Sometimes we are interested in picking out observations that seem to be much larger or much smaller than most of the other observations. That is, we are looking for outliers. Tukey suggested a method for defining outliers that is resistant to the effect of one or two extreme values and makes use of the IQR. In a box-and-whisker diagram, construct **inner fences** to the left and right of the box at a distance of 1.5 times the IQR. **Outer fences** are constructed in the same way at a distance of 3 times the IQR. Observations that lie between the inner and outer fences are called **suspected outliers**. Observations that lie beyond the outer fences are called **outliers**. The observations beyond the inner fences are denoted with a circle (•), and the whiskers are drawn only to the extreme values within or on the inner fences. When you are analyzing a set of data, suspected outliers deserve a closer look and outliers should be looked at very carefully. It does not follow that suspected outliers should be removed from the data, unless some error (such as a recording error) has been made. Moreover, it is sometimes important to determine the cause of extreme values, because outliers can often provide useful insights into the situation under consideration (such as a better way of doing things).

STATISTICAL COMMENTS There is a story that statisticians tell about Ralph Sampson, who was an excellent basketball player at the University of Virginia in the early 1980s and later was drafted by the Houston Rockets. He supposedly majored in communication studies at Virginia, and it is reported that the department there said that the average starting salary of their majors was much higher than those in the sciences; that happened because of Sampson's high starting salary with the Rockets. If this story is true, it would have been much more appropriate to report the median starting salary of majors and this median salary would have been much lower than the median starting salaries in the sciences. ∎

Example 6.2-5

Continuing with Example 6.2-4, we find that the interquartile range is IQR = 5.31 − 5.25 = 0.06. Thus, the inner fences would be constructed at a distance of 1.5(0.06) = 0.09 to the left and right of the box, and the outer fences would be constructed at a distance of 3(0.06) = 0.18 to the left and right of the box. Figure 6.2-3 shows a box plot with the fences. Of course, since the maximum is 0.06 greater than $\widetilde{q}_3$, there are no fences to the right. From this box plot, we see that there are three suspected outliers and two outliers. (You may speculate as to why there are outliers with these data and why they fall to the left — that is, they are lighter than expected.) Note that many computer programs use an asterisk to plot outliers and suspected outliers, and do not print fences. ∎

Some functions of two or more order statistics are quite important in modern statistics. We mention and illustrate one more, along with the range and the IQR, using the 100 fluoride concentrations shown in Table 6.2-6.

(a) **Midrange** = average of the extremes

$$= \frac{y_1 + y_n}{2} = \frac{0.85 + 1.06}{2} = 0.955.$$

(b) **Range** = difference of the extremes.

(c) **Interquartile range** = difference of third and first quartiles

$$= \widetilde{q}_3 - \widetilde{q}_1 = 0.97 - 0.89 = 0.08.$$

Thus, we see that the mean, the median, and the midrange are measures of the middle of the sample. In some sense, the standard deviation, the range, and the interquartile range provide measures of spread of the sample.

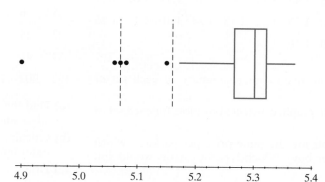

Figure 6.2-3 Box plot for weights of 39 gold coins with fences and outliers

Exercises

6.2-1. In Exercise 6.1-3, measurements for 96 plungers are given. Use those measurements to

(a) Construct a stem-and-leaf diagram using integer stems.

(b) Find the five-number summary of the data.

(c) Construct a box-and-whisker diagram. Are there any outliers?

6.2-2. When you purchase "1-pound bags" of carrots, you can buy either "baby" carrots or regular carrots. We shall compare the weights of 75 bags of each of these types of carrots. The following table gives the weights of the bags of baby carrots:

1.03	1.03	1.06	1.02	1.03	1.03	1.03	1.02	1.03	1.03
1.06	1.04	1.05	1.03	1.04	1.03	1.05	1.06	1.04	1.04
1.03	1.04	1.04	1.06	1.03	1.04	1.05	1.04	1.04	1.02
1.03	1.05	1.05	1.03	1.04	1.03	1.04	1.04	1.03	1.04
1.03	1.04	1.04	1.04	1.05	1.04	1.04	1.03	1.03	1.05
1.04	1.04	1.05	1.04	1.03	1.03	1.05	1.03	1.04	1.05
1.04	1.04	1.04	1.05	1.03	1.04	1.04	1.04	1.04	1.03
1.05	1.05	1.05	1.03	1.04					

This table gives the weights of the regular-sized carrots:

1.29	1.10	1.28	1.29	1.23	1.20	1.31	1.25	1.13	1.26
1.19	1.33	1.24	1.20	1.26	1.24	1.11	1.14	1.15	1.15
1.19	1.26	1.14	1.20	1.20	1.20	1.24	1.25	1.28	1.24
1.26	1.20	1.30	1.23	1.26	1.16	1.34	1.10	1.22	1.27
1.21	1.09	1.23	1.03	1.32	1.21	1.23	1.34	1.19	1.18
1.20	1.20	1.13	1.43	1.19	1.05	1.16	1.19	1.07	1.21
1.36	1.21	1.00	1.23	1.22	1.13	1.24	1.10	1.18	1.26
1.12	1.10	1.19	1.10	1.24					

(a) Calculate the five-number summary of each set of weights.

(b) On the same graph, construct box plots for each set of weights.

(c) If the carrots are the same price per package, which is the better buy? Which type of carrots would you select?

6.2-3. Here are underwater weights in kilograms for 82 male students:

3.7	3.6	4.0	4.3	3.8	3.4	4.1	4.0	3.7	3.4	3.5	3.8	3.7	4.9
3.5	3.8	3.3	4.8	3.4	4.6	3.5	5.3	4.4	4.2	2.5	3.1	5.2	3.8
3.3	3.4	4.1	4.6	4.0	1.4	4.3	3.8	4.7	4.4	5.0	3.2	3.1	4.2
4.9	4.5	3.8	4.2	2.7	3.8	3.8	2.0	3.4	4.9	3.3	4.3	5.6	3.2
4.7	4.5	5.2	5.0	5.0	4.0	3.8	5.3	4.5	3.8	3.8	3.4	3.6	3.3
4.2	5.1	4.0	4.7	6.5	4.4	3.6	4.7	4.5	2.3	4.0	3.7		

Here are underwater weights in kilograms for 100 female students:

2.0	2.0	2.1	1.6	1.9	2.0	2.0	1.3	1.3	1.2	2.3	1.9
2.1	1.2	2.0	1.6	1.1	2.2	2.2	1.4	1.7	2.4	1.8	1.7
2.0	2.1	1.6	1.7	1.8	0.7	1.9	1.7	1.7	1.1	2.0	2.3
0.5	1.3	2.7	1.8	2.0	1.7	1.2	0.7	1.1	1.1	1.7	1.7
1.2	1.2	0.7	2.3	1.7	2.4	1.0	2.4	1.4	1.9	2.5	2.2
2.1	1.4	2.4	1.8	2.5	1.3	0.5	1.7	1.9	1.8	1.3	2.0
2.2	1.7	2.0	2.5	1.2	1.4	1.4	1.2	2.2	2.0	1.8	1.4
1.9	1.4	1.3	2.5	1.2	1.5	0.8	2.0	2.2	1.8	2.0	1.6
1.5	1.6	1.5	2.6								

(a) Group each set of data into classes with a class width of 0.5 kilograms and in which the class marks are $0.5, 1.0, 1.5, \ldots$.

(b) Draw histograms of the grouped data.

(c) Construct box-and-whisker diagrams of the data and draw them on the same graph. Describe what this graph shows.

6.2-4. An insurance company experienced the following mobile home losses in 10,000's of dollars for 50 catastrophic events:

1	2	2	3	3	4	4	5	5	5
5	6	7	7	9	9	9	10	11	12
22	24	28	29	31	33	36	38	38	38
39	41	48	49	53	55	74	82	117	134
192	207	224	225	236	280	301	308	351	527

(a) Find the five-number summary of the data and draw a box-and-whisker diagram.

(b) Calculate the IQR and the locations of the inner and outer fences.

(c) Draw a box plot that shows the fences, suspected outliers, and outliers.

(d) Describe the distribution of losses. (See Exercise 6.1-6.)

6.2-5. In Exercise 6.1-5, data are given for the number of $1 bets a player can make in roulette before losing $5. Use those data to respond to the following:

(a) Determine the order statistics.

(b) Find the five-number summary of the data.

(c) Draw a box-and-whisker diagram.

(d) Find the locations of the inner and outer fences, and draw a box plot that shows the fences, the suspected outliers, and the outliers.

(e) In your opinion, does the median or sample mean give a better measure of the center of the data?

6.2-6. In the casino game roulette, if a player bets $1 on red (or on black or on odd or on even), the probability of winning $1 is 18/38 and the probability of losing $1 is 20/38. Suppose that a player begins with $5 and makes successive $1 bets. Let Y equal the player's maximum capital before losing the $5. One hundred observations of Y were simulated on a computer, yielding the following data:

25	9	5	5	5	9	6	5	15	45
55	6	5	6	24	21	16	5	8	7
7	5	5	35	13	9	5	18	6	10
19	16	21	8	13	5	9	10	10	6
23	8	5	10	15	7	5	5	24	9
11	34	12	11	17	11	16	5	15	5
12	6	5	5	7	6	17	20	7	8
8	6	10	11	6	7	5	12	11	18
6	21	6	5	24	7	16	21	23	15
11	8	6	8	14	11	6	9	6	10

(a) Construct an ordered stem-and-leaf display.

(b) Find the five-number summary of the data and draw a box-and-whisker diagram.

(c) Calculate the IQR and the locations of the inner and outer fences.

(d) Draw a box plot that shows the fences, suspected outliers, and outliers.

(e) Find the 90th percentile.

6.2-7. Let X denote the concentration of calcium carbonate ($CaCO_3$) in milligrams per liter. Following are 20 observations of X:

130.8	129.9	131.5	131.2	129.5
132.7	131.5	127.8	133.7	132.2
134.8	131.7	133.9	129.8	131.4
128.8	132.7	132.8	131.4	131.3

(a) Construct an ordered stem-and-leaf display, using stems of 127, 128, ..., 134.

(b) Find the midrange, range, interquartile range, median, sample mean, and sample variance.

(c) Draw a box-and-whisker diagram.

6.2-8. The weights (in grams) of 25 indicator housings used on gauges are as follows:

102.0	106.3	106.6	108.8	107.7
106.1	105.9	106.7	106.8	110.2
101.7	106.6	106.3	110.2	109.9
102.0	105.8	109.1	106.7	107.3
102.0	106.8	110.0	107.9	109.3

(a) Construct an ordered stem-and-leaf display, using integers as the stems and tenths as the leaves.

(b) Find the five-number summary of the data and draw a box plot.

(c) Are there any suspected outliers? Are there any outliers?

6.2-9. In Exercise 6.1-4, the melting points of a firm's alloy filaments are given for a sample of 50 filaments.

(a) Construct a stem-and-leaf diagram of those melting points, using as stems $30f, 30s, \ldots, 33f$.

(b) Find the five-number summary for these melting points.

(c) Construct a box-and-whisker diagram.

(d) Describe the symmetry of the data.

6.2-10. In Exercise 6.1-7, lead concentrations near the San Diego Freeway in 1976 are given. During the fall of 1977, the weekday afternoon lead concentrations (in $\mu g/m^3$) at the measurement station near the San Diego Freeway in Los Angeles were as follows:

9.5 10.7 8.3 9.8 9.1 9.4 9.6 11.9 9.5 12.6 10.5

8.9 11.4 12.0 12.4 9.9 10.9 12.3 11.0 9.2 9.3 9.3

10.5 9.4 9.4 8.2 10.4 9.3 8.7 9.8 9.1 2.9 9.8

5.7 8.2 8.1 8.8 9.7 8.1 8.8 10.3 8.6 10.2 9.4

14.8 9.9 9.3 8.2 9.9 11.6 8.7 5.0 9.9 6.3 6.5

10.2 8.8 8.0 8.7 8.9 6.8 6.6 7.3 16.7

(a) Construct a frequency distribution and display the results in the form of a histogram. Is this distribution symmetric?

(b) Calculate the sample mean and sample standard deviation.

(c) Locate $\bar{x}$, $\bar{x} \pm s$ on your histogram. How many observations lie within one standard deviation of the mean? How many lie within two standard deviations of the mean?

(d) Using the data from Exercise 6.1-7 and the data from this exercise, construct a back-to-back stem-and-leaf diagram with integer stems in the center and the leaves for 1976 going to the left and those for 1977 going to the right.

(e) Construct box-and-whisker displays of both sets of data on the same graph.

(f) Use your numerical and graphical results to interpret what you see.

REMARK   In the spring of 1977, a new traffic lane was added to the freeway. This lane reduced traffic congestion but increased traffic speed.   ∎

6.3 ORDER STATISTICS

Order statistics are the observations of the random sample, arranged, or ordered, in magnitude from the smallest to the largest. In recent years, the importance of order statistics has increased owing to the more frequent use of nonparametric inferences and robust procedures. However, order statistics have always been prominent because, among other things, they are needed to determine rather simple statistics such as the sample median, the sample range, and the empirical cdf. Recall that in Section 6.2 we discussed observed order statistics in connection with descriptive and exploratory statistical methods. We will consider certain interesting aspects about their distributions in this section.

In most of our discussions about order statistics, we will assume that the n independent observations come from a continuous-type distribution. This means, among other things, that the probability of any two observations being equal is zero. That is, the probability is 1 that the observations can be ordered from smallest to largest without having two equal values. Of course, in practice, we do frequently observe *ties*; but if the probability of a tie is small, the distribution theory that follows will hold approximately. Thus, in the discussion here, we are assuming that the probability of a tie is zero.

Example 6.3-1   The values $x_1 = 0.62, x_2 = 0.98, x_3 = 0.31, x_4 = 0.81$, and $x_5 = 0.53$ are the $n = 5$ observed values of five independent trials of an experiment with pdf $f(x) = 2x$, $0 < x < 1$. The observed order statistics are

$$y_1 = 0.31 < y_2 = 0.53 < y_3 = 0.62 < y_4 = 0.81 < y_5 = 0.98.$$

Recall that the middle observation in the ordered arrangement, here $y_3 = 0.62$, is called the *sample median* and the difference of the largest and the smallest, here

$$y_5 - y_1 = 0.98 - 0.31 = 0.67,$$

is called the *sample range*.   ∎

If $X_1, X_2, \ldots, X_n$ are observations of a random sample of size n from a continuous-type distribution, we let the random variables

$$Y_1 < Y_2 < \cdots < Y_n$$

denote the order statistics of that sample. That is,

$$Y_1 = \text{smallest of } X_1, X_2, \ldots, X_n,$$
$$Y_2 = \text{second smallest of } X_1, X_2, \ldots, X_n,$$
$$\vdots$$
$$Y_n = \text{largest of } X_1, X_2, \ldots, X_n.$$

There is a very simple procedure for determining the cdf of the rth order statistic, Y_r. This procedure depends on the binomial distribution and is illustrated in Example 6.3-2.

Example 6.3-2 Let $Y_1 < Y_2 < Y_3 < Y_4 < Y_5$ be the order statistics associated with n independent observations X_1, X_2, X_3, X_4, X_5, each from the distribution with pdf $f(x) = 2x$, $0 < x < 1$. Consider $P(Y_4 < 1/2)$. For the event $\{Y_4 < 1/2\}$ to occur, at least four of the random variables X_1, X_2, X_3, X_4, X_5 must be less than $1/2$, because Y_4 is the fourth smallest among the five observations. Thus, if the event $\{X_i < 1/2\}$, $i = 1, 2, \ldots, 5$, is called "success," we must have at least four successes in the five mutually independent trials, each of which has probability of success

$$P\left(X_i \leq \frac{1}{2}\right) = \int_0^{1/2} 2x\, dx = \left(\frac{1}{2}\right)^2 = \frac{1}{4}.$$

Hence,

$$P\left(Y_4 \leq \frac{1}{2}\right) = \binom{5}{4}\left(\frac{1}{4}\right)^4\left(\frac{3}{4}\right) + \left(\frac{1}{4}\right)^5 = 0.0156.$$

In general, if $0 < y < 1$, then the cdf of Y_4 is

$$G(y) = P(Y_4 < y) = \binom{5}{4}(y^2)^4(1 - y^2) + (y^2)^5,$$

since this represents the probability of at least four "successes" in five independent trials, each of which has probability of success

$$P(X_i < y) = \int_0^y 2x\, dx = y^2.$$

For $0 < y < 1$, the pdf of Y_4 is therefore

$$g(y) = G'(y) = \binom{5}{4}4(y^2)^3(2y)(1 - y^2) + \binom{5}{4}(y^2)^4(-2y) + 5(y^2)^4(2y)$$

$$= \frac{5!}{3!\,1!}(y^2)^3(1 - y^2)(2y), \qquad 0 < y < 1.$$

Note that in this example the cdf of each X is $F(x) = x^2$ when $0 < x < 1$. Thus,

$$g(y) = \frac{5!}{3!\,1!}[F(y)]^3[1 - F(y)]\,f(y), \qquad 0 < y < 1. \qquad \blacksquare$$

The preceding example should make the following generalization easier to read: Let $Y_1 < Y_2 < \cdots < Y_n$ be the order statistics of n independent observations from a distribution of the continuous type with cdf $F(x)$ and pdf $F'(x) = f(x)$, where $0 < F(x) < 1$ for $a < x < b$ and $F(a) = 0$, $F(b) = 1$. (It is possible that $a = -\infty$ and/or $b = +\infty$.) The event that the rth order statistic Y_r is at most y, $\{Y_r \leq y\}$, can occur if and only if at least r of the n observations are less than or equal to y. That is, here the probability of "success" on each trial is $F(y)$, and we must have at least r successes. Thus,

$$G_r(y) = P(Y_r \leq y) = \sum_{k=r}^{n} \binom{n}{k}[F(y)]^k[1 - F(y)]^{n-k}.$$

Rewriting this slightly, we have

$$G_r(y) = \sum_{k=r}^{n-1} \binom{n}{k}[F(y)]^k[1 - F(y)]^{n-k} + [F(y)]^n.$$

Hence, the pdf of Y_r is

$$g_r(y) = G_r'(y) = \sum_{k=r}^{n-1} \binom{n}{k}(k)[F(y)]^{k-1}f(y)[1 - F(y)]^{n-k}$$

$$+ \sum_{k=r}^{n-1} \binom{n}{k}[F(y)]^k(n-k)[1 - F(y)]^{n-k-1}[-f(y)]$$

$$+ n[F(y)]^{n-1}f(y). \tag{6.3-1}$$

However, since

$$\binom{n}{k}k = \frac{n!}{(k-1)!\,(n-k)!} \quad \text{and} \quad \binom{n}{k}(n-k) = \frac{n!}{k!\,(n-k-1)!},$$

it follows that the pdf of Y_r is

$$g_r(y) = \frac{n!}{(r-1)!\,(n-r)!}\,[F(y)]^{r-1}[1 - F(y)]^{n-r}f(y), \qquad a < y < b,$$

which is the first term of the first summation in $g_r(y) = G_r'(y)$, Equation 6.3-1. The remaining terms in $g_r(y) = G_r'(y)$ sum to zero because the second term of the first summation (when $k = r + 1$) equals the negative of the first term in the second summation (when $k = r$), and so on. Finally, the last term of the second summation equals the negative of $n[F(y)]^{n-1}f(y)$. To see this clearly, the student is urged to write out a number of terms in these summations. (See Exercise 6.3-4.)

It is worth noting that the pdf of the smallest order statistic is

$$g_1(y) = n[1 - F(y)]^{n-1}f(y), \qquad a < y < b,$$

and the pdf of the largest order statistic is

$$g_n(y) = n[F(y)]^{n-1}f(y), \qquad a < y < b.$$

REMARK There is one quite satisfactory way to construct heuristically the expression for the pdf of Y_r. To do this, we must recall the multinomial probability and then consider the probability element $g_r(y)(\Delta y)$ of Y_r. If the length Δy is *very* small, $g_r(y)(\Delta y)$ represents approximately the probability

$$P(y < Y_r \leq y + \Delta y).$$

Thus, we want the probability, $g_r(y)(\Delta y)$, that $(r-1)$ items fall less than y, that $(n-r)$ items are greater than $y + \Delta y$, and that one item falls between y and $y + \Delta y$. Recall that the probabilities on a single trial are

$$P(X \le y) = F(y),$$

$$P(X > y + \Delta y) = 1 - F(y + \Delta y) \approx 1 - F(y),$$

$$P(y < X \le y + \Delta y) \approx f(y)(\Delta y).$$

Thus, the multinomial probability is approximately

$$g_r(y)(\Delta y) = \frac{n!}{(r-1)!\,1!\,(n-r)!} [F(y)]^{r-1}[1 - F(y)]^{n-r}[f(y)(\Delta y)].$$

If we divide both sides by the length Δy, the formula for $g_r(y)$ results. ∎

Example 6.3-3

Returning to Example 6.3-2, we shall now graph the pdfs of the order statistics $Y_1 < Y_2 < Y_3 < Y_4 < Y_5$ when sampling from a distribution with pdf $f(x) = 2x, 0 < x < 1$, and cdf $F(x) = x^2, 0 < x < 1$. These graphs are given in Figure 6.3-1. The respective pdfs and their means are as follows:

$$g_1(y) = 10y(1 - y^2)^4, \quad 0 < y < 1; \quad \mu_1 = \frac{256}{693},$$

$$g_2(y) = 40y^3(1 - y^2)^3, \quad 0 < y < 1; \quad \mu_2 = \frac{128}{231},$$

$$g_3(y) = 60y^5(1 - y^2)^2, \quad 0 < y < 1; \quad \mu_3 = \frac{160}{231},$$

$$g_4(y) = 40y^7(1 - y^2), \quad 0 < y < 1; \quad \mu_4 = \frac{80}{99},$$

$$g_5(y) = 10y^9, \quad 0 < y < 1; \quad \mu_5 = \frac{10}{11}.$$

∎

Recall that in Theorem 5.1-2 we proved that if X has a cdf $F(x)$ of the continuous type, then $F(X)$ has a uniform distribution on the interval from 0 to 1. If $Y_1 < Y_2 < \cdots < Y_n$ are the order statistics of n independent observations $X_1, X_2, \ldots, X_n$, then

$$F(Y_1) < F(Y_2) < \cdots < F(Y_n),$$

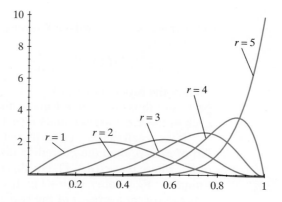

Figure 6.3-1 pdfs of order statistics, $f(x) = 2x, 0 < x < 1$

because F is a nondecreasing function and the probability of an equality is again zero. Note that this ordering could be looked upon as an ordering of the mutually independent random variables $F(X_1), F(X_2), \ldots, F(X_n)$, each of which is $U(0,1)$. That is,

$$W_1 = F(Y_1) < W_2 = F(Y_2) < \cdots < W_n = F(Y_n)$$

can be thought of as the order statistics of n independent observations from that uniform distribution. Since the cdf of $U(0,1)$ is $G(w) = w, 0 < w < 1$, the pdf of the rth order statistic, $W_r = F(Y_r)$, is

$$h_r(w) = \frac{n!}{(r-1)!\,(n-r)!}\, w^{r-1}(1-w)^{n-r}, \qquad 0 < w < 1.$$

Of course, the mean, $E(W_r) = E[F(Y_r)]$ of $W_r = F(Y_r)$, is given by the integral

$$E(W_r) = \int_0^1 w\, \frac{n!}{(r-1)!\,(n-r)!}\, w^{r-1}(1-w)^{n-r}\, dw.$$

This integral can be evaluated by integrating by parts several times, but it is easier to obtain the answer if we rewrite the integration as follows:

$$E(W_r) = \left(\frac{r}{n+1}\right) \int_0^1 \frac{(n+1)!}{r!\,(n-r)!}\, w^{r}(1-w)^{n-r}\, dw.$$

The integrand in this last expression can be thought of as the pdf of the $(r+1)$st order statistic of $n+1$ independent observations from a $U(0,1)$ distribution. This is a beta pdf with $\alpha = r+1$ and $\beta = n-r+1$; hence, the integral must equal 1, and it follows that

$$E(W_r) = \frac{r}{n+1}, \qquad r = 1, 2, \ldots, n.$$

There is an extremely interesting interpretation of $W_r = F(Y_r)$. Note that $F(Y_r)$ is the cumulated probability up to and including Y_r or, equivalently, the area under $f(x) = F'(x)$ but less than Y_r. Consequently, $F(Y_r)$ can be treated as a random area. Since $F(Y_{r-1})$ is also a random area, $F(Y_r) - F(Y_{r-1})$ is the random area under $f(x)$ between Y_{r-1} and Y_r. The expected value of the random area between any two adjacent order statistics is then

$$E[F(Y_r) - F(Y_{r-1})] = E[F(Y_r)] - E[F(Y_{r-1})]$$

$$= \frac{r}{n+1} - \frac{r-1}{n+1} = \frac{1}{n+1}.$$

Also, it is easy to show (see Exercise 6.3-6) that

$$E[F(Y_1)] = \frac{1}{n+1} \qquad \text{and} \qquad E[1 - F(Y_n)] = \frac{1}{n+1}.$$

That is, the order statistics $Y_1 < Y_2 < \cdots < Y_n$ partition the support of X into $n+1$ parts and thus create $n+1$ areas under $f(x)$ and above the x-axis. On the average, each of the $n+1$ areas equals $1/(n+1)$.

If we recall that the $(100p)$th percentile π_p is such that the area under $f(x)$ to the left of π_p is p, then the preceding discussion suggests that we let Y_r be an estimator of π_p, where $p = r/(n+1)$. For this reason, we define the $(100p)$**th percentile of the sample** as Y_r, where $r = (n+1)p$. In case $(n+1)p$ is not an integer, we use a weighted average (or an average) of the two adjacent order statistics Y_r and Y_{r+1}, where r is the greatest integer $[(n+1)p]$ (or, $\lfloor (n+1)p \rfloor$) in $(n+1)p$. In particular, the sample median is

$$\widetilde{m} = \begin{cases} Y_{(n+1)/2}, & \text{when } n \text{ is odd,} \\ \dfrac{Y_{n/2} + Y_{(n/2)+1}}{2}, & \text{when } n \text{ is even.} \end{cases}$$

Example 6.3-4

Let X equal the weight of soap in a "1000-gram" bottle. A random sample of $n = 12$ observations of X yielded the following weights, which have been ordered:

$$1013 \quad 1019 \quad 1021 \quad 1024 \quad 1026 \quad 1028$$
$$1033 \quad 1035 \quad 1039 \quad 1040 \quad 1043 \quad 1047$$

Since $n = 12$ is even, the sample median is

$$\widetilde{m} = \frac{y_6 + y_7}{2} = \frac{1028 + 1033}{2} = 1030.5.$$

The location of the 25th percentile (or first quartile) is

$$(n + 1)(0.25) = (12 + 1)(0.25) = 3.25.$$

Thus, using a weighted average, we find that the first quartile is

$$\widetilde{q}_1 = y_3 + (0.25)(y_4 - y_3) = (0.75)y_3 + (0.25)y_4$$
$$= (0.75)(1021) + (0.25)(1024) = 1021.75.$$

Similarly, the 75th percentile (or third quartile) is

$$\widetilde{q}_3 = y_9 + (0.75)(y_{10} - y_9) = (0.25)y_9 + (0.75)y_{10}$$
$$= (0.25)(1039) + (0.75)(1040) = 1039.75,$$

because $(12 + 1)(0.75) = 9.75$. Since $(12 + 1)(0.60) = 7.8$, the 60th percentile is

$$\widetilde{\pi}_{0.60} = (0.2)y_7 + (0.8)y_8 = (0.2)(1033) + (0.8)(1035) = 1034.6.$$ ∎

The $(100p)$th percentile of a distribution is often called the quantile of order p. So if $y_1 \leq y_2 \leq \cdots \leq y_n$ are the order statistics associated with the sample $x_1, x_2, \ldots, x_n$, then y_r is called the **sample quantile of order $r/(n+1)$** as well as the **$100r/(n+1)$th sample percentile**. Also, the percentile π_p of a theoretical distribution is the quantile of order p. Now, suppose the theoretical distribution is a good model for the observations. Then we plot (y_r, π_p), where $p = r/(n+1)$, for several values of r (possibly even for all r values, $r = 1, 2, \ldots, n$); we would expect these points (y_r, π_p) to lie close to a line through the origin with slope equal to 1 because $y_r \approx \pi_p$. If they are not close to that line, then we would doubt that the theoretical distribution is a good model for the observations. The plot of (y_r, π_p) for several values of r is called the **quantile–quantile plot** or, more simply, the **q–q plot**.

Given a set of observations of a random variable X, how can we decide, for example, whether or not X has an approximate normal distribution? If we have a large number of observations of X, a stem-and-leaf diagram or a histogram of the observations can often be helpful. (See Exercises 6.2-1 and 6.1-3, respectively.) For small samples, a q–q plot can be used to check on whether the sample arises from a normal distribution. For example, suppose the quantiles of a sample were plotted against the corresponding quantiles of a certain normal distribution and the pairs of points generated were on a straight line with slope 1 and intercept 0. Of course, we would then believe that we have an ideal sample from that normal distribution with that certain mean and standard deviation. Such a plot, however, requires that we know the mean and the standard deviation of this normal distribution, and we usually do not. However, since the quantile, q_p, of $N(\mu, \sigma^2)$ is related to the corresponding one, z_{1-p}, of $N(0,1)$ by $q_p = \mu + \sigma z_{1-p}$, we can always plot the quantiles of the sample against the corresponding ones of $N(0,1)$ and get the needed information. That is, if the sample quantiles are plotted as the x-coordinates of the pairs and the $N(0,1)$ quantiles as the y-coordinates, and if the graph is almost a straight line, then it is reasonable to assume that the sample arises from a normal distribution. Moreover, the reciprocal of the slope of that straight line is a good estimate of the standard deviation σ because $z_{1-p} = (q_p - \mu)/\sigma$.

Example 6.3-5

In researching groundwater it is often important to know the characteristics of the soil at a certain site. Many of these characteristics, such as porosity, are at least partially dependent upon the grain size. The diameter of individual grains of soil can be measured. Here are the diameters (in mm) of 30 randomly selected grains:

1.24	1.36	1.28	1.31	1.35	1.20	1.39	1.35	1.41	1.31
1.28	1.26	1.37	1.49	1.32	1.40	1.33	1.28	1.25	1.39
1.38	1.34	1.40	1.27	1.33	1.36	1.43	1.33	1.29	1.34

For these data, $\bar{x} = 1.33$ and $s^2 = 0.0040$. May we assume that these are observations of a random variable X that is $N(1.33, 0.0040)$? To help answer this question, we shall construct a q–q plot of the standard normal quantiles that correspond to $p = 1/31, 2/31, \ldots, 30/31$ versus the ordered observations. To find these quantiles, it is helpful to use the computer.

k	Diameters in mm (x)	$p = k/31$	z_{1-p}	k	Diameters in mm (x)	$p = k/31$	z_{1-p}
1	1.20	0.0323	−1.85	16	1.34	0.5161	0.04
2	1.24	0.0645	−1.52	17	1.34	0.5484	0.12
3	1.25	0.0968	−1.30	18	1.35	0.5806	0.20
4	1.26	0.1290	−1.13	19	1.35	0.6129	0.29
5	1.27	0.1613	−0.99	20	1.36	0.6452	0.37
6	1.28	0.1935	−0.86	21	1.36	0.6774	0.46
7	1.28	0.2258	−0.75	22	1.37	0.7097	0.55
8	1.28	0.2581	−0.65	23	1.38	0.7419	0.65
9	1.29	0.2903	−0.55	24	1.39	0.7742	0.75
10	1.31	0.3226	−0.46	25	1.39	0.8065	0.86
11	1.31	0.3548	−0.37	26	1.40	0.8387	0.99
12	1.32	0.3871	−0.29	27	1.40	0.8710	1.13
13	1.33	0.4194	−0.20	28	1.41	0.9032	1.30
14	1.33	0.4516	−0.12	29	1.43	0.9355	1.52
15	1.33	0.4839	−0.04	30	1.49	0.9677	1.85

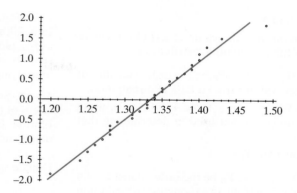

Figure 6.3-2 q–q plot, $N(0, 1)$ quantiles versus grain diameters

A q–q plot of these data is shown in Figure 6.3-2. Note that the points do fall close to a straight line, so the normal probability model seems to be appropriate on the basis of these few data. ∎

Exercises

6.3-1. Some biology students were interested in analyzing the amount of time that bees spend gathering nectar in flower patches. Thirty-nine bees visited a high-density flower patch and spent the following times (in seconds) gathering nectar:

235	210	95	146	195	840	185	610	680	990
146	404	119	47	9	4	10	169	270	95
329	151	211	127	154	35	225	140	158	116
46	113	149	420	120	45	10	18	105	

(a) Find the order statistics.

(b) Find the median and 80th percentile of the sample.

(c) Determine the first and third quartiles (i.e., 25th and 75th percentiles) of the sample.

6.3-2. Let X equal the forced vital capacity (the volume of air a person can expel from his or her lungs) of a male athlete. Seventeen observations of X, which have been ordered, are

3.4	3.6	4.1	4.3	4.5	4.9	5.2	5.4	
5.5	5.7	5.8	6.0	6.1	6.1	6.9	6.9	7.5

(a) Find the median, the first quartile, and the third quartile.

(b) Find the 40th and 60th percentiles.

6.3-3. Let $Y_1 < Y_2 < Y_3 < Y_4 < Y_5$ be the order statistics of five independent observations from an exponential distribution that has a mean of $\theta = 3$.

(a) Find the pdf of the sample median Y_3.

(b) Compute the probability that Y_4 is less than 5.

(c) Determine $P(1 < Y_1)$.

6.3-4. In the expression for $g_r(y) = G'_r(y)$ in Equation 6.3-1, let $n = 6$, and $r = 3$, and write out the summations, showing that the "telescoping" suggested in the text is achieved.

6.3-5. Let $Y_1 < Y_2 < \cdots < Y_8$ be the order statistics of eight independent observations from a continuous-type distribution with 70th percentile $\pi_{0.7} = 27.3$.

(a) Determine $P(Y_7 < 27.3)$.

(b) Find $P(Y_5 < 27.3 < Y_8)$.

6.3-6. Let $W_1 < W_2 < \cdots < W_n$ be the order statistics of n independent observations from a $U(0, 1)$ distribution.

(a) Find the pdf of W_1 and that of W_n.

(b) Use the results of (a) to verify that $E(W_1) = 1/(n+1)$ and $E(W_n) = n/(n+1)$.

(c) Show that the pdf of W_r is beta.

6.3-7. Let $Y_1 < Y_2 < \cdots < Y_{19}$ be the order statistics of $n = 19$ independent observations from the exponential distribution with mean θ.

(a) What is the pdf of Y_1?

(b) Using integration, find the value of $E[F(Y_1)]$, where F is the cdf of the exponential distribution.

6.3-8. Let $W_1 < W_2 < \cdots < W_n$ be the order statistics of n independent observations from a $U(0,1)$ distribution.

(a) Show that $E(W_r^2) = r(r+1)/(n+1)(n+2)$, using a technique similar to that used in determining that $E(W_r) = r/(n+1)$.

(b) Find the variance of W_r.

6.3-9. Let $Y_1 < Y_2 < \cdots < Y_n$ be the order statistics of a random sample of size n from an exponential distribution with pdf $f(x) = e^{-x}$, $0 < x < \infty$.

(a) Find the pdf of Y_r.

(b) Determine the pdf of $U = e^{-Y_r}$.

6.3-10. Use the heuristic argument to show that the joint pdf of the two order statistics $Y_i < Y_j$ is

$$g(y_i, y_j) = \frac{n!}{(i-1)!(j-i-1)!(n-j)!}$$
$$\times [F(y_i)]^{i-1}[F(y_j) - F(y_i)]^{j-i-1}$$
$$\times [1 - F(y_j)]^{n-j} f(y_i) f(y_j), \quad -\infty < y_i < y_j < \infty.$$

6.3-11. Use the result of Exercise 6.3-10.

(a) Find the joint pdf of Y_1 and Y_n, the first and the nth order statistics of a random sample of size n from the $U(0,1)$ distribution.

(b) Find the joint and the marginal pdfs of $W_1 = Y_1/Y_n$ and $W_2 = Y_n$.

(c) Are W_1 and W_2 independent?

(d) Use simulation to confirm your theoretical results.

6.3-12. Nine measurements are taken on the strength of a certain metal. In order, they are 7.2, 8.9, 9.7, 10.5, 10.9,

11.7, 12.9, 13.9, 15.3, and these values correspond to the 10th, 20th, ..., 90th percentiles of this sample. Construct a q–q plot of the measurements against the same percentiles of $N(0, 1)$. Does it seem reasonable that the underlying distribution of strengths could be normal?

6.3-13. Some measurements (in mm) were made on specimens of the spider *Sosippus floridanus*, which is native to Florida. Here are the lengths of nine female spiders and nine male spiders.

Female spiders	11.06	13.87	12.93	15.08	17.82
	14.14	12.26	17.82	20.17	
Male spiders	12.26	11.66	12.53	13.00	11.79
	12.46	10.65	10.39	12.26	

(a) Construct a q–q plot of the female spider lengths. Do they appear to be normally distributed?

(b) Construct a q–q plot of the male spider lengths. Do they appear to be normally distributed?

6.3-14. An interior automotive supplier places several electrical wires in a harness. A pull test measures the force required to pull spliced wires apart. A customer requires that each wire that is spliced into the harness withstand a pull force of 20 pounds. Let X equal the pull force required to pull a spliced wire apart. The following data give the values of a random sample of $n = 20$ observations of X:

28.8 24.4 30.1 25.6 26.4 23.9 22.1 22.5 27.6 28.1

20.8 27.7 24.4 25.1 24.6 26.3 28.2 22.2 26.3 24.4

(a) Construct a q–q plot, using the ordered array and the corresponding quantiles of $N(0,1)$.

(b) Does X appear to have a normal distribution?

6.4 MAXIMUM LIKELIHOOD ESTIMATION

In earlier chapters, we alluded to estimating characteristics of the distribution from the corresponding ones of the sample, hoping that the latter would be reasonably close to the former. For example, the sample mean $\bar{x}$ can be thought of as an estimate of the distribution mean μ, and the sample variance s^2 can be used as an estimate of the distribution variance σ^2. Even the relative frequency histogram associated with a sample can be taken as an estimate of the pdf of the underlying distribution. But how good are these estimates? What makes an estimate good? Can we say anything about the closeness of an estimate to an unknown parameter?

In this section, we consider random variables for which the functional form of the pmf or pdf is known, but the distribution depends on an unknown parameter (say, θ) that may have any value in a set (say, Ω) called the **parameter space**. For example, perhaps it is known that $f(x; \theta) = (1/\theta)e^{-x/\theta}$, $0 < x < \infty$, and that $\theta \in \Omega = \{\theta : 0 < \theta < \infty\}$. In certain instances, it might be necessary for the experimenter to select precisely one member of the family $\{f(x, \theta), \theta \in \Omega\}$ as the most likely pdf of the random variable. That is, the experimenter needs a point estimate of the parameter θ, namely, the value of the parameter that corresponds to the selected pdf.

In one common estimation scenario, we take a random sample from the distribution to elicit some information about the unknown parameter θ. That is, we repeat the experiment n independent times, observe the sample, $X_1, X_2, \ldots, X_n$, and try to estimate the value of θ by using the observations $x_1, x_2, \ldots, x_n$. The function of $X_1, X_2, \ldots, X_n$ used to estimate θ—say, the statistic $u(X_1, X_2, \ldots, X_n)$—is called an **estimator** of θ. We want it to be such that the computed **estimate** $u(x_1, x_2, \ldots, x_n)$ is usually close to θ. Since we are estimating one member of $\theta \in \Omega$, such an estimator is often called a **point estimator**.

The following example should help motivate one principle that is often used in finding point estimates: Suppose that X is $b(1, p)$, so that the pmf of X is

$$f(x; p) = p^x (1-p)^{1-x}, \qquad x = 0, 1, \qquad 0 \le p \le 1.$$

We note that $p \in \Omega = \{p : 0 \le p \le 1\}$, where Ω represents the parameter space—that is, the space of all possible values of the parameter p. Given a random sample $X_1, X_2, \ldots, X_n$, the problem is to find an estimator $u(X_1, X_2, \ldots, X_n)$ such that $u(x_1, x_2, \ldots, x_n)$ is a good point estimate of p, where $x_1, x_2, \ldots, x_n$ are the observed values of the random sample. Now, the probability that $X_1, X_2, \ldots, X_n$ takes these particular values is (with Σx_i denoting $\sum_{i=1}^{n} x_i$)

$$P(X_1 = x_1, \ldots, X_n = x_n) = \prod_{i=1}^{n} p^{x_i}(1-p)^{1-x_i} = p^{\Sigma x_i}(1-p)^{n-\Sigma x_i},$$

which is the joint pmf of $X_1, X_2, \ldots, X_n$ evaluated at the observed values. One reasonable way to proceed toward finding a good estimate of p is to regard this probability (or joint pmf) as a function of p and find the value of p that maximizes it. That is, we find the p value most likely to have produced these sample values. The joint pmf, when regarded as a function of p, is frequently called the **likelihood function**. Thus, here the likelihood function is

$$
\begin{aligned}
L(p) &= L(p; x_1, x_2, \ldots, x_n) \\
&= f(x_1; p) f(x_2; p) \cdots f(x_n; p) \\
&= p^{\Sigma x_i}(1-p)^{n-\Sigma x_i}, \qquad 0 \le p \le 1.
\end{aligned}
$$

If $\sum_{i=1}^{n} x_i = 0$, then $L(p) = (1-p)^n$, which is maximized over $p \in [0, 1]$ by taking $\widehat{p} = 0$. If, on the other hand, $\sum_{i=1}^{n} x_i = n$, then $L(p) = p^n$ and this is maximized over $p \in [0, 1]$ by taking $\widehat{p} = 1$. If $\sum_{i=1}^{n} x_i$ equals neither 0 nor n, then $L(0) = L(1) = 0$ while $L(p) > 0$ for all $p \in (0, 1)$; thus, in this case it suffices to maximize $L(p)$ for $0 < p < 1$, which we do by standard methods of calculus. The derivative of $L(p)$ is

$$L'(p) = (\Sigma x_i) p^{\Sigma x_i - 1}(1-p)^{n-\Sigma x_i} - (n - \Sigma x_i) p^{\Sigma x_i}(1-p)^{n-\Sigma x_i - 1}.$$

Setting this first derivative equal to zero gives us, with the restriction that $0 < p < 1$,

$$p^{\Sigma x_i}(1-p)^{n-\Sigma x_i}\left(\frac{\Sigma x_i}{p} - \frac{n - \Sigma x_i}{1-p}\right) = 0.$$

Since $0 < p < 1$, the preceding equation equals zero when

$$\frac{\Sigma x_i}{p} - \frac{n - \Sigma x_i}{1-p} = 0. \tag{6.4-1}$$

Multiplying each member of Equation 6.4-1 by $p(1 - p)$ and simplifying, we obtain

$$\sum_{i=1}^{n} x_i - np = 0$$

or, equivalently,

$$p = \frac{\sum_{i=1}^{n} x_i}{n} = \bar{x}.$$

It can be shown that $L''(\bar{x}) < 0$, so that $L(\bar{x})$ is a maximum. The corresponding statistic, namely, $(\sum_{i=1}^{n} X_i)/n = \bar{X}$, is called the **maximum likelihood estimator** and is denoted by $\hat{p}$; that is,

$$\hat{p} = \frac{1}{n} \sum_{i=1}^{n} X_i = \bar{X}.$$

When finding a maximum likelihood estimator, it is often easier to find the value of the parameter that maximizes the natural logarithm of the likelihood function rather than the value of the parameter that maximizes the likelihood function itself. Because the natural logarithm function is a strictly increasing function, the solutions will be the same. To see this, note that for $0 < p < 1$, the example we have been considering gives us

$$\ln L(p) = \left(\sum_{i=1}^{n} x_i \right) \ln p + \left(n - \sum_{i=1}^{n} x_i \right) \ln(1 - p).$$

To find the maximum, we set the first derivative equal to zero to obtain

$$\frac{d\left[\ln L(p)\right]}{dp} = \left(\sum_{i=1}^{n} x_i \right) \left(\frac{1}{p} \right) + \left(n - \sum_{i=1}^{n} x_i \right) \left(\frac{-1}{1-p} \right) = 0,$$

which is the same as Equation 6.4-1. Thus, the solution is $p = \bar{x}$ and the maximum likelihood estimator for p is $\hat{p} = \bar{X}$.

Motivated by the preceding example, we present the formal definition of maximum likelihood estimators (this definition is used in both the discrete and continuous cases).

Let $X_1, X_2, \ldots, X_n$ be a random sample from a distribution that depends on one or more unknown parameters $\theta_1, \theta_2, \ldots, \theta_m$ with pmf or pdf that is denoted by $f(x; \theta_1, \theta_2, \ldots, \theta_m)$. Suppose that $(\theta_1, \theta_2, \ldots, \theta_m)$ is restricted to a given parameter space Ω. Then the joint pmf or pdf of $X_1, X_2, \ldots, X_n$, namely,

$$L(\theta_1, \theta_2, \ldots, \theta_m) = f(x_1; \theta_1, \ldots, \theta_m) f(x_2; \theta_1, \ldots, \theta_m)$$
$$\cdots f(x_n; \theta_1, \ldots, \theta_m), \qquad (\theta_1, \theta_2, \ldots, \theta_m) \in \Omega,$$

when regarded as a function of $\theta_1, \theta_2, \ldots, \theta_m$, is called the **likelihood function**. Say

$$[u_1(x_1, \ldots, x_n), u_2(x_1, \ldots, x_n), \ldots, u_m(x_1, \ldots, x_n)]$$

is that m-tuple in Ω that maximizes $L(\theta_1, \theta_2, \ldots, \theta_m)$. Then

$$\widehat{\theta}_1 = u_1(X_1, \ldots, X_n),$$
$$\widehat{\theta}_2 = u_2(X_1, \ldots, X_n),$$
$$\vdots$$
$$\widehat{\theta}_m = u_m(X_1, \ldots, X_n)$$

are **maximum likelihood estimators** of $\theta_1, \theta_2, \ldots, \theta_m$, respectively; and the corresponding observed values of these statistics, namely,

$$u_1(x_1, \ldots, x_n), u_2(x_1, \ldots, x_n), \ldots, u_m(x_1, \ldots, x_n),$$

are called **maximum likelihood estimates**. In many practical cases, these estimators (and estimates) are unique.

For many applications, there is just one unknown parameter. In these cases, the likelihood function is given by

$$L(\theta) = \prod_{i=1}^{n} f(x_i; \theta).$$

Some additional examples will help clarify these definitions.

Example 6.4-1   Let $X_1, X_2, \ldots, X_n$ be a random sample from the exponential distribution with pdf

$$f(x; \theta) = \frac{1}{\theta} e^{-x/\theta}, \qquad 0 < x < \infty, \qquad \theta \in \Omega = \{\theta : 0 < \theta < \infty\}.$$

The likelihood function is given by

$$L(\theta) = L(\theta; x_1, x_2, \ldots, x_n)$$
$$= \left(\frac{1}{\theta} e^{-x_1/\theta}\right)\left(\frac{1}{\theta} e^{-x_2/\theta}\right) \cdots \left(\frac{1}{\theta} e^{-x_n/\theta}\right)$$
$$= \frac{1}{\theta^n} \exp\left(\frac{-\sum_{i=1}^{n} x_i}{\theta}\right), \qquad 0 < \theta < \infty.$$

The natural logarithm of $L(\theta)$ is

$$\ln L(\theta) = -(n) \ln(\theta) - \frac{1}{\theta} \sum_{i=1}^{n} x_i, \qquad 0 < \theta < \infty.$$

Thus,

$$\frac{d\left[\ln L(\theta)\right]}{d\theta} = \frac{-n}{\theta} + \frac{\sum_{i=1}^{n} x_i}{\theta^2} = 0.$$

The solution of this equation for θ is

$$\theta = \frac{1}{n} \sum_{i=1}^{n} x_i = \overline{x}.$$

Note that

$$\frac{d\,[\ln L(\theta)]}{d\theta} = \frac{1}{\theta}\left(-n + \frac{n\bar{x}}{\theta}\right) \begin{cases} > 0, & \theta < \bar{x}, \\ = 0, & \theta = \bar{x}, \\ < 0, & \theta > \bar{x}. \end{cases}$$

Hence, $\ln L(\theta)$ does have a maximum at $\bar{x}$, and it follows that the maximum likelihood estimator for θ is

$$\widehat{\theta} = \overline{X} = \frac{1}{n}\sum_{i=1}^{n} X_i.$$

■

Example 6.4-2

Let $X_1, X_2, \ldots, X_n$ be a random sample from the geometric distribution with pmf $f(x;p) = (1-p)^{x-1}p$, $x = 1, 2, 3, \ldots$. The likelihood function is given by

$$L(p) = (1-p)^{x_1-1}p(1-p)^{x_2-1}p\cdots(1-p)^{x_n-1}p$$

$$= p^n(1-p)^{\sum x_i - n}, \qquad 0 \le p \le 1.$$

The natural logarithm of $L(p)$ is

$$\ln L(p) = n \ln p + \left(\sum_{i=1}^{n} x_i - n\right)\ln(1-p), \qquad 0 < p < 1.$$

Thus, restricting p to $0 < p < 1$, so as to be able to take the derivative, we have

$$\frac{d \ln L(p)}{dp} = \frac{n}{p} - \frac{\sum_{i=1}^{n} x_i - n}{1-p} = 0.$$

Solving for p, we obtain

$$p = \frac{n}{\sum_{i=1}^{n} x_i} = \frac{1}{\bar{x}},$$

and, by the second derivative test, this solution provides a maximum. So the maximum likelihood estimator of p is

$$\widehat{p} = \frac{n}{\sum_{i=1}^{n} X_i} = \frac{1}{\overline{X}}.$$

This estimator agrees with our intuition because, in n observations of a geometric random variable, there are n successes in the $\sum_{i=1}^{n} x_i$ trials. Thus, the estimate of p is the number of successes divided by the total number of trials.

■

In the following important example, we find the maximum likelihood estimators of the parameters associated with the normal distribution.

Example 6.4-3

Let $X_1, X_2, \ldots, X_n$ be a random sample from $N(\theta_1, \theta_2)$, where

$$\Omega = \{(\theta_1, \theta_2): -\infty < \theta_1 < \infty, 0 < \theta_2 < \infty\}.$$

That is, here we let $\theta_1 = \mu$ and $\theta_2 = \sigma^2$. Then

$$L(\theta_1, \theta_2) = \prod_{i=1}^{n} \frac{1}{\sqrt{2\pi\theta_2}} \exp\left[-\frac{(x_i - \theta_1)^2}{2\theta_2}\right]$$

or, equivalently,

$$L(\theta_1,\theta_2) = \left(\frac{1}{\sqrt{2\pi\theta_2}}\right)^n \exp\left[\frac{-\sum_{i=1}^n(x_i-\theta_1)^2}{2\theta_2}\right], \qquad (\theta_1,\theta_2)\in\Omega.$$

The natural logarithm of the likelihood function is

$$\ln L(\theta_1,\theta_2) = -\frac{n}{2}\ln(2\pi\theta_2) - \frac{\sum_{i=1}^n(x_i-\theta_1)^2}{2\theta_2}.$$

The partial derivatives with respect to θ_1 and θ_2 are

$$\frac{\partial(\ln L)}{\partial\theta_1} = \frac{1}{\theta_2}\sum_{i=1}^n(x_i-\theta_1)$$

and

$$\frac{\partial(\ln L)}{\partial\theta_2} = \frac{-n}{2\theta_2} + \frac{1}{2\theta_2^2}\sum_{i=1}^n(x_i-\theta_1)^2.$$

The equation $\partial(\ln L)/\partial\theta_1 = 0$ has the solution $\theta_1 = \bar{x}$. Setting $\partial(\ln L)/\partial\theta_2 = 0$ and replacing θ_1 by $\bar{x}$ yields

$$\theta_2 = \frac{1}{n}\sum_{i=1}^n(x_i-\bar{x})^2.$$

By considering the usual condition on the second-order partial derivatives, we see that these solutions do provide a maximum. Thus, the maximum likelihood estimators of $\mu = \theta_1$ and $\sigma^2 = \theta_2$ are

$$\widehat{\theta_1} = \overline{X} \qquad\text{and}\qquad \widehat{\theta_2} = \frac{1}{n}\sum_{i=1}^n(X_i-\overline{X})^2 = V. \qquad\blacksquare$$

It is interesting to note that in our first illustration, where $\widehat{p} = \overline{X}$, and in Example 6.4-1, where $\widehat{\theta} = \overline{X}$, the expected value of the estimator is equal to the corresponding parameter. This observation leads to the following definition.

Definition 6.4-1
If $E[u(X_1,X_2,\ldots,X_n)] = \theta$, then the statistic $u(X_1,X_2,\ldots,X_n)$ is called an **unbiased estimator** of θ. Otherwise, it is said to be **biased**.

Example 6.4-4 Let $Y_1 < Y_2 < Y_3 < Y_4$ be the order statistics of a random sample X_1,X_2,X_3,X_4 from a uniform distribution with pdf $f(x;\theta) = 1/\theta$, $0 < x \leq \theta$. The likelihood function is

$$L(\theta) = \left(\frac{1}{\theta}\right)^4, \qquad 0 < x_i \leq \theta, \ i = 1,2,3,4,$$

and equals zero if $\theta < x_i$ or if $x_i \leq 0$. To maximize $L(\theta)$, we must make θ as small as possible; hence, the maximum likelihood estimator is

$$\widehat{\theta} = \max(X_i) = Y_4$$

because θ cannot be less than any X_i. Since $F(x; \theta) = x/\theta$, $0 < x \leq \theta$, the pdf of Y_4 is

$$g_4(y_4) = \frac{4!}{3!1!} \left(\frac{y_4}{\theta}\right)^3 \left(\frac{1}{\theta}\right) = 4\frac{y_4^3}{\theta^4}, \qquad 0 < y_4 \leq \theta.$$

Accordingly,

$$E(Y_4) = \int_0^{\theta} y_4 \cdot 4\frac{y_4^3}{\theta^4}\, dy_4 = \frac{4}{5}\theta$$

and Y_4 is a biased estimator of θ. However, $5Y_4/4$ is unbiased. ∎

Example 6.4-5

We have shown that when sampling from $N(\theta_1 = \mu, \theta_2 = \sigma^2)$, one finds that the maximum likelihood estimators of μ and σ^2 are

$$\widehat{\theta_1} = \widehat{\mu} = \overline{X} \qquad \text{and} \qquad \widehat{\theta_2} = \widehat{\sigma^2} = \frac{(n-1)S^2}{n}.$$

Recalling that the distribution of $\overline{X}$ is $N(\mu, \sigma^2/n)$, we see that $E(\overline{X}) = \mu$; thus, $\overline{X}$ is an unbiased estimator of μ.

In Theorem 5.5-2, we showed that the distribution of $(n-1)S^2/\sigma^2$ is $\chi^2(n-1)$. Hence,

$$E(S^2) = E\left[\frac{\sigma^2}{n-1} \frac{(n-1)S^2}{\sigma^2}\right] = \frac{\sigma^2}{n-1}(n-1) = \sigma^2.$$

That is, the sample variance

$$S^2 = \frac{1}{n-1}\sum_{i=1}^{n}(X_i - \overline{X})^2$$

is an unbiased estimator of σ^2. Consequently, since

$$E(\widehat{\theta_2}) = \frac{n-1}{n}E(S^2) = \frac{n-1}{n}\sigma^2,$$

$\widehat{\theta_2}$ is a biased estimator of $\theta_2 = \sigma^2$. ∎

Sometimes it is impossible to find maximum likelihood estimators in a convenient closed form, and numerical methods must be used to maximize the likelihood function. For example, suppose that $X_1, X_2, \ldots, X_n$ is a random sample from a gamma distribution with parameters $\alpha = \theta_1$ and $\beta = \theta_2$, where $\theta_1 > 0, \theta_2 > 0$. It is difficult to maximize

$$L(\theta_1, \theta_2; x_1, \ldots, x_n) = \left[\frac{1}{\Gamma(\theta_1)\theta_2^{\theta_1}}\right]^n (x_1 x_2 \cdots x_n)^{\theta_1 - 1} \exp\left(-\sum_{i=1}^{n} x_i/\theta_2\right)$$

with respect to θ_1 and θ_2, owing to the presence of the gamma function $\Gamma(\theta_1)$. Thus, numerical methods must be used to maximize L once $x_1, x_2, \ldots, x_n$ are observed.

There are other ways, however, to easily obtain point estimates of θ_1 and θ_2. One of the early methods was to simply equate the first sample moment to the first theoretical moment. Next, if needed, the two second moments are equated, then the third moments, and so on, until we have enough equations to solve for the parameters. As an illustration, in the gamma distribution situation, let us simply equate the first two moments of the distribution to the corresponding moments of the empirical distribution. This seems like a reasonable way in which to find estimators, since the empirical distribution converges in some sense to the probability distribution, and hence corresponding moments should be about equal. In this situation, we have

$$\theta_1 \theta_2 = \overline{X}, \qquad \theta_1 \theta_2^2 = V,$$

the solutions of which are

$$\widetilde{\theta}_1 = \frac{\overline{X}^2}{V} \qquad \text{and} \qquad \widetilde{\theta}_2 = \frac{V}{\overline{X}}.$$

We say that these latter two statistics, $\widetilde{\theta}_1$ and $\widetilde{\theta}_2$, are respective estimators of θ_1 and θ_2 found by the **method of moments**.

To generalize this discussion, let $X_1, X_2, \ldots, X_n$ be a random sample of size n from a distribution with pdf $f(x; \theta_1, \theta_2, \ldots, \theta_r)$, $(\theta_1, \ldots, \theta_r) \in \Omega$. The expectation $E(X^k)$ is frequently called the kth moment of the distribution, $k = 1, 2, 3, \ldots$. The sum $M_k = \sum_{i=1}^{n} X_i^k / n$ is the kth moment of the sample, $k = 1, 2, 3, \ldots$. The method of moments can be described as follows. Equate $E(X^k)$ to M_k, beginning with $k = 1$ and continuing until there are enough equations to provide unique solutions for $\theta_1, \theta_2, \ldots, \theta_r$ — say, $h_i(M_1, M_2, \ldots)$, $i = 1, 2, \ldots, r$, respectively. Note that this could be done in an equivalent manner by equating $\mu = E(X)$ to $\overline{X}$ and $E[(X - \mu)^k]$ to $\sum_{i=1}^{n} (X_i - \overline{X})^k / n$, $k = 2, 3$, and so on, until unique solutions for $\theta_1, \theta_2, \ldots, \theta_r$ are obtained. This alternative procedure was used in the preceding illustration. In most practical cases, the estimator $\widetilde{\theta}_i = h_i(M_1, M_2, \ldots)$ of θ_i, found by the method of moments, is an estimator of θ_i that in some sense gets close to that parameter when n is large, $i = 1, 2, \ldots, r$.

The next two examples—the first for a one-parameter family and the second for a two-parameter family—illustrate the method-of-moments technique for finding estimators.

Example 6.4-6 Let $X_1, X_2, \ldots, X_n$ be a random sample of size n from the distribution with pdf $f(x; \theta) = \theta x^{\theta - 1}$, $0 < x < 1$, $0 < \theta < \infty$. Sketch the graphs of this pdf for $\theta = 1/4$, 1, and 4. Note that sets of observations for these three values of θ would look very different. How do we estimate the value of θ? The mean of this distribution is given by

$$E(X) = \int_0^1 x \theta x^{\theta - 1} \, dx = \frac{\theta}{\theta + 1}.$$

We shall set the distribution mean equal to the sample mean and solve for θ. We have

$$\overline{x} = \frac{\theta}{\theta + 1}.$$

Solving for θ, we obtain the method-of-moments estimator,

$$\widetilde{\theta} = \frac{\overline{X}}{1 - \overline{X}}.$$

Thus, an estimate of θ by the method of moments is $\overline{x}/(1 - \overline{x})$. ■

Recall that in the method of moments, if two parameters have to be estimated, the first two sample moments are set equal to the first two distribution moments that are given in terms of the unknown parameters. These two equations are then solved simultaneously for the unknown parameters.

Example 6.4-7

Let the distribution of X be $N(\mu, \sigma^2)$. Then

$$E(X) = \mu \qquad \text{and} \qquad E(X^2) = \sigma^2 + \mu^2.$$

For a random sample of size n, the first two moments are given by

$$m_1 = \frac{1}{n} \sum_{i=1}^{n} x_i \qquad \text{and} \qquad m_2 = \frac{1}{n} \sum_{i=1}^{n} x_i^2.$$

We set $m_1 = E(X)$ and $m_2 = E(X^2)$ and solve for μ and σ^2. That is,

$$\frac{1}{n} \sum_{i=1}^{n} x_i = \mu \qquad \text{and} \qquad \frac{1}{n} \sum_{i=1}^{n} x_i^2 = \sigma^2 + \mu^2.$$

The first equation yields $\bar{x}$ as the estimate of μ. Replacing μ^2 with $\bar{x}^2$ in the second equation and solving for σ^2, we obtain

$$\frac{1}{n} \sum_{i=1}^{n} x_i^2 - \bar{x}^2 = \sum_{i=1}^{n} \frac{(x_i - \bar{x})^2}{n} = v$$

as the solution of σ^2. Thus, the method-of-moments estimators for μ and σ^2 are $\tilde{\mu} = \overline{X}$ and $\widetilde{\sigma^2} = V$, which are the same as the maximum likelihood estimators. Of course, $\tilde{\mu} = \overline{X}$ is unbiased, whereas $\widetilde{\sigma^2} = V$ is biased. ∎

In Example 6.4-5, we showed that $\overline{X}$ and S^2 are unbiased estimators of μ and σ^2, respectively, when one is sampling from a normal distribution. This is also true when one is sampling from any distribution with a finite variance σ^2. That is, $E(\overline{X}) = \mu$ and $E(S^2) = \sigma^2$, provided that the sample arises from a distribution with variance $\sigma^2 < \infty$. (See Exercise 6.4-11.) Although S^2 is an unbiased estimator of σ^2, S is a biased estimator of σ. In Exercise 6.4-14, you are asked to show that, when one is sampling from a normal distribution, cS is an unbiased estimator of σ, where

$$c = \frac{\sqrt{n-1}\, \Gamma\left(\dfrac{n-1}{2}\right)}{\sqrt{2}\, \Gamma\left(\dfrac{n}{2}\right)}.$$

REMARK Later we show that S^2 is an unbiased estimator of σ^2, provided it exists, for every distribution, not just the normal. ∎

Exercises

6.4-1. Let $X_1, X_2, \ldots, X_n$ be a random sample from $N(\mu, \sigma^2)$, where the mean $\theta = \mu$ is such that $-\infty < \theta < \infty$ and σ^2 is a known positive number. Show that the maximum likelihood estimator for θ is $\widehat{\theta} = \overline{X}$.

6.4-2. A random sample $X_1, X_2, \ldots, X_n$ of size n is taken from $N(\mu, \sigma^2)$, where the variance $\theta = \sigma^2$ is such that $0 < \theta < \infty$ and μ is a known real number. Show that the maximum likelihood estimator for θ is $\widehat{\theta} = (1/n)\sum_{i=1}^n (X_i - \mu)^2$ and that this estimator is an unbiased estimator of θ.

6.4-3. A random sample $X_1, X_2, \ldots, X_n$ of size n is taken from a Poisson distribution with a mean of λ, $0 < \lambda < \infty$.

(a) Show that the maximum likelihood estimator for λ is $\widehat{\lambda} = \overline{X}$.

(b) Let X equal the number of flaws per 100 feet of a used computer tape. Assume that X has a Poisson distribution with a mean of λ. If 40 observations of X yielded 5 zeros, 7 ones, 12 twos, 9 threes, 5 fours, 1 five, and 1 six, find the maximum likelihood estimate of λ.

6.4-4. For determining half-lives of radioactive isotopes, it is important to know what the background radiation is in a given detector over a specific period. The following data were taken in a γ-ray detection experiment over 98 ten-second intervals:

58	50	57	58	64	63	54	64	59	41	43	56	60	50
46	59	54	60	59	60	67	52	65	63	55	61	68	58
63	36	42	54	58	54	40	60	64	56	61	51	48	50
60	42	62	67	58	49	66	58	57	59	52	54	53	53
57	43	73	65	45	43	57	55	73	62	68	55	51	55
53	68	58	53	51	73	44	50	53	62	58	47	63	59
59	56	60	59	50	52	62	51	66	51	56	53	59	57

Assume that these data are observations of a Poisson random variable with mean λ.

(a) Find the values of $\overline{x}$ and s^2.

(b) What is the value of the maximum likelihood estimator of λ?

(c) Is S^2 an unbiased estimator of λ?

(d) Which of $\overline{x}$ and s^2 would you recommend for estimating λ? Why? You could compare the variance of $\overline{X}$ with the variance of S^2, which is

$$\text{Var}(S^2) = \frac{\lambda(2\lambda n + n - 1)}{n(n-1)}.$$

6.4-5. Let $X_1, X_2, \ldots, X_n$ be a random sample from distributions with the given probability density functions. In each case, find the maximum likelihood estimator $\widehat{\theta}$.

(a) $f(x; \theta) = (1/\theta^2)\, x\, e^{-x/\theta}$, $\quad 0 < x < \infty$, $\quad 0 < \theta < \infty$.

(b) $f(x; \theta) = (1/2\theta^3)\, x^2\, e^{-x/\theta}$, $\quad 0 < x < \infty$, $\quad 0 < \theta < \infty$.

(c) $f(x; \theta) = (1/2)\, e^{-|x-\theta|}$, $\quad -\infty < x < \infty$, $\quad -\infty < \theta < \infty$.

HINT: Finding θ involves minimizing $\sum |x_i - \theta|$, which is a difficult problem. When $n = 5$, do it for $x_1 = 6.1$, $x_2 = -1.1$, $x_3 = 3.2$, $x_4 = 0.7$, and $x_5 = 1.7$, and you will see the answer. (See also Exercise 2.2-8.)

6.4-6. Find the maximum likelihood estimates for $\theta_1 = \mu$ and $\theta_2 = \sigma^2$ if a random sample of size 12 from $N(\mu, \sigma^2)$ yielded the following values:

26.7	25.2	27.1	19.8	22.8
26.5	24.6	27.2	25.0	26.4
23.4	27.4			

6.4-7. Let $f(x; \theta) = \theta x^{\theta-1}$, $0 < x < 1$, $\theta \in \Omega = \{\theta : 0 < \theta < \infty\}$. Let $X_1, X_2, \ldots, X_n$ denote a random sample of size n from this distribution.

(a) Sketch the pdf of X for **(i)** $\theta = 1/2$, **(ii)** $\theta = 1$, and **(iii)** $\theta = 2$.

(b) Show that $\widehat{\theta} = -n/\ln\left(\prod_{i=1}^n X_i\right)$ is the maximum likelihood estimator of θ.

(c) For each of the following three sets of 10 observations from the given distribution, calculate the values of the maximum likelihood estimate and the method-of-moments estimate of θ:

(i)	0.0256	0.3051	0.0278	0.8971	0.0739
	0.3191	0.7379	0.3671	0.9763	0.0102
(ii)	0.9960	0.3125	0.4374	0.7464	0.8278
	0.9518	0.9924	0.7112	0.2228	0.8609
(iii)	0.4698	0.3675	0.5991	0.9513	0.6049
	0.9917	0.1551	0.0710	0.2110	0.2154

6.4-8. Let $f(x; \theta) = (1/\theta)x^{(1-\theta)/\theta}$, $0 < x < 1$, $0 < \theta < \infty$.

(a) Show that the maximum likelihood estimator of θ is $\widehat{\theta} = -(1/n)\sum_{i=1}^n \ln X_i$.

(b) Show that $E(\widehat{\theta}) = \theta$ and thus that $\widehat{\theta}$ is an unbiased estimator of θ.

6.4-9. Let $X_1, X_2, \ldots, X_n$ be a random sample of size n from the exponential distribution whose pdf is $f(x; \theta) = (1/\theta)e^{-x/\theta}$, $0 < x < \infty$, $0 < \theta < \infty$.

(a) Show that $\overline{X}$ is an unbiased estimator of θ.

(b) Show that the variance of $\overline{X}$ is θ^2/n.

(c) What is a good estimate of θ if a random sample of size 5 yielded the sample values 3.5, 8.1, 0.9, 4.4, and 0.5?

6.4-10. Let $X_1, X_2, \ldots, X_n$ be a random sample of size n from a geometric distribution for which p is the probability of success.

(a) Use the method of moments to find a point estimate for p.

(b) Explain intuitively why your estimate makes good sense.

(c) Use the following data to give a point estimate of p:

3   34   7   4   19   2   1   19   43   2

22   4   19   11   7   1   2   21   15   16

6.4-11. Let $X_1, X_2, \ldots, X_n$ be a random sample from a distribution having finite variance σ^2. Show that

$$S^2 = \sum_{i=1}^{n} \frac{(X_i - \overline{X})^2}{n-1}$$

is an unbiased estimator of σ^2. HINT: Write

$$S^2 = \frac{1}{n-1}\left(\sum_{i=1}^{n} X_i^2 - n\overline{X}^2\right)$$

and compute $E(S^2)$.

6.4-12. Let $X_1, X_2, \ldots, X_n$ be a random sample from $b(1, p)$ (i.e., n Bernoulli trials). Thus,

$$Y = \sum_{i=1}^{n} X_i \text{ is } b(n, p).$$

(a) Show that $\overline{X} = Y/n$ is an unbiased estimator of p.

(b) Show that $\mathrm{Var}(\overline{X}) = p(1-p)/n$.

(c) Show that $E[\overline{X}(1-\overline{X})/n] = (n-1)[p(1-p)/n^2]$.

(d) Find the value of c so that $c\overline{X}(1-\overline{X})$ is an unbiased estimator of $\mathrm{Var}(\overline{X}) = p(1-p)/n$.

6.4-13. Let $X_1, X_2, \ldots, X_n$ be a random sample from a uniform distribution on the interval $(\theta - 1, \theta + 1)$.

(a) Find the method-of-moments estimator of θ.

(b) Is your estimator in part (a) an unbiased estimator of θ?

(c) Given the following $n = 5$ observations of X, give a point estimate of θ:

6.61   7.70   6.98   8.36   7.26

(d) The method-of-moments estimator actually has greater variance than the maximum likelihood estimator of θ, namely $[\min(X_i) + \max(X_i)]/2$. Compute the value of the latter estimator for the $n = 5$ observations in (c).

6.4-14. Let $X_1, X_2, \ldots, X_n$ be a random sample of size n from a normal distribution.

(a) Show that an unbiased estimator of σ is cS, where

$$c = \frac{\sqrt{n-1}\,\Gamma\left(\dfrac{n-1}{2}\right)}{\sqrt{2}\,\Gamma\left(\dfrac{n}{2}\right)}.$$

HINT: Recall that the distribution of $(n-1)S^2/\sigma^2$ is $\chi^2(n-1)$.

(b) Find the value of c when $n = 5$; when $n = 6$.

(c) Graph c as a function of n. What is the limit of c as n increases without bound?

6.4-15. Given the following 25 observations from a gamma distribution with mean $\mu = \alpha\theta$ and variance $\sigma^2 = \alpha\theta^2$, use the method-of-moments estimators to find point estimates of α and θ:

6.9 7.3 6.7 6.4 6.3 5.9 7.0 7.1 6.5 7.6 7.2 7.1 6.1

7.3 7.6 7.6 6.7 6.3 5.7 6.7 7.5 5.3 5.4 7.4 6.9

6.4-16. An urn contains 36 balls, of which N_1 are red and N_2 are blue. A random sample of $n = 6$ balls is selected from the urn without replacement, and X is equal to the number of red balls in the sample. This experiment was repeated 20 times (the 6 balls being returned to the urn before each repetition), yielding the following data:

2   1   0   1   1   1   4   2   3   1

0   2   1   0   0   3   2   1   0   0

Using these data, guess the value of N_1 and give a reason for your guess.

6.4-17. Let the pdf of X be defined by

$$f(x) = \begin{cases} \left(\dfrac{4}{\theta^2}\right)x, & 0 < x \le \dfrac{\theta}{2}, \\[2mm] -\left(\dfrac{4}{\theta^2}\right)x + \dfrac{4}{\theta}, & \dfrac{\theta}{2} < x \le \theta, \\[2mm] 0, & \text{elsewhere,} \end{cases}$$

where $\theta \in \Omega = \{\theta : 0 < \theta \le 2\}$.

(a) Sketch the graph of this pdf when $\theta = 1/2, \theta = 1$, and $\theta = 2$.

(b) Find an estimator of θ by the method of moments.

(c) For the following observations of X, give a point estimate of θ:

0.3206   0.2408   0.2577   0.3557   0.4188

0.5601   0.0240   0.5422   0.4532   0.5592

6.4-18. Let independent random samples, each of size n, be taken from the k normal distributions with means $\mu_j = c + d[j - (k+1)/2]$, $j = 1, 2, \ldots, k$, respectively, and common variance σ^2. Find the maximum likelihood estimators of c and d.

6.4-19. Let the independent normal random variables $Y_1, Y_2, \ldots, Y_n$ have the respective distributions $N(\mu, \gamma^2 x_i^2)$, $i = 1, 2, \ldots, n$, where $x_1, x_2, \ldots, x_n$ are known but not all the same and no one of which is equal to zero. Find the maximum likelihood estimators for μ and γ^2.

6.5 A SIMPLE REGRESSION PROBLEM

There is often interest in the relation between two variables—for example, the temperature at which a certain chemical reaction is performed and the yield of a chemical compound resulting from the reaction. Frequently, one of these variables, say, x, is known in advance of the other, so there is interest in predicting a future random variable Y. Since Y is a random variable, we cannot predict its future observed value $Y = y$ with certainty. Let us first concentrate on the problem of estimating the mean of Y—that is, $E(Y \mid x)$. Now, $E(Y \mid x)$ is usually a function of x. For example, in our illustration with the yield, say Y, of the chemical reaction, we might expect $E(Y \mid x)$ to increase with increasing temperature x. Sometimes $E(Y \mid x) = \mu(x)$ is assumed to be of a given form, such as linear, quadratic, or exponential; that is, $\mu(x)$ could be assumed to be equal to $\alpha + \beta x$, $\alpha + \beta x + \gamma x^2$, or $\alpha e^{\beta x}$. To estimate $E(Y \mid x) = \mu(x)$, or, equivalently, the parameters α, β, and γ, we observe the random variable Y for each of n possibly different values of x—say, $x_1, x_2, \ldots, x_n$. Once the n independent experiments have been performed, we have n pairs of known numbers $(x_1, y_1), (x_2, y_2), \ldots, (x_n, y_n)$. These pairs are then used to estimate the mean $E(Y \mid x)$. Problems like this are often classified under **regression** because $E(Y \mid x) = \mu(x)$ is frequently called a regression curve.

REMARK A model for the mean that is of the form $\alpha + \beta x + \gamma x^2$ is called a linear model because it is linear in the parameters, α, β, and γ. Note, however, that a plot of this model versus x is not a straight line unless $\gamma = 0$. Thus, a linear model may be nonlinear in x. On the other hand, $\alpha e^{\beta x}$ is not a linear model, because it is not linear in α and β. ∎

Let us begin with the case in which $E(Y \mid x) = \mu(x)$ is a linear function of x. The data points are $(x_1, y_1), (x_2, y_2), \ldots, (x_n, y_n)$, so the first problem is that of fitting a straight line to the set of data. (See Figure 6.5-1.) In addition to assuming that the mean of Y is a linear function, we assume that, for a particular value of x, the value of Y will differ from its mean by a random amount ε. We further assume that the distribution of ε is $N(0, \sigma^2)$. So we have, for our linear model,

$$Y_i = \alpha_1 + \beta x_i + \varepsilon_i,$$

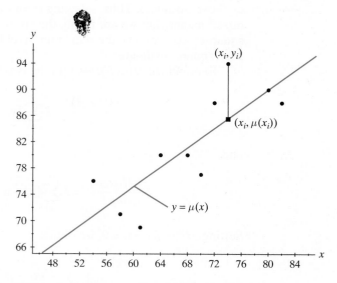

Figure 6.5-1 Scatter plot and the line $y = \mu(x)$

where ε_i, for $i = 1, 2, \ldots, n$, are independent and $N(0, \sigma^2)$. The unknown parameters α_1 and β are the Y-intercept and slope, respectively, of the line $\mu(x) = \alpha_1 + \beta x$.

We shall now find point estimates, specifically maximum likelihood estimates, for α_1, β, and σ^2. For convenience, we let $\alpha_1 = \alpha - \beta \bar{x}$, so that

$$Y_i = \alpha + \beta(x_i - \bar{x}) + \varepsilon_i, \text{ where } \bar{x} = \frac{1}{n} \sum_{i=1}^{n} x_i.$$

Then Y_i is equal to a nonrandom quantity, $\alpha + \beta(x_i - \bar{x})$, plus a mean-zero normal random variable ε_i. Hence, $Y_1, Y_2, \ldots, Y_n$ are mutually independent normal variables with respective means $\alpha + \beta(x_i - \bar{x})$, $i = 1, 2, \ldots, n$, and unknown variance σ^2. Their joint pdf is therefore the product of the individual probability density functions; that is, the likelihood function equals

$$L(\alpha, \beta, \sigma^2) = \prod_{i=1}^{n} \frac{1}{\sqrt{2\pi\sigma^2}} \exp\left\{-\frac{[y_i - \alpha - \beta(x_i - \bar{x})]^2}{2\sigma^2}\right\}$$

$$= \left(\frac{1}{2\pi\sigma^2}\right)^{n/2} \exp\left\{-\frac{\sum_{i=1}^{n} [y_i - \alpha - \beta(x_i - \bar{x})]^2}{2\sigma^2}\right\}.$$

To maximize $L(\alpha, \beta, \sigma^2)$ or, equivalently, to minimize

$$-\ln L(\alpha, \beta, \sigma^2) = \frac{n}{2} \ln(2\pi\sigma^2) + \frac{\sum_{i=1}^{n} [y_i - \alpha - \beta(x_i - \bar{x})]^2}{2\sigma^2},$$

we must select α and β to minimize

$$H(\alpha, \beta) = \sum_{i=1}^{n} [y_i - \alpha - \beta(x_i - \bar{x})]^2.$$

Since $|y_i - \alpha - \beta(x_i - \bar{x})| = |y_i - \mu(x_i)|$ is the vertical distance from the point (x_i, y_i) to the line $y = \mu(x)$, we note that $H(\alpha, \beta)$ represents the sum of the squares of those distances. Thus, selecting α and β so that the sum of the squares is minimized means that we are fitting the straight line to the data by the **method of least squares**. Accordingly, the maximum likelihood estimates of α and β are also called **least squares estimates.**

To minimize $H(\alpha, \beta)$, we find the two first-order partial derivatives

$$\frac{\partial H(\alpha, \beta)}{\partial \alpha} = 2 \sum_{i=1}^{n} [y_i - \alpha - \beta(x_i - \bar{x})](-1)$$

and

$$\frac{\partial H(\alpha, \beta)}{\partial \beta} = 2 \sum_{i=1}^{n} [y_i - \alpha - \beta(x_i - \bar{x})][-(x_i - \bar{x})].$$

Setting $\partial H(\alpha, \beta)/\partial \alpha = 0$, we obtain

$$\sum_{i=1}^{n} y_i - n\alpha - \beta \sum_{i=1}^{n} (x_i - \bar{x}) = 0.$$

Since

$$\sum_{i=1}^{n} (x_i - \bar{x}) = 0,$$

we have

$$\sum_{i=1}^{n} y_i - n\alpha = 0;$$

thus,

$$\widehat{\alpha} = \bar{Y}.$$

With α replaced by $\bar{y}$, the equation $\partial H(\alpha, \beta)/\partial \beta = 0$ yields

$$\sum_{i=1}^{n} (y_i - \bar{y})(x_i - \bar{x}) - \beta \sum_{i=1}^{n} (x_i - \bar{x})^2 = 0$$

or, equivalently,

$$\widehat{\beta} = \frac{\sum_{i=1}^{n} (Y_i - \bar{Y})(x_i - \bar{x})}{\sum_{i=1}^{n} (x_i - \bar{x})^2} = \frac{\sum_{i=1}^{n} Y_i(x_i - \bar{x})}{\sum_{i=1}^{n} (x_i - \bar{x})^2}.$$

Standard methods of multivariate calculus can be used to show that this solution obtained by equating the first-order partial derivatives of $H(\alpha, \beta)$ to zero is indeed a point of minimum. Hence, the line that best estimates the mean line, $\mu(x) = \alpha + \beta(x_i - \bar{x})$, is $\widehat{\alpha} + \widehat{\beta}(x_i - \bar{x})$, where

$$\widehat{\alpha} = \bar{y} \tag{6.5-1}$$

and

$$\widehat{\beta} = \frac{\sum_{i=1}^{n} y_i(x_i - \bar{x})}{\sum_{i=1}^{n} (x_i - \bar{x})^2} = \frac{\sum_{i=1}^{n} x_i y_i - \left(\frac{1}{n}\right)\left(\sum_{i=1}^{n} x_i\right)\left(\sum_{i=1}^{n} y_i\right)}{\sum_{i=1}^{n} x_i^2 - \left(\frac{1}{n}\right)\left(\sum_{i=1}^{n} x_i\right)^2}. \tag{6.5-2}$$

To find the maximum likelihood estimator of σ^2, consider the partial derivative

$$\frac{\partial[-\ln L(\alpha, \beta, \sigma^2)]}{\partial(\sigma^2)} = \frac{n}{2\sigma^2} - \frac{\sum_{i=1}^{n} [y_i - \alpha - \beta(x_i - \bar{x})]^2}{2(\sigma^2)^2}.$$

Setting this equal to zero and replacing α and β by their solutions $\widehat{\alpha}$ and $\widehat{\beta}$, we obtain

$$\widehat{\sigma^2} = \frac{1}{n} \sum_{i=1}^{n} [Y_i - \widehat{\alpha} - \widehat{\beta}(x_i - \bar{x})]^2. \tag{6.5-3}$$

A formula that is useful in calculating $n\widehat{\sigma^2}$ is

$$n\widehat{\sigma^2} = \sum_{i=1}^{n} y_i^2 - \frac{1}{n}\left(\sum_{i=1}^{n} y_i\right)^2 - \widehat{\beta}\sum_{i=1}^{n} x_i y_i + \widehat{\beta}\left(\frac{1}{n}\right)\left(\sum_{i=1}^{n} x_i\right)\left(\sum_{i=1}^{n} y_i\right). \tag{6.5-4}$$

Note that the summand in Equation 6.5-3 for $\widehat{\sigma^2}$ is the square of the difference between the value of Y_i and the estimated mean of Y_i. Let $\widehat{Y}_i = \widehat{\alpha} + \widehat{\beta}(x_i - \bar{x})$, the estimated mean value of Y_i, given x. The difference

$$Y_i - \widehat{Y}_i = Y_i - \widehat{\alpha} - \widehat{\beta}(x_i - \bar{x})$$

Table 6.5-1 Calculations for test score data

x	y	x^2	xy	y^2	$\widehat{y}$	$y - \widehat{y}$	$(y - \widehat{y})^2$
70	77	4,900	5,390	5,929	82.561566	−5.561566	30.931016
74	94	5,476	6,956	8,836	85.529956	8.470044	71.741645
72	88	5,184	6,336	7,744	84.045761	3.954239	15.636006
68	80	4,624	5,440	6,400	81.077371	−1.077371	1.160728
58	71	3,364	4,118	5,041	73.656395	−2.656395	7.056434
54	76	2,916	4,104	5,776	70.688004	5.311996	28.217302
82	88	6,724	7,216	7,744	91.466737	−3.466737	12.018265
64	80	4,096	5,120	6,400	78.108980	1.891020	3.575957
80	90	6,400	7,200	8,100	89.982542	0.017458	0.000305
61	69	3,721	4,209	4,761	75.882687	−6.882687	47.371380
683	813	47,405	56,089	66,731	812.999999	0.000001	217.709038

is called the ith **residual**, $i = 1, 2, \ldots, n$. The maximum likelihood estimate of σ^2 is then the sum of the squares of the residuals divided by n. It should always be true that the sum of the residuals is equal to zero. However, in practice, due to rounding off, the sum of the observed residuals, $y_i - \widehat{y}_i$, sometimes differs slightly from zero. A graph of the residuals plotted as a scatter plot of the points $x_i, y_i - \widehat{y}_i, i = 1, 2, \ldots, n$, can show whether or not linear regression provides the best fit.

Example 6.5-1

The data plotted in Figure 6.5-1 are 10 pairs of test scores of 10 students in a psychology class, x being the score on a preliminary test and y the score on the final examination. The values of x and y are shown in Table 6.5-1. The sums that are needed to calculate estimates of the parameters are also given. Of course, the estimates of α and β have to be found before the residuals can be calculated.

Thus, $\widehat{\alpha} = 813/10 = 81.3$, and

$$\widehat{\beta} = \frac{56,089 - (683)(813)/10}{47,405 - (683)(683)/10} = \frac{561.1}{756.1} = 0.742.$$

Since $\bar{x} = 683/10 = 68.3$, the least squares regression line is

$$\widehat{y} = 81.3 + (0.742)(x - 68.3).$$

The maximum likelihood estimate of σ^2 is

$$\widehat{\sigma^2} = \frac{217.709038}{10} = 21.7709.$$

A plot of the residuals for these data is shown in Figure 6.5-2. ∎

We shall now consider the problem of finding the distributions of $\widehat{\alpha}$, $\widehat{\beta}$, and $\widehat{\sigma^2}$ (or distributions of functions of these estimators). We would like to be able to

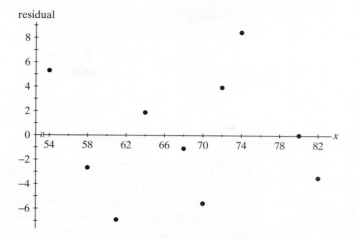

Figure 6.5-2 Residuals plot for data in Table 6.5-1

say something about the error of the estimates to find confidence intervals for the parameters.

The preceding discussion treated $x_1, x_2, \ldots, x_n$ as nonrandom constants. Of course, many times they can be set by the experimenter; for example, an experimental chemist might produce a compound at many different temperatures. But these numbers might instead be observations on an earlier random variable, such as an SAT score or a preliminary test grade (as in Example 6.5-1). Nevertheless, we consider the problem on the condition that the x-values are given in either case. Thus, in finding the distributions of $\widehat{\alpha}$, $\widehat{\beta}$, and $\widehat{\sigma^2}$, the only random variables are $Y_1, Y_2, \ldots, Y_n$.

Since $\widehat{\alpha}$ is a linear function of independent and normally distributed random variables, $\widehat{\alpha}$ has a normal distribution with mean

$$E(\widehat{\alpha}) = E\left(\frac{1}{n} \sum_{i=1}^{n} Y_i\right) = \frac{1}{n} \sum_{i=1}^{n} E(Y_i)$$

$$= \frac{1}{n} \sum_{i=1}^{n} [\alpha + \beta(x_i - \overline{x})] = \alpha$$

and variance

$$\mathrm{Var}(\widehat{\alpha}) = \left(\frac{1}{n}\right)^2 \sum_{i=1}^{n} \mathrm{Var}(Y_i) = \frac{\sigma^2}{n}.$$

The estimator $\widehat{\beta}$ is also a linear function of $Y_1, Y_2, \ldots, Y_n$ and hence has a normal distribution with mean

$$E(\widehat{\beta}) = \frac{\sum_{i=1}^{n}(x_i - \overline{x})E(Y_i)}{\sum_{i=1}^{n}(x_i - \overline{x})^2}$$

$$= \frac{\sum_{i=1}^{n}(x_i - \overline{x})[\alpha + \beta(x_i - \overline{x})]}{\sum_{i=1}^{n}(x_i - \overline{x})^2}$$

$$= \frac{\alpha \sum_{i=1}^{n}(x_i - \overline{x}) + \beta \sum_{i=1}^{n}(x_i - \overline{x})^2}{\sum_{i=1}^{n}(x_i - \overline{x})^2} = \beta$$

and variance

$$Var(\widehat{\beta}) = \sum_{i=1}^{n} \left[\frac{x_i - \bar{x}}{\sum_{j=1}^{n} (x_j - \bar{x})^2} \right]^2 Var(Y_i)$$

$$= \frac{\sum_{i=1}^{n} (x_i - \bar{x})^2}{\left[\sum_{i=1}^{n} (x_i - \bar{x})^2 \right]^2} \sigma^2 = \frac{\sigma^2}{\sum_{i=1}^{n} (x_i - \bar{x})^2}.$$

STATISTICAL COMMENTS We now give an illustration (see Ledolter and Hogg in the References) using data from the *Challenger* explosion on January 28, 1986. It would not be appropriate to actually carry out an analysis of these data using the regression methods introduced in this section, for they require the variables to be continuous while in this case the *Y* variable is discrete. Rather, we present the illustration to make the point that it can be very important to examine the relationship between two variables, and to do so using all available data.

The *Challenger* space shuttle was launched from Cape Kennedy in Florida on a very cold January morning. Meteorologists had forecasted temperatures (as of January 27) in the range of 26°–29° Fahrenheit. The night before the launch there was much debate among engineers and NASA officials whether a launch under such low-temperature conditions would be advisable. Several engineers advised against a launch because they thought that O-ring failures were related to temperature. Data on O-ring failures experienced in previous launches were available and were studied the night before the launch. There were seven previous incidents of known distressed O-rings. Figure 6.5-3(a) displays this information; it is a simple scatter plot of the number of distressed rings per launch against temperature at launch.

From this plot alone, there does not seem to be a strong relationship between the number of O-ring failures and temperature. On the basis of this information, along with many other technical and political considerations, it was decided to launch the *Challenger* space shuttle. As you all know, the launch resulted in disaster: the loss of seven lives and billions of dollars, and a serious setback to the space program.

One may argue that engineers looked at the scatter plot of the number of failures against temperature but could not see a relationship. However, this argument misses the fact that engineers did not display *all the data that were relevant to the question*. They looked only at instances in which there were failures; they ignored

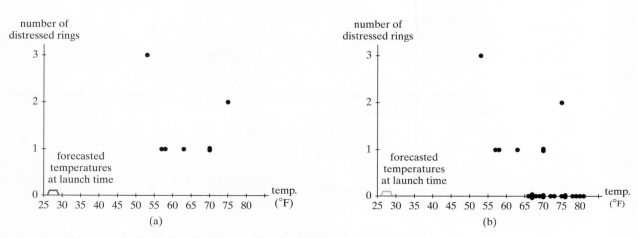

Figure 6.5-3 Number of distressed rings per launch versus temperature

the cases where there were no failures. In fact, there were 17 previous launches in which no failures occurred. A scatter plot of the number of distressed O-rings per launch against temperature using data from all previous shuttle launches is given in Figure 6.5-3(b).

It is difficult to look at these data and not see a relationship between failures and temperature. Moreover, one recognizes that an extrapolation is required and that an inference about the number of failures outside the observed range of temperature is needed. The actual temperature at launch was 31°F, while the lowest temperature recorded at a previous launch was 53°F. It is always very dangerous to extrapolate inferences to a region for which one does not have data. If NASA officials had looked at this plot, certainly the launch would have been delayed. This example shows why it is important to have statistically minded engineers involved in important decisions.

These comments raise two interesting points: (1) It is important to produce a scatter plot of one variable against another. (2) It is also important to plot *relevant data*. Yes, it is true that some data were used in making the decision to launch the *Challenger*. But not all the relevant data were utilized. To make good decisions, it takes knowledge of statistics as well as subject knowledge, common sense, and an ability to question the relevance of information. ■

Exercises

6.5-1. Show that the residuals, $Y_i - \widehat{Y}_i$ $(i = 1, 2, \ldots, n)$, from the least squares fit of the simple linear regression model sum to zero.

6.5-2. In some situations where the regression model is useful, it is known that the mean of Y when $X = 0$ is equal to 0, i.e., $Y_i = \beta x_i + \varepsilon_i$ where ε_i for $i = 1, 2, \ldots, n$ are independent and $N(0, \sigma^2)$.

(a) Obtain the maximum likelihood estimators, $\widehat{\beta}$ and $\widehat{\sigma^2}$, of β and σ^2 under this model.

(b) Find the distributions of $\widehat{\beta}$ and $\widehat{\sigma^2}$. (You may use, without proof, the fact that $\widehat{\beta}$ and $\widehat{\sigma^2}$ are independent, together with Theorem 9.3-1.)

6.5-3. The midterm and final exam scores of 10 students in a statistics course are tabulated as shown.

(a) Calculate the least squares regression line for these data.

(b) Plot the points and the least squares regression line on the same graph.

(c) Find the value of $\widehat{\sigma^2}$.

Midterm	Final	Midterm	Final
70	87	67	73
74	79	70	83
80	88	64	79
84	98	74	91
80	96	82	94

6.5-4. The final grade in a calculus course was predicted on the basis of the student's high school grade point average in mathematics, Scholastic Aptitude Test (SAT) score in mathematics, and score on a mathematics entrance examination. The predicted grades x and the earned grades y for 10 students are given (2.0 represents a C, 2.3 a C+, 2.7 a B−, etc.).

(a) Calculate the least squares regression line for these data.

(b) Plot the points and the least squares regression line on the same graph.

(c) Find the value of $\widehat{\sigma^2}$.

x	y	x	y
2.0	1.3	2.7	3.0
3.3	3.3	4.0	4.0
3.7	3.3	3.7	3.0
2.0	2.0	3.0	2.7
2.3	1.7	2.3	3.0

6.5-5. A student who considered himself to be a "car guy" was interested in how the horsepower and weight of a car affected the time that it takes the car to go from 0 to 60 mph. The following table gives, for each of 14 cars, the horsepower, the time in seconds to go from 0 to 60 mph, and the weight in pounds:

Horsepower	0–60	Weight	Horsepower	0–60	Weight
230	8.1	3516	282	6.2	3627
225	7.8	3690	300	6.4	3892
375	4.7	2976	220	7.7	3377
322	6.6	4215	250	7.0	3625
190	8.4	3761	315	5.3	3230
150	8.4	2940	200	6.2	2657
178	7.2	2818	300	5.5	3518

(a) Calculate the least squares regression line for "0–60" versus horsepower.

(b) Plot the points and the least squares regression line on the same graph.

(c) Calculate the least squares regression line for "0–60" versus weight.

(d) Plot the points and the least squares regression line on the same graph.

(e) Which of the two variables, horsepower or weight, has the most effect on the "0–60" time?

6.5-6. Let x and y equal the ACT scores in social science and natural science, respectively, for a student who is applying for admission to a small liberal arts college. A sample of $n = 15$ such students yielded the following data:

x	y	x	y	x	y
32	28	30	27	26	32
23	25	17	23	16	22
23	24	20	30	21	28
23	32	17	18	24	31
26	31	18	18	30	26

(a) Calculate the least squares regression line for these data.

(b) Plot the points and the least squares regression line on the same graph.

(c) Find point estimates for α, β, and σ^2.

6.5-7. The Federal Trade Commission measured the number of milligrams of tar and carbon monoxide (CO) per cigarette for all domestic cigarettes. Let x and y equal the measurements of tar and CO, respectively, for 100-millimeter filtered and mentholated cigarettes. A sample of 12 brands yielded the following data:

Brand	x	y	Brand	x	y
Capri	9	6	Now	3	4
Carlton	4	6	Salem	17	18
Kent	14	14	Triumph	6	8
Kool Milds	12	12	True	7	8
Marlboro Lights	10	12	Vantage	8	13
Merit Ultras	5	7	Virginia Slims	15	13

(a) Calculate the least squares regression line for these data.

(b) Plot the points and the least squares regression line on the same graph.

(c) Find point estimates for α, β, and σ^2.

6.5-8. The data in the following table, part of a set of data collected by Ledolter and Hogg (see References), provide the number of miles per gallon (mpg) for city and highway driving of 2007 midsize-model cars, as well as the curb weight of the cars:

Type	mpg City	mpg Hwy	Curb Weight
Ford Fusion V6 SE	20	28	3230
Chevrolet Sebring Sedan Base	24	32	3287
Toyota Camry Solara SE	24	34	3240
Honda Accord Sedan	20	29	3344
Audi A6 3.2	21	29	3825
BMW 5-series 525i Sedan	20	29	3450
Chrysler PT Cruiser Base	22	29	3076
Mercedes E-Class E350 Sedan	19	26	3740
Volkswagen Passat Sedan 2.0T	23	32	3305
Nissan Altima 2.5	26	35	3055
Kia Optima LX	24	34	3142

(a) Find the least squares regression line for highway mpg (y) and city mpg (x).

(b) Plot the points and the least squares regression line on the same graph.

(c) Repeat parts (a) and (b) for the regression of highway mpg (y) on curb weight (x).

6.5-9. Using an Instron 4204, rectangular strips of Plexiglas® were stretched to failure in a tensile test. The following data give the change in length, in millimeters (mm), before breaking (x) and the cross–sectional area in square millimeters (mm^2) (y):

(5.28, 52.36) (5.40, 52.58) (4.65, 51.07) (4.76, 52.28) (5.55, 53.02)

(5.73, 52.10) (5.84, 52.61) (4.97, 52.21) (5.50, 52.39) (6.24, 53.77)

(a) Find the equation of the least squares regression line.

(b) Plot the points and the line on the same graph.

(c) Interpret your output.

6.5-10. The "golden ratio" is $\phi = (1 + \sqrt{5})/2$. John Putz, a mathematician who was interested in music, analyzed Mozart's sonata movements, which are divided into two distinct sections, both of which are repeated in performance (see References). The length of the "Exposition" in measures is represented by a and the length of the "Development and Recapitulation" is represented by b. Putz's conjecture was that Mozart divided his movements close to the golden ratio. That is, Putz was interested in studying whether a scatter plot of $a + b$ against b not only would be linear, but also would actually fall along the line $y = \phi x$. Here are the data in tabular form, in which the first column identifies the piece and movement by the Köchel cataloging system:

(a) Make a scatter plot of the points $a + b$ against the points b. Is this plot linear?

(b) Find the equation of the least squares regression line. Superimpose it on the scatter plot.

Köchel	a	b	$a + b$	Köchel	a	b	$a + b$
279, I	38	62	100	279, II	28	46	74
279, III	56	102	158	280, I	56	88	144
280, II	24	36	60	280, III	77	113	190
281, I	40	69	109	281, II	46	60	106
282, I	15	18	33	282, III	39	63	102
283, I	53	67	120	283, II	14	23	37
283, III	102	171	273	284, I	51	76	127
309, I	58	97	155	311, I	39	73	112
310, I	49	84	133	330, I	58	92	150
330, III	68	103	171	332, I	93	136	229
332, III	90	155	245	333, I	63	102	165
333, II	31	50	81	457, I	74	93	167
533, I	102	137	239	533, II	46	76	122
545, I	28	45	73	547a, I	78	118	196
570, I	79	130	209				

(c) On the scatter plot, superimpose the line $y = \phi x$. Compare this line with the least squares regression line (graphically if you wish).

(d) Find the sample mean of the points $(a + b)/b$. Is the mean close to ϕ?

6.6* ASYMPTOTIC DISTRIBUTIONS OF MAXIMUM LIKELIHOOD ESTIMATORS

Let us consider a distribution of the continuous type with pdf $f(x; \theta)$ such that the parameter θ is not involved in the support of the distribution. Moreover, we want $f(x; \theta)$ to possess a number of mathematical properties that we do not list here. However, in particular, we want to be able to find the maximum likelihood estimator $\widehat{\theta}$ by solving

$$\frac{\partial [\ln L(\theta)]}{\partial \theta} = 0,$$

where here we use a partial derivative sign because $L(\theta)$ involves $x_1, x_2, \ldots, x_n$, too. That is,

$$\frac{\partial [\ln L(\widehat{\theta})]}{\partial \theta} = 0,$$

where now, with $\widehat{\theta}$ in this expression, $L(\widehat{\theta}) = f(X_1; \widehat{\theta})f(X_2; \widehat{\theta}) \cdots f(X_n; \widehat{\theta})$. We can approximate the left-hand member of this latter equation by a linear function found from the first two terms of a Taylor's series expanded about θ, namely,

$$\frac{\partial[\ln L(\theta)]}{\partial \theta} + (\widehat{\theta} - \theta)\frac{\partial^2[\ln L(\theta)]}{\partial \theta^2} \approx 0,$$

when $L(\theta) = f(X_1;\theta)f(X_2;\theta)\cdots f(X_n;\theta)$.

Obviously, this approximation is good enough only if $\widehat{\theta}$ is close to θ, and an adequate mathematical proof involves those conditions, which we have not given here. (See Hogg, McKean, and Craig, 2013.) But a heuristic argument can be made by solving for $\widehat{\theta} - \theta$ to obtain

$$\widehat{\theta} - \theta = \frac{\dfrac{\partial[\ln L(\theta)]}{\partial \theta}}{-\dfrac{\partial^2[\ln L(\theta)]}{\partial \theta^2}}. \tag{6.6-1}$$

Recall that

$$\ln L(\theta) = \ln f(X_1;\theta) + \ln f(X_2;\theta) + \cdots + \ln f(X_n;\theta)$$

and

$$\frac{\partial \ln L(\theta)}{\partial \theta} = \sum_{i=1}^{n} \frac{\partial[\ln f(X_i;\theta)]}{\partial \theta}, \tag{6.6-2}$$

which is the numerator in Equation 6.6-1. However, Equation 6.6-2 gives the sum of the n independent and identically distributed random variables

$$Y_i = \frac{\partial[\ln f(X_i;\theta)]}{\partial \theta}, \qquad i = 1,2,\ldots,n,$$

and thus, by the central limit theorem, has an approximate normal distribution with mean (in the continuous case) equal to

$$\int_{-\infty}^{\infty} \frac{\partial[\ln f(x;\theta)]}{\partial \theta} f(x;\theta)\,dx = \int_{-\infty}^{\infty} \frac{\partial[f(x;\theta)]}{\partial \theta} \frac{f(x;\theta)}{f(x;\theta)}\,dx$$

$$= \int_{-\infty}^{\infty} \frac{\partial[f(x;\theta)]}{\partial \theta}\,dx$$

$$= \frac{\partial}{\partial \theta}\left[\int_{-\infty}^{\infty} f(x;\theta)\,dx\right]$$

$$= \frac{\partial}{\partial \theta}[1]$$

$$= 0.$$

Clearly, we need a certain mathematical condition that makes it permissible to interchange the operations of integration and differentiation in those last steps. Of course, the integral of $f(x;\theta)$ is equal to 1 because it is a pdf.

Since we now know that the mean of each Y is

$$\int_{-\infty}^{\infty} \frac{\partial[\ln f(x;\theta)]}{\partial \theta} f(x;\theta)\,dx = 0,$$

let us take derivatives of each member of this equation with respect to θ, obtaining

$$\int_{-\infty}^{\infty} \left\{ \frac{\partial^2[\ln f(x;\theta)]}{\partial \theta^2} f(x;\theta) + \frac{\partial[\ln f(x;\theta)]}{\partial \theta} \frac{\partial[f(x;\theta)]}{\partial \theta} \right\} dx = 0.$$

However,

$$\frac{\partial[f(x;\theta)]}{\partial\theta} = \frac{\partial[\ln f(x;\theta)]}{\partial\theta} f(x;\theta);$$

so

$$\int_{-\infty}^{\infty} \left\{\frac{\partial[\ln f(x;\theta)]}{\partial\theta}\right\}^2 f(x;\theta)\, dx = -\int_{-\infty}^{\infty} \frac{\partial^2[\ln f(x;\theta)]}{\partial\theta^2} f(x;\theta)\, dx.$$

Since $E(Y) = 0$, this last expression provides the variance of $Y = \partial[\ln f(X;\theta)]/\partial\theta$. Then the variance of the sum in Equation 6.6-2 is n times this value, namely,

$$-nE\left\{\frac{\partial^2[\ln f(X;\theta)]}{\partial\theta^2}\right\}.$$

Let us rewrite Equation 6.6-1 as

$$\frac{\sqrt{n}\,(\widehat{\theta}-\theta)}{\left(\dfrac{1}{\sqrt{-E\{\partial^2[\ln f(X;\theta)]/\partial\theta^2\}}}\right)} = \frac{\left(\dfrac{\partial[\ln L(\theta)]/\partial\theta}{\sqrt{-nE\{\partial^2[\ln f(X;\theta)]/\partial\theta^2\}}}\right)}{\left(\dfrac{-\dfrac{1}{n}\dfrac{\partial^2[\ln L(\theta)]}{\partial\theta^2}}{E\{-\partial^2[\ln f(X;\theta)]/\partial\theta^2\}}\right)}. \tag{6.6-3}$$

Since it is the sum of n independent random variables (see Equation 6.6-2),

$$\partial[\ln f(X_i;\theta)]/\partial\theta, \qquad i = 1, 2, \ldots, n,$$

the numerator of the right-hand member of Equation 6.6-3 has an approximate $N(0,1)$ distribution, and the aforementioned unstated mathematical conditions require, in some sense, that

$$-\frac{1}{n}\frac{\partial^2[\ln L(\theta)]}{\partial\theta^2} \qquad \text{converge to} \qquad E\{-\partial^2[\ln f(X;\theta)]/\partial\theta^2\}.$$

Accordingly, the ratios given in Equation 6.6-3 must be approximately $N(0,1)$. That is, $\widehat{\theta}$ has an approximate normal distribution with mean θ and standard deviation

$$\frac{1}{\sqrt{-nE\{\partial^2[\ln f(X;\theta)]/\partial\theta^2\}}}.$$

Example 6.6-1 (Continuation of Example 6.4-1.) With the underlying exponential pdf

$$f(x;\theta) = \frac{1}{\theta}e^{-x/\theta}, \qquad 0 < x < \infty, \qquad \theta \in \Omega = \{\theta : 0 < \theta < \infty\},$$

$\overline{X}$ is the maximum likelihood estimator. Since

$$\ln f(x;\theta) = -\ln\theta - \frac{x}{\theta}$$

and

$$\frac{\partial[\ln f(x;\theta)]}{\partial\theta} = -\frac{1}{\theta} + \frac{x}{\theta^2} \qquad \text{and} \qquad \frac{\partial^2[\ln f(x;\theta)]}{\partial\theta} = \frac{1}{\theta^2} - \frac{2x}{\theta^3},$$

we have

$$-E\left[\frac{1}{\theta^2} - \frac{2X}{\theta^3}\right] = -\frac{1}{\theta^2} + \frac{2\theta}{\theta^3} = \frac{1}{\theta^2},$$

because $E(X) = \theta$. That is, $\overline{X}$ has an approximate normal distribution with mean θ and standard deviation $\theta/\sqrt{n}$. Thus, the random interval $\overline{X} \pm 1.96(\theta/\sqrt{n})$ has an approximate probability of 0.95 that it covers θ. Substituting the observed $\overline{x}$ for θ, as well as for $\overline{X}$, we say that $\overline{x} \pm 1.96\overline{x}/\sqrt{n}$ is an approximate 95% confidence interval for θ. ∎

While the development of the preceding result used a continuous-type distribution, the result holds for the discrete type also, as long as the support does not involve the parameter. This is illustrated in the next example.

Example 6.6-2 (Continuation of Exercise 6.4-3.) If the random sample arises from a Poisson distribution with pmf

$$f(x; \lambda) = \frac{\lambda^x e^{-\lambda}}{x!}, \qquad x = 0, 1, 2, \ldots; \qquad \lambda \in \Omega = \{\lambda : 0 < \lambda < \infty\},$$

then the maximum likelihood estimator for λ is $\widehat{\lambda} = \overline{X}$. Now,

$$\ln f(x; \lambda) = x \ln \lambda - \lambda - \ln x!.$$

Also,

$$\frac{\partial[\ln f(x; \lambda)]}{\partial \lambda} = \frac{x}{\lambda} - 1 \qquad \text{and} \qquad \frac{\partial^2[\ln f(x; \lambda)]}{\partial \lambda^2} = -\frac{x}{\lambda^2}.$$

Thus,

$$-E\left(-\frac{X}{\lambda^2}\right) = \frac{\lambda}{\lambda^2} = \frac{1}{\lambda},$$

and $\widehat{\lambda} = \overline{X}$ has an approximate normal distribution with mean λ and standard deviation $\sqrt{\lambda/n}$. Finally, $\overline{x} \pm 1.645\sqrt{\overline{x}/n}$ serves as an approximate 90% confidence interval for λ. With the data in Exercise 6.4-3, $\overline{x} = 2.225$, and it follows that this interval ranges from 1.837 to 2.613. ∎

It is interesting that there is another theorem which is somewhat related to the preceding result in that the variance of $\widehat{\theta}$ serves as a lower bound for the variance of every unbiased estimator of θ. Thus, we know that if a certain unbiased estimator has a variance equal to that lower bound, we cannot find a better one, and hence that estimator is the best in the sense of being the minimum-variance unbiased estimator. So, in the limit, the maximum likelihood estimator is this type of best estimator.

We describe this **Rao–Cramér inequality** here without proof. Let $X_1, X_2, \ldots, X_n$ be a random sample from a distribution of the continuous type with pdf $f(x; \theta)$, $\theta \in \Omega = \{\theta : c < \theta < d\}$, where the support of X does not depend upon θ, so that we can differentiate, with respect to θ, under integral signs like that in the following integral:

$$\int_{-\infty}^{\infty} f(x; \theta)\, dx = 1.$$

If $Y = u(X_1, X_2, \ldots, X_n)$ is an unbiased estimator of θ, then

$$\text{Var}(Y) \geq \frac{1}{n \int_{-\infty}^{\infty} \{[\partial \ln f(x;\theta)/\partial \theta]\}^2 f(x;\theta)\, dx}$$

$$= \frac{-1}{n \int_{-\infty}^{\infty} [\partial^2 \ln f(x;\theta)/\partial \theta^2]\, f(x;\theta)\, dx}.$$

Note that the integrals in the denominators are, respectively, the expectations

$$E\left\{\left[\frac{\partial \ln f(X;\theta)}{\partial \theta}\right]^2\right\} \qquad \text{and} \qquad E\left[\frac{\partial^2 \ln f(X;\theta)}{\partial \theta^2}\right];$$

sometimes one is easier to compute than the other. Note also that although the Rao–Cramér lower bound has been stated only for a continuous-type distribution, it is also true for a discrete-type distribution, with summations replacing integrals.

We have computed this lower bound for each of two distributions: exponential with mean θ and Poisson with mean λ. Those respective lower bounds were θ^2/n and λ/n. (See Examples 6.6-1 and 6.6-2.) Since, in each case, the variance of $\overline{X}$ equals the lower bound, then $\overline{X}$ is the minimum-variance unbiased estimator.

Let us consider another example.

Example 6.6-3

(Continuation of Exercise 6.4-7.) Let the pdf of X be given by

$$f(x;\theta) = \theta x^{\theta-1}, \qquad 0 < x < 1, \qquad \theta \in \Omega = \{\theta : 0 < \theta < \infty\}.$$

We then have

$$\ln f(x;\theta) = \ln \theta + (\theta - 1) \ln x,$$

$$\frac{\partial \ln f(x;\theta)}{\partial \theta} = \frac{1}{\theta} + \ln x,$$

and

$$\frac{\partial^2 \ln f(x;\theta)}{\partial \theta^2} = -\frac{1}{\theta^2}.$$

Since $E(-1/\theta^2) = -1/\theta^2$, the greatest lower bound of the variance of every unbiased estimator of θ is θ^2/n. Moreover, the maximum likelihood estimator $\widehat{\theta} = -n/\ln \prod_{i=1}^{n} X_i$ has an approximate normal distribution with mean θ and variance θ^2/n. Thus, in a limiting sense, $\widehat{\theta}$ is the minimum variance unbiased estimator of θ. ∎

To measure the value of estimators, their variances are compared with the Rao–Cramér lower bound. The ratio of the Rao–Cramér lower bound to the actual variance of any unbiased estimator is called the **efficiency** of that estimator. An estimator with an efficiency of, say, 50%, means that $1/0.5 = 2$ times as many sample observations are needed to do as well in estimation as can be done with the minimum variance unbiased estimator (the 100% efficient estimator).

Exercises

6.6-1. Let $X_1, X_2, \ldots, X_{10}$ be a random sample from $N(\theta, 20)$.

(a) Show that $Y = (X_1 + X_2)/2$ is an unbiased estimator of θ.

(b) Find the Rao–Cramér lower bound for the variance of an unbiased estimator of θ for $n = 10$.

(c) What is the efficiency of Y in part (a)?

6.6-2. Let $X_1, X_2, \ldots, X_{20}$ denote a random sample from $b(2, p)$. We know that $\overline{X}$ is an unbiased estimator of p and that $\mathrm{Var}(\overline{X}) = p(1 - p)/n$. (See Exercise 6.4-12.)

(a) Find the Rao–Cramér lower bound for the variance of every unbiased estimator of p.

(b) What is the efficiency of $\overline{X}$ as an estimator of p?

6.6-3. (Continuation of Exercise 6.4-2.) In sampling from a normal distribution with known mean μ, the maximum likelihood estimator of $\theta = \sigma^2$ is $\widehat{\theta} = \sum_{i=1}^{n}(X_i - \mu)^2/n$.

(a) Determine the Rao–Cramér lower bound.

(b) What is the approximate distribution of $\widehat{\theta}$?

(c) What is the exact distribution of $n\widehat{\theta}/\theta$, where $\theta = \sigma^2$?

6.6-4. Find the Rao–Cramér lower bound, and thus the asymptotic variance of the maximum likelihood estimator $\widehat{\theta}$, if the random sample $X_1, X_2, \ldots, X_n$ is taken from each of the distributions having the following pdfs:

(a) $f(x; \theta) = (1/\theta^2)\, x\, e^{-x/\theta}$, $\quad 0 < x < \infty$, $\quad 0 < \theta < \infty$.

(b) $f(x; \theta) = (1/2\theta^3)\, x^2\, e^{-x/\theta}$, $\quad 0 < x < \infty$, $\quad 0 < \theta < \infty$.

(c) $f(x; \theta) = (1/\theta)\, x^{(1-\theta)/\theta}$, $\quad 0 < x < 1$, $\quad 0 < \theta < \infty$.

6.7 SUFFICIENT STATISTICS

We first define a sufficient statistic $Y = u(X_1, X_2, \ldots, X_n)$ for a parameter, using a statement that, in most books, is given as a necessary and sufficient condition for sufficiency, namely, the well-known Fisher–Neyman factorization theorem. We do this because we find that readers at the introductory level can apply such a definition easily. However, using this definition, we shall note, by examples, its implications, one of which is also sometimes used as the definition of sufficiency. An understanding of Example 6.7-3 is most important in an appreciation of the value of sufficient statistics.

> **Definition 6.7-1**
> **(Factorization Theorem)** Let $X_1, X_2, \ldots, X_n$ denote random variables with joint pdf or pmf $f(x_1, x_2, \ldots, x_n; \theta)$, which depends on the parameter θ. The statistic $Y = u(X_1, X_2, \ldots, X_n)$ is sufficient for θ if and only if
>
> $$f(x_1, x_2, \ldots, x_n; \theta) = \phi[u(x_1, x_2, \ldots, x_n); \theta]h(x_1, x_2, \ldots, x_n),$$
>
> where ϕ depends on $x_1, x_2, \ldots, x_n$ only through $u(x_1, \ldots, x_n)$ and $h(x_1, \ldots, x_n)$ does not depend on θ.

Let us consider several important examples and consequences of this definition. We first note, however, that in all instances in this book the random variables $X_1, X_2, \ldots, X_n$ will be of a random sample, and hence their joint pdf or pmf will be of the form

$$f(x_1; \theta)f(x_2; \theta) \cdots f(x_n; \theta).$$

Example 6.7-1 Let $X_1, X_2, \ldots, X_n$ denote a random sample from a Poisson distribution with parameter $\lambda > 0$. Then

$$f(x_1;\lambda)f(x_2;\lambda)\cdots f(x_n;\lambda) = \frac{\lambda^{\Sigma x_i}e^{-n\lambda}}{x_1!x_2!\cdots x_n!} = (\lambda^{n\bar{x}}e^{-n\lambda})\left(\frac{1}{x_1!x_2!\cdots x_n!}\right),$$

where $\bar{x} = (1/n)\sum_{i=1}^{n} x_i$. Thus, from the factorization theorem (Definition 6.7-1), it is clear that the sample mean $\bar{X}$ is a sufficient statistic for λ. It can easily be shown that the maximum likelihood estimator for λ is also $\bar{X}$, so here the maximum likelihood estimator is a function of a sufficient statistic. ∎

In Example 6.7-1, if we replace $n\bar{x}$ by $\sum_{i=1}^{n} x_i$, it is quite obvious that the sum $\sum_{i=1}^{n} X_i$ is also a sufficient statistic for λ. This certainly agrees with our intuition, because if we know one of the statistics $\bar{X}$ and $\sum_{i=1}^{n} X_i$, we can easily find the other. If we generalize this idea, we see that if Y is sufficient for a parameter θ, then every single-valued function of Y not involving θ, but with a single-valued inverse, is also a sufficient statistic for θ. The reason is that if we know either Y or that function of Y, we know the other. More formally, if $W = v(Y) = v[u(X_1, X_2, \ldots, X_n)]$ is that function and $Y = v^{-1}(W)$ is the single-valued inverse, then the factorization theorem can be written as

$$f(x_1, x_2, \ldots, x_n; \theta) = \phi[v^{-1}\{v[u(x_1, x_2, \ldots, x_n)]\}; \theta]\, h(x_1, x_2, \ldots, x_n).$$

The first factor of the right-hand member of this equation depends on $x_1, x_2, \ldots, x_n$ through $v[u(x_1, x_2, \ldots, x_n)]$, so $W = v[u(X_1, X_2, \ldots, X_n)]$ is a sufficient statistic for θ. We illustrate this fact and the factorization theorem with an underlying distribution of the continuous type.

Example 6.7-2 Let $X_1, X_2, \ldots, X_n$ be a random sample from $N(\mu, 1)$, $-\infty < \mu < \infty$. The joint pdf of these random variables is

$$\frac{1}{(2\pi)^{n/2}} \exp\left[-\frac{1}{2}\sum_{i=1}^{n}(x_i - \mu)^2\right]$$

$$= \frac{1}{(2\pi)^{n/2}} \exp\left[-\frac{1}{2}\sum_{i=1}^{n}[(x_i - \bar{x}) + (\bar{x} - \mu)]^2\right]$$

$$= \left\{\exp\left[-\frac{n}{2}(\bar{x} - \mu)^2\right]\right\}\left\{\frac{1}{(2\pi)^{n/2}} \exp\left[-\frac{1}{2}\sum_{i=1}^{n}(x_i - \bar{x})^2\right]\right\}.$$

From the factorization theorem, we see that $\bar{X}$ is sufficient for μ. Now, $\bar{X}^3$ is also sufficient for μ, because knowing $\bar{X}^3$ is equivalent to having knowledge of the value of $\bar{X}$. However, $\bar{X}^2$ does not have this property, and it is not sufficient for μ. ∎

One extremely important consequence of the sufficiency of a statistic Y is that the conditional probability of any given event A in the support of $X_1, X_2, \ldots, X_n$, given that $Y = y$, does not depend on θ. This consequence is sometimes used as the definition of sufficiency and is illustrated in the next example.

Example 6.7-3 Let $X_1, X_2, \ldots, X_n$ be a random sample from a distribution with pmf

$$f(x; p) = p^x(1 - p)^{1-x}, \qquad x = 0, 1,$$

where the parameter p is between 0 and 1. We know that

$$Y = X_1 + X_2 + \cdots + X_n$$

is $b(n,p)$ and Y is sufficient for p because the joint pmf of $X_1, X_2, \ldots, X_n$ is

$$p^{x_1}(1-p)^{1-x_1} \cdots p^{x_n}(1-p)^{1-x_n} = [p^{\Sigma x_i}(1-p)^{n-\Sigma x_i}](1),$$

where $\phi(y;p) = p^y(1-p)^{n-y}$ and $h(x_1, x_2, \ldots, x_n) = 1$. What, then, is the conditional probability $P(X_1 = x_1, \ldots, X_n = x_n \mid Y = y)$, where $y = 0, 1, \ldots, n-1$, or n? Unless the sum of the nonnegative integers $x_1, x_2, \ldots, x_n$ equals y, this conditional probability is obviously equal to zero, which does not depend on p. Hence, it is interesting to consider the solution only when $y = x_1 + \cdots + x_n$. From the definition of conditional probability, we have

$$
\begin{aligned}
P(X_1 = x_1, \ldots, X_n = x_n \mid Y = y) &= \frac{P(X_1 = x_1, \ldots, X_n = x_n)}{P(Y = y)} \\
&= \frac{p^{x_1}(1-p)^{1-x_1} \cdots p^{x_n}(1-p)^{1-x_n}}{\binom{n}{y} p^y(1-p)^{n-y}} \\
&= \frac{1}{\binom{n}{y}},
\end{aligned}
$$

where $y = x_1 + \cdots + x_n$. Since y equals the number of ones in the collection $x_1, x_2, \ldots, x_n$, this answer is only the probability of selecting a particular arrangement, namely, $x_1, x_2, \ldots, x_n$, of y ones and $n - y$ zeros, and does not depend on the parameter p. That is, given that the sufficient statistic $Y = y$, the conditional probability of $X_1 = x_1, X_2 = x_2, \ldots, X_n = x_n$ does not depend on the parameter p. ∎

It is interesting to observe that the underlying pdf or pmf in Examples 6.7-1, 6.7-2, and 6.7-3 can be written in the exponential form

$$f(x;\theta) = \exp[K(x)p(\theta) + S(x) + q(\theta)],$$

where the support is free of θ. That is, we have, respectively,

$$\frac{e^{-\lambda}\lambda^x}{x!} = \exp\{x \ln \lambda - \ln x! - \lambda\}, \qquad x = 0, 1, 2, \ldots,$$

$$\frac{1}{\sqrt{2\pi}} e^{-(x-\mu)^2/2} = \exp\left\{ x\mu - \frac{x^2}{2} - \frac{\mu^2}{2} - \frac{1}{2} \ln(2\pi) \right\}, \qquad -\infty < x < \infty,$$

and

$$p^x(1-p)^{1-x} = \exp\left\{ x \ln\left(\frac{p}{1-p}\right) + \ln(1-p) \right\}, \qquad x = 0, 1.$$

In each of these examples, the sum $\sum_{i=1}^{n} X_i$ of the observations of the random sample is a sufficient statistic for the parameter. This idea is generalized by Theorem 6.7-1.

Theorem 6.7-1

Let $X_1, X_2, \ldots, X_n$ be a random sample from a distribution with a pdf or pmf of the exponential form

$$f(x; \theta) = \exp[K(x)p(\theta) + S(x) + q(\theta)]$$

on a support free of θ. Then the statistic $\sum_{i=1}^{n} K(X_i)$ is sufficient for θ.

Proof The joint pdf (pmf) of $X_1, X_2, \ldots, X_n$ is

$$\exp\left[p(\theta) \sum_{i=1}^{n} K(x_i) + \sum_{i=1}^{n} S(x_i) + nq(\theta)\right]$$

$$= \left\{\exp\left[p(\theta) \sum_{i=1}^{n} K(x_i) + nq(\theta)\right]\right\} \left\{\exp\left[\sum_{i=1}^{n} S(x_i)\right]\right\}.$$

In accordance with the factorization theorem, the statistic $\sum_{i=1}^{n} K(X_i)$ is sufficient for θ. $\qquad\square$

In many cases, Theorem 6.7-1 permits the student to find a sufficient statistic for a parameter with very little effort, as shown in the next example.

Example 6.7-4

Let $X_1, X_2, \ldots, X_n$ be a random sample from an exponential distribution with pdf

$$f(x; \theta) = \frac{1}{\theta} e^{-x/\theta} = \exp\left[x\left(-\frac{1}{\theta}\right) - \ln\theta\right], \qquad 0 < x < \infty,$$

provided that $0 < \theta < \infty$. Here, $K(x) = x$. Thus, $\sum_{i=1}^{n} X_i$ is sufficient for θ; of course, $\overline{X} = \sum_{i=1}^{n} X_i/n$ is also sufficient. ∎

Note that if there is a sufficient statistic for the parameter under consideration and if the maximum likelihood estimator of this parameter is unique, then the maximum likelihood estimator is a function of the sufficient statistic. To see this heuristically, consider the following: If a sufficient statistic exists, then the likelihood function is

$$L(\theta) = f(x_1, x_2, \ldots, x_n; \theta) = \phi[u(x_1, x_2, \ldots, x_n); \theta] h(x_1, x_2, \ldots, x_n).$$

Since $h(x_1, x_2, \ldots, x_n)$ does not depend on θ, we maximize $L(\theta)$ by maximizing $\phi[u(x_1, x_2, \ldots, x_n); \theta]$. But ϕ is a function of $x_1, x_2, \ldots, x_n$ only through the statistic $u(x_1, x_2, \ldots, x_n)$. Thus, if there is a unique value of θ that maximizes ϕ, then it must be a function of $u(x_1, x_2, \ldots, x_n)$. That is, $\widehat{\theta}$ is a function of the sufficient statistic $u(X_1, X_2, \ldots, X_n)$. This fact was alluded to in Example 6.7-1, but it could be checked with the use of other examples and exercises.

In many cases, we have two (or more) parameters—say, θ_1 and θ_2. All of the preceding concepts can be extended to these situations. For example, Definition 6.7-1 (the factorization theorem) becomes the following in the case of two parameters: If

$$f(x_1, \ldots, x_n; \theta_1, \theta_2) = \phi[u_1(x_1, \ldots, x_n), u_2(x_1, \ldots, x_n); \theta_1, \theta_2] h(x_1, \ldots, x_n),$$

where ϕ depends on $x_1, x_2, \ldots, x_n$ only through $u_1(x_1, \ldots, x_n)$, $u_2(x_1, \ldots, x_n)$, and $h(x_1, x_2, \ldots, x_n)$ does not depend upon θ_1 or θ_2, then $Y_1 = u_1(X_1, X_2, \ldots, X_n)$ and $Y_2 = u_2(X_1, X_2, \ldots, X_n)$ are **jointly sufficient statistics** for θ_1 and θ_2.

Example 6.7-5

Let $X_1, X_2, \ldots, X_n$ denote a random sample from a normal distribution $N(\theta_1 = \mu, \theta_2 = \sigma^2)$. Then

$$\prod_{i=1}^{n} f(x_i; \theta_1, \theta_2) = \left(\frac{1}{\sqrt{2\pi\theta_2}} \right)^n \exp\left[-\sum_{i=1}^{n} (x_i - \theta_1)^2 \Big/ 2\theta_2 \right]$$

$$= \exp\left[\left(-\frac{1}{2\theta_2} \right) \sum_{i=1}^{n} x_i^2 + \left(\frac{\theta_1}{\theta_2} \right) \sum_{i=1}^{n} x_i - \frac{n\theta_1^2}{2\theta_2} - n \ln \sqrt{2\pi\theta_2} \right] \cdot (1).$$

Thus,

$$Y_1 = \sum_{i=1}^{n} X_i^2 \quad \text{and} \quad Y_2 = \sum_{i=1}^{n} X_i$$

are joint sufficient statistics for θ_1 and θ_2. Of course, the single-valued functions of Y_1 and Y_2, namely,

$$\overline{X} = \frac{Y_2}{n} \quad \text{and} \quad S^2 = \frac{Y_1 - Y_2^2/n}{n-1},$$

are also joint sufficient statistics for θ_1 and θ_2. ∎

Actually, we can see from Definition 6.7-1 and Example 6.7-5 that if we can write the pdf in the exponential form, it is easy to find joint sufficient statistics. In that example,

$$f(x; \theta_1, \theta_2) = \exp\left(\frac{-1}{2\theta_2} x^2 + \frac{\theta_1}{\theta_2} x - \frac{\theta_1^2}{2\theta_2} - \ln \sqrt{2\pi\theta_2} \right);$$

so

$$Y_1 = \sum_{i=1}^{n} X_i^2 \quad \text{and} \quad Y_2 = \sum_{i=1}^{n} X_i$$

are joint sufficient statistics for θ_1 and θ_2. A much more complicated illustration is given if we take a random sample $(X_1, Y_1), (X_2, Y_2), \ldots, (X_n, Y_n)$ from a bivariate normal distribution with parameters $\theta_1 = \mu_X$, $\theta_2 = \mu_Y$, $\theta_3 = \sigma_X^2$, $\theta_4 = \sigma_Y^2$, and $\theta_5 = \rho$. In Exercise 6.7-3, we write the bivariate normal pdf $f(x, y; \theta_1, \theta_2, \theta_3, \theta_4, \theta_5)$ in exponential form and see that $Z_1 = \sum_{i=1}^{n} X_i^2$, $Z_2 = \sum_{i=1}^{n} Y_i^2$, $Z_3 = \sum_{i=1}^{n} X_i Y_i$, $Z_4 = \sum_{i=1}^{n} X_i$, and $Z_5 = \sum_{i=1}^{n} Y_i$ are joint sufficient statistics for θ_1, θ_2, θ_3, θ_4, and θ_5. Of course, the single-valued functions

$$\overline{X} = \frac{Z_4}{n}, \quad \overline{Y} = \frac{Z_5}{n}, \quad S_X^2 = \frac{Z_1 - Z_4^2/n}{n-1},$$

$$S_Y^2 = \frac{Z_2 - Z_5^2/n}{n-1}, \quad R = \frac{(Z_3 - Z_4 Z_5/n)/(n-1)}{S_X S_Y}$$

are also joint sufficient statistics for those parameters.

The important point to stress for cases in which sufficient statistics exist is that once the sufficient statistics are given, there is no additional information about the parameters left in the remaining (conditional) distribution. That is, all statistical inferences should be based upon the sufficient statistics. To help convince the reader of this in point estimation, we state and prove the well-known **Rao–Blackwell theorem**.

Theorem 6.7-2

Let $X_1, X_2, \ldots, X_n$ be a random sample from a distribution with pdf or pmf $f(x; \theta)$, $\theta \in \Omega$. Let $Y_1 = u_1(X_1, X_2, \ldots, X_n)$ be a sufficient statistic for θ, and let $Y_2 = u_2(X_1, X_2, \ldots, X_n)$ be an unbiased estimator of θ, where Y_2 is not a function of Y_1 alone. Then $E(Y_2 \mid y_1) = u(y_1)$ defines a statistic $u(Y_1)$, a function of the sufficient statistic Y_1, which is an unbiased estimator of θ, and its variance is less than that of Y_2.

Proof Let $g(y_1, y_2; \theta)$ be the joint pdf or pmf of Y_1 and Y_2. Let $g_1(y_1; \theta)$ be the marginal of Y_1; thus,

$$\frac{g(y_1, y_2; \theta)}{g_1(y_1; \theta)} = h(y_2 \mid y_1)$$

is the conditional pdf or pmf of Y_2, given that $Y_1 = y_1$. This equation does not depend upon θ, since Y_1 is a sufficient statistic for θ. Of course, in the continuous case,

$$u(y_1) = \int_{S_2} y_2 h(y_2 \mid y_1)\, dy_2 = \int_{S_2} y_2 \frac{g(y_1, y_2; \theta)}{g_1(y_1; \theta)}\, dy_2$$

and

$$E[u(Y_1)] = \int_{S_1} \left(\int_{S_2} y_2 \frac{g(y_1, y_2; \theta)}{g_1(y_1; \theta)}\, dy_2 \right) g_1(y_1; \theta)\, dy_1$$

$$= \int_{S_1} \int_{S_2} y_2\, g(y_1, y_2; \theta)\, dy_2\, dy_1 = \theta,$$

because Y_2 is an unbiased estimator of θ. Thus, $u(Y_1)$ is also an unbiased estimator of θ.

Now, consider

$$\mathrm{Var}(Y_2) = E[(Y_2 - \theta)^2] = E[\{Y_2 - u(Y_1) + u(Y_1) - \theta\}^2]$$

$$= E[\{Y_2 - u(Y_1)\}^2] + E[\{u(Y_1) - \theta\}^2] + 2E[\{Y_2 - u(Y_1)\}\{u(Y_1) - \theta\}].$$

But the latter expression (i.e., the third term) is equal to

$$2 \int_{S_1} [u(y_1) - \theta] \left\{ \int_{S_2} [y_2 - u(y_1)] h(y_2 \mid y_1)\, dy_2 \right\} g(y_1; \theta)\, dy_1 = 0,$$

because $u(y_1)$ is the mean $E(Y_2 \mid y_1)$ of Y_2 in the conditional distribution given by $h(y_2 \mid y_1)$. Thus,

$$\mathrm{Var}(Y_2) = E[\{Y_2 - u(Y_1)\}^2] + \mathrm{Var}[u(Y_1)].$$

However, $E[\{(Y_2 - u(Y_1))\}^2 \geq 0$, as it is the expected value of a positive expression. Therefore,

$$\mathrm{Var}(Y_2) \geq \mathrm{Var}[u(Y_1)]. \qquad \square$$

The importance of this theorem is that it shows that for every other unbiased estimator of θ, we can always find an unbiased estimator based on the sufficient statistic that has a variance at least as small as the first unbiased estimator. Hence, in that sense, the one based upon the sufficient statistic is at least as good as the first one. More importantly, we might as well begin our search for an unbiased estimator with the smallest variance by considering only those unbiased estimators based upon the sufficient statistics. Moreover, in an advanced course we show that if the underlying distribution is described by a pdf or pmf of the exponential form, then, if an unbiased estimator exists, there is only one function of the sufficient statistic that is unbiased. That is, that unbiased estimator is unique. (See Hogg, McKean, and Craig, 2013.)

There is one other useful result involving a sufficient statistic Y for a parameter θ, particularly with a pdf of the exponential form. It is that if another statistic Z has a distribution that is free of θ, then Y and Z are independent. This is the reason $Z = (n - 1)S^2$ is independent of $Y = \overline{X}$ when the sample arises from a distribution that is $N(\theta, \sigma^2)$. The sample mean is a sufficient statistic for θ, and

$$Z = (n - 1)S^2 = \sum_{i=1}^{n} (X_i - \overline{X})^2$$

has a distribution that is free of θ. To see this, we note that the mgf of Z, namely, $E(e^{tZ})$, is

$$\int_{-\infty}^{\infty} \int_{-\infty}^{\infty} \cdots \int_{-\infty}^{\infty} \exp\left[t \sum_{i=1}^{n} (x_i - \overline{x})^2 \right] \left(\frac{1}{\sqrt{2\pi}\sigma} \right)^n \exp\left[-\frac{\sum(x_i - \theta)^2}{2\sigma^2} \right] dx_1 dx_2 \ldots dx_n.$$

Changing variables by letting $x_i - \theta = w_i$, $i = 1, 2, \ldots, n$, the preceding expression becomes

$$\int_{-\infty}^{\infty} \int_{-\infty}^{\infty} \cdots \int_{-\infty}^{\infty} \exp\left[t \sum_{i=1}^{n} (w_i - \overline{w})^2 \right] \left(\frac{1}{\sqrt{2\pi}\sigma} \right)^n \exp\left[-\frac{\sum w_i^2}{2\sigma^2} \right] dw_1 dw_2 \ldots dw_n,$$

which is free of θ.

An outline of the proof of this result is given by noting that

$$\int_y [h(z \mid y) - g_2(z)] \, g_1(y; \theta) \, dy = g_2(z) - g_2(z) = 0$$

for all $\theta \in \Omega$. However, $h(z \mid y)$ is free of θ due to the hypothesis of sufficiency; so $h(z \mid y) - g_2(z)$ is free of θ, since Z has a distribution that is free of θ. Since $N(\theta, \sigma^2)$ is of the exponential form, $Y = \overline{X}$ has a pdf $g_1(y \mid \theta)$ that requires $h(z \mid y) - g_2(z)$ to be equal to zero. That is,

$$h(z \mid y) = g_2(z),$$

which means that Z and Y are independent. This proves the independence of $\overline{X}$ and S^2, which was stated in Theorem 5.5-2.

Example 6.7-6   Let $X_1, X_2, \ldots, X_n$ be a random sample from a gamma distribution with α (given) and $\theta > 0$, which is of exponential form. Now, $Y = \sum_{i=1}^{n} X_i$ is a sufficient statistic for θ, since the gamma pdf is of the exponential form. Clearly, then,

$$Z = \frac{\sum_{i=1}^{n} a_i X_i}{\sum_{i=1}^{n} X_i},$$

where not all constants $a_1, a_2, \ldots, a_n$ are equal, has a distribution that is free of the spread parameter θ because the mgf of Z, namely,

$$E(e^{tZ}) = \int_0^{\infty} \int_0^{\infty} \cdots \int_0^{\infty} \frac{e^{t\Sigma a_i X_i / \Sigma X_i}}{[\Gamma(\alpha)]^n \theta^{n\alpha}} (x_1 x_2 \cdots x_n)^{\alpha - 1} e^{-\Sigma x_i / \theta} \, dx_1 dx_2 \ldots dx_n,$$

and does not depend upon θ, as is seen by the transformation $w_i = x_i/\theta$, $i = 1, 2, \ldots, n$. So Y and Z are independent statistics. ∎

This special case of the independence of Y and Z concerning one sufficient statistic Y and one parameter θ was first observed by Hogg (1953) and then generalized to several sufficient statistics for more than one parameter by Basu (1955) and is usually called **Basu's theorem**.

Due to these results, sufficient statistics are extremely important and estimation problems are based upon them when they exist.

Exercises

6.7-1. Let $X_1, X_2, \ldots, X_n$ be a random sample from $N(0, \sigma^2)$.

(a) Find a sufficient statistic Y for σ^2.

(b) Show that the maximum likelihood estimator for σ^2 is a function of Y.

(c) Is the maximum likelihood estimator for σ^2 unbiased?

6.7-2. Let $X_1, X_2, \ldots, X_n$ be a random sample from a Poisson distribution with mean $\lambda > 0$. Find the conditional probability $P(X_1 = x_1, \ldots, X_n = x_n | Y = y)$, where $Y = X_1 + \cdots + X_n$ and the nonnegative integers $x_1, x_2, \ldots, x_n$ sum to y. Note that this probability does not depend on λ.

6.7-3. Write the bivariate normal pdf $f(x, y; \theta_1, \theta_2, \theta_3, \theta_4, \theta_5)$ in exponential form and show that $Z_1 = \sum_{i=1}^{n} X_i^2$, $Z_2 = \sum_{i=1}^{n} Y_i^2$, $Z_3 = \sum_{i=1}^{n} X_i Y_i$, $Z_4 = \sum_{i=1}^{n} X_i$, and $Z_5 = \sum_{i=1}^{n} Y_i$ are joint sufficient statistics for $\theta_1, \theta_2, \theta_3, \theta_4$, and θ_5.

6.7-4. Let $X_1, X_2, \ldots, X_n$ be a random sample from a distribution with pdf $f(x; \theta) = \theta x^{\theta-1}$, $0 < x < 1$, where $0 < \theta$.

(a) Find a sufficient statistic Y for θ.

(b) Show that the maximum likelihood estimator $\widehat{\theta}$ is a function of Y.

(c) Argue that $\widehat{\theta}$ is also sufficient for θ.

6.7-5. Let $X_1, X_2, \ldots, X_n$ be a random sample from a gamma distribution with $\alpha = 1$ and $1/\theta > 0$. Show that $Y = \sum_{i=1}^{n} X_i$ is a sufficient statistic, Y has a gamma distribution with parameters n and $1/\theta$, and $(n-1)/Y$ is an unbiased estimator of θ.

6.7-6. Let $X_1, X_2, \ldots, X_n$ be a random sample from a gamma distribution with known parameter α and unknown parameter $\theta > 0$.

(a) Show that $Y = \sum_{i=1}^{n} X_i$ is a sufficient statistic for θ.

(b) Show that the maximum likelihood estimator of θ is a function of Y and is an unbiased estimator of θ.

6.7-7. Let $X_1, X_2, \ldots, X_n$ be a random sample from the distribution with pmf $f(x; p) = p(1-p)^{x-1}$, $x = 1, 2, 3, \ldots$, where $0 < p \le 1$.

(a) Show that $Y = \sum_{i=1}^{n} X_i$ is a sufficient statistic for p.

(b) Find a function of Y that is an unbiased estimator of $\theta = 1/p$.

6.7-8. Let $X_1, X_2, \ldots, X_n$ be a random sample from $N(0, \theta)$, where $\sigma^2 = \theta > 0$ is unknown. Argue that the sufficient statistic $Y = \sum_{i=1}^{n} X_i^2$ for θ and $Z = \sum_{i=1}^{n} a_i X_i / \sum_{i=1}^{n} X_i$ are independent. HINT: Let $x_i = \theta w_i$, $i = 1, 2, \ldots, n$, in the multivariate integral representing $E[e^{tZ}]$.

6.7-9. Let $X_1, X_2, \ldots, X_n$ be a random sample from $N(\theta_1, \theta_2)$. Show that the sufficient statistics $Y_1 = \overline{X}$ and $Y_2 = S^2$ are independent of the statistic

$$Z = \sum_{i=1}^{n-1} \frac{(X_{i+1} - X_i)^2}{S^2}$$

because Z has a distribution that is free of θ_1 and θ_2.

HINT: Let $w_i = (x_i - \theta_1)/\sqrt{\theta_2}$, $i = 1, 2, \ldots, n$, in the multivariate integral representing $E[e^{tZ}]$.

6.7-10. Find a sufficient statistic for θ, given a random sample, $X_1, X_2, \ldots, X_n$, from a distribution with pdf $f(x; \theta) = \{\Gamma(2\theta)/[\Gamma(\theta)]^2\} x^{\theta-1}(1-x)^{\theta-1}$, $0 < x < 1$.

6.7-11. Let $X_1, X_2, \ldots, X_{15}$ be a random sample from a distribution with pdf $(x; \theta) = (1/2)\theta^3 x^2 e^{-\theta x}$, $0 < x < \infty$. Show that $Y = \sum_{i=1}^{15} X_i$ and $Z = (X_1 + X_2)/Y$ are independent.

6.7-12. Let $X_1, X_2, \ldots, X_n$ be a random sample from $N(0, \sigma^2)$, where n is odd. Let Y and Z be the mean and median of the sample. Argue that Y and $Z - Y$ are independent so that the variance of Z is $\text{Var}(Y) + \text{Var}(Z - Y)$. We know that $\text{Var}(Y) = \sigma^2/n$, so that we could estimate the $\text{Var}(Z - Y)$ by Monte Carlo. This might be more efficient than estimating $\text{Var}(Z)$ directly since $\text{Var}(Z - Y) \le \text{Var}(Z)$. This scheme is often called the **Monte Carlo Swindle**.

6.8 BAYESIAN ESTIMATION

We now describe another approach to estimation that is used by a group of statisticians who call themselves Bayesians. To understand their approach fully would require more text than we can allocate to this topic, but let us begin this brief introduction by considering a simple application of the theorem of the Reverend Thomas Bayes. (See Section 1.5.)

Example 6.8-1

Suppose we know that we are going to select an observation from a Poisson distribution with mean λ equal to 2 or 4. Moreover, prior to performing the experiment, we believe that $\lambda = 2$ has about four times as much chance of being the parameter as does $\lambda = 4$; that is, the prior probabilities are $P(\lambda = 2) = 0.8$ and $P(\lambda = 4) = 0.2$. The experiment is now performed and we observe that $x = 6$. At this point, our intuition tells us that $\lambda = 2$ seems less likely than before, as the observation $x = 6$ is much more probable with $\lambda = 4$ than with $\lambda = 2$, because, in an obvious notation,

$$P(X = 6 \mid \lambda = 2) = 0.995 - 0.983 = 0.012$$

and

$$P(X = 6 \mid \lambda = 4) = 0.889 - 0.785 = 0.104,$$

from Table III in Appendix B. Our intuition can be supported by computing the conditional probability of $\lambda = 2$, given that $X = 6$:

$$P(\lambda = 2 \mid X = 6) = \frac{P(\lambda = 2, X = 6)}{P(X = 6)}$$

$$= \frac{P(\lambda = 2)P(X = 6 \mid \lambda = 2)}{P(\lambda = 2)P(X = 6 \mid \lambda = 2) + P(\lambda = 4)P(X = 6 \mid \lambda = 4)}$$

$$= \frac{(0.8)(0.012)}{(0.8)(0.012) + (0.2)(0.104)} = 0.316.$$

This conditional probability is called the posterior probability of $\lambda = 2$, given the single data point (here, $x = 6$). In a similar fashion, the posterior probability of $\lambda = 4$ is found to be 0.684. Thus, we see that the probability of $\lambda = 2$ has decreased from 0.8 (the prior probability) to 0.316 (the posterior probability) with the observation of $x = 6$. ∎

In a more practical application, the parameter, say, θ can take many more than two values as in Example 6.8-1. Somehow Bayesians must assign prior probabilities to this total parameter space through a prior pdf $h(\theta)$. They have developed procedures for assessing these prior probabilities, and we simply cannot do justice to these methods here. Somehow $h(\theta)$ reflects the prior weights that the Bayesian wants to assign to the various possible values of θ. In some instances, if $h(\theta)$ is a constant and thus θ has the uniform prior distribution, we say that the Bayesian has a **noninformative** prior. If, in fact, some knowledge of θ exists in advance of experimentation, noninformative priors should be avoided if at all possible.

Also, in more practical examples, we usually take several observations, not just one. That is, we take a random sample, and there is frequently a good statistic, say, Y, for the parameter θ. Suppose we are considering a continuous case and the pdf of Y, say, $g(y; \theta)$, can be thought of as the conditional pdf of Y, given θ. [Henceforth in this section, we write $g(y; \theta) = g(y \mid \theta)$.] Thus, we can treat

$$g(y \mid \theta)h(\theta) = k(y, \theta)$$

as the joint pdf of the statistic Y and the parameter. Of course, the marginal pdf of Y is

$$k_1(y) = \int_{-\infty}^{\infty} h(\theta)g(y \mid \theta)\, d\theta.$$

Consequently,

$$\frac{k(y, \theta)}{k_1(y)} = \frac{g(y \mid \theta)h(\theta)}{k_1(y)} = k(\theta \mid y)$$

would serve as the conditional pdf of the parameter, given that $Y = y$. This formula is essentially Bayes' theorem, and $k(\theta \mid y)$ is called the **posterior pdf of** θ, given that $Y = y$.

Bayesians believe that everything which needs to be known about the parameter is summarized in this posterior pdf $k(\theta \mid y)$. Suppose, for example, that they were pressed into making a point estimate of the parameter θ. They would note that they would be guessing the value of a random variable, here θ, given its pdf $k(\theta \mid y)$. There are many ways that this could be done: The mean, the median, or the mode of that distribution would be reasonable guesses. However, in the final analysis, the best guess would clearly depend upon the penalties for various errors created by incorrect guesses. For instance, if we were penalized by taking the square of the error between the guess, say, $w(y)$, and the real value of the parameter θ, clearly we would use the conditional mean

$$w(y) = \int_{-\infty}^{\infty} \theta k(\theta \mid y)\, d\theta$$

as our Bayes estimate of θ. The reason is that, in general, if Z is a random variable, then the function of b, $E[(Z - b)^2]$, is minimized by $b = E(Z)$. (See Example 2.2-4.) Likewise, if the penalty (loss) function is the absolute value of the error, $|\theta - w(y)|$, then we use the median of the distribution, because with any random variable Z, $E[|Z - b|]$ is minimized when b equals the median of the distribution of Z. (See Exercise 2.2-8.)

Example 6.8-2

Suppose that Y has a binomial distribution with parameters n and $p = \theta$. Then the pmf of Y, given θ, is

$$g(y \mid \theta) = \binom{n}{y}\theta^y(1 - \theta)^{n-y}, \qquad y = 0, 1, 2, \ldots, n.$$

Let us take the prior pdf of the parameter to be the beta pdf:

$$h(\theta) = \frac{\Gamma(\alpha + \beta)}{\Gamma(\alpha)\Gamma(\beta)}\,\theta^{\alpha-1}(1 - \theta)^{\beta-1}, \qquad 0 < \theta < 1.$$

Such a prior pdf provides a Bayesian a great deal of flexibility through the selection of the parameters α and β. Thus, the joint probabilities can be described by a product of a binomial pmf with parameters n and θ and this beta pdf, namely,

$$k(y, \theta) = \binom{n}{y}\frac{\Gamma(\alpha + \beta)}{\Gamma(\alpha)\Gamma(\beta)}\,\theta^{y+\alpha-1}(1 - \theta)^{n-y+\beta-1},$$

on the support given by $y = 0, 1, 2, \ldots, n$ and $0 < \theta < 1$. We find

$$k_1(y) = \int_0^1 k(y, \theta) \, d\theta$$

$$= \binom{n}{y} \frac{\Gamma(\alpha + \beta)}{\Gamma(\alpha)\Gamma(\beta)} \frac{\Gamma(\alpha + y)\Gamma(n + \beta - y)}{\Gamma(n + \alpha + \beta)}$$

on the support $y = 0, 1, 2, \ldots, n$ by comparing the integral with one involving a beta pdf with parameters $y + \alpha$ and $n - y + \beta$. Therefore,

$$k(\theta \mid y) = \frac{k(y, \theta)}{k_1(y)}$$

$$= \frac{\Gamma(n + \alpha + \beta)}{\Gamma(\alpha + y)\Gamma(n + \beta - y)} \theta^{y+\alpha-1}(1 - \theta)^{n-y+\beta-1}, \qquad 0 < \theta < 1,$$

which is a beta pdf with parameters $y + \alpha$ and $n - y + \beta$. With the squared error loss function we must minimize, with respect to $w(y)$, the integral

$$\int_0^1 [\theta - w(y)]^2 \, k(\theta \mid y) \, d\theta,$$

to obtain the Bayes estimator. But, as noted earlier, if Z is a random variable with a second moment, then $E[(Z - b)^2]$ is minimized by $b = E(Z)$. In the preceding integration, θ is like the Z with pdf $k(\theta \mid y)$, and $w(y)$ is like the b, so the minimization is accomplished by taking

$$w(y) = E(\theta \mid y) = \frac{\alpha + y}{\alpha + \beta + n},$$

which is the mean of the beta distribution with parameters $y + \alpha$ and $n - y + \beta$. (See Exercise 5.2-8.) It is instructive to note that this Bayes estimator can be written as

$$w(y) = \left(\frac{n}{\alpha + \beta + n}\right)\left(\frac{y}{n}\right) + \left(\frac{\alpha + \beta}{\alpha + \beta + n}\right)\left(\frac{\alpha}{\alpha + \beta}\right),$$

which is a weighted average of the maximum likelihood estimate y/n of θ and the mean $\alpha/(\alpha + \beta)$ of the prior pdf of the parameter. Moreover, the respective weights are $n/(\alpha + \beta + n)$ and $(\alpha + \beta)/(\alpha + \beta + n)$. Thus, we see that α and β should be selected so that not only is $\alpha/(\alpha + \beta)$ the desired prior mean, but also the sum $\alpha + \beta$ plays a role corresponding to a sample size. That is, if we want our prior opinion to have as much weight as a sample size of 20, we would take $\alpha + \beta = 20$. So if our prior mean is 3/4, we select $\alpha = 15$ and $\beta = 5$. That is, the prior pdf of θ is beta(15, 5). If we observe $n = 40$ and $y = 28$, then the posterior pdf is beta($28 + 15 = 43$, $12 + 5 = 17$). The prior and posterior pdfs are shown in Figure 6.8-1. ∎

In Example 6.8-2, it is quite convenient to note that it is not really necessary to determine $k_1(y)$ to find $k(\theta \mid y)$. If we divide $k(y, \theta)$ by $k_1(y)$, we get the product of a factor that depends on y but does *not* depend on θ — say, $c(y)$ — and we have

$$\theta^{y+\alpha-1}(1 - \theta)^{n-y+\beta-1}.$$

That is,

$$k(\theta \mid y) = c(y)\,\theta^{y+\alpha-1}(1 - \theta)^{n-y+\beta-1}, \qquad 0 < \theta < 1.$$

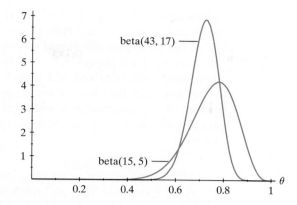

Figure 6.8-1 Beta prior and posterior pdfs

However, $c(y)$ must be that "constant" needed to make $k(\theta \mid y)$ a pdf, namely,

$$c(y) = \frac{\Gamma(n + \alpha + \beta)}{\Gamma(y + \alpha)\Gamma(n - y + \beta)}.$$

Accordingly, Bayesians frequently write that $k(\theta \mid y)$ is proportional to $k(y, \theta) = g(y \mid \theta)h(\theta)$; that is,

$$k(\theta \mid y) \propto g(y \mid \theta)\, h(\theta).$$

Then, to actually form the pdf $k(\theta \mid y)$, they simply find the "constant" (which is, of course, actually some function of y) such that the expression integrates to 1.

Example 6.8-3

Suppose that $Y = \overline{X}$ is the mean of a random sample of size n that arises from the normal distribution $N(\theta, \sigma^2)$, where σ^2 is known. Then $g(y \mid \theta)$ is $N(\theta, \sigma^2/n)$. Suppose further that we are able to assign prior weights to θ through a prior pdf $h(\theta)$ that is $N(\theta_0, \sigma_0^2)$. Then we have

$$k(\theta \mid y) \propto \frac{1}{\sqrt{2\pi}\,(\sigma/\sqrt{n})} \frac{1}{\sqrt{2\pi}\,\sigma_0} \exp\left[-\frac{(y - \theta)^2}{2(\sigma^2/n)} - \frac{(\theta - \theta_0)^2}{2\sigma_0^2}\right].$$

If we eliminate all constant factors (including factors involving y only), then

$$k(\theta \mid y) \propto \exp\left[-\frac{(\sigma_0^2 + \sigma^2/n)\theta^2 - 2(y\sigma_0^2 + \theta_0\sigma^2/n)\theta}{2(\sigma^2/n)\sigma_0^2}\right].$$

This expression can be simplified by completing the square, to read (after eliminating factors not involving θ)

$$k(\theta \mid y) \propto \exp\left\{-\frac{[\theta - (y\sigma_0^2 + \theta_0\sigma^2/n)/(\sigma_0^2 + \sigma^2/n)]^2}{[2(\sigma^2/n)\sigma_0^2]/[\sigma_0^2 + (\sigma^2/n)]}\right\}.$$

That is, the posterior pdf of the parameter is obviously normal with mean

$$\frac{y\sigma_0^2 + \theta_0\sigma^2/n}{\sigma_0^2 + \sigma^2/n} = \left(\frac{\sigma_0^2}{\sigma_0^2 + \sigma^2/n}\right)y + \left(\frac{\sigma^2/n}{\sigma_0^2 + \sigma^2/n}\right)\theta_0$$

and variance $(\sigma^2/n)\sigma_0^2/(\sigma_0^2 + \sigma^2/n)$. If the squared error loss function is used, then this posterior mean is the Bayes estimator. Again, note that it is a weighted average of the maximum likelihood estimate $y = \bar{x}$ and the prior mean θ_0. The Bayes estimator $w(y)$ will always be a value between the prior judgment and the usual estimate. Note also, here and in Example 6.8-2, that the Bayes estimator gets closer to the maximum likelihood estimate as n increases. Thus, the Bayesian procedures permit the decision maker to enter his or her prior opinions into the solution in a very formal way so that the influence of those prior notions will be less and less as n increases. ■

In Bayesian statistics, all the information is contained in the posterior pdf $k(\theta \mid y)$. In Examples 6.8-2 and 6.8-3, we found Bayesian point estimates with the use of the squared error loss function. Note that if the loss function is the absolute value of the error, $|w(y) - \theta|$, then the Bayes estimator would be the median of the posterior distribution of the parameter, which is given by $k(\theta \mid y)$. Hence, the Bayes estimator changes—as it should—with different loss functions.

Finally, if an interval estimate of θ is desired, we would find two functions of y—say, $u(y)$ and $v(y)$—such that

$$\int_{u(y)}^{v(y)} k(\theta \mid y)\, d\theta = 1 - \alpha,$$

where α is small—say, $\alpha = 0.05$. Then the observed interval from $u(y)$ to $v(y)$ would serve as an interval estimate for the parameter in the sense that the posterior probability of the parameter's being in that interval is $1 - \alpha$. In Example 6.8-3, where the posterior pdf of the parameter was normal, the interval

$$\frac{y\sigma_0^2 + \theta_0\sigma^2/n}{\sigma_0^2 + \sigma^2/n} \pm 1.96 \sqrt{\frac{(\sigma^2/n)\sigma_0^2}{\sigma_0^2 + \sigma^2/n}}$$

serves as an interval estimate for θ with posterior probability of 0.95.

In closing this short section on Bayesian estimation, note that we could have begun with the sample observations $X_1, X_2, \ldots, X_n$, rather than some statistic Y. Then, in our discussion, we would replace $g(y \mid \theta)$ by the likelihood function

$$L(\theta) = f(x_1 \mid \theta)f(x_2 \mid \theta) \cdots f(x_n \mid \theta),$$

which is the joint pdf of $X_1, X_2, \ldots, X_n$, given θ. Thus, we find that

$$k(\theta \mid x_1, x_2, \ldots, x_n) \propto h(\theta)f(x_1 \mid \theta)f(x_2 \mid \theta) \cdots f(x_n \mid \theta) = h(\theta)L(\theta).$$

Now, $k(\theta \mid x_1, x_2, \ldots, x_n)$ contains all the information about θ, given the data. Thus, depending on the loss function, we would choose our Bayes estimate of θ as some characteristic of this posterior distribution, such as the mean or the median. It is interesting to observe that if the loss function is zero for some small neighborhood about the true parameter θ and is some large positive constant otherwise, then the Bayes estimate, $w(x_1, x_2, \ldots, x_n)$, is essentially the mode of this conditional pdf, $k(\theta \mid x_1, x_2, \ldots, x_n)$. The reason for this is that we want to take the estimate so that it has as much posterior probability as possible in a small neighborhood around it. Finally, note that if $h(\theta)$ is a constant (a noninformative prior), then this Bayes estimate using the mode is exactly the same as the maximum likelihood estimate. More generally, if $h(\theta)$ is not a constant, then the Bayes estimate using the mode can be thought of as a weighted maximum likelihood estimate in which the weights reflect prior opinion about θ. That is, that value of θ which maximizes $h(\theta)L(\theta)$ is the mode of the posterior distribution of the parameter given the data and can be used as the Bayes estimate associated with the appropriate loss function.

Example 6.8-4

Let us consider again Example 6.8-2, but now say that $X_1, X_2, \ldots, X_n$ is a random sample from the Bernoulli distribution with pmf

$$f(x \mid \theta) = \theta^x (1 - \theta)^{1-x}, \qquad x = 0, 1.$$

With the same prior pdf of θ, the joint distribution of $X_1, X_2, \ldots, X_n$ and θ is given by

$$\frac{\Gamma(\alpha + \beta)}{\Gamma(\alpha)\Gamma(\beta)} \theta^{\alpha-1}(1 - \theta)^{\beta-1} \theta^{\sum_{i=1}^{n} x_i}(1 - \theta)^{n-\sum_{i=1}^{n} x_i}, \qquad 0 < \theta < 1, \ x_i = 0, 1.$$

Of course, the posterior pdf of θ, given that $X_1 = x_1, X_2 = x_2, \ldots, X_n = x_n$, is such that

$$k(\theta \mid x_1, x_2, \ldots, x_n) \propto \theta^{\sum_{i=1}^{n} x_i + \alpha - 1}(1 - \theta)^{n - \sum_{i=1}^{n} x_i + \beta - 1}, \qquad 0 < \theta < 1,$$

which is beta with $\alpha^* = \sum x_i + \alpha$, $\beta^* = n - \sum x_i + \beta$. The conditional mean of θ is

$$\frac{\sum_{i=1}^{n} x_i + \alpha}{n + \alpha + \beta} = \left(\frac{n}{n + \alpha + \beta}\right)\left(\frac{\sum_{i=1}^{n} x_i}{n}\right) + \left(\frac{\alpha + \beta}{n + \alpha + \beta}\right)\left(\frac{\alpha}{\alpha + \beta}\right),$$

which, with $y = \sum x_i$, is exactly the same result as that of Example 6.8-2. ∎

Exercises

6.8-1. Let Y be the sum of the observations of a random sample from a Poisson distribution with mean θ. Let the prior pdf of θ be gamma with parameters $\alpha = 3$ and $\beta = 2$.

(a) Find the posterior pdf of θ, given that $Y = y$.

(b) If the loss function is $[w(y) - \theta]^2$, find the Bayesian point estimate $w(y)$.

(c) Show that $w(y)$ found in (b) is a weighted average of the maximum likelihood estimate y/n and the prior mean 6, with respective weights of $n/(n + 1/2)$ and $(1/2)/(n + 1/2)$.

6.8-2. Let $X_1, X_2, \ldots, X_n$ be a random sample from a gamma distribution with known α and with $\theta = 1/\tau$. Say τ has a prior pdf that is gamma with parameters α_0 and θ_0, so that the prior mean is $\alpha_0\theta_0$.

(a) Find the posterior pdf of τ, given that $X_1 = x_1, X_2 = x_2, \ldots, X_n = x_n$.

(b) Find the mean of the posterior distribution found in part (a), and write it as a function of the sample mean $\overline{X}$ and $\alpha_0\theta_0$.

(c) Explain how you would find a 95% interval estimate of τ if $n = 10$, $\alpha = 3$, $\alpha_0 = 10$, and $\theta_0 = 2$.

6.8-3. In Example 6.8-2, take $n = 30$, $\alpha = 15$, and $\beta = 5$.

(a) Using the squared error loss, compute the expected loss (risk function) associated with the Bayes estimator $w(Y)$.

(b) The risk function associated with the usual estimator Y/n is, of course, $\theta(1 - \theta)/30$. Find those values of θ for which the risk function in part (a) is less than $\theta(1-\theta)/30$. In particular, if the prior mean $\alpha/(\alpha+\beta) = 3/4$ is a reasonable guess, then the risk function in part (a) is the better of the two (i.e., is smaller in a neighborhood of $\theta = 3/4$) for what values of θ?

6.8-4. Consider a random sample $X_1, X_2, \ldots, X_6$ from a distribution with pdf

$$f(x \mid \theta) = 3\theta x^2 e^{-\theta x^3}, \qquad 0 < x < \infty.$$

Let θ have a prior pdf that is gamma with $\alpha = 5$ and the usual $\theta = 1/4$. Find the conditional mean of θ, given that $X_1 = 1$, $X_2 = 0$, $X_3 = 2$, $X_4 = 3$, $X_5 = 4$, and $X_6 = 5$.

6.8-5. In Example 6.8-3, suppose the loss function $|\theta - w(Y)|$ is used. What is the Bayes estimator $w(Y)$?

6.8-6. Let Y be the largest order statistic of a random sample of size 4 from a distribution with pdf $f(x|\theta) = 1/\theta$, $0 < x < \theta$. Say θ has the prior pdf

$$h(\theta) = 2/\theta^3, \qquad 1 < \theta < \infty.$$

(a) If $w(Y)$ is the Bayes estimator of θ and $[\theta - w(Y)]^2$ is the loss function, find $w(Y)$.

(b) Find the Bayesian estimator w(Y) if the loss function is $[\theta - w(Y)]$.

6.8-7. Refer to Example 6.8-3. Suppose we select $\sigma_0^2 = d\sigma^2$, where σ^2 is known in that example. What value do we assign to d so that the variance of the posterior pdf of the parameter is two thirds of the variance of $Y = \overline{X}$, namely, σ^2/n?

6.8-8. Consider the likelihood function $L(\alpha, \beta, \sigma^2)$ of Section 6.5. Let α and β be independent with priors $N(\alpha_1, \sigma_1^2)$ and $N(\beta_0, \sigma_0^2)$. Determine the posterior mean of $\alpha + \beta(x - \overline{x})$.

6.9* MORE BAYESIAN CONCEPTS

Let $X_1, X_2, \ldots, X_n$ be a random sample from a distribution with pdf (pmf) $f(x \mid \theta)$, and let $h(\theta)$ be the prior pdf. Then the distribution associated with the marginal pdf of $X_1, X_2, \ldots, X_n$, namely,

$$k_1(x_1, x_2, \ldots, x_n) = \int_{-\infty}^{\infty} f(x_1 \mid \theta) f(x_2 \mid \theta) \cdots f(x_n \mid \theta) h(\theta) \, d\theta,$$

is called the **predictive distribution** because it provides the best description of the probabilities on $X_1, X_2, \ldots, X_n$. Often this creates some interesting distributions. For example, suppose there is only one X with the normal pdf

$$f(x \mid \theta) = \frac{\sqrt{\theta}}{\sqrt{2\pi}} e^{-(\theta x^2)/2}, \qquad -\infty < x < \infty.$$

Here, $\theta = 1/\sigma^2$, the inverse of the variance, is called the **precision** of X. Say this precision has the gamma pdf

$$h(\theta) = \frac{1}{\Gamma(\alpha)\beta^\alpha} \theta^{\alpha-1} e^{-\theta/\beta}, \qquad 0 < \theta < \infty.$$

Then the predictive pdf is

$$k_1(x) = \int_0^{\infty} \frac{\theta^{\alpha+\frac{1}{2}-1} e^{-\left(\frac{x^2}{2} + \frac{1}{\beta}\right)\theta}}{\Gamma(\alpha)\beta^\alpha \sqrt{2\pi}} \, d\theta$$

$$= \frac{\Gamma(\alpha + 1/2)}{\Gamma(\alpha)\beta^\alpha \sqrt{2\pi}} \frac{1}{(1/\beta + x^2/2)^{\alpha+1/2}}, \qquad -\infty < x < \infty.$$

Note that if $\alpha = r/2$ and $\beta = 2/r$, where r is a positive integer, then

$$k_1(x) \propto \frac{1}{(1 + x^2/r)^{(r+1)/2}}, \qquad -\infty < x < \infty,$$

which is a t pdf with r degrees of freedom. So if the inverse of the variance—or precision θ—of a normal distribution varies as a gamma random variable, a generalization of a t distribution has been created that has heavier tails than the normal distribution. This **mixture** of normals (different from a mixed distribution) is attained by weighing with the gamma distribution in a process often called **compounding**.

Another illustration of compounding is given in the next example.

Example 6.9-1 Suppose X has a gamma distribution with the two parameters k and θ^{-1}. (That is, the usual α is replaced by k and θ by its reciprocal.) Say $h(\theta)$ is gamma with parameters α and β, so that

$$k_1(x) = \int_0^\infty \frac{\theta^k x^{k-1} e^{-\theta x}}{\Gamma(k)} \frac{1}{\Gamma(\alpha)\beta^\alpha} \theta^{\alpha-1} e^{-\theta/\beta} \, d\theta$$

$$= \int_0^\infty \frac{x^{k-1}\theta^{k+\alpha-1}e^{-\theta(x+1/\beta)}}{\Gamma(k)\Gamma(\alpha)\beta^\alpha} \, d\theta$$

$$= \frac{\Gamma(k+\alpha)x^{k-1}}{\Gamma(k)\Gamma(\alpha)\beta^\alpha} \frac{1}{(x+1/\beta)^{k+\alpha}}$$

$$= \frac{\Gamma(k+x)\beta^k x^{k-1}}{\Gamma(k)\Gamma(\alpha)(1+\beta x)^{k+\alpha}}, \qquad 0 < x < \infty.$$

Of course, this is a generalization of the F distribution, which we obtain by letting $\alpha = r_2/2$, $k = r_1/2$, and $\beta = r_1/r_2$. ∎

Note how well the prior $h(\theta)$ "fits" with $f(x \mid \theta)$ or $f(x_1 \mid \theta)f(x_2 \mid \theta)\cdots f(x_n \mid \theta)$ in all of our examples, and the posterior distribution is of exactly the same form as the prior. In Example 6.8-2, both the prior and the posterior were beta. In Example 6.8-3, both the prior and posterior were normal. In Example 6.9-1, both the prior and the posterior (if we had found it) were gamma. When this type of pairing occurs, we say that that class of prior pdfs (pmfs) is a **conjugate family of priors**. Obviously, this makes the mathematics easier, and usually the parameters in the prior distribution give us enough flexibility to obtain good fits.

Example 6.9-2

(Berry, 1996) This example deals with *predictive probabilities*, and it concerns the breakage of glass panels in high-rise buildings. One such case involved 39 panels, and of the 39 panels that broke, it was known that 3 broke due to nickel sulfide (NiS) stones found in them. Loss of evidence prevented the causes of breakage of the other 36 panels from being known. So the court wanted to know whether the manufacturer of the panels or the builder was at fault for the breakage of these 36 panels.

From expert testimony, it was thought that usually about 5% breakage is caused by NiS stones. That is, if this value of p is selected from a beta distribution, we have

$$\frac{\alpha}{\alpha+\beta} = 0.05. \tag{6.9-1}$$

Moreover, the expert thought that if two panels from the same lot break and one breakage was caused by NiS stones, then, due to the pervasive nature of the manufacturing process, the probability of the second panel breaking due to NiS stones increases to about 95%. Thus, the posterior estimate of p (see Example 6.8-2) with one "success" after one trial is

$$\frac{\alpha+1}{\alpha+\beta+1} = 0.95. \tag{6.9-2}$$

Solving Equations 6.9-1 and 6.9-2 for α and β, we obtain

$$\alpha = \frac{1}{360} \qquad \text{and} \qquad \beta = \frac{19}{360}.$$

Now updating the posterior probability with 3 "successes" out of 3 trials, we obtain the posterior estimate of p:

$$\frac{\alpha+3}{\alpha+\beta+3} = \frac{1/360+3}{20/360+3} = \frac{1081}{1100} = 0.983.$$

Of course, the court that heard the case wanted to know the expert's opinion about the probability that all of the remaining 36 panels broke because of NiS stones. Using updated probabilities after the third break, then the fourth, and so on, we obtain the product

$$\left(\frac{1/360+3}{20/360+3}\right)\left(\frac{1/360+4}{20/360+4}\right)\left(\frac{1/360+5}{20/360+5}\right)\cdots\left(\frac{1/360+38}{20/360+38}\right) = 0.8664.$$

That is, the expert held that the probability that all 36 breakages were caused by NiS stones was about 87%, which is the needed value in the court's decision. ∎

We now look at a situation in which we have two unknown parameters; we will use, for convenience, what is called a noninformative prior, which usually puts uniform distributions on the parameters. Let us begin with a random sample $X_1, X_2, \ldots, X_n$ from the normal distribution $N(\theta_1, \theta_2)$, and suppose we have little prior knowledge about θ_1 and θ_2. We then use the noninformative prior that θ_1 and $\ln \theta_2$ are uniform and independent; that is,

$$h_1(\theta_1)h_2(\theta_2) \propto \frac{1}{\theta_2}, \qquad -\infty < \theta_1 < \infty, \ 0 < \theta_2 < \infty.$$

Of course, we immediately note that we cannot find a constant c such that c/θ_2 is a joint pdf on that support. That is, this noninformative prior pdf is not a pdf at all; hence, it is called an **improper** prior. However, we use it anyway, because it will be satisfactory when multiplied by the joint pdf of $X_1, X_2, \ldots, X_n$. We have the product

$$\left(\frac{1}{\theta_2}\right)\left(\frac{1}{\sqrt{2\pi\theta_2}}\right)^n \exp\left[-\sum_{i=1}^{n}\frac{(x_i-\theta_1)^2}{2\theta_2}\right].$$

Thus,

$$k_{12}(\theta_1, \theta_2 \mid x_1, x_2, \ldots, x_n) \propto \left(\frac{1}{\theta_2}\right)^{\frac{n}{2}+1} \exp\left[-\frac{1}{2}\left\{(n-1)s^2 + n(\bar{x}-\theta_1)^2\right\}/\theta_2\right]$$

since $\sum_{i=1}^{n}(x_i-\theta_1)^2 = (n-1)s^2 + n(\bar{x}-\theta_1)^2 = D$. It then follows that

$$k_1(\theta_1 \mid x_1, x_2, \ldots, x_n) \propto \int_0^\infty k_{12}(\theta_1, \theta_2 \mid x_1, x_2, \ldots, x_n)\, d\theta_2.$$

Changing variables by letting $z = 1/\theta_2$, we obtain

$$k_1(\theta_1 \mid x_1, x_2, \ldots, x_n) \propto \int_0^\infty \frac{z^{n/2+1}}{z^2} e^{-\frac{1}{2}Dz}\, dz$$

$$\propto D^{-n/2} = \left[(n-1)s^2 + n(\bar{x}-\theta_1)^2\right]^{-n/2}.$$

To get this pdf in a more familiar form, let $t = (\theta_1 - \bar{x})/(s/\sqrt{n})$, with Jacobian $s/\sqrt{n}$, to yield

$$k(t \mid x_1, x_2, \ldots, x_n) \propto \frac{1}{[1+t^2/(n-1)]^{[(n-1)+1]/2}}, \qquad -\infty < t < \infty.$$

That is, the conditional pdf of t, given $x_1, x_2, \ldots, x_n$, is Student's t with $n-1$ degrees of freedom. Thus, a $(1-\alpha)$ **probability interval** for θ_1 is given by

$$-t_{\alpha/2} < \frac{\theta_1 - \bar{x}}{s/\sqrt{n}} < t_{\alpha/2},$$

or

$$\bar{x} - t_{\alpha/2}\, s/\sqrt{n} < \theta_1 < \bar{x} + t_{\alpha/2}\, s/\sqrt{n}.$$

The reason we get the same answer in this case is that we use a noninformative prior. Bayesians do not like to use a noninformative prior if they really know something about the parameters. For example, say they believe that the precision $1/\theta_2$ has a gamma distribution with parameters α and β instead of the noninformative prior. Then finding the conditional pdf of θ_1 becomes a much more difficult integration. However, it can be done, but we leave it to a more advanced course. (See Hogg, McKean, and Craig, 2013.)

Example 6.9-3

(Johnson and Albert, 1999) The data in this example, a sample of $n = 13$ measurements of the National Oceanographic and Atmospheric Administration (NOAA)/Environmental Protection Agency (EPA) ultraviolet (UV) index taken in Los Angeles, were collected from archival data of every Sunday in October during the years 1995–1997 in a database maintained by NOAA. The 13 UV readings are

$$7,\ 6,\ 5,\ 5,\ 3,\ 6,\ 5,\ 5,\ 3,\ 5,\ 5,\ 4,\ 4,$$

and, although they are integer values, we assume that they are taken from a $N(\mu, \sigma^2)$ distribution.

The Bayesian analysis, using a noninformative prior in the preceding discussion, implies that, with $\mu = \theta_1$,

$$\frac{\mu - 4.846}{0.317}, \qquad \text{where} \qquad \bar{x} = 4.846 \qquad \text{and} \qquad \frac{s}{\sqrt{n}} = 0.317,$$

has a posterior t distribution with $n - 1 = 12$ degrees of freedom. For example, a posterior 95% probability interval for μ is

$$(4.846 - [t_{0.025}(12)][0.317],\ 4.846 + [t_{0.025}(12)][0.317]) = (4.155,\ 5.537). \quad \blacksquare$$

Example 6.9-4

Tsutakawa et. al. (1985) discuss mortality rates from stomach cancer over the period 1972–1981 in males aged 45–64 in 84 cities in Missouri. Ten-year observed mortality rates in 20 of these cities are listed in Table 6.9-1, where y_i represents the number of deaths due to stomach cancer among this subpopulation in city i from 1972–1981, and n_i is the estimated size of this subpopulation in city i at the beginning of 1977 (estimated by linear interpolation from the 1970 and 1980 U.S. Census figures). Let p_i, $i = 1, 2, \ldots, 20$, represent the corresponding probabilities of death due to stomach cancer, and assume that $p_1, p_2, \ldots, p_{20}$ are taken independently from a beta distribution with parameters α and β. Then the posterior mean of p_i is

$$\widehat{p_i}\left(\frac{n_i}{n_i + \alpha + \beta}\right) + \left(\frac{\alpha}{\alpha + \beta}\right)\left(\frac{\alpha + \beta}{n_i + \alpha + \beta}\right), \quad i = 1, 2, \ldots, 20,$$

where $\widehat{p_i} = y_i/n_i$. Of course, the parameters α and β are unknown, but we have assumed that $p_1, p_2, \ldots, p_{20}$ arose from a similar distribution for these cities in Missouri; that is, we assume that our prior knowledge concerning the proportions is *exchangeable*. So it would be reasonable to estimate $\alpha/(\alpha + \beta)$, the prior mean of a proportion, with the formula

$$\bar{y} = \frac{y_1 + y_2 + \cdots + y_{20}}{n_1 + n_2 + \cdots + n_{20}} = \frac{71}{71,478} = 0.000993,$$

Table 6.9-1 Cancer mortality rates

y_i	n_i	$\widehat{p}_i$	Posterior Estimate	y_i	n_i	$\widehat{p}_i$	Posterior Estimate
0	1083	0	0.00073	0	855	0	0.00077
2	3461	0.00058	0.00077	0	657	0	0.00081
1	1208	0.00083	0.00095	1	1025	0.00098	0.00099
0	527	0	0.00084	2	1668	0.00120	0.00107
1	583	0.00172	0.00111	3	582	0.00515	0.00167
0	917	0	0.00076	1	857	0.00117	0.00103
1	680	0.00147	0.00108	1	917	0.00109	0.00102
54	53637	0.00101	0.00101	0	874	0	0.00077
0	395	0	0.00088	1	581	0.00172	0.00111
3	588	0.00510	0.00167	0	383	0	0.00088

for the data given in Table 6.9-1. Thus, the posterior estimate of p_i is found by *shrinking* $\widehat{p}_i$ toward the pooled estimate of the mean $\alpha/(\alpha + \beta)$—namely, $\bar{y}$. That is, the posterior estimate is

$$\widehat{p}_i\left(\frac{n_i}{n_i + \alpha + \beta}\right) + \bar{y}\left(\frac{\alpha + \beta}{n_i + \alpha + \beta}\right).$$

The only question remaining is how much weight should be given to the prior, represented by $\alpha + \beta$, relative to $n_1, n_2, \ldots, n_{20}$. Considering the sizes of the samples from the various cities, we selected $\alpha + \beta = 3000$ (which means that the prior is worth about a sample of size 3000), which resulted in the posterior probabilities given in Table 6.9-1. Note how this type of shrinkage tends to pull the posterior estimates much closer to the average, particularly those associated with small sample sizes. Baseball fans might try this type of shrinkage in predicting some of the final batting averages of the better batters about a quarter of the way through the season. ∎

It is clear that difficult integration caused Bayesians great problems until very recent times, in which advances in computer methods "solved" many of these problems. As a simple illustration, suppose the pdf of a statistic Y is $f(y \mid \theta)$ and the prior pdf $h(\theta)$ is such that

$$k(\theta \mid y) = \frac{f(y \mid \theta)\,h(\theta)}{\int_{-\infty}^{\infty} f(y \mid \tau)\,h(\tau)\,d\tau}$$

is not a nice pdf with which to deal. In particular, say that we have a squared error loss and we wish to determine $E(\theta \mid y)$, namely,

$$\delta(y) = \frac{\int_{-\infty}^{\infty} \theta f(y \mid \theta)\,h(\theta)\,d\theta}{\int_{-\infty}^{\infty} f(y \mid \theta)\,h(\theta)\,d\theta},$$

but cannot do it easily. Let $f(y \mid \theta) = w(\theta)$. Then we wish to evaluate the ratio

$$\frac{E[\theta\,w(\theta)]}{E[w(\theta)]},$$

where y is given and the expected values are taken with respect to θ. To do so, we simply generate a number of θ values, say, $\theta_1, \theta_2, \ldots, \theta_m$ (where m is large), from the distribution given by $h(\theta)$. Then we estimate the numerator and denominator of the desired ratio by

$$\sum_{i=1}^{m} \frac{\theta_i\, w(\theta_i)}{m} \quad \text{and} \quad \sum_{i=1}^{m} \frac{w(\theta_i)}{m},$$

respectively, to obtain

$$\tau = \frac{\sum_{i=1}^{m} \theta_i\, w(\theta_i)/m}{\sum_{i=1}^{m} w(\theta_i)/m}.$$

In addition to this simple Monte Carlo procedure, there are additional ones that are extremely useful in Bayesian inferences. Two of these are the **Gibbs sampler** and the **Markov chain Monte Carlo (MCMC)**. The latter is used in **hierarchical Bayes models** in which the prior has another parameter that has its own prior (called the **hyperprior**). That is, we have

$$f(y \mid \theta), \qquad h(\theta \mid \tau), \qquad \text{and} \qquad g(\tau).$$

Hence,

$$k(\theta, \tau \mid y) = \frac{f(y \mid \theta)\, h(\theta \mid \tau)\, g(\tau)}{\int_{-\infty}^{\infty} \int_{-\infty}^{\infty} f(y \mid \eta)\, h(\eta \mid v)\, g(v)\, d\eta\, dv}$$

and

$$k_1(\theta \mid y) = \int_{-\infty}^{\infty} k(\theta, \tau \mid y)\, d\tau.$$

Thus, a Bayes estimator, for a squared error loss, is

$$\int_{-\infty}^{\infty} \theta\, k_1(\theta \mid y)\, d\theta.$$

Using the Gibbs sampler, we can generate a stream of values $(\theta_1, \tau_1), (\theta_2, \tau_2), \ldots$ that allows us to estimate $k(\theta, \tau \mid y)$ and $\int_{-\infty}^{\infty} \theta\, k_1(\theta \mid y)\, d\theta$. These procedures are the MCMC procedures. (For additional references, see Hogg, McKean, and Craig, 2013.)

Exercises

6.9-1. Let X have a Poisson distribution with parameter θ. Let θ be $\Gamma(3, 2)$. Show that the marginal pmf of X (the compound distribution) is

$$k_1(x) = \frac{\Gamma(3+x)2^x}{\Gamma(3)x!(3)^{3+x}}, \qquad x = 0, 1, 2, 3, \ldots,$$

which is a generalization of the negative binomial distribution.

6.9-2. Suppose X is $b(n, \theta)$ and θ is beta(α, β). Show that the marginal pdf of X (the compound distribution) is

$$k_1(x) = \frac{n!\,\Gamma(\alpha + \beta)\,\Gamma(x + \alpha)\,\Gamma(n - x + \beta)}{x!\,(n - x)!\,\Gamma(\alpha)\,\Gamma(\beta)\,\Gamma(n + \alpha + \beta)},$$

for $x = 0, 1, 2, \ldots, n$.

6.9-3. Let X have the geometric pmf $\theta(1 - \theta)^{x-1}$, $x = 1, 2, 3, \ldots$, where θ is beta with parameters α and β. Show that the compound pmf is

$$\frac{\Gamma(\alpha + \beta)\,\Gamma(\alpha + 1)\,\Gamma(\beta + x - 1)}{\Gamma(\alpha)\,\Gamma(\beta)\,\Gamma(\alpha + \beta + x)}, \qquad x = 1, 2, 3, \ldots.$$

With $\alpha = 1$, this is one form of **Zipf's law**,

$$\frac{\beta}{(\beta + x)(\beta + x - 1)}, \qquad x = 1, 2, 3, \ldots.$$

6.9-4. Let X have the pdf

$$f(x \mid \theta) = \theta \tau x^{\tau - 1} e^{-\theta x^{\tau}}, \qquad 0 < x < \infty,$$

where the distribution of θ is $\Gamma(\alpha, \beta)$. Find the compound distribution of X, which is called the **Burr distribution**.

6.9-5. Let $X_1, X_2, \ldots, X_n$ be a random sample from a gamma distribution with $\alpha = 1, \theta$. Let $h(\theta) \propto 1/\theta$, $0 < \theta < \infty$, be an improper noninformative prior.

(a) Find the posterior pdf of θ.

(b) Change variables by letting $z = 1/\theta$, and show that the posterior distribution of Z is $\Gamma(n, 1/y)$, where $y = \sum_{i=1}^{n} x_i$.

(c) Use $2yz$ to obtain a $(1 - \alpha)$ probability interval for z and, of course, for θ.

6.9-6. Let X_1, X_2 be a random sample from the Cauchy distribution with pdf

$$f(x \mid \theta_1, \theta_2) = \frac{1}{\pi} \frac{\theta_2}{\theta_2^2 + (x - \theta_1)^2},$$

$$-\infty < x < \infty, \quad -\infty < \theta_1 < \infty, \quad 0 < \theta_2 < \infty.$$

Consider the noninformative prior $h(\theta_1, \theta_2) \propto 1$ on that support. Obtain the posterior pdf (except for constants) of θ_1, θ_2 if $x_1 = 3$ and $x_2 = 7$. For estimates, find θ_1, θ_2 that maximizes this posterior pdf; that is, find the mode of that posterior. (This might require some reasonable "trial and error" or an advanced method of maximizing a function of two variables.)

HISTORICAL COMMENTS When a statistician thinks of estimation, he or she recalls R. A. Fisher's contributions to many aspects of the subject: maximum likelihood, estimation, efficiency, and sufficiency. Of course, many more statisticians have contributed to that discipline since the 1920s. It would be an interesting exercise for the reader to go through the tables of contents of the *Journal of the American Statistical Association*, the *Annals of Statistics*, and related journals to observe how many articles are about estimation. Often our friends ask, "What is there left to do in mathematics?" University libraries are full of expanding journals of new mathematics, including statistics.

We must observe that most maximum likelihood estimators have approximate normal distributions for large sample sizes, and we give a heuristic proof of it in this chapter. These estimators are of what is called the *regular cases*—in particular, those cases in which the parameters are not in the endpoints of the support of X. Abraham de Moivre proved this theorem for $\widehat{p}$ of the binomial distribution, and Laplace and Gauss did so for $\overline{X}$ in a number of other distributions. This is the real reason the normal distribution is so important: Most estimators of parameters have approximate normal distributions, allowing us to construct confidence intervals (see Chapter 7) and perform tests (see Chapter 8) with such estimates.

The Neo-Bayesian movement in America really started with J. Savage in the 1950s. Initially, Bayesians were limited in their work because it was extremely difficult to compute certain distributions, such as the conditional one, $k(\theta \mid x_1, x_2, \ldots, x_n)$. However, toward the end of the 1970s, computers were becoming more useful and thus computing was much easier. In particular, the Bayesians developed Gibbs sampling and Markov chain Monte Carlo (MCMC). It is our opinion that the Bayesians will continue to expand and Bayes methods will be a major approach to statistical inferences, possibly even dominating professional applications. This is difficult for three fairly classical (non-Bayesian) statisticians (as we are) to admit, but, in all fairness, we cannot ignore the strong trend toward Bayesian methods.

INTERVAL ESTIMATION

7.1 CONFIDENCE INTERVALS FOR MEANS

Given a random sample $X_1, X_2, \ldots, X_n$ from a normal distribution $N(\mu, \sigma^2)$, we shall now consider the closeness of $\overline{X}$, the unbiased estimator of μ, to the unknown mean μ. To do this, we use the error structure (distribution) of $\overline{X}$, namely, that $\overline{X}$ is $N(\mu, \sigma^2/n)$ (see Corollary 5.5-1), to construct what is called a confidence interval for the unknown parameter μ when the variance σ^2 is known. For the probability $1 - \alpha$, we can find a number $z_{\alpha/2}$ from Table V in Appendix B such that

$$P\left(-z_{\alpha/2} \le \frac{\overline{X} - \mu}{\sigma/\sqrt{n}} \le z_{\alpha/2}\right) = 1 - \alpha.$$

For example, if $1 - \alpha = 0.95$, then $z_{\alpha/2} = z_{0.025} = 1.96$, and if $1 - \alpha = 0.90$, then $z_{\alpha/2} = z_{0.05} = 1.645$. Now, recalling that $\sigma > 0$, we see that the following inequalities are equivalent:

$$-z_{\alpha/2} \le \frac{\overline{X} - \mu}{\sigma/\sqrt{n}} \le z_{\alpha/2},$$

$$-z_{\alpha/2}\left(\frac{\sigma}{\sqrt{n}}\right) \le \overline{X} - \mu \le z_{\alpha/2}\left(\frac{\sigma}{\sqrt{n}}\right),$$

$$-\overline{X} - z_{\alpha/2}\left(\frac{\sigma}{\sqrt{n}}\right) \le -\mu \le -\overline{X} + z_{\alpha/2}\left(\frac{\sigma}{\sqrt{n}}\right),$$

$$\overline{X} + z_{\alpha/2}\left(\frac{\sigma}{\sqrt{n}}\right) \ge \mu \ge \overline{X} - z_{\alpha/2}\left(\frac{\sigma}{\sqrt{n}}\right).$$

Thus, since the probability of the first of these is $1 - \alpha$, the probability of the last must also be $1 - \alpha$, because the latter is true if and only if the former is true. That is, we have

$$P\left[\overline{X} - z_{\alpha/2}\left(\frac{\sigma}{\sqrt{n}}\right) \le \mu \le \overline{X} + z_{\alpha/2}\left(\frac{\sigma}{\sqrt{n}}\right)\right] = 1 - \alpha.$$

So the probability that the random interval

$$\left[\overline{X} - z_{\alpha/2}\left(\frac{\sigma}{\sqrt{n}}\right), \overline{X} + z_{\alpha/2}\left(\frac{\sigma}{\sqrt{n}}\right)\right]$$

includes the unknown mean μ is $1 - \alpha$.

Once the sample is observed and the sample mean computed to equal $\overline{x}$, the interval $[\overline{x} - z_{\alpha/2}(\sigma/\sqrt{n}), \overline{x} + z_{\alpha/2}(\sigma/\sqrt{n})]$ becomes known. Since the probability that the random interval covers μ before the sample is drawn is equal to $1 - \alpha$, we now call the computed interval, $\overline{x} \pm z_{\alpha/2}(\sigma/\sqrt{n})$ (for brevity), a $100(1 - \alpha)\%$ **confidence interval** for the unknown mean μ. For example, $\overline{x} \pm 1.96(\sigma/\sqrt{n})$ is a 95% confidence interval for μ. The number $100(1 - \alpha)\%$, or equivalently, $1 - \alpha$, is called the **confidence coefficient**.

We see that the confidence interval for μ is centered at the point estimate $\overline{x}$ and is completed by subtracting and adding the quantity $z_{\alpha/2}(\sigma/\sqrt{n})$. Note that as n increases, $z_{\alpha/2}(\sigma/\sqrt{n})$ decreases, resulting in a shorter confidence interval with the same confidence coefficient $1 - \alpha$. A shorter confidence interval gives a more precise estimate of μ, regardless of the confidence we have in the estimate of μ. Statisticians who are not restricted by time, money, effort, or the availability of observations can obviously make the confidence interval as short as they like by increasing the sample size n. For a fixed sample size n, the length of the confidence interval can also be shortened by decreasing the confidence coefficient $1 - \alpha$. But if this is done, we achieve a shorter confidence interval at the expense of losing some confidence.

Example 7.1-1

Let X equal the length of life of a 60-watt light bulb marketed by a certain manufacturer. Assume that the distribution of X is $N(\mu, 1296)$. If a random sample of $n = 27$ bulbs is tested until they burn out, yielding a sample mean of $\overline{x} = 1478$ hours, then a 95% confidence interval for μ is

$$\left[\overline{x} - z_{0.025}\left(\frac{\sigma}{\sqrt{n}}\right), \overline{x} + z_{0.025}\left(\frac{\sigma}{\sqrt{n}}\right)\right] = \left[1478 - 1.96\left(\frac{36}{\sqrt{27}}\right), 1478 + 1.96\left(\frac{36}{\sqrt{27}}\right)\right]$$

$$= [1478 - 13.58, 1478 + 13.58]$$

$$= [1464.42, 1491.58].$$ ■

The next example will help to give a better intuitive feeling for the interpretation of a confidence interval.

Example 7.1-2

Let $\overline{x}$ be the observed sample mean of five observations of a random sample from the normal distribution $N(\mu, 16)$. A 90% confidence interval for the unknown mean μ is

$$\left[\overline{x} - 1.645\sqrt{\frac{16}{5}}, \overline{x} + 1.645\sqrt{\frac{16}{5}}\right].$$

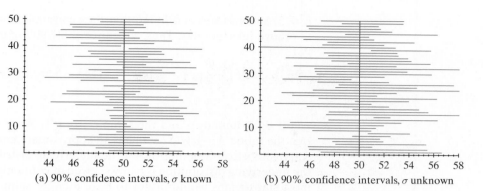

(a) 90% confidence intervals, σ known   (b) 90% confidence intervals, σ unknown

Figure 7.1-1 Confidence intervals using z and t

For a particular sample, this interval either does or does not contain the mean μ. However, if many such intervals were calculated, about 90% of them should contain the mean μ. Fifty random samples of size 5 from the normal distribution $N(50, 16)$ were simulated on a computer. A 90% confidence interval was calculated for each random sample, as if the mean were unknown. Figure 7.1-1(a) depicts each of these 50 intervals as a line segment. Note that 45 (or 90%) of them contain the mean, $\mu = 50$. In other simulations of 50 confidence intervals, the number of 90% confidence intervals containing the mean could be larger or smaller. [In fact, if W is a random variable that counts the number of 90% confidence intervals containing the mean, then the distribution of W is $b(50, 0.90)$.] ■

If we cannot assume that the distribution from which the sample arose is normal, we can still obtain an approximate confidence interval for μ. By the central limit theorem, provided that n is large enough, the ratio $(\overline{X} - \mu)/(\sigma/\sqrt{n})$ has the approximate normal distribution $N(0, 1)$ when the underlying distribution is not normal. In this case,

$$P\left(-z_{\alpha/2} \leq \frac{\overline{X} - \mu}{\sigma/\sqrt{n}} \leq z_{\alpha/2}\right) \approx 1 - \alpha,$$

and

$$\left[\overline{x} - z_{\alpha/2}\left(\frac{\sigma}{\sqrt{n}}\right), \overline{x} + z_{\alpha/2}\left(\frac{\sigma}{\sqrt{n}}\right)\right]$$

is an approximate $100(1 - \alpha)\%$ confidence interval for μ.

The closeness of the approximate probability $1 - \alpha$ to the exact probability depends on both the underlying distribution and the sample size. When the underlying distribution is unimodal (has only one mode), symmetric, and continuous, the approximation is usually quite good even for small n, such as $n = 5$. As the underlying distribution becomes "less normal" (i.e., badly skewed or discrete), a larger sample size might be required to keep a reasonably accurate approximation. But, in almost all cases, an n of at least 30 is usually adequate.

Example 7.1-3   Let X equal the amount of orange juice (in grams per day) consumed by an American. Suppose it is known that the standard deviation of X is $\sigma = 96$. To estimate the mean μ of X, an orange growers' association took a random sample of

$n = 576$ Americans and found that they consumed, on the average, $\bar{x} = 133$ grams of orange juice per day. Thus, an approximate 90% confidence interval for μ is

$$133 \pm 1.645\left(\frac{96}{\sqrt{576}}\right), \qquad \text{or} \qquad [133 - 6.58, \, 133 + 6.58] = [126.42, \, 139.58]. \quad \blacksquare$$

If σ^2 is unknown and the sample size n is 30 or greater, we shall use the fact that the ratio $(\overline{X} - \mu)/(S/\sqrt{n})$ has an approximate normal distribution $N(0,1)$. This statement is true whether or not the underlying distribution is normal. However, if the underlying distribution is badly skewed or contaminated with occasional outliers, most statisticians would prefer to have a larger sample size—say, 50 or more—and even that might not produce good results. After this next example, we consider what to do when n is small.

Example 7.1-4

Lake Macatawa, an inlet lake on the east side of Lake Michigan, is divided into an east basin and a west basin. To measure the effect on the lake of salting city streets in the winter, students took 32 samples of water from the west basin and measured the amount of sodium in parts per million in order to make a statistical inference about the unknown mean μ. They obtained the following data:

13.0	18.5	16.4	14.8	19.4	17.3	23.2	24.9
20.8	19.3	18.8	23.1	15.2	19.9	19.1	18.1
25.1	16.8	20.4	17.4	25.2	23.1	15.3	19.4
16.0	21.7	15.2	21.3	21.5	16.8	15.6	17.6

For these data, $\bar{x} = 19.07$ and $s^2 = 10.60$. Thus, an approximate 95% confidence interval for μ is

$$\bar{x} \pm 1.96\left(\frac{s}{\sqrt{n}}\right), \qquad \text{or} \qquad 19.07 \pm 1.96\sqrt{\frac{10.60}{32}}, \qquad \text{or} \qquad [17.94, \, 20.20]. \quad \blacksquare$$

So we have found a confidence interval for the mean μ of a normal distribution, assuming that the value of the standard deviation σ is known or assuming that σ is unknown but the sample size is large. However, in many applications, the sample sizes are small and we do not know the value of the standard deviation, although in some cases we might have a very good idea about its value. For example, a manufacturer of light bulbs probably has a good notion from past experience of the value of the standard deviation of the length of life of different types of light bulbs. But certainly, most of the time, the investigator will not have any more idea about the standard deviation than about the mean—and frequently less. Let us consider how to proceed under these circumstances.

If the random sample arises from a normal distribution, we use the fact that

$$T = \frac{\overline{X} - \mu}{S/\sqrt{n}}$$

has a t distribution with $r = n - 1$ degrees of freedom (see Equation 5.5-2), where S^2 is the usual unbiased estimator of σ^2. Select $t_{\alpha/2}(n-1)$ so that $P[T \geq t_{\alpha/2}(n-1)] = \alpha/2$. [See Figure 5.5-2(b) and Table VI in Appendix B.] Then

$$1 - \alpha = P\left[-t_{\alpha/2}(n{-}1) \le \frac{\overline{X} - \mu}{S/\sqrt{n}} \le t_{\alpha/2}(n{-}1) \right]$$

$$= P\left[-t_{\alpha/2}(n{-}1)\left(\frac{S}{\sqrt{n}}\right) \le \overline{X} - \mu \le t_{\alpha/2}(n{-}1)\left(\frac{S}{\sqrt{n}}\right) \right]$$

$$= P\left[-\overline{X} - t_{\alpha/2}(n{-}1)\left(\frac{S}{\sqrt{n}}\right) \le -\mu \le -\overline{X} + t_{\alpha/2}(n{-}1)\left(\frac{S}{\sqrt{n}}\right) \right]$$

$$= P\left[\overline{X} - t_{\alpha/2}(n{-}1)\left(\frac{S}{\sqrt{n}}\right) \le \mu \le \overline{X} + t_{\alpha/2}(n{-}1)\left(\frac{S}{\sqrt{n}}\right) \right].$$

Thus, the observations of a random sample provide $\overline{x}$ and s^2, and

$$\left[\overline{x} - t_{\alpha/2}(n{-}1)\left(\frac{s}{\sqrt{n}}\right), \ \overline{x} + t_{\alpha/2}(n{-}1)\left(\frac{s}{\sqrt{n}}\right) \right]$$

is a $100(1 - \alpha)\%$ confidence interval for μ.

Example 7.1-5

Let X equal the amount of butterfat in pounds produced by a typical cow during a 305-day milk production period between her first and second calves. Assume that the distribution of X is $N(\mu, \sigma^2)$. To estimate μ, a farmer measured the butterfat production for $n = 20$ cows and obtained the following data:

| 481 | 537 | 513 | 583 | 453 | 510 | 570 | 500 | 457 | 555 |
| 618 | 327 | 350 | 643 | 499 | 421 | 505 | 637 | 599 | 392 |

For these data, $\overline{x} = 507.50$ and $s = 89.75$. Thus, a point estimate of μ is $\overline{x} = 507.50$. Since $t_{0.05}(19) = 1.729$, a 90% confidence interval for μ is

$$507.50 \pm 1.729\left(\frac{89.75}{\sqrt{20}}\right) \quad \text{or}$$

$$507.50 \pm 34.70, \qquad \text{or equivalently,} \qquad [472.80, 542.20]. \qquad \blacksquare$$

Let T have a t distribution with $n-1$ degrees of freedom. Then $t_{\alpha/2}(n{-}1) > z_{\alpha/2}$. Consequently, we would expect the interval $\overline{x} \pm z_{\alpha/2}(\sigma/\sqrt{n})$ to be shorter than the interval $\overline{x} \pm t_{\alpha/2}(n{-}1)(s/\sqrt{n})$. After all, we have more information, namely, the value of σ, in constructing the first interval. However, the length of the second interval is very much dependent on the value of s. If the observed s is smaller than σ, a shorter confidence interval could result by the second procedure. But on the average, $\overline{x} \pm z_{\alpha/2}(\sigma/\sqrt{n})$ is the shorter of the two confidence intervals (Exercise 7.1-14).

Example 7.1-6

In Example 7.1-2, 50 confidence intervals were simulated for the mean of a normal distribution, assuming that the variance was known. For those same data, since $t_{0.05}(4) = 2.132$, $\overline{x} \pm 2.132(s/\sqrt{5})$ was used to calculate a 90% confidence interval for μ. For those particular 50 intervals, 46 contained the mean $\mu = 50$. These 50 intervals are depicted in Figure 7.1-1(b). Note the different lengths of the intervals. Some are longer and some are shorter than the corresponding z intervals. The average length of the 50 t intervals is 7.137, which is quite close to the expected length of such an interval: 7.169. (See Exercise 7.1-14.) The length of the intervals that use z and $\sigma = 4$ is 5.885. $\qquad \blacksquare$

If we are not able to assume that the underlying distribution is normal, but μ and σ are both unknown, approximate confidence intervals for μ can still be constructed with the formula

$$T = \frac{\overline{X} - \mu}{S/\sqrt{n}},$$

which now only has an approximate t distribution. Generally, this approximation is quite good (i.e., it is robust) for many nonnormal distributions; in particular, it works well if the underlying distribution is symmetric, unimodal, and of the continuous type. However, if the distribution is highly skewed, there is great danger in using that approximation. In such a situation, it would be safer to use certain nonparametric methods for finding a confidence interval for the median of the distribution, one of which is given in Section 7.5.

There is one other aspect of confidence intervals that should be mentioned. So far, we have created only what are called **two-sided confidence intervals** for the mean μ. Sometimes, however, we might want only a lower (or upper) bound on μ. We proceed as follows.

Say $\overline{X}$ is the mean of a random sample of size n from the normal distribution $N(\mu, \sigma^2)$, where, for the moment, assume that σ^2 is known. Then

$$P\left(\frac{\overline{X} - \mu}{\sigma/\sqrt{n}} \leq z_\alpha\right) = 1 - \alpha,$$

or equivalently,

$$P\left[\overline{X} - z_\alpha\left(\frac{\sigma}{\sqrt{n}}\right) \leq \mu\right] = 1 - \alpha.$$

Once $\overline{X}$ is observed to be equal to $\overline{x}$, it follows that $[\overline{x} - z_\alpha(\sigma/\sqrt{n}), \infty)$ is a $100(1-\alpha)\%$ **one-sided confidence interval** for μ. That is, with the confidence coefficient $1 - \alpha$, $\overline{x} - z_\alpha(\sigma/\sqrt{n})$ is a lower bound for μ. Similarly, $(-\infty, \overline{x} + z_\alpha(\sigma/\sqrt{n})]$ is a one-sided confidence interval for μ and $\overline{x} + z_\alpha(\sigma/\sqrt{n})$ provides an upper bound for μ with confidence coefficient $1 - \alpha$.

When σ is unknown, we would use $T = (\overline{X} - \mu)/(S/\sqrt{n})$ to find the corresponding lower or upper bounds for μ, namely,

$$\overline{x} - t_\alpha(n-1)(s/\sqrt{n}) \quad \text{and} \quad \overline{x} + t_\alpha(n-1)(s/\sqrt{n}).$$

Exercises

7.1-1. A random sample of size 16 from the normal distribution $N(\mu, 25)$ yielded $\overline{x} = 73.8$. Find a 95% confidence interval for μ.

7.1-2. A random sample of size 8 from $N(\mu, 72)$ yielded $\overline{x} = 85$. Find the following confidence intervals for μ:

(a) 99%.   (b) 95%.   (c) 90%.   (d) 80%.

7.1-3. To determine the effect of 100% nitrate on the growth of pea plants, several specimens were planted and then watered with 100% nitrate every day. At the end of

two weeks, the plants were measured. Here are data on seven of them:

17.5   14.5   15.2   14.0   17.3   18.0   13.8

Assume that these data are a random sample from a normal distribution $N(\mu, \sigma^2)$.

(a) Find the value of a point estimate of μ.

(b) Find the value of a point estimate of σ.

(c) Give the endpoints for a 90% confidence interval for μ.

7.1-4. Let X equal the weight in grams of a "52-gram" snack pack of candies. Assume that the distribution of X is $N(\mu, 4)$. A random sample of $n = 10$ observations of X yielded the following data:

55.95	56.54	57.58	55.13	57.48
56.06	59.93	58.30	52.57	58.46

(a) Give a point estimate for μ.

(b) Find the endpoints for a 95% confidence interval for μ.

(c) On the basis of these very limited data, what is the probability that an individual snack pack selected at random is filled with less than 52 grams of candy?

7.1-5. As a clue to the amount of organic waste in Lake Macatawa (see Example 7.1-4), a count was made of the number of bacteria colonies in 100 milliliters of water. The number of colonies, in hundreds, for $n = 30$ samples of water from the east basin yielded

93	140	8	120	3	120	33	70	91	61
7	100	19	98	110	23	14	94	57	9
66	53	28	76	58	9	73	49	37	92

Find an approximate 90% confidence interval for the mean number (say, μ_E) of colonies in 100 milliliters of water in the east basin.

7.1-6. To determine whether the bacteria count was lower in the west basin of Lake Macatawa than in the east basin, $n = 37$ samples of water were taken from the west basin and the number of bacteria colonies in 100 milliliters of water was counted. The sample characteristics were $\bar{x} = 11.95$ and $s = 11.80$, measured in hundreds of colonies. Find an approximate 95% confidence interval for the mean number of colonies (say, μ_W) in 100 milliliters of water in the west basin.

7.1-7. Thirteen tons of cheese, including "22-pound" wheels (label weight), is stored in some old gypsum mines. A random sample of $n = 9$ of these wheels yielded the following weights in pounds:

21.50	18.95	18.55	19.40	19.15
22.35	22.90	22.20	23.10	

Assuming that the distribution of the weights of the wheels of cheese is $N(\mu, \sigma^2)$, find a 95% confidence interval for μ.

7.1-8. Assume that the yield per acre for a particular variety of soybeans is $N(\mu, \sigma^2)$. For a random sample of $n = 5$ plots, the yields in bushels per acre were 37.4, 48.8, 46.9, 55.0, and 44.0.

(a) Give a point estimate for μ.

(b) Find a 90% confidence interval for μ.

7.1-9. During the Friday night shift, $n = 28$ mints were selected at random from a production line and weighed. They had an average weight of $\bar{x} = 21.45$ grams and a standard deviation of $s = 0.31$ grams. Give the lower endpoint of a 90% one-sided confidence interval for μ, the mean weight of all the mints.

7.1-10. A leakage test was conducted to determine the effectiveness of a seal designed to keep the inside of a plug airtight. An air needle was inserted into the plug, and the plug and needle were placed under water. The pressure was then increased until leakage was observed. Let X equal the pressure in pounds per square inch. Assume that the distribution of X is $N(\mu, \sigma^2)$. The following $n = 10$ observations of X were obtained:

$$3.1 \quad 3.3 \quad 4.5 \quad 2.8 \quad 3.5 \quad 3.5 \quad 3.7 \quad 4.2 \quad 3.9 \quad 3.3$$

Use the observations to

(a) Find a point estimate of μ.

(b) Find a point estimate of σ.

(c) Find a 95% one-sided confidence interval for μ that provides an upper bound for μ.

7.1-11. Students took $n = 35$ samples of water from the east basin of Lake Macatawa (see Example 7.1-4) and measured the amount of sodium in parts per million. For their data, they calculated $\bar{x} = 24.11$ and $s^2 = 24.44$. Find an approximate 90% confidence interval for μ, the mean of the amount of sodium in parts per million.

7.1-12. In nuclear physics, detectors are often used to measure the energy of a particle. To calibrate a detector, particles of known energy are directed into it. The values of signals from 15 different detectors, for the same energy, are

260	216	259	206	265	284	291	229
232	250	225	242	240	252	236	

(a) Find a 95% confidence interval for μ, assuming that these are observations from a $N(\mu, \sigma^2)$ distribution.

(b) Construct a box-and-whisker diagram of the data.

(c) Are these detectors doing a good job or a poor job of putting out the same signal for the same input energy?

7.1-13. A study was conducted to measure (1) the amount of cervical spine movement induced by different methods of gaining access to the mouth and nose to begin resuscitation of a football player who is wearing a helmet and (2) the time it takes to complete each method. One method involves using a manual screwdriver to remove the side clips holding the face mask in place and then flipping the mask up. Twelve measured times in seconds for the manual screwdriver are

33.8 31.6 28.5 29.9 29.8 26.0 35.7 27.2 29.1 32.1 26.1 24.1

Assume that these are independent observations of a normally distributed random variable that is $N(\mu, \sigma^2)$.

(a) Find point estimates of μ and σ.

(b) Find a 95% one-sided confidence interval for μ that provides an upper bound for μ.

(c) Does the assumption of normality seem to be justified? Why?

7.1-14. Let $X_1, X_2, \ldots, X_n$ be a random sample of size n from the normal distribution $N(\mu, \sigma^2)$. Calculate the expected length of a 95% confidence interval for μ, assuming that $n = 5$ and the variance is

(a) known.

(b) unknown.

HINT: To find $E(S)$, first determine $E[\sqrt{(n-1)S^2/\sigma^2}]$, recalling that $(n-1)S^2/\sigma^2$ is $\chi^2(n-1)$. (See Exercise 6.4-14.)

7.1-15. An automotive supplier of interior parts places several electrical wires in a harness. A pull test measures the force required to pull spliced wires apart. A customer requires that each wire spliced into the harness must withstand a pull force of 20 pounds. Let X equal the pull force required to pull 20 gauge wires apart. Assume that the

distribution of X is $N(\mu, \sigma^2)$. The following data give 20 observations of X:

28.8 24.4 30.1 25.6 26.4 23.9 22.1 22.5 27.6 28.1

20.8 27.7 24.4 25.1 24.6 26.3 28.2 22.2 26.3 24.4

(a) Find point estimates for μ and σ.

(b) Find a 99% one-sided confidence interval for μ that provides a lower bound for μ.

7.1-16. Let S^2 be the variance of a random sample of size n from $N(\mu, \sigma^2)$. Using the fact that $(n-1)S^2/\sigma^2$ is $\chi^2(n-1)$, note that the probability

$$P\left[a \le \frac{(n-1)S^2}{\sigma^2} \le b\right] = 1 - \alpha,$$

where $a = \chi^2_{1-\alpha/2}(n-1)$ and $b = \chi^2_{\alpha/2}(n-1)$. Rewrite the inequalities to obtain

$$P\left[\frac{(n-1)S^2}{b} \le \sigma^2 \le \frac{(n-1)S^2}{a}\right] = 1 - \alpha.$$

If $n = 13$ and $12S^2 = \sum_{i=1}^{13}(x_i - \bar{x})^2 = 128.41$, show that $[6.11, 24.57]$ is a 90% **confidence interval for the variance** σ^2. Accordingly, $[2.47, 4.96]$ is a 90% confidence interval for σ.

7.1-17. Let $\bar{X}$ be the mean of a random sample of size n from $N(\mu, 9)$. Find n so that $P(\bar{X}-1 < \mu < \bar{X}+1) = 0.90$.

7.2 CONFIDENCE INTERVALS FOR THE DIFFERENCE OF TWO MEANS

Suppose that we are interested in comparing the means of two normal distributions. Let $X_1, X_2, \ldots, X_n$ and $Y_1, Y_2, \ldots, Y_m$ be, respectively, two independent random samples of sizes n and m from the two normal distributions $N(\mu_X, \sigma_X^2)$ and $N(\mu_Y, \sigma_Y^2)$. Suppose, for now, that σ_X^2 and σ_Y^2 are known. The random samples are independent; thus, the respective sample means $\bar{X}$ and $\bar{Y}$ are also independent and have distributions $N(\mu_X, \sigma_X^2/n)$ and $N(\mu_Y, \sigma_Y^2/m)$. Consequently, the distribution of $W = \bar{X} - \bar{Y}$ is $N(\mu_X - \mu_Y, \sigma_X^2/n + \sigma_Y^2/m)$ and

$$P\left(-z_{\alpha/2} \le \frac{(\bar{X} - \bar{Y}) - (\mu_X - \mu_Y)}{\sqrt{\sigma_X^2/n + \sigma_Y^2/m}} \le z_{\alpha/2}\right) = 1 - \alpha,$$

which can be rewritten as

$$P[(\bar{X} - \bar{Y}) - z_{\alpha/2}\sigma_W \le \mu_X - \mu_Y \le (\bar{X} - \bar{Y}) + z_{\alpha/2}\sigma_W] = 1 - \alpha,$$

where $\sigma_W = \sqrt{\sigma_X^2/n + \sigma_Y^2/m}$ is the standard deviation of $\bar{X} - \bar{Y}$. Once the experiments have been performed and the means $\bar{x}$ and $\bar{y}$ computed, the interval

$$[\bar{x} - \bar{y} - z_{\alpha/2}\sigma_W, \bar{x} - \bar{y} + z_{\alpha/2}\sigma_W]$$

or, equivalently, $\bar{x}-\bar{y}\pm z_{\alpha/2}\sigma_W$ provides a $100(1-\alpha)\%$ confidence interval for $\mu_X-\mu_Y$. Note that this interval is centered at the point estimate $\bar{x}-\bar{y}$ of $\mu_X-\mu_Y$ and is completed by subtracting and adding the product of $z_{\alpha/2}$ and the standard deviation of the point estimator.

Example 7.2-1

In the preceding discussion, let $n=15$, $m=8$, $\bar{x}=70.1$, $\bar{y}=75.3$, $\sigma_X^2=60$, $\sigma_Y^2=40$, and $1-\alpha=0.90$. Thus, $1-\alpha/2=0.95=\Phi(1.645)$. Hence,

$$1.645\sigma_W = 1.645\sqrt{\frac{60}{15}+\frac{40}{8}} = 4.935,$$

and, since $\bar{x}-\bar{y}=-5.2$, it follows that

$$[-5.2-4.935, -5.2+4.935] = [-10.135, -0.265]$$

is a 90% confidence interval for $\mu_X-\mu_Y$. Because the confidence interval does not include zero, we suspect that μ_Y is greater than μ_X. ∎

If the sample sizes are large and σ_X and σ_Y are unknown, we can replace σ_X^2 and σ_Y^2 with s_x^2 and s_y^2, where s_x^2 and s_y^2 are the values of the respective unbiased estimates of the variances. This means that

$$\bar{x}-\bar{y}\pm z_{\alpha/2}\sqrt{\frac{s_x^2}{n}+\frac{s_y^2}{m}}$$

serves as an approximate $100(1-\alpha)\%$ confidence interval for $\mu_X-\mu_Y$.

Now consider the problem of constructing confidence intervals for the difference of the means of two normal distributions when the variances are unknown but the sample sizes are small. Let $X_1, X_2, \ldots, X_n$ and $Y_1, Y_2, \ldots, Y_m$ be two independent random samples from the distributions $N(\mu_X, \sigma_X^2)$ and $N(\mu_Y, \sigma_Y^2)$, respectively. If the sample sizes are not large (say, considerably smaller than 30), this problem can be a difficult one. However, even in these cases, if we can assume common, but unknown, variances (say, $\sigma_X^2=\sigma_Y^2=\sigma^2$), there is a way out of our difficulty.

We know that

$$Z = \frac{\overline{X}-\overline{Y}-(\mu_X-\mu_Y)}{\sqrt{\sigma^2/n+\sigma^2/m}}$$

is $N(0,1)$. Moreover, since the random samples are independent,

$$U = \frac{(n-1)S_X^2}{\sigma^2}+\frac{(m-1)S_Y^2}{\sigma^2}$$

is the sum of two independent chi-square random variables; thus, the distribution of U is $\chi^2(n+m-2)$. In addition, the independence of the sample means and sample variances implies that Z and U are independent. According to the definition of a T random variable,

$$T = \frac{Z}{\sqrt{U/(n+m-2)}}$$

has a t distribution with $n + m - 2$ degrees of freedom. That is,

$$T = \frac{\dfrac{\overline{X} - \overline{Y} - (\mu_X - \mu_Y)}{\sqrt{\sigma^2/n + \sigma^2/m}}}{\sqrt{\left[\dfrac{(n-1)S_X^2}{\sigma^2} + \dfrac{(m-1)S_Y^2}{\sigma^2}\right] \Big/ (n+m-2)}}$$

$$= \frac{\overline{X} - \overline{Y} - (\mu_X - \mu_Y)}{\sqrt{\left[\dfrac{(n-1)S_X^2 + (m-1)S_Y^2}{n+m-2}\right]\left[\dfrac{1}{n} + \dfrac{1}{m}\right]}}$$

has a t distribution with $r = n + m - 2$ degrees of freedom. Thus, with $t_0 = t_{\alpha/2}(n+m-2)$, we have

$$P(-t_0 \leq T \leq t_0) = 1 - \alpha.$$

Solving the inequalities for $\mu_X - \mu_Y$ yields

$$P\left(\overline{X} - \overline{Y} - t_0 S_P \sqrt{\frac{1}{n} + \frac{1}{m}} \leq \mu_X - \mu_Y \leq \overline{X} - \overline{Y} + t_0 S_P \sqrt{\frac{1}{n} + \frac{1}{m}}\right),$$

where the pooled estimator of the common standard deviation is

$$S_P = \sqrt{\frac{(n-1)S_X^2 + (m-1)S_Y^2}{n+m-2}}.$$

If $\overline{x}$, $\overline{y}$, and s_p are the observed values of $\overline{X}$, $\overline{Y}$, and S_P, then

$$\left[\overline{x} - \overline{y} - t_0 s_p \sqrt{\frac{1}{n} + \frac{1}{m}}, \overline{x} - \overline{y} + t_0 s_p \sqrt{\frac{1}{n} + \frac{1}{m}}\right]$$

is a $100(1 - \alpha)\%$ confidence interval for $\mu_X - \mu_Y$.

Example 7.2-2

Suppose that scores on a standardized test in mathematics taken by students from large and small high schools are $N(\mu_X, \sigma^2)$ and $N(\mu_Y, \sigma^2)$, respectively, where σ^2 is unknown. If a random sample of $n = 9$ students from large high schools yielded $\overline{x} = 81.31$, $s_x^2 = 60.76$, and a random sample of $m = 15$ students from small high schools yielded $\overline{y} = 78.61$, $s_y^2 = 48.24$, then the endpoints for a 95% confidence interval for $\mu_X - \mu_Y$ are given by

$$81.31 - 78.61 \pm 2.074 \sqrt{\frac{8(60.76) + 14(48.24)}{22}} \sqrt{\frac{1}{9} + \frac{1}{15}}$$

because $t_{0.025}(22) = 2.074$. The 95% confidence interval is $[-3.65, 9.05]$. ∎

REMARKS The assumption of equal variances, namely, $\sigma_X^2 = \sigma_Y^2$, can be modified somewhat so that we are still able to find a confidence interval for $\mu_X - \mu_Y$. That is, if we know the ratio σ_X^2/σ_Y^2 of the variances, we can still make this type of statistical inference by using a random variable with a t distribution. (See Exercise 7.2-8.) However, if we do not know the ratio of the variances and yet suspect that the unknown σ_X^2 and σ_Y^2 differ by a great deal, what do we do? It is safest to return to

$$\frac{\overline{X} - \overline{Y} - (\mu_X - \mu_Y)}{\sqrt{\sigma_X^2/n + \sigma_Y^2/m}}$$

for the inference about $\mu_X - \mu_Y$ but replacing σ_X^2 and σ_Y^2 by their respective estimators S_X^2 and S_Y^2. That is, consider

$$W = \frac{\overline{X} - \overline{Y} - (\mu_X - \mu_Y)}{\sqrt{S_X^2/n + S_Y^2/m}}.$$

What is the distribution of W? As before, we note that if n and m are large enough and the underlying distributions are close to normal (or at least not badly skewed), then W has an approximate normal distribution and a confidence interval for $\mu_X - \mu_Y$ can be found by considering

$$P(-z_{\alpha/2} \leq W \leq z_{\alpha/2}) \approx 1 - \alpha.$$

However, for smaller n and m, Welch has proposed a Student's t distribution as the approximating one for W. Welch's proposal was later modified by Aspin. [See A. A. Aspin, "Tables for Use in Comparisons Whose Accuracy Involves Two Variances, Separately Estimated," *Biometrika*, **36** (1949), pp. 290–296, with an appendix by B. L. Welch in which he makes the suggestion used here.] The approximating Student's t distribution has r degrees of freedom, where

$$\frac{1}{r} = \frac{c^2}{n-1} + \frac{(1-c)^2}{m-1} \qquad \text{and} \qquad c = \frac{s_x^2/n}{s_x^2/n + s_y^2/m}.$$

An equivalent formula for r is

$$r = \frac{\left(\dfrac{s_x^2}{n} + \dfrac{s_y^2}{m}\right)^2}{\dfrac{1}{n-1}\left(\dfrac{s_x^2}{n}\right)^2 + \dfrac{1}{m-1}\left(\dfrac{s_y^2}{m}\right)^2}. \tag{7.2-1}$$

In particular, the assignment of r by this rule provides protection in the case in which the smaller sample size is associated with the larger variance by greatly reducing the number of degrees of freedom from the usual $n + m - 2$. Of course, this reduction increases the value of $t_{\alpha/2}$. If r is not an integer, then use the greatest integer in r; that is, use $[r]$ as the number of degrees of freedom associated with the approximating Student's t distribution. An approximate $100(1-\alpha)\%$ confidence interval for $\mu_X - \mu_Y$ is given by

$$\overline{x} - \overline{y} \pm t_{\alpha/2}(r)\sqrt{\frac{s_x^2}{n} + \frac{s_y^2}{m}}.$$

It is interesting to consider the two-sample T in more detail. It is

$$T = \frac{\overline{X} - \overline{Y} - (\mu_X - \mu_Y)}{\sqrt{\dfrac{(n-1)S_X^2 + (m-1)S_Y^2}{n+m-2}\left(\dfrac{1}{n} + \dfrac{1}{m}\right)}} \tag{7.2-2}$$

$$= \frac{\overline{X} - \overline{Y} - (\mu_X - \mu_Y)}{\sqrt{\left[\dfrac{(n-1)S_X^2}{nm} + \dfrac{(m-1)S_Y^2}{nm}\right]\left[\dfrac{n+m}{n+m-2}\right]}}.$$

Now, since $(n-1)/n \approx 1$, $(m-1)/m \approx 1$, and $(n+m)/(n+m-2) \approx 1$, we have

$$T \approx \frac{\overline{X} - \overline{Y} - (\mu_X - \mu_Y)}{\sqrt{\dfrac{S_X^2}{m} + \dfrac{S_Y^2}{n}}}.$$

We note that, in this form, each variance is divided by the wrong sample size! That is, if the sample sizes are large or the variances known, we would like

$$\sqrt{\frac{S_X^2}{n} + \frac{S_Y^2}{m}} \qquad \text{or} \qquad \sqrt{\frac{\sigma_X^2}{n} + \frac{\sigma_Y^2}{m}}$$

in the denominator; so T seems to change the sample sizes. Thus, using this T is particularly bad when the sample sizes and the variances are unequal; hence, caution must be taken in using that T to construct a confidence interval for $\mu_X - \mu_Y$. That is, if $n < m$ and $\sigma_X^2 < \sigma_Y^2$, then T does not have a distribution which is close to that of a Student t-distribution with $n + m - 2$ degrees of freedom: Instead, its spread is much less than the Student t's as the term s_y^2/n in the denominator is much larger than it should be. By contrast, if $m < n$ and $\sigma_X^2 < \sigma_Y^2$, then $s_x^2/m + s_y^2/n$ is generally smaller than it should be and the distribution of T is spread out more than that of the Student t.

There is a way out of this difficulty, however: When the underlying distributions are close to normal, but the sample sizes and the variances are seemingly much different, we suggest the use of

$$W = \frac{\overline{X} - \overline{Y} - (\mu_X - \mu_Y)}{\sqrt{\dfrac{S_X^2}{n} + \dfrac{S_Y^2}{m}}}, \tag{7.2-3}$$

where Welch proved that W has an approximate t distribution with $[r]$ degrees of freedom, with the number of degrees of freedom given by Equation 7.2-1. ∎

Example 7.2-3   To help understand the preceding remarks, a simulation was done with *Maple*. In order to obtain a q–q plot of the quantiles of a t distribution, a CAS or some type of computer program is very important because of the challenge in finding these quantiles.

Maple was used to simulate $N = 500$ observations of T (Equation 7.2-2) and $N = 500$ observations of W (Equation 7.2-3). In Figure 7.2-1, $n = 6$, $m = 18$, the

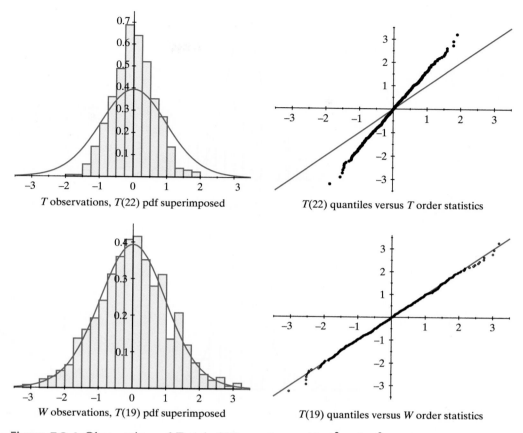

Figure 7.2-1 Observations of T and of W, $n = 6$, $m = 18$, $\sigma_X^2 = 1$, $\sigma_Y^2 = 36$

X observations were generated from the $N(0, 1)$ distribution, and the Y observations were generated from the $N(0, 36)$ distribution. For the value of r for Welch's approximate t distribution, we used the distribution variances rather than the sample variances so that we could use the same r for each of the 500 values of W.

For the simulation results shown in Figure 7.2-2, $n = 18$, $m = 6$, the X observations were generated from the $N(0, 1)$ distribution, and the Y observations were generated from the $N(0, 36)$ distribution. In both cases, Welch's W with a corrected number of r degrees of freedom is much better than the usual T when the variances and sample sizes are unequal, as they are in these examples. ∎

In some applications, two measurements—say, X and Y—are taken on the same subject. In these cases, X and Y may be dependent random variables. Many times these are "before" and "after" measurements, such as weight before and after participating in a diet-and-exercise program. To compare the means of X and Y, it is not permissible to use the t statistics and confidence intervals that we just developed, because in that situation X and Y are independent. Instead, we proceed as follows.

Let $(X_1, Y_1), (X_2, Y_2), \ldots, (X_n, Y_n)$ be n pairs of dependent measurements. Let $D_i = X_i - Y_i$, $i = 1, 2, \ldots, n$. Suppose that $D_1, D_2, \ldots, D_n$ can be thought of as

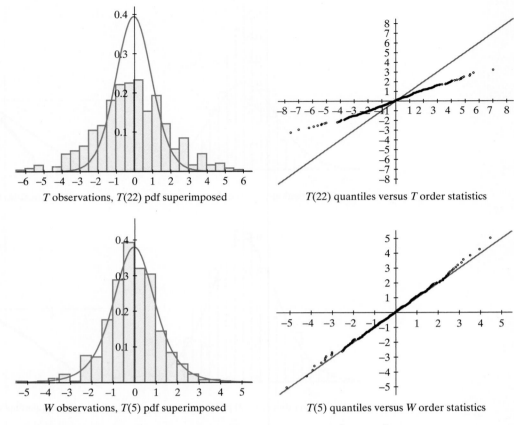

Figure 7.2-2 Observations of T and of W, $n = 18$, $m = 6$, $\sigma_X^2 = 1$, $\sigma_Y^2 = 36$

a random sample from $N(\mu_D, \sigma_D^2)$, where μ_D and σ_D are the mean and standard deviation of each difference. To form a confidence interval for $\mu_X - \mu_Y$, use

$$T = \frac{\overline{D} - \mu_D}{S_D/\sqrt{n}},$$

where $\overline{D}$ and S_D are, respectively, the sample mean and sample standard deviation of the n differences. Thus, T is a t statistic with $n - 1$ degrees of freedom. The endpoints for a $100(1 - \alpha)\%$ confidence interval for $\mu_D = \mu_X - \mu_Y$ are then

$$\overline{d} \pm t_{\alpha/2}(n-1) \frac{s_d}{\sqrt{n}},$$

where $\overline{d}$ and s_d are the observed mean and standard deviation of the sample of the D values. Of course, this is like the confidence interval for a single mean, presented in the last section.

Example 7.2-4

An experiment was conducted to compare people's reaction times to a red light versus a green light. When signaled with either the red or the green light, the subject was asked to hit a switch to turn off the light. When the switch was hit, a clock was turned off and the reaction time in seconds was recorded. The following results give the reaction times for eight subjects:

Subject	Red (x)	Green (y)	$d = x - y$
1	0.30	0.43	−0.13
2	0.23	0.32	−0.09
3	0.41	0.58	−0.17
4	0.53	0.46	0.07
5	0.24	0.27	−0.03
6	0.36	0.41	−0.05
7	0.38	0.38	0.00
8	0.51	0.61	−0.10

For these data, $\bar{d} = -0.0625$ and $s_d = 0.0765$. To form a 95% confidence interval for $\mu_D = \mu_X - \mu_Y$, we find, from Table VI in Appendix B, that $t_{0.025}(7) = 2.365$. Thus, the endpoints for the confidence interval are

$$-0.0625 \pm 2.365 \frac{0.0765}{\sqrt{8}}, \qquad \text{or} \qquad [-0.1265, 0.0015].$$

In this very limited data set, zero is included in the confidence interval but is close to the endpoint 0.0015. We suspect that if more data were taken, zero might not be included in the confidence interval. If that actually were to happen, it would seem that people react faster to a red light. ∎

Of course, we can find one-sided confidence intervals for the difference of the means, $\mu_X - \mu_Y$. Suppose we believe that we have changed some characteristic of the X distribution and created a Y distribution such that we think that $\mu_X > \mu_Y$. Let us find a one-sided 95% confidence interval that is a lower bound for $\mu_X - \mu_Y$. Say this lower bound is greater than zero. Then we would feel 95% confident that the mean μ_X is larger than the mean μ_Y. That is, the change that was made seemed to decrease the mean; this would be good in some cases, such as golf or racing. In other cases, in which we hope the change would be such that $\mu_X < \mu_Y$, we would find a one-sided confidence interval which is an upper bound for $\mu_X - \mu_Y$, and we would hope that it would be less than zero. These ideas are illustrated in Exercises 7.2-5, 7.2-10, and 7.2-11.

Exercises

7.2-1. The length of life of brand X light bulbs is assumed to be $N(\mu_X, 784)$. The length of life of brand Y light bulbs is assumed to be $N(\mu_Y, 627)$ and independent of X. If a random sample of $n = 56$ brand X light bulbs yielded a mean of $\bar{x} = 937.4$ hours and a random sample of size $m = 57$ brand Y light bulbs yielded a mean of $\bar{y} = 988.9$ hours, find a 90% confidence interval for $\mu_X - \mu_Y$.

7.2-2. Let $X_1, X_2, \ldots, X_5$ be a random sample of SAT mathematics scores, assumed to be $N(\mu_X, \sigma^2)$, and let $Y_1, Y_2, \ldots, Y_8$ be an independent random sample of SAT verbal scores, assumed to be $N(\mu_Y, \sigma^2)$. If the following data are observed, find a 90% confidence interval for $\mu_X - \mu_Y$:

$$x_1 = 644 \quad x_2 = 493 \quad x_3 = 532 \quad x_4 = 462 \quad x_5 = 565$$

$$y_1 = 623 \quad y_2 = 472 \quad y_3 = 492 \quad y_4 = 661 \quad y_5 = 540$$

$$y_6 = 502 \quad y_7 = 549 \quad y_8 = 518$$

7.2-3. Independent random samples of the heights of adult males living in two countries yielded the following results: $n = 12$, $\overline{x} = 65.7$ inches, $s_x = 4$ inches and $m = 15$, $\overline{y} = 68.2$ inches, $s_y = 3$ inches. Find an approximate 98% confidence interval for the difference $\mu_X - \mu_Y$ of the means of the populations of heights. Assume that $\sigma_X^2 = \sigma_Y^2$.

7.2-4. [*Medicine and Science in Sports and Exercise* (January 1990).] Let X and Y equal, respectively, the blood volumes in milliliters for a male who is a paraplegic and participates in vigorous physical activities and for a male who is able-bodied and participates in everyday, ordinary activities. Assume that X is $N(\mu_X, \sigma_X^2)$ and Y is $N(\mu_Y, \sigma_Y^2)$. Following are $n = 7$ observations of X:

$$1612 \quad 1352 \quad 1456 \quad 1222 \quad 1560 \quad 1456 \quad 1924$$

Following are $m = 10$ observations of Y:

1082	1300	1092	1040	910
1248	1092	1040	1092	1288

Use the observations of X and Y to

(a) Give a point estimate for $\mu_X - \mu_Y$.

(b) Find a 95% confidence interval for $\mu_X - \mu_Y$. Since the variances σ_X^2 and σ_Y^2 might not be equal, use Welch's T.

7.2-5. A biologist who studies spiders was interested in comparing the lengths of female and male green lynx spiders. Assume that the length X of the male spider is approximately $N(\mu_X, \sigma_X^2)$ and the length Y of the female spider is approximately $N(\mu_Y, \sigma_Y^2)$. Following are $n = 30$ observations of X:

5.20	4.70	5.75	7.50	6.45	6.55
4.70	4.80	5.95	5.20	6.35	6.95
5.70	6.20	5.40	6.20	5.85	6.80
5.65	5.50	5.65	5.85	5.75	6.35
5.75	5.95	5.90	7.00	6.10	5.80

Following are $m = 30$ observations of Y:

8.25	9.95	5.90	7.05	8.45	7.55
9.80	10.80	6.60	7.55	8.10	9.10
6.10	9.30	8.75	7.00	7.80	8.00
9.00	6.30	8.35	8.70	8.00	7.50
9.50	8.30	7.05	8.30	7.95	9.60

The units of measurement for both sets of observations are millimeters. Find an approximate one-sided 95% confidence interval that is an upper bound for $\mu_X - \mu_Y$.

7.2-6. A test was conducted to determine whether a wedge on the end of a plug fitting designed to hold a seal onto the plug was doing its job. The data taken were in the form of measurements of the force required to remove a seal from the plug with the wedge in place (say, X) and the force required without the plug (say, Y). Assume that the distributions of X and Y are $N(\mu_X, \sigma^2)$ and $N(\mu_Y, \sigma^2)$, respectively. Ten independent observations of X are

$$3.26 \quad 2.26 \quad 2.62 \quad 2.62 \quad 2.36 \quad 3.00 \quad 2.62 \quad 2.40 \quad 2.30 \quad 2.40$$

Ten independent observations of Y are

$$1.80 \quad 1.46 \quad 1.54 \quad 1.42 \quad 1.32 \quad 1.56 \quad 1.36 \quad 1.64 \quad 2.00 \quad 1.54$$

(a) Find a 95% confidence interval for $\mu_X - \mu_Y$.

(b) Construct box-and-whisker diagrams of these data on the same figure.

(c) Is the wedge necessary?

7.2-7. An automotive supplier is considering changing its electrical wire harness to save money. The idea is to replace a current 20-gauge wire with a 22-gauge wire. Since not all wires in the harness can be changed, the new wire must work with the current wire splice process. To determine whether the new wire is compatible, random samples were selected and measured with a pull test. A pull test measures the force required to pull the spliced wires apart. The minimum pull force required by the customer is 20 pounds. Twenty observations of the forces needed for the current wire are

$$28.8 \quad 24.4 \quad 30.1 \quad 25.6 \quad 26.4 \quad 23.9 \quad 22.1 \quad 22.5 \quad 27.6 \quad 28.1$$
$$20.8 \quad 27.7 \quad 24.4 \quad 25.1 \quad 24.6 \quad 26.3 \quad 28.2 \quad 22.2 \quad 26.3 \quad 24.4$$

Twenty observations of the forces needed for the new wire are

$$14.1 \quad 12.2 \quad 14.0 \quad 14.6 \quad 8.5 \quad 12.6 \quad 13.7 \quad 14.8 \quad 14.1 \quad 13.2$$
$$12.1 \quad 11.4 \quad 10.1 \quad 14.2 \quad 13.6 \quad 13.1 \quad 11.9 \quad 14.8 \quad 11.1 \quad 13.5$$

(a) Does the current wire meet the customer's specifications?

(b) Find a 90% confidence interval for the difference of the means for these two sets of wire.

(c) Construct box-and-whisker diagrams of the two sets of data on the same figure.

(d) What is your recommendation for this company?

7.2-8. Let $\overline{X}$, $\overline{Y}$, S_X^2, and S_Y^2 be the respective sample means and unbiased estimates of the variances obtained from independent samples of sizes n and m from the normal distributions $N(\mu_X, \sigma_X^2)$ and $N(\mu_Y, \sigma_Y^2)$, where μ_X, μ_Y, σ_X^2, and σ_Y^2 are unknown. If $\sigma_X^2/\sigma_Y^2 = d$, a known constant,

(a) Argue that $\dfrac{(\overline{X} - \overline{Y}) - (\mu_X - \mu_Y)}{\sqrt{d\sigma_Y^2/n + \sigma_Y^2/m}}$ is $N(0, 1)$.

(b) Argue that $\dfrac{(n-1)S_X^2}{d\sigma_Y^2} + \dfrac{(m-1)S_Y^2}{\sigma_Y^2}$ is $\chi^2(n+m-2)$.

(c) Argue that the two random variables in (a) and (b) are independent.

(d) With these results, construct a random variable (not depending upon σ_Y^2) that has a t distribution and that can be used to construct a confidence interval for $\mu_X - \mu_Y$.

7.2-9. Students in a semester-long health-fitness program have their percentage of body fat measured at the beginning of the semester and at the end of the semester. The following measurements give these percentages for 10 men and for 10 women:

Males		Females	
Pre-program %	Post-program %	Pre-program %	Post-program %
11.10	9.97	22.90	22.89
19.50	15.80	31.60	33.47
14.00	13.02	27.70	25.75
8.30	9.28	21.70	19.80
12.40	11.51	19.36	18.00
7.89	7.40	25.03	22.33
12.10	10.70	26.90	25.26
8.30	10.40	25.75	24.90
12.31	11.40	23.63	21.80
10.00	11.95	25.06	24.28

(a) Find a 90% confidence interval for the mean of the difference in the percentages for the males.

(b) Find a 90% confidence interval for the mean of the difference in the percentages for the females.

(c) On the basis of these data, have these percentages decreased?

(d) If possible, check whether each set of differences comes from a normal distribution.

7.2-10. Twenty-four 9th- and 10th-grade high school girls were put on an ultraheavy rope-jumping program. The following data give the time difference for each girl ("before program time" minus "after program time") for the 40-yard dash:

0.28	0.01	0.13	0.33	−0.03	0.07	−0.18	−0.14
−0.33	0.01	0.22	0.29	−0.08	0.23	0.08	0.04
−0.30	−0.08	0.09	0.70	0.33	−0.34	0.50	0.06

(a) Give a point estimate of μ_D, the mean of the difference in race times.

(b) Find a one-sided 95% confidence interval that is a lower bound for μ_D.

(c) Does it look like the rope-jumping program was effective?

7.2-11. The Biomechanics Lab at Hope College tested healthy old women and healthy young women to discover whether or not lower extremity response time to a stimulus is a function of age. Let X and Y respectively equal the independent response times for these two groups when taking steps in the anterior direction. Find a one-sided 95% confidence interval that is a lower bound for $\mu_X - \mu_Y$ if $n = 60$ observations of X yielded $\overline{x} = 671$ and $s_x = 129$, while $m = 60$ observations of Y yielded $\overline{y} = 480$ and $s_y = 93$.

7.2-12. Let X and Y equal the hardness of the hot and cold water, respectively, in a campus building. Hardness is measured in terms of the calcium ion concentration (in ppm). The following data were collected ($n = 12$ observations of X and $m = 10$ observations of Y):

x:	133.5	137.2	136.3	133.3	137.5	135.4
	138.4	137.1	136.5	139.4	137.9	136.8
y:	134.0	134.7	136.0	132.7	134.6	135.2
	135.9	135.6	135.8	134.2		

(a) Calculate the sample means and the sample variances of these data.

(b) Construct a 95% confidence interval for $\mu_X - \mu_Y$, assuming that the distributions of X and Y are $N(\mu_X, \sigma_X^2)$ and $N(\mu_Y, \sigma_Y^2)$, respectively.

(c) Construct box plots of the two sets of data on the same graph.

(d) Do the means seem to be equal or different?

7.2-13. Ledolter and Hogg (see References) report that two rubber compounds were tested for tensile strength. Rectangular materials were prepared and pulled in a longitudinal direction. A sample of 14 specimens, 7 from compound A and 7 from compound B, was prepared, but it was later found that two B specimens were defective and they had to be removed from the test. The tensile strength (in units of 100 pounds per square inch) of the remaining specimens are as follows:

A:	32	30	33	32	29	34	32
B:	33	35	36	37	35		

Calculate a 95% confidence interval for the difference of the mean tensile strengths of the two rubber compounds. State your assumptions.

7.2-14. Let S_X^2 and S_Y^2 be the respective variances of two independent random samples of sizes n and m from $N(\mu_X, \sigma_X^2)$ and $N(\mu_Y, \sigma_Y^2)$. Use the fact that $F = [S_Y^2/\sigma_Y^2]/[S_X^2/\sigma_X^2]$ has an F distribution, with parameters $r_1 = m-1$ and $r_2 = n-1$, to rewrite $P(c \le F \le d) = 1-\alpha$, where $c = F_{1-\alpha/2}(r_1, r_2)$ and $d = F_{\alpha/2}(r_1, r_2)$, so that

$$P\left(c\frac{S_X^2}{S_Y^2} \le \frac{\sigma_X^2}{\sigma_Y^2} \le d\frac{S_X^2}{S_Y^2}\right) = 1 - \alpha.$$

If the observed values are $n = 13$, $m = 9$, $12s_x^2 = 128.41$, and $8s_y^2 = 36.72$, show that a 98% **confidence interval for the ratio of the two variances**, σ_X^2/σ_Y^2, is $[0.41, 10.49]$, so that $[0.64, 3.24]$ is a 98% confidence interval for σ_X/σ_Y.

7.3 CONFIDENCE INTERVALS FOR PROPORTIONS

We have suggested that the histogram is a good description of how the observations of a random sample are distributed. We might naturally inquire about the accuracy of those relative frequencies (or percentages) associated with the various classes. To illustrate, in Example 6.1-1 concerning the weights of $n = 40$ candy bars, we found that the relative frequency of the class interval (22.25, 23.15) was 8/40 = 0.20, or 20%. If we think of this collection of 40 weights as a random sample observed from a larger population of candy bar weights, how close is 20% to the true percentage (or 0.20 to the true proportion) of weights in that class interval for the entire population of weights for this type of candy bar?

In considering this problem, we generalize it somewhat by treating the class interval (22.25, 23.15) as "success." That is, there is some true probability of success, p—namely, the proportion of the population in that interval. Let Y equal the frequency of measurements in the interval out of the n observations, so that (under the assumptions of independence and constant probability p) Y has the binomial distribution $b(n, p)$. Thus, the problem is to determine the accuracy of the relative frequency Y/n as an estimator of p. We solve this problem by finding, for the unknown p, a confidence interval based on Y/n.

In general, when observing n Bernoulli trials with probability p of success on each trial, we shall find a confidence interval for p based on Y/n, where Y is the number of successes and Y/n is an unbiased point estimator for p.

In Section 5.7, we noted that

$$\frac{Y - np}{\sqrt{np(1-p)}} = \frac{(Y/n) - p}{\sqrt{p(1-p)/n}}$$

has an approximate normal distribution $N(0, 1)$, provided that n is large enough. This means that, for a given probability $1-\alpha$, we can find a $z_{\alpha/2}$ in Table V in Appendix B such that

$$P\left[-z_{\alpha/2} \leq \frac{(Y/n)-p}{\sqrt{p(1-p)/n}} \leq z_{\alpha/2}\right] \approx 1-\alpha. \tag{7.3-1}$$

If we proceed as we did when we found a confidence interval for μ in Section 7.1, we would obtain

$$P\left[\frac{Y}{n} - z_{\alpha/2}\sqrt{\frac{p(1-p)}{n}} \leq p \leq \frac{Y}{n} + z_{\alpha/2}\sqrt{\frac{p(1-p)}{n}}\right] \approx 1-\alpha.$$

Unfortunately, the unknown parameter p appears in the endpoints of this inequality. There are two ways out of this dilemma. First, we could make an additional approximation, namely, replacing p with Y/n in $p(1-p)/n$ in the endpoints. That is, if n is large enough, it is still true that

$$P\left[\frac{Y}{n} - z_{\alpha/2}\sqrt{\frac{(Y/n)(1-Y/n)}{n}} \leq p \leq \frac{Y}{n} + z_{\alpha/2}\sqrt{\frac{(Y/n)(1-Y/n)}{n}}\right] \approx 1-\alpha.$$

Thus, for large n, if the observed Y equals y, then the interval

$$\left[\frac{y}{n} - z_{\alpha/2}\sqrt{\frac{(y/n)(1-y/n)}{n}}, \frac{y}{n} + z_{\alpha/2}\sqrt{\frac{(y/n)(1-y/n)}{n}}\right]$$

serves as an approximate $100(1-\alpha)\%$ confidence interval for p. Frequently, this interval is written as

$$\frac{y}{n} \pm z_{\alpha/2}\sqrt{\frac{(y/n)(1-y/n)}{n}} \tag{7.3-2}$$

for brevity. This formulation clearly notes, as does $\bar{x} \pm z_{\alpha/2}(\sigma/\sqrt{n})$ in Section 7.1, the reliability of the estimate y/n, namely, that we are $100(1-\alpha)\%$ confident that p is within $z_{\alpha/2}\sqrt{(y/n)(1-y/n)/n}$ of $\widehat{p} = y/n$.

A second way to solve for p in the inequality in Equation 7.3-1 is to note that

$$\frac{|Y/n - p|}{\sqrt{p(1-p)/n}} \leq z_{\alpha/2}$$

is equivalent to

$$H(p) = \left(\frac{Y}{n} - p\right)^2 - \frac{z_{\alpha/2}^2 p(1-p)}{n} \leq 0. \tag{7.3-3}$$

But $H(p)$ is a quadratic expression in p. Thus, we can find those values of p for which $H(p) \leq 0$ by finding the two zeros of $H(p)$. Letting $\widehat{p} = Y/n$ and $z_0 = z_{\alpha/2}$ in Equation 7.3-3, we have

$$H(p) = \left(1 + \frac{z_0^2}{n}\right)p^2 - \left(2\widehat{p} + \frac{z_0^2}{n}\right)p + \widehat{p}^2.$$

By the quadratic formula, the zeros of $H(p)$ are, after simplifications,

$$\frac{\widehat{p} + z_0^2/(2n) \pm z_0\sqrt{\widehat{p}(1-\widehat{p})/n + z_0^2/(4n^2)}}{1 + z_0^2/n}, \tag{7.3-4}$$

and these zeros give the endpoints for an approximate $100(1 - \alpha)\%$ confidence interval for p. If n is large, $z_0^2/(2n)$, $z_0^2/(4n^2)$, and z_0^2/n are small. Thus, the confidence intervals given by Equations 7.3-2 and 7.3-4 are approximately equal when n is large.

Example 7.3-1

Let us return to the example of the histogram of the candy bar weights, Example 6.1-1, with $n = 40$ and $y/n = 8/40 = 0.20$. If $1 - \alpha = 0.90$, so that $z_{\alpha/2} = 1.645$, then, using Equation 7.3-2, we find that the endpoints

$$0.20 \pm 1.645\sqrt{\frac{(0.20)(0.80)}{40}}$$

serve as an approximate 90% confidence interval for the true fraction p. That is, $[0.096, 0.304]$, which is the same as $[9.6\%, 30.4\%]$, is an approximate 90% confidence interval for the percentage of weights of the entire population in the interval $(22.25, 23.15)$. If we had used the endpoints given by Equation 7.3-4, the confidence interval would be $[0.117, 0.321]$. Because of the small sample size, there is a non-negligible difference in these intervals. If the sample size had been $n = 400$ and $y = 80$, so that $y/n = 80/400 = 0.20$, the two 90% confidence intervals would have been $[0.167, 0.233]$ and $[0.169, 0.235]$, respectively, which differ very little. ∎

Example 7.3-2

In a certain political campaign, one candidate has a poll taken at random among the voting population. The results are that $y = 185$ out of $n = 351$ voters favor this candidate. Even though $y/n = 185/351 = 0.527$, should the candidate feel very confident of winning? From Equation 7.3-2, an approximate 95% confidence interval for the fraction p of the voting population who favor the candidate is

$$0.527 \pm 1.96\sqrt{\frac{(0.527)(0.473)}{351}}$$

or, equivalently, $[0.475, 0.579]$. Thus, there is a good possibility that p is less than 50%, and the candidate should certainly take this possibility into account in campaigning. ∎

One-sided confidence intervals are sometimes appropriate for p. For example, we may be interested in an upper bound on the proportion of defectives in manufacturing some item. Or we may be interested in a lower bound on the proportion of voters who favor a particular candidate. The one-sided confidence interval for p given by

$$\left[0, \frac{y}{n} + z_\alpha\sqrt{\frac{(y/n)[1 - (y/n)]}{n}}\right]$$

provides an upper bound for p, while

$$\left[\frac{y}{n} - z_\alpha\sqrt{\frac{(y/n)[1 - (y/n)]}{n}}, 1\right]$$

provides a lower bound for p.

REMARK Sometimes the confidence intervals suggested here are not very close to having the stated confidence coefficient. This is particularly true if n is small or if one of Y or $n - Y$ is close to zero. It is obvious that something is wrong if $Y = 0$ or $n - Y = 0$, because the radical is then equal to zero.

It has been suggested (see, e.g., Agresti and Coull, 1998) that we use $\widetilde{p} = (Y + 2)/(n + 4)$ as an estimator for p in those cases because the results are usually much better. It is true that $\widetilde{p}$ is a biased estimator of p, but it is a Bayes shrinkage estimator if we use the beta prior pdf with parameters $\alpha = 2$, $\beta = 2$. In those cases in which n is small or Y or $n - Y$ is close to zero,

$$\widetilde{p} \pm z_{\alpha/2}\sqrt{\widetilde{p}(1 - \widetilde{p})/(n + 4)} \tag{7.3-5}$$

provides a much better $100(1 - \alpha)\%$ confidence interval for p. A similar statement can be made about one-sided confidence intervals.

Look again at Equation 7.3-4. If we form a 95% confidence interval using this equation, we find that $z_0 = 1.96 \approx 2$. Thus, a 95% confidence interval is centered approximately at

$$\frac{\widehat{p} + z_0^2/(2n)}{1 + z_0^2/n} = \frac{y + z_0^2/2}{n + z_0^2} \approx \frac{y + 2}{n + 4}.$$

This result is consistent with Equation 7.3-5 for 95% confidence intervals. ∎

Example 7.3-3 Returning to the data in Example 7.3-1, and using Equation 7.3-5, we have $\widetilde{p} = (8 + 2)/(40 + 4) = 0.227$. Thus, a 90% confidence interval is

$$0.227 \pm 1.645\sqrt{\frac{(0.227)(0.773)}{44}},$$

or $[0.123, 0.331]$. If it had been true that $y = 80$ and $n = 400$, the confidence interval given by Equation 7.3-5 would have been $[0.170, 0.236]$. ∎

Frequently, there are two (or more) possible independent ways of performing an experiment; suppose these have probabilities of success p_1 and p_2, respectively. Let n_1 and n_2 be the number of independent trials associated with these two methods, and let us say that they result in Y_1 and Y_2 successes, respectively. In order to make a statistical inference about the difference $p_1 - p_2$, we proceed as follows.

Since the independent random variables Y_1/n_1 and Y_2/n_2 have respective means p_1 and p_2 and variances $p_1(1 - p_1)/n_1$ and $p_2(1 - p_2)/n_2$, we know from Section 5.4 that the difference $Y_1/n_1 - Y_2/n_2$ must have mean $p_1 - p_2$ and variance

$$\frac{p_1(1 - p_1)}{n_1} + \frac{p_2(1 - p_2)}{n_2}.$$

(Recall that the variances are added to get the variance of a difference of two independent random variables.) Moreover, the fact that Y_1/n_1 and Y_2/n_2 have approximate normal distributions would suggest that the difference

$$\frac{Y_1}{n_1} - \frac{Y_2}{n_2}$$

would have an approximate normal distribution with the above mean and variance. (See Theorem 5.5-1.) That is,

$$\frac{(Y_1/n_1) - (Y_2/n_2) - (p_1 - p_2)}{\sqrt{p_1(1-p_1)/n_1 + p_2(1-p_2)/n_2}}$$

has an approximate normal distribution $N(0,1)$. If we now replace p_1 and p_2 in the denominator of this ratio by Y_1/n_1 and Y_2/n_2, respectively, it is still true for large enough n_1 and n_2 that the new ratio will be approximately $N(0,1)$. Thus, for a given $1 - \alpha$, we can find $z_{\alpha/2}$ from Table V in Appendix B, so that

$$P\left[-z_{\alpha/2} \leq \frac{(Y_1/n_1) - (Y_2/n_2) - (p_1 - p_2)}{\sqrt{(Y_1/n_1)(1 - Y_1/n_1)/n_1 + (Y_2/n_2)(1 - Y_2/n_2)/n_2}} \leq z_{\alpha/2}\right] \approx 1 - \alpha.$$

Once Y_1 and Y_2 are observed to be y_1 and y_2, respectively, this approximation can be solved to obtain an approximate $100(1 - \alpha)\%$ confidence interval

$$\frac{y_1}{n_1} - \frac{y_2}{n_2} \pm z_{\alpha/2}\sqrt{\frac{(y_1/n_1)(1 - y_1/n_1)}{n_1} + \frac{(y_2/n_2)(1 - y_2/n_2)}{n_2}}$$

for the unknown difference $p_1 - p_2$. Note again how this form indicates the reliability of the estimate $y_1/n_1 - y_2/n_2$ of the difference $p_1 - p_2$.

Example 7.3-4

Two detergents were tested for their ability to remove stains of a certain type. An inspector judged the first one to be successful on 63 out of 91 independent trials and the second one to be successful on 42 out of 79 independent trials. The respective relative frequencies of success are $63/91 = 0.692$ and $42/79 = 0.532$. An approximate 90% confidence interval for the difference $p_1 - p_2$ of the two detergents is

$$\left(\frac{63}{91} - \frac{42}{79}\right) \pm 1.645\sqrt{\frac{(63/91)(28/91)}{91} + \frac{(42/79)(37/79)}{79}}$$

or, equivalently, $[0.039, 0.283]$. Accordingly, since this interval does not include zero, it seems that the first detergent is probably better than the second one for removing the type of stains in question. ∎

Exercises

7.3-1. A machine shop manufactures toggle levers. A lever is flawed if a standard nut cannot be screwed onto the threads. Let p equal the proportion of flawed toggle levers that the shop manufactures. If there were 24 flawed levers out of a sample of 642 that were selected randomly from the production line,

(a) Give a point estimate of p.

(b) Use Equation 7.3-2 to find an approximate 95% confidence interval for p.

(c) Use Equation 7.3-4 to find an approximate 95% confidence interval for p.

(d) Use Equation 7.3-5 to find an approximate 95% confidence interval for p.

(e) Find a one-sided 95% confidence interval for p that provides an upper bound for p.

7.3-2. Let p equal the proportion of letters mailed in the Netherlands that are delivered the next day. Suppose that

$y = 142$ out of a random sample of $n = 200$ letters were delivered the day after they were mailed.

(a) Give a point estimate of p.

(b) Use Equation 7.3-2 to find an approximate 90% confidence interval for p.

(c) Use Equation 7.3-4 to find an approximate 90% confidence interval for p.

(d) Use Equation 7.3-5 to find an approximate 90% confidence interval for p.

(e) Find a one-sided 90% confidence interval for p that provides a lower bound for p.

7.3-3. Let p equal the proportion of triathletes who suffered a training-related overuse injury during the past year. Out of 330 triathletes who responded to a survey, 167 indicated that they had suffered such an injury during the past year.

(a) Use these data to give a point estimate of p.

(b) Use these data to find an approximate 90% confidence interval for p.

(c) Do you think that the 330 triathletes who responded to the survey may be considered a random sample from the population of triathletes?

7.3-4. Let p equal the proportion of Americans who favor the death penalty. If a random sample of $n = 1234$ Americans yielded $y = 864$ who favored the death penalty, find an approximate 95% confidence interval for p.

7.3-5. In order to estimate the proportion, p, of a large class of college freshmen that had high school GPAs from 3.2 to 3.6, inclusive, a sample of $n = 50$ students was taken. It was found that $y = 9$ students fell into this interval.

(a) Give a point estimate of p.

(b) Use Equation 7.3-2 to find an approximate 95% confidence interval for p.

(c) Use Equation 7.3-4 to find an approximate 95% confidence interval for p.

(d) Use Equation 7.3-5 to find an approximate 95% confidence interval for p.

7.3-6. Let p equal the proportion of Americans who select jogging as one of their recreational activities. If 1497 out of a random sample of 5757 selected jogging, find an approximate 98% confidence interval for p.

7.3-7. In developing countries in Africa and the Americas, let p_1 and p_2 be the respective proportions of women with nutritional anemia. Find an approxi-

mate 90% confidence interval for $p_1 - p_2$, given that a random sample of $n_1 = 2100$ African women yielded $y_1 = 840$ with nutritional anemia and a random sample of $n_2 = 1900$ women from the Americas yielded $y_2 = 323$ women with nutritional anemia.

7.3-8. A proportion, p, that many public opinion polls estimate is the number of Americans who would say yes to the question, "If something were to happen to the president of the United States, do you think that the vice president would be qualified to take over as president?" In one such random sample of 1022 adults, 388 said yes.

(a) On the basis of the given data, find a point estimate of p.

(b) Find an approximate 90% confidence interval for p.

(c) Give updated answers to this question if new poll results are available.

7.3-9. Consider the following two groups of women: Group 1 consists of women who spend less than $500 annually on clothes; Group 2 comprises women who spend over $1000 annually on clothes. Let p_1 and p_2 equal the proportions of women in these two groups, respectively, who believe that clothes are too expensive. If 1009 out of a random sample of 1230 women from group 1 and 207 out of a random sample 340 from group 2 believe that clothes are too expensive,

(a) Give a point estimate of $p_1 - p_2$.

(b) Find an approximate 95% confidence interval for $p_1 - p_2$.

7.3-10. A candy manufacturer selects mints at random from the production line and weighs them. For one week, the day shift weighed $n_1 = 194$ mints and the night shift weighed $n_2 = 162$ mints. The numbers of these mints that weighed at most 21 grams was $y_1 = 28$ for the day shift and $y_2 = 11$ for the night shift. Let p_1 and p_2 denote the proportions of mints that weigh at most 21 grams for the day and night shifts, respectively.

(a) Give a point estimate of p_1.

(b) Give the endpoints for a 95% confidence interval for p_1.

(c) Give a point estimate of $p_1 - p_2$.

(d) Find a one-sided 95% confidence interval that gives a lower bound for $p_1 - p_2$.

7.3-11. For developing countries in Asia (excluding China) and Africa, let p_1 and p_2 be the respective proportions of preschool children with chronic malnutrition (stunting). If respective random samples of $n_1 = 1300$ and $n_2 = 1100$ yielded $y_1 = 520$ and $y_2 = 385$ children with chronic malnutrition, find an approximate 95% confidence interval for $p_1 - p_2$.

7.3-12. An environmental survey contained a question asking what respondents thought was the major cause of air pollution in this country, giving the choices "automobiles," "factories," and "incinerators." Two versions of the test, A and B, were used. Let p_A and p_B be the respective proportions of people using forms A and B who select "factories." If 170 out of 460 people who used version A chose "factories" and 141 out of 440 people who used version B chose "factories,"

(a) Find a 95% confidence interval for $p_A - p_B$.

(b) Do the versions seem to be consistent concerning this answer? Why or why not?

7.4 SAMPLE SIZE

In statistical consulting, the first question frequently asked is, "How large should the sample size be to estimate a mean?" In order to convince the inquirer that the answer will depend on the variation associated with the random variable under observation, the statistician could correctly respond, "Only one observation is needed, provided that the standard deviation of the distribution is zero." That is, if σ equals zero, then the value of that one observation would necessarily equal the unknown mean of the distribution. This, of course, is an extreme case and one that is not met in practice; however, it should help convince people that the smaller the variance, the smaller is the sample size needed to achieve a given degree of accuracy. This assertion will become clearer as we consider several examples. Let us begin with a problem that involves a statistical inference about the unknown mean of a distribution.

Example 7.4-1 A mathematics department wishes to evaluate a new method of teaching calculus with a computer. At the end of the course, the evaluation will be made on the basis of scores of the participating students on a standard test. There is particular interest in estimating μ, the mean score for students taking the course. Thus, there is a desire to determine the number of students, n, who are to be selected at random from a larger group of students to take the course. Since new computing equipment must be purchased, the department cannot afford to let all of the school's students take calculus the new way. In addition, some of the staff question the value of this approach and hence do not want to expose every student to this new procedure. So, let us find the sample size n such that we are fairly confident that $\bar{x} \pm 1$ contains the unknown test mean μ. From past experience, it is believed that the standard deviation associated with this type of test is about 15. (The mean is also known when students take the standard calculus course.) Accordingly, using the fact that the sample mean of the test scores, $\overline{X}$, is approximately $N(\mu, \sigma^2/n)$, we see that the interval given by $\bar{x} \pm 1.96(15/\sqrt{n}\,)$ will serve as an approximate 95% confidence interval for μ. That is, we want

$$1.96\left(\frac{15}{\sqrt{n}}\right) = 1$$

or, equivalently,

$$\sqrt{n} = 29.4, \qquad \text{and thus} \qquad n \approx 864.36,$$

or $n = 865$ because n must be an integer. ■

It is quite likely that, in the preceding example, it had not been anticipated that as many as 865 students would be needed in this study. If that is the case, the statistician must discuss with those involved in the experiment whether or not the accuracy and the confidence level could be relaxed some. For example, rather than requiring $\bar{x} \pm 1$ to be a 95% confidence interval for μ, possibly $\bar{x} \pm 2$ would be a satisfactory 80% one. If this modification is acceptable, we now have

$$1.282\left(\frac{15}{\sqrt{n}}\right) = 2$$

or, equivalently,

$$\sqrt{n} = 9.615, \qquad \text{so that} \qquad n \approx 92.4.$$

Since n must be an integer, we would probably use 93 in practice. Most likely, the persons involved in the project would find that a more reasonable sample size. Of course, any sample size greater than 93 could be used. Then either the length of the confidence interval could be decreased from $\bar{x} \pm 2$ or the confidence coefficient could be increased from 80%, or a combination of both approaches could be taken. Also, since there might be some question as to whether the standard deviation σ actually equals 15, the sample standard deviation s would no doubt be used in the construction of the interval. For instance, suppose that the sample characteristics observed are

$$n = 145, \qquad \bar{x} = 77.2, \qquad s = 13.2;$$

then

$$\bar{x} \pm \frac{1.282s}{\sqrt{n}}, \qquad \text{or} \qquad 77.2 \pm 1.41,$$

provides an approximate 80% confidence interval for μ.

In general, if we want the $100(1-\alpha)\%$ confidence interval for μ, $\bar{x} \pm z_{\alpha/2}(\sigma/\sqrt{n})$, to be no longer than that given by $\bar{x} \pm \varepsilon$, then the sample size n is the solution of

$$\varepsilon = \frac{z_{\alpha/2}\sigma}{\sqrt{n}}, \qquad \text{where} \quad \Phi(z_{\alpha/2}) = 1 - \frac{\alpha}{2}.$$

That is,

$$n = \frac{z_{\alpha/2}^2 \sigma^2}{\varepsilon^2}, \tag{7.4-1}$$

where it is assumed that σ^2 is known. We sometimes call $\varepsilon = z_{\alpha/2}(\sigma/\sqrt{n})$ the **maximum error of the estimate**. If the experimenter has no idea about the value of σ^2, it may be necessary to first take a preliminary sample to estimate σ^2.

The type of statistic we see most often in newspapers and magazines is an estimate of a proportion p. We might, for example, want to know the percentage of the labor force that is unemployed or the percentage of voters favoring a certain candidate. Sometimes extremely important decisions are made on the basis of these estimates. If this is the case, we would most certainly desire short confidence intervals for p with large confidence coefficients. We recognize that these conditions will require a large sample size. If, to the contrary, the fraction p being estimated is not too important, an estimate associated with a longer confidence interval with a smaller confidence coefficient is satisfactory, and in that case a smaller sample size can be used.

Example 7.4-2 Suppose we know that the unemployment rate has been about 8% (0.08). However, we wish to update our estimate in order to make an important decision about the national economic policy. Accordingly, let us say we wish to be 99% confident that the new estimate of p is within 0.001 of the true p. If we assume Bernoulli trials (an assumption that might be questioned), the relative frequency y/n, based upon a large sample size n, provides the approximate 99% confidence interval:

$$\frac{y}{n} \pm 2.576\sqrt{\frac{(y/n)(1-y/n)}{n}}.$$

Although we do not know y/n exactly before sampling, since y/n will be near 0.08, we do know that

$$2.576\sqrt{\frac{(y/n)(1-y/n)}{n}} \approx 2.576\sqrt{\frac{(0.08)(0.92)}{n}},$$

and we want this number to equal 0.001. That is,

$$2.576\sqrt{\frac{(0.08)(0.92)}{n}} = 0.001$$

or, equivalently,

$$\sqrt{n} = 2576\sqrt{0.0736}, \qquad \text{and then} \qquad n \approx 488{,}394.$$

That is, under our assumptions, such a sample size is needed in order to achieve the reliability and the accuracy desired. Because n is so large, we would probably be willing to increase the error, say, to 0.01, and perhaps reduce the confidence level to 98%. In that case,

$$\sqrt{n} = (2.326/0.01)\sqrt{0.0736} \qquad \text{and} \qquad n \approx 3{,}982,$$

which is a more reasonable sample size. ∎

From the preceding example, we hope that the student will recognize how important it is to know the sample size (or the length of the confidence interval and the confidence coefficient) before he or she can place much weight on a statement such as "Fifty-one percent of the voters seem to favor candidate A, 46% favor candidate B, and 3% are undecided." Is this statement based on a sample of 100 or 2000 or 10,000 voters? If we assume Bernoulli trials, the approximate 95% confidence intervals for the fraction of voters favoring candidate A in these cases are, respectively, $[0.41, 0.61]$, $[0.49, 0.53]$, and $[0.50, 0.52]$. Quite obviously, the first interval, with $n = 100$, does not assure candidate A of the support of at least half the voters, whereas the interval with $n = 10{,}000$ is more convincing.

In general, to find the required sample size to estimate p, recall that the point estimate of p is $\widehat{p} = y/n$ and an approximate $1 - \alpha$ confidence interval for p is

$$\widehat{p} \pm z_{\alpha/2}\sqrt{\frac{\widehat{p}(1-\widehat{p})}{n}}.$$

Suppose we want an estimate of p that is within ε of the unknown p with $100(1-\alpha)\%$ confidence, where $\varepsilon = z_{\alpha/2}\sqrt{\widehat{p}(1-\widehat{p})/n}$ is the **maximum error of the point estimate** $\widehat{p} = y/n$. Since $\widehat{p}$ is unknown before the experiment is run, we cannot use the value of $\widehat{p}$ in our determination of n. However, if it is known that p is about equal to p^*, the necessary sample size n is the solution of

$$\varepsilon = \frac{z_{\alpha/2}\sqrt{p^*(1-p^*)}}{\sqrt{n}}.$$

That is,

$$n = \frac{z_{\alpha/2}^2 p^*(1 - p^*)}{\varepsilon^2}. \tag{7.4-2}$$

Often, however, we do not have a strong prior idea about p, as we did in Example 7.4-2 about the rate of unemployment. It is interesting to observe that no matter what value p takes between 0 and 1, it is always true that $p^*(1 - p^*) \leq 1/4$. Hence,

$$n = \frac{z_{\alpha/2}^2 p^*(1 - p^*)}{\varepsilon^2} \leq \frac{z_{\alpha/2}^2}{4\varepsilon^2}.$$

Thus, if we want the $100(1 - \alpha)\%$ confidence interval for p to be no longer than $y/n \pm \varepsilon$, a solution for n that provides this protection is

$$n = \frac{z_{\alpha/2}^2}{4\varepsilon^2}. \tag{7.4-3}$$

REMARK Up to this point in the text, we have used the "hat" ($\frown$) notation to indicate an estimator, as in $\widehat{p} = Y/n$ and $\widehat{\mu} = \overline{X}$. Note, however, that in the previous discussion we used $\widehat{p} = y/n$, an estimate of p. Occasionally, statisticians find it convenient to use the "hat" notation for an estimate as well as an estimator. It is usually clear from the context which is being used. ■

Example 7.4-3 A possible gubernatorial candidate wants to assess initial support among the voters before making an announcement about her candidacy. If the fraction p of voters who are favorable, without any advance publicity, is around 0.15, the candidate will enter the race. From a poll of n voters selected at random, the candidate would like the estimate y/n to be within 0.03 of p. That is, the decision will be based on a 95% confidence interval of the form $y/n \pm 0.03$. Since the candidate has no idea about the magnitude of p, a consulting statistician formulates the equation

$$n = \frac{(1.96)^2}{4(0.03)^2} = 1067.11.$$

Thus, the sample size should be around 1068 to achieve the desired reliability and accuracy. Suppose that 1068 voters around the state were selected at random and interviewed and $y = 214$ express support for the candidate. Then $\widehat{p} = 214/1068 = 0.20$ is a point estimate of p, and an approximate 95% confidence interval for p is

$$0.20 \pm 1.96\sqrt{(0.20)(0.80)/n}, \qquad \text{or} \qquad 0.20 \pm 0.024.$$

That is, we are 95% confident that p belongs to the interval $[0.176, 0.224]$. On the basis of this sample, the candidate decided to run for office. Note that, for a confidence coefficient of 95%, we found a sample size so that the maximum error of the estimate would be 0.03. From the data that were collected, the maximum error of the estimate is only 0.024. We ended up with a smaller error because we found the sample size assuming that $p = 0.50$, while, in fact, p is closer to 0.20. ■

Suppose that you want to estimate the proportion p of a student body that favors a new policy. How large should the sample be? If p is close to $1/2$ and you want to be 95% confident that the maximum error of the estimate is $\varepsilon = 0.02$, then

$$n = \frac{(1.96)^2}{4(0.02)^2} = 2401.$$

Such a sample size makes sense at a large university. However, if you are a student at a small college, the entire enrollment could be less than 2401. Thus, we now give a procedure that can be used to determine the sample size when the population is not so large relative to the desired sample size.

Let N equal the size of a population, and assume that N_1 individuals in the population have a certain characteristic C (e.g., favor a new policy). Let $p = N_1/N$, the proportion with this characteristic. Then $1 - p = 1 - N_1/N$. If we take a sample of size n without replacement, then X, the number of observations with the characteristic C, has a hypergeometric distribution. The mean and variance of X are, respectively,

$$\mu = n\left(\frac{N_1}{N}\right) = np$$

and

$$\sigma^2 = n\left(\frac{N_1}{N}\right)\left(1 - \frac{N_1}{N}\right)\left(\frac{N-n}{N-1}\right) = np(1-p)\left(\frac{N-n}{N-1}\right).$$

The mean and variance of X/n are, respectively,

$$E\left(\frac{X}{n}\right) = \frac{\mu}{n} = p$$

and

$$\operatorname{Var}\left(\frac{X}{n}\right) = \frac{\sigma^2}{n^2} = \frac{p(1-p)}{n}\left(\frac{N-n}{N-1}\right).$$

To find an approximate confidence interval for p, we can use the normal approximation:

$$P\left[-z_{\alpha/2} \le \frac{\dfrac{X}{n} - p}{\sqrt{\dfrac{p(1-p)}{n}\left(\dfrac{N-n}{N-1}\right)}} \le z_{\alpha/2}\right] \approx 1 - \alpha.$$

Thus,

$$1 - \alpha \approx P\left[\frac{X}{n} - z_{\alpha/2}\sqrt{\frac{p(1-p)}{n}\left(\frac{N-n}{N-1}\right)} \le p \le \frac{X}{n} + z_{\alpha/2}\sqrt{\frac{p(1-p)}{n}\left(\frac{N-n}{N-1}\right)}\right].$$

Replacing p under the radical with $\widehat{p} = x/n$, we find that an approximate $1 - \alpha$ confidence interval for p is

$$\widehat{p} \pm z_{\alpha/2}\sqrt{\frac{\widehat{p}(1-\widehat{p})}{n}\left(\frac{N-n}{N-1}\right)}.$$

This is similar to the confidence interval for p when the distribution of X is $b(n,p)$. If N is large relative to n, then

$$\frac{N-n}{N-1} = \frac{1-n/N}{1-1/N} \approx 1,$$

so in this case the two intervals are essentially equal.

Suppose now that we are interested in determining the sample size n that is required to have $1 - \alpha$ confidence that the maximum error of the estimate of p is ε. We let

$$\varepsilon = z_{\alpha/2}\sqrt{\frac{p(1-p)}{n}\left(\frac{N-n}{N-1}\right)}$$

and solve for n. After some simplification, we obtain

$$n = \frac{Nz_{\alpha/2}^2 p(1-p)}{(N-1)\varepsilon^2 + z_{\alpha/2}^2 p(1-p)}$$

$$= \frac{z_{\alpha/2}^2 p(1-p)/\varepsilon^2}{\dfrac{N-1}{N} + \dfrac{z_{\alpha/2}^2 p(1-p)/\varepsilon^2}{N}}.$$

If we let

$$m = \frac{z_{\alpha/2}^2 p^*(1-p^*)}{\varepsilon^2},$$

which is the n value given by Equation 7.4-2, then we choose

$$n = \frac{m}{1 + \dfrac{m-1}{N}}$$

for our sample size n.

If we know nothing about p, we set $p^* = 1/2$ to determine m. For example, if the size of the student body is $N = 4000$ and $1 - \alpha = 0.95$, $\varepsilon = 0.02$, and we let $p^* = 1/2$, then $m = 2401$ and

$$n = \frac{2401}{1 + 2400/4000} = 1501,$$

rounded up to the nearest integer. Thus, we would sample approximately 37.5% of the student body.

Example 7.4-4

Suppose that a college of $N = 3000$ students is interested in assessing student support for a new form for teacher evaluation. To estimate the proportion p in favor of the new form, how large a sample is required so that the maximum error of the estimate of p is $\varepsilon = 0.03$ with 95% confidence? If we assume that p is completely unknown, we use $p^* = 1/2$ to obtain

$$m = \frac{(1.96)^2}{4(0.03)^2} = 1068,$$

rounded up to the nearest integer. Thus, the desired sample size is

$$n = \frac{1068}{1 + 1067/3000} = 788,$$

rounded up to the nearest integer. ∎

Exercises

7.4-1. Let X equal the tarsus length for a male grackle. Assume that the distribution of X is $N(\mu, 4.84)$. Find the sample size n that is needed so that we are 95% confident that the maximum error of the estimate of μ is 0.4.

7.4-2. Let X equal the excess weight of soap in a "1000-gram" bottle. Assume that the distribution of X is $N(\mu, 169)$. What sample size is required so that we have 95% confidence that the maximum error of the estimate of μ is 1.5?

7.4-3. A company packages powdered soap in "6-pound" boxes. The sample mean and standard deviation of the soap in these boxes are currently 6.09 pounds and 0.02 pound, respectively. If the mean fill can be lowered by 0.01 pound, $14,000 would be saved per year. Adjustments were made in the filling equipment, but it can be assumed that the standard deviation remains unchanged.

(a) How large a sample is needed so that the maximum error of the estimate of the new μ is $\varepsilon = 0.001$ with 90% confidence?

(b) A random sample of size $n = 1219$ yielded $\bar{x} = 6.048$ and $s = 0.022$. Calculate a 90% confidence interval for μ.

(c) Estimate the savings per year with these new adjustments.

(d) Estimate the proportion of boxes that will now weigh less than 6 pounds.

7.4-4. Measurements of the length in centimeters of $n = 29$ fish yielded an average length of $\bar{x} = 16.82$ and $s^2 = 34.9$. Determine the size of a new sample so that $\bar{x} \pm 0.5$ is an approximate 95% confidence interval for μ.

7.4-5. A quality engineer wanted to be 98% confident that the maximum error of the estimate of the mean strength, μ, of the left hinge on a vanity cover molded by a machine is 0.25. A preliminary sample of size $n = 32$ parts yielded a sample mean of $\bar{x} = 35.68$ and a standard deviation of $s = 1.723$.

(a) How large a sample is required?

(b) Does this seem to be a reasonable sample size? (Note that destructive testing is needed to obtain the data.)

7.4-6. A manufacturer sells a light bulb that has a mean life of 1450 hours with a standard deviation of 33.7 hours. A new manufacturing process is being tested, and there is interest in knowing the mean life μ of the new bulbs. How large a sample is required so that $\bar{x} \pm 5$ is a 95% confidence interval for μ? You may assume that the change in the standard deviation is minimal.

7.4-7. For a public opinion poll for a close presidential election, let p denote the proportion of voters who favor candidate A. How large a sample should be taken if we want the maximum error of the estimate of p to be equal to

(a) 0.03 with 95% confidence?

(b) 0.02 with 95% confidence?

(c) 0.03 with 90% confidence?

7.4-8. Some college professors and students examined 137 Canadian geese for patent schistosome in the year they hatched. Of these 137 birds, 54 were infected. The professors and students were interested in estimating p, the proportion of infected birds of this type. For future studies, determine the sample size n so that the estimate of p is within $\varepsilon = 0.04$ of the unknown p with 90% confidence.

7.4-9. A die has been loaded to change the probability of rolling a 6. In order to estimate p, the new probability of rolling a 6, how many times must the die be rolled so that we are 99% confident that the maximum error of the estimate of p is $\varepsilon = 0.02$?

7.4-10. A seed distributor claims that 80% of its beet seeds will germinate. How many seeds must be tested for germination in order to estimate p, the true proportion that will germinate, so that the maximum error of the estimate is $\varepsilon = 0.03$ with 90% confidence?

7.4-11. Some dentists were interested in studying the fusion of embryonic rat palates by a standard transplantation technique. When no treatment is used, the probability of fusion equals approximately 0.89. The dentists would like to estimate p, the probability of fusion, when vitamin A is lacking.

(a) How large a sample n of rat embryos is needed for $y/n \pm 0.10$ to be a 95% confidence interval for p?

(b) If $y = 44$ out of $n = 60$ palates showed fusion, give a 95% confidence interval for p.

7.4-12. Let p equal the proportion of college students who favor a new policy for alcohol consumption on campus. How large a sample is required to estimate p so that the maximum error of the estimate of p is 0.04 with 95% confidence when the size of the student body is

(a) $N = 1500$?

(b) $N = 15,000$?

(c) $N = 25,000$?

7.4-13. Out of 1000 welds that have been made on a tower, it is suspected that 15% are defective. To estimate p, the proportion of defective welds, how many welds

must be inspected to have approximately 95% confidence that the maximum error of the estimate of p is 0.04?

7.4-14. If Y_1/n and Y_2/n are the respective independent relative frequencies of success associated with the two binomial distributions $b(n, p_1)$ and $b(n, p_2)$, compute n such that the approximate probability that the random interval $(Y_1/n - Y_2/n) \pm 0.05$ covers $p_1 - p_2$ is at least 0.80. HINT: Take $p_1^* = p_2^* = 1/2$ to provide an upper bound for n.

7.4-15. If $\overline{X}$ and $\overline{Y}$ are the respective means of two independent random samples of the same size n, find n if we want $\overline{x} - \overline{y} \pm 4$ to be a 90% confidence interval for $\mu_X - \mu_Y$. Assume that the standard deviations are known to be $\sigma_X = 15$ and $\sigma_Y = 25$.

7.5 DISTRIBUTION-FREE CONFIDENCE INTERVALS FOR PERCENTILES

In Section 6.3, we defined sample percentiles in terms of order statistics and noted that the sample percentiles can be used to estimate corresponding distribution percentiles. In this section, we use order statistics to construct confidence intervals for unknown distribution percentiles. Since little is assumed about the underlying distribution (except that it is of the continuous type) in the construction of these confidence intervals, they are often called **distribution-free confidence intervals**.

If $Y_1 < Y_2 < Y_3 < Y_4 < Y_5$ are the order statistics of a random sample of size $n = 5$ from a continuous-type distribution, then the sample median Y_3 could be thought of as an estimator of the distribution median $\pi_{0.5}$. We shall let $m = \pi_{0.5}$. We could simply use the sample median Y_3 as an estimator of the distribution median m. However, we are certain that all of us recognize that, with only a sample of size 5, we would be quite lucky if the observed $Y_3 = y_3$ were very close to m. Thus, we now describe how a confidence interval can be constructed for m.

Instead of simply using Y_3 as an estimator of m, let us also compute the probability that the random interval (Y_1, Y_5) includes m. That is, let us determine $P(Y_1 < m < Y_5)$. Doing this is easy if we say that we have success if an individual observation—say, X—is less than m; then the probability of success on one of the independent trials is $P(X < m) = 0.5$. In order for the first order statistic Y_1 to be less than m and the last order statistic Y_5 to be greater than m, we must have at least one success, but not five successes. That is,

$$P(Y_1 < m < Y_5) = \sum_{k=1}^{4} \binom{5}{k} \left(\frac{1}{2}\right)^k \left(\frac{1}{2}\right)^{5-k}$$

$$= 1 - \left(\frac{1}{2}\right)^5 - \left(\frac{1}{2}\right)^5 = \frac{15}{16}.$$

So the probability that the random interval (Y_1, Y_5) includes m is $15/16 \approx 0.94$. Suppose now that this random sample is actually taken and the order statistics are observed to equal $y_1 < y_2 < y_3 < y_4 < y_5$, respectively. Then (y_1, y_5) is a 94% confidence interval for m.

It is interesting to note what happens as the sample size increases. Let $Y_1 < Y_2 < \cdots < Y_n$ be the order statistics of a random sample of size n from a distribution of the continuous type. Then $P(Y_1 < m < Y_n)$ is the probability that there is at least one "success" but not n successes, where the probability of success on each trial is $P(X < m) = 0.5$. Consequently,

$$P(Y_1 < m < Y_n) = \sum_{k=1}^{n-1} \binom{n}{k} \left(\frac{1}{2}\right)^k \left(\frac{1}{2}\right)^{n-k}$$

$$= 1 - \left(\frac{1}{2}\right)^n - \left(\frac{1}{2}\right)^n = 1 - \left(\frac{1}{2}\right)^{n-1}.$$

This probability increases as n increases, so that the corresponding confidence interval (y_1, y_n) would have the very large confidence coefficient $1 - (1/2)^{n-1}$. Unfortunately, the interval (y_1, y_n) tends to get wider as n increases; thus, we are not "pinning down" m very well. However, if we used the interval (y_2, y_{n-1}) or (y_3, y_{n-2}), we would obtain shorter intervals, but also smaller confidence coefficients. Let us investigate this possibility further.

With the order statistics $Y_1 < Y_2 < \cdots < Y_n$ associated with a random sample of size n from a continuous-type distribution, consider $P(Y_i < m < Y_j)$, where $i < j$. For example, we might want

$$P(Y_2 < m < Y_{n-1}) \qquad \text{or} \qquad P(Y_3 < m < Y_{n-2}).$$

On each of the n independent trials, we say that we have success if that X is less than m; thus, the probability of success on each trial is $P(X < m) = 0.5$. Consequently, to have the ith order statistic Y_i less than m and the jth order statistic greater than m, we must have at least i successes but fewer than j successes (or else $Y_j < m$). That is,

$$P(Y_i < m < Y_j) = \sum_{k=i}^{j-1} \binom{n}{k} \left(\frac{1}{2}\right)^k \left(\frac{1}{2}\right)^{n-k} = 1 - \alpha.$$

For particular values of n, i, and j, this probability—say, $1 - \alpha$—which is the sum of probabilities from a binomial distribution, can be calculated directly or approximated by an area under the normal pdf, provided that n is large enough. The observed interval (y_i, y_j) could then serve as a $100(1 - \alpha)\%$ confidence interval for the unknown distribution median.

Example 7.5-1

The lengths in centimeters of $n = 9$ fish of a particular species captured off the New England coast were 32.5, 27.6, 29.3, 30.1, 15.5, 21.7, 22.8, 21.2, and 19.0. Thus, the observed order statistics are

$$15.5 < 19.0 < 21.2 < 21.7 < 22.8 < 27.6 < 29.3 < 30.1 < 32.5.$$

Before the sample is drawn, we know that

$$P(Y_2 < m < Y_8) = \sum_{k=2}^{7} \binom{9}{k} \left(\frac{1}{2}\right)^k \left(\frac{1}{2}\right)^{9-k}$$

$$= 0.9805 - 0.0195 = 0.9610,$$

from Table II in Appendix B. Thus, the confidence interval $(y_2 = 19.0, y_8 = 30.1)$ for m, the median of the lengths of all fish of this species, has a 96.1% confidence coefficient. ∎

So that the student need not compute many of these probabilities, Table 7.5-1 lists the necessary information for constructing confidence intervals of the form (y_i, y_{n+1-i}) for the unknown m for sample sizes $n = 5, 6, \ldots, 20$. The subscript i is selected so that the confidence coefficient $P(Y_i < m < Y_{n+1-i})$ is greater than 90% and as close to 95% as possible.

For sample sizes larger than 20, we approximate those binomial probabilities with areas under the normal curve. To illustrate how good these approximations are, we compute the probability corresponding to $n = 16$ in Table 7.5-1. Here, using Table II, we have

$$1 - \alpha = P(Y_5 < m < Y_{12}) = \sum_{k=5}^{11} \binom{16}{k}\left(\frac{1}{2}\right)^k \left(\frac{1}{2}\right)^{16-k}$$

$$= P(W = 5, 6, \ldots, 11)$$

$$= 0.9616 - 0.0384 = 0.9232,$$

where W is $b(16, 1/2)$. The normal approximation gives

$$1 - \alpha = P(4.5 < W < 11.5) = P\left(\frac{4.5 - 8}{2} < \frac{W - 8}{2} < \frac{11.5 - 8}{2}\right),$$

because W has mean $np = 8$ and variance $np(1 - p) = 4$. The standardized variable $Z = (W - 8)/2$ has an approximate normal distribution. Thus,

$$1 - \alpha \approx \Phi\left(\frac{3.5}{2}\right) - \Phi\left(\frac{-3.5}{2}\right) = \Phi(1.75) - \Phi(-1.75)$$

$$= 0.9599 - 0.0401 = 0.9198.$$

This value compares very favorably with the probability 0.9232 recorded in Table 7.5-1. (Note that Minitab or some other computer program can also be used.)

Table 7.5-1 Information for confidence intervals for m

n	$(i, n+1-i)$	$P(Y_i < m < Y_{n+1-i})$	n	$(i, n+1-i)$	$P(Y_i < m < Y_{n+1-i})$
5	$(1, 5)$	0.9376	13	$(3, 11)$	0.9776
6	$(1, 6)$	0.9688	14	$(4, 11)$	0.9426
7	$(1, 7)$	0.9844	15	$(4, 12)$	0.9648
8	$(2, 7)$	0.9296	16	$(5, 12)$	0.9232
9	$(2, 8)$	0.9610	17	$(5, 13)$	0.9510
10	$(2, 9)$	0.9786	18	$(5, 14)$	0.9692
11	$(3, 9)$	0.9346	19	$(6, 14)$	0.9364
12	$(3, 10)$	0.9614	20	$(6, 15)$	0.9586

The argument used to find a confidence interval for the median m of a distribution of the continuous type can be applied to any percentile π_p. In this case, we say that we have success on a single trial if that X is less than π_p. Thus, the probability of success on each of the independent trials is $P(X < \pi_p) = p$. Accordingly, with $i < j$, $1 - \alpha = P(Y_i < \pi_p < Y_j)$ is the probability that we have at least i successes but fewer than j successes. Hence,

$$1 - \alpha = P(Y_i < \pi_p < Y_j) = \sum_{k=i}^{j-1} \binom{n}{k} p^k (1-p)^{n-k}.$$

Once the sample is observed and the order statistics determined, the known interval (y_i, y_j) could serve as a $100(1-\alpha)\%$ confidence interval for the unknown distribution percentile π_p.

Example 7.5-2

Let the following numbers represent the order statistics of the $n = 27$ observations obtained in a random sample from a certain population of incomes (measured in hundreds of dollars):

261	269	271	274	279	280	283	284	286
287	292	293	296	300	304	305	313	321
322	329	341	343	356	364	391	417	476

Say we are interested in estimating the 25th percentile, $\pi_{0.25}$, of the population. Since $(n + 1)p = 28(1/4) = 7$, the seventh order statistic, namely, $y_7 = 283$, would be a point estimate of $\pi_{0.25}$. To find a confidence interval for $\pi_{0.25}$, let us move down and up a few order statistics from y_7—say, to y_4 and y_{10}. What is the confidence coefficient associated with the interval (y_4, y_{10})? Before the sample was drawn, we had

$$1 - \alpha = P(Y_4 < \pi_{0.25} < Y_{10}) = \sum_{k=4}^{9} \binom{27}{k} (0.25)^k (0.75)^{27-k} = 0.8201.$$

For the normal approximation, we use W, which is $b(27, 1/4)$ with mean $27/4 = 6.75$ and variance $81/16$. Hence,

$$1 - \alpha = P(4 \le W \le 9) = P(3.5 < W < 9.5)$$

$$\approx \Phi\left(\frac{9.5 - 6.75}{9/4}\right) - \Phi\left(\frac{3.5 - 6.75}{9/4}\right)$$

$$= \Phi\left(\frac{11}{9}\right) - \Phi\left(-\frac{13}{9}\right) = 0.8149.$$

Thus, $(y_4 = 274, y_{10} = 287)$ is an 82.01% (or approximate 81.49%) confidence interval for $\pi_{0.25}$. Note that we could choose other intervals, such as $(y_3 = 271, y_{11} = 292)$, and these would have different confidence coefficients. The persons involved in the study must select the desired confidence coefficient, and then the appropriate order statistics are taken, usually quite symmetrically about the $(n + 1)p$th order statistic. ∎

When the number of observations is large, it is important to be able to determine the order statistics rather easily. As illustrated in the next example, a stem-and-leaf diagram, as introduced in Section 6.2, can be helpful in determining the needed order statistics.

Example 7.5-3

The measurements of butterfat produced by $n = 90$ cows during a 305-day milk production period following their first calf are summarized in Table 7.5-2, in which each leaf consists of two digits. From this display, it is quite easy to see that $y_8 = 392$.

Table 7.5-2 Ordered stem-and-leaf diagram of butterfat production

Stems	Leaves								
2s	74								
2•									
3∗									
3t	27	39							
3f	45	50							
3s									
3•	80	88	92	94	95				
4∗	17	18							
4t	21	22	27	34	37	39			
4f	44	52	53	53	57	58			
4s	60	64	66	70	70	72	75	78	
4•	81	86	89	91	92	94	96	97	99
5∗	00	00	01	02	05	09	10	13	13	16
5t	24	26	31	32	32	37	37	39	
5f	40	41	44	55					
5s	61	70	73	74					
5•	83	83	86	93	99				
6∗	07	08	11	12	13	17	18	19	
6t	27	28	35	37					
6f	43	43	45						
6s	72								
6•	91	96							

It takes a little more work to show that $y_{38} = 494$ and $y_{53} = 526$ creates an interval $(494, 526)$ which serves as a confidence interval for the unknown median m of all butterfat production for the given breed of cows. Its confidence coefficient is

$$P(Y_{38} < m < Y_{53}) = \sum_{k=38}^{52} \binom{90}{k} \left(\frac{1}{2}\right)^k \left(\frac{1}{2}\right)^{90-k}$$

$$\approx \Phi\left(\frac{52.5 - 45}{\sqrt{22.5}}\right) - \Phi\left(\frac{37.5 - 45}{\sqrt{22.5}}\right)$$

$$= \Phi(1.58) - \Phi(-1.58) = 0.8858.$$

Similarly, $(y_{17} = 437, y_{29} = 470)$ is a confidence interval for the first quartile, $\pi_{0.25}$, with confidence coefficient

$$P(Y_{17} < \pi_{0.25} < Y_{29}) \approx \Phi\left(\frac{28.5 - 22.5}{\sqrt{16.875}}\right) - \Phi\left(\frac{16.5 - 22.5}{\sqrt{16.875}}\right)$$

$$= \Phi(1.46) - \Phi(-1.46) = 0.8558.$$

Using the binomial distribution, the confidence coefficients are 0.8867 and 0.8569, respectively. ■

It is interesting to compare the length of a confidence interval for the mean μ obtained with $\bar{x} \pm t_{\alpha/2}(n-1)(s/\sqrt{n})$ against the length of a $100(1-\alpha)\%$ confidence interval for the median m obtained with the distribution-free techniques of this section. Usually, if the sample arises from a distribution that does not deviate too much from the normal, the confidence interval based upon $\bar{x}$ is much shorter. After all, we assume much more when we create that confidence interval. With the distribution-free method, all we assume is that the distribution is of the continuous type. So if the distribution is highly skewed or heavy-tailed so that outliers could exist, a distribution-free technique is safer and much more robust. Moreover, the distribution-free technique provides a way to get confidence intervals for various percentiles, and investigators are often interested in such intervals.

Exercises

7.5-1. Let $Y_1 < Y_2 < Y_3 < Y_4 < Y_5 < Y_6$ be the order statistics of a random sample of size $n = 6$ from a distribution of the continuous type having $(100p)$th percentile π_p. Compute

(a) $P(Y_2 < \pi_{0.5} < Y_5)$.

(b) $P(Y_1 < \pi_{0.25} < Y_4)$.

(c) $P(Y_4 < \pi_{0.9} < Y_6)$.

7.5-2. For $n = 12$ year-2007 model sedans whose horsepower is between 290 and 390, the following measurements give the time in seconds for the car to go from 0 to 60 mph:

6.0 6.3 5.0 6.0 5.7 5.9 6.8 5.5 5.4 4.8 5.4 5.8

(a) Find a 96.14% confidence interval for the median, m.

(b) The interval (y_1, y_7) could serve as a confidence interval for $\pi_{0.3}$. Find it and give its confidence coefficient.

7.5-3. A sample of $n = 9$ electrochromic mirrors was used to measure the following low-end reflectivity percentages:

7.12 7.22 6.78 6.31 5.99 6.58 7.80 7.40 7.05

(a) Find the endpoints for an approximate 95% confidence interval for the median, m.

(b) The interval (y_3, y_7) could serve as a confidence interval for m. Find it and give its confidence coefficient.

7.5-4. Let m denote the median weight of "80-pound" bags of water softener pellets. Use the following random sample of $n = 14$ weights to find an approximate 95% confidence interval for m:

80.51	80.28	80.40	80.35	80.38	80.28	80.27
80.16	80.59	80.56	80.32	80.27	80.53	80.32

(a) Find a 94.26% confidence interval for m.

(b) The interval (y_6, y_{12}) could serve as a confidence interval for $\pi_{0.6}$. What is its confidence coefficient?

7.5-5. A biologist who studies spiders selected a random sample of 20 male green lynx spiders (a spider that does not weave a web, but chases and leaps on its prey) and measured the lengths (in millimeters) of one of the front legs of the 20 spiders. Use the following measurements to construct a confidence interval for m that has a confidence coefficient about equal to 0.95:

15.10	13.55	15.75	20.00	15.45
13.60	16.45	14.05	16.95	19.05
16.40	17.05	15.25	16.65	16.25
17.75	15.40	16.80	17.55	19.05

7.5-6. A company manufactures mints that have a label weight of 20.4 grams. The company regularly samples from the production line and weighs the selected mints. During two mornings of production it sampled 81 mints, obtaining the following weights:

21.8	21.7	21.7	21.6	21.3	21.6	21.5	21.3	21.2
21.0	21.6	21.6	21.6	21.5	21.4	21.8	21.7	21.6
21.6	21.3	21.9	21.9	21.6	21.0	20.7	21.8	21.7
21.7	21.4	20.9	22.0	21.3	21.2	21.0	21.0	21.9
21.7	21.5	21.5	21.1	21.3	21.3	21.2	21.0	20.8
21.6	21.6	21.5	21.5	21.2	21.5	21.4	21.4	21.3
21.2	21.8	21.7	21.7	21.6	20.5	21.8	21.7	21.5
21.4	21.4	21.9	21.8	21.7	21.4	21.3	20.9	21.9
20.7	21.1	20.8	20.6	20.6	22.0	22.0	21.7	21.6

(a) Construct an ordered stem-and-leaf display using stems of $20f$, $20s$, $20\bullet$, $21*, \ldots, 22*$.

(b) Find **(i)** the three quartiles, **(ii)** the 60th percentile, and **(iii)** the 15th percentile.

(c) Find approximate 95% confidence intervals for **(i)** $\pi_{0.25}$, **(ii)** $m = \pi_{0.5}$, and **(iii)** $\pi_{0.75}$.

7.5-7. Here are the weights (in grams) of 25 indicator housings used on gauges (see Exercise 6.2-8):

102.0	106.3	106.6	108.8	107.7
106.1	105.9	106.7	106.8	110.2
101.7	106.6	106.3	110.2	109.9
102.0	105.8	109.1	106.7	107.3
102.0	106.8	110.0	107.9	109.3

(a) List the observations in order of magnitude.

(b) Give point estimates of $\pi_{0.25}$, m, and $\pi_{0.75}$.

(c) Find the following confidence intervals and, from Table II in Appendix B, state the associated confidence coefficient:

   (i) (y_3, y_{10}), a confidence interval for $\pi_{0.25}$.

   (ii) (y_9, y_{17}), a confidence interval for the median m.

   (iii) (y_{16}, y_{23}), a confidence interval for $\pi_{0.75}$.

(d) Use $\bar{x} \pm t_{\alpha/2}(24)(s/\sqrt{25})$ to find a confidence interval for μ, whose confidence coefficient corresponds to that of (c), part (ii). Compare these two confidence intervals of the middles.

7.5-8. The biologist of Exercise 7.5-5 also selected a random sample of 20 female green lynx spiders and measured the length (again in millimeters) of one of their front legs. Use the following data to construct a confidence interval for m that has a confidence coefficient about equal to 0.95:

15.85	18.00	11.45	15.60	16.10
18.80	12.85	15.15	13.30	16.65
16.25	16.15	15.25	12.10	16.20
14.80	14.60	17.05	14.15	15.85

7.5-9. Let X equal the amount of fluoride in a certain brand of toothpaste. The specifications are 0.85–1.10 mg/g. Table 6.1-3 lists 100 such measurements.

(a) Give a point estimate of the median $m = \pi_{0.50}$.

(b) Find an approximate 95% confidence interval for the median m. If possible, use a computer to find the exact confidence level.

(c) Give a point estimate for the first quartile.

(d) Find an approximate 95% confidence interval for the first quartile and, if possible, give the exact confidence coefficient.

(e) Give a point estimate for the third quartile.

(f) Find an approximate 95% confidence interval for the third quartile and, if possible, give the exact confidence coefficient.

7.5-10. When placed in solutions of varying ionic strength, paramecia grow blisters in order to counteract the flow of water. The following 60 measurements in microns are blister lengths:

7.42	5.73	3.80	5.20	11.66	8.51	6.31	8.49
10.31	6.92	7.36	5.92	6.74	8.93	9.61	11.38
12.78	11.43	6.57	13.50	10.58	8.03	10.07	8.71
10.09	11.16	7.22	10.10	6.32	10.30	10.75	11.51
11.55	11.41	9.40	4.74	6.52	12.10	6.01	5.73
7.57	7.80	6.84	6.95	8.93	8.92	5.51	6.71
10.40	13.44	9.33	8.57	7.08	8.11	13.34	6.58
8.82	7.70	12.22	7.46				

(a) Construct an ordered stem-and-leaf diagram.

(b) Give a point estimate of the median $m = \pi_{0.50}$.

(c) Find an approximate 95% confidence interval for m.

(d) Give a point estimate for the 40th percentile, $\pi_{0.40}$.

(e) Find an approximate 90% confidence interval for $\pi_{0.40}$.

7.5-11. Using the weights of Verica's 39 gold coins given in Example 6.2-4, find approximate 95% confidence intervals for $\pi_{0.25}$, $\pi_{0.5}$, and $\pi_{0.75}$. Give the exact confidence coefficients for the intervals.

7.5-12. Let $Y_1 < Y_2 < \cdots < Y_8$ be the order statistics of eight independent observations from a continuous-type distribution with 70th percentile $\pi_{0.7} = 27.3$.

(a) Determine $P(Y_7 < 27.3)$.

(b) Find $P(Y_5 < 27.3 < Y_8)$.

7.6* MORE REGRESSION

In this section, we develop confidence intervals for important quantities in the linear regression model using the notation and assumptions of Section 6.5. It can be shown (Exercise 7.6-13) that

$$\sum_{i=1}^{n}[Y_i - \alpha - \beta(x_i - \bar{x})]^2 = \sum_{i=1}^{n}\{(\widehat{\alpha} - \alpha) + (\widehat{\beta} - \beta)(x_i - \bar{x})$$
$$+ [Y_i - \widehat{\alpha} - \widehat{\beta}(x_i - \bar{x})]\}^2$$
$$= n(\widehat{\alpha} - \alpha)^2 + (\widehat{\beta} - \beta)^2 \sum_{i=1}^{n}(x_i - \bar{x})^2$$
$$+ \sum_{i=1}^{n}[Y_i - \widehat{\alpha} - \widehat{\beta}(x_i - \bar{x})]^2. \quad (7.6\text{-}1)$$

From the fact that Y_i, $\widehat{\alpha}$, and $\widehat{\beta}$ have normal distributions, it follows that each of

$$\frac{[Y_i - \alpha - \beta(x_i - \bar{x})]^2}{\sigma^2}, \quad \frac{(\widehat{\alpha} - \alpha)^2}{\left[\dfrac{\sigma^2}{n}\right]}, \quad \text{and} \quad \frac{(\widehat{\beta} - \beta)^2}{\left[\dfrac{\sigma^2}{\sum_{i=1}^{n}(x_i - \bar{x})^2}\right]}$$

has a chi-square distribution with one degree of freedom. Since $Y_1, Y_2, \ldots, Y_n$ are mutually independent,

$$\frac{\sum_{i=1}^{n}[Y_i - \alpha - \beta(x_i - \bar{x})]^2}{\sigma^2}$$

is $\chi^2(n)$. That is, the left-hand member of Equation 7.6-1 divided by σ^2 is $\chi^2(n)$ and is equal to the sum of two $\chi^2(1)$ variables and

$$\frac{\sum_{i=1}^{n}[Y_i - \widehat{\alpha} - \widehat{\beta}(x_i - \bar{x})]^2}{\sigma^2} = \frac{n\widehat{\sigma^2}}{\sigma^2} \geq 0.$$

Thus, we might guess that $n\widehat{\sigma^2}/\sigma^2$ is $\chi^2(n-2)$. This is true, and moreover, $\widehat{\alpha}$, $\widehat{\beta}$, and $\widehat{\sigma^2}$ are mutually independent. [For a proof, see Hogg, McKean, and Craig, *Introduction to Mathematical Statistics*, 7th ed. (Upper Saddle River, NJ: Prentice Hall, 2013).]

Suppose now that we are interested in forming a confidence interval for β, the slope of the line. We can use the fact that

$$T_1 = \frac{\sqrt{\sum_{i=1}^{n} (x_i - \bar{x})^2} \left(\dfrac{\widehat{\beta} - \beta}{\sigma} \right)}{\sqrt{\dfrac{n\widehat{\sigma}^2}{\sigma^2(n-2)}}} = \frac{\widehat{\beta} - \beta}{\sqrt{\dfrac{n\widehat{\sigma}^2}{(n-2)\sum_{i=1}^{n} (x_i - \bar{x})^2}}}$$

has a t distribution with $n - 2$ degrees of freedom. Therefore,

$$P\left[-t_{\gamma/2}(n-2) \leq \frac{\widehat{\beta} - \beta}{\sqrt{\dfrac{n\widehat{\sigma}^2}{(n-2)\sum_{i=1}^{n} (x_i - \bar{x})^2}}} \leq t_{\gamma/2}(n-2) \right] = 1 - \gamma,$$

and it follows that

$$\left[\widehat{\beta} - t_{\gamma/2}(n-2)\sqrt{\frac{n\widehat{\sigma}^2}{(n-2)\sum_{i=1}^{n} (x_i - \bar{x})^2}} \, , \right.$$

$$\left. \widehat{\beta} + t_{\gamma/2}(n-2)\sqrt{\frac{n\widehat{\sigma}^2}{(n-2)\sum_{i=1}^{n} (x_i - \bar{x})^2}} \right]$$

is a $100(1 - \gamma)\%$ confidence interval for β.

Similarly,

$$T_2 = \frac{\dfrac{\sqrt{n}(\widehat{\alpha} - \alpha)}{\sigma}}{\sqrt{\dfrac{n\widehat{\sigma}^2}{\sigma^2(n-2)}}} = \frac{\widehat{\alpha} - \alpha}{\sqrt{\dfrac{\widehat{\sigma}^2}{n-2}}}$$

has a t distribution with $n - 2$ degrees of freedom. Thus, T_2 can be used to make inferences about α. (See Exercise 7.6-14.) The fact that $n\widehat{\sigma}^2/\sigma^2$ has a chi-square distribution with $n - 2$ degrees of freedom can be used to make inferences about the variance σ^2. (See Exercise 7.6-15.)

We have noted that $\widehat{Y} = \widehat{\alpha} + \widehat{\beta}(x - \bar{x})$ is a point estimate for the mean of Y for some given x, or we could think of this as a prediction of the value of Y for this given x. But how close is $\widehat{Y}$ to the mean of Y or to Y itself? We shall now find a confidence interval for $\alpha + \beta(x - \bar{x})$ and a prediction interval for Y, given a particular value of x.

To find a confidence interval for

$$E(Y) = \mu(x) = \alpha + \beta(x - \bar{x}),$$

let

$$\widehat{Y} = \widehat{\alpha} + \widehat{\beta}(x - \bar{x}).$$

Recall that $\widehat{Y}$ is a linear combination of normally and independently distributed random variables $\widehat{\alpha}$ and $\widehat{\beta}$, so $\widehat{Y}$ has a normal distribution. Furthermore,

$$E(\widehat{Y}) = E[\widehat{\alpha} + \widehat{\beta}(x - \bar{x})]$$

$$= \alpha + \beta(x - \bar{x})$$

and

$$\text{Var}(\widehat{Y}) = \text{Var}[\widehat{\alpha} + \widehat{\beta}(x - \bar{x})]$$

$$= \frac{\sigma^2}{n} + \frac{\sigma^2}{\sum_{i=1}^{n}(x_i - \bar{x})^2}(x - \bar{x})^2$$

$$= \sigma^2\left[\frac{1}{n} + \frac{(x - \bar{x})^2}{\sum_{i=1}^{n}(x_i - \bar{x})^2}\right].$$

Recall that the distribution of $n\widehat{\sigma^2}/\sigma^2$ is $\chi^2(n-2)$. Since $\widehat{\alpha}$ and $\widehat{\beta}$ are independent of $\widehat{\sigma^2}$, we can form the t statistic

$$T = \frac{\dfrac{\widehat{\alpha} + \widehat{\beta}(x - \bar{x}) - [\alpha + \beta(x - \bar{x})]}{\sigma\sqrt{\dfrac{1}{n} + \dfrac{(x - \bar{x})^2}{\sum_{i=1}^{n}(x_i - \bar{x})^2}}}}{\sqrt{\dfrac{n\widehat{\sigma^2}}{(n-2)\sigma^2}}},$$

which has a t distribution with $r = n-2$ degrees of freedom. Next we select $t_{\gamma/2}(n-2)$ from Table VI in Appendix B so that

$$P[-t_{\gamma/2}(n-2) \le T \le t_{\gamma/2}(n-2)] = 1 - \gamma.$$

This becomes

$$P[\widehat{\alpha} + \widehat{\beta}(x - \bar{x}) - ct_{\gamma/2}(n-2) \le \alpha + \beta(x - \bar{x})$$
$$\le \widehat{\alpha} + \widehat{\beta}(x - \bar{x}) + ct_{\gamma/2}(n-2)]$$
$$= 1 - \gamma,$$

where

$$c = \sqrt{\frac{n\widehat{\sigma^2}}{n-2}}\sqrt{\frac{1}{n} + \frac{(x - \bar{x})^2}{\sum_{i=1}^{n}(x_i - \bar{x})^2}}.$$

Thus, the endpoints for a $100(1 - \gamma)\%$ confidence interval for $\mu(x) = \alpha + \beta(x - \bar{x})$ are

$$\widehat{\alpha} + \widehat{\beta}(x - \bar{x}) \pm ct_{\gamma/2}(n-2).$$

Note that the width of this interval depends on the particular value of x, because c depends on x. (See Example 7.6-1.)

We have used $(x_1, y_1), (x_2, y_2), \ldots, (x_n, y_n)$ to estimate α and β. Suppose that we are given a value of x, say, x_{n+1}. A point estimate of the corresponding value of Y is

$$\widehat{y}_{n+1} = \widehat{\alpha} + \widehat{\beta}(x_{n+1} - \bar{x}).$$

However, $\widehat{y}_{n+1}$ is just one possible value of the random variable

$$Y_{n+1} = \alpha + \beta(x_{n+1} - \bar{x}) + \varepsilon_{n+1}.$$

What can we say about possible values for Y_{n+1}? We shall now obtain a **prediction interval** for Y_{n+1} when $x = x_{n+1}$ that is similar to the confidence interval for the mean of Y when $x = x_{n+1}$.

We have

$$Y_{n+1} = \alpha + \beta(x_{n+1} - \bar{x}) + \varepsilon_{n+1},$$

where ε_{n+1} is $N(0, \sigma^2)$. Now,

$$W = Y_{n+1} - \widehat{\alpha} - \widehat{\beta}(x_{n+1} - \overline{x})$$

is a linear combination of normally and independently distributed random variables, so W has a normal distribution. The mean of W is

$$E(W) = E[Y_{n+1} - \widehat{\alpha} - \widehat{\beta}(x_{n+1} - \overline{x})]$$
$$= \alpha + \beta(x_{n+1} - \overline{x}) - \alpha - \beta(x_{n+1} - \overline{x}) = 0.$$

Since $Y_{n+1}, \widehat{\alpha}$ and $\widehat{\beta}$ are independent, the variance of W is

$$\mathrm{Var}(W) = \sigma^2 + \frac{\sigma^2}{n} + \frac{\sigma^2}{\sum_{i=1}^{n}(x_i - \overline{x})^2}(x_{n+1} - \overline{x})^2$$

$$= \sigma^2 \left[1 + \frac{1}{n} + \frac{(x_{n+1} - \overline{x})^2}{\sum_{i=1}^{n}(x_i - \overline{x})^2} \right].$$

Recall that $n\widehat{\sigma^2}/[(n-2)\sigma^2]$ is $\chi^2(n-2)$. Since $Y_{n+1}, \widehat{\alpha}$, and $\widehat{\beta}$ are independent of $\widehat{\sigma^2}$, we can form the t statistic

$$T = \frac{\dfrac{Y_{n+1} - \widehat{\alpha} - \widehat{\beta}(x_{n+1} - \overline{x})}{\sigma\sqrt{1 + \dfrac{1}{n} + \dfrac{(x_{n+1} - \overline{x})^2}{\sum_{i=1}^{n}(x_i - \overline{x})^2}}}}{\sqrt{\dfrac{n\widehat{\sigma^2}}{(n-2)\sigma^2}}},$$

which has a t distribution with $r = n - 2$ degrees of freedom. Now we select a constant $t_{\gamma/2}(n-2)$ from Table VI in Appendix B so that

$$P[-t_{\gamma/2}(n-2) \leq T \leq t_{\gamma/2}(n-2)] = 1 - \gamma.$$

Solving this inequality for Y_{n+1}, we have

$$P[\widehat{\alpha} + \widehat{\beta}(x_{n+1} - \overline{x}) - d\,t_{\gamma/2}(n-2) \leq Y_{n+1}$$
$$\leq \widehat{\alpha} + \widehat{\beta}(x_{n+1} - \overline{x}) + d t_{\gamma/2}(n-2)]$$
$$= 1 - \gamma,$$

where

$$d = \sqrt{\frac{n\widehat{\sigma^2}}{n-2}}\sqrt{1 + \frac{1}{n} + \frac{(x_{n+1} - \overline{x})^2}{\sum_{i=1}^{n}(x_i - \overline{x})^2}}.$$

Thus, the endpoints for a $100(1 - \gamma)\%$ prediction interval for Y_{n+1} are

$$\widehat{\alpha} + \widehat{\beta}(x_{n+1} - \overline{x}) \pm d t_{\gamma/2}(n-2).$$

Observe that

$$d^2 = c^2 + \frac{n\widehat{\sigma^2}}{n-2}$$

when $x_{n+1} = x$, implying that the $100(1 - \gamma)\%$ prediction interval for Y at $X = x$ is somewhat wider than the $100(1-\gamma)\%$ prediction interval for $\mu(x)$. This makes sense, since the difference between one observation of Y (at a given X) and its predictor tends to vary more than the difference between the mean of the entire population of Y values (at the same X) and its estimator.

The collection of all $100(1-\gamma)\%$ confidence intervals for $\{\mu(x) : -\infty < x < \infty\}$ is called a **pointwise** $100(1-\gamma)\%$ **confidence band** for $\mu(x)$. Similarly, the collection of all $100(1-\gamma)\%$ prediction intervals for $\{Y(x) = \alpha + \beta x + \varepsilon : -\infty < x < \infty\}$ is called a **pointwise** $100(1-\gamma)\%$ **prediction band** for Y. Note, from the expressions for c and d in the confidence and prediction intervals, respectively, that these bands are narrowest at $x = \bar{x}$.

We shall now use the data in Example 6.5-1 to illustrate a 95% confidence interval for $\mu(x)$ and a 95% prediction interval for Y for a given value of x. To find such intervals, we use Equations 6.5-1, 6.5-2, and 6.5-4.

Example 7.6-1 To find a 95% confidence interval for $\mu(x)$ using the data in Example 6.5-1, note that we have already found that $\bar{x} = 68.3$, $\widehat{\alpha} = 81.3$, $\widehat{\beta} = 561.1/756.1 = 0.7421$, and $\widehat{\sigma^2} = 21.7709$. We also need

$$\sum_{i=1}^{n}(x_i - \bar{x})^2 = \sum_{i=1}^{n} x_i^2 - \left(\frac{1}{n}\right)\left(\sum_{i=1}^{n} x_i\right)^2$$

$$= 47,405 - \frac{683^2}{10} = 756.1.$$

For 95% confidence, $t_{0.025}(8) = 2.306$. When $x = 60$, the endpoints for a 95% confidence interval for $\mu(60)$ are

$$81.3 + 0.7421(60 - 68.3) \pm \left[\sqrt{\frac{10(21.7709)}{8}}\sqrt{\frac{1}{10} + \frac{(60 - 68.3)^2}{756.1}}\right](2.306),$$

or

$$75.1406 \pm 5.2589.$$

Similarly, when $x = 70$, the endpoints for a 95% confidence interval for $\mu(70)$ are

$$82.5616 \pm 3.8761.$$

Note that the lengths of these intervals depend on the particular value of x. A pointwise 95% confidence band for $\mu(x)$ is graphed in Figure 7.6-1(a) along with the scatter diagram and $\widehat{y} = \widehat{\alpha} + \widehat{\beta}(x - \bar{x})$.

The endpoints for a 95% prediction interval for Y when $x = 60$ are

$$81.3 + 0.7421(60 - 68.3) \pm \left[\sqrt{\frac{10(21.7709)}{8}}\sqrt{1.1 + \frac{(60 - 68.3)^2}{756.1}}\right](2.306),$$

or

$$75.1406 \pm 13.1289.$$

Note that this interval is much wider than the confidence interval for $\mu(60)$. In Figure 7.6-1(b), the pointwise 95% prediction band for Y is graphed along with the scatter diagram and the least squares regression line. ∎

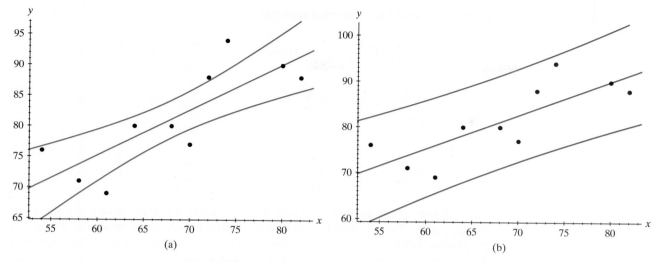

Figure 7.6-1 A pointwise 95% (a) confidence band for $\mu(x)$ and (b) prediction band for Y

We now generalize the simple regression model to the **multiple regression** case. Suppose we observe several x-values—say, $x_1, x_2, \ldots, x_k$—along with the y-value. For example, suppose that x_1 equals the student's ACT composite score, x_2 equals the student's high school class rank, and y equals the student's first-year GPA in college. We want to estimate a regression function $E(Y) = \mu(x_1, x_2, \ldots, x_k)$ from some observed data. If

$$\mu(x_1, x_2, \ldots, x_k) = \beta_1 x_1 + \beta_2 x_2 + \cdots + \beta_k x_k,$$

then we say that we have a **linear model** because this expression is linear in the coefficients $\beta_1, \beta_2, \ldots, \beta_k$.

To illustrate, note that the model in Section 6.5 is linear in $\alpha = \beta_1$ and $\beta = \beta_2$, with $x_1 = 1$ and $x_2 = x$, giving the mean $\alpha + \beta x$. (For convenience, there the mean of the x-values was subtracted from x.) Suppose, however, that we had wished to use the cubic function $\beta_1 + \beta_2 x + \beta_3 x^2 + \beta_4 x^3$ as the mean. This cubic expression still provides a linear model (i.e., linear in the β-values), and we would take $x_1 = 1$, $x_2 = x$, $x_3 = x^2$, and $x_4 = x^3$.

Say our n observation points are

$$(x_{1j}, x_{2j}, \ldots, x_{kj}, y_j), \qquad j = 1, 2, \ldots, n.$$

To fit the linear model $\beta_1 x_1 + \beta_2 x_2 + \cdots + \beta_k x_k$ by the method of least squares, we minimize

$$G = \sum_{j=1}^{n} (y_j - \beta_1 x_{1j} - \beta_2 x_{2j} - \cdots - \beta_k x_{kj})^2.$$

If we equate the k first order partial derivatives

$$\frac{\partial G}{\partial \beta_i} = \sum_{j=1}^{n} (-2)(y_j - \beta_1 x_{1j} - \beta_2 x_{2j} - \cdots - \beta_k x_{kj})(x_{ij}), \qquad i = 1, 2, \ldots, k,$$

to zero, we obtain the k **normal equations**

$$\beta_1 \sum_{j=1}^{n} x_{1j}^2 + \beta_2 \sum_{j=1}^{n} x_{1j}x_{2j} + \cdots + \beta_k \sum_{j=1}^{n} x_{1j}x_{kj} = \sum_{j=1}^{n} x_{1j}y_j,$$

$$\beta_1 \sum_{j=1}^{n} x_{2j}x_{1j} + \beta_2 \sum_{j=1}^{n} x_{2j}^2 + \cdots + \beta_k \sum_{j=1}^{n} x_{2j}x_{kj} = \sum_{j=1}^{n} x_{2j}y_j,$$

$$\vdots \qquad \vdots \qquad \ddots \qquad \vdots \qquad \vdots$$

$$\beta_1 \sum_{j=1}^{n} x_{kj}x_{1j} + \beta_2 \sum_{j=1}^{n} x_{kj}x_{2j} + \cdots + \beta_k \sum_{j=1}^{n} x_{kj}^2 = \sum_{j=1}^{n} x_{kj}y_j.$$

The solution of the preceding k equations provides the least squares estimates of $\beta_1, \beta_2, \ldots, \beta_k$. These estimates are also maximum likelihood estimates of $\beta_1, \beta_2, \ldots, \beta_k$, provided that the random variables $Y_1, Y_2, \ldots, Y_n$ are mutually independent and Y_j is $N(\beta_1 x_{1j} + \beta_2 x_{2j} + \cdots + \beta_k x_{kj}, \sigma^2)$, $j = 1, 2, \ldots, n$.

Example 7.6-2 By the method of least squares, we fit $y = \beta_1 x_1 + \beta_2 x_2 + \beta_3 x_3$ to the five observed points (x_1, x_2, x_3, y):

$$(1, 1, 0, 4), \quad (1, 0, 1, 3), \quad (1, 2, 3, 2), \quad (1, 3, 0, 6), \quad (1, 0, 0, 1).$$

Note that $x_1 = 1$ in each point, so we are really fitting $y = \beta_1 + \beta_2 x_2 + \beta_3 x_3$. Since

$$\sum_{j=1}^{5} x_{1j}^2 = 5, \quad \sum_{j=1}^{5} x_{1j}x_{2j} = 6, \quad \sum_{j=1}^{5} x_{1j}x_{3j} = 4, \quad \sum_{j=1}^{5} x_{1j}y_j = 16,$$

$$\sum_{j=1}^{5} x_{2j}x_{1j} = 6, \quad \sum_{j=1}^{5} x_{2j}^2 = 14, \quad \sum_{j=1}^{5} x_{2j}x_{3j} = 6, \quad \sum_{j=1}^{5} x_{2j}y_j = 26,$$

$$\sum_{j=1}^{5} x_{3j}x_{1j} = 4, \quad \sum_{j=1}^{5} x_{3j}x_{2j} = 6, \quad \sum_{j=1}^{5} x_{3j}^2 = 10, \quad \sum_{j=1}^{5} x_{3j}y_j = 9,$$

the normal equations are

$$5\beta_1 + 6\beta_2 + 4\beta_3 = 16,$$

$$6\beta_1 + 14\beta_2 + 6\beta_3 = 26,$$

$$4\beta_1 + 6\beta_2 + 10\beta_3 = 9.$$

Solving these three linear equations in three unknowns, we obtain

$$\widehat{\beta}_1 = \frac{274}{112}, \qquad \widehat{\beta}_2 = \frac{127}{112}, \qquad \widehat{\beta}_3 = -\frac{85}{112}.$$

Thus, the least squares fit is

$$y = \frac{274x_1 + 127x_2 - 85x_3}{112}.$$

If x_1 always equals 1, then the equation reads

$$y = \frac{274 + 127x_2 - 85x_3}{112}.$$

It is interesting to observe that the usual two-sample problem is actually a linear model. Let $\beta_1 = \mu_1$ and $\beta_2 = \mu_2$, and consider n pairs of (x_1, x_2) that equal $(1, 0)$ and m pairs that equal $(0, 1)$. This would require each of the first n variables $Y_1, Y_2, \ldots, Y_n$ to have the mean

$$\beta_1 \cdot 1 + \beta_2 \cdot 0 = \beta_1 = \mu_1$$

and the next m variables $Y_{n+1}, Y_{n+2}, \ldots, Y_{n+m}$ to have the mean

$$\beta_1 \cdot 0 + \beta_2 \cdot 1 = \beta_2 = \mu_2.$$

This is the background of the two-sample problem, but with the usual $X_1, X_2, \ldots, X_n$ and $Y_1, Y_2, \ldots, Y_m$ replaced by $Y_1, Y_2, \ldots, Y_n$ and $Y_{n+1}, Y_{n+2}, \ldots, Y_{n+m}$, respectively.

Exercises

7.6-1. The mean of Y when $x = 0$ in the simple linear regression model is $\alpha - \beta \bar{x} = \alpha_1$. The least squares estimator of α_1 is $\hat{\alpha} - \hat{\beta} \bar{x} = \hat{\alpha}_1$.

(a) Find the distribution of $\hat{\alpha}_1$ under the usual model assumptions.

(b) Obtain an expression for a $100(1 - \gamma)\%$ two-sided confidence interval for α_1.

7.6-2. Obtain a two-sided $100(1 - \gamma)\%$ prediction interval for the average of m future independent observations taken at the same X-value, x^*.

7.6-3. For the data given in Exercise 6.5-3, with the usual assumptions,

(a) Find a 95% confidence interval for $\mu(x)$ when $x = 68, 75$, and 82.

(b) Find a 95% prediction interval for Y when $x = 68, 75$, and 82.

7.6-4. For the data given in Exercise 6.5-4, with the usual assumptions,

(a) Find a 95% confidence interval for $\mu(x)$ when $x = 2, 3$, and 4.

(b) Find a 95% prediction interval for Y when $x = 2, 3$, and 4.

7.6-5. For the cigarette data in Exercise 6.5-7, with the usual assumptions,

(a) Find a 95% confidence interval for $\mu(x)$ when $x = 5, 10$, and 15.

(b) Determine a 95% prediction interval for Y when $x = 5, 10$, and 15.

7.6-6. A computer center recorded the number of programs it maintained during each of 10 consecutive years.

(a) Calculate the least squares regression line for the data shown.

(b) Plot the points and the line on the same graph.

(c) Find a 95% prediction interval for the number of programs in year 11 under the usual assumptions.

Year	Number of Programs
1	430
2	480
3	565
4	790
5	885
6	960
7	1200
8	1380
9	1530
10	1591

7.6-7. For the ACT scores in Exercise 6.5-6, with the usual assumptions,

(a) Find a 95% confidence interval for $\mu(x)$ when $x = 17, 20, 23, 26$, and 29.

(b) Determine a 90% prediction interval for Y when $x = 17, 20, 23, 26$, and 29.

7.6-8. By the method of least squares, fit the regression plane $y = \beta_1 + \beta_2 x_1 + \beta_3 x_2$ to the following 12 observations of (x_1, x_2, y): (1, 1, 6), (0, 2, 3), (3, 0, 10),

$(-2, 0, -4)$, $(-1, 2, 0)$, $(0, 0, 1)$, $(2, 1, 8)$, $(-1, -1, -2)$, $(0, -3, -3)$, $(2, 1, 5)$, $(1, 1, 1)$, $(-1, 0, -2)$.

7.6-9. By the method of least squares, fit the cubic equation $y = \beta_1 + \beta_2 x + \beta_3 x^2 + \beta_4 x^3$ to the following 10 observed data points (x, y): $(0, 1)$, $(-1, -3)$, $(0, 3)$, $(1, 3)$, $(-1, -1)$, $(2, 10)$, $(0, 0)$, $(-2, -9)$, $(-1, -2)$, $(2, 8)$.

7.6-10. We would like to fit the quadratic curve $y = \beta_1 + \beta_2 x + \beta_3 x^2$ to a set of points $(x_1, y_1), (x_2, y_2), \ldots, (x_n, y_n)$ by the method of least squares. To do this, let

$$h(\beta_1, \beta_2, \beta_3) = \sum_{i=1}^{n} (y_i - \beta_1 - \beta_2 x_i - \beta_3 x_i^2)^2.$$

(a) By setting the three first partial derivatives of h with respect to β_1, β_2, and β_3 equal to 0, show that β_1, β_2, and β_3 satisfy the following set of equations (called normal equations), all of which are sums going from 1 to n:

$$\beta_1 n + \beta_2 \sum x_i + \beta_3 \sum x_i^2 = \sum y_i;$$

$$\beta_1 \sum x_i + \beta_2 \sum x_i^2 + \beta_3 \sum x_i^3 = \sum x_i y_i;$$

$$\beta_1 \sum x_i^2 + \beta_2 \sum x_i^3 + \beta_3 \sum x_i^4 = \sum x_i^2 y_i.$$

(b) For the data

$(6.91, 17.52)$ $(4.32, 22.69)$ $(2.38, 17.61)$ $(7.98, 14.29)$

$(8.26, 10.77)$ $(2.00, 12.87)$ $(3.10, 18.63)$ $(7.69, 16.77)$

$(2.21, 14.97)$ $(3.42, 19.16)$ $(8.18, 11.15)$ $(5.39, 22.41)$

$(1.19, 7.50)$ $(3.21, 19.06)$ $(5.47, 23.89)$ $(7.35, 16.63)$

$(2.32, 15.09)$ $(7.54, 14.75)$ $(1.27, 10.75)$ $(7.33, 17.42)$

$(8.41, 9.40)$ $(8.72, 9.83)$ $(6.09, 22.33)$ $(5.30, 21.37)$

$(7.30, 17.36)$

$n = 25$, $\sum x_i = 133.34$, $\sum x_i^2 = 867.75$, $\sum x_i^3 = 6197.21$, $\sum x_i^4 = 46{,}318.88$, $\sum y_i = 404.22$, $\sum x_i y_i = 2138.38$, and $\sum x_i^2 y_i = 13{,}380.30$. Show that $a = -1.88$, $b = 9.86$, and $c = -0.995$.

(c) Plot the points and the linear regression line for these data.

(d) Calculate and plot the residuals. Does linear regression seem to be appropriate?

(e) Show that the least squares quadratic regression line is $\widehat{y} = -1.88 + 9.86x - 0.995x^2$.

(f) Plot the points and this least squares quadratic regression curve on the same graph.

(g) Plot the residuals for quadratic regression and compare this plot with that in part (d).

7.6-11. (The information presented in this exercise comes from the Westview Blueberry Farm and National Oceanic and Atmospheric Administration Reports [NOAA].) For the following paired data, (x, y), x gives the Holland, Michigan, rainfall for June, and y gives the blueberry production in thousands of pounds from the Westview Blueberry Farm:

$(4.11, 56.2)$	$(5.49, 45.3)$	$(5.35, 31.0)$	$(6.53, 30.1)$
$(5.18, 40.0)$	$(4.89, 38.5)$	$(2.09, 50.0)$	$(1.40, 45.8)$
$(4.52, 45.9)$	$(1.11, 32.4)$	$(0.60, 18.2)$	$(3.80, 56.1)$

The data are from 1971 to 1989 for those years in which the last frost occurred May 10 or earlier.

(a) Find the correlation coefficient for these data.

(b) Find the least squares regression line.

(c) Make a scatter plot of the data with the least squares regression line on the plot.

(d) Calculate and plot the residuals. Does linear regression seem to be appropriate?

(e) Find the least squares quadratic regression curve.

(f) Calculate and plot the residuals. Does quadratic regression seem to be appropriate?

(g) Give a short interpretation of your results.

7.6-12. Explain why the model $\mu(x) = \beta_1 e^{\beta_2 x}$ is not a linear model. Would taking the logarithms of both sides yield a linear model for $\ln \mu(x)$?

7.6-13. Show that

$$\sum_{i=1}^{n} [Y_i - \alpha - \beta(x_i - \bar{x})]^2$$

$$= n(\widehat{\alpha} - \alpha)^2 + (\widehat{\beta} - \beta)^2 \sum_{i=1}^{n} (x_i - \bar{x})^2$$

$$+ \sum_{i=1}^{n} [Y_i - \widehat{\alpha} - \widehat{\beta}(x_i - \bar{x})]^2.$$

7.6-14. Show that the endpoints for a $100(1 - \gamma)\%$ confidence interval for α are

$$\widehat{\alpha} \pm t_{\gamma/2}(n-2)\sqrt{\frac{\widehat{\sigma}^2}{n-2}}.$$

7.6-15. Show that a $100(1 - \gamma)\%$ confidence interval for σ^2 is

$$\left[\frac{n\widehat{\sigma}^2}{\chi_{\gamma/2}^2(n-2)}, \frac{n\widehat{\sigma}^2}{\chi_{1-\gamma/2}^2(n-2)} \right].$$

7.6-16. Find 95% confidence intervals for α, β, and σ^2 for the predicted and earned grades data in Exercise 6.5-4.

7.6-17. Find 95% confidence intervals for α, β, and σ^2 for the midterm and final exam scores data in Exercise 6.5-3.

7.6-18. Using the cigarette data in Exercise 6.5-7, find 95% confidence intervals for α, β, and σ^2 under the usual assumptions.

7.6-19. Using the data in Exercise 6.5-8(a), find 95% confidence intervals for α, β, and σ^2.

7.6-20. Using the ACT scores in Exercise 6.5-6, find 95% confidence intervals for α, β, and σ^2 under the usual assumptions.

7.7* RESAMPLING METHODS

Sampling and resampling methods have become more useful in recent years due to the power of computers. These methods are even used in introductory courses to convince students that statistics have distributions—that is, that statistics are random variables with distributions. At this stage in the book, the reader should be convinced that this is true, although we did use some sampling in Section 5.6 to help sell the idea that the sample mean has an approximate normal distribution.

Resampling methods, however, are used for more than showing that statistics have certain distributions. Rather, they are needed in finding approximate distributions of certain statistics that are used to make statistical inferences. We already know a great deal about the distribution of $\overline{X}$, and resampling methods are not needed for $\overline{X}$. In particular, $\overline{X}$ has an approximate normal distribution with mean μ and standard deviation $\sigma/\sqrt{n}$. Of course, if the latter is unknown, we can estimate it by $s/\sqrt{n}$ and note that $(\overline{X} - \mu)/(s/\sqrt{n})$ has an approximate $N(0, 1)$ distribution, provided that the sample size is large enough and the underlying distribution is not too badly skewed with a long, heavy tail.

We know something about the distribution of S^2 *if the random sample arises from a normal distribution* or one fairly close to it. However, the statistic S^2 is not very robust, in that its distribution changes a great deal as the underlying distribution changes. It is not like $\overline{X}$, which always has an approximate normal distribution, provided that the mean μ and variance σ^2 of the underlying distribution exist. So what do we do about distributions of statistics like the sample variance S^2, whose distribution depends so much on having a given underlying distribution? We use resampling methods that essentially substitute computation for theory. We need to have some idea about the distributions of these various estimators to find confidence intervals for the corresponding parameters.

Let us now explain resampling. Suppose that we need to find the distribution of some statistic, such as S^2, but we do not believe that we are sampling from a normal distribution. We observe the values of $X_1, X_2, \ldots, X_n$ to be $x_1, x_2, \ldots, x_n$. Actually, if we know nothing about the underlying distribution, then the empirical distribution found by placing the weight $1/n$ on each x_i is the best estimate of that distribution. Therefore, to get some idea about the distribution of S^2, let us take a random sample of size n from this empirical distribution; then we are sampling from the n values with replacement. We compute S^2 for that sample; say it is s_1^2. We then do it again, getting s_2^2. And again, we compute s_3^2. We continue to do this a large number of times, say, N, where N might be 1000, 2000, or even 10,000. Once we have these N values of S^2, we can construct a histogram, a stem-and-leaf display, or a q–q plot—anything to help us get some information about the distribution of S^2 when the sample arises from this empirical distribution, which is an estimate of the real underlying distribution. Clearly, we must use the computer for all of this sampling. We illustrate the resampling procedure by using, not S^2, but a statistic called the *trimmed mean*.

Although we usually do not know the underlying distribution, we state that, in this illustration, it is of the Cauchy type, because there are certain basic ideas we want to review or introduce for the first time. The pdf of the Cauchy is

$$f(x) = \frac{1}{\pi(1+x^2)}, \qquad -\infty < x < \infty.$$

The cdf is

$$F(x) = \int_{-\infty}^{x} \frac{1}{\pi(1+w^2)} \, dw = \frac{1}{\pi} \arctan x + \frac{1}{2}, \qquad -\infty < x < \infty.$$

If we want to generate some X-values that have this distribution, we let Y have the uniform distribution $U(0, 1)$ and define X by

$$Y = F(X) = \frac{1}{\pi} \arctan X + \frac{1}{2}$$

or, equivalently,

$$X = \tan\left[\pi\left(Y - \frac{1}{2}\right)\right].$$

We can generate 40 values of Y on the computer and then calculate the 40 values of X. Let us now add $\theta = 5$ to each X-value to create a sample from a Cauchy distribution with a median of 5. That is, we have a random sample of 40 W-values, where $W = X + 5$. We will consider some statistics used to estimate the median, θ, of this distribution. Of course, usually the value of the median is unknown, but here we know that it is equal to $\theta = 5$, and our statistics are estimates of this known number. These 40 values of W are as follows, after ordering:

−7.34	−5.92	−2.98	0.19	0.77	0.95	2.86	3.17	3.76	4.20
4.20	4.27	4.31	4.42	4.60	4.73	4.84	4.87	4.90	4.96
4.98	5.00	5.09	5.09	5.14	5.22	5.23	5.42	5.50	5.83
5.94	5.95	6.00	6.01	6.24	6.82	9.62	10.03	18.27	93.62

It is interesting to observe that many of these 40 values are between 3 and 7 and hence are close to $\theta = 5$; it is almost as if they had arisen from a normal distribution with mean $\mu = 5$ and $\sigma^2 = 1$. But then we note the outliers; these very large or small values occur because of the heavy and long tails of the Cauchy distribution and suggest that the sample mean $\overline{X}$ is not a very good estimator of the middle. And it is not in this sample, because $\bar{x} = 6.67$. In a more theoretical course, it can be shown that, due to the fact that the mean μ and the variance σ^2 do not exist for a Cauchy distribution, $\overline{X}$ is not any better than a single observation X_i in estimating the median θ. The sample median $\widetilde{m}$ is a much better estimate of θ, as it is not influenced by the outliers. Here the median equals 4.97, which is fairly close to 5. Actually, the maximum likelihood estimator found by maximizing

$$L(\theta) = \prod_{i=1}^{40} \frac{1}{\pi[1 + (x_i - \theta)^2]}$$

is extremely good but requires difficult numerical methods to compute. Then advanced theory shows that, in the case of a Cauchy distribution, a **trimmed mean**, found by ordering the sample, discarding the smallest and largest $3/8 = 37.5\%$ of the sample, and averaging the middle 25%, is almost as good as the maximum likelihood estimator but is much easier to compute. This trimmed mean is usually denoted by $\overline{X}_{0.375}$; we use $\overline{X}_t$ for brevity, and here $\overline{x}_t = 4.96$. For this sample, it is not quite as good as the median; but, for most samples, it is better. Trimmed means are often very useful and many times are used with a smaller trimming percentage. For example, in sporting events such as skating and diving, often the smallest and largest of the judges' scores are discarded.

For this Cauchy example, let us resample from the empirical distribution created by placing the "probability" 1/40 on each of our 40 observations. With each of these samples, we find our trimmed mean $\overline{X}_t$. That is, we order the observations of each resample and average the middle 25% of the order statistics—namely, the middle 10 order statistics. We do this $N = 1000$ times, thus obtaining $N = 1000$ values of $\overline{X}_t$. These values are summarized with the histogram in Figure 7.7-1(a).

From this resampling procedure, which is called **bootstrapping**, we have some idea about the distribution if the sample arises from the empirical distribution and, hopefully, from the underlying distribution, which is approximated by the empirical distribution. While the distribution of the sample mean $\overline{X}$ is not normal if the sample arises from a Cauchy-type distribution, the approximate distribution of $\overline{X}_t$ is normal. From the histogram of trimmed mean values in Figure 7.7-1(a), that looks to be the case. This observation is supported by the q–q plot in Figure 7.7-1(b) of the quantiles of a standard normal distribution versus those of the 1000 $\overline{x}_t$-values: The plot is very close to being a straight line.

How do we find a confidence interval for θ? Recall that the middle of the distribution of $\overline{X}_t - \theta$ is zero. So a guess at θ would be the amount needed to move the histogram of $\overline{X}_t$-values over so that zero is more or less in the middle of the translated histogram. We recognize that this histogram was generated from the original sample $X_1, X_2, \ldots, X_{40}$ and thus is really only an estimate of the distribution of $\overline{X}_t$.

We could get a point estimate of θ by moving it over until its median (or mean) is at zero. Clearly, however, some error is incurred in doing so—and we really want some bounds for θ as given by a confidence interval.

To find that confidence interval, let us proceed as follows: In the $N = 1000$ resampled values of $\overline{X}_t$, we find two points—say, c and d—such that about 25 values are less than c and about 25 are greater than d. That is, c and d are about on

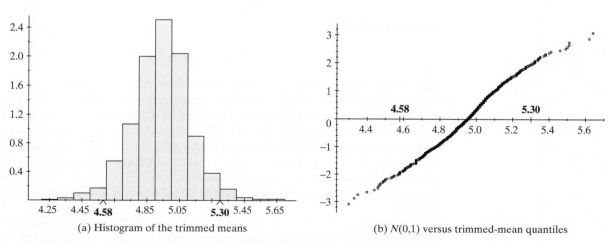

(a) Histogram of the trimmed means   (b) $N(0,1)$ versus trimmed-mean quantiles

Figure 7.7-1 $N = 1000$ observations of trimmed means

the respective 2.5th and 97.5th percentiles of the empirical distribution of these $N = 1000$ resampled $\overline{X}_t$-values. Thus, θ should be big enough so that over 2.5% of the $\overline{X}_t$-values are less than c and small enough so that over 2.5% of the $\overline{X}_t$-values are greater than d. This requires that $c < \theta$ and $\theta < d$; thus, $[c, d]$ serves as an approximate 95% confidence interval for θ as found by the **percentile method**. With our bootstrapped distribution of $N = 1000$ $\overline{X}_t$-values, this 95% confidence interval for θ runs from 4.58 to 5.30, and these two points are marked on the histogram and the q–q plot. Clearly, we could change this percentage to other values, such as 90%.

This percentile method, associated with the bootstrap method, is a nonparametric procedure, as we make no assumptions about the underlying distribution. It is interesting to compare the answer it produces with that obtained by using the order statistics $Y_1 < Y_2 < \cdots < Y_{40}$. If the sample arises from a continuous-type distribution, then, with the use of a calculator or computer, we have, when θ is the median,

$$P(Y_{14} < \theta < Y_{27}) = \sum_{k=14}^{26} \binom{40}{k} \left(\frac{1}{2}\right)^{40} = 0.9615.$$

(See Section 7.5.) Since, in our illustration, $Y_{14} = 4.42$ and $Y_{27} = 5.23$, the interval $[4.42, 5.23]$ is an approximate 96% confidence interval for θ. Of course, $\theta = 5$ is included in each of the two confidence intervals. In this case, the bootstrap confidence interval is a little more symmetric about $\theta = 5$ and somewhat shorter, but it did require much more work.

We have now illustrated bootstrapping, which allows us to substitute computation for theory to make statistical inferences about characteristics of the underlying distribution. This method is becoming more important as we encounter complicated data sets that clearly do not satisfy certain underlying assumptions. For example, consider the distribution of $T = (\overline{X} - \mu)/(S/\sqrt{n})$ when the random sample arises from an exponential distribution that has pdf $f(x) = e^{-x}$, $0 < x < \infty$, with mean $\mu = 1$. First, we will *not* use resampling, but we will simulate the distribution of T when the sample size $n = 16$ by taking $N = 1000$ random samples from this known exponential distribution. Here

$$F(x) = \int_0^x e^{-w} \, dw = 1 - e^{-x}, \qquad 0 < x < \infty.$$

So $Y = F(X)$ means

$$X = -\ln(1 - Y)$$

and X has that given exponential distribution with $\mu = 1$, provided that Y has the uniform distribution $U(0, 1)$. With the computer, we select $n = 16$ values of Y, determine the corresponding $n = 16$ values of X, and, finally, compute the value of $T = (\overline{X} - 1)/(S/\sqrt{16})$—say, T_1. We repeat this process over and over again, obtaining not only T_1, but also the values of $T_2, T_3, \ldots, T_{1000}$. We have done this and display the histogram of the 1000 T-values in Figure 7.7-2(a). Moreover the q–q plot with quantiles of $N(0, 1)$ on the y-axis is displayed in Figure 7.7-2(b). Both the histogram and the q–q plot show that the distribution of T in this case is skewed to the left.

In the preceding illustration, we knew the underlying distribution. Let us now sample from the exponential distribution with mean $\mu = 1$, but add a value θ to each X. Thus, we will try to estimate the new mean $\theta + 1$. The authors know the

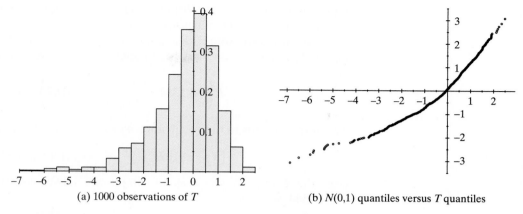

(a) 1000 observations of T (b) $N(0,1)$ quantiles versus T quantiles

Figure 7.7-2 T observations from an exponential distribution

value of θ, but the readers do not know it at this time. The observed 16 values of this random sample are

11.9776	9.3889	9.9798	13.4676	9.2895	10.1242	9.5798	9.3148
9.0605	9.1680	11.0394	9.1083	10.3720	9.0523	13.2969	10.5852

At this point we are trying to find a confidence interval for $\mu = \theta + 1$, and we pretend that we do not know that the underlying distribution is exponential. Actually, this is the case in practice: We do not know the underlying distribution. So we use the empirical distribution as the best guess of the underlying distribution; it is found by placing the weight 1/16 on each of the observations. The mean of this empirical distribution is $\bar{x} = 10.3003$. Therefore, we obtain some idea about the distribution of T by now simulating

$$T = \frac{\overline{X} - 10.3003}{S/\sqrt{16}}$$

with $N = 1000$ random samples from the empirical distribution.

We obtain $t_1, t_2, \ldots, t_{1000}$, and these values are used to construct a histogram, shown in Figure 7.7-3(a), and a q–q plot, illustrated in Figure 7.7-3(b). These two

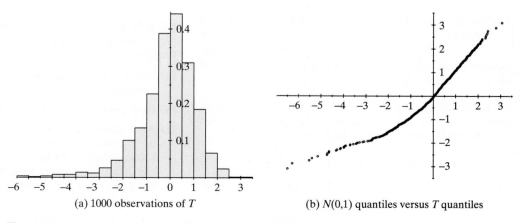

(a) 1000 observations of T (b) $N(0,1)$ quantiles versus T quantiles

Figure 7.7-3 T observations from an empirical distribution

figures look somewhat like those in Figure 7.7-2. Moreover, the 0.025th and 0.975th quantiles of the 1000 t-values are $c = -3.1384$ and $d = 1.8167$, respectively.

Now we have some idea about the 2.5th and 97.5th percentiles of the T distribution. Hence, as a very rough approximation, we can write

$$P\left(-3.1384 \leq \frac{\overline{X} - \mu}{S/\sqrt{16}} \leq 1.8167\right) \approx 0.95.$$

This formula leads to the rough approximate 95% confidence interval

$$[\overline{x} - 1.8167s/\sqrt{16}, \ \overline{x} - (-3.1384)s/\sqrt{16}]$$

once the $\overline{x}$ and s of the *original* sample are substituted. With $\overline{x} = 10.3003$ and $s = 1.4544$, we have

$$[10.3003 - 1.8167(1.4544)/4, \ 10.3003 + 3.1384(1.4544)/4] = [9.6397, 11.4414]$$

as a 95% approximate confidence interval for $\mu = \theta + 1$. Note that, because we added $\theta = 9$ to each x-value, the interval does cover $\theta + 1 = 10$.

It is easy to see how this procedure gets its name, because it is like "pulling yourself up by your own bootstraps," with the empirical distribution acting as the bootstraps.

Exercises

7.7-1. If time and computing facilities are available, consider the following 40 losses, due to wind-related catastrophes, that were recorded to the nearest \$1 million (these data include only those losses of \$2 million or more, and, for convenience, they have been ordered and recorded in millions of dollars):

2	2	2	2	2	2	2	2	2	2
2	2	3	3	3	3	4	4	4	5
5	5	5	6	6	6	6	8	8	9
15	17	22	23	24	24	25	27	32	43

To illustrate bootstrapping, take resamples of size $n = 40$ as many as $N = 100$ times, computing the value of $T = (\overline{X} - 5)/(S/\sqrt{40})$ each time. Here the value 5 is the median of the original sample. Construct a histogram of the bootstrapped values of T.

7.7-2. Consider the following 16 observed values, rounded to the nearest tenth, from the exponential distribution that was given in this section:

12.0	9.4	10.0	13.5	9.3	10.1	9.6	9.3
9.1	9.2	11.0	9.1	10.4	9.1	13.3	10.6

(a) Take resamples of size $n = 16$ from these observations about $N = 200$, times and compute s^2 each time. Construct a histogram of these 200 bootstrapped values of S^2.

(b) Simulate $N = 200$ random samples of size $n = 16$ from an exponential distribution with θ equal to the mean of the data in part (a) minus 9. For each sample, calculate the value of s^2. Construct a histogram of these 200 values of S^2.

(c) Construct a q–q plot of the two sets of sample variances and compare these two empirical distributions of S^2.

7.7-3. Refer to the data in Example 7.5-1 and take resamples of size $n = 9$ exactly $N = 1000$ times and compute the fifth order statistic, y_5, each time.

(a) Construct a histogram of these $N = 1000$ fifth order statistics.

(b) Find a point estimate of the median, $\pi_{0.50}$.

(c) Also, calculate a 96% confidence interval for $\pi_{0.50}$ by finding two numbers, the first of which has $(1000)(0.02) = 20$ values less than it and the second has 20 values greater than it. How does this interval compare to the one given in that example?

7.7-4. Refer to the data in Example 7.5-2 and take resamples of size $n = 27$ exactly $N = 500$ times and compute the seventh order statistic, y_7, each time.

(a) Construct a histogram of these $N = 500$ seventh order statistics.

(b) Give a point estimate of $\pi_{0.25}$.

(c) Find an 82% confidence interval for $\pi_{0.25}$ by finding two numbers, the first of which has $(500)(0.09) = 45$ values less than it and the second has 205 values greater than it.

(d) How does this interval compare to the one given in that example?

7.7-5. Let $X_1, X_2, \ldots, X_{21}$ and $Y_1, Y_2, \ldots, X_{21}$ be independent random samples of sizes $n = 21$ and $m = 21$ from $N(0,1)$ distributions. Then $F = S_X^2/S_Y^2$ has an F distribution with 20 and 20 degrees of freedom.

(a) Illustrate this situation empirically by simulating 100 observations of F.

   (i) Plot a relative frequency histogram with the $F(20, 20)$ pdf superimposed.

   (ii) Construct a q–q plot of the quantiles of $F(20, 20)$ versus the order statistics of your simulated data. Is the plot linear?

(b) Consider the following 21 observations of the $N(0,1)$ random variable X:

   0.1616 −0.8593 0.3105 0.3932 −0.2357 0.9697 1.3633

  −0.4166 0.7540 −1.0570 −0.1287 −0.6172 0.3208 0.9637

   0.2494 −1.1907 −2.4699 −0.1931 1.2274 −1.2826 −1.1532

Consider also the following 21 observations of the $N(0,1)$ random variable Y:

   0.4419 −0.2313 0.9233 −0.1203 1.7659 −0.2022 0.9036

  −0.4996 −0.8778 −0.8574 2.7574 1.1033 0.7066 1.3595

  −0.0056 −0.5545 −0.1491 −0.9774 −0.0868 1.7462 −0.2636

Sampling with replacement, resample with a sample of size 21 from each of these sets of observations. Calculate the value of $w = s_X^2/s_Y^2$. Repeat in order to simulate 100 observations of W from these two empirical distributions. Use the same graphical comparisons that you used in part (a) to see if the 100 observations represent observations from an approximate $F(20, 20)$ distribution.

(c) Consider the following 21 observations of the exponential random variable X with mean 1:

   0.6958 1.6394 0.2464 1.5827 0.0201 0.4544 0.8427

   0.6385 0.1307 1.0223 1.3423 1.6653 0.0081 5.2150

   0.5453 0.08440 1.2346 0.5721 1.5167 0.4843 0.9145

Consider also the following 21 observations of the exponential random variable Y with mean 1:

   1.1921 0.3708 0.0874 0.5696 0.1192 0.0164 1.6482

   0.2453 0.4522 3.2312 1.4745 0.8870 2.8097 0.8533

   0.1466 0.9494 0.0485 4.4379 1.1244 0.2624 1.3655

Sampling with replacement, resample with a sample of size 21 from each of these sets of observations. Calculate the value of $w = s_X^2/s_Y^2$. Repeat in order to simulate 100 observations of W from these two empirical distributions. Use the same graphical comparisons that you used in part (a) to see if the 100 observations represent observations from an approximate $F(20, 20)$ distribution.

7.7-6. The following 54 pairs of data give, for Old Faithful geyser, the duration in minutes of an eruption and the time in minutes until the next eruption:

(2.500, 72) (4.467, 88) (2.333, 62) (5.000, 87) (1.683, 57) (4.500, 94)

(4.500, 91) (2.083, 51) (4.367, 98) (1.583, 59) (4.500, 93) (4.550, 86)

(1.733, 70) (2.150, 63) (4.400, 91) (3.983, 82) (1.767, 58) (4.317, 97)

(1.917, 59) (4.583, 90) (1.833, 58) (4.767, 98) (1.917, 55) (4.433, 107)

(1.750, 61) (4.583, 82) (3.767, 91) (1.833, 65) (4.817, 97) (1.900, 52)

(4.517, 94) (2.000, 60) (4.650, 84) (1.817, 63) (4.917, 91) (4.000, 83)

(4.317, 84) (2.133, 71) (4.783, 83) (4.217, 70) (4.733, 81) (2.000, 60)

(4.717, 91) (1.917, 51) (4.233, 85) (1.567, 55) (4.567, 98) (2.133, 49)

(4.500, 85) (1.717, 65) (4.783, 102) (1.850, 56) (4.583, 86) (1.733, 62)

(a) Calculate the correlation coefficient, and construct a scatterplot, of these data.

(b) To estimate the distribution of the correlation coefficient, R, resample 500 samples of size 54 from the empirical distribution, and for each sample, calculate the value of R.

(c) Construct a histogram of these 500 observations of R.

(d) Simulate 500 samples of size 54 from a bivariate normal distribution with correlation coefficient equal to the correlation coefficient of the geyser data. For each sample of 54, calculate the correlation coefficient.

(e) Construct a histogram of the 500 observations of the correlation coefficient.

(f) Construct a q–q plot of the 500 observations of R from the bivariate normal distribution of part (d) versus the 500 observations in part (b). Do the two distributions of R appear to be about equal?

HISTORICAL COMMENTS One topic among many important ones in this chapter is regression, a technique that leads to a mathematical model of the result of some process in terms of some associated (explanatory) variables. We create such models to give us some idea of the value of a response variable if we know the values of certain explanatory variables. If we have an idea of the form of the equation relating these variables, then we can "fit" this model to the data; that is, we can determine approximate values for the unknown parameters in the model from the data. Now, no model is exactly correct; but, as the well-known statistician George Box observed, "Some are useful." That is, while models may be wrong and we should check them as best we can, they may be good enough approximations to shed some light on the issues of interest.

Once satisfactory models are found, they may be used

1. to determine the effect of each explanatory variable (some may have very little effect and can be dropped),
2. to estimate the response variable for given values of important explanatory variables,
3. to predict the future, such as upcoming sales (although this sometimes should be done with great care),
4. to often substitute a cheaper explanatory variable for an expensive one that is difficult to obtain [such as chemical oxygen demand (COD) for biological oxygen demand (BOD)].

The name *bootstrap* and the resulting technique were first used by Brad Efron of Stanford University. Efron knew that the expression "to pull oneself up by his or her own bootstraps" seems to come from *The Surprising Adventures of Baron Munchausen* by Rudolph Erich Raspe. The baron had fallen from the sky and found himself in a hole 9 fathoms deep and had no idea how to get out. He comments as follows: "Looking down I observed that I had on a pair of boots with exceptionally sturdy straps. Grasping them firmly, I pulled with all my might. Soon I had hoisted myself to the top and stepped out on terra firma without further ado."

Of course, in statistical *bootstrapping*, statisticians pull themselves up by their bootstraps (the empirical distributions) by recognizing that the empirical distribution is the best estimate of the underlying distribution without a lot of other assumptions. So they use the empirical distribution as if it is the underlying distribution to find approximate distributions of statistics of interest.

TESTS OF STATISTICAL HYPOTHESES

8.1 TESTS ABOUT ONE MEAN

We begin this chapter on tests of statistical hypotheses with an application in which we define many of the terms associated with testing.

Example 8.1-1

Let X equal the breaking strength of a steel bar. If the bar is manufactured by process I, X is $N(50, 36)$, i.e., X is normally distributed with $\mu = 50$ and $\sigma^2 = 36$. It is hoped that if process II (a new process) is used, X will be $N(55, 36)$. Given a large number of steel bars manufactured by process II, how could we test whether the five-unit increase in the mean breaking strength was realized?

In this problem, we are assuming that X is $N(\mu, 36)$ and μ is equal to 50 or 55. We want to test the **simple null hypothesis** H_0: $\mu = 50$ against the **simple alternative hypothesis** H_1: $\mu = 55$. Note that each of these hypotheses completely specifies the distribution of X. That is, H_0 states that X is $N(50, 36)$ and H_1 states that X is $N(55, 36)$. (If the alternative hypothesis had been H_1: $\mu > 50$, it would be a **composite hypothesis**, because it is composed of all normal distributions with $= 36$ and means greater than 50.) In order to test which of the two hypotheses, H_0 or H_1, is true, we shall set up a rule based on the breaking strengths $x_1, x_2, \ldots, x_n$ of n bars (the observed values of a random sample of size n from this new normal distribution). The rule leads to a decision to accept or reject H_0; hence, it is necessary to partition the sample space into two parts—say, C and C'—so that if $(x_1, x_2, \ldots, x_n) \in C$, H_0 is rejected, and if $(x_1, x_2, \ldots, x_n) \in C'$, H_0 is accepted (not rejected). The rejection region C for H_0 is called the **critical region** for the test. Often, the partitioning of the sample space is specified in terms of the values of a statistic called the **test statistic**. In this example, we could let $\overline{X}$ be the test statistic and, say, take $C = \{(x_1, x_2, \ldots, x_n) : \overline{x} \geq 53\}$; that is, we will reject H_0 if $\overline{x} \geq 53$. If $(x_1, x_2, \ldots, x_n) \in C$ when H_0 is true, H_0 would be rejected when it is true, a **Type I error**. If $(x_1, x_2, \ldots, x_n) \in C'$ when H_1 is true, H_0 would be accepted (i.e., not rejected) when in fact H_1 is true, a **Type II error**. The probability of a Type I error is called the **significance level** of the test and is denoted by α. That is, $\alpha = P[(X_1, X_2, \ldots, X_n) \in C; H_0]$ is the probability that $(X_1, X_2, \ldots, X_n)$

falls into C when H_0 is true. The probability of a Type II error is denoted by β; that is, $\beta = P[(X_1, X_2, \ldots, X_n) \in C'; H_1]$ is the probability of accepting (failing to reject) H_0 when it is false.

As an illustration, suppose $n = 16$ bars were tested and $C = \{\overline{x} : \overline{x} \geq 53\}$. Then $\overline{X}$ is $N(50, 36/16)$ when H_0 is true and is $N(55, 36/16)$ when H_1 is true. Thus,

$$\alpha = P(\overline{X} \geq 53; H_0) = P\left(\frac{\overline{X} - 50}{6/4} \geq \frac{53 - 50}{6/4}; H_0\right)$$

$$= 1 - \Phi(2) = 0.0228$$

and

$$\beta = P(\overline{X} < 53; H_1) = P\left(\frac{\overline{X} - 55}{6/4} < \frac{53 - 55}{6/4}; H_1\right)$$

$$= \Phi\left(-\frac{4}{3}\right) = 1 - 0.9087 = 0.0913.$$

Figure 8.1-1 shows the graphs of the probability density functions of $\overline{X}$ when H_0 and H_1, respectively, are true. Note that by changing the critical region, C, it is possible to decrease (increase) the size of α but this leads to an increase (decrease) in the size of β. Both α and β can be decreased if the sample size n is increased.   ■

Through another example, we define a *p*-value obtained in testing a hypothesis about a mean.

Example 8.1-2   Assume that the underlying distribution is normal with unknown mean μ but known variance $\sigma^2 = 100$. Say we are testing the simple null hypothesis H_0: $\mu = 60$ against the composite alternative hypothesis H_1: $\mu > 60$ with a sample mean $\overline{X}$ based on $n = 52$ observations. Suppose that we obtain the observed sample mean of $\overline{x} = 62.75$. If we compute the probability of obtaining an $\overline{X}$ of that value of 62.75 or greater when $\mu = 60$, then we obtain the *p*-**value** associated with $\overline{x} = 62.75$. That is,

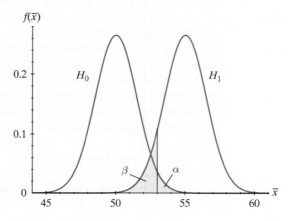

Figure 8.1-1 pdf of $\overline{X}$ under H_0 and H_1

$$p\text{-value} = P(\overline{X} \geq 62.75; \mu = 60)$$

$$= P\left(\frac{\overline{X} - 60}{10/\sqrt{52}} \geq \frac{62.75 - 60}{10/\sqrt{52}}; \mu = 60\right)$$

$$= 1 - \Phi\left(\frac{62.75 - 60}{10/\sqrt{52}}\right) = 1 - \Phi(1.983) = 0.0237.$$

If this p-value is small, we tend to reject the hypothesis $H_0: \mu = 60$. For example, rejecting $H_0: \mu = 60$ if the p-value is less than or equal to $\alpha = 0.05$ is exactly the same as rejecting H_0 if

$$\bar{x} \geq 60 + (1.645)\left(\frac{10}{\sqrt{52}}\right) = 62.281.$$

Here

$$p\text{-value} = 0.0237 < \alpha = 0.05 \qquad \text{and} \qquad \bar{x} = 62.75 > 62.281.$$

To help the reader keep the definition of p-value in mind, we note that it can be thought of as that **tail-end probability**, under H_0, of the distribution of the statistic (here $\overline{X}$) beyond the observed value of the statistic. (See Figure 8.1-2 for the p-value associated with $\bar{x} = 62.75$.)

If the alternative were the two-sided $H_1: \mu \neq 60$, then the p-value would have been double 0.0237; that is, then the p-value $= 2(0.0237) = 0.0474$ because we include both tails. ∎

When we sample from a normal distribution, the null hypothesis is generally of the form $H_0: \mu = \mu_0$. There are three possibilities of interest for a composite alternative hypothesis: (i) that μ has increased, or $H_1: \mu > \mu_0$; (ii) that μ has decreased, or $H_1: \mu < \mu_0$; and (iii) that μ has changed, but it is not known whether it has increased or decreased, which leads to the two-sided alternative hypothesis, or $H_1: \mu \neq \mu_0$.

To test $H_0: \mu = \mu_0$ against one of these three alternative hypotheses, a random sample is taken from the distribution and an observed sample mean, $\bar{x}$, that is close to μ_0 supports H_0. The closeness of $\bar{x}$ to μ_0 is measured in terms of standard deviations of $\overline{X}$, $\sigma/\sqrt{n}$, when σ is known, a measure that is sometimes called the **standard error of the mean**. Thus, the test statistic could be defined by

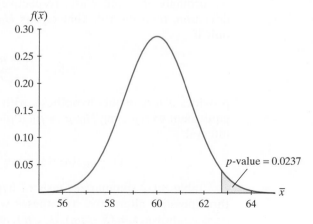

Figure 8.1-2 Illustration of p-value

Table 8.1-1 Tests of hypotheses about one mean, variance known

H_0	H_1	Critical Region
$\mu = \mu_0$	$\mu > \mu_0$	$z \geq z_\alpha$ or $\overline{x} \geq \mu_0 + z_\alpha \sigma / \sqrt{n}$
$\mu = \mu_0$	$\mu < \mu_0$	$z \leq -z_\alpha$ or $\overline{x} \leq \mu_0 - z_\alpha \sigma / \sqrt{n}$
$\mu = \mu_0$	$\mu \neq \mu_0$	$|z| \geq z_{\alpha/2}$ or $|\overline{x} - \mu_0| \geq z_{\alpha/2} \sigma / \sqrt{n}$

$$Z = \frac{\overline{X} - \mu_0}{\sqrt{\sigma^2/n}} = \frac{\overline{X} - \mu_0}{\sigma/\sqrt{n}}, \tag{8.1-1}$$

and the critical regions, at a significance level α, for the three respective alternative hypotheses would be (i) $z \geq z_\alpha$, (ii) $z \leq -z_\alpha$, and (iii) $|z| \geq z_{\alpha/2}$. In terms of $\overline{x}$, these three critical regions become (i) $\overline{x} \geq \mu_0 + z_\alpha(\sigma/\sqrt{n})$, (ii) $\overline{x} \leq \mu_0 - z_\alpha(\sigma/\sqrt{n})$, and (iii) $|\overline{x} - \mu_0| \geq z_{\alpha/2}(\sigma/\sqrt{n})$.

The three tests and critical regions are summarized in Table 8.1-1. The underlying assumption is that the distribution is $N(\mu, \sigma^2)$ and σ^2 is known.

It is usually the case that the variance σ^2 is not known. Accordingly, we now take a more realistic position and assume that the variance is unknown. Suppose our null hypothesis is $H_0: \mu = \mu_0$ and the two-sided alternative hypothesis is $H_1: \mu \neq \mu_0$. Recall from Section 7.1, for a random sample $X_1, X_2, \ldots, X_n$ taken from a normal distribution $N(\mu, \sigma^2)$, a confidence interval for μ is based on

$$T = \frac{\overline{X} - \mu}{\sqrt{S^2/n}} = \frac{\overline{X} - \mu}{S/\sqrt{n}}.$$

This suggests that T might be a good statistic to use for the test of $H_0: \mu = \mu_0$ with μ replaced by μ_0. In addition, it is the natural statistic to use if we replace σ^2/n by its unbiased estimator S^2/n in $(\overline{X} - \mu_0)/\sqrt{\sigma^2/n}$ in Equation 8.1-1. If $\mu = \mu_0$, we know that T has a t distribution with $n - 1$ degrees of freedom. Thus, with $\mu = \mu_0$,

$$P[\,|T| \geq t_{\alpha/2}(n-1)] = P\left[\frac{|\overline{X} - \mu_0|}{S/\sqrt{n}} \geq t_{\alpha/2}(n-1)\right] = \alpha.$$

Accordingly, if $\overline{x}$ and s are, respectively, the sample mean and sample standard deviation, then the rule that rejects $H_0: \mu = \mu_0$ and accepts $H_1: \mu \neq \mu_0$ if and only if

$$|t| = \frac{|\overline{x} - \mu_0|}{s/\sqrt{n}} \geq t_{\alpha/2}(n-1)$$

provides a test of this hypothesis with significance level α. Note that this rule is equivalent to rejecting $H_0: \mu = \mu_0$ if μ_0 is not in the open $100(1-\alpha)\%$ confidence interval

$$\left(\overline{x} - t_{\alpha/2}(n-1)\left[s/\sqrt{n}\,\right], \overline{x} + t_{\alpha/2}(n-1)\left[s/\sqrt{n}\,\right]\right).$$

Table 8.1-2 summarizes tests of hypotheses for a single mean, along with the three possible alternative hypotheses, when the underlying distribution is $N(\mu, \sigma^2)$, σ^2 is unknown, $t = (\overline{x} - \mu_0)/(s/\sqrt{n})$, and $n \leq 30$. If $n > 30$, we use Table 8.1-1 for approximate tests, with σ replaced by s.

Table 8.1-2 Tests of hypotheses for one mean, variance unknown		
H_0	H_1	Critical Region
$\mu = \mu_0$	$\mu > \mu_0$	$t \geq t_\alpha(n-1)$ or $\bar{x} \geq \mu_0 + t_\alpha(n-1)s/\sqrt{n}$
$\mu = \mu_0$	$\mu < \mu_0$	$t \leq -t_\alpha(n-1)$ or $\bar{x} \leq \mu_0 - t_\alpha(n-1)s/\sqrt{n}$
$\mu = \mu_0$	$\mu \neq \mu_0$	$|t| \geq t_{\alpha/2}(n-1)$ or $|\bar{x} - \mu_0| \geq t_{\alpha/2}(n-1)s/\sqrt{n}$

Example 8.1-3

Let X (in millimeters) equal the growth in 15 days of a tumor induced in a mouse. Assume that the distribution of X is $N(\mu, \sigma^2)$. We shall test the null hypothesis H_0: $\mu = \mu_0 = 4.0$ mm against the two-sided alternative hypothesis H_1: $\mu \neq 4.0$. If we use $n = 9$ observations and a significance level of $\alpha = 0.10$, the critical region is

$$|t| = \frac{|\bar{x} - 4.0|}{s/\sqrt{9}} \geq t_{\alpha/2}(8) = 1.860.$$

If we are given that $n = 9$, $\bar{x} = 4.3$, and $s = 1.2$, we see that

$$t = \frac{4.3 - 4.0}{1.2/\sqrt{9}} = \frac{0.3}{0.4} = 0.75.$$

Thus,

$$|t| = |0.75| < 1.860,$$

and we accept (do not reject) H_0: $\mu = 4.0$ at the $\alpha = 10\%$ significance level. (See Figure 8.1-3.) The p-value is the two-sided probability of $|T| \geq 0.75$, namely,

$$p\text{-value} = P(|T| \geq 0.75) = 2P(T \geq 0.75).$$

With our t tables with eight degrees of freedom, we cannot find this p-value exactly. It is about 0.50, because

$$P(|T| \geq 0.706) = 2P(T \geq 0.706) = 0.50.$$

However, Minitab gives a p-value of 0.4747. (See Figure 8.1-3.) ∎

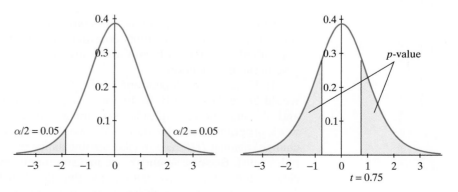

Figure 8.1-3 Test about mean of tumor growths

REMARK In discussing the test of a statistical hypothesis, the word *accept* H_0 might better be replaced by *do not reject* H_0. That is, if, in Example 8.1-3, $\bar{x}$ is close enough to 4.0 so that we accept $\mu = 4.0$, we do not want that acceptance to imply that μ is actually equal to 4.0. We want to say that the data do not deviate enough from $\mu = 4.0$ for us to reject that hypothesis; that is, we do not reject $\mu = 4.0$ with these observed data. With this understanding, we sometimes use *accept*, and sometimes *fail to reject* or *do not reject*, the null hypothesis. ∎

The next example illustrates the use of the t statistic with a one-sided alternative hypothesis.

Example 8.1-4

In attempting to control the strength of the wastes discharged into a nearby river, a paper firm has taken a number of measures. Members of the firm believe that they have reduced the oxygen-consuming power of their wastes from a previous mean μ of 500 (measured in parts per million of permanganate). They plan to test H_0: $\mu = 500$ against H_1: $\mu < 500$, using readings taken on $n = 25$ consecutive days. If these 25 values can be treated as a random sample, then the critical region, for a significance level of $\alpha = 0.01$, is

$$t = \frac{\bar{x} - 500}{s/\sqrt{25}} \leq -t_{0.01}(24) = -2.492.$$

The observed values of the sample mean and sample standard deviation were $\bar{x} = 308.8$ and $s = 115.15$. Since

$$t = \frac{308.8 - 500}{115.15/\sqrt{25}} = -8.30 < -2.492,$$

we clearly reject the null hypothesis and accept H_1: $\mu < 500$. Note, however, that although an improvement has been made, there still might exist the question of whether the improvement is adequate. The one-sided 99% confidence interval for μ, namely,

$$[0, 308.8 + 2.492(115.25/\sqrt{25})] = [0, 366.191],$$

provides an upper bound for μ and may help the company answer this question. ∎

Oftentimes, there is interest in comparing the means of two different distributions or populations. We must consider two situations: that in which X and Y are dependent and that in which X and Y are independent. We consider the independent case in the next section.

If X and Y are dependent, let $W = X - Y$, and the hypothesis that $\mu_X = \mu_Y$ would be replaced with the hypothesis H_0: $\mu_W = 0$. For example, suppose that X and Y equal the resting pulse rate for a person before and after taking an eight-week program in aerobic dance. We would be interested in testing H_0: $\mu_W = 0$ (no change) against H_1: $\mu_W > 0$ (the aerobic dance program decreased the mean resting pulse rate). Because X and Y are measurements on the same person, X and Y are clearly dependent. If we can assume that the distribution of W is (approximately) $N(\mu_W, \sigma^2)$, then we can choose to use the appropriate t test for a single mean from Table 8.1-2. This is often called a **paired t test**.

Example 8.1-5

Twenty-four girls in the 9th and 10th grades were put on an ultraheavy rope-jumping program. Someone thought that such a program would increase their speed in the 40-yard dash. Let W equal the difference in time to run the 40-yard dash—the "before-program time" minus the "after-program time." Assume that the distribution of W is approximately $N(\mu_W, \sigma_W^2)$. We shall test the null hypothesis $H_0\colon \mu_W = 0$ against the alternative hypothesis $H_1\colon \mu_W > 0$. The test statistic and the critical region that has an $\alpha = 0.05$ significance level are given by

$$t = \frac{\overline{w} - 0}{s_w/\sqrt{24}} \geq t_{0.05}(23) = 1.714.$$

The following data give the difference in time that it took each girl to run the 40-yard dash, with positive numbers indicating a faster time after the program:

0.28	0.01	0.13	0.33	−0.03	0.07	−0.18	−0.14
−0.33	0.01	0.22	0.29	−0.08	0.23	0.08	0.04
−0.30	−0.08	0.09	0.70	0.33	−0.34	0.50	0.06

For these data, $\overline{w} = 0.0788$ and $s_w = 0.2549$. Thus, the observed value of the test statistic is

$$t = \frac{0.0788 - 0}{0.2549/\sqrt{24}} = 1.514.$$

Since $1.514 < 1.714$, the null hypothesis is not rejected. Note, however, that $t_{0.10}(23) = 1.319$ and $t = 1.514 > 1.319$. Hence, the null hypothesis would be rejected at an $\alpha = 0.10$ significance level. Another way of saying this is that

$$0.05 < p\text{-value} < 0.10.$$

It would be instructive to draw a figure illustrating this double inequality. ∎

There are two ways of viewing a statistical test. One of these is through the p-value of the test; this approach is becoming more popular and is included in most computer printouts, so we mention it again. After observing the test statistic, we can say that the p-value is the probability, under the hypothesis H_0, of the test statistic being at least as extreme (in the direction of rejection of H_0) as the observed one. That is, the p-value is the tail-end probability. As an illustration, say a golfer averages about 90 for an 18-hole round, with a standard deviation of 3, and she takes some lessons to improve. To test her possible improvement, namely, $H_0\colon \mu = 90$, against $H_1\colon \mu < 90$, she plays $n = 16$ rounds of golf. Assume a normal distribution with $\sigma = 3$. If the golfer averaged $\overline{x} = 87.9375$, then

$$p\text{-value} = P(\overline{X} \leq 87.9375) = P\left(\frac{\overline{X} - 90}{3/4} \leq \frac{87.9375 - 90}{3/4}\right) = 0.0030.$$

The fact that the p-value is less than 0.05 is equivalent to the fact that $\overline{x} < 88.77$, because $P(\overline{X} \leq 88.77; \mu = 90) = 0.05$. Since $\overline{x} = 87.9375$ is an observed value of a random variable, namely, $\overline{X}$, it follows that the p-value, a function of $\overline{x}$, is also an observed value of a random variable. That is, before the random experiment is performed, the probability that the p-value is less than or equal to α is approximately equal to α when the null hypothesis is true. Many statisticians believe that the observed p-value provides an understandable measure of the truth of H_0: The smaller the p-value, the less they believe in H_0.

Two additional examples of the p-value may be based on Examples 8.1-3 and 8.1-4. In two-sided tests for means and proportions, the p-value is the probability of the extreme values in both directions. With the mouse data (Example 8.1-3), the p-value is

$$p\text{-value} = P(|T| \geq 0.75).$$

In Table VI in Appendix B, we see that if T has a t distribution with eight degrees of freedom, then $P(T \geq 0.706) = 0.25$. Thus, $P(|T| \geq 0.706) = 0.50$ and the p-value will be a little smaller than 0.50. In fact, $P(|T| \geq 0.75) = 0.4747$ (a probability that was found with Minitab), which is not less than $\alpha = 0.10$; hence, we do not reject H_0 at that significance level. In the example concerned with waste (Example 8.1-4), the p-value is essentially zero, since $P(T \leq -8.30) \approx 0$, where T has a t distribution with 24 degrees of freedom. Consequently, we reject H_0.

The other way of looking at tests of hypotheses is through the consideration of confidence intervals, particularly for two-sided alternatives and the corresponding tests. For example, with the mouse data (Example 8.1-3), a 90% confidence interval for the unknown mean is

$$4.3 \pm (1.86)(1.2)/\sqrt{9}, \qquad \text{or} \qquad [3.56, 5.04],$$

since $t_{0.05}(8) = 1.86$. Note that this confidence interval covers the hypothesized value $\mu = 4.0$ and we do not reject H_0: $\mu = 4.0$. If the confidence interval did not cover $\mu = 4.0$, then we would have rejected H_0: $\mu = 4.0$. Many statisticians believe that estimation is much more important than tests of hypotheses and accordingly approach statistical tests through confidence intervals. For one-sided tests, we use one-sided confidence intervals.

Exercises

8.1-1. Assume that IQ scores for a certain population are approximately $N(\mu, 100)$. To test H_0: $\mu = 110$ against the one-sided alternative hypothesis H_1: $\mu > 110$, we take a random sample of size $n = 16$ from this population and observe $\bar{x} = 113.5$.

(a) Do we accept or reject H_0 at the 5% significance level?

(b) Do we accept or reject H_0 at the 10% significance level?

(c) What is the p-value of this test?

8.1-2. Assume that the weight of cereal in a "12.6-ounce box" is $N(\mu, 0.2^2)$. The Food and Drug Association (FDA) allows only a small percentage of boxes to contain less than 12.6 ounces. We shall test the null hypothesis H_0: $\mu = 13$ against the alternative hypothesis H_1: $\mu < 13$.

(a) Use a random sample of $n = 25$ to define the test statistic and the critical region that has a significance level of $\alpha = 0.025$.

(b) If $\bar{x} = 12.9$, what is your conclusion?

(c) What is the p-value of this test?

8.1-3. Let X equal the Brinell hardness measurement of ductile iron subcritically annealed. Assume that the distribution of X is $N(\mu, 100)$. We shall test the null hypothesis H_0: $\mu = 170$ against the alternative hypothesis H_1: $\mu > 170$, using $n = 25$ observations of X.

(a) Define the test statistic and a critical region that has a significance level of $\alpha = 0.05$. Sketch a figure showing this critical region.

(b) A random sample of $n = 25$ observations of X yielded the following measurements:

170	167	174	179	179	156	163	156	187
156	183	179	174	179	170	156	187	
179	183	174	187	167	159	170	179	

Calculate the value of the test statistic and state your conclusion clearly.

(c) Give the approximate p-value of this test.

8.1-4. Let X equal the thickness of spearmint gum manufactured for vending machines. Assume that the distribution of X is $N(\mu, \sigma^2)$. The target thickness is 7.5

hundredths of an inch. We shall test the null hypothesis H_0: $\mu = 7.5$ against a two-sided alternative hypothesis, using 10 observations.

(a) Define the test statistic and critical region for an $\alpha = 0.05$ significance level. Sketch a figure illustrating this critical region.

(b) Calculate the value of the test statistic and state your decision clearly, using the following $n = 10$ thicknesses in hundredths of an inch for pieces of gum that were selected randomly from the production line:

$$7.65 \quad 7.60 \quad 7.65 \quad 7.70 \quad 7.55$$

$$7.55 \quad 7.40 \quad 7.40 \quad 7.50 \quad 7.50$$

(c) Is $\mu = 7.50$ contained in a 95% confidence interval for μ?

8.1-5. The mean birth weight of infants in the United States is $\mu = 3315$ grams. Let X be the birth weight (in grams) of a randomly selected infant in Jerusalem. Assume that the distribution of X is $N(\mu, \sigma^2)$, where μ and σ^2 are unknown. We shall test the null hypothesis H_0: $\mu = 3315$ against the alternative hypothesis H_1: $\mu < 3315$, using $n = 30$ randomly selected Jerusalem infants.

(a) Define a critical region that has a significance level of $\alpha = 0.05$.

(b) If the random sample of $n = 30$ yielded $\bar{x} = 3189$ and $s = 488$, what would be your conclusion?

(c) What is the approximate p-value of your test?

8.1-6. Let X equal the forced vital capacity (FVC) in liters for a female college student. (The FVC is the amount of air that a student can force out of her lungs.) Assume that the distribution of X is approximately $N(\mu, \sigma^2)$. Suppose it is known that $\mu = 3.4$ liters. A volleyball coach claims that the FVC of volleyball players is greater than 3.4. She plans to test her claim with a random sample of size $n = 9$.

(a) Define the null hypothesis.

(b) Define the alternative (coach's) hypothesis.

(c) Define the test statistic.

(d) Define a critical region for which $\alpha = 0.05$. Draw a figure illustrating your critical region.

(e) Calculate the value of the test statistic given that the random sample yielded the following FVCs:

$$3.4 \quad 3.6 \quad 3.8 \quad 3.3 \quad 3.4 \quad 3.5 \quad 3.7 \quad 3.6 \quad 3.7$$

(f) What is your conclusion?

(g) What is the approximate p-value of this test?

8.1-7. Vitamin B_6 is one of the vitamins in a multiple vitamin pill manufactured by a pharmaceutical company. The pills are produced with a mean of 50 mg of vitamin B_6 per pill. The company believes that there is a deterioration of 1 mg/month, so that after 3 months it expects that $\mu = 47$. A consumer group suspects that $\mu < 47$ after 3 months.

(a) Define a critical region to test H_0: $\mu = 47$ against H_1: $\mu < 47$ at an $\alpha = 0.05$ significance level based on a random sample of size $n = 20$.

(b) If the 20 pills yielded a mean of $\bar{x} = 46.94$ with a standard deviation of $s = 0.15$, what is your conclusion?

(c) What is the approximate p-value of this test?

8.1-8. A company that manufactures brackets for an automaker regularly selects brackets from the production line and performs a torque test. The goal is for mean torque to equal 125. Let X equal the torque and assume that X is $N(\mu, \sigma^2)$. We shall use a sample of size $n = 15$ to test H_0: $\mu = 125$ against a two-sided alternative hypothesis.

(a) Give the test statistic and a critical region with significance level $\alpha = 0.05$. Sketch a figure illustrating the critical region.

(b) Use the following observations to calculate the value of the test statistic and state your conclusion:

$$128 \quad 149 \quad 136 \quad 114 \quad 126 \quad 142 \quad 124 \quad 136$$

$$122 \quad 118 \quad 122 \quad 129 \quad 118 \quad 122 \quad 129$$

8.1-9. The ornamental ground cover *Vinca minor* is spreading rapidly through the Hope College Biology Field Station because it can outcompete the small, native woody vegetation. In an attempt to discover whether *Vinca minor* utilized natural chemical weapons to inhibit the growth of the native vegetation, Hope biology students conducted an experiment in which they treated 33 sunflower seedlings with extracts taken from *Vinca minor* roots for several weeks and then measured the heights of the seedlings. Let X equal the height of one of these seedlings and assume that the distribution of X is $N(\mu, \sigma^2)$. The observed growths (in cm) were

$$11.5 \quad 11.8 \quad 15.7 \quad 16.1 \quad 14.1 \quad 10.5 \quad 15.2 \quad 19.0 \quad 12.8 \quad 12.4 \quad 19.2$$

$$13.5 \quad 16.5 \quad 13.5 \quad 14.4 \quad 16.7 \quad 10.9 \quad 13.0 \quad 15.1 \quad 17.1 \quad 13.3 \quad 12.4$$

$$8.5 \quad 14.3 \quad 12.9 \quad 11.1 \quad 15.0 \quad 13.3 \quad 15.8 \quad 13.5 \quad 9.3 \quad 12.2 \quad 10.3$$

The students also planted some control sunflower seedlings that had a mean height of 15.7 cm. We shall test the null hypothesis H_0: $\mu = 15.7$ against the alternative hypothesis H_1: $\mu < 15.7$.

(a) Calculate the value of the test statistic and give limits for the p-value of this test.

(b) What is your conclusion?

(c) Find an approximate 98% one-sided confidence interval that gives an upper bound for μ.

8.1-10. In a mechanical testing lab, Plexiglass® strips are stretched to failure. Let X equal the change in length in mm before breaking. Assume that the distribution of X is $N(\mu, \sigma^2)$. We shall test the null hypothesis H_0: $\mu = 5.70$ against the alternative hypothesis H_1: $\mu > 5.70$, using $n = 8$ observations of X.

(a) Define the test statistic and a critical region that has a significance level of $\alpha = 0.05$. Sketch a figure showing this critical region.

(b) A random sample of eight observations of X yielded the following data:

$$5.71 \quad 5.80 \quad 6.03 \quad 5.87 \quad 6.22 \quad 5.92 \quad 5.57 \quad 5.83$$

Calculate the value of the test statistic and state your conclusion clearly.

(c) Give the approximate value of or bounds for the p-value of this test.

8.1-11. A vendor of milk products produces and sells low-fat dry milk to a company that uses it to produce baby formula. In order to determine the fat content of the milk, both the company and the vendor take an observation from each lot and test it for fat content in percent. Ten sets of paired test results are as follows:

Lot Number	Company Test Results (x)	Vendor Test Results (y)
1	0.50	0.79
2	0.58	0.71
3	0.90	0.82
4	1.17	0.82
5	1.14	0.73
6	1.25	0.77
7	0.75	0.72
8	1.22	0.79
9	0.74	0.72
10	0.80	0.91

Let μ_D denote the mean of the difference $x - y$. Test H_0: $\mu_D = 0$ against H_1: $\mu_D > 0$, using a paired t test with the differences. Let $\alpha = 0.05$.

8.1-12. To test whether a golf ball of brand A can be hit a greater distance off the tee than a golf ball of brand B, each of 17 golfers hit a ball of each brand, 8 hitting ball A before ball B and 9 hitting ball B before ball A. The results in yards are as follows:

Golfer	Distance for Ball A	Distance for Ball B	Golfer	Distance for Ball A	Distance for Ball B
1	265	252	10	274	260
2	272	276	11	274	267
3	246	243	12	269	267
4	260	246	13	244	251
5	274	275	14	212	222
6	263	246	15	235	235
7	255	244	16	254	255
8	258	245	17	224	231
9	276	259			

Assume that the differences of the paired A distance and B distance are approximately normally distributed, and test the null hypothesis H_0: $\mu_D = 0$ against the alternative hypothesis H_1: $\mu_D > 0$, using a paired t test with the 17 differences. Let $\alpha = 0.05$.

8.1-13. A company that manufactures motors receives reels of 10,000 terminals per reel. Before using a reel of terminals, 20 terminals are randomly selected to be tested. The test is the amount of pressure needed to pull the terminal apart from its mate. This amount of pressure should continue to increase from test to test as the terminal is "roughed up." (Since this kind of testing is destructive testing, a terminal that is tested cannot be used in a motor.) Let W equal the difference of the pressures: "test No. 1 pressure" minus "test No. 2 pressure." Assume that the distribution of W is $N(\mu_W, \sigma_W^2)$. We shall test the null hypothesis H_0: $\mu_W = 0$ against the alternative hypothesis H_1: $\mu_W < 0$, using 20 pairs of observations.

(a) Give the test statistic and a critical region that has a significance level of $\alpha = 0.05$. Sketch a figure illustrating this critical region.

(b) Use the following data to calculate the value of the test statistic, and state your conclusion clearly:

Terminal	Test 1	Test 2	Terminal	Test 1	Test 2
1	2.5	3.8	11	7.3	8.2
2	4.0	3.9	12	7.2	6.6
3	5.2	4.7	13	5.9	6.8
4	4.9	6.0	14	7.5	6.6
5	5.2	5.7	15	7.1	7.5
6	6.0	5.7	16	7.2	7.5
7	5.2	5.0	17	6.1	7.3
8	6.6	6.2	18	6.3	7.1
9	6.7	7.3	19	6.5	7.2
10	6.6	6.5	20	6.5	6.7

(c) What would the conclusion be if $\alpha = 0.01$?

(d) What is the approximate p-value of this test?

8.1-14. A researcher claims that she can reduce the variance of $N(\mu, 100)$ by a new manufacturing process. If S^2 is the variance of a random sample of size n from this new distribution, she tests $H_0: \sigma^2 = 100$ against $H_1: \sigma^2 < 100$ by rejecting H_0 if $(n-1)S^2/100 \leq \chi^2_{1-\alpha}(n-1)$ since $(n-1)S^2/100$ is $\chi^2(n-1)$ when H_0 is true.

(a) If $n = 23$, $s^2 = 32.52$, and $\alpha = 0.025$, would she reject H_0?

(b) Based on the same distributional result, what would be a reasonable test of $H_0: \sigma^2 = 100$ against a two-sided alternative hypothesis $H_1: \sigma^2 \neq 100$ when $\alpha = 0.05$?

8.1-15. Let $X_1, X_2, \ldots, X_{19}$ be a random sample of size $n = 19$ from the normal distribution $N(\mu, \sigma^2)$.

(a) Find a critical region, C, of size $\alpha = 0.05$ for testing $H_0: \sigma^2 = 30$ against $H_1: \sigma^2 = 80$.

(b) Find the approximate value of β, the probability of a Type II error, for the critical region C of part (a).

8.2 TESTS OF THE EQUALITY OF TWO MEANS

Let independent random variables X and Y have normal distributions $N(\mu_X, \sigma_X^2)$ and $N(\mu_Y, \sigma_Y^2)$, respectively. There are times when we are interested in testing whether the distributions of X and Y are the same. So if the assumption of normality is valid, we would be interested in testing whether the two means are equal. (A test for the equality of the two variances is given in the next section.)

When X and Y are independent and normally distributed, we can test hypotheses about their means with the same t statistic that we used to construct a confidence interval for $\mu_X - \mu_Y$ in Section 7.2. Recall that the t statistic used to construct the confidence interval assumed that the variances of X and Y were equal. (That is why we shall consider a test for the equality of two variances in the next section.)

We begin with an example and then give a table that lists some hypotheses and critical regions. A botanist is interested in comparing the growth response of dwarf pea stems against two different levels of the hormone indoleacetic acid (IAA). Using 16-day-old pea plants, the botanist obtains 5-mm sections and floats these sections on solutions with different hormone concentrations to observe the effect of the hormone on the growth of the pea stem. Let X and Y denote, respectively, the independent growths that can be attributed to the hormone during the first 26 hours after sectioning for $(0.5)(10)^{-4}$ and 10^{-4} levels of concentration of IAA. The botanist would like to test the null hypothesis $H_0: \mu_X - \mu_Y = 0$ against the alternative hypothesis $H_1: \mu_X - \mu_Y < 0$. If we can assume that X and Y are independent and normally distributed with a common variance, and if we assume respective random samples of sizes n and m, then we can find a test based on the statistic

$$T = \frac{\overline{X} - \overline{Y}}{\sqrt{\{[(n-1)S_X^2 + (m-1)S_Y^2]/(n+m-2)\}(1/n + 1/m)}} \tag{8.2-1}$$

$$= \frac{\overline{X} - \overline{Y}}{S_P\sqrt{1/n + 1/m}},$$

where

$$S_P = \sqrt{\frac{(n-1)S_X^2 + (m-1)S_Y^2}{n+m-2}}. \qquad (8.2\text{-}2)$$

Now, T has a t distribution with $r = n + m - 2$ degrees of freedom when H_0 is true and the variances are equal. Thus, the hypothesis H_0 will be rejected in favor of H_1 if the observed value of T is less than $-t_\alpha(n+m-2)$.

Example 8.2-1   In the preceding discussion, the botanist measured the growths of pea stem segments, in millimeters, for $n = 11$ observations of X:

$$0.8 \quad 1.8 \quad 1.0 \quad 0.1 \quad 0.9 \quad 1.7 \quad 1.0 \quad 1.4 \quad 0.9 \quad 1.2 \quad 0.5$$

She did the same with $m = 13$ observations of Y:

$$1.0 \quad 0.8 \quad 1.6 \quad 2.6 \quad 1.3 \quad 1.1 \quad 2.4$$

$$1.8 \quad 2.5 \quad 1.4 \quad 1.9 \quad 2.0 \quad 1.2$$

For these data, $\bar{x} = 1.03$, $s_x^2 = 0.24$, $\bar{y} = 1.66$, and $s_y^2 = 0.35$. The critical region for testing $H_0: \mu_X - \mu_Y = 0$ against $H_1: \mu_X - \mu_Y < 0$ is $t \leq -t_{0.05}(22) = -1.717$, where t is the two-sample t found in Equation 8.2-1. Since

$$t = \frac{1.03 - 1.66}{\sqrt{\{[10(0.24) + 12(0.35)]/(11 + 13 - 2)\}(1/11 + 1/13)}}$$

$$= -2.81 \; < \; -1.717,$$

H_0 is clearly rejected at an $\alpha = 0.05$ significance level. Notice that the approximate p-value of this test is 0.005, because $-t_{0.005}(22) = -2.819$. (See Figure 8.2-1.) Notice also that the sample variances do not differ too much; thus, most statisticians would use this two-sample t test.

It is instructive to construct box-and-whisker diagrams to gain a visual comparison of the two samples. For these two sets of data, the five-number summaries (minimum, three quartiles, maximum) are

$$0.1 \quad 0.8 \quad 1.0 \quad 1.4 \quad 1.8$$

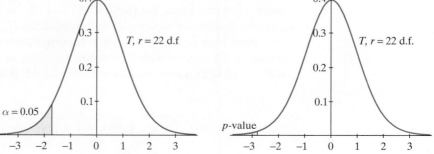

Figure 8.2-1 Critical region and p-value for pea stem growths

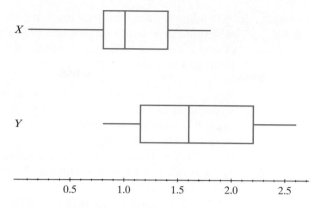

Figure 8.2-2 Box plots for pea stem growths

for the X sample and

$$0.8 \qquad 1.15 \qquad 1.6 \qquad 2.2 \qquad 2.6$$

for the Y sample. The two box plots are shown in Figure 8.2-2. ∎

Assuming independent random samples of sizes n and m, let $\bar{x}$, $\bar{y}$, and s_p^2 represent the observed unbiased estimates of the respective parameters μ_X, μ_Y, and $\sigma_X^2 = \sigma_Y^2$ of two normal distributions with a common variance. Then α-level tests of certain hypotheses are given in Table 8.2-1 when $\sigma_X^2 = \sigma_Y^2$. If the common-variance assumption is violated, but not too badly, the test is satisfactory, but the significance levels are only approximate. The t statistic and s_p are given in Equations 8.2-1 and 8.2-2, respectively.

REMARK Again, to emphasize the relationship between confidence intervals and tests of hypotheses, we note that each of the tests in Table 8.2-1 has a corresponding confidence interval. For example, the first one-sided test is equivalent to saying that we reject H_0: $\mu_X - \mu_Y = 0$ if zero is not in the one-sided confidence interval with lower bound

$$\bar{x} - \bar{y} - t_\alpha(n+m-2)s_p\sqrt{1/n + 1/m}.$$ ∎

Table 8.2-1 Tests of hypotheses for equality of two means

H_0	H_1	Critical Region
$\mu_X = \mu_Y$	$\mu_X > \mu_Y$	$t \geq t_\alpha(n+m-2)$ or
		$\bar{x} - \bar{y} \geq t_\alpha(n+m-2)s_p\sqrt{1/n + 1/m}$
$\mu_X = \mu_Y$	$\mu_X < \mu_Y$	$t \leq -t_\alpha(n+m-2)$ or
		$\bar{x} - \bar{y} \leq -t_\alpha(n+m-2)s_p\sqrt{1/n + 1/m}$
$\mu_X = \mu_Y$	$\mu_X \neq \mu_Y$	$|t| \geq t_{\alpha/2}(n+m-2)$ or
		$|\bar{x} - \bar{y}| \geq t_{\alpha/2}(n+m-2)s_p\sqrt{1/n + 1/m}$

Example 8.2-2

A product is packaged by a machine with 24 filler heads numbered 1 to 24, with the odd-numbered heads on one side of the machine and the even on the other side. Let X and Y equal the fill weights in grams when a package is filled by an odd-numbered head and an even-numbered head, respectively. Assume that the distributions of X and Y are $N(\mu_X, \sigma^2)$ and $N(\mu_Y, \sigma^2)$, respectively, and that X and Y are independent. We would like to test the null hypothesis H_0: $\mu_X - \mu_Y = 0$ against the alternative hypothesis H_1: $\mu_X - \mu_Y \neq 0$. To perform the test, after the machine has been set up and is running, we shall select one package at random from each filler head and weigh it. The test statistic is that given by Equation 8.2-1 with $n = m = 12$. At an $\alpha = 0.10$ significance level, the critical region is $|t| \geq t_{0.05}(22) = 1.717$.

For the $n = 12$ observations of X, namely,

1071	1076	1070	1083	1082	1067
1078	1080	1075	1084	1075	1080

$\bar{x} = 1076.75$ and $s_x^2 = 29.30$. For the $m = 12$ observations of Y, namely,

1074	1069	1075	1067	1068	1079
1082	1064	1070	1073	1072	1075

$\bar{y} = 1072.33$ and $s_y^2 = 26.24$. The calculated value of the test statistic is

$$t = \frac{1076.75 - 1072.33}{\sqrt{\dfrac{11(29.30) + 11(26.24)}{22}\left(\dfrac{1}{12} + \dfrac{1}{12}\right)}} = 2.05.$$

Since

$$|t| = |2.05| = 2.05 > 1.717,$$

the null hypothesis is rejected at an $\alpha = 0.10$ significance level. Note, however, that

$$|t| = 2.05 < 2.074 = t_{0.025}(22),$$

so that the null hypothesis would not be rejected at an $\alpha = 0.05$ significance level. That is, the p-value is between 0.05 and 0.10.

Again, it is instructive to construct box plots on the same graph for these two sets of data. The box plots in Figure 8.2-3 were constructed with the use of the five-number summary for the observations of X (1067, 1072, 1077, 1081.5, and 1084) and the five-number summary for the observations of Y (1064, 1068.25, 1072.5, 1075, and 1082). It looks like additional sampling would be advisable to test that the filler heads on the two sides of the machine are filling in a similar manner. If not, some corrective action needs to be taken. ∎

We would like to give two modifications of tests about two means. First, if we are able to assume that we know the variances of X and Y, then the appropriate test statistic to use for testing H_0: $\mu_X = \mu_Y$ is

$$Z = \frac{\overline{X} - \overline{Y}}{\sqrt{\dfrac{\sigma_X^2}{n} + \dfrac{\sigma_Y^2}{m}}}, \tag{8.2-3}$$

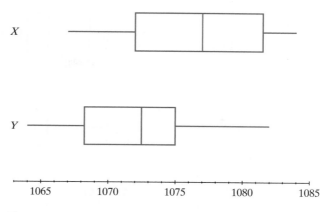

Figure 8.2-3 Box plots for fill weights

which has a standard normal distribution when the null hypothesis is true and, of course, when the populations are normally distributed. Second, if the variances are unknown and the sample sizes are large, replace σ_X^2 with S_X^2 and σ_Y^2 with S_Y^2 in Equation 8.2-3. The resulting statistic will have an approximate $N(0,1)$ distribution.

Example 8.2-3

The target thickness for Fruit Flavored Gum and for Fruit Flavored Bubble Gum is 6.7 hundredths of an inch. Let the independent random variables X and Y equal the respective thicknesses of these gums in hundredths of an inch, and assume that their distributions are $N(\mu_X, \sigma_X^2)$ and $N(\mu_Y, \sigma_Y^2)$, respectively. Because bubble gum has more elasticity than regular gum, it seems as if it would be harder to roll it out to the correct thickness. Thus, we shall test the null hypothesis H_0: $\mu_X = \mu_Y$ against the alternative hypothesis H_1: $\mu_X < \mu_Y$, using samples of sizes $n = 50$ and $m = 40$.

Because the variances are unknown and the sample sizes are large, the test statistic that is used is

$$Z = \frac{\overline{X} - \overline{Y}}{\sqrt{\dfrac{S_X^2}{50} + \dfrac{S_Y^2}{40}}}.$$

At an approximate significance level of $\alpha = 0.01$, the critical region is

$$z \leq -z_{0.01} = -2.326.$$

The observed values of X were

6.85 6.60 6.70 6.75 6.75 6.90 6.85 6.90 6.70 6.85

6.60 6.70 6.75 6.70 6.70 6.70 6.55 6.60 6.95 6.95

6.80 6.80 6.70 6.75 6.60 6.70 6.65 6.55 6.55 6.60

6.60 6.70 6.80 6.75 6.60 6.75 6.50 6.75 6.70 6.65

6.70 6.70 6.55 6.65 6.60 6.65 6.60 6.65 6.80 6.60

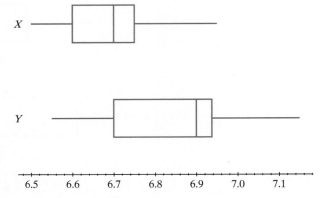

Figure 8.2-4 Box plots for gum thicknesses

for which $\bar{x} = 6.701$ and $s_x = 0.108$. The observed values of Y were

7.10	7.05	6.70	6.75	6.90	6.90	6.65	6.60	6.55	6.55
6.85	6.90	6.60	6.85	6.95	7.10	6.95	6.90	7.15	7.05
6.70	6.90	6.85	6.95	7.05	6.75	6.90	6.80	6.70	6.75
6.90	6.90	6.70	6.70	6.90	6.90	6.70	6.70	6.90	6.95

for which $\bar{y} = 6.841$ and $s_y = 0.155$. Since the calculated value of the test statistic is

$$z = \frac{6.701 - 6.841}{\sqrt{0.108^2/50 + 0.155^2/40}} = -4.848 < -2.326,$$

the null hypothesis is clearly rejected.

The box-and-whisker diagrams in Figure 8.2-4 were constructed with the use of the five-number summary of the observations of X (6.50, 6.60, 6.70, 6.75, and 6.95) and the five-number summary of the observations of Y (6.55, 6.70, 6.90, 6.94, and 7.15). This graphical display also confirms our conclusion. ■

REMARK To have satisfactory tests, our assumptions must be satisfied reasonably well. As long as the underlying distributions have finite means and variances and are not highly skewed, the normal assumptions are not too critical, as $\overline{X}$ and $\overline{Y}$ have approximate normal distributions by the central limit theorem. As distributions become nonnormal and highly skewed, the sample mean and sample variance become more dependent, and that causes problems in using the Student's t as an approximating distribution for T. In these cases, some of the nonparametric methods described later could be used. (See Section 8.4.)

When the distributions are close to normal, but the variances seem to differ by a great deal, the t statistic should again be avoided, particularly if the sample sizes are also different. In that case, use Z or the modification produced by substituting the sample variances for the distribution variances. In the latter situation, if n and m are large enough, there is no problem. With small n and m, most statisticians would use Welch's suggestion (or other modifications of it); that is, they would use an approximating Student's t distribution with r degrees of freedom, where r is given by Equation 7.2-1. We actually give a test for the equality of variances that could be employed to decide whether to use T or a modification of Z. However, most statisticians do not place much confidence in this test of $\sigma_X^2 = \sigma_Y^2$ and would use a modification of Z (possibly Welch's) if they suspected that the variances differed greatly. Alternatively, nonparametric methods described in Section 8.4 could be used. ∎

Exercises

(In some of the exercises that follow, we must make assumptions such as the existence of normal distributions with equal variances.)

8.2-1. The botanist in Example 8.2-1 is really interested in testing for synergistic interaction. That is, given the two hormones gibberellin (GA$_3$) and indoleacetic acid (IAA), let X_1 and X_2 equal the growth responses (in mm) of dwarf pea stem segments to GA$_3$ and IAA, respectively and separately. Also, let $X = X_1 + X_2$ and let Y equal the growth response when both hormones are present. Assuming that X is $N(\mu_X, \sigma^2)$ and Y is $N(\mu_Y, \sigma^2)$, the botanist is interested in testing the hypothesis H_0: $\mu_X = \mu_Y$ against the alternative hypothesis of synergistic interaction H_1: $\mu_X < \mu_Y$.

(a) Using $n = m = 10$ observations of X and Y, define the test statistic and the critical region. Sketch a figure of the t pdf and show the critical region on your figure. Let $\alpha = 0.05$.

(b) Given $n = 10$ observations of X, namely,

  2.1 2.6 2.6 3.4 2.1 1.7 2.6 2.6 2.2 1.2

and $m = 10$ observations of Y, namely,

  3.5 3.9 3.0 2.3 2.1 3.1 3.6 1.8 2.9 3.3

calculate the value of the test statistic and state your conclusion. Locate the test statistic on your figure.

(c) Construct two box plots on the same figure. Does this confirm your conclusion?

8.2-2. Let X and Y denote the weights in grams of male and female common gallinules, respectively. Assume that X is $N(\mu_X, \sigma_X^2)$ and Y is $N(\mu_Y, \sigma_Y^2)$.

(a) Given $n = 16$ observations of X and $m = 13$ observations of Y, define a test statistic and a critical region for testing the null hypothesis H_0: $\mu_X = \mu_Y$ against the one-sided alternative hypothesis H_1: $\mu_X > \mu_Y$. Let $\alpha = 0.01$. (Assume that the variances are equal.)

(b) Given that $\bar{x} = 415.16$, $s_x^2 = 1356.75$, $\bar{y} = 347.40$, and $s_y^2 = 692.21$, calculate the value of the test statistic and state your conclusion.

(c) Although we assumed that $\sigma_X^2 = \sigma_Y^2$, let us say we suspect that that equality is not valid. Thus, use the test proposed by Welch.

8.2-3. Let X equal the weight in grams of a Low-Fat Strawberry Kudo and Y the weight of a Low-Fat Blueberry Kudo. Assume that the distributions of X and Y are $N(\mu_X, \sigma_X^2)$ and $N(\mu_Y, \sigma_Y^2)$, respectively. Let

  21.7 21.0 21.2 20.7 20.4 21.9 20.2 21.6 20.6

be $n = 9$ observations of X, and let

  21.5   20.5   20.3   21.6   21.7   21.3   23.0

  21.3   18.9   20.0   20.4   20.8   20.3

be $m = 13$ observations of Y. Use these observations to answer the following questions:

(a) Test the null hypothesis H_0: $\mu_X = \mu_Y$ against a two-sided alternative hypothesis. You may select the significance level. Assume that the variances are equal.

(b) Construct and interpret box-and-whisker diagrams to support your conclusions.

8.2-4. Among the data collected for the World Health Organization air quality monitoring project is a measure of suspended particles, in $\mu g/m^3$. Let X and Y equal the concentration of suspended particles in $\mu g/m^3$ in the city centers (commercial districts), of Melbourne and Houston, respectively. Using $n = 13$ observations of X and $m = 16$ observations of Y, we shall test H_0: $\mu_X = \mu_Y$ against H_1: $\mu_X < \mu_Y$.

(a) Define the test statistic and critical region, assuming that the variances are equal. Let $\alpha = 0.05$.

(b) If $\bar{x} = 72.9$, $s_x = 25.6$, $\bar{y} = 81.7$, and $s_y = 28.3$, calculate the value of the test statistic and state your conclusion.

(c) Give bounds for the p-value of this test.

8.2-5. Some nurses in county public health conducted a survey of women who had received inadequate prenatal care. They used information from birth certificates to select mothers for the survey. The mothers selected were divided into two groups: 14 mothers who said they had five or fewer prenatal visits and 14 mothers who said they had six or more prenatal visits. Let X and Y equal the respective birth weights of the babies from these two sets of mothers, and assume that the distribution of X is $N(\mu_X, \sigma^2)$ and the distribution of Y is $N(\mu_Y, \sigma^2)$.

(a) Define the test statistic and critical region for testing H_0: $\mu_X - \mu_Y = 0$ against H_1: $\mu_X - \mu_Y < 0$. Let $\alpha = 0.05$.

(b) Given that the observations of X were

49	108	110	82	93	114	134
114	96	52	101	114	120	116

and the observations of Y were

133	108	93	119	119	98	106
131	87	153	116	129	97	110

calculate the value of the test statistic and state your conclusion.

(c) Approximate the p-value.

(d) Construct box plots on the same figure for these two sets of data. Do the box plots support your conclusion?

8.2-6. Let X and Y equal the forces required to pull stud No. 3 and stud No. 4 out of a window that has been manufactured for an automobile. Assume that the distributions of X and Y are $N(\mu_X, \sigma_X^2)$ and $N(\mu_Y, \sigma_Y^2)$, respectively.

(a) If $m = n = 10$ observations are selected randomly, define a test statistic and a critical region for testing H_0: $\mu_X - \mu_Y = 0$ against a two-sided alternative hypothesis. Let $\alpha = 0.05$. Assume that the variances are equal.

(b) Given $n = 10$ observations of X, namely,

111 120 139 136 138 149 143 145 111 123

and $m = 10$ observations of Y, namely,

152 155 133 134 119 155 142 146 157 149

calculate the value of the test statistic and state your conclusion clearly.

(c) What is the approximate p-value of this test?

(d) Construct box plots on the same figure for these two sets of data. Do the box plots confirm your decision in part (b)?

8.2-7. Let X and Y equal the number of milligrams of tar in filtered and nonfiltered cigarettes, respectively. Assume that the distributions of X and Y are $N(\mu_X, \sigma_X^2)$ and $N(\mu_Y, \sigma_Y^2)$, respectively. We shall test the null hypothesis H_0: $\mu_X - \mu_Y = 0$ against the alternative hypothesis H_1: $\mu_X - \mu_Y < 0$, using random samples of sizes $n = 9$ and $m = 11$ observations of X and Y, respectively.

(a) Define the test statistic and a critical region that has an $\alpha = 0.01$ significance level. Sketch a figure illustrating this critical region.

(b) Given $n = 9$ observations of X, namely,

0.9 1.1 0.1 0.7 0.4 0.9 0.8 1.0 0.4

and $m = 11$ observations of Y, namely,

1.5 0.9 1.6 0.5 1.4 1.9 1.0 1.2 1.3 1.6 2.1

calculate the value of the test statistic and state your conclusion clearly. Locate the value of the test statistic on your figure.

8.2-8. Let X and Y denote the tarsus lengths of male and female grackles, respectively. Assume that X is $N(\mu_X, \sigma_X^2)$ and Y is $N(\mu_Y, \sigma_Y^2)$. Given that $n = 25$, $\bar{x} = 33.80$, $s_x^2 = 4.88$, $m = 29$, $\bar{y} = 31.66$, and $s_y^2 = 5.81$, test the null hypothesis H_0: $\mu_X = \mu_Y$ against H_1: $\mu_X > \mu_Y$ with $\alpha = 0.01$.

8.2-9. When a stream is turbid, it is not completely clear due to suspended solids in the water. The higher the turbidity, the less clear is the water. A stream was studied on 26 days, half during dry weather (say, observations of X) and the other half immediately after a significant rainfall (say, observations of Y). Assume that the distributions of X and Y are $N(\mu_X, \sigma^2)$ and $N(\mu_Y, \sigma^2)$, respectively. The following turbidities were recorded in units of NTUs (nephelometric turbidity units):

x:	2.9	14.9	1.0	12.6	9.4	7.6	3.6
	3.1	2.7	4.8	3.4	7.1	7.2	

y:	7.8	4.2	2.4	12.9	17.3	10.4	5.9
	4.9	5.1	8.4	10.8	23.4	9.7	

(a) Test the null hypothesis H_0: $\mu_X = \mu_Y$ against H_1: $\mu_X < \mu_Y$. Give bounds for the p-value and state your conclusion.

(b) Draw box-and-whisker diagrams on the same graph. Does this figure confirm your answer?

8.2-10. Plants convert carbon dioxide (CO_2) in the atmosphere, along with water and energy from sunlight, into the energy they need for growth and reproduction. Experiments were performed under normal atmospheric air conditions and in air with enriched CO_2 concentrations to determine the effect on plant growth. The plants were given the same amount of water and light for a four-week period. The following table gives the plant growths in grams:

Normal Air 4.67 4.21 2.18 3.91 4.09 5.24 2.94 4.71
4.04 5.79 3.80 4.38

Enriched Air 5.04 4.52 6.18 7.01 4.36 1.81 6.22 5.70

On the basis of these data, determine whether CO_2-enriched atmosphere increases plant growth.

8.2-11. Let X equal the fill weight in April and Y the fill weight in June for an 8-pound box of bleach. We shall test the null hypothesis $H_0: \mu_X - \mu_Y = 0$ against the alternative hypothesis $H_1: \mu_X - \mu_Y > 0$ given that $n = 90$ observations of X yielded $\bar{x} = 8.10$ and $s_x = 0.117$ and $m = 110$ observations of Y yielded $\bar{y} = 8.07$ and $s_y = 0.054$.

(a) What is your conclusion if $\alpha = 0.05$?

HINT: Do the variances seem to be equal?

(b) What is the approximate p-value of this test?

8.2-12. Let X and Y denote the respective lengths of male and female green lynx spiders. Assume that the distributions of X and Y are $N(\mu_X, \sigma_X^2)$ and $N(\mu_Y, \sigma_Y^2)$, respectively, and that $\sigma_Y^2 > \sigma_X^2$. Thus, use the modification of Z to test the hypothesis $H_0: \mu_X - \mu_Y = 0$ against the alternative hypothesis $H_1: \mu_X - \mu_Y < 0$.

(a) Define the test statistic and a critical region that has a significance level of $\alpha = 0.025$.

(b) Using the data given in Exercise 7.2-5, calculate the value of the test statistic and state your conclusion.

(c) Draw two box-and-whisker diagrams on the same figure. Does your figure confirm the conclusion of this exercise?

8.2-13. Students looked at the effect of a certain fertilizer on plant growth. The students tested this fertilizer on one group of plants (Group A) and did not give fertilizer to a second group (Group B). The growths of the plants, in mm, over six weeks were as follows:

Group A: 55 61 33 57 17 46 50 42 71 51 63

Group B: 31 27 12 44 9 25 34 53 33 21 32

(a) Test the null hypothesis that the mean growths are equal against the alternative that the fertilizer enhanced growth. Assume that the variances are equal.

(b) Construct box plots of the two sets of growths on the same graph. Does this confirm your answer to part (a)?

8.2-14. An ecology laboratory studied tree dispersion patterns for the sugar maple, whose seeds are dispersed by the wind, and the American beech, whose seeds are dispersed by mammals. In a plot of area 50 m by 50 m, they measured distances between like trees, yielding the following distances in meters for 19 American beech trees and 19 sugar maple trees:

American beech:	5.00	5.00	6.50	4.25	4.25	8.80	6.50
	7.15	6.15	2.70	2.70	11.40	9.70	
	6.10	9.35	2.85	4.50	4.50	6.50	
sugar maple:	6.00	4.00	6.00	6.45	5.00	5.00	5.50
	2.35	2.35	3.90	3.90	5.35	3.15	
	2.10	4.80	3.10	5.15	3.10	6.25	

(a) Test the null hypothesis that the means are equal against the one-sided alternative that the mean for the distances between beech trees is greater than that between maple trees.

(b) Construct two box plots to confirm your answer.

8.2-15. Say X and Y are independent random variables with distributions that are $N(\mu_X, \sigma_X^2)$ and $N(\mu_Y, \sigma_Y^2)$. We wish to test $H_0: \sigma_X^2 = \sigma_Y^2$ against $H_1: \sigma_X^2 > \sigma_Y^2$.

(a) Argue that, if H_0 is true, the ratio of the two variances of the samples of sizes n and m, S_X^2/S_Y^2, has an $F(n-1, m-1)$ distribution.

(b) If $n = m = 31$, $\bar{x} = 8.153$, $s_x^2 = 1.410$, $\bar{y} = 5.917$, $s_y^2 = 0.4399$, $s_x^2/s_y^2 = 3.2053$, and $\alpha = 0.01$, show that H_0 is rejected and H_1 is accepted since $3.2053 > 2.39$.

(c) Where did the 2.39 come from?

8.2-16. To measure air pollution in a home, let X and Y equal the amount of suspended particulate matter (in $\mu g/m^3$) measured during a 24-hour period in a home in which there is no smoker and a home in which there is a smoker, respectively. We shall test the null hypothesis $H_0: \sigma_X^2/\sigma_Y^2 = 1$ against the one-sided alternative hypothesis $H_1: \sigma_X^2/\sigma_Y^2 > 1$.

(a) If a random sample of size $n = 9$ yielded $\bar{x} = 93$ and $s_x = 12.9$ while a random sample of size $m = 11$ yielded $\bar{y} = 132$ and $s_y = 7.1$, define a critical region and give your conclusion if $\alpha = 0.05$.

(b) Now test $H_0: \mu_X = \mu_Y$ against $H_1: \mu_X < \mu_Y$ if $\alpha = 0.05$.

8.2-17. Consider the distributions $N(\mu_X, 400)$ and $N(\mu_Y, 225)$. Let $\theta = \mu_X - \mu_Y$. Say $\bar{x}$ and $\bar{y}$ denote the observed means of two independent random samples, each of size n, from the respective distributions. Say we reject $H_0: \theta = 0$ and accept $H_1: \theta > 0$ if $\bar{x} - \bar{y} \geq c$. Let $K(\theta)$ be the power function of the test. Find n and c so that $K(0) = 0.05$ and $K(10) = 0.90$, approximately.

8.3 TESTS ABOUT PROPORTIONS

Suppose a manufacturer of a certain printed circuit observes that approximately a proportion $p = 0.06$ of the circuits fail. An engineer and statistician working together suggest some changes that might improve the design of the product. To test this new procedure, it was agreed that $n = 200$ circuits would be produced by the proposed method and then checked. Let Y equal the number of these 200 circuits that fail. Clearly, if the number of failures, Y, is such that $Y/200$ is about equal to 0.06, then it seems that the new procedure has not resulted in an improvement. Also, on the one hand, if Y is small, so that $Y/200$ is about 0.02 or 0.03, we might believe that the new method is better than the old. On the other hand, if $Y/200$ is 0.09 or 0.10, the proposed method has perhaps caused a greater proportion of failures.

What we need to establish is a formal rule that tells us when to accept the new procedure as an improvement. In addition, we must know the consequences of this rule. As an example of such a rule, we could accept the new procedure as an improvement if $Y \leq 7$ or $Y/n \leq 0.035$. We do note, however, that the probability of failure could still be about $p = 0.06$ even with the new procedure, and yet we could observe 7 or fewer failures in $n = 200$ trials. That is, we could erroneously accept the new method as being an improvement when, in fact, it was not. This decision is a mistake we call a Type I error. By contrast, the new procedure might actually improve the product so that p is much smaller, say, $p = 0.03$, and yet we could observe $y = 9$ failures, so that $y/200 = 0.045$. Thus, we could, again erroneously, not accept the new method as resulting in an improvement when, in fact, it had. This decision is a mistake we call a Type II error. We must study the probabilities of these two types of errors to understand fully the consequences of our rule.

Let us begin by modeling the situation. If we believe that these trials, conducted under the new procedure, are independent, and that each trial has about the same probability of failure, then Y is binomial $b(200, p)$. We wish to make a statistical inference about p using the unbiased estimator $\widehat{p} = Y/200$. Of course, we could construct a one-sided confidence interval—say, one that has 95% confidence of providing an upper bound for p—and obtain

$$\left[0, \widehat{p} + 1.645 \sqrt{\frac{\widehat{p}(1-\widehat{p})}{200}} \right].$$

This inference is appropriate and many statisticians simply make it. If the limits of this confidence interval contain 0.06, they would not say that the new procedure is necessarily better, at least until more data are taken. If, however, the upper limit of the confidence interval is less than 0.06, then those same statisticians would feel 95% confident that the true p is now less than 0.06. Hence, they would support the conclusion that the new procedure has improved the manufacturing of the printed circuits in question.

While this use of confidence intervals is highly appropriate, and later we indicate the relationship of confidence intervals to tests of hypotheses, every student of statistics should also have some understanding of the basic concepts in the latter area. Here, in our illustration, we are testing whether the probability of failure has or has not decreased from 0.06 when the new manufacturing procedure is used. The null hypothesis is $H_0: p = 0.06$ and the alternative hypothesis is $H_1: p < 0.06$. Since, in our illustration, we make a Type I error if $Y \leq 7$ when, in fact, $p = 0.06$, we can calculate the probability of this error. We denote that probability by α and call it the significance level of the test. Under our assumptions, it is

$$\alpha = P(Y \leq 7; p = 0.06) = \sum_{y=0}^{7} \binom{200}{y} (0.06)^y (0.94)^{200-y}.$$

Since n is rather large and p is small, these binomial probabilities can be approximated very well by Poisson probabilities with $\lambda = 200(0.06) = 12$. That is, from the Poisson table, the probability of a Type I error is

$$\alpha \approx \sum_{y=0}^{7} \frac{12^y e^{-12}}{y!} = 0.090.$$

Thus, the approximate significance level of this test is $\alpha = 0.090$. (Using the binomial distribution, we find that the exact value of α is 0.0829, which you can easily verify with Minitab.)

This value of α is reasonably small. However, what about the probability of a Type II error in case p has been improved to, say, 0.03? This error occurs if $Y > 7$ when, in fact, $p = 0.03$; hence, its probability, denoted by β, is

$$\beta = P(Y > 7; p = 0.03) = \sum_{y=8}^{200} \binom{200}{y}(0.03)^y(0.97)^{200-y}.$$

Again, we use the Poisson approximation, here with $\lambda = 200(0.03) = 6$, to obtain

$$\beta \approx 1 - \sum_{y=0}^{7} \frac{6^y e^{-6}}{y!} = 1 - 0.744 = 0.256.$$

(The binomial distribution tells us that the exact probability is 0.2539, so the approximation is very good.) The engineer and the statistician who created the new procedure probably are not too pleased with this answer. That is, they might note that if their new procedure of manufacturing circuits has actually decreased the probability of failure to 0.03 from 0.06 (*a big improvement*), there is still a good chance, 0.256, that $H_0: p = 0.06$ is accepted and their improvement rejected. In Section 8.5, more will be said about modifying tests so that satisfactory values of the probabilities of the two types of errors, namely, α and β, can be obtained; however, to decrease both of them, we need larger sample sizes.

Without worrying more about the probability of the Type II error here, we present a frequently used procedure for testing $H_0: p = p_0$, where p_0 is some specified probability of success. This test is based upon the fact that the number of successes Y in n independent Bernoulli trials is such that Y/n has an approximate normal distribution $N[p_0, p_0(1 - p_0)/n]$, provided that $H_0: p = p_0$ is true and n is large. Suppose the alternative hypothesis is $H_1: p > p_0$; that is, it has been hypothesized by a research worker that something has been done to increase the probability of success. Consider the test of $H_0: p = p_0$ against $H_1: p > p_0$ that rejects H_0 and accepts H_1 if and only if

$$Z = \frac{Y/n - p_0}{\sqrt{p_0(1 - p_0)/n}} \geq z_\alpha.$$

That is, if Y/n exceeds p_0 by z_α standard deviations of Y/n, we reject H_0 and accept the hypothesis $H_1: p > p_0$. Since, under H_0, Z is approximately $N(0, 1)$, the approximate probability of this occurring when $H_0: p = p_0$ is true is α. So the significance level of this test is approximately α.

If the alternative is $H_1: p < p_0$ instead of $H_1: p > p_0$, then the appropriate α-level test is given by $Z \leq -z_\alpha$. Hence, if Y/n is smaller than p_0 by z_α standard deviations of Y/n, we accept $H_1: p < p_0$.

Example 8.3-1 It was claimed that many commercially manufactured dice are not fair because the "spots" are really indentations, so that, for example, the 6-side is lighter than the 1-side. Let p equal the probability of rolling a 6 with one of these dice. To test H_0: $p = 1/6$ against the alternative hypothesis $H_1: p > 1/6$, several such dice will be rolled to yield a total of $n = 8000$ observations. Let Y equal the number of times that 6 resulted in the 8000 trials. The test statistic is

$$Z = \frac{Y/n - 1/6}{\sqrt{(1/6)(5/6)/n}} = \frac{Y/8000 - 1/6}{\sqrt{(1/6)(5/6)/8000}}.$$

If we use a significance level of $\alpha = 0.05$, the critical region is

$$z \geq z_{0.05} = 1.645.$$

The results of the experiment yielded $y = 1389$, so the calculated value of the test statistic is

$$z = \frac{1389/8000 - 1/6}{\sqrt{(1/6)(5/6)/8000}} = 1.67.$$

Since

$$z = 1.67 > 1.645,$$

the null hypothesis is rejected, and the experimental results indicate that these dice favor a 6 more than a fair die would. (You could perform your own experiment to check out other dice.) ∎

There are times when a two-sided alternative is appropriate; that is, here we test $H_0: p = p_0$ against $H_1: p \neq p_0$. For example, suppose that the pass rate in the usual beginning statistics course is p_0. There has been an intervention (say, some new teaching method) and it is not known whether the pass rate will increase, decrease, or stay about the same. Thus, we test the null (no-change) hypothesis $H_0: p = p_0$ against the two-sided alternative $H_1: p \neq p_0$. A test with the approximate significance level α for doing this is to reject $H_0: p = p_0$ if

$$|Z| = \frac{|Y/n - p_0|}{\sqrt{p_0(1 - p_0)/n}} \geq z_{\alpha/2},$$

since, under H_0, $P(|Z| \geq z_{\alpha/2}) \approx \alpha$. These tests of approximate significance level α are summarized in Table 8.3-1. The rejection region for H_0 is often called the critical region of the test, and we use that terminology in the table.

The p-value associated with a test is the probability, under the null hypothesis H_0, that the test statistic (a random variable) is equal to or exceeds the observed value (a constant) of the test statistic in the direction of the alternative hypothesis.

Table 8.3-1 Tests of hypotheses for one proportion

H_0	H_1	Critical Region
$p = p_0$	$p > p_0$	$z = \dfrac{y/n - p_0}{\sqrt{p_0(1 - p_0)/n}} \geq z_\alpha$
$p = p_0$	$p < p_0$	$z = \dfrac{y/n - p_0}{\sqrt{p_0(1 - p_0)/n}} \leq -z_\alpha$
$p = p_0$	$p \neq p_0$	$|z| = \dfrac{|y/n - p_0|}{\sqrt{p_0(1 - p_0)/n}} \geq z_{\alpha/2}$

Rather than select the critical region ahead of time, the p-value of a test can be reported and the reader then makes a decision. In Example 8.3-1, the value of the test statistic was $z = 1.67$. Because the alternative hypothesis was $H_1: p > 1/6$, the p-value is

$$P(Z \geq 1.67) = 0.0475.$$

Note that this p-value is less than $\alpha = 0.05$, which would lead to the rejection of H_0 at an $\alpha = 0.05$ significance level. If the alternative hypothesis were two sided, $H_1: p \neq 1/6$, then the p-value would be $P(|Z| \geq 1.67) = 0.095$ and would not lead to the rejection of H_0 at $\alpha = 0.05$.

Often there is interest in tests about p_1 and p_2, the probabilities of success for two different distributions or the proportions of two different populations having a certain characteristic. For example, if p_1 and p_2 denote the respective proportions of homeowners and renters who vote in favor of a proposal to reduce property taxes, a politician might be interested in testing $H_0: p_1 = p_2$ against the one-sided alternative hypothesis $H_1: p_1 > p_2$.

Let Y_1 and Y_2 represent, respectively, the numbers of observed successes in n_1 and n_2 independent trials with probabilities of success p_1 and p_2. Recall that the distribution of $\widehat{p}_1 = Y_1/n_1$ is approximately $N[p_1, p_1(1 - p_1)/n_1]$ and the distribution of $\widehat{p}_2 = Y_2/n_2$ is approximately $N[p_2, p_2(1 - p_2)/n_2]$. Thus, the distribution of $\widehat{p}_1 - \widehat{p}_2 = Y_1/n_1 - Y_2/n_2$ is approximately $N[p_1 - p_2, p_1(1 - p_1)/n_1 + p_2(1 - p_2)/n_2]$. It follows that the distribution of

$$Z = \frac{Y_1/n_1 - Y_2/n_2 - (p_1 - p_2)}{\sqrt{p_1(1 - p_1)/n_1 + p_2(1 - p_2)/n_2}} \qquad (8.3\text{-}1)$$

is approximately $N(0, 1)$. To test $H_0: p_1 - p_2 = 0$ or, equivalently, $H_0: p_1 = p_2$, let $p = p_1 = p_2$ be the common value under H_0. We shall estimate p with $\widehat{p} = (Y_1 + Y_2)/(n_1 + n_2)$. Replacing p_1 and p_2 in the denominator of Equation 8.3-1 with this estimate, we obtain the test statistic

$$Z = \frac{\widehat{p}_1 - \widehat{p}_2 - 0}{\sqrt{\widehat{p}(1 - \widehat{p})(1/n_1 + 1/n_2)}},$$

which has an approximate $N(0, 1)$ distribution for large sample sizes when the null hypothesis is true.

The three possible alternative hypotheses and their critical regions are summarized in Table 8.3-2.

Table 8.3-2 Tests of Hypotheses for two proportions

H_0	H_1	Critical Region
$p_1 = p_2$	$p_1 > p_2$	$z = \dfrac{\widehat{p}_1 - \widehat{p}_2}{\sqrt{\widehat{p}(1 - \widehat{p})(1/n_1 + 1/n_2)}} \geq z_\alpha$
$p_1 = p_2$	$p_1 < p_2$	$z = \dfrac{\widehat{p}_1 - \widehat{p}_2}{\sqrt{\widehat{p}(1 - \widehat{p})(1/n_1 + 1/n_2)}} \leq -z_\alpha$
$p_1 = p_2$	$p_1 \neq p_2$	$|z| = \dfrac{|\widehat{p}_1 - \widehat{p}_2|}{\sqrt{\widehat{p}(1 - \widehat{p})(1/n_1 + 1/n_2)}} \geq z_{\alpha/2}$

REMARK   In testing both $H_0: p = p_0$ and $H_0: p_1 = p_2$, statisticians sometimes use different denominators for z. For tests of single proportions, $\sqrt{p_0(1 - p_0)/n}$ can be replaced by $\sqrt{(y/n)(1 - y/n)/n}$, and for tests of the equality of two proportions, the following denominator can be used:

$$\sqrt{\frac{\widehat{p}_1(1 - \widehat{p}_1)}{n_1} + \frac{\widehat{p}_2(1 - \widehat{p}_2)}{n_2}}.$$

We do not have a strong preference one way or the other since the two methods provide about the same numerical result. The substitutions do provide better estimates of the standard deviations of the numerators when the null hypotheses are clearly false. There is some advantage to this result if the null hypothesis is likely to be false. In addition, the substitutions tie together the use of confidence intervals and tests of hypotheses. For example, if the null hypothesis is $H_0: p = p_0$, then the alternative hypothesis $H_1: p < p_0$ is accepted if

$$z = \frac{\widehat{p} - p_0}{\sqrt{\dfrac{\widehat{p}(1 - \widehat{p})}{n}}} \leq -z_\alpha.$$

This formula is equivalent to the statement that

$$p_0 \notin \left[0, \widehat{p} + z_\alpha \sqrt{\frac{\widehat{p}(1 - \widehat{p})}{n}} \right),$$

where the latter is a one-sided confidence interval providing an upper bound for p. Or if the alternative hypothesis is $H_1: p \neq p_0$, then H_0 is rejected if

$$\frac{|\widehat{p} - p_0|}{\sqrt{\dfrac{\widehat{p}(1 - \widehat{p})}{n}}} \geq z_{\alpha/2}.$$

This inequality is equivalent to

$$p_0 \notin \left(\widehat{p} - z_{\alpha/2} \sqrt{\frac{\widehat{p}(1 - \widehat{p})}{n}}, \widehat{p} + z_{\alpha/2} \sqrt{\frac{\widehat{p}(1 - \widehat{p})}{n}} \right),$$

where the latter is a confidence interval for p. However, using the forms given in Tables 8.3-1 and 8.3-2, we do get better approximations to α-level significance tests. Thus, there are trade-offs, and it is difficult to say that one is better than the other. Fortunately, the numerical answers are about the same.

In the second situation in which the estimates of p_1 and p_2 are the observed $\widehat{p}_1 = y_1/n_1$ and $\widehat{p}_2 = y_2/n_2$, we have, with large values of n_1 and n_2, an approximate 95% confidence interval for $p_1 - p_2$ given by

$$\frac{y_1}{n_1} - \frac{y_2}{n_2} \pm 1.96 \sqrt{\frac{(y_1/n_1)(1 - y_1/n_1)}{n_1} + \frac{(y_2/n_2)(1 - y_2/n_2)}{n_2}}.$$

If $p_1 - p_2 = 0$ is not in this interval, we reject H_0: $p_1 - p_2 = 0$ at the $\alpha = 0.05$ significance level. This is equivalent to saying that we reject H_0: $p_1 - p_2 = 0$ if

$$\frac{\left|\dfrac{y_1}{n_1} - \dfrac{y_2}{n_2}\right|}{\sqrt{\dfrac{(y_1/n_1)(1 - y_1/n_1)}{n_1} + \dfrac{(y_2/n_2)(1 - y_2/n_2)}{n_2}}} \geq 1.96.$$

In general, if the estimator $\widehat{\theta}$ (often, the maximum likelihood estimator) of θ has an approximate (sometimes exact) normal distribution $N(\theta, \sigma_{\widehat{\theta}}^2)$, then H_0: $\theta = \theta_0$ is rejected in favor of H_1: $\theta \neq \theta_0$ at the approximate (sometimes exact) α significance level if

$$\theta_0 \notin (\widehat{\theta} - z_{\alpha/2}\, \sigma_{\widehat{\theta}}, \widehat{\theta} + z_{\alpha/2}\, \sigma_{\widehat{\theta}})$$

or, equivalently,

$$\frac{|\widehat{\theta} - \theta_0|}{\sigma_{\widehat{\theta}}} \geq z_{\alpha/2}.$$

Note that $\sigma_{\widehat{\theta}}$ often depends upon some unknown parameter that must be estimated and substituted in $\sigma_{\widehat{\theta}}$ to obtain $\widehat{\sigma}_{\widehat{\theta}}$. Sometimes $\sigma_{\widehat{\theta}}$ or its estimate is called the **standard error** of $\widehat{\theta}$. This was the case in our last illustration when, with $\theta = p_1 - p_2$ and $\widehat{\theta} = \widehat{p_1} - \widehat{p_2}$, we substituted y_1/n_1 for p_1 and y_2/n_2 for p_2 in

$$\sqrt{\frac{p_1(1 - p_1)}{n_1} + \frac{p_2(1 - p_2)}{n_2}}$$

to obtain the standard error of $\widehat{p_1} - \widehat{p_2} = \widehat{\theta}$. ∎

Exercises

8.3-1. Let Y be $b(100, p)$. To test H_0: $p = 0.08$ against H_1: $p < 0.08$, we reject H_0 and accept H_1 if and only if $Y \leq 6$.

(a) Determine the significance level α of the test.

(b) Find the probability of the Type II error if, in fact, $p = 0.04$.

8.3-2. A bowl contains two red balls, two white balls, and a fifth ball that is either red or white. Let p denote the probability of drawing a red ball from the bowl. We shall test the simple null hypothesis H_0: $p = 3/5$ against the simple alternative hypothesis H_1: $p = 2/5$. Draw four balls at random from the bowl, one at a time and with replacement. Let X equal the number of red balls drawn.

(a) Define a critical region C for this test in terms of X.

(b) For the critical region C defined in part (a), find the values of α and β.

8.3-3. Let Y be $b(192, p)$. We reject H_0: $p = 0.75$ and accept H_1: $p > 0.75$ if and only if $Y \geq 152$. Use the normal approximation to determine

(a) $\alpha = P(Y \geq 152; p = 0.75)$.

(b) $\beta = P(Y < 152)$ when $p = 0.80$.

8.3-4. Let p denote the probability that, for a particular tennis player, the first serve is good. Since $p = 0.40$, this player decided to take lessons in order to increase p. When the lessons are completed, the hypothesis H_0: $p = 0.40$ will be tested against H_1: $p > 0.40$ on the basis of $n = 25$ trials. Let y equal the number of first serves that are good, and let the critical region be defined by $C = \{y : y \geq 13\}$.

(a) Determine $\alpha = P(Y \geq 13; p = 0.40)$. Use Table II in the appendix.

(b) Find $\beta = P(Y < 13)$ when $p = 0.60$; that is, $\beta = P(Y \leq 12; p = 0.60)$. Use Table II.

8.3-5. If a newborn baby has a birth weight that is less than 2500 grams (5.5 pounds), we say that the baby has a low birth weight. The proportion of babies with a low birth weight is an indicator of lack of nutrition for the mothers. For the United States, approximately 7% of babies have a low birth weight. Let p equal the proportion of babies born in the Sudan who weigh less than 2500 grams. We shall test the null hypothesis $H_0: p = 0.07$ against the alternative hypothesis $H_1: p > 0.07$. In a random sample of $n = 209$ babies, $y = 23$ weighed less than 2500 grams.

(a) What is your conclusion at a significance level of $\alpha = 0.05$?

(b) What is your conclusion at a significance level of $\alpha = 0.01$?

(c) Find the p-value for this test.

8.3-6. It was claimed that 75% of all dentists recommend a certain brand of gum for their gum-chewing patients. A consumer group doubted this claim and decided to test $H_0: p = 0.75$ against the alternative hypothesis $H_1: p < 0.75$, where p is the proportion of dentists who recommend that brand of gum. A survey of 390 dentists found that 273 recommended the given brand of gum.

(a) Which hypothesis would you accept if the significance level is $\alpha = 0.05$?

(b) Which hypothesis would you accept if the significance level is $\alpha = 0.01$?

(c) Find the p-value for this test.

8.3-7. The management of the Tigers baseball team decided to sell only low-alcohol beer in their ballpark to help combat rowdy fan conduct. They claimed that more than 40% of the fans would approve of this decision. Let p equal the proportion of Tiger fans on opening day who approved of the decision. We shall test the null hypothesis $H_0: p = 0.40$ against the alternative hypothesis $H_1: p > 0.40$.

(a) Define a critical region that has an $\alpha = 0.05$ significance level.

(b) If, out of a random sample of $n = 1278$ fans, $y = 550$ said that they approved of the new policy, what is your conclusion?

8.3-8. Let p equal the proportion of drivers who use a seat belt in a state that does not have a mandatory seat belt law. It was claimed that $p = 0.14$. An advertising campaign was conducted to increase this proportion. Two months after the campaign, $y = 104$ out of a random sample of $n = 590$ drivers were wearing their seat belts. Was the campaign successful?

(a) Define the null and alternative hypotheses.

(b) Define a critical region with an $\alpha = 0.01$ significance level.

(c) What is your conclusion?

8.3-9. According to a population census in 1986, the percentage of males who are 18 or 19 years old and are married was 3.7%. We shall test whether this percentage increased from 1986 to 1988.

(a) Define the null and alternative hypotheses.

(b) Define a critical region that has an approximate significance level of $\alpha = 0.01$. Sketch a standard normal pdf to illustrate this critical region.

(c) If $y = 20$ out of a random sample of $n = 300$ males, each 18 or 19 years old, were married (*U.S. Bureau of the Census, Statistical Abstract of the United States: 1988*), what is your conclusion? Show the calculated value of the test statistic on your figure in part (b).

8.3-10. Because of tourism in the state, it was proposed that public schools in Michigan begin after Labor Day. To determine whether support for this change was greater than 65%, a public poll was taken. Let p equal the proportion of Michigan adults who favor a post–Labor Day start. We shall test $H_0: p = 0.65$ against $H_1: p > 0.65$.

(a) Define a test statistic and an $\alpha = 0.025$ critical region.

(b) Given that 414 out of a sample of 600 favor a post–Labor Day start, calculate the value of the test statistic.

(c) Find the p-value and state your conclusion.

(d) Find a 95% one-sided confidence interval that gives a lower bound for p.

8.3-11. A machine shop that manufactures toggle levers has both a day and a night shift. A toggle lever is defective if a standard nut cannot be screwed onto the threads. Let p_1 and p_2 be the proportion of defective levers among those manufactured by the day and night shifts, respectively. We shall test the null hypothesis, $H_0: p_1 = p_2$, against a two-sided alternative hypothesis based on two random samples, each of 1000 levers taken from the production of the respective shifts.

(a) Define the test statistic and a critical region that has an $\alpha = 0.05$ significance level. Sketch a standard normal pdf illustrating this critical region.

(b) If $y_1 = 37$ and $y_2 = 53$ defectives were observed for the day and night shifts, respectively, calculate the value of the test statistic. Locate the calculated test statistic on your figure in part (a) and state your conclusion.

8.3-12. Let p equal the proportion of yellow candies in a package of mixed colors. It is claimed that $p = 0.20$.

(a) Define a test statistic and critical region with a significance level of $\alpha = 0.05$ for testing $H_0: p = 0.20$ against a two-sided alternative hypothesis.

(b) To perform the test, each of 20 students counted the number of yellow candies, y, and the total number of candies, n, in a 48.1-gram package, yielding the following ratios, y/n: 8/56, 13/55, 12/58, 13/56, 14/57, 5/54, 14/56, 15/57, 11/54, 13/55, 10/57, 8/59, 10/54, 11/55, 12/56, 11/57, 6/54, 7/58, 12/58, 14/58. If each individual tests $H_0: p = 0.20$, what proportion of the students rejected the null hypothesis?

(c) If we may assume that the null hypothesis is true, what proportion of the students would you have expected to reject the null hypothesis?

(d) For each of the 20 ratios in part (b), a 95% confidence interval for p can be calculated. What proportion of these 95% confidence intervals contain $p = 0.20$?

(e) If the 20 results are pooled so that $\sum_{i=1}^{20} y_i$ equals the number of yellow candies and $\sum_{i=1}^{20} n_i$ equals the total sample size, do we reject $H_0: p = 0.20$?

8.3-13. Let p_m and p_f be the respective proportions of male and female white-crowned sparrows that return to their hatching site. Give the endpoints for a 95% confidence interval for $p_m - p_f$ if 124 out of 894 males and 70 out of 700 females returned (*The Condor*, 1992, pp. 117–133). Does your result agree with the conclusion of a test of $H_0: p_1 = p_2$ against $H_1: p_1 \neq p_2$ with $\alpha = 0.05$?

8.3-14. For developing countries in Africa and the Americas, let p_1 and p_2 be the respective proportions of babies with a low birth weight (below 2500 grams). We shall test $H_0: p_1 = p_2$ against the alternative hypothesis $H_1: p_1 > p_2$.

(a) Define a critical region that has an $\alpha = 0.05$ significance level.

(b) If respective random samples of sizes $n_1 = 900$ and $n_2 = 700$ yielded $y_1 = 135$ and $y_2 = 77$ babies with a low birth weight, what is your conclusion?

(c) What would your decision be with a significance level of $\alpha = 0.01$?

(d) What is the p-value of your test?

8.3-15. Each of six students has a deck of cards and selects a card randomly from his or her deck.

(a) Show that the probability of at least one match is equal to 0.259.

(b) Now let each of the students randomly select an integer from 1–52, inclusive. Let p equal the probability of at least one match. Test the null hypothesis $H_0: p = 0.259$ against an appropriate alternative hypothesis. Give a reason for your alternative.

(c) Perform this experiment a large number of times. What is your conclusion?

8.3-16. Let p be the fraction of engineers who do not understand certain basic statistical concepts. Unfortunately, in the past, this number has been high, about $p = 0.73$. A new program to improve the knowledge of statistical methods has been implemented, and it is expected that under this program p would decrease from the aforesaid 0.73 value. To test $H_0: p = 0.73$ against $H_1: p < 0.73$, 300 engineers in the new program were tested and 204 (i.e., 68%) did not comprehend certain basic statistical concepts. Compute the p-value to determine whether this result indicates progress. That is, can we reject H_0 is favor of H_1? Use $\alpha = 0.05$.

8.4 THE WILCOXON TESTS

As mentioned earlier in the text, at times it is clear that the normality assumptions are not met and that other procedures, sometimes referred to as **nonparametric** or **distribution-free** methods, should be considered. For example, suppose some hypothesis, say, $H_0: m = m_0$, against $H_1: m \neq m_0$, is made about the unknown median, m, of a continuous-type distribution. From the data, we could construct a $100(1 - \alpha)\%$ confidence interval for m, and if m_0 is not in that interval, we would reject H_0 at the α significance level.

Now let X be a continuous-type random variable and let m denote the median of X. To test the hypothesis $H_0: m = m_0$ against an appropriate alternative hypothesis, we could also use a **sign test**. That is, if $X_1, X_2, \ldots, X_n$ denote the observations of a random sample from this distribution, and if we let Y equal the number of negative differences among $X_1 - m_0, X_2 - m_0, \ldots, X_n - m_0$, then Y has the binomial distribution $b(n, 1/2)$ under H_0 and is the test statistic for the sign test. If Y is too large or too small, we reject $H_0: m = m_0$.

Example 8.4-1

Let X denote the length of time in seconds between two calls entering a call center. Let m be the unique median of this continuous-type distribution. We test the null hypothesis H_0: $m = 6.2$ against the alternative hypothesis H_1: $m < 6.2$. Table II in Appendix B tells us that if Y is the number of lengths of time between calls in a random sample of size 20 that are less than 6.2, then the critical region $C = \{y : y \geq 14\}$ has a significance level of $\alpha = 0.0577$. A random sample of size 20 yielded the following data:

$$6.8 \quad 5.7 \quad 6.9 \quad 5.3 \quad 4.1 \quad 9.8 \quad 1.7 \quad 7.0$$

$$2.1 \quad 19.0 \quad 18.9 \quad 16.9 \quad 10.4 \quad 44.1 \quad 2.9 \quad 2.4$$

$$4.8 \quad 18.9 \quad 4.8 \quad 7.9$$

Since $y = 9$, the null hypothesis is not rejected. ∎

The sign test can also be used to test the hypothesis that two possibly dependent continuous-type random variables X and Y are such that $p = P(X > Y) = 1/2$. To test the hypothesis H_0: $p = 1/2$ against an appropriate alternative hypothesis, consider the independent pairs $(X_1, Y_1), (X_2, Y_2), \ldots, (X_n, Y_n)$. Let W denote the number of pairs for which $X_k - Y_k > 0$. When H_0 is true, W is $b(n, 1/2)$, and the test can be based upon the statistic W. For example, say X is the length of the right foot of a person and Y the length of the corresponding left foot. Thus, there is a natural pairing, and here H_0: $p = P(X > Y) = 1/2$ suggests that either foot of a particular individual is equally likely to be longer.

One major objection to the sign test is that it does not take into account the magnitude of the differences $X_1 - m_0, \ldots, X_n - m_0$. We now discuss a **test of Wilcoxon** that does take into account the magnitude of the differences $|X_k - m_0|$, $k = 1, 2, \ldots, n$. However, in addition to assuming that the random variable X is of the continuous type, we must also assume that the pdf of X is symmetric about the median in order to find the distribution of this new statistic. Because of the continuity assumption, we assume, in the discussion which follows, that no two observations are equal and that no observation is equal to the median.

We are interested in testing the hypothesis H_0: $m = m_0$, where m_0 is some given constant. With our random sample $X_1, X_2, \ldots, X_n$, we rank the absolute values $|X_1 - m_0|, |X_2 - m_0|, \ldots, |X_n - m_0|$ in ascending order according to magnitude. That is, for $k = 1, 2, \ldots, n$, we let R_k denote the rank of $|X_k - m_0|$ among $|X_1 - m_0|, |X_2 - m_0|, \ldots, |X_n - m_0|$. Note that $R_1, R_2, \ldots, R_n$ is a permutation of the first n positive integers, $1, 2, \ldots, n$. Now, with each R_k, we associate the sign of the difference $X_k - m_0$; that is, if $X_k - m_0 > 0$, we use R_k, but if $X_k - m_0 < 0$, we use $-R_k$. The Wilcoxon statistic W is the sum of these n signed ranks, and therefore is often called the **Wilcoxon signed rank statistic**.

Example 8.4-2

Suppose the lengths of $n = 10$ sunfish are

$$x_i: \ 5.0 \ \ 3.9 \ \ 5.2 \ \ 5.5 \ \ 2.8 \ \ 6.1 \ \ 6.4 \ \ 2.6 \ \ 1.7 \ \ 4.3$$

We shall test H_0: $m = 3.7$ against the alternative hypothesis H_1: $m > 3.7$. Thus, we have

$x_k - m_0$:	1.3,	0.2,	1.5,	1.8,	−0.9,	2.4,	2.7,	−1.1,	−2.0,	0.6
$\|x_k - m_0\|$:	1.3,	0.2,	1.5,	1.8,	0.9,	2.4,	2.7,	1.1,	2.0,	0.6
Ranks:	5,	1,	6,	7,	3,	9,	10,	4,	8,	2
Signed Ranks:	5,	1,	6,	7,	−3,	9,	10,	−4,	−8,	2

Therefore, the Wilcoxon statistic is equal to

$$W = 5 + 1 + 6 + 7 - 3 + 9 + 10 - 4 - 8 + 2 = 25.$$

Incidentally, the positive answer seems reasonable because the number of the 10 lengths that are less than 3.7 is 3, which is the statistic used in the sign test. ∎

If the hypothesis H_0: $m = m_0$ is true, about one half of the differences would be negative and thus about one half of the signs would be negative. Hence, it seems that the hypothesis H_0: $m = m_0$ is supported if the observed value of W is close to zero. If the alternative hypothesis is H_1: $m > m_0$, we would reject H_0 if the observed $W = w$ is too large, since, in this case, the larger deviations $|X_k - m_0|$ would usually be associated with observations for which $x_k - m_0 > 0$. That is, the critical region would be of the form $\{w : w \geq c_1\}$. If the alternative hypothesis is H_1: $m < m_0$, the critical region would be of the form $\{w : w \leq c_2\}$. Of course, the critical region would be of the form $\{w : w \leq c_3 \text{ or } w \geq c_4\}$ for a two-sided alternative hypothesis H_1: $m \neq m_0$. In order to find the values of $c_1, c_2, c_3,$ and c_4 that yield desired significance levels, it is necessary to determine the distribution of W under H_0. Accordingly, we consider certain characteristics of this distribution.

When H_0: $m = m_0$ is true,

$$P(X_k < m_0) = P(X_k > m_0) = \frac{1}{2}, \qquad k = 1, 2, \ldots, n.$$

Hence, the probability is 1/2 that a negative sign is associated with the rank R_k of $|X_k - m_0|$. Moreover, the assignments of these n signs are independent because $X_1, X_2, \ldots, X_n$ are mutually independent. In addition, W is a sum that contains the integers $1, 2, \ldots, n$, each with a positive or negative sign. Since the underlying distribution is symmetric, it seems intuitively obvious that W has the same distribution as the random variable

$$V = \sum_{k=1}^{n} V_k,$$

where $V_1, V_2, \ldots, V_n$ are independent and

$$P(V_k = k) = P(V_k = -k) = \frac{1}{2}, \qquad k = 1, 2, \ldots, n.$$

That is, V is a sum that contains the integers $1, 2, \ldots, n$, and these integers receive their algebraic signs by independent assignments.

Since W and V have the same distribution, their means and variances are equal, and we can easily find those of V. Now, the mean of V_k is

$$E(V_k) = -k\left(\frac{1}{2}\right) + k\left(\frac{1}{2}\right) = 0;$$

thus,

$$E(W) = E(V) = \sum_{k=1}^{n} E(V_k) = 0.$$

The variance of V_k is

$$\text{Var}(V_k) = E(V_k^2) = (-k)^2\left(\frac{1}{2}\right) + (k)^2\left(\frac{1}{2}\right) = k^2.$$

Hence,

$$\text{Var}(W) = \text{Var}(V) = \sum_{k=1}^{n} \text{Var}(V_k) = \sum_{k=1}^{n} k^2 = \frac{n(n+1)(2n+1)}{6}.$$

We shall not try to find the distribution of W in general, since that pmf does not have a convenient expression. However, we demonstrate how we could find the distribution of W (or V) with enough patience and computer support. Recall that the moment-generating function of V_i is

$$M_k(t) = e^{t(-k)}\left(\frac{1}{2}\right) + e^{t(+k)}\left(\frac{1}{2}\right) = \frac{e^{-kt} + e^{kt}}{2}, \qquad k = 1, 2, \dots, n.$$

Let $n = 2$; then the moment-generating function of $V_1 + V_2$ is

$$M(t) = E[e^{t(V_1+V_2)}].$$

From the independence of V_1 and V_2, we obtain

$$M(t) = E(e^{tV_1})E(e^{tV_2})$$

$$= \left(\frac{e^{-t} + e^{t}}{2}\right)\left(\frac{e^{-2t} + e^{2t}}{2}\right)$$

$$= \frac{e^{-3t} + e^{-t} + e^{t} + e^{3t}}{4}.$$

This means that each of the points $-3, -1, 1, 3$ in the support of $V_1 + V_2$ has probability $1/4$.

Next let $n = 3$; then the moment-generating function of $V_1 + V_2 + V_3$ is

$$M(t) = E[e^{t(V_1+V_2+V_3)}]$$

$$= E[e^{t(V_1+V_2)}]E(e^{tV_3})$$

$$= \left(\frac{e^{-3t} + e^{-t} + e^{t} + e^{3t}}{4}\right)\left(\frac{e^{-3t} + e^{3t}}{2}\right)$$

$$= \frac{e^{-6t} + e^{-4t} + e^{-2t} + 2e^{0} + e^{2t} + e^{4t} + e^{6t}}{8}.$$

Thus, the points $-6, -4, -2, 0, 2, 4$, and 6 in the support of $V_1 + V_2 + V_3$ have the respective probabilities $1/8, 1/8, 1/8, 2/8, 1/8, 1/8$, and $1/8$. Obviously, this procedure can be continued for $n = 4, 5, 6, \dots$, but it is rather tedious. Fortunately, however, even though $V_1, V_2, \dots, V_n$ are not identically distributed random variables, the sum V of them still has an approximate normal distribution for large samples. To obtain this normal approximation for V (or W), a more general form of the central limit theorem, due to Liapounov, can be used which allows us to say that the standardized random variable

$$Z = \frac{W - 0}{\sqrt{n(n+1)(2n+1)/6}}$$

is approximately $N(0, 1)$ when H_0 is true. We accept this theorem without proof, so that we can use this normal distribution to approximate probabilities such as

$P(W \geq c; H_0) \approx P(Z \geq z_\alpha; H_0)$ when the sample size n is sufficiently large. The next example illustrates this approximation.

Example 8.4-3

The moment-generating function of W or of V is given by

$$M(t) = \prod_{i=1}^{n} \frac{e^{-kt} + e^{kt}}{2}.$$

Using a computer algebra system such as *Maple*, we can expand $M(t)$ and find the coefficients of e^{kt}, which is equal to $P(W = k)$. In Figure 8.4-1, we have drawn a probability histogram for the distribution of W along with the approximating $N[0, n(n + 1)(2n + 1)/6]$ pdf for $n = 4$ (a poor approximation) and for $n = 10$. It is important to note that the widths of the rectangles in the probability histogram are equal to 2, so the "half-unit correction for continuity" mentioned in Section 5.7 now is equal to 1. ∎

Example 8.4-4

Let m be the median of a symmetric distribution of the continuous type. To test the hypothesis $H_0: m = 160$ against the alternative hypothesis $H_1: m > 160$, we take a random sample of size $n = 16$. For an approximate significance level of $\alpha = 0.05$, H_0 is rejected if the computed $W = w$ is such that

$$z = \frac{w}{\sqrt{16(17)(33)/6}} \geq 1.645,$$

or

$$w \geq 1.645\sqrt{\frac{16(17)(33)}{6}} = 63.626.$$

Say the observed values of a random sample are 176.9, 158.3, 152.1, 158.8, 172.4, 169.8, 159.7, 162.7, 156.6, 174.5, 184.4, 165.2, 147.8, 177.8, 160.1, and 160.5. In Table 8.4-1, the magnitudes of the differences $|x_k - 160|$ have been ordered and ranked. Those differences $x_k - 160$ which were negative have been underlined, and the ranks are under the ordered values. For this set of data,

$$w = 1 - 2 + 3 - 4 - 5 + 6 + \cdots + 16 = 60.$$

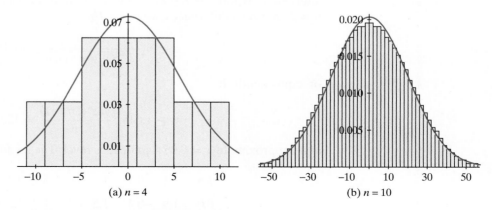

Figure 8.4-1 The Wilcoxon distribution

Table 8.4-1 Ordered absolute differences from 160							
0.1	0.3	0.5	1.2	1.7	2.7	3.4	5.2
1	2	3	4	5	6	7	8
7.9	9.8	12.2	12.4	14.5	16.9	17.8	24.4
9	10	11	12	13	14	15	16

Since $60 < 63.626$, H_0 is not rejected at the 0.05 significance level. It is interesting to note that H_0 would have been rejected at $\alpha = 0.10$, since, with a unit correction made for continuity, the approximate p-value is

$$p\text{-value} = P(W \geq 60)$$

$$= P\left(\frac{W - 0}{\sqrt{(16)(17)(33)/6}} \geq \frac{59 - 0}{\sqrt{(16)(17)(33)/6}}\right)$$

$$\approx P(Z \geq 1.525) = 0.0636.$$

(*Maple* produces a p-value equal to $4{,}251/65{,}536 = 0.0649$.) Such a p-value would indicate that the data are too few to reject H_0, but if the pattern continues, we shall most certainly reject H_0 with a larger sample size. ∎

Although theoretically we could ignore the possibilities that $x_k = m_0$ for some k and that $|x_k - m_0| = |x_j - m_0|$ for some $k \neq j$, these situations do occur in applications. Usually, in practice, if $x_k = m_0$ for some k, that observation is deleted and the test is performed with a reduced sample size. If the absolute values of the differences from m_0 of two or more observations are equal, each observation is assigned the average of the corresponding ranks. The change this causes in the distribution of W is not very great, provided that the number of ties is relatively small; thus, we continue using the same normal approximation.

We now give an example that has some tied observations.

Example 8.4-5

We consider some paired data for percentage of body fat measured at the beginning and the end of a semester. Let m equal the median of the differences, $x - y$. We shall use the Wilcoxon statistic to test the null hypothesis H_0: $m = 0$ against the alternative hypothesis H_1: $m > 0$ with the differences given below. Since there are $n = 25$ nonzero differences, we reject H_0 if

$$z = \frac{w - 0}{\sqrt{(25)(26)(51)/6}} \geq 1.645$$

or, equivalently, if

$$w \geq 1.645\sqrt{\frac{(25)(26)(51)}{6}} = 122.27$$

at an approximate $\alpha = 0.05$ significance level. The 26 differences are

$$1.8 \quad -3.1 \quad 0.1 \quad 1.1 \quad 0.6 \quad -5.1 \quad 9.2 \quad 0.2 \quad 0.4$$

$$0.0 \quad 1.9 \quad -0.4 \quad -1.5 \quad 1.4 \quad -1.0 \quad 2.2 \quad 0.8 \quad -0.4$$

$$2.0 \quad -5.8 \quad -3.4 \quad -2.3 \quad 3.0 \quad 2.7 \quad 0.2 \quad 3.2$$

Table 8.4-2 Ordered absolute values, changes in percentage of body fat												
0.1	0.2	0.2	0.4	<u>0.4</u>	<u>0.4</u>	0.6	0.8	<u>1.0</u>	1.1	1.4	<u>1.5</u>	1.8
1	2.5	2.5	5	5	5	7	8	9	10	11	12	13
1.9	2.0	2.2	<u>2.3</u>	2.7	3.0	<u>3.1</u>	3.2	<u>3.4</u>	<u>5.1</u>	<u>5.8</u>	9.2	
14	15	16	17	18	19	20	21	22	23	24	25	

Table 8.4-2 lists the ordered nonzero absolute values, with those that were originally negative underlined. The rank is under each observation. Note that in the case of ties, the average of the ranks of the tied measurements is given.

The value of the Wilcoxon statistic is

$$w = 1 + 2.5 + 2.5 + 5 - 5 - 5 + \cdots + 25 = 51.$$

Since $51 < 122.27$, we fail to reject the null hypothesis. The approximate p-value of this test, using the continuity correction, is

$$p\text{-value} = P(W \geq 51)$$

$$\approx P\left(Z \geq \frac{50 - 0}{\sqrt{(25)(26)(51)/6}}\right) = P(Z \geq 0.673) = 0.2505. \quad \blacksquare$$

Another method due to Wilcoxon for testing the equality of two distributions of the continuous type uses the magnitudes of the observations. For this test, it is assumed that the respective cdfs F and G have the same shape and spread but possibly different locations; that is, there exists a constant c such that $F(x) = G(x + c)$ for all x. To proceed with the test, place the combined sample of $x_1, x_2, \ldots, x_{n_1}$ and $y_1, y_2, \ldots, y_{n_2}$ in increasing order of magnitude. Assign the ranks $1, 2, 3, \ldots, n_1 + n_2$ to the ordered values. In the case of ties, assign the average of the ranks associated with the tied values. Let w equal the sum of the ranks of $y_1, y_2, \ldots, y_{n_2}$. If the distribution of Y is shifted to the right of that of X, the values of Y would tend to be larger than the values of X and w would usually be larger than expected when $F(z) = G(z)$. If m_X and m_Y are the respective medians, the critical region for testing $H_0: m_X = m_Y$ against $H_1: m_X < m_Y$ would be of the form $w \geq c$. Similarly, if the alternative hypothesis is $m_X > m_Y$, the critical region would be of the form $w \leq c$.

We shall not derive the distribution of W. However, if n_1 and n_2 are both greater than 7, and there are no ties, a normal approximation can be used. With $F(z) = G(z)$, the mean and variance of W are

$$\mu_W = \frac{n_2(n_1 + n_2 + 1)}{2}$$

and

$$\text{Var}(W) = \frac{n_1 n_2(n_1 + n_2 + 1)}{12},$$

and the statistic

$$Z = \frac{W - n_2(n_1 + n_2 + 1)/2}{\sqrt{n_1 n_2(n_1 + n_2 + 1)/12}}$$

is approximately $N(0, 1)$.

Example 8.4-6

The weights of of the contents of $n_1 = 8$ and $n_2 = 8$ tins of cinnamon packaged by companies A and B, respectively, selected at random, yielded the following observations of X and Y:

x: 117.1 121.3 127.8 121.9 117.4 124.5 119.5 115.1

y: 123.5 125.3 126.5 127.9 122.1 125.6 129.8 117.2

The critical region for testing $H_0: m_X = m_Y$ against $H_1: m_X < m_Y$ is of the form $w \geq c$. Since $n_1 = n_2 = 8$, at an approximate $\alpha = 0.05$ significance level H_0 is rejected if

$$z = \frac{w - 8(8 + 8 + 1)/2}{\sqrt{[(8)(8)(8 + 8 + 1)]/12}} \geq 1.645,$$

or

$$w \geq 1.645\sqrt{\frac{(8)(8)(17)}{12}} + 4(17) = 83.66.$$

To calculate the value of W, it is sometimes helpful to construct a **back-to-back stem-and-leaf display**. In such a display, the stems are put in the center and the leaves go to the left and the right. (See Table 8.4-3.)

Reading from this two-sided stem-and-leaf display, we show the combined sample in Table 8.4-4, with the Company B (y) weights underlined. The ranks are given beneath the values.

From Table 8.4-4, the computed W is

$$w = 3 + 8 + 9 + 11 + 12 + 13 + 15 + 16 = 87 > 83.66.$$

Table 8.4-3 Back-to-back stem-and-leaf diagram of weights of cinnamon

x Leaves		Stems	y Leaves	
	51	$11f$		
74	71	$11s$	72	
	95	$11\bullet$		
19	13	$12*$		
		$12t$	21	35
	45	$12f$	53	56
	78	$12s$	65	79
		$12\bullet$	98	

Multiply numbers by 10^{-1}.

Table 8.4-4	Combined ordered samples						
115.1	117.1	117.2	117.4	119.5	121.3	121.9	122.1
1	2	3	4	5	6	7	8
123.5	124.5	125.3	125.6	126.5	127.8	127.9	129.8
9	10	11	12	13	14	15	16

Thus, H_0 is rejected. Finally, making a half-unit correction for continuity, we see that the p-value of this test is

$$p\text{-value} = P(W \geq 87)$$

$$= P\left(\frac{W - 68}{\sqrt{90.667}} \geq \frac{86.5 - 68}{\sqrt{90.667}} \right)$$

$$\approx P(Z \geq 1.943) = 0.0260. \quad \blacksquare$$

Exercises

8.4-1. It is claimed that the median weight m of certain loads of candy is 40,000 pounds.

(a) Use the following 13 observations and the Wilcoxon statistic to test the null hypothesis H_0: $m = 40,000$ against the one-sided alternative hypothesis H_1: $m < 40,000$ at an approximate significance level of $\alpha = 0.05$:

41,195 39,485 41,229 36,840 38,050 40,890 38,345

34,930 39,245 31,031 40,780 38,050 30,906

(b) What is the approximate p-value of this test?

(c) Use the sign test to test the same hypothesis.

(d) Calculate the p-value from the sign test and compare it with the p-value obtained from the Wilcoxon test.

8.4-2. A course in economics was taught to two groups of students, one in a classroom situation and the other online. There were 24 students in each group. The students were first paired according to cumulative grade point averages and background in economics, and then assigned to the courses by a flip of a coin. (The procedure was repeated 24 times.) At the end of the course each class was given the same final examination. Use the Wilcoxon test to test the hypothesis that the two methods of teaching are equally effective against a two-sided alternative. The differences in the final scores for each pair of students were as follows (the online student's score was subtracted from the corresponding classroom student's score):

14	−4	−6	−2	−1	18
6	12	8	−4	13	7
2	6	21	7	−2	11
−3	−14	−2	17	−4	−5

8.4-3. Let X equal the weight (in grams) of a Hershey's grape-flavored Jolly Rancher. Denote the median of X by m. We shall test H_0: $m = 5.900$ against H_1: $m > 5.900$. A random sample of size $n = 25$ yielded the following ordered data:

5.625 5.665 5.697 5.837 5.863 5.870 5.878 5.884 5.908

5.967 6.019 6.020 6.029 6.032 6.037 6.045 6.049

6.050 6.079 6.116 6.159 6.186 6.199 6.307 6.387

(a) Use the sign test to test the hypothesis.

(b) Use the Wilcoxon test statistic to test the hypothesis.

(c) Use a t test to test the hypothesis.

(d) Write a short comparison of the three tests.

8.4-4. The outcomes on $n = 10$ simulations of a Cauchy random variable were $-1.9415, 0.5901, -5.9848, -0.0790,$ $-0.7757, -1.0962, 9.3820, -74.0216, -3.0678,$ and 3.8545. For the Cauchy distribution, the mean does not exist, but for this one, the median is believed to equal zero. Use the Wilcoxon test and these data to test H_0: $m = 0$ against the alternative hypothesis H_1: $m \neq 0$. Let $\alpha \approx 0.05$.

8.4-5. Let x equal a student's GPA in the fall semester and y the same student's GPA in the spring semester. Let m equal the median of the differences, $x - y$. We shall test the null hypothesis H_0: $m = 0$ against an appropriate alternative hypothesis that you select on the basis of your past experience. Use a Wilcoxon test and the following 15 observations of paired data to test H_0:

x	y	x	y
2.88	3.22	3.98	3.76
3.67	3.49	4.00	3.96
2.76	2.54	3.39	3.52
2.34	2.17	2.59	2.36
2.46	2.53	2.78	2.62
3.20	2.98	2.85	3.06
3.17	2.98	3.25	3.16
2.90	2.84		

8.4-6. Let m equal the median of the posttest grip strengths in the right arms of male freshmen in a study of health dynamics. We shall use observations on $n = 15$ such students to test the null hypothesis H_0: $m = 50$ against the alternative hypothesis H_1: $m > 50$.

(a) Using the Wilcoxon statistic, define a critical region that has an approximate significance level of $\alpha = 0.05$.

(b) Given the observed values

58.0 52.5 46.0 57.5 52.0 45.5 65.5 71.0

57.0 54.0 48.0 58.0 35.5 44.0 53.0

what is your conclusion?

(c) What is the p-value of this test?

8.4-7. Let X equal the weight in pounds of a "1-pound" bag of carrots. Let m equal the median weight of a population of these bags. Test the null hypothesis H_0: $m = 1.14$ against the alternative hypothesis H_1: $m > 1.14$.

(a) With a sample of size $n = 14$, use the Wilcoxon statistic to define a critical region. Use $\alpha \approx 0.10$.

(b) What would be your conclusion if the observed weights were

1.12 1.13 1.19 1.25 1.06 1.31 1.12

1.23 1.29 1.17 1.20 1.11 1.18 1.23

(c) What is the p-value of your test?

8.4-8. A pharmaceutical company is interested in testing the effect of humidity on the weight of pills that are sold in aluminum packaging. Let X and Y denote the respective

weights of pills and their packaging (in grams), when the packaging is good and when it is defective, after the pill has spent 1 week in a chamber containing 100% humidity and heated to 30 °C.

(a) Use the Wilcoxon test to test H_0: $m_X = m_Y$ against H_0: $m_X - m_Y < 0$ on the following random samples of $n_1 = 12$ observations of X and $n_2 = 12$ observations of Y:

x: 0.7565 0.7720 0.7776 0.7750 0.7494 0.7615
    0.7741 0.7701 0.7712 0.7719 0.7546 0.7719

y: 0.7870 0.7750 0.7720 0.7876 0.7795 0.7972
    0.7815 0.7811 0.7731 0.7613 0.7816 0.7851

What is the p-value?

(b) Construct and interpret a q–q plot of these data. HINT: This is a q–q plot of the empirical distribution of X against that of Y.

8.4-9. Let us compare the failure times of a certain type of light bulb produced by two different manufacturers, X and Y, by testing 10 bulbs selected at random from each of the outputs. The data, in hundreds of hours used before failure, are

x: 5.6 4.6 6.8 4.9 6.1 5.3 4.5 5.8 5.4 4.7

y: 7.2 8.1 5.1 7.3 6.9 7.8 5.9 6.7 6.5 7.1

(a) Use the Wilcoxon test to test the equality of medians of the failure times at the approximate 5% significance level. What is the p-value?

(b) Construct and interpret a q–q plot of these data. HINT: This is a q–q plot of the empirical distribution of X against that of Y.

8.4-10. Let X and Y denote the heights of blue spruce trees, measured in centimeters, growing in two large fields. We shall compare these heights by measuring 12 trees selected at random from each of the fields. Take $\alpha \approx 0.05$, and use the statistic W—the sum of the ranks of the observations of Y in the combined sample—to test the hypothesis H_0: $m_X = m_Y$ against the alternative hypothesis H_1: $m_X < m_Y$ on the basis of the following $n_1 = 12$ observations of X and $n_2 = 12$ observations of Y.

x: 90.4 77.2 75.9 83.2 84.0 90.2
    87.6 67.4 77.6 69.3 83.3 72.7

y: 92.7 78.9 82.5 88.6 95.0 94.4
    73.1 88.3 90.4 86.5 84.7 87.5

8.4-11. Let X and Y equal the sizes of grocery orders from, respectively, a south-side and a north-side food store of the same chain. We shall test the null hypothesis

$H_0: m_X = m_Y$ against a two-sided alternative, using the following ordered observations:

x:	5.13	8.22	11.81	13.77	15.36
	23.71	31.39	34.65	40.17	75.58
y:	4.42	6.47	7.12	10.50	12.12
	12.57	21.29	33.14	62.84	72.05

(a) Use the Wilcoxon test when $\alpha = 0.05$. What is the p-value of this two-sided test?

(b) Construct a q–q plot and interpret it. HINT: This is a q–q plot of the empirical distribution of X against that of Y.

8.4-12. A charter bus line has 48-passenger and 38-passenger buses. Let m_{48} and m_{38} denote the median number of miles traveled per day by the respective buses. With $\alpha = 0.05$, use the Wilcoxon statistic to test $H_0: m_{48} = m_{38}$ against the one-sided alternative H_1: $m_{48} > m_{38}$. Use the following data, which give the numbers of miles traveled per day for respective random samples of sizes 9 and 11:

48-passenger buses:	331	308	300	414	253
	323	452	396	104	

38-passenger buses:	248	393	260	355	279	184
	386	450	432	196	197	

8.4-13. A company manufactures and packages soap powder in 6-pound boxes. The quality assurance department was interested in comparing the fill weights of packages from the east and west lines. Taking random samples from the two lines, the department obtained the following weights:

East line (x):	6.06	6.04	6.11	6.06	6.06
	6.07	6.06	6.08	6.05	6.09
West line (y):	6.08	6.03	6.04	6.07	6.11
	6.08	6.08	6.10	6.06	6.04

(a) Let m_X and m_Y denote the median weights for the east and west lines, respectively. Test $H_0: m_X = m_Y$ against a two-sided alternative hypothesis, using the Wilcoxon test with $\alpha \approx 0.05$. Find the p-value of this two-sided test.

(b) Construct and interpret a q–q plot of these data.

8.4-14. In Exercise 8.2-13, data are given that show the effect of a certain fertilizer on plant growth. The growths of the plants in mm over six weeks are repeated here, where Group A received fertilizer and Group B did not:

Group A: 55 61 33 57 17 46 50 42 71 51 63

Group B: 31 27 12 44 9 25 34 53 33 21 32

We shall test the hypothesis that fertilizer enhanced the growth of the plants.

(a) Construct a back-to-back stem-and-leaf display in which the stems are put down the center of the diagram and the Group A leaves go to the left while the Group B leaves go to the right.

(b) Calculate the value of the Wilcoxon statistic and give your conclusion.

(c) How does this result compare with that using the t test in Exercise 8.2-13?

8.4-15. With $\alpha = 0.05$, use the Wilcoxon statistic to test $H_0: m_X = m_Y$ against a two-sided alternative. Use the following observations of X and Y, which have been ordered for your convenience:

x:	−2.3864	−2.2171	−1.9148	−1.9097	−1.4883
	−1.2007	−1.1077	−0.3601	0.4325	1.0598
	1.3035	1.5241	1.7133	1.7656	2.4912
y:	−1.7613	−0.9391	−0.7437	−0.5530	−0.2469
	0.0647	0.2031	0.3219	0.3579	0.6431
	0.6557	0.6724	0.6762	0.9041	1.3571

8.4-16. Data were collected during a step-direction experiment in the biomechanics laboratory at Hope College. The goal of the study is to establish differences in stepping responses between healthy young and healthy older adults. In one part of the experiment, the subjects are told in what direction they should take a step. Then, when given a signal, the subject takes a step in that direction, and the time it takes for them to lift their foot to take the step is measured. The direction is repeated a few times throughout the testing, and for each subject, a mean of all the "liftoff" times in a certain direction is calculated. The mean liftoff times (in thousandths of a second) for the anterior direction, ordered for your convenience, are as follows:

Young Subjects 397 433 450 468 485 488 498 504 561
565 569 576 577 579 581 586 696

Older Subjects 463 538 549 573 588 590 594 626 627
653 674 728 818 835 863 888 936

(a) Construct a back-to-back stem-and-leaf display. Use stems $3\bullet, 4*, \ldots, 9*$.

(b) Use the Wilcoxon statistic to test the null hypothesis that the response times are equal against the alternative that the times for the young subjects are less than that for the older subjects.

(c) What outcome does a t test give?

8.4-17. Some measurements in mm were made on a species of spiders, named *Sosippus floridanus*, that are native to Florida. There are 10 female spiders and 10 male spiders. The body lengths and the lengths of their front and back legs are repeated here:

Female body lengths	Female front legs	Female back legs	Male body lengths	Male front legs	Male back legs
11.06	15.03	19.29	12.26	21.22	25.54
13.87	17.96	22.74	11.66	18.62	23.94
12.93	17.56	21.28	12.53	18.62	23.94
15.08	21.22	25.54	13.00	19.95	25.80
17.82	22.61	28.86	11.79	19.15	25.40
14.14	20.08	25.14	12.46	19.02	25.27
12.26	16.49	20.22	10.65	17.29	22.21
17.82	18.75	24.61	10.39	17.02	21.81
20.17	23.01	28.46	12.26	18.49	23.41
16.88	22.48	28.59	14.07	22.61	28.86

In this exercise, we shall use the Wilcoxon statistic to compare the sizes of the female and male spiders. For each of the following instructions, construct back-to-back stem-and-leaf displays:

(a) Test the null hypothesis that the body lengths of female and male spiders are equal against the alternative hypothesis that female spiders are longer.

(b) Test the null hypothesis that the lengths of the front legs of the female and male spiders are equal against a two-sided alternative.

(c) Test the null hypothesis that the lengths of the back legs of the female and male spiders are equal against a two-sided alternative.

8.4-18. In Exercise 8.2-10, growth data are given for plants in normal air and for plants in CO_2-enriched air. Those data are repeated here:

Normal Air (x) 4.67 4.21 2.18 3.91 4.09 5.24 2.94 4.71 4.04 5.79 3.80 4.38

Enriched Air (y) 5.04 4.52 6.18 7.01 4.36 1.81 6.22 5.70

In this exercise, we shall test the null hypothesis that the medians are equal, namely, H_0: $m_X = m_Y$, against the alternative hypothesis H_1: $m_X < m_Y$. You may select the significance level. However, give the approximate p-value or state clearly why you arrived at a particular conclusion, for each of the tests. Show your work.

(a) What is your conclusion from the Wilcoxon test?

(b) What was your conclusion from the t test in Exercise 8.2-10?

(c) Write a comparison of these two tests.

8.5 POWER OF A STATISTICAL TEST

In Chapter 8, we gave several tests of fairly common statistical hypotheses in such a way that we described the significance level α and the p-values of each. Of course, those tests were based on good (sufficient) statistics of the parameters, when the latter exist. In this section, we consider the probability of making the other type of error: accepting the null hypothesis H_0 when the alternative hypothesis H_1 is true. This consideration leads to ways to find most powerful tests of the null hypothesis H_0 against the alternative hypothesis H_1.

The first example introduces a new concept using a test about p, the probability of success. The sample size is kept small so that Table II in Appendix B can be used to find probabilities. The application is one that you can actually perform.

Example 8.5-1 Assume that when given a name tag, a person puts it on either the right or left side. Let p equal the probability that the name tag is placed on the right side. We shall test the null hypothesis H_0: $p = 1/2$ against the composite alternative hypothesis H_1: $p < 1/2$. (Included with the null hypothesis are those values of p which are greater than 1/2; that is, we could think of H_0 as H_0: $p \geq 1/2$.) We shall give name tags to a random sample of $n = 20$ people, denoting the placements of their name tags with Bernoulli random variables, $X_1, X_2, \ldots, X_{20}$, where $X_{i=1}$ if a person places the name tag on the right and $X_i = 0$ if a person places the name tag on the left. For our test statistic, we can then use $Y = \sum_{i=1}^{20} X_i$, which has the binomial distribution $b(20, p)$. Say the critical region is defined by $C = \{y : y \leq 6\}$ or, equivalently, by $\{(x_1, x_2, \ldots, x_{20}) : \sum_{i=1}^{20} x_i \leq 6\}$. Since Y is $b(20, 1/2)$ if $p = 1/2$, the significance level of the corresponding test is

$$\alpha = P\left(Y \leq 6; p = \frac{1}{2}\right) = \sum_{y=0}^{6} \binom{20}{y}\left(\frac{1}{2}\right)^{20} = 0.0577,$$

from Table II in Appendix B. Of course, the probability β of a Type II error has different values, with different values of p selected from the composite alternative hypothesis $H_1 : p < 1/2$. For example, with $p = 1/4$,

$$\beta = P\left(7 \leq Y \leq 20; p = \frac{1}{4}\right) = \sum_{y=7}^{20} \binom{20}{y}\left(\frac{1}{4}\right)^{y}\left(\frac{3}{4}\right)^{20-y} = 0.2142,$$

whereas with $p = 1/10$,

$$\beta = P\left(7 \leq Y \leq 20; p = \frac{1}{10}\right) = \sum_{y=7}^{20} \binom{20}{y}\left(\frac{1}{10}\right)^{y}\left(\frac{9}{10}\right)^{20-y} = 0.0024.$$

Instead of considering the probability β of accepting H_0 when H_1 is true, we could compute the probability K of rejecting H_0 when H_1 is true. After all, β and $K = 1 - \beta$ provide the same information. Since K is a function of p, we denote this explicitly by writing $K(p)$. The probability

$$K(p) = \sum_{y=0}^{6} \binom{20}{y} p^y (1-p)^{20-y}, \qquad 0 < p \leq \frac{1}{2},$$

is called the **power function of the test**. Of course, $\alpha = K(1/2) = 0.0577$, $1 - K(1/4) = 0.2142$, and $1 - K(1/10) = 0.0024$. The value of the power function at a specified p is called the **power** of the test at that point. For instance, $K(1/4) = 0.7858$ and $K(1/10) = 0.9976$ are the powers at $p = 1/4$ and $p = 1/10$, respectively. An acceptable power function assumes small values when H_0 is true and larger values when p differs much from $p = 1/2$. (See Figure 8.5-1 for a graph of this power function.) ∎

In Example 8.5-1, we introduced the new concept of the power function of a test. We now show how the sample size can be selected so as to create a test with appropriate power.

Example 8.5-2 Let $X_1, X_2, \ldots, X_n$ be a random sample of size n from the normal distribution $N(\mu, 100)$, which we can suppose is a possible distribution of scores of students in a statistics course that uses a new method of teaching (e.g., computer-related materials). We wish to decide between $H_0 : \mu = 60$ (the *no-change* hypothesis because, let us say, this was the mean by the previous method of teaching) and the research worker's hypothesis $H_1 : \mu > 60$. Let us consider a sample of size $n = 25$. Of course, the sample mean $\overline{X}$ is the maximum likelihood estimator of μ; thus, it seems reasonable to base our decision on this statistic. Initially, we use the rule to reject H_0 and accept H_1 if and only if $\overline{x} \geq 62$. What are the consequences of this test? These are summarized in the power function of the test.

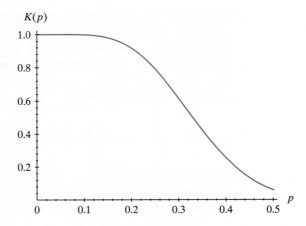

Figure 8.5-1 Power function: $K(p) = P(Y \leq 6; p)$, where Y is $b(20, p)$

We first find the probability of rejecting H_0: $\mu = 60$ for various values of $\mu \geq 60$. The probability of rejecting H_0 is given by

$$K(\mu) = P(\overline{X} \geq 62; \mu),$$

because this test calls for the rejection of H_0: $\mu = 60$ when $\bar{x} \geq 62$. When the new process has the general mean μ, $\overline{X}$ has the normal distribution $N(\mu, 100/25 = 4)$. Accordingly,

$$K(\mu) = P\left(\frac{\overline{X} - \mu}{2} \geq \frac{62 - \mu}{2}; \mu\right)$$

$$= 1 - \Phi\left(\frac{62 - \mu}{2}\right), \qquad 60 \leq \mu,$$

is the probability of rejecting H_0: $\mu = 60$ by using this particular test. Several values of $K(\mu)$ are given in Table 8.5-1. Figure 8.5-2 depicts the graph of the function $K(\mu)$.

Table 8.5-1 Values of the power function

μ	$K(\mu)$
60	0.1587
61	0.3085
62	0.5000
63	0.6915
64	0.8413
65	0.9332
66	0.9772

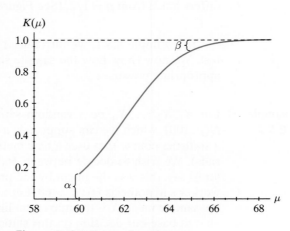

Figure 8.5-2 Power function
$K(\mu) = 1 - \Phi([62 - \mu]/2)$

The probability $K(\mu)$ of rejecting H_0: $\mu = 60$ is called the *power function* of the test. At the value μ_1 of the parameter, $K(\mu_1)$ is the power at μ_1. The power at $\mu = 60$ is $K(60) = 0.1587$, and this is the probability of rejecting H_0: $\mu = 60$ when H_0 is true. That is, $K(60) = 0.1587 = \alpha$ is the probability of a Type I error and is called the *significance level* of the test.

The power at $\mu = 65$ is $K(65) = 0.9332$, and this is the probability of making the correct decision (namely, rejecting H_0: $\mu = 60$ when $\mu = 65$). Hence, we are pleased that here it is large. When $\mu = 65$, $1 - K(65) = 0.0668$ is the probability of not rejecting H_0: $\mu = 60$ when $\mu = 65$; that is, it is the probability of a Type II error and is denoted by $\beta = 0.0668$. These α- and β-values are displayed in Figure 8.5-2. Clearly, the probability $\beta = 1 - K(\mu_1)$ of a Type II error depends on which value— say, μ_1—is taken in the alternative hypothesis H_1: $\mu > 60$. Thus, while $\beta = 0.0668$ when $\mu = 65$, β is equal to $1 - K(63) = 0.3085$ when $\mu = 63$.

Frequently, statisticians like to have the significance level α smaller than 0.1587—say, around 0.05 or less—because it is a probability of an error, namely, a Type I error. Thus, if we would like $\alpha = 0.05$, then, with $n = 25$, we can no longer use the critical region $\bar{x} \geq 62$; rather, we use $\bar{x} \geq c$, where c is selected such that

$$K(60) = P(\overline{X} \geq c; \mu = 60) = 0.05.$$

However, when $\mu = 60$, $\overline{X}$ is $N(60, 4)$, and it follows that

$$K(60) = P\left(\frac{\overline{X} - 60}{2} \geq \frac{c - 60}{2}; \mu = 60\right)$$

$$= 1 - \Phi\left(\frac{c - 60}{2}\right) = 0.05.$$

From Table Va in Appendix B, we have

$$\frac{c - 60}{2} = 1.645 = z_{0.05} \qquad \text{and} \qquad c = 60 + 3.29 = 63.29.$$

Although this change reduces α from 0.1587 to 0.05, it increases β at $\mu = 65$ from 0.0668 to

$$\beta = 1 - P(\overline{X} \geq 63.29; \mu = 65)$$

$$= 1 - P\left(\frac{\overline{X} - 65}{2} \geq \frac{63.29 - 65}{2}; \mu = 65\right)$$

$$= \Phi(-0.855) = 0.1963.$$

In general, without changing the sample size or the type of test of the hypothesis, a decrease in α causes an increase in β, and a decrease in β causes an increase in α. *Both* probabilities α and β of the two types of errors can be decreased only by increasing the sample size or, in some way, constructing a better test of the hypothesis.

For example, if $n = 100$ and we desire a test with significance level $\alpha = 0.05$, then, since $\overline{X}$ is $N(\mu, 100/100 = 1)$,

$$\alpha = P(\overline{X} \geq c; \mu = 60) = 0.05$$

means that

$$P\left(\frac{\overline{X} - 60}{1} \geq \frac{c - 60}{1}; \mu = 60\right) = 0.05$$

and $c - 60 = 1.645$. Thus, $c = 61.645$. The power function is

$$K(\mu) = P(\overline{X} \geq 61.645; \mu)$$

$$= P\left(\frac{\overline{X} - \mu}{1} \geq \frac{61.645 - \mu}{1}; \mu\right) = 1 - \Phi(61.645 - \mu).$$

In particular, this means that at $\mu = 65$,

$$\beta = 1 - K(\mu) = \Phi(61.645 - 65) = \Phi(-3.355) = 0.0004;$$

so, with $n = 100$, both α and β have decreased from their respective original values of 0.1587 and 0.0668 when $n = 25$.

Rather than guess at the value of n, an ideal power function determines the sample size. Let us use a critical region of the form $\overline{x} \geq c$. Further, suppose that we want $\alpha = 0.025$ and, when $\mu = 65$, $\beta = 0.05$. Thus, since $\overline{X}$ is $N(\mu, 100/n)$, it follows that

$$0.025 = P(\overline{X} \geq c; \mu = 60) = 1 - \Phi\left(\frac{c - 60}{10/\sqrt{n}}\right)$$

and

$$0.05 = 1 - P(\overline{X} \geq c; \mu = 65) = \Phi\left(\frac{c - 65}{10/\sqrt{n}}\right).$$

That is,

$$\frac{c - 60}{10/\sqrt{n}} = 1.96 \qquad \text{and} \qquad \frac{c - 65}{10/\sqrt{n}} = -1.645.$$

Solving these equations simultaneously for c and $10/\sqrt{n}$, we obtain

$$c = 60 + 1.96 \frac{5}{3.605} = 62.718;$$

$$\frac{10}{\sqrt{n}} = \frac{5}{3.605}.$$

Hence,

$$\sqrt{n} = 7.21 \qquad \text{and} \qquad n = 51.98.$$

Since n must be an integer, we would use $n = 52$ and thus obtain $\alpha \approx 0.025$ and $\beta \approx 0.05$. ∎

The next example is an extension of Example 8.5-1.

Example 8.5-3
To test H_0: $p = 1/2$ against H_1: $p < 1/2$, we take a random sample of Bernoulli trials, $X_1, X_2, \ldots, X_n$, and use for our test statistic $Y = \sum_{i=1}^{n} X_i$, which has a binomial distribution $b(n, p)$. Let the critical region be defined by $C = \{y: y \leq c\}$. The power function for this test is defined by $K(p) = P(Y \leq c; p)$. We shall find the values of n and c so that $K(1/2) \approx 0.05$ and $K(1/4) \approx 0.90$. That is, we would like the significance level to be $\alpha = K(1/2) = 0.05$ and the power at $p = 1/4$ to equal 0.90. We proceed as follows: Since

$$0.05 = P\left(Y \leq c; p = \frac{1}{2}\right) = P\left(\frac{Y - n/2}{\sqrt{n(1/2)(1/2)}} \leq \frac{c - n/2}{\sqrt{n(1/2)(1/2)}}\right),$$

it follows that

$$(c - n/2)/\sqrt{n/4} \approx -1.645;$$

and since

$$0.90 = P\left(Y \le c; p = \frac{1}{4}\right) = P\left(\frac{Y - n/4}{\sqrt{n(1/4)(3/4)}} \le \frac{c - n/4}{\sqrt{n(1/4)(3/4)}}\right),$$

it follows that

$$(c - n/4)/\sqrt{3n/16} \approx 1.282.$$

Therefore,

$$\frac{n}{4} \approx 1.645\sqrt{\frac{n}{4}} + 1.282\sqrt{\frac{3n}{16}} \qquad \text{and} \qquad \sqrt{n} \approx 4(1.378) = 5.512.$$

Thus, n is approximately 30.4, and we round upward to 31. From either of the first two approximate equalities, we find that c is about equal to 10.9. Using $n = 31$ and $c = 10.9$ means that $K(1/2) = 0.05$ and $K(1/4) = 0.90$ are only approximate. In fact, since Y must be an integer, we could let $c = 10.5$. Then, with $n = 31$,

$$\alpha = K\left(\frac{1}{2}\right) = P\left(Y \le 10.5; p = \frac{1}{2}\right) \approx 0.0362;$$

$$K\left(\frac{1}{4}\right) = P\left(Y \le 10.5; p = \frac{1}{4}\right) \approx 0.8730.$$

Or we could let $c = 11.5$ and $n = 32$, in which case

$$\alpha = K\left(\frac{1}{2}\right) = P\left(Y \le 11.5; p = \frac{1}{2}\right) \approx 0.0558;$$

$$K\left(\frac{1}{4}\right) = P\left(Y \le 11.5; p = \frac{1}{4}\right) \approx 0.9235. \qquad \blacksquare$$

Exercises

8.5-1. A certain size of bag is designed to hold 25 pounds of potatoes. A farmer fills such bags in the field. Assume that the weight X of potatoes in a bag is $N(\mu, 9)$. We shall test the null hypothesis $H_0: \mu = 25$ against the alternative hypothesis $H_1: \mu < 25$. Let X_1, X_2, X_3, X_4 be a random sample of size 4 from this distribution, and let the critical region C for this test be defined by $\bar{x} \le 22.5$, where $\bar{x}$ is the observed value of $\overline{X}$.

(a) What is the power function $K(\mu)$ of this test? In particular, what is the significance level $\alpha = K(25)$ for your test?

(b) If the random sample of four bags of potatoes yielded the values $x_1 = 21.24$, $x_2 = 24.81$, $x_3 = 23.62$, and $x_4 = 26.82$, would your test lead you to accept or reject H_0?

(c) What is the p-value associated with $\bar{x}$ in part (b)?

8.5-2. Let X equal the number of milliliters of a liquid in a bottle that has a label volume of 350 ml. Assume that the distribution of X is $N(\mu, 4)$. To test the null hypothesis H_0: $\mu = 355$ against the alternative hypothesis $H_1: \mu < 355$, let the critical region be defined by $C = \{\bar{x}: \bar{x} \le 354.05\}$, where $\bar{x}$ is the sample mean of the contents of a random sample of $n = 12$ bottles.

(a) Find the power function $K(\mu)$ for this test.

(b) What is the (approximate) significance level of the test?

(c) Find the values of $K(354.05)$ and $K(353.1)$, and sketch the graph of the power function.

(d) Use the following 12 observations to state your conclusion from this test:

350	353	354	356	353	352
354	355	357	353	354	355

(e) What is the approximate p-value of the test?

8.5-3. Assume that SAT mathematics scores of students who attend small liberal arts colleges are $N(\mu, 8100)$. We shall test H_0: $\mu = 530$ against the alternative hypothesis H_1: $\mu < 530$. Given a random sample of size $n = 36$ SAT mathematics scores, let the critical region be defined by $C = \{\bar{x} : \bar{x} \le 510.77\}$, where $\bar{x}$ is the observed mean of the sample.

(a) Find the power function, $K(\mu)$, for this test.

(b) What is the value of the significance level of the test?

(c) What is the value of $K(510.77)$?

(d) Sketch the graph of the power function.

(e) What is the p-value associated with **(i)** $\bar{x} = 507.35$; **(ii)** $\bar{x} = 497.45$?

8.5-4. Let X be $N(\mu, 100)$. To test H_0: $\mu = 80$ against H_1: $\mu > 80$, let the critical region be defined by $C = \{(x_1, x_2, \ldots, x_{25}) : \bar{x} \ge 83\}$, where $\bar{x}$ is the sample mean of a random sample of size $n = 25$ from this distribution.

(a) What is the power function $K(\mu)$ for this test?

(b) What is the significance level of the test?

(c) What are the values of $K(80)$, $K(83)$, and $K(86)$?

(d) Sketch the graph of the power function.

(e) What is the p-value corresponding to $\bar{x} = 83.41$?

8.5-5. Let X equal the yield of alfalfa in tons per acre per year. Assume that X is $N(1.5, 0.09)$. It is hoped that a new fertilizer will increase the average yield. We shall test the null hypothesis H_0: $\mu = 1.5$ against the alternative hypothesis H_1: $\mu > 1.5$. Assume that the variance continues to equal $\sigma^2 = 0.09$ with the new fertilizer. Using $\bar{X}$, the mean of a random sample of size n, as the test statistic, reject H_0 if $\bar{x} \ge c$. Find n and c so that the power function $K(\mu) = P(\bar{X} \ge c : \mu)$ is such that $\alpha = K(1.5) = 0.05$ and $K(1.7) = 0.95$.

8.5-6. Let X equal the butterfat production (in pounds) of a Holstein cow during the 305-day milking period following the birth of a calf. Assume that the distribution of X is $N(\mu, 140^2)$. To test the null hypothesis H_0: $\mu = 715$

against the alternative hypothesis H_1: $\mu < 715$, let the critical region be defined by $C = \{\bar{x} : \bar{x} \le 668.94\}$, where $\bar{x}$ is the sample mean of $n = 25$ butterfat weights from 25 cows selected at random.

(a) Find the power function $K(\mu)$ for this test.

(b) What is the significance level of the test?

(c) What are the values of $K(668.94)$ and $K(622.88)$?

(d) Sketch a graph of the power function.

(e) What conclusion do you draw from the following 25 observations of X?

425	710	661	664	732	714	934	761	744
653	725	657	421	573	535	602	537	405
874	791	721	849	567	468	975		

(f) What is the approximate p-value of the test?

8.5-7. In Exercise 8.5-6, let $C = \{\bar{x} : \bar{x} \le c\}$ be the critical region. Find values for n and c so that the significance level of this test is $\alpha = 0.05$ and the power at $\mu = 650$ is 0.90.

8.5-8. Let X have a Bernoulli distribution with pmf

$$f(x; p) = p^x (1 - p)^{1-x}, \qquad x = 0, 1, \qquad 0 \le p \le 1.$$

We would like to test the null hypothesis H_0: $p \le 0.4$ against the alternative hypothesis H_1: $p > 0.4$. For the test statistic, use $Y = \sum_{i=1}^{n} X_i$, where $X_1, X_2, \ldots, X_n$ is a random sample of size n from this Bernoulli distribution. Let the critical region be of the form $C = \{y : y \ge c\}$.

(a) Let $n = 100$. On the same set of axes, sketch the graphs of the power functions corresponding to the three critical regions, $C_1 = \{y : y \ge 40\}$, $C_2 = \{y : y \ge 50\}$, and $C_3 = \{y : y \ge 60\}$. Use the normal approximation to compute the probabilities.

(b) Let $C = \{y : y \ge 0.45n\}$. On the same set of axes, sketch the graphs of the power functions corresponding to the three samples of sizes 10, 100, and 1000.

8.5-9. Let p denote the probability that, for a particular tennis player, the first serve is good. Since $p = 0.40$, this player decided to take lessons in order to increase p. When the lessons are completed, the hypothesis H_0: $p = 0.40$ will be tested against H_1: $p > 0.40$ on the basis of $n = 25$ trials. Let y equal the number of first serves that are good, and let the critical region be defined by $C = \{y : y \ge 14\}$.

(a) Find the power function $K(p)$ for this test.

(b) What is the value of the significance level, $\alpha = K(0.40)$? Use Table II in Appendix B.

(c) Evaluate $K(p)$ at $p = 0.45, 0.50, 0.60, 0.70, 0.80$, and 0.90. Use Table II.

(d) Sketch the graph of the power function.

(e) If $y = 15$ following the lessons, would H_0 be rejected?

(f) What is the p-value associated with $y = 15$?

8.5-10. Let $X_1, X_2, \ldots, X_8$ be a random sample of size $n = 8$ from a Poisson distribution with mean λ. Reject the simple null hypothesis H_0: $\lambda = 0.5$, and accept H_1: $\lambda > 0.5$, if the observed sum $\sum_{i=1}^{8} x_i \geq 8$.

(a) Compute the significance level α of the test.

(b) Find the power function $K(\lambda)$ of the test as a sum of Poisson probabilities.

(c) Using Table III in Appendix B, determine $K(0.75)$, $K(1)$, and $K(1.25)$.

8.5-11. Let p equal the fraction defective of a certain manufactured item. To test H_0: $p = 1/26$ against H_1: $p > 1/26$, we inspect n items selected at random and

let Y be the number of defective items in this sample. We reject H_0 if the observed $y \geq c$. Find n and c so that $\alpha = K(1/26) \approx 0.05$ and $K(1/10) \approx 0.90$, where $K(p) = P(Y \geq c; p)$. HINT: Use either the normal or Poisson approximation to help solve this exercise.

8.5-12. Let X_1, X_2, X_3 be a random sample of size $n = 3$ from an exponential distribution with mean $\theta > 0$. Reject the simple null hypothesis H_0: $\theta = 2$, and accept the composite alternative hypothesis H_1: $\theta < 2$, if the observed sum $\sum_{i=1}^{3} x_i \leq 2$.

(a) What is the power function $K(\theta)$, written as an integral?

(b) Using integration by parts, define the power function as a summation.

(c) With the help of Table III in Appendix B, determine $\alpha = K(2)$, $K(1)$, $K(1/2)$, and $K(1/4)$.

8.6 BEST CRITICAL REGIONS

In this section, we consider the properties a satisfactory hypothesis test (or critical region) should possess. To introduce our investigation, we begin with a nonstatistical example.

Example 8.6-1

Say that you have α dollars with which to buy books. Further, suppose that you are not interested in the books themselves, but only in filling as much of your bookshelves as possible. How do you decide which books to buy? Does the following approach seem reasonable? First of all, take all the available free books. Then start choosing those books for which the cost of filling an inch of bookshelf is smallest. That is, choose those books for which the ratio c/w is a minimum, where w is the width of the book in inches and c is the cost of the book. Continue choosing books this way until you have spent the α dollars. ∎

To see how Example 8.6-1 provides the background for selecting a good critical region of size α, let us consider a test of the simple hypothesis H_0: $\theta = \theta_0$ against a simple alternative hypothesis H_1: $\theta = \theta_1$. In this discussion, we assume that the random variables $X_1, X_2, \ldots, X_n$ under consideration have a joint pmf of the discrete type, which we here denote by $L(\theta; x_1, x_2, \ldots, x_n)$. That is,

$$P(X_1 = x_1, X_2 = x_2, \ldots, X_n = x_n) = L(\theta; x_1, x_2, \ldots, x_n).$$

A critical region C of size α is a set of points $(x_1, x_2, \ldots, x_n)$ with probability α when $\theta = \theta_0$. For a good test, this set C of points should have a large probability when $\theta = \theta_1$, because, under H_1: $\theta = \theta_1$, we wish to reject H_0: $\theta = \theta_0$. Accordingly, the first point we would place in the critical region C is the one with the smallest ratio:

$$\frac{L(\theta_0; x_1, x_2, \ldots, x_n)}{L(\theta_1; x_1, x_2, \ldots, x_n)}.$$

That is, the "cost" in terms of probability under H_0: $\theta = \theta_0$ is small compared with the probability that we can "buy" if $\theta = \theta_1$. The next point to add to C would be the one with the next-smallest ratio. We would continue to add points to C in this manner until the probability of C, under H_0: $\theta = \theta_0$, equals α. In this way, for the given significance level α, we have achieved the region C with the largest probability when H_1: $\theta = \theta_1$ is true. We now formalize this discussion by defining a best critical region and proving the well-known Neyman–Pearson lemma.

Definition 8.6-1
Consider the test of the simple null hypothesis H_0: $\theta = \theta_0$ against the simple alternative hypothesis H_1: $\theta = \theta_1$. Let C be a critical region of size α; that is, $\alpha = P(C; \theta_0)$. Then C is a **best critical region of size** α if, for every other critical region D of size $\alpha = P(D; \theta_0)$, we have

$$P(C; \theta_1) \geq P(D; \theta_1).$$

That is, when H_1: $\theta = \theta_1$ is true, the probability of rejecting H_0: $\theta = \theta_0$ with the use of the critical region C is at least as great as the corresponding probability with the use of any other critical region D of size α.

Thus, a best critical region of size α is the critical region that has the greatest power among all critical regions of size α. The Neyman–Pearson lemma gives sufficient conditions for a best critical region of size α.

Theorem 8.6-1

(Neyman–Pearson Lemma) Let $X_1, X_2, \ldots, X_n$ be a random sample of size n from a distribution with pdf or pmf $f(x; \theta)$, where θ_0 and θ_1 are two possible values of θ. Denote the joint pdf or pmf of $X_1, X_2, \ldots, X_n$ by the likelihood function

$$L(\theta) = L(\theta; x_1, x_2, \ldots, x_n) = f(x_1; \theta)f(x_2; \theta) \cdots f(x_n; \theta).$$

If there exist a positive constant k and a subset C of the sample space such that

(a) $P[(X_1, X_2, \ldots, X_n) \in C; \theta_0] = \alpha$,

(b) $\dfrac{L(\theta_0)}{L(\theta_1)} \leq k$ for $(x_1, x_2, \ldots, x_n) \in C$, and

(c) $\dfrac{L(\theta_0)}{L(\theta_1)} \geq k$ for $(x_1, x_2, \ldots, x_n) \in C'$,

then C is a best critical region of size α for testing the simple null hypothesis H_0: $\theta = \theta_0$ against the simple alternative hypothesis H_1: $\theta = \theta_1$.

Proof We prove the theorem when the random variables are of the continuous type; for discrete-type random variables, replace the integral signs by summation signs. To simplify the exposition, we shall use the following notation:

$$\int_B L(\theta) = \int \cdots \int_B L(\theta; x_1, x_2, \ldots, x_n) \, dx_1 \, dx_2 \cdots dx_n.$$

Assume that there exists another critical region of size α—say, D—such that, in this new notation,

$$\alpha = \int_C L(\theta_0) = \int_D L(\theta_0).$$

Then we have

$$0 = \int_C L(\theta_0) - \int_D L(\theta_0)$$

$$= \int_{C\cap D'} L(\theta_0) + \int_{C\cap D} L(\theta_0) - \int_{C\cap D} L(\theta_0) - \int_{C'\cap D} L(\theta_0).$$

Hence,

$$0 = \int_{C\cap D'} L(\theta_0) - \int_{C'\cap D} L(\theta_0).$$

By hypothesis (b), $kL(\theta_1) \geq L(\theta_0)$ at each point in C and therefore in $C \cap D'$; thus,

$$k\int_{C\cap D'} L(\theta_1) \geq \int_{C\cap D'} L(\theta_0).$$

By hypothesis (c), $kL(\theta_1) \leq L(\theta_0)$ at each point in C' and therefore in $C' \cap D$; thus, we obtain

$$k\int_{C'\cap D} L(\theta_1) \leq \int_{C'\cap D} L(\theta_0).$$

Consequently,

$$0 = \int_{C\cap D'} L(\theta_0) - \int_{C'\cap D} L(\theta_0) \leq (k)\left\{ \int_{C\cap D'} L(\theta_1) - \int_{C'\cap D} L(\theta_1)\right\}.$$

That is,

$$0 \leq (k)\left\{ \int_{C\cap D'} L(\theta_1) + \int_{C\cap D} L(\theta_1) - \int_{C\cap D} L(\theta_1) - \int_{C'\cap D} L(\theta_1)\right\}$$

or, equivalently,

$$0 \leq (k)\left\{ \int_C L(\theta_1) - \int_D L(\theta_1)\right\}.$$

Thus,

$$\int_C L(\theta_1) \geq \int_D L(\theta_1);$$

that is, $P(C; \theta_1) \geq P(D; \theta_1)$. Since that is true for every critical region D of size α, C is a best critical region of size α. □

For a realistic application of the Neyman–Pearson lemma, consider the next example, in which the test is based on a random sample from a normal distribution.

Example 8.6-2 Let $X_1, X_2, \ldots, X_n$ be a random sample from the normal distribution $N(\mu, 36)$. We shall find the best critical region for testing the simple hypothesis H_0: $\mu = 50$ against the simple alternative hypothesis H_1: $\mu = 55$. Using the ratio of the likelihood functions, namely, $L(50)/L(55)$, we shall find those points in the sample space for which this ratio is less than or equal to some positive constant k. That is, we shall solve the following inequality:

$$\frac{L(50)}{L(55)} = \frac{(72\pi)^{-n/2}\exp\left[-\left(\frac{1}{72}\right)\sum_{i=1}^{n}(x_i-50)^2\right]}{(72\pi)^{-n/2}\exp\left[-\left(\frac{1}{72}\right)\sum_{i=1}^{n}(x_i-55)^2\right]}$$

$$= \exp\left[-\left(\frac{1}{72}\right)\left(10\sum_{i=1}^{n}x_i+n50^2-n55^2\right)\right] \le k.$$

If we take the natural logarithm of each member of the inequality, we find that

$$-10\sum_{i=1}^{n}x_i-n50^2+n55^2 \le (72)\ln k.$$

Thus,

$$\frac{1}{n}\sum_{i=1}^{n}x_i \ge -\frac{1}{10n}[n50^2-n55^2+(72)\ln k]$$

or, equivalently,

$$\bar{x} \ge c,$$

where $c = -(1/10n)[n50^2-n55^2+(72)\ln k]$. Hence, $L(50)/L(55) \le k$ is equivalent to $\bar{x} \ge c$. According to the Neyman–Pearson lemma, a best critical region is

$$C = \{(x_1,x_2,\ldots,x_n): \bar{x} \ge c\},$$

where c is selected so that the size of the critical region is α. Say $n = 16$ and $c = 53$. Then, since $\overline{X}$ is $N(50,36/16)$ under H_0, we have

$$\alpha = P(\overline{X} \ge 53; \mu = 50)$$

$$= P\left(\frac{\overline{X}-50}{6/4} \ge \frac{3}{6/4}; \mu = 50\right) = 1-\Phi(2) = 0.0228. \qquad \blacksquare$$

This last example illustrates what is often true, namely, that the inequality

$$L(\theta_0)/L(\theta_1) \le k$$

can be expressed in terms of a function $u(x_1,x_2,\ldots,x_n)$, say,

$$u(x_1,\ldots,x_n) \le c_1$$

or

$$u(x_1,\ldots,x_n) \ge c_2,$$

where c_1 or c_2 is selected so that the size of the critical region is α. Thus, the test can be based on the statistic $u(x_1,\ldots,x_n)$. As an example, if we want α to be a given value—say, 0.05—we could then choose our c_1 or c_2. In Example 8.6-2, with $\alpha = 0.05$, we want

$$0.05 = P(\overline{X} \geq c; \mu = 50)$$

$$= P\left(\frac{\overline{X} - 50}{6/4} \geq \frac{c - 50}{6/4}; \mu = 50\right) = 1 - \Phi\left(\frac{c - 50}{6/4}\right).$$

Hence, it must be true that $(c - 50)/(3/2) = 1.645$, or, equivalently,

$$c = 50 + \frac{3}{2}(1.645) \approx 52.47.$$

Example 8.6-3 Let $X_1, X_2, \ldots, X_n$ denote a random sample of size n from a Poisson distribution with mean λ. A best critical region for testing $H_0: \lambda = 2$ against $H_1: \lambda = 5$ is given by

$$\frac{L(2)}{L(5)} = \frac{2^{\Sigma x_i} e^{-2n}}{x_1! x_2! \cdots x_n!} \frac{x_1! x_2! \cdots x_n!}{5^{\Sigma x_i} e^{-5n}} \leq k.$$

This inequality can be written as

$$\left(\frac{2}{5}\right)^{\Sigma x_i} e^{3n} \leq k, \qquad \text{or} \qquad (\Sigma x_i) \ln\left(\frac{2}{5}\right) + 3n \leq \ln k.$$

Since $\ln(2/5) < 0$, the latter inequality is the same as

$$\sum_{i=1}^{n} x_i \geq \frac{\ln k - 3n}{\ln(2/5)} = c.$$

If $n = 4$ and $c = 13$, then

$$\alpha = P\left(\sum_{i=1}^{4} X_i \geq 13; \lambda = 2\right) = 1 - 0.936 = 0.064,$$

from Table III in Appendix B, since $\sum_{i=1}^{4} X_i$ has a Poisson distribution with mean 8 when $\lambda = 2$. ∎

When $H_0: \theta = \theta_0$ and $H_1: \theta = \theta_1$ are both simple hypotheses, a critical region of size α is a best critical region if the probability of rejecting H_0 when H_1 is true is a maximum compared with all other critical regions of size α. The test using the best critical region is called a **most powerful test**, because it has the greatest value of the power function at $\theta = \theta_1$ compared with that of other tests with significance level α. If H_1 is a composite hypothesis, the power of a test depends on each simple alternative in H_1.

Definition 8.6-2
A test defined by a critical region C of size α is a **uniformly most powerful test** if it is a most powerful test against each simple alternative in H_1. The critical region C is called a **uniformly most powerful critical region of size α**.

Let us consider again Example 8.6-2 when the alternative hypothesis is composite.

Example 8.6-4

Let $X_1, X_2, \ldots, X_n$ be a random sample from $N(\mu, 36)$. We have seen that, in testing $H_0: \mu = 50$ against $H_1: \mu = 55$, a best critical region C is defined by $C = \{(x_1, x_2, \ldots, x_n): \bar{x} \geq c\}$, where c is selected so that the significance level is α. Now consider testing $H_0: \mu = 50$ against the one-sided composite alternative hypothesis $H_1: \mu > 50$. For each simple hypothesis in H_1—say, $\mu = \mu_1$—the quotient of the likelihood functions is

$$\frac{L(50)}{L(\mu_1)} = \frac{(72\pi)^{-n/2} \exp\left[-\left(\frac{1}{72}\right)\sum_{i=1}^{n}(x_i - 50)^2\right]}{(72\pi)^{-n/2} \exp\left[-\left(\frac{1}{72}\right)\sum_{i=1}^{n}(x_i - \mu_1)^2\right]}$$

$$= \exp\left[-\frac{1}{72}\left\{2(\mu_1 - 50)\sum_{i=1}^{n} x_i + n(50^2 - \mu_1^2)\right\}\right].$$

Now, $L(50)/L(\mu_1) \leq k$ if and only if

$$\bar{x} \geq \frac{(-72)\ln(k)}{2n(\mu_1 - 50)} + \frac{50 + \mu_1}{2} = c.$$

Thus, the best critical region of size α for testing $H_0: \mu = 50$ against $H_1: \mu = \mu_1$, where $\mu_1 > 50$, is given by $C = \{(x_1, x_2, \ldots, x_n): \bar{x} \geq c\}$, where c is selected such that $P(\bar{X} \geq c; H_0: \mu = 50) = \alpha$. Note that the same value of c can be used for each $\mu_1 > 50$, but (of course) k does not remain the same. Since the critical region C defines a test that is most powerful against each simple alternative $\mu_1 > 50$, this is a uniformly most powerful test, and C is a uniformly most powerful critical region of size α. Again, if $\alpha = 0.05$, then $c \approx 52.47$. ∎

Example 8.6-5

Let Y have the binomial distribution $b(n, p)$. To find a uniformly most powerful test of the simple null hypothesis $H_0: p = p_0$ against the one-sided alternative hypothesis $H_1: p > p_0$, consider, with $p_1 > p_0$,

$$\frac{L(p_0)}{L(p_1)} = \frac{\binom{n}{y} p_0^y (1 - p_0)^{n-y}}{\binom{n}{y} p_1^y (1 - p_1)^{n-y}} \leq k.$$

This is equivalent to

$$\left[\frac{p_0(1 - p_1)}{p_1(1 - p_0)}\right]^y \left[\frac{1 - p_0}{1 - p_1}\right]^n \leq k$$

and

$$y \ln\left[\frac{p_0(1 - p_1)}{p_1(1 - p_0)}\right] \leq \ln k - n \ln\left[\frac{1 - p_0}{1 - p_1}\right].$$

Since $p_0 < p_1$, we have $p_0(1 - p_1) < p_1(1 - p_0)$. Thus, $\ln[p_0(1 - p_1)/p_1(1 - p_0)] < 0$. It follows that

$$\frac{y}{n} \geq \frac{\ln k - n \ln[(1 - p_0)/(1 - p_1)]}{n \ln[p_0(1 - p_1)/p_1(1 - p_0)]} = c$$

for each $p_1 > p_0$.

It is interesting to note that if the alternative hypothesis is the one-sided H_1: $p < p_0$, then a uniformly most powerful test is of the form $(y/n) \leq c$. Thus, the tests of H_0: $p = p_0$ against the one-sided alternatives given in Table 8.3-1 are uniformly most powerful. ■

Exercise 8.6-5 will demonstrate that uniformly most powerful tests do not always exist; in particular, they usually do not exist when the composite alternative hypothesis is two sided.

REMARK We close this section with one easy but important observation: If a sufficient statistic $Y = u(X_1, X_2, \ldots, X_n)$ exists for θ, then, by the factorization theorem (Definition 6.7-1),

$$\frac{L(\theta_0)}{L(\theta_1)} = \frac{\phi[u(x_1, x_2, \ldots, x_n); \theta_0] \, h(x_1, x_2, \ldots, x_n)}{\phi[u(x_1, x_2, \ldots, x_n); \theta_1] \, h(x_1, x_2, \ldots, x_n)}$$

$$= \frac{\phi[u(x_1, x_2, \ldots, x_n); \theta_0)}{\phi[u(x_1, x_2, \ldots, x_n); \theta_1]}.$$

Thus, $L(\theta_0)/L(\theta_1) \leq k$ provides a critical region that is a function of the observations $x_1, x_2, \ldots x_n$ only through the observed value of the sufficient statistic $y = u(x_1, x_2, \ldots, x_n)$. Hence, best critical and uniformly most powerful critical regions are based upon sufficient statistics when they exist. ■

Exercises

8.6-1. Let $X_1, X_2, \ldots, X_n$ be a random sample from a normal distribution $N(\mu, 64)$.

(a) Show that $C = \{(x_1, x_2, \ldots, x_n) : \bar{x} \leq c\}$ is a best critical region for testing H_0: $\mu = 80$ against H_1: $\mu = 76$.

(b) Find n and c so that $\alpha \approx 0.05$ and $\beta \approx 0.05$.

8.6-2. Let $X_1, X_2, \ldots, X_n$ be a random sample from $N(0, \sigma^2)$.

(a) Show that $C = \{(x_1, x_2, \ldots, x_n) : \sum_{i=1}^{n} x_i^2 \geq c\}$ is a best critical region for testing H_0: $\sigma^2 = 4$ against H_1: $\sigma^2 = 16$.

(b) If $n = 15$, find the value of c so that $\alpha = 0.05$. HINT: Recall that $\sum_{i=1}^{n} X_i^2 / \sigma^2$ is $\chi^2(n)$.

(c) If $n = 15$ and c is the value found in part (b), find the approximate value of $\beta = P(\sum_{i=1}^{n} X_i^2 < c$; $\sigma^2 = 16)$.

8.6-3. Let X have an exponential distribution with a mean of θ; that is, the pdf of X is $f(x; \theta) = (1/\theta)e^{-x/\theta}$, $0 < x < \infty$. Let $X_1, X_2, \ldots, X_n$ be a random sample from this distribution.

(a) Show that a best critical region for testing H_0: $\theta = 3$ against H_1: $\theta = 5$ can be based on the statistic $\sum_{i=1}^{n} X_i$.

(b) If $n = 12$, use the fact that $(2/\theta) \sum_{i=1}^{12} X_i$ is $\chi^2(24)$ to find a best critical region of size $\alpha = 0.10$.

(c) If $n = 12$, find a best critical region of size $\alpha = 0.10$ for testing H_0: $\theta = 3$ against H_1: $\theta = 7$.

(d) If H_1: $\theta > 3$, is the common region found in parts (b) and (c) a uniformly most powerful critical region of size $\alpha = 0.10$?

8.6-4. Let $X_1, X_2, \ldots, X_n$ be a random sample of Bernoulli trials $b(1, p)$.

(a) Show that a best critical region for testing H_0: $p = 0.9$ against H_1: $p = 0.8$ can be based on the statistic $Y = \sum_{i=1}^{n} X_i$, which is $b(n, p)$.

(b) If $C = \{(x_1, x_2, \ldots, x_n) : \sum_{i=1}^{n} x_i \leq n(0.85)\}$ and $Y = \sum_{i=1}^{n} X_i$, find the value of n such that $\alpha = P[Y \leq n(0.85); p = 0.9] \approx 0.10$. HINT: Use the normal approximation for the binomial distribution.

(c) What is the approximate value of $\beta = P[Y > n(0.85); p = 0.8]$ for the test given in part (b)?

(d) Is the test of part (b) a uniformly most powerful test when the alternative hypothesis is $H_1: p < 0.9$?

8.6-5. Let $X_1, X_2, \ldots, X_n$ be a random sample from the normal distribution $N(\mu, 36)$.

(a) Show that a uniformly most powerful critical region for testing $H_0: \mu = 50$ against $H_1: \mu < 50$ is given by $C_2 = \{\bar{x}: \bar{x} \leq c\}$.

(b) With this result and that of Example 8.6-4, argue that a uniformly most powerful test for testing $H_0: \mu = 50$ against $H_1: \mu \neq 50$ does not exist.

8.6-6. Let $X_1, X_2, \ldots, X_n$ be a random sample from the normal distribution $N(\mu, 9)$. To test the hypothesis $H_0: \mu = 80$ against $H_1: \mu \neq 80$, consider the following three critical regions: $C_1 = \{\bar{x}: \bar{x} \geq c_1\}$, $C_2 = \{\bar{x}: \bar{x} \leq c_2\}$, and $C_3 = \{\bar{x}: |\bar{x} - 80| \geq c_3\}$.

(a) If $n = 16$, find the values of c_1, c_2, c_3 such that the size of each critical region is 0.05. That is, find c_1, c_2, c_3 such that

$$0.05 = P(\overline{X} \in C_1; \mu = 80) = P(\overline{X} \in C_2; \mu = 80)$$

$$= P(\overline{X} \in C_3; \mu = 80).$$

(b) On the same graph paper, sketch the power functions for these three critical regions.

8.6-7. Let $X_1, X_2, \ldots, X_{10}$ be a random sample of size 10 from a Poisson distribution with mean μ.

(a) Show that a uniformly most powerful critical region for testing $H_0: \mu = 0.5$ against $H_1: \mu > 0.5$ can be defined with the use of the statistic $\sum_{i=1}^{10} X_i$.

(b) What is a uniformly most powerful critical region of size $\alpha = 0.068$? Recall that $\sum_{i=1}^{10} X_i$ has a Poisson distribution with mean 10μ.

(c) Sketch the power function of this test.

8.6-8. Consider a random sample $X_1, X_2, \ldots, X_n$ from a distribution with pdf $f(x; \theta) = \theta(1 - x)^{\theta - 1}$, $0 < x < 1$, where $0 < \theta$. Find the form of the uniformly most powerful test of $H_0: \theta = 1$ against $H_1: \theta > 1$.

8.6-9. Let $X_1, X_2, \ldots, X_5$ be a random sample from the Bernoulli distribution $p(x; \theta) = \theta^x (1 - \theta)^{1-x}$. We reject $H_0: \theta = 1/2$ and accept $H_1: \theta < 1/2$ if $Y = \sum_{i=1}^{5} X_i \leq c$. Show that this is a uniformly most powerful test and find the power function $K(\theta)$ if $c = 1$.

8.7* LIKELIHOOD RATIO TESTS

In this section, we consider a general test-construction method that is applicable when either or both of the null and alternative hypotheses—say, H_0 and H_1—are composite. We continue to assume that the functional form of the pdf is known, but that it depends on one or more unknown parameters. That is, we assume that the pdf of X is $f(x; \theta)$, where θ represents one or more unknown parameters. We let Ω denote the total parameter space—that is, the set of all possible values of the parameter θ given by either H_0 or H_1. These hypotheses will be stated as

$$H_0: \theta \in \omega, \qquad H_1: \theta \in \omega',$$

where ω is a subset of Ω and ω' is the complement of ω with respect to Ω. The test will be constructed with the use of a ratio of likelihood functions that have been maximized in ω and Ω, respectively. In a sense, this is a natural generalization of the ratio appearing in the Neyman–Pearson lemma when the two hypotheses were simple.

Definition 8.7-1

The **likelihood ratio** is the quotient

$$\lambda = \frac{L(\hat{\omega})}{L(\hat{\Omega})},$$

where $L(\hat{\omega})$ is the maximum of the likelihood function with respect to θ when $\theta \in \omega$ and $L(\hat{\Omega})$ is the maximum of the likelihood function with respect to θ when $\theta \in \Omega$.

Because λ is the quotient of nonnegative functions, $\lambda \geq 0$. In addition, since $\omega \subset \Omega$, it follows that $L(\widehat{\omega}) \leq L(\widehat{\Omega})$ and hence $\lambda \leq 1$. Thus, $0 \leq \lambda \leq 1$. If the maximum of L in ω is much smaller than that in Ω, it would seem that the data $x_1, x_2, \ldots, x_n$ do not support the hypothesis $H_0: \theta \in \omega$. That is, a small value of the ratio $\lambda = L(\widehat{\omega})/L(\widehat{\Omega})$ would lead to the rejection of H_0. In contrast, a value of the ratio λ that is close to 1 would support the null hypothesis H_0. This reasoning leads us to the next definition.

Definition 8.7-2
To test $H_0: \theta \in \omega$ against $H_1: \theta \in \omega'$, the **critical region for the likelihood ratio test** is the set of points in the sample space for which

$$\lambda = \frac{L(\widehat{\omega})}{L(\widehat{\Omega})} \leq k,$$

where $0 < k < 1$ and k is selected so that the test has a desired significance level α.

The next example illustrates these definitions.

Example 8.7-1

Assume that the weight X in ounces of a "10-pound" bag of sugar is $N(\mu, 5)$. We shall test the hypothesis $H_0: \mu = 162$ against the alternative hypothesis $H_1: \mu \neq 162$. Thus, $\Omega = \{\mu : -\infty < \mu < \infty\}$ and $\omega = \{162\}$. To find the likelihood ratio, we need $L(\widehat{\omega})$ and $L(\widehat{\Omega})$. When H_0 is true, μ can take on only one value, namely, $\mu = 162$. Hence, $L(\widehat{\omega}) = L(162)$. To find $L(\widehat{\Omega})$, we must find the value of μ that maximizes $L(\mu)$. Recall that $\widehat{\mu} = \overline{x}$ is the maximum likelihood estimate of μ. Then $L(\widehat{\Omega}) = L(\overline{x})$, and the likelihood ratio $\lambda = L(\widehat{\omega})/L(\widehat{\Omega})$ is given by

$$\lambda = \frac{(10\pi)^{-n/2} \exp\left[-\left(\dfrac{1}{10}\right)\displaystyle\sum_{i=1}^{n}(x_i - 162)^2\right]}{(10\pi)^{-n/2} \exp\left[-\left(\dfrac{1}{10}\right)\displaystyle\sum_{i=1}^{n}(x_i - \overline{x})^2\right]}$$

$$= \frac{\exp\left[-\left(\dfrac{1}{10}\right)\displaystyle\sum_{i=1}^{n}(x_i - \overline{x})^2 - \left(\dfrac{n}{10}\right)(\overline{x} - 162)^2\right]}{\exp\left[-\left(\dfrac{1}{10}\right)\displaystyle\sum_{i=1}^{n}(x_i - \overline{x})^2\right]}$$

$$= \exp\left[-\frac{n}{10}(\overline{x} - 162)^2\right].$$

On the one hand, a value of $\overline{x}$ close to 162 would tend to support H_0, and in that case λ is close to 1. On the other hand, an $\overline{x}$ that differs from 162 by too much would tend to support H_1. (See Figure 8.7-1 for the graph of this likelihood ratio when $n = 5$.)

A critical region for a likelihood ratio is given by $\lambda \leq k$, where k is selected so that the significance level of the test is α. Using this criterion and simplifying the

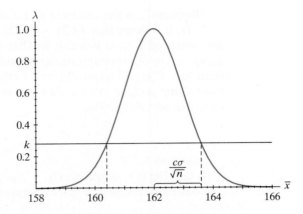

Figure 8.7-1 The likelihood ratio for testing $H_0\colon \mu = 162$

inequality as we do when we use the Neyman–Pearson lemma, we find that $\lambda \leq k$ is equivalent to each of the following inequalities:

$$-\left(\frac{n}{10}\right)(\bar{x} - 162)^2 \leq \ln k,$$

$$(\bar{x} - 162)^2 \geq -\left(\frac{10}{n}\right)\ln k,$$

$$\frac{|\bar{x} - 162|}{\sqrt{5}/\sqrt{n}} \geq \frac{\sqrt{-(10/n)\ln k}}{\sqrt{5}/\sqrt{n}} = c.$$

Since $Z = (\overline{X} - 162)/(\sqrt{5}/\sqrt{n})$ is $N(0,1)$ when $H_0\colon \mu = 162$ is true, let $c = z_{\alpha/2}$. Thus, the critical region is

$$C = \left\{ \bar{x}\colon \frac{|\bar{x} - 162|}{\sqrt{5}/\sqrt{n}} \geq z_{\alpha/2} \right\}.$$

To illustrate, if $\alpha = 0.05$, then $z_{0.025} = 1.96$. ∎

As illustrated in Example 8.7-1, the inequality $\lambda \leq k$ can often be expressed in terms of a statistic whose distribution is known. Also, note that although the likelihood ratio test is an intuitive test, it leads to the same critical region as that given by the Neyman–Pearson lemma when H_0 and H_1 are both simple hypotheses.

Suppose now that the random sample $X_1, X_2, \ldots, X_n$ arises from the normal population $N(\mu, \sigma^2)$, where both μ and σ^2 are unknown. Let us consider the likelihood ratio test of the null hypothesis $H_0\colon \mu = \mu_0$ against the two-sided alternative hypothesis $H_1\colon \mu \neq \mu_0$. For this test,

$$\omega = \{(\mu, \sigma^2)\colon \mu = \mu_0,\ 0 < \sigma^2 < \infty\}$$

and

$$\Omega = \{(\mu, \sigma^2)\colon -\infty < \mu < \infty,\ 0 < \sigma^2 < \infty\}.$$

If $(\mu, \sigma^2) \in \Omega$, then the observed maximum likelihood estimates are $\widehat{\mu} = \bar{x}$ and $\widehat{\sigma^2} = (1/n) \sum_{i=1}^{n} (x_i - \bar{x})^2$.
Thus,

$$L(\widehat{\Omega}) = \left[\frac{1}{2\pi \left(\dfrac{1}{n} \right) \sum_{i=1}^{n} (x_i - \bar{x})^2} \right]^{n/2} \exp\left[-\frac{\sum_{i=1}^{n} (x_i - \bar{x})^2}{\left(\dfrac{2}{n} \right) \sum_{i=1}^{n} (x_i - \bar{x})^2} \right]$$

$$= \left[\frac{ne^{-1}}{2\pi \sum_{i=1}^{n} (x_i - \bar{x})^2} \right]^{n/2}.$$

Similarly, if $(\mu, \sigma^2) \in \omega$, then the observed maximum likelihood estimates are $\widehat{\mu} = \mu_0$ and $\widehat{\sigma^2} = (1/n) \sum_{i=1}^{n} (x_i - \mu_0)^2$. Hence,

$$L(\widehat{\omega}) = \left[\frac{1}{2\pi \left(\dfrac{1}{n} \right) \sum_{i=1}^{n} (x_i - \mu_0)^2} \right]^{n/2} \exp\left[-\frac{\sum_{i=1}^{n} (x_i - \mu_0)^2}{\left(\dfrac{2}{n} \right) \sum_{i=1}^{n} (x_i - \mu_0)^2} \right]$$

$$= \left[\frac{ne^{-1}}{2\pi \sum_{i=1}^{n} (x_i - \mu_0)^2} \right]^{n/2}.$$

The likelihood ratio $\lambda = L(\widehat{\omega})/L(\widehat{\Omega})$ for this test is

$$\lambda = \frac{\left[\dfrac{ne^{-1}}{2\pi \sum_{i=1}^{n} (x_i - \mu_0)^2} \right]^{n/2}}{\left[\dfrac{ne^{-1}}{2\pi \sum_{i=1}^{n} (x_i - \bar{x})^2} \right]^{n/2}} = \left[\frac{\sum_{i=1}^{n} (x_i - \bar{x})^2}{\sum_{i=1}^{n} (x_i - \mu_0)^2} \right]^{n/2}.$$

However, note that

$$\sum_{i=1}^{n} (x_i - \mu_0)^2 = \sum_{i=1}^{n} (x_i - \bar{x} + \bar{x} - \mu_0)^2 = \sum_{i=1}^{n} (x_i - \bar{x})^2 + n(\bar{x} - \mu_0)^2.$$

If this substitution is made in the denominator of λ, we have

$$\lambda = \left[\frac{\sum_{i=1}^{n} (x_i - \bar{x})^2}{\sum_{i=1}^{n} (x_i - \bar{x})^2 + n(\bar{x} - \mu_0)^2} \right]^{n/2}$$

$$= \left[\frac{1}{1 + \dfrac{n(\bar{x} - \mu_0)^2}{\sum_{i=1}^{n} (x_i - \bar{x})^2}} \right]^{n/2}.$$

Note that λ is close to 1 when $\bar{x}$ is close to μ_0 and λ is small when $\bar{x}$ and μ_0 differ by a great deal. The likelihood ratio test, given by the inequality $\lambda \leq k$, is the same as

$$\frac{1}{1 + \dfrac{n(\bar{x} - \mu_0)^2}{\sum_{i=1}^{n}(x_i - \bar{x})^2}} \leq k^{2/n}$$

or, equivalently,

$$\frac{n(\bar{x} - \mu_0)^2}{\dfrac{1}{n-1}\sum_{i=1}^{n}(x_i - \bar{x})^2} \geq (n-1)(k^{-2/n} - 1).$$

When H_0 is true, $\sqrt{n}(\overline{X} - \mu_0)/\sigma$ is $N(0, 1)$ and $\sum_{i=1}^{n}(X_i - \overline{X})^2/\sigma^2$ has an independent chi-square distribution $\chi^2(n-1)$. Hence, under H_0,

$$T = \frac{\sqrt{n}(\overline{X} - \mu_0)/\sigma}{\sqrt{\sum_{i=1}^{n}(X_i - \overline{X})^2/[\sigma^2(n-1)]}} = \frac{\sqrt{n}(\overline{X} - \mu_0)}{\sqrt{\sum_{i=1}^{n}(X_i - \overline{X})^2/(n-1)}} = \frac{\overline{X} - \mu_0}{S/\sqrt{n}}$$

has a t distribution with $r = n - 1$ degrees of freedom. In accordance with the likelihood ratio test criterion, H_0 is rejected if the observed

$$T^2 \geq (n-1)(k^{-2/n} - 1).$$

That is, we reject H_0: $\mu = \mu_0$ and accept H_1: $\mu \neq \mu_0$ at the α level of significance if the observed $|T| \geq t_{\alpha/2}(n - 1)$.

Note that this test is exactly the same as that listed in Table 8.1-2 for testing H_0: $\mu = \mu_0$ against H_1: $\mu \neq \mu_0$. That is, the test listed there is a likelihood ratio test. As a matter of fact, all six of the tests given in Tables 8.1-2 and 8.2-1 are likelihood ratio tests. Thus, the examples and exercises associated with those tables are illustrations of the use of such tests.

The final development of this section concerns a test about the variance of a normal population. Let $X_1, X_2, \ldots, X_n$ be a random sample from $N(\mu, \sigma^2)$, where μ and σ^2 are unknown. We wish to test H_0: $\sigma^2 = \sigma_0^2$ against H_1: $\sigma^2 \neq \sigma_0^2$. For this purpose, we have

$$\omega = \{(\mu, \sigma^2): -\infty < \mu < \infty, \ \sigma^2 = \sigma_0^2\}$$

and

$$\Omega = \{(\mu, \sigma^2): -\infty < \mu < \infty, \ 0 < \sigma^2 < \infty\}.$$

As in the test concerning the mean, we obtain

$$L(\widehat{\Omega}) = \left[\frac{ne^{-1}}{2\pi \sum_{i=1}^{n}(x_i - \bar{x})^2}\right]^{n/2}.$$

If $(\mu, \sigma^2) \in \omega$, then $\widehat{\mu} = \bar{x}$ and $\widehat{\sigma^2} = \sigma_0^2$; thus,

$$L(\widehat{\omega}) = \left(\frac{1}{2\pi\sigma_0^2}\right)^{n/2} \exp\left[-\frac{\sum_{i=1}^{n}(x_i - \bar{x})^2}{2\sigma_0^2}\right].$$

Accordingly, the likelihood ratio test $\lambda = L(\hat{\omega})/L(\hat{\Omega})$ is

$$\lambda = \left(\frac{w}{n}\right)^{n/2} \exp\left(-\frac{w}{2} + \frac{n}{2}\right) \leq k,$$

where $w = \sum_{i=1}^{n}(x_i - \bar{x})^2/\sigma_0^2$. Solving this inequality for w, we obtain a solution of the form $w \leq c_1$ or $w \geq c_2$, where the constants c_1 and c_2 are appropriate functions of the constants k and n so as to achieve the desired significance level α. However these values of c_1 and c_2 do not place probability $\alpha/2$ in each of the two regions $w \leq c_1$ and $w \geq c_2$. Since $W = \sum_{i=1}^{n}(X_i - \bar{X})^2/\sigma_0^2$ is $\chi^2(n-1)$ if $H_0: \sigma^2 = \sigma_0^2$ is true, most statisticians modify this test slightly by taking $c_1 = \chi_{1-\alpha/2}^2(n-1)$ and $c_2 = \chi_{\alpha/2}^2(n-1)$. As a matter of fact, most tests involving normal assumptions are likelihood ratio tests or modifications of them; included are tests involving regression and analysis of variance (see Chapter 9).

REMARK Note that likelihood ratio tests are based on sufficient statistics when they exist, as was also true of best critical and uniformly most powerful critical regions. ∎

Exercises

8.7-1. In Example 8.7-1, suppose that $n = 20$ and $\bar{x} = 161.1$.

(a) Is H_0 accepted if $\alpha = 0.10$?

(b) Is H_0 accepted if $\alpha = 0.05$?

(c) What is the p-value of this test?

8.7-2. Assume that the weight X in ounces of a "10-ounce" box of cornflakes is $N(\mu, 0.03)$. Let $X_1, X_2, \ldots, X_n$ be a random sample from this distribution.

(a) To test the hypothesis $H_0: \mu \geq 10.35$ against the alternative hypothesis $H_1: \mu < 10.35$, what is the critical region of size $\alpha = 0.05$ specified by the likelihood ratio test criterion? HINT: Note that if $\mu \geq 10.35$ and $\bar{x} < 10.35$, then $\hat{\mu} = 10.35$.

(b) If a random sample of $n = 50$ boxes yielded a sample mean of $\bar{x} = 10.31$, is H_0 rejected? HINT: Find the critical value z_α when H_0 is true by taking $\mu = 10.35$, which is the extreme value in $\mu \geq 10.35$.

(c) What is the p-value of this test?

8.7-3. Let $X_1, X_2, \ldots, X_n$ be a random sample of size n from the normal distribution $N(\mu, 100)$.

(a) To test $H_0: \mu = 230$ against $H_1: \mu > 230$, what is the critical region specified by the likelihood ratio test criterion?

(b) Is this test uniformly most powerful?

(c) If a random sample of $n = 16$ yielded $\bar{x} = 232.6$, is H_0 accepted at a significance level of $\alpha = 0.10$?

(d) What is the p-value of this test?

8.7-4. Let $X_1, X_2, \ldots, X_n$ be a random sample of size n from the normal distribution $N(\mu, \sigma_0^2)$, where σ_0^2 is known but μ is unknown.

(a) Find the likelihood ratio test for $H_0: \mu = \mu_0$ against $H_1: \mu \neq \mu_0$. Show that this critical region for a test with significance level α is given by $|\bar{X} - \mu_0| > z_{\alpha/2}\sigma_0/\sqrt{n}$.

(b) Test $H_0: \mu = 59$ against $H_1: \mu \neq 59$ when $\sigma^2 = 225$ and a sample of size $n = 100$ yielded $\bar{x} = 56.13$. Let $\alpha = 0.05$.

(c) What is the p-value of this test? Note that H_1 is a two-sided alternative.

8.7-5. It is desired to test the hypothesis $H_0: \mu = 30$ against the alternative hypothesis $H_1: \mu \neq 30$, where μ is the mean of a normal distribution and σ^2 is unknown. If a random sample of size $n = 9$ has $\bar{x} = 32.8$ and $s = 4$, is H_0 accepted at an $\alpha = 0.05$ significance level? What is the approximate p-value of this test?

8.7-6. To test $H_0: \mu = 335$ against $H_1: \mu < 335$ under normal assumptions, a random sample of size 17 yielded $\bar{x} = 324.8$ and $s = 40$. Is H_0 accepted at an $\alpha = 0.10$ significance level?

8.7-7. Let X have a normal distribution in which μ and σ^2 are both unknown. It is desired to test $H_0: \mu = 1.80$ against $H_1: \mu > 1.80$ at an $\alpha = 0.10$ significance level. If a random sample of size $n = 121$ yielded $\bar{x} = 1.84$ and $s = 0.20$, is H_0 accepted or rejected? What is the p-value of this test?

8.7-8. Let $X_1, X_2, \ldots, X_n$ be a random sample from an exponential distribution with mean θ. Show that the likelihood ratio test of $H_0: \theta = \theta_0$ against $H_1: \theta \neq \theta_0$ has a critical region of the form $\sum_{i=1}^{n} x_i \leq c_1$ or $\sum_{i=1}^{n} x_i \geq c_2$. How would you modify this test so that chi-square tables can be used easily?

8.7-9. Let independent random samples of sizes n and m be taken respectively from two normal distributions with unknown means μ_X and μ_Y and unknown variances σ_X^2 and σ_Y^2.

(a) Show that when $\sigma_X^2 = \sigma_Y^2$, the likelihood ratio for testing $H_0: \mu_X = \mu_Y$ against $H_1: \mu_X \neq \mu_Y$ is a function of the usual two-sample t statistic.

(b) Show that the likelihood ratio for testing $H_0: \sigma_X^2 = \sigma_Y^2$ against $H_1: \sigma_X^2 \neq \sigma_Y^2$ is a function of the usual two-sample F statistic.

8.7-10. Referring back to Exercise 6.4-19, find the likelihood ratio test of $H_0: \gamma = 1$, μ unspecified, against all alternatives.

8.7-11. Let $Y_1, Y_2, \ldots, Y_n$ be n independent random variables with normal distributions $N(\beta x_i, \sigma^2)$, where $x_1, x_2, \ldots, x_n$ are known and not all equal and β and σ^2 are unknown parameters.

(a) Find the likelihood ratio test for $H_0: \beta = 0$ against $H_1: \beta \neq 0$.

(b) Can this test be based on a statistic with a well-known distribution?

HISTORICAL COMMENTS Most of the tests presented in this section result from the use of methods found in the theories of Jerzy Neyman and Egon Pearson, a son of Karl Pearson. Neyman and Pearson formed a team, particularly in the 1920s and 1930s, which produced theoretical results that were important contributions to the area of testing statistical hypotheses.

Neyman and Pearson knew, in testing hypotheses, that they needed a critical region that had high probability when the alternative was true, but they did not have a procedure to find the best one. Neyman was thinking about this late one day when his wife told him they had to attend a concert. He kept thinking of this problem during the concert and finally, in the middle of the concert, the solution came to him: Select points in the critical region for which the ratio of the pdf under the alternative hypothesis to that under the null hypothesis is as large as possible. Hence, the Neyman–Pearson lemma was born. Sometimes solutions occur at the strangest times.

Shortly after Wilcoxon proposed his two-sample test, Mann and Whitney suggested a test based on the estimate of the probability $P(X < Y)$. In this test, they let U equal the number of times that $X_i < Y_j$, $i = 1, 2, \ldots, n_1$ and $j = 1, 2, \ldots, n_2$. Using the data in Example 8.4-6, we find that the computed U is $u = 51$ among all $n_1 n_2 = (8)(8) = 64$ pairs of (X, Y). Thus, the estimate of $P(X < Y)$ is $51/64$ or, in general, $u/n_1 n_2$. At the time of the Mann–Whitney suggestion, it was noted that U was just a linear function of Wilcoxon's W and hence really provided the same test. That relationship is

$$U = W - \frac{n_2(n_2 + 1)}{2},$$

which in our special case is

$$51 = 87 - \frac{8(9)}{2} = 87 - 36.$$

Thus, we often read about the test of Mann, Whitney, and Wilcoxon. From this observation, this test could be thought of as one testing $H_0: P(X < Y) = 1/2$ against the alternative $H_1: P(X < Y) > 1/2$ with critical region of the form $w \geq c$.

Note that the two-sample Wilcoxon test is much less sensitive to extreme values than is the Student's t test based on $\overline{X} - \overline{Y}$. Therefore, if there is considerable skewness or contamination, these proposed distribution-free tests are much safer. In particular, that of Wilcoxon is quite good and does not lose too much power in case the distributions are close to normal ones. It is important to note that the one-sample Wilcoxon test requires symmetry of the underlying distribution, but the two-sample Wilcoxon test does not and thus can be used for skewed distributions.

From theoretical developments beyond the scope of this text, the two Wilcoxon tests are strong competitors of the usual one- and two-sample tests based upon normality assumptions, so the Wilcoxon tests should be considered if those assumptions are questioned.

Computer programs, including Minitab, will calculate the value of the Wilcoxon or Mann–Whitney statistic. However, it is instructive to do these tests by hand so that you can see what is being calculated!

Chapter

MORE TESTS

9

9.1 CHI-SQUARE GOODNESS-OF-FIT TESTS

We now consider applications of the very important chi-square statistic, first proposed by Karl Pearson in 1900. As the reader will see, it is a very adaptable test statistic and can be used for many different types of tests. In particular, one application allows us to test the appropriateness of different probabilistic models.

So that the reader can get some idea as to why Pearson first proposed his chi-square statistic, we begin with the binomial case. That is, let Y_1 be $b(n, p_1)$, where $0 < p_1 < 1$. According to the central limit theorem,

$$Z = \frac{Y_1 - np_1}{\sqrt{np_1(1 - p_1)}}$$

has a distribution that is approximately $N(0,1)$ for large n, particularly when $np_1 \geq 5$ and $n(1 - p_1) \geq 5$. Thus, it is not surprising that $Q_1 = Z^2$ is approximately $\chi^2(1)$. If we let $Y_2 = n - Y_1$ and $p_2 = 1 - p_1$, we see that Q_1 may be written as

$$Q_1 = \frac{(Y_1 - np_1)^2}{np_1(1 - p_1)} = \frac{(Y_1 - np_1)^2}{np_1} + \frac{(Y_1 - np_1)^2}{n(1 - p_1)}.$$

Since

$$(Y_1 - np_1)^2 = (n - Y_1 - n[1 - p_1])^2 = (Y_2 - np_2)^2,$$

we have

$$Q_1 = \frac{(Y_1 - np_1)^2}{np_1} + \frac{(Y_2 - np_2)^2}{np_2}.$$

Let us now carefully consider each term in this last expression for Q_1. Of course, Y_1 is the number of "successes," and np_1 is the expected number of "successes"; that is, $E(Y_1) = np_1$. Likewise, Y_2 and np_2 are, respectively, the number and the expected number of "failures." So each numerator consists of the square of the difference of an observed number and an expected number. Note that Q_1 can be written as

$$Q_1 = \sum_{i=1}^{2} \frac{(Y_i - np_i)^2}{np_i},$$

(9.1-1)

and we have seen intuitively that it has an approximate chi-square distribution with one degree of freedom. In a sense, Q_1 measures the "closeness" of the observed numbers to the corresponding expected numbers. For example, if the observed values of Y_1 and Y_2 equal their expected values, then the computed Q_1 is equal to $q_1 = 0$; but if they differ much from them, then the computed $Q_1 = q_1$ is relatively large.

To generalize, we let an experiment have k (instead of only two) mutually exclusive and exhaustive outcomes, say, $A_1, A_2, \ldots, A_k$. Let $p_i = P(A_i)$, and thus $\sum_{i=1}^{k} p_i = 1$. The experiment is repeated n independent times, and we let Y_i represent the number of times the experiment results in A_i, $i = 1, 2, \ldots, k$. This joint distribution of $Y_1, Y_2, \ldots, Y_{k-1}$ is a straightforward generalization of the binomial distribution, as follows.

In considering the joint pmf, we see that

$$f(y_1, y_2, \ldots, y_{k-1}) = P(Y_1 = y_1, Y_2 = y_2, \ldots, Y_{k-1} = y_{k-1}),$$

where $y_1, y_2, \ldots, y_{k-1}$ are nonnegative integers such that $y_1 + y_2 + \cdots + y_{k-1} \leq n$. Note that we do not need to consider Y_k, since, once the other $k-1$ random variables are observed to equal $y_1, y_2, \ldots, y_{k-1}$, respectively, we know that

$$Y_k = n - y_1 - y_2 - \cdots - y_{k-1} = y_k, \quad \text{say.}$$

From the independence of the trials, the probability of each particular arrangement of y_1 A_1s, y_2 A_2s, $\ldots, y_k$ A_ks is

$$p_1^{y_1} p_2^{y_2} \cdots p_k^{y_k}.$$

The number of such arrangements is the multinomial coefficient

$$\binom{n}{y_1, y_2, \ldots, y_k} = \frac{n!}{y_1! \, y_2! \cdots y_k!}.$$

Hence, the product of these two expressions gives the joint pmf of $Y_1, Y_2, \ldots, Y_{k-1}$:

$$f(y_1, y_2, \ldots, y_{k-1}) = \frac{n!}{y_1! \, y_2! \cdots y_k!} p_1^{y_1} p_2^{y_2} \cdots p_k^{y_k}.$$

(Recall that $y_k = n - y_1 - y_2 - \cdots - y_{k-1}$.)

Pearson then constructed an expression similar to Q_1 (Equation 9.1-1), which involves Y_1 and $Y_2 = n - Y_1$, that we denote by Q_{k-1}, which involves $Y_1, Y_2, \ldots, Y_{k-1}$, and $Y_k = n - Y_1 - Y_2 - \cdots - Y_{k-1}$, namely,

$$Q_{k-1} = \sum_{i=1}^{k} \frac{(Y_i - np_i)^2}{np_i}.$$

He argued that Q_{k-1} has an approximate chi-square distribution with $k-1$ degrees of freedom in much the same way we argued that Q_1 is approximately $\chi^2(1)$. We accept Pearson's conclusion, as the proof is beyond the level of this text.

Some writers suggest that n should be large enough so that $np_i \geq 5$, $i = 1, 2, \ldots, k$, to be certain that the approximating distribution is adequate. This is probably good advice for the beginner to follow, although we have seen the approximation work very well when $np_i \geq 1$, $i = 1, 2, \ldots, k$. The important thing to guard against is allowing some particular np_i to become so small that the corresponding term in Q_{k-1}, namely, $(Y_i - np_i)^2/np_i$, tends to dominate the others because of its small denominator. In any case, it is important to realize that Q_{k-1} has only an approximate chi-square distribution.

We shall now show how we can use the fact that Q_{k-1} is approximately $\chi^2(k-1)$ to test hypotheses about probabilities of various outcomes. Let an experiment have k mutually exclusive and exhaustive outcomes, $A_1, A_2, \ldots, A_k$. We would like to test whether $p_i = P(A_i)$ is equal to a known number p_{i0}, $i = 1, 2, \ldots, k$. That is, we shall test the hypothesis

$$H_0: \ p_i = p_{i0}, \qquad i = 1, 2, \ldots, k.$$

In order to test such a hypothesis, we shall take a sample of size n; that is, we repeat the experiment n independent times. We tend to favor H_0 if the observed number of times that A_i occurred, say, y_i, and the number of times A_i was expected to occur if H_0 were true, namely, np_{i0}, are approximately equal. That is, if

$$q_{k-1} = \sum_{i=1}^{k} \frac{(y_i - np_{i0})^2}{np_{i0}}$$

is "small," we tend to favor H_0. Since the distribution of Q_{k-1} is approximately $\chi^2(k-1)$, we shall reject H_0 if $q_{k-1} \geq \chi_\alpha^2(k-1)$, where α is the desired significance level of the test.

Example 9.1-1

If persons are asked to record a string of random digits, such as

$$3 \quad 7 \quad 2 \quad 4 \quad 1 \quad 9 \quad 7 \quad 2 \quad 1 \quad 5 \quad 0 \quad 8 \ldots,$$

we usually find that they are reluctant to record the same or even the two closest numbers in adjacent positions. And yet, in true random-digit generation, the probability of the next digit being the same as the preceding one is $p_{10} = 1/10$, the probability of the next being only one away from the preceding (assuming that 0 is one away from 9) is $p_{20} = 2/10$, and the probability of all other possibilities is $p_{30} = 7/10$. We shall test one person's concept of a random sequence by asking her to record a string of 51 digits that seems to represent a random-digit generation. Thus, we shall test

$$H_0: \ p_1 = p_{10} = \frac{1}{10}, \ p_2 = p_{20} = \frac{2}{10}, \ p_3 = p_{30} = \frac{7}{10}.$$

The critical region for an $\alpha = 0.05$ significance level is $q_2 \geq \chi_{0.05}^2(2) = 5.991$. The sequence of digits was as follows:

5	8	3	1	9	4	6	7	9	2	6	3	0
8	7	5	1	3	6	2	1	9	5	4	8	0
3	7	1	4	6	0	4	3	8	2	7	3	9
8	5	6	1	8	7	0	3	5	2	5	2	

We went through this listing and observed how many times the next digit was the same as or was one away from the preceding one:

	Frequency	Expected Number
Same	0	$50(1/10) = 5$
One away	8	$50(2/10) = 10$
Other	42	$50(7/10) = 35$
Total	50	50

The computed chi-square statistic is

$$\frac{(0-5)^2}{5} + \frac{(8-10)^2}{10} + \frac{(42-35)^2}{35} = 6.8 > 5.991 = \chi_{0.05}^2(2).$$

Thus, we would say that this string of 51 digits does not seem to be random. ■

One major disadvantage in the use of the chi-square test is that it is a many-sided test. That is, the alternative hypothesis is very general, and it would be difficult to restrict alternatives to situations such as $H_1: p_1 > p_{10}, p_2 > p_{20}, p_3 < p_{30}$ (with $k = 3$). As a matter of fact, some statisticians would probably test H_0 against this particular alternative H_1 by using a linear function of Y_1, Y_2, and Y_3. However, that sort of discussion is beyond the scope of the book because it involves knowing more about the distributions of linear functions of the dependent random variables Y_1, Y_2, and Y_3. In any case, the student who truly recognizes that this chi-square statistic tests $H_0: p_i = p_{i0}, i = 1, 2, \ldots, k$, against all alternatives can usually appreciate the fact that it is more difficult to reject H_0 at a given significance level α when the chi-square statistic is used than it would be if some appropriate "one-sided" test statistic were available.

Many experiments yield a set of data, say, $x_1, x_2, \ldots, x_n$, and the experimenter is often interested in determining whether these data can be treated as the observed values of a random sample $X_1, X_2, \ldots, X_n$ from a given distribution. That is, would this proposed distribution be a reasonable probabilistic model for these sample items? To see how the chi-square test can help us answer questions of this sort, consider a very simple example.

Example 9.1-2 Let X denote the number of heads that occur when four coins are tossed at random. Under the assumption that the four coins are independent and the probability of heads on each coin is $1/2$, X is $b(4, 1/2)$. One hundred repetitions of this experiment resulted in 0, 1, 2, 3, and 4 heads being observed on 7, 18, 40, 31, and 4 trials, respectively. Do these results support the assumptions? That is, is $b(4, 1/2)$ a reasonable model for the distribution of X? To answer this, we begin by letting $A_1 = \{0\}$, $A_2 = \{1\}$, $A_3 = \{2\}$, $A_4 = \{3\}$, and $A_5 = \{4\}$. If $p_{i0} = P(X \in A_i)$ when X is $b(4, 1/2)$, then

$$p_{10} = p_{50} = \binom{4}{0}\left(\frac{1}{2}\right)^4 = \frac{1}{16} = 0.0625,$$

$$p_{20} = p_{40} = \binom{4}{1}\left(\frac{1}{2}\right)^4 = \frac{4}{16} = 0.25,$$

$$p_{30} = \binom{4}{2}\left(\frac{1}{2}\right)^4 = \frac{6}{16} = 0.375.$$

At an approximate $\alpha = 0.05$ significance level, the null hypothesis

$$H_0: \ p_i = p_{i0}, \qquad i = 1, 2, \ldots, 5,$$

is rejected if the observed value of Q_4 is greater than $\chi^2_{0.05}(4) = 9.488$. If we use the 100 repetitions of this experiment that resulted in the observed values $y_1 = 7$, $y_2 = 18$, $y_3 = 40$, $y_4 = 31$, and $y_5 = 4$, of $Y_1, Y_2, \ldots, Y_5$, respectively, then the computed value of Q_4 is

$$q_4 = \frac{(7 - 6.25)^2}{6.25} + \frac{(18 - 25)^2}{25} + \frac{(40 - 37.5)^2}{37.5} + \frac{(31 - 25)^2}{25} + \frac{(4 - 6.25)^2}{6.25}$$
$$= 4.47.$$

Since $4.47 < 9.488$, the hypothesis is not rejected. That is, the data support the hypothesis that $b(4, 1/2)$ is a reasonable probabilistic model for X. Recall that the mean of a chi-square random variable is its number of degrees of freedom. In this example, the mean is 4 and the observed value of Q_4 is 4.47, just a little greater than the mean. ∎

Thus far, all the hypotheses H_0 tested with the chi-square statistic Q_{k-1} have been simple ones (i.e., completely specified—namely, in $H_0: p_i = p_{i0}, i = 1, 2, \ldots, k$, each p_{i0} has been known). This is not always the case, and it frequently happens that $p_{10}, p_{20}, \ldots, p_{k0}$ are functions of one or more unknown parameters. For example, suppose that the hypothesized model for X in Example 9.1-2 was $H_0: X$ is $b(4, p)$, $0 < p < 1$. Then

$$p_{i0} = P(X \in A_i) = \frac{4!}{(i-1)!(5-i)!}p^{i-1}(1-p)^{5-i}, \qquad i = 1, 2, \ldots, 5,$$

which is a function of the unknown parameter p. Of course, if $H_0: p_i = p_{i0}$, $i = 1, 2, \ldots, 5$, is true, then, for large n,

$$Q_4 = \sum_{i=1}^{5} \frac{(Y_i - np_{i0})^2}{np_{i0}}$$

still has an approximate chi-square distribution with four degrees of freedom. The difficulty is that when $Y_1, Y_2, \ldots, Y_5$ are observed to be equal to $y_1, y_2, \ldots, y_5$, Q_4 cannot be computed, since $p_{10}, p_{20}, \ldots, p_{50}$ (and hence Q_4) are functions of the unknown parameter p.

One way out of the difficulty would be to estimate p from the data and then carry out the computations with the use of this estimate. It is interesting to note the following: Say the estimation of p is carried out by minimizing Q_4 with respect to p, yielding $\tilde{p}$. This $\tilde{p}$ is sometimes called a **minimum chi-square estimator** of p. If, then, this $\tilde{p}$ is used in Q_4, the statistic Q_4 still has an approximate chi-square distribution, but with only $4 - 1 = 3$ degrees of freedom. That is, the number of degrees of freedom of the approximating chi-square distribution is reduced by one for each parameter estimated by the minimum chi-square technique. We accept this result without proof (as it is a rather difficult one). Although we have considered it when p_{i0}, $i = 1, 2, \ldots, k$, is a function of only one parameter, it holds when there is more than one unknown parameter, say, d. Hence, in a more general situation, the test would be completed by computing Q_{k-1}, using Y_i and the estimated p_{i0}, $i = 1, 2, \ldots, k$, to obtain q_{k-1} (i.e., q_{k-1} is the minimized chi-square). This value q_{k-1} would then be compared with a critical value $\chi_\alpha^2(k - 1 - d)$. In our special case, the computed (minimized) chi-square q_4 would be compared with $\chi_\alpha^2(3)$.

There is still one trouble with all of this: It is usually very difficult to find minimum chi-square estimators. Hence, most statisticians usually use some reasonable method of estimating the parameters. (Maximum likelihood is satisfactory.) They then compute q_{k-1}, recognizing that it is somewhat larger than the minimized chi-square, and compare it with $\chi_\alpha^2(k - 1 - d)$. Note that this approach provides a slightly larger probability of rejecting H_0 than would the scheme in which the minimized chi-square were used because the computed q_{k-1} is larger than the minimum q_{k-1}.

Example 9.1-3

Let X denote the number of alpha particles emitted by barium-133 in one tenth of a second. The following 50 observations of X were taken with a Geiger counter in a fixed position:

7	4	3	6	4	4	5	3	5	3
5	5	3	2	5	4	3	3	7	6
6	4	3	11	9	6	7	4	5	4
7	3	2	8	6	7	4	1	9	8
4	8	9	3	9	7	7	9	3	10

The experimenter is interested in determining whether X has a Poisson distribution. To test H_0: X is Poisson, we first estimate the mean of X—say, λ—with the sample mean, $\bar{x} = 5.4$, of these 50 observations. We then partition the set of outcomes for this experiment into the sets $A_1 = \{0, 1, 2, 3\}$, $A_2 = \{4\}$, $A_3 = \{5\}$, $A_4 = \{6\}$, $A_5 = \{7\}$, and $A_6 = \{8, 9, 10, \ldots\}$. (Note that we combined $\{0, 1, 2, 3\}$ into one set A_1 and $\{8, 9, 10, \ldots\}$ into another A_6 so that the expected number of outcomes for each set would be at least five when H_0 is true.) In Table 9.1-1, the data are grouped and the estimated probabilities specified by the hypothesis that X has a Poisson distribution

Table 9.1-1 Grouped Geiger counter data

	Outcome					
	A_1	A_2	A_3	A_4	A_5	A_6
Frequency	13	9	6	5	7	10
Probability	0.213	0.160	0.173	0.156	0.120	0.178
Expected ($50p_i$)	10.65	8.00	8.65	7.80	6.00	8.90

with an estimated $\widehat{\lambda} = \bar{x} = 5.4$ are given. Since one parameter was estimated, Q_{6-1} has an approximate chi-square distribution with $r = 6-1-1 = 4$ degrees of freedom. Also, since

$$q_5 = \frac{[13 - 50(0.213)]^2}{50(0.213)} + \cdots + \frac{[10 - 50(0.178)]^2}{50(0.178)}$$

$$= 2.763 < 9.488 = \chi^2_{0.05}(4),$$

H_0 is not rejected at the 5% significance level. That is, with only these data, we are quite willing to accept the model that X has a Poisson distribution. ∎

Let us now consider the problem of testing a model for the distribution of a random variable W of the continuous type. That is, if $F(w)$ is the distribution function of W, we wish to test

$$H_0: \ F(w) = F_0(w),$$

where $F_0(w)$ is some known distribution function of the continuous type. Recall that we have considered problems of this type in which we used q–q plots. In order to use the chi-square statistic, we must partition the set of possible values of W into k sets. One way this can be done is as follows: Partition the interval $[0, 1]$ into k sets with the points $b_0, b_1, b_2, \ldots, b_k$, where

$$0 = b_0 < b_1 < b_2 < \cdots < b_k = 1.$$

Let $a_i = F_0^{-1}(b_i)$, $i = 1, 2, \ldots, k-1$; $A_1 = (-\infty, a_1]$, $A_i = (a_{i-1}, a_i]$ for $i = 2, 3, \ldots, k-1$, and $A_k = (a_{k-1}, \infty)$; and $p_i = P(W \in A_i)$, $i = 1, 2, \ldots, k$. Let Y_i denote the number of times the observed value of W belongs to A_i, $i = 1, 2, \ldots, k$, in n independent repetitions of the experiment. Then $Y_1, Y_2, \ldots, Y_k$ have a multinomial distribution with parameters $n, p_1, p_2, \ldots, p_{k-1}$. Also, let $p_{i0} = P(W \in A_i)$ when the distribution function of W is $F_0(w)$. The hypothesis that we actually test is a modification of H_0, namely,

$$H_0': \ p_i = p_{i0}, \qquad i = 1, 2, \ldots, k.$$

This hypothesis is rejected if the observed value of the chi-square statistic

$$Q_{k-1} = \sum_{i=1}^{k} \frac{(Y_i - np_{i0})^2}{np_{i0}}$$

is at least as great as $\chi^2_\alpha(k-1)$. If the hypothesis $H_0': p_i = p_{i0}$, $i = 1, 2, \ldots, k$, is not rejected, we do not reject the hypothesis $H_0: F(w) = F_0(w)$.

Example 9.1-4 Example 6.1-5 gives 105 observations of the times in minutes between calls to 911. Also given is a histogram of these data, with the exponential pdf with $\theta = 20$ superimposed. We shall now use a chi-square goodness-of-fit test to see whether or not this is an appropriate model for the data. That is, if X is equal to the time between calls to 911, we shall test the null hypothesis that the distribution of X is exponential with a mean of $\theta = 20$. Table 9.1-2 groups the data into nine classes and gives the probabilities and expected values of these classes. Using the frequencies and expected values, the chi-square goodness-of-fit statistic is

$$q_8 = \frac{(41 - 38.0520)^2}{38.0520} + \frac{(22 - 24.2655)^2}{24.2655} + \cdots + \frac{(2 - 2.8665)^2}{2.8665} = 4.6861.$$

Table 9.1-2 Summary of times between calls to 911

Class	Frequency	Probability	Expected
$A_1 = [0, 9]$	41	0.3624	38.0520
$A_2 = (9, 18]$	22	0.2311	24.2655
$A_3 = (18, 27]$	11	0.1473	15.4665
$A_4 = (27, 36]$	10	0.0939	9.8595
$A_5 = (36, 45]$	9	0.0599	6.2895
$A_6 = (45, 54]$	5	0.0382	4.0110
$A_7 = (54, 63]$	2	0.0244	2.5620
$A_8 = (63, 72]$	3	0.0155	1.6275
$A_9 = (72, \infty)$	2	0.0273	2.8665

The p-value associated with this test is 0.7905, which means that it is an extremely good fit.

Note that we assumed that we knew $\theta = 20$. We could also have run this test letting $\theta = \bar{x}$, remembering that we then lose one degree of freedom. For this example, the outcome would be about the same. ∎

It is also true, in dealing with models of random variables of the continuous type, that we must frequently estimate unknown parameters. For example, let H_0 be that W is $N(\mu, \sigma^2)$, where μ and σ^2 are unknown. With a random sample $W_1, W_2, \ldots, W_n$, we first can estimate μ and σ^2, possibly with $\bar{w}$ and s_w^2. We partition the space $\{w : -\infty < w < \infty\}$ into k mutually disjoint sets $A_1, A_2, \ldots, A_k$. We then use the estimates of μ and σ^2—say, $\bar{w}$ and $s^2 = s_w^2$, respectively, to estimate

$$\widehat{p}_{i0} = \int_{A_i} \frac{1}{s\sqrt{2\pi}} \exp\left[-\frac{(w - \bar{w})^2}{2s^2}\right] dw,$$

$i = 1, 2, \ldots, k$. Using the observed frequencies $y_1, y_2, \ldots, y_k$ of $A_1, A_2, \ldots, A_k$, respectively, from the observed random sample $w_1, w_2, \ldots, w_n$, and $\widehat{p}_{10}, \widehat{p}_{20}, \ldots, \widehat{p}_{k0}$ estimated with $\bar{w}$ and $s^2 = s_w^2$, we compare the computed

$$q_{k-1} = \sum_{i=1}^{k} \frac{(y_i - n\widehat{p}_{i0})^2}{n\widehat{p}_{i0}}$$

with $\chi_\alpha^2(k-1-2)$. This value q_{k-1} will again be somewhat larger than that which would be found using minimum chi-square estimation, and certain caution should be observed. Several exercises illustrate the procedure in which one or more parameters must be estimated. Finally, note that the methods given in this section frequently are classified under the more general title of goodness-of-fit tests. In particular, then, the tests in this section would be **chi-square goodness-of-fit tests**.

Exercises

9.1-1. A 1-pound bag of candy-coated chocolate-covered peanuts contained 240 pieces of candy, each colored brown, orange, green, or yellow. Test the null hypothesis that the machine filling these bags treats the four colors of candy equally likely; that is, test

$$H_0: \ p_B = p_O = p_G = p_Y = \frac{1}{4}.$$

The observed values were 52 brown, 65 orange, 60 green, and 63 yellow candies. You may select the significance level or give an approximate p-value.

9.1-2. A particular brand of candy-coated chocolate comes in six different colors that we shall denote as $A_1 = $ {brown}, $A_2 = $ {pink}, $A_3 = $ {yellow}, $A_4 = $ {orange}, $A_5 = $ {green}, and $A_6 = $ {coffee}. Let p_i equal the probability that the color of a piece of candy selected at random belongs to A_i, $i = 1, 2, \ldots, 6$. Test the null hypothesis

$$H_0 : p_1 = 0.3, p_2 = 0.2, p_3 = 0.1, p_4 = 0.1, p_5 = 0.2, p_6 = 0.1,$$

using a random sample of $n = 600$ pieces of candy whose colors yielded the respective frequencies 95, 105, 100, 102, 96, and 102. You may select the significance level or give an approximate p-value.

9.1-3. In the Michigan Lottery Daily3 Game, twice a day a three-digit integer is generated one digit at a time. Let p_i denote the probability of generating digit i, $i = 0, 1, \ldots, 9$. Let $\alpha = 0.05$, and use the following 50 digits to test $H_0: p_0 = p_1 = \cdots = p_9 = 1/10$:

1	6	9	9	3	8	5	0	6	7
4	7	5	9	4	6	5	6	4	4
4	8	0	9	3	2	1	5	4	5
7	3	2	1	4	6	7	1	3	4
4	8	8	6	1	6	1	2	8	8

9.1-4. In a biology laboratory, students use corn to test the Mendelian theory of inheritance. The theory claims that frequencies of the four categories "smooth and yellow," "wrinkled and yellow," "smooth and purple," and "wrinkled and purple" will occur in the ratio 9:3:3:1. If a student counted 124, 30, 43, and 11, respectively, for these four categories, would these data support the Mendelian theory? Let $\alpha = 0.05$.

9.1-5. Let X equal the number of male children in a three-child family. We shall use a chi-square goodness-of-fit statistic to test the null hypothesis that the distribution of X is $b(3, 0.5)$.

(a) Define the test statistic and critical region, using an $\alpha = 0.05$ significance level.

(b) Among students who were taking statistics, 64 came from families with three children. For these families,

$X = 0, 1, 2,$ and 3 for 14, 18, 15, and 17 families, respectively. Calculate the value of the test statistic and state your conclusion, considering how the sample was selected.

9.1-6. It has been claimed that, for a penny minted in 1999 or earlier, the probability of observing heads upon spinning the penny is $p = 0.30$. Three students got together, and they would each spin a penny and record the number X of heads out of the three spins. They repeated this experiment $n = 200$ times, observing 0, 1, 2, and 3 heads 57, 95, 38, and 10 times, respectively. Use these data to test the hypotheses that X is $b(3, 0.30)$. Give limits for the p-value of this test. In addition, out of the 600 spins, calculate the number of heads occurring and then a 95% confidence interval for p.

9.1-7. A rare type of heredity change causes the bacterium in *E. coli* to become resistant to the drug streptomycin. This type of change, called *mutation*, can be detected by plating many bacteria on petri dishes containing an antibiotic medium. Any colonies that grow on this medium result from a single mutant cell. A sample of $n = 150$ petri dishes of streptomycin agar were each plated with 10^6 bacteria, and the numbers of colonies were counted on each dish. The observed results were that 92 dishes had 0 colonies, 46 had 1, 8 had 2, 3 had 3, and 1 dish had 4 colonies. Let X equal the number of colonies per dish. Test the hypothesis that X has a Poisson distribution. Use $\bar{x} = 0.5$ as an estimate of λ. Let $\alpha = 0.01$.

9.1-8. For determining the half-lives of radioactive isotopes, it is important to know what the background radiation is for a given detector over a certain period. A γ-ray detection experiment over 300 one-second intervals yielded the following data:

0	2	4	6	6	1	7	4	6	1	1	2	3	6	4	2	7	4	4	2
2	5	4	4	4	1	2	4	3	2	2	5	0	3	1	1	0	0	5	2
7	1	3	3	3	2	3	1	4	1	3	5	3	5	1	3	3	0	3	2
6	1	1	4	6	3	6	4	4	2	2	4	3	3	6	1	6	2	5	0
6	3	4	3	1	1	4	6	1	5	1	1	4	1	4	1	1	1	3	3
4	3	3	2	5	2	1	3	5	3	2	7	0	4	2	3	3	5	6	1
4	2	6	4	2	0	4	4	7	3	5	2	2	3	1	3	1	3	6	5
4	8	2	2	4	2	2	1	4	7	5	2	1	1	4	1	4	3	6	2
1	1	2	2	2	2	3	5	4	3	2	2	3	3	2	4	4	3	2	2
3	6	1	1	3	3	2	1	4	5	5	1	2	3	3	1	3	7	2	5
4	2	0	6	2	3	2	3	0	4	4	5	2	5	3	0	4	6	2	2
2	2	2	5	2	2	3	4	2	3	7	1	1	7	1	3	6	0	5	3
0	0	3	3	0	2	4	3	1	2	3	3	3	4	3	2	2	7	5	3
5	1	1	2	2	6	1	3	1	4	4	2	3	4	5	1	3	4	3	1
0	3	7	4	0	5	2	5	4	4	2	2	3	2	4	6	5	5	3	4

Do these look like observations of a Poisson random variable with mean $\lambda = 3$? To answer this question, do the following:

(a) Find the frequencies of $0, 1, 2, \ldots, 8$.

(b) Calculate the sample mean and sample variance. Are they approximately equal to each other?

(c) Construct a probability histogram with $\lambda = 3$ and a relative frequency histogram on the same graph.

(d) Use $\alpha = 0.05$ and a chi-square goodness-of-fit test to answer this question.

9.1-9. Let X equal the amount of butterfat (in pounds) produced by 90 cows during a 305-day milk production period following the birth of their first calf. Test the hypothesis that the distribution of X is $N(\mu, \sigma^2)$, using $k = 10$ classes of equal probability. You may take $\bar{x} = 511.633$ and $s_x = 87.576$ as estimates of μ and σ, respectively. The data are as follows:

486	537	513	583	453	510	570	500	458	555
618	327	350	643	500	497	421	505	637	599
392	574	492	635	460	696	593	422	499	524
539	339	472	427	532	470	417	437	388	481
537	489	418	434	466	464	544	475	608	444
573	611	586	613	645	540	494	532	691	478
513	583	457	612	628	516	452	501	453	643
541	439	627	619	617	394	607	502	395	470
531	526	496	561	491	380	345	274	672	509

9.1-10. A biologist is studying the life cycle of the avian schistosome that causes swimmer's itch. His study uses Menganser ducks for the adult parasites and aquatic snails as intermediate hosts for the larval stages. The life history is cyclic. (For more information, see http://swimmersitch.org/.) As a part of this study, the biologist and his students used snails from a natural population to measure the distances (in cm) that snails travel per day. The conjecture is that snails that had a patent infection would not travel as far as those without without such an infection.

Here are the measurements in cm that snails traveled per day. There are 39 in the infected group and 31 in the control group.

Distances for Infected Snail Group (ordered):

263	238	226	220	170	155	139	123	119	107	107	97	90
90	90	79	75	74	71	66	60	55	47	47	47	45
43	41	40	39	38	38	35	32	32	28	19	10	10

Distances for Control Snail Group (ordered):

314	300	274	246	190	186	185	182	180	141	132
129	110	100	95	95	93	83	55	52	50	48
48	44	40	32	30	25	24	18	7		

(a) Find the sample means and sample standard deviations for the two groups of snails.

(b) Make box plots of the two groups of snails on the same graph.

(c) For the control snail group, test the hypothesis that the distances come from an exponential distribution. Use $\bar{x}$ as an estimate of θ. Group the data into 5 or 10 classes, with equal probabilities for each class. Thus, the expected value will be either 6.2 or 3.1, respectively.

(d) For the infected snail group, test the hypothesis that the distances come from a gamma distribution with $\alpha = 2$ and $\theta = 42$. Use 10 classes with equal probabilities so that the expected value of each class is 3.9. Use Minitab or some other computer program to calculate the boundaries of the classes.

9.1-11. In Exercise 6.1-4, data are given for the melting points for 50 metal alloy filaments. Here the data are repeated:

320	326	325	318	322	320	329	317	316	331
320	320	317	329	316	308	321	319	322	335
318	313	327	314	329	323	327	323	324	314
308	305	328	330	322	310	324	314	312	318
313	320	324	311	317	325	328	319	310	324

Test the hypothesis that these are observations of a normally distributed random variable. Note that you must estimate two parameters: μ and σ.

9.2 CONTINGENCY TABLES

In this section, we demonstrate the flexibility of the chi-square test. We first look at a method for testing whether two or more multinomial distributions are equal, sometimes called a *test for homogeneity*. Then we consider a *test for independence of attributes of classification*. Both of these lead to a similar test statistic.

Suppose that each of two independent experiments can end in one of the k mutually exclusive and exhaustive events $A_1, A_2, \ldots, A_k$. Let

$$p_{ij} = P(A_i), \qquad i = 1, 2, \ldots, k, \qquad j = 1, 2.$$

That is, $p_{11}, p_{21}, \ldots, p_{k1}$ are the probabilities of the events in the first experiment, and $p_{12}, p_{22}, \ldots, p_{k2}$ are those associated with the second experiment. Let the experiments be repeated n_1 and n_2 independent times, respectively. Also, let $Y_{11}, Y_{21}, \ldots, Y_{k1}$ be the frequencies of $A_1, A_2, \ldots, A_k$ associated with the n_1 independent trials of the first experiment. Similarly, let $Y_{12}, Y_{22}, \ldots, Y_{k2}$ be the respective frequencies associated with the n_2 trials of the second experiment. Of course, $\sum_{i=1}^{k} Y_{ij} = n_j, j = 1, 2$. From the sampling distribution theory corresponding to the basic chi-square test, we know that each of

$$\sum_{i=1}^{k} \frac{(Y_{ij} - n_j p_{ij})^2}{n_j p_{ij}}, \qquad j = 1, 2,$$

has an approximate chi-square distribution with $k - 1$ degrees of freedom. Since the two experiments are independent (and thus the two chi-square statistics are independent), the sum

$$\sum_{j=1}^{2} \sum_{i=1}^{k} \frac{(Y_{ij} - n_j p_{ij})^2}{n_j p_{ij}}$$

is approximately chi-square with $k - 1 + k - 1 = 2k - 2$ degrees of freedom.

Usually, the $p_{ij}, i = 1, 2, \ldots, k, j = 1, 2$, are unknown, and frequently we wish to test the hypothesis

$$H_0: \ p_{11} = p_{12}, \ p_{21} = p_{22}, \ \ldots, \ p_{k1} = p_{k2};$$

that is, H_0 is the hypothesis that the corresponding probabilities associated with the two independent experiments are equal. Under H_0, we can estimate the unknown

$$p_{i1} = p_{i2}, \qquad i = 1, 2, \ldots, k,$$

by using the relative frequency $(Y_{i1} + Y_{i2})/(n_1 + n_2), i = 1, 2, \ldots, k$. That is, if H_0 is true, we can say that the two experiments are actually parts of a larger one in which $Y_{i1} + Y_{i2}$ is the frequency of the event $A_i, i = 1, 2, \ldots, k$. Note that we have to estimate only the $k - 1$ probabilities $p_{i1} = p_{i2}$, using

$$\frac{Y_{i1} + Y_{i2}}{n_1 + n_2}, \qquad i = 1, 2, \ldots, k - 1,$$

since the sum of the k probabilities must equal 1. That is, the estimator of $p_{k1} = p_{k2}$ is

$$1 - \frac{Y_{11} + Y_{12}}{n_1 + n_2} - \cdots - \frac{Y_{k-1,1} + Y_{k-1,2}}{n_1 + n_2} = \frac{Y_{k1} + Y_{k2}}{n_1 + n_2}.$$

Substituting these estimators, we find that

$$Q = \sum_{j=1}^{2} \sum_{i=1}^{k} \frac{[Y_{ij} - n_j(Y_{i1} + Y_{i2})/(n_1 + n_2)]^2}{n_j(Y_{i1} + Y_{i2})/(n_1 + n_2)}$$

has an approximate chi-square distribution with $2k - 2 - (k - 1) = k - 1$ degrees of freedom. Here $k - 1$ is subtracted from $2k - 2$, because that is the number of estimated parameters. The critical region for testing H_0 is of the form

$$q \geq \chi_\alpha^2(k-1).$$

Example 9.2-1

To test two methods of instruction, 50 students are selected at random from each of two groups. At the end of the instruction period, each student is assigned a grade (A, B, C, D, or F) by an evaluating team. The data are recorded as follows:

	Grade					
	A	B	C	D	F	Totals
Group I	8	13	16	10	3	50
Group II	4	9	14	16	7	50

Accordingly, if the hypothesis H_0 that the corresponding probabilities are equal is true, then the respective estimates of the probabilities are

$$\frac{8+4}{100} = 0.12, \ 0.22, \ 0.30, \ 0.26, \ \frac{3+7}{100} = 0.10.$$

Thus, the estimates of $n_1 p_{i1} = n_2 p_{i2}$ are 6, 11, 15, 13, and 5, respectively. Hence, the computed value of Q is

$$q = \frac{(8-6)^2}{6} + \frac{(13-11)^2}{11} + \frac{(16-15)^2}{15} + \frac{(10-13)^2}{13} + \frac{(3-5)^2}{5}$$

$$+ \frac{(4-6)^2}{6} + \frac{(9-11)^2}{11} + \frac{(14-15)^2}{15} + \frac{(16-13)^2}{13} + \frac{(7-5)^2}{5}$$

$$= \frac{4}{6} + \frac{4}{11} + \frac{1}{15} + \frac{9}{13} + \frac{4}{5} + \frac{4}{6} + \frac{4}{11} + \frac{1}{15} + \frac{9}{13} + \frac{4}{5} = 5.18.$$

Now, under H_0, Q has an approximate chi-square distribution with $k - 1 = 4$ degrees of freedom, so the $\alpha = 0.05$ critical region is $q \geq 9.488 = \chi_{0.05}^2(4)$. Here $q = 5.18 < 9.488$, and hence H_0 is not rejected at the 5% significance level. Furthermore, the p-value for $q = 5.18$ is 0.269, which is greater than most significance levels. Thus, with these data, we cannot say that there is a difference between the two methods of instruction. ∎

It is fairly obvious how this procedure can be extended to testing the equality of h independent multinomial distributions. That is, let

$$p_{ij} = P(A_i), \quad i = 1, 2, \ldots, k, \quad j = 1, 2, \ldots, h,$$

and test

$$H_0: p_{i1} = p_{i2} = \cdots = p_{ih} = p_i, \quad i = 1, 2, \ldots, k.$$

Repeat the jth experiment n_j independent times, and let $Y_{1j}, Y_{2j}, \ldots, Y_{kj}$ denote the frequencies of the respective events $A_1, A_2, \ldots, A_k$. Now,

$$Q = \sum_{j=1}^{h} \sum_{i=1}^{k} \frac{(Y_{ij} - n_j p_{ij})^2}{n_j p_{ij}}$$

has an approximate chi-square distribution with $h(k-1)$ degrees of freedom. Under H_0, we must estimate $k-1$ probabilities, using

$$\widehat{p}_i = \frac{\sum_{j=1}^{h} Y_{ij}}{\sum_{j=1}^{h} n_j}, \qquad i = 1, 2, \ldots, k-1,$$

because the estimate of p_k follows from $\widehat{p}_k = 1 - \widehat{p}_1 - \widehat{p}_2 - \cdots - \widehat{p}_{k-1}$. We use these estimates to obtain

$$Q = \sum_{j=1}^{h} \sum_{i=1}^{k} \frac{(Y_{ij} - n_j \widehat{p}_i)^2}{n_j \widehat{p}_i},$$

which has an approximate chi-square distribution, with its degrees of freedom given by $h(k-1) - (k-1) = (h-1)(k-1)$.

Let us see how we can use the preceding procedures to test the equality of two or more independent distributions that are not necessarily multinomial. Suppose first that we are given random variables U and V with distribution functions $F(u)$ and $G(v)$, respectively. It is sometimes of interest to test the hypothesis H_0: $F(x) = G(x)$ for all x. Previously, we considered tests of $\mu_U = \mu_V$, $\sigma_U^2 = \sigma_V^2$. In Section 8.4, we will look at the two-sample Wilcoxon test. Now we shall assume only that the distributions are independent and of the continuous type.

We are interested in testing the hypothesis H_0: $F(x) = G(x)$ for all x. This hypothesis will be replaced by another one. Partition the real line into k mutually disjoint sets $A_1, A_2, \ldots, A_k$. Let

$$p_{i1} = P(U \in A_i), \qquad i = 1, 2, \ldots, k,$$

and

$$p_{i2} = P(V \in A_i), \qquad i = 1, 2, \ldots, k.$$

We observe that if $F(x) = G(x)$ for all x, then $p_{i1} = p_{i2}$, $i = 1, 2, \ldots, k$. We replace the hypothesis H_0: $F(x) = G(x)$ with the less restrictive hypothesis H_0': $p_{i1} = p_{i2}$, $i = 1, 2, \ldots, k$. That is, we are now essentially interested in testing the equality of two multinomial distributions.

Let n_1 and n_2 denote the number of independent observations of U and V, respectively. For $i = 1, 2, \ldots, k$, let Y_{ij} denote the number of these observations of U and V, $j = 1, 2$, respectively, that fall into a set A_i. At this point, we proceed to make the test of H_0' as described earlier. Of course, if H_0' is rejected at the (approximate) significance level α, then H_0 is rejected with the same probability. However, if H_0' is true, H_0 is not necessarily true. Thus, if H_0' is not rejected, then we do not reject H_0.

In applications, the question of how to select $A_1, A_2, \ldots, A_k$ is frequently raised. Obviously, there is no single choice for k or for the dividing marks of the partition. But it is interesting to observe that the combined sample can be used in this selection without upsetting the approximate distribution of Q. For example, suppose that $n_1 = n_2 = 20$. Then we could easily select the dividing marks of the partition so that $k = 4$, and one fourth of the combined sample falls into each of the four sets.

Example 9.2-2

Select, at random, 20 cars of each of two comparable major-brand models. All 40 cars are submitted to accelerated life testing; that is, they are driven many miles over very poor roads in a short time, and their failure times (in weeks) are recorded as follows:

Brand U:	25	31	20	42	39	19	35	36	44	26
	38	31	29	41	43	36	28	31	25	38

Brand V:	28	17	33	25	31	21	16	19	31	27
	23	19	25	22	29	32	24	20	34	26

If we use 23.5, 28.5, and 34.5 as dividing marks, we note that exactly one fourth of the 40 cars fall into each of the resulting four sets. Thus, the data can be summarized as follows:

	A_1	A_2	A_3	A_4	Totals
Brand U	2	4	4	10	20
Brand V	8	6	6	0	20

The estimate of each p_i is $10/40 = 1/4$, which, multiplied by $n_j = 20$, gives 5. Hence, the computed Q is

$$q = \frac{(2-5)^2}{5} + \frac{(4-5)^2}{5} + \frac{(4-5)^2}{5} + \frac{(10-5)^2}{5} + \frac{(8-5)^2}{5}$$

$$+ \frac{(6-5)^2}{5} + \frac{(6-5)^2}{5} + \frac{(0-5)^2}{5}$$

$$= \frac{72}{5} = 14.4 > 7.815 = \chi^2_{0.05}(3).$$

Also, the p-value is 0.0024. Thus, it seems that the two brands of cars have different distributions for the length of life under accelerated life testing. Brand U seems better than brand V. ∎

Again, it should be clear how this approach can be extended to more than two distributions, and this extension will be illustrated in the exercises.

Now let us suppose that a random experiment results in an outcome that can be classified by two different attributes, such as height and weight. Assume that the first attribute is assigned to one and only one of k mutually exclusive and exhaustive event—say $A_1, A_2, \ldots, A_k$—and the second attribute falls into one and only one of h mutually exclusive and exhaustive events—say $B_1, B_2, \ldots, B_h$. Let the probability of $A_i \cap B_j$ be defined by

$$p_{ij} = P(A_i \cap B_j), \qquad i = 1, 2, \ldots, k, \qquad j = 1, 2, \ldots, h.$$

The random experiment is to be repeated n independent times, and Y_{ij} will denote the frequency of the event $A_i \cap B_j$. Since there are kh such events as $A_i \cap B_j$, the random variable

$$Q_{kh-1} = \sum_{j=1}^{h} \sum_{i=1}^{k} \frac{(Y_{ij} - np_{ij})^2}{np_{ij}}$$

has an approximate chi-square distribution with $kh-1$ degrees of freedom, provided that n is large.

Suppose that we wish to test the hypothesis of the independence of the A and B attributes, namely,

$$H_0: \; P(A_i \cap B_j) = P(A_i)P(B_j), \qquad i = 1, 2, \ldots, k, \qquad j = 1, 2, \ldots, h.$$

Let us denote $P(A_i)$ by $p_{i\cdot}$ and $P(B_j)$ by $p_{\cdot j}$; that is,

$$p_{i\cdot} = \sum_{j=1}^{h} p_{ij} = P(A_i) \qquad \text{and} \qquad p_{\cdot j} = \sum_{i=1}^{k} p_{ij} = P(B_j).$$

Of course,

$$1 = \sum_{j=1}^{h}\sum_{i=1}^{k} p_{ij} = \sum_{j=1}^{h} p_{\cdot j} = \sum_{i=1}^{k} p_{i\cdot}.$$

Then the hypothesis can be formulated as

$$H_0: \; p_{ij} = p_{i\cdot}p_{\cdot j}, \qquad i = 1, 2, \ldots, k, \qquad j = 1, 2, \ldots, h.$$

To test H_0, we can use Q_{kh-1} with p_{ij} replaced by $p_{i\cdot}p_{\cdot j}$. But if $p_{i\cdot}$, $i = 1, 2, \ldots, k$, and $p_{\cdot j}$, $j = 1, 2, \ldots, h$, are unknown, as they usually are in applications, we cannot compute Q_{kh-1} once the frequencies are observed. In such a case, we estimate these unknown parameters by

$$\widehat{p}_{i\cdot} = \frac{y_{i\cdot}}{n}, \qquad \text{where} \quad y_{i\cdot} = \sum_{j=1}^{h} y_{ij}$$

is the observed frequency of A_i, $i = 1, 2, \ldots, k$; and

$$\widehat{p}_{\cdot j} = \frac{y_{\cdot j}}{n}, \qquad \text{where} \quad y_{\cdot j} = \sum_{i=1}^{k} y_{ij}$$

is the observed frequency of B_j, $j = 1, 2, \ldots, h$. Since $\sum_{i=1}^{k} p_{i\cdot} = \sum_{j=1}^{h} p_{\cdot j} = 1$, we actually estimate only $k - 1 + h - 1 = k + h - 2$ parameters. So if these estimates are used in Q_{kh-1}, with $p_{ij} = p_{i\cdot}p_{\cdot j}$, then, according to the rule stated earlier, the random variable

$$Q = \sum_{j=1}^{h}\sum_{i=1}^{k} \frac{[Y_{ij} - n(Y_{i\cdot}/n)(Y_{\cdot j}/n)]^2}{n(Y_{i\cdot}/n)(Y_{\cdot j}/n)}$$

has an approximate chi-square distribution with $kh - 1 - (k + h - 2) = (k-1)(h-1)$ degrees of freedom, provided that H_0 is true. The hypothesis H_0 is rejected if the computed value of this statistic exceeds $\chi_\alpha^2[(k-1)(h-1)]$.

Example 9.2-3

The 400 undergraduate students in a random sample at the University of Iowa were classified according to the college in which the students were enrolled and according to their gender. The results are recorded in Table 9.2-1, called a $k \times h$ **contingency table**, where, in this case, $k = 2$ and $h = 5$. (Do not be concerned about the numbers in parentheses at this point.) Incidentally, these data do actually reflect the composition of the undergraduate colleges at Iowa, but they were modified a little to make the computations easier in this example.

Table 9.2-1 Undergraduates at the University of Iowa

Gender	College Business	Engineering	Liberal Arts	Nursing	Pharmacy	Totals
Male	21	16	145	2	6	190
	(16.625)	(9.5)	(152)	(7.125)	(4.75)	
Female	14	4	175	13	4	210
	(18.375)	(10.5)	(168)	(7.875)	(5.25)	
Totals	35	20	320	15	10	400

We desire to test the null hypothesis H_0: $p_{ij} = p_{i.}p_{.j}$, $i = 1, 2$ and $j = 1, 2, 3, 4, 5$, that the college in which a student enrolls is independent of the gender of that student. Under H_0, estimates of the probabilities are

$$\widehat{p}_{1.} = \frac{190}{400} = 0.475 \quad \text{and} \quad \widehat{p}_{2.} = \frac{210}{400} = 0.525$$

and

$$\widehat{p}_{.1} = \frac{35}{400} = 0.0875, \quad \widehat{p}_{.2} = 0.05, \quad \widehat{p}_{.3} = 0.8, \quad \widehat{p}_{.4} = 0.0375, \quad \widehat{p}_{.5} = 0.025.$$

The expected numbers $n(y_{i.}/n)(y_{.j}/n)$ are computed as follows:

$$400(0.475)(0.0875) = 16.625,$$

$$400(0.525)(0.0875) = 18.375,$$

$$400(0.475)(0.05) = 9.5,$$

and so on. These are the values recorded in parentheses in Table 9.2-1. The computed chi-square statistic is

$$q = \frac{(21 - 16.625)^2}{16.625} + \frac{(14 - 18.375)^2}{18.375} + \cdots + \frac{(4 - 5.25)^2}{5.25}$$

$$= 1.15 + 1.04 + 4.45 + 4.02 + 0.32 + 0.29 + 3.69$$

$$+ 3.34 + 0.33 + 0.30 = 18.93.$$

Since the number of degrees of freedom equals $(k - 1)(h - 1) = 4$, this $q = 18.93 > 13.28 = \chi^2_{0.01}(4)$, and we reject H_0 at the $\alpha = 0.01$ significance level. Moreover, since the first two terms of q come from the business college, the next two from engineering, and so on, it is clear that the enrollments in engineering and nursing are more highly dependent on gender than in the other colleges, because they have contributed the most to the value of the chi-square statistic. It is also interesting to note that one expected number is less than 5, namely, 4.75. However, as the associated term in q does not contribute an unusual amount to the chi-square value, it does not concern us. ∎

It is fairly obvious how to extend the preceding testing procedure to more than two attributes. For example, if the third attribute falls into one and only one of m mutually exclusive and exhaustive events—say, $C_1, C_2, \ldots, C_m$—then we test the independence of the three attributes by using

$$Q = \sum_{r=1}^{m} \sum_{j=1}^{h} \sum_{i=1}^{k} \frac{[Y_{ijr} - n(Y_{i..}/n)(Y_{.j.}/n)(Y_{..r}/n)]^2}{n(Y_{i..}/n)(Y_{.j.}/n)(Y_{..r}/n)},$$

where Y_{ijr}, $Y_{i..}$, $Y_{.j.}$, and $Y_{..r}$ are the respective observed frequencies of the events $A_i \cap B_j \cap C_r$, A_i, B_j, and C_r in n independent trials of the experiment. If n is large and if the three attributes are independent, then Q has an approximate chi-square distribution with $khm - 1 - (k - 1) - (h - 1) - (m - 1) = khm - k - h - m + 2$ degrees of freedom.

Rather than explore this extension further, it is more instructive to note some interesting uses of contingency tables.

Example 9.2-4

Say we observed 30 values $x_1, x_2, \ldots, x_{30}$ that are claimed to be the values of a random sample. That is, the corresponding random variables $X_1, X_2, \ldots, X_{30}$ were supposed to be mutually independent and each of these random variables is supposed to have the same distribution. Say, however, by looking at the 30 values, we detect an upward trend which indicates that there might have been some dependence and/or the random variables did not actually have the same distribution. One simple way to test whether they could be thought of as being observed values of a random sample is the following: Mark each x high (H) or low (L), depending on whether it is above or below the sample median. Then divide the x values into three groups: $x_1, \ldots, x_{10}$; $x_{11}, \ldots, x_{20}$; and $x_{21}, \ldots, x_{30}$. Certainly, if the observations are those of a random sample, we would expect five H's and five L's in each group. That is, the attribute classified as H or L should be independent of the group number. The summary of these data provides a 3×2 contingency table. For example, say the 30 values are

5.6	8.2	7.8	4.8	5.5	8.1	6.7	7.7	9.3	6.9
8.2	10.1	7.5	6.9	11.1	9.2	8.7	10.3	10.7	10.0
9.2	11.6	10.3	11.7	9.9	10.6	10.0	11.4	10.9	11.1

The median can be taken to be the average of the two middle observations in magnitude, namely, 9.2 and 9.3. Marking each item H or L after comparing it with this median, we obtain the following 3×2 contingency table:

Group	L	H	Totals
1	9	1	10
2	5	5	10
3	1	9	10
Totals	15	15	30

Here each $n(y_{i.}/n)(y_{.j}/n) = 30(10/30)(15/30) = 5$, so that the computed value of Q is

$$q = \frac{(9-5)^2}{5} + \frac{(1-5)^2}{5} + \frac{(5-5)^2}{5} + \frac{(5-5)^2}{5} + \frac{(1-5)^2}{5} + \frac{(9-5)^2}{5}$$

$$= 12.8 > 5.991 = \chi^2_{0.05}(2),$$

since in this instance $(k-1)(h-1) = 2$ degrees of freedom. (The p-value is 0.0017.) Hence, we reject the conjecture that these 30 values could be the observations of a random sample. Obviously, modifications could be made to this scheme: dividing the sample into more (or fewer) than three groups and rating items differently, such as low (L), middle (M), and high (H). ∎

It cannot be emphasized enough that the chi-square statistic can be used fairly effectively in almost any situation in which there should be independence. For example, suppose that we have a group of workers who have essentially the same qualifications (training, experience, etc.). Many believe that the salary and gender of the workers should be independent attributes, yet there have been several claims in special cases that there is a dependence—or discrimination—in attributes associated with such a problem.

Example 9.2-5

Two groups of workers have the same qualifications for a particular type of work. Their experience in salaries is summarized by the following 2×5 contingency table, in which the upper bound of each salary range is not included in that listing:

	Salary (Thousands of Dollars)					
Group	27–29	29–31	31–33	33–35	35 and over	Totals
1	6	11	16	14	13	60
2	5	9	8	6	2	30
Totals	11	20	24	20	15	90

To test whether the group assignment and the salaries seem to be independent with these data at the $\alpha = 0.05$ significance level, we compute

$$q = \frac{[6 - 90(60/90)(11/90)]^2}{90(60/90)(11/90)} + \cdots + \frac{[2 - 90(30/90)(15/90)]^2}{90(30/90)(15/90)}$$

$$= 4.752 < 9.488 = \chi^2_{0.05}(4).$$

Also, the p-value is 0.314. Hence, with these limited data, group assignment and salaries seem to be independent. ∎

Before turning to the exercises, note that we could have thought of the last two examples in this section as testing the equality of two or more multinomial distributions. In Example 9.2-4, the three groups define three binomial distributions, and in Example 9.2-5, the two groups define two multinomial distributions. What would have happened if we had used the computations outlined earlier in the section? It is interesting to note that we obtain exactly the same value of chi-square and in each case the number of degrees of freedom is equal to $(k-1)(h-1)$. Hence, it makes no difference whether we think of it as a test of independence or a test of the equality of several multinomial distributions. Our advice is to use the terminology that seems most natural for the particular situation.

Exercises

9.2-1. We wish to see if two groups of nurses distribute their time in six different categories about the same way. That is, the hypothesis under consideration is H_0: $p_{i1} = p_{i2}$, $i = 1, 2, \ldots, 6$. To test this hypothesis, nurses are observed at random throughout several days, each observation resulting in a mark in one of the six categories. A summary of the results is given by the following frequency table:

| | Category | | | | | | |
	1	2	3	4	5	6	Totals
Group I	95	36	71	21	45	32	300
Group II	53	26	43	18	32	28	200

Use a chi-square test with $\alpha = 0.05$.

9.2-2. Suppose that a third group of nurses was observed along with groups I and II of Exercise 9.2-1, resulting in the respective frequencies 130, 75, 136, 33, 61, and 65. Test H_0: $p_{i1} = p_{i2} = p_{i3}$, $i = 1, 2, \ldots, 6$, at the $\alpha = 0.025$ significance level.

9.2-3. Each of two comparable classes of 15 students responded to two different methods of instructions, giving the following scores on a standardized test:

| Class U: | 91 | 42 | 39 | 62 | 55 | 82 | 67 | 44 |
| | 51 | 77 | 61 | 52 | 76 | 41 | 59 | |

| Class V: | 80 | 71 | 55 | 67 | 61 | 93 | 49 | 78 |
| | 57 | 88 | 79 | 81 | 63 | 51 | 75 | |

Use a chi-square test with $\alpha = 0.05$ to test the equality of the distributions of test scores by dividing the combined sample into three equal parts (low, middle, high).

9.2-4. Suppose that a third class (W) of 15 students was observed along with classes U and V of Exercise 9.2-3, resulting in scores of

| 91 | 73 | 67 | 83 | 59 | 98 | 87 | 69 |
| 78 | 80 | 65 | 94 | 82 | 74 | 85 | |

Again, use a chi-square test with $\alpha = 0.05$ to test the equality of the three distributions by dividing the combined sample into three equal parts.

9.2-5. In the following contingency table, 2000 individuals are classified by the area of the district they inhabit and by whether they favor or oppose the election of Candidate A as the manager of the district:

| | Election of Candidate A as District Manager | | |
Area	Favor	Oppose	Totals
Rural	620	380	1000
Urban	550	450	1000
Totals	1170	830	2000

Test the null hypothesis that area and opinion on the election of Candidate A are independent. Use a 5% significance level for this test.

9.2-6. A random survey of 100 students asked each student to select the most preferred form of recreational activity from four choices. Following are the results of the survey:

| | Recreational Choice | | | | |
Gender	Jogging	Swimming	Tennis	Basketball	Totals
Male	10	5	18	20	53
Female	20	10	5	12	47
Totals	30	15	23	32	100

Test whether the choice is independent of the gender of the respondents. Approximate the p-value of the test. Would we reject the null hypothesis at $\alpha = 0.05$?

9.2-7. Three hundred music majors in a random sample were classified as follows by gender and by the kind of instrument that they played:

Gender	Instrument				Totals
	Piano	Brass	String	Woodwind	
Male	86	60	44	10	200
Female	40	33	25	2	100
Totals	126	93	69	12	300

Test whether the selection of instrument is independent of the gender of the respondent. Approximate the p-value of this test.

9.2-8. A student who uses a certain college's recreational facilities was interested in whether there is a difference between the facilities used by men and those used by women. Use $\alpha = 0.05$ and the following data to test the null hypothesis that facility and gender are independent attributes:

Gender	Facility		Totals
	Racquetball Court	Track	
Male	51	30	81
Female	43	48	91
Totals	94	78	172

9.2-9. A survey of high school girls classified them by two attributes: whether or not they participated in sports and whether or not they had one or more older brothers. Use the following data to test the null hypothesis that these two attributes of classification are independent:

Older Brother(s)	Participated in Sports		Totals
	Yes	No	
Yes	12	8	20
No	13	27	40
Totals	25	35	60

Approximate the p-value of this test. Do we reject the null hypothesis if $\alpha = 0.05$?

9.2-10. A random sample of 60 men who were tested for cholesterol was classified according to age and cholesterol level and grouped into the following contingency table:

Age	Cholesterol Level			Totals
	<180	180–220	>220	
<40	5	12	13	30
≥ 40	7	4	19	30
Totals	12	16	32	60

Test the null hypothesis H_0: Age and cholesterol level are independent attributes of classification. What is your conclusion if $\alpha = 0.05$?

9.2-11. Although high school grades and testing scores, such as SAT or ACT, can be used to predict first-year college grade-point average (GPA), many educators claim that a more important factor influencing GPA is the living conditions of students. In particular, it is claimed that the roommate of the student will have a great influence on his or her grades. To test this hypothesis, suppose we selected at random 200 students and classified each according to the following two attributes:

(a) Ranking of the student's roommate on a scale from 1 to 5, with 1 denoting a person who was difficult to live with and discouraged scholarship, and 5 signifying a person who was congenial and encouraged scholarship.

(b) The student's first-year GPA.

Say this classification gives the following 5×4 contingency table:

Rank of Roommate	Grade-Point Average				Totals
	Under 2.00	2.00–2.69	2.70–3.19	3.20–4.00	
1	8	9	10	4	31
2	5	11	15	11	42
3	6	7	20	14	47
4	3	5	22	23	53
5	1	3	11	12	27
Totals	23	35	78	64	200

Compute the chi-square statistic used to test the independence of the two attributes, and compare it with the critical value associated with $\alpha = 0.05$.

9.2-12. In a psychology experiment, 200 students were divided into majors emphasizing left-hemisphere brain skills (e.g., philosophy, physics, and mathematics) and majors emphasizing right-hemisphere skills (e.g., art, music, theater, and dance). They were also classified into one of three groups on the basis of hand posture (right noninverted, left inverted, and left noninverted). The data are as follows:

	LH	RH
RN	120	30
LI	10	10
LN	10	20

Do these data show sufficient evidence to accept the claim that the choice of college major is related to hand posture? Let $\alpha = 0.01$.

9.2-13. A study was conducted to determine the media credibility for reporting news. Those surveyed were asked to give their age, gender, education, and the most credible medium. The results of the survey are as follows:

Age	Most Credible Medium			
	Newspaper	Television	Radio	Totals
Under 35	30	68	10	108
35–54	61	79	20	160
Over 54	98	43	21	162
Totals	189	190	51	430

Gender	Most Credible Medium			
	Newspaper	Television	Radio	Totals
Male	92	108	19	219
Female	97	81	32	210
Totals	189	189	51	429

Education	Most Credible Medium			
	Newspaper	Television	Radio	Totals
Grade School	45	22	6	73
High School	94	115	30	239
College	49	52	13	114
Totals	188	189	49	426

(a) Test whether media credibility and age are independent.

(b) Test whether media credibility and gender are independent.

(c) Test whether media credibility and education are independent.

(d) Give the approximate p-value for each test.

9.3 ONE-FACTOR ANALYSIS OF VARIANCE

Frequently, experimenters want to compare more than two treatments: yields of several different corn hybrids; results due to three or more teaching techniques; or miles per gallon obtained from many different types of compact cars. Sometimes the different treatment distributions of the resulting observations are due to changing the level of a certain factor (e.g., different doses of a given drug). Thus, the consideration of the equality of the different means of the various distributions comes under the analysis of a **one-factor experiment**.

In Section 8.2, we discussed how to compare the means of two normal distributions. More generally, let us now consider m normal distributions with unknown means $\mu_1, \mu_2, \ldots, \mu_m$ and an unknown, but common, variance σ^2. One inference that we wish to consider is a test of the equality of the m means, namely, H_0: $\mu_1 = \mu_2 = \cdots = \mu_m = \mu$, with μ unspecified, against all possible alternative hypotheses H_1. In order to test this hypothesis, we shall take independent random samples from these distributions. Let $X_{i1}, X_{i2}, \ldots, X_{in_i}$ represent a random sample of size n_i from the normal distribution $N(\mu_i, \sigma^2)$, $i = 1, 2, \ldots, m$. In Table 9.3-1, we have

Table 9.3-1 One-factor random samples					
					Means
X_1:	X_{11}	X_{12}	$\cdots$	X_{1n_1}	$\overline{X}_{1.}$
X_2:	X_{21}	X_{22}	$\cdots$	X_{2n_2}	$\overline{X}_{2.}$
$\vdots$	$\vdots$	$\vdots$	$\vdots$	$\vdots$	$\vdots$
X_m:	X_{m1}	X_{m2}	$\cdots$	X_{mn_m}	$\overline{X}_{m.}$
Grand Mean:					$\overline{X}_{..}$

indicated these random samples along with the row means (sample means), where, with $n = n_1 + n_2 + \cdots + n_m$,

$$\overline{X}_{..} = \frac{1}{n} \sum_{i=1}^{m} \sum_{j=1}^{n_i} X_{ij} \quad \text{and} \quad \overline{X}_{i.} = \frac{1}{n_i} \sum_{j=1}^{n_i} X_{ij}, \quad i = 1, 2, \ldots, m.$$

The dot in the notation for the means, $\overline{X}_{..}$ and $\overline{X}_{i.}$, indicates the index over which the average is taken. Here $\overline{X}_{..}$ is an average taken over both indices, while $\overline{X}_{i.}$ is taken over just the index j.

To determine a critical region for a test of H_0, we shall first partition the sum of squares associated with the variance of the combined samples into two parts. This sum of squares is given by

$$\text{SS(TO)} = \sum_{i=1}^{m} \sum_{j=1}^{n_i} (X_{ij} - \overline{X}_{..})^2$$

$$= \sum_{i=1}^{m} \sum_{j=1}^{n_i} (X_{ij} - \overline{X}_{i.} + \overline{X}_{i.} - \overline{X}_{..})^2$$

$$= \sum_{i=1}^{m} \sum_{j=1}^{n_i} (X_{ij} - \overline{X}_{i.})^2 + \sum_{i=1}^{m} \sum_{j=1}^{n_i} (\overline{X}_{i.} - \overline{X}_{..})^2$$

$$+ 2 \sum_{i=1}^{m} \sum_{j=1}^{n_i} (X_{ij} - \overline{X}_{i.})(\overline{X}_{i.} - \overline{X}_{..}).$$

The last term of the right-hand member of this identity may be written as

$$2 \sum_{i=1}^{m} \left[(\overline{X}_{i.} - \overline{X}_{..}) \sum_{j=1}^{n_i} (X_{ij} - \overline{X}_{i.}) \right] = 2 \sum_{i=1}^{m} (\overline{X}_{i.} - \overline{X}_{..})(n_i \overline{X}_{i.} - n_i \overline{X}_{i.}) = 0,$$

and the preceding term may be written as

$$\sum_{i=1}^{m} \sum_{j=1}^{n_i} (\overline{X}_{i.} - \overline{X}_{..})^2 = \sum_{i=1}^{m} n_i (\overline{X}_{i.} - \overline{X}_{..})^2.$$

Thus,

$$SS(TO) = \sum_{i=1}^{m} \sum_{j=1}^{n_i} (X_{ij} - \overline{X}_{i \cdot})^2 + \sum_{i=1}^{m} n_i (\overline{X}_{i \cdot} - \overline{X}_{\cdot \cdot})^2.$$

For notation, let

$$SS(TO) = \sum_{i=1}^{m} \sum_{j=1}^{n_i} (X_{ij} - \overline{X}_{\cdot \cdot})^2, \text{ the total sum of squares;}$$

$$SS(E) = \sum_{i=1}^{m} \sum_{j=1}^{n_i} (X_{ij} - \overline{X}_{i \cdot})^2, \quad \text{the sum of squares within treatments, groups, or classes, often called the error sum of squares;}$$

$$SS(T) = \sum_{i=1}^{m} n_i (\overline{X}_{i \cdot} - \overline{X}_{\cdot \cdot})^2, \quad \text{the sum of squares among the different treatments, groups, or classes, often called the between-treatment sum of squares.}$$

Hence,

$$SS(TO) = SS(E) + SS(T).$$

When H_0 is true, we may regard $X_{ij}, i = 1, 2, \ldots, m, j = 1, 2, \ldots, n_i$, as a random sample of size $n = n_1 + n_2 + \cdots + n_m$ from the normal distribution $N(\mu, \sigma^2)$. Then $SS(TO)/(n - 1)$ is an unbiased estimator of σ^2 because $SS(TO)/\sigma^2$ is $\chi^2(n-1)$, so that $E[SS(TO)/\sigma^2] = n - 1$ and $E[SS(TO)/(n - 1)] = \sigma^2$. An unbiased estimator of σ^2 based only on the sample from the ith distribution is

$$W_i = \frac{\sum_{j=1}^{n_i} (X_{ij} - \overline{X}_{i \cdot})^2}{n_i - 1} \qquad \text{for } i = 1, 2, \ldots, m,$$

because $(n_i - 1)W_i/\sigma^2$ is $\chi^2(n_i-1)$. Thus,

$$E \left[\frac{(n_i - 1)W_i}{\sigma^2} \right] = n_i - 1,$$

and so

$$E(W_i) = \sigma^2, \qquad i = 1, 2, \ldots, m.$$

It follows that the sum of m of these independent chi-square random variables, namely,

$$\sum_{i=1}^{m} \frac{(n_i - 1)W_i}{\sigma^2} = \frac{SS(E)}{\sigma^2},$$

is also chi-square with $(n_1 - 1) + (n_2 - 1) + \cdots + (n_m - 1) = n - m$ degrees of freedom. Hence, $SS(E)/(n - m)$ is an unbiased estimator of σ^2. We now have

$$\frac{SS(TO)}{\sigma^2} = \frac{SS(E)}{\sigma^2} + \frac{SS(T)}{\sigma^2},$$

where

$$\frac{SS(TO)}{\sigma^2} \text{ is } \chi^2(n-1) \quad \text{and} \quad \frac{SS(E)}{\sigma^2} \text{ is } \chi^2(n-m).$$

Because $SS(T) \geq 0$, there is a theorem (see subsequent remark) which states that $SS(E)$ and $SS(T)$ are independent and the distribution of $SS(T)/\sigma^2$ is $\chi^2(m-1)$.

REMARK The sums of squares, $SS(T)$, $SS(E)$, and $SS(TO)$, are examples of **quadratic forms** in the variables X_{ij}, $i = 1, 2, \ldots, m$, $j = 1, 2, \ldots, n_i$. That is, each term in these sums of squares is of second degree in X_{ij}. Furthermore, the coefficients of the variables are real numbers, so these sums of squares are called **real quadratic forms**. The next theorem, stated without proof, is used in this chapter. [For a proof, see Hogg, McKean, and Craig, *Introduction to Mathematical Statistics*, 7th ed. (Upper Saddle River: Prentice Hall, 2013).] ∎

Theorem 9.3-1

Let $Q = Q_1 + Q_2 + \cdots + Q_k$, where $Q, Q_1, \ldots, Q_k$ are $k + 1$ real quadratic forms in n mutually independent random variables normally distributed with the same variance σ^2. Let $Q/\sigma^2, Q_1/\sigma^2, \ldots, Q_{k-1}/\sigma^2$ have chi-square distributions with $r, r_1, \ldots, r_{k-1}$ degrees of freedom, respectively. If Q_k is nonnegative, then

(a) $Q_1, \ldots, Q_k$ are mutually independent, and hence,

(b) Q_k/σ^2 has a chi-square distribution with $r - (r_1 + \cdots + r_{k-1}) = r_k$ degrees of freedom.

Since, under H_0, $SS(T)/\sigma^2$ is $\chi^2(m-1)$, we have $E[SS(T)/\sigma^2] = m - 1$ and it follows that $E[SS(T)/(m - 1)] = \sigma^2$. Now, the estimator of σ^2, namely, $SS(E)/(n - m)$, which is based on $SS(E)$, is always unbiased, whether H_0 is true or false. However, if the means $\mu_1, \mu_2, \ldots, \mu_m$ are not equal, the expected value of the estimator that is based on $SS(T)$ will be greater than σ^2. To make this last statement clear, we have

$$E[SS(T)] = E\left[\sum_{i=1}^{m} n_i(\overline{X}_{i.} - \overline{X}_{..})^2\right] = E\left[\sum_{i=1}^{m} n_i\overline{X}_{i.}^2 - n\overline{X}_{..}^2\right]$$

$$= \sum_{i=1}^{m} n_i\{\text{Var}(\overline{X}_{i.}) + [E(\overline{X}_{i.})]^2\} - n\{\text{Var}(\overline{X}_{..}) + [E(\overline{X}_{..})]^2\}$$

$$= \sum_{i=1}^{m} n_i\left\{\frac{\sigma^2}{n_i} + \mu_i^2\right\} - n\left\{\frac{\sigma^2}{n} + \overline{\mu}^2\right\}$$

$$= (m - 1)\sigma^2 + \sum_{i=1}^{m} n_i(\mu_i - \overline{\mu})^2,$$

where $\overline{\mu} = (1/n)\sum_{i=1}^{m} n_i\mu_i$. If $\mu_1 = \mu_2 = \cdots = \mu_m = \mu$, then

$$E\left(\frac{SS(T)}{m - 1}\right) = \sigma^2.$$

If the means are not all equal, then

$$E\left[\frac{SS(T)}{m-1}\right] = \sigma^2 + \sum_{i=1}^{m} n_i \frac{(\mu_i - \overline{\mu})^2}{m-1} > \sigma^2.$$

We can base our test of H_0 on the ratio of $SS(T)/(m-1)$ and $SS(E)/(n-m)$, both of which are unbiased estimators of σ^2, provided that $H_0: \mu_1 = \mu_2 = \cdots = \mu_m$ is true, so that, under H_0, the ratio would assume values near 1. However, in the case that the means $\mu_1, \mu_2, \ldots, \mu_m$ begin to differ, this ratio tends to become large, since $E[SS(T)/(m-1)]$ gets larger. Under H_0, the ratio

$$\frac{SS(T)/(m-1)}{SS(E)/(n-m)} = \frac{[SS(T)/\sigma^2]/(m-1)}{[SS(E)/\sigma^2]/(n-m)} = F$$

has an F distribution with $m-1$ and $n-m$ degrees of freedom because $SS(T)/\sigma^2$ and $SS(E)/\sigma^2$ are independent chi-square variables. We would reject H_0 if the observed value of F is too large because this would indicate that we have a relatively large $SS(T)$, suggesting that the means are unequal. Thus, the critical region is of the form $F \geq F_\alpha(m-1, n-m)$.

The information used for tests of the equality of several means is often summarized in an **analysis-of-variance table**, or **ANOVA** table, like that given in Table 9.3-2, where the mean square (MS) is the sum of squares (SS) divided by its degrees of freedom.

Example 9.3-1 Let X_1, X_2, X_3, X_4 be independent random variables that have normal distributions $N(\mu_i, \sigma^2)$, $i = 1, 2, 3, 4$. We shall test

$$H_0: \mu_1 = \mu_2 = \mu_3 = \mu_4 = \mu$$

against all alternatives on the basis of a random sample of size $n_i = 3$ from each of the four distributions. A critical region of size $\alpha = 0.05$ is given by

$$F = \frac{SS(T)/(4-1)}{SS(E)/(12-4)} \geq 4.07 = F_{0.05}(3, 8).$$

The observed data are shown in Table 9.3-3. (Clearly, these data are not observations from normal distributions; they were selected to illustrate the calculations.)

Table 9.3-2 Analysis-of-variance table

Source	Sum of Squares (SS)	Degrees of Freedom	Mean Square (MS)	F Ratio
Treatment	SS(T)	$m-1$	$MS(T) = \dfrac{SS(T)}{m-1}$	$\dfrac{MS(T)}{MS(E)}$
Error	SS(E)	$n-m$	$MS(E) = \dfrac{SS(E)}{n-m}$	
Total	SS(TO)	$n-1$		

Table 9.3-3 Illustrative data

	Observations			$\overline{X}_{i\cdot}$
x_1:	13	8	9	10
x_2:	15	11	13	13
x_3:	8	12	7	9
x_4:	11	15	10	12
$\overline{x}_{\cdot\cdot}$				11

For the given data, the calculated SS(TO), SS(E), and SS(T) are

$$\text{SS(TO)} = (13 - 11)^2 + (8 - 11)^2 + \cdots + (15 - 11)^2 + (10 - 11)^2 = 80,$$

$$\text{SS(E)} = (13 - 10)^2 + (8 - 10)^2 + \cdots + (15 - 12)^2 + (10 - 12)^2 = 50,$$

$$\text{SS(T)} = 3[(10 - 11)^2 + (13 - 11)^2 + (9 - 11)^2 + (12 - 11)^2] = 30.$$

Note that since SS(TO) = SS(E) + SS(T), only two of the three values need to be calculated directly from the data. Here the computed value of F is

$$\frac{30/3}{50/8} = 1.6 < 4.07,$$

and H_0 is not rejected. The p-value is the probability, under H_0, of obtaining an F that is at least as large as this computed value of F. It is often given by computer programs.

The information for this example is summarized in Table 9.3-4. Again, we note that (here and elsewhere) the F statistic is the ratio of two appropriate mean squares. ■

Formulas that sometimes simplify the calculations of SS(TO), SS(T), and SS(E) (and also reduce roundoff errors created by subtracting the averages from the observations) are

Table 9.3-4 ANOVA table for illustrative data

Source	Sum of Squares (SS)	Degrees of Freedom	Mean Square (MS)	F Ratio	p-value
Treatment	30	3	30/3	1.6	0.264
Error	50	8	50/8		
Total	80	11			

$$\text{SS(TO)} = \sum_{i=1}^{m} \sum_{j=1}^{n_i} X_{ij}^2 - \frac{1}{n}\left[\sum_{i=1}^{m} \sum_{j=1}^{n_i} X_{ij}\right]^2,$$

$$\text{SS(T)} = \sum_{i=1}^{m} \frac{1}{n_i}\left[\sum_{j=1}^{n_i} X_{ij}\right]^2 - \frac{1}{n}\left[\sum_{i=1}^{m} \sum_{j=1}^{n_i} X_{ij}\right]^2,$$

and

$$\text{SS(E)} = \text{SS(TO)} - \text{SS(T)}.$$

It is interesting to note that in these formulas each square is divided by the number of observations in the sum being squared: X_{ij}^2 by 1, $(\sum_{j=1}^{n_i} X_{ij})^2$ by n_i, and $(\sum_{i=1}^{m} \sum_{j=1}^{n_i} X_{ij})^2$ by n. The preceding formulas are used in Example 9.3-2. Although they are useful, you are encouraged to use appropriate statistical packages on a computer to aid you with these calculations.

If the sample sizes are all at least equal to 7, insight can be gained by plotting box-and-whisker diagrams on the same figure, for each of the samples. This technique is also illustrated in Example 9.3-2.

Example 9.3-2

A window that is manufactured for an automobile has five studs for attaching it. A company that manufactures these windows performs "pullout tests" to determine the force needed to pull a stud out of the window. Let X_i, $i = 1, 2, 3, 4, 5$, equal the force required at position i, and assume that the distribution of X_i is $N(\mu_i, \sigma^2)$. We shall test the null hypothesis $H_0: \mu_1 = \mu_2 = \mu_3 = \mu_4 = \mu_5$, using seven independent observations at each position. At an $\alpha = 0.01$ significance level, H_0 is rejected if the computed

$$F = \frac{\text{SS(T)}/(5-1)}{\text{SS(E)}/(35-5)} \geq 4.02 = F_{0.01}(4, 30).$$

The observed data, along with certain sums, are given in Table 9.3-5. For these data,

Table 9.3-5 Pullout test data								$\sum_{j=1}^{7} x_{ij}$	$\sum_{j=1}^{7} x_{ij}^2$
			Observations						
x_1:	92	90	87	105	86	83	102	645	59,847
x_2:	100	108	98	110	114	97	94	721	74,609
x_3:	143	149	138	136	139	120	145	970	134,936
x_4:	147	144	160	149	152	131	134	1017	148,367
x_5:	142	155	119	134	133	146	152	981	138,415
Totals								4334	556,174

$$SS(TO) = 556,174 - \frac{1}{35}(4334)^2 = 19{,}500.97,$$

$$SS(T) = \frac{1}{7}(645^2 + 721^2 + 970^2 + 1017^2 + 981^2)$$

$$- \frac{1}{35}(4334)^2 = 16{,}672.11,$$

$$SS(E) = 19{,}500.97 - 16{,}672.11 = 2828.86.$$

Since the computed F is

$$F = \frac{16{,}672.11/4}{2828.86/30} = 44.20,$$

the null hypothesis is clearly rejected. This information obtained from the equations is summarized in Table 9.3-6.

But why is H_0 rejected? The box-and-whisker diagrams shown in Figure 9.3-1 help to answer this question. It looks like the forces required to pull out studs in positions 1 and 2 are similar, and those in positions 3, 4, and 5 are quite similar, but different from, positions 1 and 2. (See Exercise 9.3-10.) An examination of the window would confirm that this is the case. ∎

As with the two-sample t test, the F test works quite well even if the underlying distributions are nonnormal, unless they are highly skewed or the variances are quite different. In these latter cases, we might need to transform the observations

Table 9.3-6 ANOVA table for pullout tests

Source	Sum of Squares (SS)	Degrees of Freedom	Mean Square (MS)	F
Treatment	16,672.11	4	4,168.03	44.20
Error	2,828.86	30	94.30	
Total	19,500.97	34		

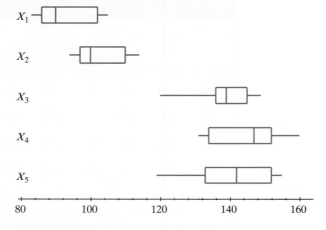

Figure 9.3-1 Box plots for pullout tests

to make the data more symmetric with about the same variances or to use certain nonparametric methods that are beyond the scope of this text.

Exercises

(In some of the exercises that follow, we must make assumptions, such as normal distributions with equal variances.)

9.3-1. Let μ_1, μ_2, μ_3 be, respectively, the means of three normal distributions with a common, but unknown, variance σ^2. In order to test, at the $\alpha = 0.05$ significance level, the hypothesis $H_0 : \mu_1 = \mu_2 = \mu_3$ against all possible alternative hypotheses, we take a random sample of size 4 from each of these distributions. Determine whether we accept or reject H_0 if the observed values from the three distributions are, respectively, as follows:

$$
\begin{array}{lcccc}
x_1: & 4 & 8 & 5 & 7 \\
x_2: & 10 & 12 & 9 & 11 \\
x_3: & 9 & 5 & 8 & 8
\end{array}
$$

9.3-2. Let μ_i be the average yield in bushels per acre of variety i of corn, $i = 1, 2, 3, 4$. In order to test the hypothesis $H_0: \mu_1 = \mu_2 = \mu_3 = \mu_4$ at the 5% significance level, four test plots for each of the four varieties of corn are planted. Determine whether we accept or reject H_0 if the yield in bushels per acre of the four varieties of corn are, respectively, as follows:

$$
\begin{array}{lcccc}
x_1: & 158.82 & 166.99 & 164.30 & 168.73 \\
x_2: & 176.84 & 165.69 & 167.87 & 166.18 \\
x_3: & 180.16 & 168.84 & 170.65 & 173.58 \\
x_4: & 151.58 & 163.51 & 164.57 & 160.75
\end{array}
$$

9.3-3. Four groups of three pigs each were fed individually four different feeds for a specified length of time to test the hypothesis $H_0: \mu_1 = \mu_2 = \mu_3 = \mu_4$, where μ_i, $i = 1, 2, 3, 4$, is the mean weight gain for each of the feeds. Determine whether the null hypothesis is accepted or rejected at a 5% significance level if the observed weight gains in pounds are, respectively, as follows:

$$
\begin{array}{lccc}
x_1: & 194.11 & 182.80 & 187.43 \\
x_2: & 216.06 & 203.50 & 216.88 \\
x_3: & 178.10 & 189.20 & 181.33 \\
x_4: & 197.11 & 202.68 & 209.18
\end{array}
$$

9.3-4. Ledolter and Hogg (see References) report that a civil engineer wishes to compare the strengths of three different types of beams, one (A) made of steel and two (B and C) made of different and more expensive alloys.

A certain deflection (in units of 0.001 inch) was measured for each beam when submitted to a given force; thus, a small deflection would indicate a beam of great strength. The order statistics for the three samples, of respective sizes $n_1 = 8$, $n_2 = 6$, and $n_3 = 6$, are as follows:

A:	79	82	83	84	85	86	86	87
B:	74	75	76	77	78	82		
C:	77	78	79	79	79	82		

(a) Use these data, $\alpha = 0.05$, and the F test to test the equality of the three means.

(b) For each set of data, construct box-and-whisker diagrams on the same figure and give an interpretation of your diagrams.

9.3-5. The female cuckoo lays her eggs in other birds' nests. The "foster parents" are usually deceived, probably because of the similarity in sizes of their own eggs and cuckoo eggs. Latter (see References) investigated this possible explanation and measured the lengths of cuckoo eggs (in mm) that were found in the nests of three species. Following are his results:

Hedge sparrow:	22.0	23.9	20.9	23.8	25.0
	24.0	21.7	23.8	22.8	23.1
	23.1	23.5	23.0	23.0	
Robin:	21.8	23.0	23.3	22.4	23.0
	23.0	23.0	22.4	23.9	22.3
	22.0	22.6	22.0	22.1	21.1
	23.0				
Wren:	19.8	22.1	21.5	20.9	22.0
	21.0	22.3	21.0	20.3	20.9
	22.0	20.0	20.8	21.2	21.0

(a) Construct an ANOVA table to test the equality of the three means.

(b) For each set of data, construct box-and-whisker diagrams on the same figure.

(c) Interpret your results.

9.3-6. Let X_1, X_2, X_3, X_4 equal the cholesterol level of a woman under the age of 50, a man under 50, a woman 50 or older, and a man 50 or older, respectively. Assume that the distribution of X_i is $N(\mu_i, \sigma^2)$, $i = 1, 2, 3, 4$. We shall test the null hypothesis H_0: $\mu_1 = \mu_2 = \mu_3 = \mu_4$, using seven observations of each X_i.

(a) Give a critical region for an $\alpha = 0.05$ significance level.

(b) Construct an ANOVA table and state your conclusion, using the following data:

x_1:	221	213	202	183	185	197	162
x_2:	271	192	189	209	227	236	142
x_3:	262	193	224	201	161	178	265
x_4:	192	253	248	278	232	267	289

(c) Give bounds on the p-value for this test.

(d) For each set of data, construct box-and-whisker diagrams on the same figure and give an interpretation of your diagram.

9.3-7. Montgomery (see References) examines the strengths of a synthetic fiber that may be affected by the percentage of cotton in the fiber. Five levels of this percentage are considered, with five observations taken at each level.

Percentage of Cotton	Tensile Strength in lb/in^2				
15	7	7	15	11	9
20	12	17	12	18	18
25	14	18	18	19	19
30	19	25	22	19	23
35	7	10	11	15	11

Use the F test, with $\alpha = 0.05$, to see if there are differences in the breaking strengths due to the percentages of cotton used.

9.3-8. Different sizes of nails are packaged in "1-pound" boxes. Let X_i equal the weight of a box with nail size $(4i)C$, $i = 1, 2, 3, 4, 5$, where $4C, 8C, 12C, 16C$, and $20C$ are the sizes of the sinkers from smallest to largest. Assume that the distribution of X_i is $N(\mu_i, \sigma^2)$. To test the null hypothesis that the mean weights of "1-pound" boxes are all equal for different sizes of nails, we shall use random samples of size 7, weighing the nails to the nearest hundredth of a pound.

(a) Give a critical region for an $\alpha = 0.05$ significance level.

(b) Construct an ANOVA table and state your conclusion, using the following data:

x_1:	1.03	1.04	1.07	1.03	1.08	1.06	1.07
x_2:	1.03	1.10	1.08	1.05	1.06	1.06	1.05
x_3:	1.03	1.08	1.06	1.02	1.04	1.04	1.07
x_4:	1.10	1.10	1.09	1.09	1.06	1.05	1.08
x_5:	1.04	1.06	1.07	1.06	1.05	1.07	1.05

(c) For each set of data, construct box-and-whisker diagrams on the same figure and give an interpretation of your diagrams.

9.3-9. Let X_i, $i = 1, 2, 3, 4$, equal the distance (in yards) that a golf ball travels when hit from a tee, where i denotes the index of the ith manufacturer. Assume that the distribution of X_i is $N(\mu_i, \sigma^2)$, $i = 1, 2, 3, 4$, when a ball is hit by a certain golfer. We shall test the null hypothesis H_0: $\mu_1 = \mu_2 = \mu_3 = \mu_4$, using three observations of each random variable.

(a) Give a critical region for an $\alpha = 0.05$ significance level.

(b) Construct an ANOVA table and state your conclusion, using the following data:

x_1:	240	221	265
x_2:	286	256	272
x_3:	259	245	232
x_4:	239	215	223

(c) What would your conclusion be if $\alpha = 0.025$?

(d) What is the approximate p-value of this test?

9.3-10. From the box-and-whisker diagrams in Figure 9.3-1, it looks like the means of X_1 and X_2 could be equal and also that the means of X_3, X_4, and X_5 could be equal but different from the first two.

(a) Using the data in Example 9.3-2, as well as a t test and an F test, test H_0: $\mu_1 = \mu_2$ against a two-sided alternative hypothesis. Let $\alpha = 0.05$. Do the F and t tests give the same result?

(b) Using the data in Example 9.3-2, test H_0: $\mu_3 = \mu_4 = \mu_5$. Let $\alpha = 0.05$.

9.3-11. The driver of a diesel-powered automobile decided to test the quality of three types of diesel fuel sold in the area. The test is to be based on miles per gallon (mpg). Make the usual assumptions, take $\alpha = 0.05$, and use the following data to test the null hypothesis that the three means are equal:

Brand A:	38.7	39.2	40.1	38.9	
Brand B:	41.9	42.3	41.3		
Brand C:	40.8	41.2	39.5	38.9	40.3

9.3-12. A particular process puts a coating on a piece of glass so that it is sensitive to touch. Randomly throughout the day, pieces of glass are selected from the production line and the resistance is measured at 12 different locations on the glass. On each of three different days, December 6, December 7, and December 22, the following data give the means of the 12 measurements on each of 11 pieces of glass:

December 6: 175.05 177.44 181.94 176.51 182.12 164.34

163.20 168.12 171.26 171.92 167.87

December 7: 175.93 176.62 171.39 173.90 178.34 172.90

174.67 174.27 177.16 184.13 167.21

December 22: 167.27 161.48 161.86 173.83 170.75 172.90

173.27 170.82 170.93 173.89 177.68

(a) Use these data to test whether the means on all three days are equal.

(b) Use box-and-whisker diagrams to confirm your answer.

9.3-13. For an aerosol product, there are three weights: the tare weight (container weight), the concentrate weight, and the propellant weight. Let X_1, X_2, X_3 denote the propellant weights on three different days. Assume that each of these independent random variables has a normal distribution with common variance and respective means μ_1, μ_2, and μ_3. We shall test the null hypothesis $H_0: \mu_1 = \mu_2 = \mu_3$, using nine observations of each of the random variables.

(a) Give a critical region for an $\alpha = 0.01$ significance level.

(b) Construct an ANOVA table and state your conclusion, using the following data:

x_1:	43.06	43.32	42.63	42.86	43.05
	42.87	42.94	42.80	42.36	
x_2:	42.33	42.81	42.13	42.41	42.39
	42.10	42.42	41.42	42.52	
x_3:	42.83	42.57	42.96	43.16	42.25
	42.24	42.20	41.97	42.61	

(c) For each set of data, construct box-and-whisker diagrams on the same figure and give an interpretation of your diagrams.

9.3-14. Ledolter and Hogg (see References) report the comparison of three workers with different experience who manufacture brake wheels for a magnetic brake. Worker A has four years of experience, worker B has seven years, and worker C has one year. The company is concerned about the product's quality, which is measured by the difference between the specified diameter and the actual diameter of the brake wheel. On a given day, the supervisor selects nine brake wheels at random from the output of each worker. The following data give the differences between the specified and actual diameters in hundredths of an inch:

Worker A:	2.0	3.0	2.3	3.5	3.0	2.0	4.0	4.5	3.0
Worker B:	1.5	3.0	4.5	3.0	3.0	2.0	2.5	1.0	2.0
Worker C:	2.5	3.0	2.0	2.5	1.5	2.5	2.5	3.0	3.5

(a) Test whether there are statistically significant differences in the quality among the three different workers.

(b) Do box plots of the data confirm your answer in part (a)?

9.3-15. Ledolter and Hogg (see References) report that an operator of a feedlot wants to compare the effectiveness of three different cattle feed supplements. He selects a random sample of 15 one-year-old heifers from his lot of over 1000 and divides them into three groups at random. Each group gets a different feed supplement. Upon noting that one heifer in group A was lost due to an accident, the operator records the gains in weight (in pounds) over a six-month period as follows:

Group A:	500	650	530	680	
Group B:	700	620	780	830	860
Group C:	500	520	400	580	410

(a) Test whether there are differences in the mean weight gains due to the three different feed supplements.

(b) Do box plots of the data confirm your answer in part (a)?

9.4 TWO-WAY ANALYSIS OF VARIANCE

The test of the equality of several means, considered in Section 9.3, is an example of a statistical inference method called the analysis of variance (ANOVA). This method derives its name from the fact that the quadratic form $SS(TO) = (n-1)S^2$—the total sum of squares about the combined sample mean—is decomposed into its components and analyzed. In this section, other problems in the analysis of variance will be investigated; here we restrict our considerations to the two-factor case, but the reader can see how it can be extended to three-factor and other cases.

Consider a situation in which it is desirable to investigate the effects of two factors that influence an outcome of an experiment. For example, a teaching method (lecture, discussion, computer assisted, television, etc.) and the size of a class might influence a student's score on a standard test; or the type of car and the grade of gasoline used might change the number of miles per gallon. In this latter example, if the number of miles per gallon is not affected by the grade of gasoline, we would no doubt use the least expensive grade.

The first analysis-of-variance model that we discuss is referred to as a **two-way classification with one observation per cell**. Assume that there are two factors (attributes), one of which has a levels and the other b levels. There are thus $n = ab$ possible combinations, each of which determines a cell. Let us think of these cells as being arranged in a rows and b columns. Here we take one observation per cell, and we denote the observation in the ith row and jth column by X_{ij}. Assume further that X_{ij} is $N(\mu_{ij}, \sigma^2)$, $i = 1, 2, \ldots, a$, and $j = 1, 2, \ldots, b$; and the $n = ab$ random variables are independent. [The assumptions of normality and homogeneous (same) variances can be somewhat relaxed in applications, with little change in the significance levels of the resulting tests.] We shall assume that the means μ_{ij} are composed of a row effect, a column effect, and an overall effect in some additive way, namely, $\mu_{ij} = \mu + \alpha_i + \beta_j$, where $\sum_{i=1}^{a} \alpha_i = 0$ and $\sum_{j=1}^{b} \beta_j = 0$. The parameter α_i represents the ith row effect, and the parameter β_j represents the jth column effect.

REMARK There is no loss in generality in assuming that

$$\sum_{i=1}^{a} \alpha_i = \sum_{j=1}^{b} \beta_j = 0.$$

To see this, let $\mu_{ij} = \mu' + \alpha_i' + \beta_j'$. Write

$$\overline{\alpha}' = \left(\frac{1}{a}\right) \sum_{i=1}^{a} \alpha_i' \quad \text{and} \quad \overline{\beta}' = \left(\frac{1}{b}\right) \sum_{j=1}^{b} \beta_j'.$$

We have

$$\mu_{ij} = (\mu' + \overline{\alpha}' + \overline{\beta}') + (\alpha_i' - \overline{\alpha}') + (\beta_j' - \overline{\beta}') = \mu + \alpha_i + \beta_j,$$

where $\sum_{i=1}^{a} \alpha_i = 0$ and $\sum_{j=1}^{b} \beta_j = 0$. The reader is asked to find μ, α_i, and β_j for one display of μ_{ij} in Exercise 9.4-2.   ■

To test the hypothesis that there is no row effect, we would test H_A: $\alpha_1 = \alpha_2 = \cdots = \alpha_a = 0$, since $\sum_{i=1}^{a} \alpha_i = 0$. Similarly, to test that there is no column effect, we would test H_B: $\beta_1 = \beta_2 = \cdots = \beta_b = 0$, since $\sum_{j=1}^{b} \beta_j = 0$. To test these hypotheses, we shall again partition the total sum of squares into several components. Letting

$$\overline{X}_{i\cdot} = \frac{1}{b} \sum_{j=1}^{b} X_{ij}, \ \overline{X}_{\cdot j} = \frac{1}{a} \sum_{i=1}^{a} X_{ij}, \ \overline{X}_{\cdot\cdot} = \frac{1}{ab} \sum_{i=1}^{a} \sum_{j=1}^{b} X_{ij},$$

we have

$$\text{SS(TO)} = \sum_{i=1}^{a} \sum_{j=1}^{b} (X_{ij} - \overline{X}_{..})^2$$

$$= \sum_{i=1}^{a} \sum_{j=1}^{b} [(\overline{X}_{i.} - \overline{X}_{..}) + (\overline{X}_{.j} - \overline{X}_{..}) + (X_{ij} - \overline{X}_{i.} - \overline{X}_{.j} + \overline{X}_{..})]^2$$

$$= b \sum_{i=1}^{a} (\overline{X}_{i.} - \overline{X}_{..})^2 + a \sum_{j=1}^{b} (\overline{X}_{.j} - \overline{X}_{..})^2$$

$$+ \sum_{i=1}^{a} \sum_{j=1}^{b} (X_{ij} - \overline{X}_{i.} - \overline{X}_{.j} + \overline{X}_{..})^2$$

$$= \text{SS(A)} + \text{SS(B)} + \text{SS(E)},$$

where SS(A) is the sum of squares among levels of factor A, or among rows; SS(B) is the sum of squares among levels of factor B, or among columns; and SS(E) is the error or residual sum of squares. In Exercise 9.4-4, the reader is asked to show that the three cross-product terms in the square of the trinomial sum to zero. The distribution of the error sum of squares does not depend on the mean μ_{ij}, provided that the additive model is correct. Hence, its distribution is the same whether H_A or H_B is true or not, and thus SS(E) acts as a "measuring stick," as did SS(E) in Section 9.3. This can be seen more clearly by writing

$$\text{SS(E)} = \sum_{i=1}^{a} \sum_{j=1}^{b} (X_{ij} - \overline{X}_{i.} - \overline{X}_{.j} + \overline{X}_{..})^2$$

$$= \sum_{i=1}^{a} \sum_{j=1}^{b} [X_{ij} - (\overline{X}_{i.} - \overline{X}_{..}) - (\overline{X}_{.j} - \overline{X}_{..}) - \overline{X}_{..}]^2$$

and noting the similarity of the summand in the right-hand member to

$$X_{ij} - \mu_{ij} = X_{ij} - \alpha_i - \beta_j - \mu.$$

We now show that $\text{SS(A)}/\sigma^2$, $\text{SS(B)}/\sigma^2$, and $\text{SS(E)}/\sigma^2$ are independent chi-square variables, provided that both H_A and H_B are true—that is, when all the means μ_{ij} have a common value μ. To do this, we first note that $\text{SS(TO)}/\sigma^2$ is $\chi^2(ab-1)$. In addition, from Section 9.3, we see that expressions such as $\text{SS(A)}/\sigma^2$ and $\text{SS(B)}/\sigma^2$ are chi-square variables, namely, $\chi^2(a-1)$ and $\chi^2(b-1)$, by replacing the n_i of Section 9.3 by a and b, respectively. Obviously, $\text{SS(E)} \geq 0$, and hence by Theorem 9.3-1, $\text{SS(A)}/\sigma^2$, $\text{SS(B)}/\sigma^2$, and $\text{SS(E)}/\sigma^2$ are independent chi-square variables with $a-1$, $b-1$, and $ab-1-(a-1)-(b-1) = (a-1)(b-1)$ degrees of freedom, respectively.

To test the hypothesis $H_A: \alpha_1 = \alpha_2 = \cdots = \alpha_a = 0$, we shall use the row sum of squares SS(A) and the residual sum of squares SS(E). When H_A is true, $\text{SS(A)}/\sigma^2$ and $\text{SS(E)}/\sigma^2$ are independent chi-square variables with $a-1$ and $(a-1)(b-1)$ degrees of freedom, respectively. Thus, $\text{SS(A)}/(a-1)$ and $\text{SS(E)}/[(a-1)(b-1)]$ are both unbiased estimators of σ^2 when H_A is true. However, $E[\text{SS(A)}/(a-1)] > \sigma^2$ when H_A is not true, and hence we would reject H_A when

$$F_A = \frac{\text{SS(A)}/[\sigma^2(a-1)]}{\text{SS(E)}/[\sigma^2(a-1)(b-1)]} = \frac{\text{SS(A)}/(a-1)}{\text{SS(E)}/[(a-1)(b-1)]}$$

is "too large." Since F_A has an F distribution with $a - 1$ and $(a - 1)(b - 1)$ degrees of freedom when H_A is true, H_A is rejected if the observed value of F_A equals or exceeds $F_\alpha[a-1, (a-1)(b-1)]$.

Similarly, the test of the hypothesis $H_B: \beta_1 = \beta_2 = \cdots = \beta_b = 0$ against all alternatives can be based on

$$F_B = \frac{\text{SS(B)}/[\sigma^2(b-1)]}{\text{SS(E)}/[\sigma^2(a-1)(b-1)]} = \frac{\text{SS(B)}/(b-1)}{\text{SS(E)}/[(a-1)(b-1)]},$$

which has an F distribution with $b-1$ and $(a-1)(b-1)$ degrees of freedom, provided that H_B is true.

Table 9.4-1 is the ANOVA table that summarizes the information needed for these tests of hypotheses. The formulas for F_A and F_B show that each of them is a ratio of two mean squares.

Example 9.4-1

Each of three cars is driven with each of four different brands of gasoline. The number of miles per gallon driven for each of the $ab = (3)(4) = 12$ different combinations is recorded in Table 9.4-2.

We would like to test whether we can expect the same mileage for each of these four brands of gasoline. In our notation, we test the hypothesis

$$H_B: \beta_1 = \beta_2 = \beta_3 = \beta_4 = 0$$

Table 9.4-1 Two-way ANOVA table, one observation per cell

Source	Sum of Squares (SS)	Degrees of Freedom	Mean Square (MS)	F
Factor A (row)	SS(A)	$a - 1$	$\text{MS(A)} = \dfrac{\text{SS(A)}}{a - 1}$	$\dfrac{\text{MS(A)}}{\text{MS(E)}}$
Factor B (column)	SS(B)	$b - 1$	$\text{MS(B)} = \dfrac{\text{SS(B)}}{b - 1}$	$\dfrac{\text{MS(B)}}{\text{MS(E)}}$
Error	SS(E)	$(a-1)(b-1)$	$\text{MS(E)} = \dfrac{\text{SS(E)}}{(a-1)(b-1)}$	
Total	SS(TO)	$ab - 1$		

Table 9.4-2 Gas mileage data

	Gasoline				
Car	1	2	3	4	$\overline{X}_{i\cdot}$
1	26	28	31	31	29
2	24	25	28	27	26
3	25	25	28	26	26
$\overline{X}_{\cdot j}$	25	26	29	28	27

against all alternatives. At a 1% significance level, we shall reject H_B if the computed F, namely,

$$\frac{\text{SS(B)}/(4-1)}{\text{SS(E)}/[(3-1)(4-1)]} \geq 9.78 = F_{0.01}(3,6).$$

We have

$$\text{SS(B)} = 3[(25-27)^2 + (26-27)^2 + (29-27)^2 + (28-27)^2] = 30;$$

$$\text{SS(E)} = (26-29-25+27)^2 + (24-26-25+27)^2 + \cdots$$

$$+ (26-26-28+27)^2 = 4.$$

Hence, the computed F is

$$\frac{30/3}{4/6} = 15 > 9.78,$$

and the hypothesis H_B is rejected. That is, the gasolines seem to give different performances (at least with these three cars).

The information for this example is summarized in Table 9.4-3. ■

In a two-way classification problem, particular combinations of the two factors might interact differently from what is expected from the additive model. For instance, in Example 9.4-1, gasoline 3 seemed to be the best gasoline and car 1 the best car; however, it sometimes happens that the two best do not "mix" well and the joint performance is poor. That is, there might be a strange interaction between this combination of car and gasoline, and accordingly, the joint performance is not as good as expected. Sometimes it happens that we get good results from a combination of some of the poorer levels of each factor. This phenomenon is called interaction, and it frequently occurs in practice (e.g., in chemistry). In order to test for possible interaction, we shall consider a two-way classification problem in which $c > 1$ independent observations per cell are taken.

Assume that X_{ijk}, $i = 1, 2, \ldots, a$; $j = 1, 2, \ldots, b$; and $k = 1, 2, \ldots, c$, are $n = abc$ random variables that are mutually independent and have normal distributions with a common, but unknown, variance σ^2. The mean of each X_{ijk}, $k = 1, 2, \ldots, c$, is $\mu_{ij} = \mu + \alpha_i + \beta_j + \gamma_{ij}$, where $\sum_{i=1}^{a} \alpha_i = 0$, $\sum_{j=1}^{b} \beta_j = 0$, $\sum_{i=1}^{a} \gamma_{ij} = 0$, and $\sum_{j=1}^{b} \gamma_{ij} = 0$. The parameter γ_{ij} is called the **interaction** associated with cell (i, j). That is, the interaction between the ith level of one classification and the jth level of

Table 9.4-3 ANOVA table for gas mileage data					
Source	Sum of Squares (SS)	Degrees of Freedom	Mean Square (MS)	F	p-value
Row (A)	24	2	12	18	0.003
Column (B)	30	3	10	15	0.003
Error	4	6	2/3		
Total	58	11			

the other classification is γ_{ij}. In Exercise 9.4-6, the reader is asked to determine μ, α_i, β_j, and γ_{ij} for some given μ_{ij}.

To test the hypotheses that (a) the row effects are equal to zero, (b) the column effects are equal to zero, and (c) there is no interaction, we shall again partition the total sum of squares into several components. Letting

$$\overline{X}_{ij\cdot} = \frac{1}{c} \sum_{k=1}^{c} X_{ijk},$$

$$\overline{X}_{i\cdot\cdot} = \frac{1}{bc} \sum_{j=1}^{b} \sum_{k=1}^{c} X_{ijk},$$

$$\overline{X}_{\cdot j\cdot} = \frac{1}{ac} \sum_{i=1}^{a} \sum_{k=1}^{c} X_{ijk},$$

$$\overline{X}_{\cdots} = \frac{1}{abc} \sum_{i=1}^{a} \sum_{j=1}^{b} \sum_{k=1}^{c} X_{ijk},$$

we have

$$\text{SS(TO)} = \sum_{i=1}^{a} \sum_{j=1}^{b} \sum_{k=1}^{c} (X_{ijk} - \overline{X}_{\cdots})^2$$

$$= bc \sum_{i=1}^{a} (\overline{X}_{i\cdot\cdot} - \overline{X}_{\cdots})^2 + ac \sum_{j=1}^{b} (\overline{X}_{\cdot j\cdot} - \overline{X}_{\cdots})^2$$

$$+ c \sum_{i=1}^{a} \sum_{j=1}^{b} (\overline{X}_{ij\cdot} - \overline{X}_{i\cdot\cdot} - \overline{X}_{\cdot j\cdot} + \overline{X}_{\cdots})^2 + \sum_{i=1}^{a} \sum_{j=1}^{b} \sum_{k=1}^{c} (X_{ijk} - \overline{X}_{ij\cdot})^2$$

$$= \text{SS(A)} + \text{SS(B)} + \text{SS(AB)} + \text{SS(E)},$$

where SS(A) is the row sum of squares, or the sum of squares among levels of factor A; SS(B) is the column sum of squares, or the sum of squares among levels of factor B; SS(AB) is the interaction sum of squares; and SS(E) is the error sum of squares. Again, we can show that the cross-product terms sum to zero.

To consider the joint distribution of SS(A), SS(B), SS(AB), and SS(E), let us assume that all the means equal the same value μ. Of course, we know that $\text{SS(TO)}/\sigma^2$ is $\chi^2(abc-1)$. Also, by letting the n_i of Section 9.3 equal bc and ac, respectively, we know that $\text{SS(A)}/\sigma^2$ and $\text{SS(B)}/\sigma^2$ are $\chi^2(a-1)$ and $\chi^2(b-1)$. Moreover,

$$\frac{\sum_{k=1}^{c} (X_{ijk} - \overline{X}_{ij\cdot})^2}{\sigma^2}$$

is $\chi^2(c-1)$; hence, $\text{SS(E)}/\sigma^2$ is the sum of ab independent chi-square variables such as this and thus is $\chi^2[ab(c-1)]$. Of course $\text{SS(AB)} \geq 0$; so, according to Theorem 9.3-1, $\text{SS(A)}/\sigma^2$, $\text{SS(B)}/\sigma^2$, $\text{SS(AB)}/\sigma^2$, and $\text{SS(E)}/\sigma^2$ are mutually independent chi-square variables with $a-1$, $b-1$, $(a-1)(b-1)$, and $ab(c-1)$ degrees of freedom, respectively.

To test the hypotheses concerning row, column, and interaction effects, we form F statistics in which the numerators are affected by deviations from the respective hypotheses, whereas the denominator is a function of SS(E), whose distribution depends only on the value of σ^2 and not on the values of the cell means. Hence, SS(E) acts as our measuring stick here.

The statistic for testing the hypothesis

$$H_{AB}: \gamma_{ij} = 0, \ i = 1, 2, \ldots, a; \ j = 1, 2, \ldots, b,$$

against all alternatives is

$$F_{AB} = \frac{c \sum_{i=1}^{a} \sum_{j=1}^{b} (\overline{X}_{ij\cdot} - \overline{X}_{i\cdot\cdot} - \overline{X}_{\cdot j\cdot} + \overline{X}_{\cdots})^2 / [\sigma^2 (a-1)(b-1)]}{\sum_{i=1}^{a} \sum_{j=1}^{b} \sum_{k=1}^{c} (X_{ijk} - \overline{X}_{ij\cdot})^2 / [\sigma^2 ab(c-1)]}$$

$$= \frac{SS(AB)/[(a-1)(b-1)]}{SS(E)/[ab(c-1)]},$$

which has an F distribution with $(a-1)(b-1)$ and $ab(c-1)$ degrees of freedom when H_{AB} is true. If the computed $F_{AB} \geq F_{\alpha}[(a-1)(b-1), ab(c-1)]$, we reject H_{AB} and say that there is a difference among the means, since there seems to be interaction. Most statisticians do *not* proceed to test row and column effects if H_{AB} is rejected.

The statistic for testing the hypothesis

$$H_A: \alpha_1 = \alpha_2 = \cdots = \alpha_a = 0$$

against all alternatives is

$$F_A = \frac{bc \sum_{i=1}^{a} (\overline{X}_{i\cdot\cdot} - \overline{X}_{\cdots})^2 / [\sigma^2 (a-1)]}{\sum_{i=1}^{a} \sum_{j=1}^{b} \sum_{k=1}^{c} (X_{ijk} - \overline{X}_{ij\cdot})^2 / [\sigma^2 ab(c-1)]} = \frac{SS(A)/(a-1)}{SS(E)/[ab(c-1)]},$$

which has an F distribution with $a-1$ and $ab(c-1)$ degrees of freedom when H_A is true. The statistic for testing the hypothesis

$$H_B: \beta_1 = \beta_2 = \cdots = \beta_b = 0$$

against all alternatives is

$$F_B = \frac{ac \sum_{j=1}^{b} (\overline{X}_{\cdot j\cdot} - \overline{X}_{\cdots})^2 / [\sigma^2 (b-1)]}{\sum_{i=1}^{a} \sum_{j=1}^{b} \sum_{k=1}^{c} (X_{ijk} - \overline{X}_{ij\cdot})^2 / [\sigma^2 ab(c-1)]} = \frac{SS(B)/(b-1)}{SS(E)/[ab(c-1)]},$$

which has an F distribution with $b-1$ and $ab(c-1)$ degrees of freedom when H_B is true. Each of these hypotheses is rejected if the observed value of F is greater than a given constant that is selected to yield the desired significance level.

Table 9.4-4 is the ANOVA table that summarizes the information needed for these tests of hypotheses.

Example 9.4-2 Consider the following experiment: One hundred eight people were randomly divided into 6 groups with 18 people in each group. Each person was given sets of three numbers to add. The three numbers were either in a "down array" or an "across array," representing the two levels of factor A. The levels of factor B are determined by the number of digits in the numbers to be added: one-digit, two-digit, or three-digit numbers. Table 9.4-5 illustrates this experiment with a sample problem for each cell; note, however, that an individual person works problems only of one of these types. Each person was placed in one of the six groups and was told to work as many problems as possible in 90 seconds. The measurement that was recorded was the average number of problems worked correctly in two trials.

Table 9.4-4 Two-way ANOVA table, c observations per cell

Source	Sum of Squares (SS)	Degrees of Freedom	Mean Square (MS)	F
Factor A (row)	SS(A)	$a - 1$	$MS(A) = \dfrac{SS(A)}{a - 1}$	$\dfrac{MS(A)}{MS(E)}$
Factor B (column)	SS(B)	$b - 1$	$MS(B) = \dfrac{SS(B)}{b - 1}$	$\dfrac{MS(B)}{MS(E)}$
Factor AB (interaction)	SS(AB)	$(a - 1)(b - 1)$	$MS(AB) = \dfrac{SS(AB)}{(a - 1)(b - 1)}$	$\dfrac{MS(AB)}{MS(E)}$
Error	SS(E)	$ab(c - 1)$	$MS(E) = \dfrac{SS(E)}{ab(c - 1)}$	
Total	SS(TO)	$abc - 1$		

Table 9.4-5 Illustration of arrays for numbers of digits

Type of Array	Number of Digits 1	2	3
Down	5	25	259
	3	69	567
	8	37	130
Across	5 + 3 + 8 =	25 + 69 + 37 =	259 + 567 + 130 =

Whenever this many subjects are used, a computer becomes an invaluable tool. A computer program provided the summary shown in Table 9.4-6 of the sample means of the rows, the columns, and the six cells. Each cell mean is the average for 18 people.

Simply considering these means, we can see clearly that there is a column effect: It is not surprising that it is easier to add one-digit than three-digit numbers.

The most interesting feature of these results is that they show the possibility of interaction. The largest cell mean occurs for those adding one-digit numbers in an across array. Note, however, that for two- and three-digit numbers, the down arrays have larger means than the across arrays.

The computer provided the ANOVA table given in Table 9.4-7. The number of degrees of freedom for SS(E) is not in our F table in Appendix B. However, the rightmost column, obtained from the computer printout, provides the p-value of each test, namely, the probability of obtaining an F as large as or larger than the calculated F ratio. Note, for example, that, to test for interaction, $F = 5.51$ and the p-value is 0.0053. Thus, the hypothesis of no interaction would be rejected at the $\alpha = 0.05$ or $\alpha = 0.01$ significance level, but it would not be rejected with $\alpha = 0.001$. ∎

Table 9.4-6 Cell, row, and column means for adding numbers

Type of Array	Number of Digits			Row Means
	1	2	3	
Down	23.806	10.694	6.278	13.593
Across	26.056	6.750	3.944	12.250
Column means	24.931	8.722	5.111	

Table 9.4-7 ANOVA table for adding numbers

Source	Sum of Squares	Degrees of Freedom	Mean Square	F	p-value
Factor A (array)	48.678	1	48.669	2.885	0.0925
Factor B (number of digits)	8022.73	2	4011.363	237.778	<0.0001
Interaction	185.92	2	92.961	5.510	0.0053
Error	1720.76	102	16.870		
Total	9978.08	107			

Exercises

(In some of the exercises that follow, we must make assumptions, such as normal distributions with equal variances.)

9.4-1. For the data given in Example 9.4-1, test the hypothesis $H_A: \alpha_1 = \alpha_2 = \alpha_3 = 0$ against all alternatives at the 5% significance level.

9.4-2. With $a = 3$ and $b = 5$, find μ, α_i, and β_j if μ_{ij}, $i = 1, 2, 3$ and $j = 1, 2, 3, 4, 5$, are given by

$$
\begin{array}{ccccc}
7 & 4 & 8 & 9 & 11 \\
11 & 8 & 12 & 13 & 15 \\
9 & 6 & 10 & 11 & 13
\end{array}
$$

Note that in an "additive" model such as this one, one row (column) can be determined by adding a constant value to each of the elements of another row (column).

9.4-3. We wish to compare compressive strengths of concrete corresponding to $a = 3$ different drying methods (treatments). Concrete is mixed in batches that are just large enough to produce three cylinders. Although care is taken to achieve uniformity, we expect some variability among the $b = 5$ batches used to obtain the following compressive strengths (there is little reason to suspect interaction; hence, only one observation is taken in each cell):

Treatment	Batch				
	B_1	B_2	B_3	B_4	B_5
A_1	52	47	44	51	42
A_2	60	55	49	52	43
A_3	56	48	45	44	38

(a) Use the 5% significance level and test H_A: $\alpha_1 = \alpha_2 = \alpha_3 = 0$ against all alternatives.

(b) Use the 5% significance level and test H_B: $\beta_1 = \beta_2 = \beta_3 = \beta_4 = \beta_5 = 0$ against all alternatives. (See Ledolter and Hogg in References.)

9.4-4. Show that the cross-product terms formed from $(\overline{X}_{i\cdot} - \overline{X}_{\cdot\cdot})$, $(\overline{X}_{\cdot j} - \overline{X}_{\cdot\cdot})$, and $(X_{ij} - \overline{X}_{i\cdot} - \overline{X}_{\cdot j} + \overline{X}_{\cdot\cdot})$ sum to zero, $i = 1, 2, \ldots a$ and $j = 1, 2, \ldots, b$. HINT: For example, write

$$\sum_{i=1}^{a} \sum_{j=1}^{b} (\overline{X}_{\cdot j} - \overline{X}_{\cdot\cdot})(X_{ij} - \overline{X}_{i\cdot} - \overline{X}_{\cdot j} + \overline{X}_{\cdot\cdot})$$

$$= \sum_{j=1}^{b} (\overline{X}_{\cdot j} - \overline{X}_{\cdot\cdot}) \sum_{i=1}^{a} [(X_{ij} - \overline{X}_{\cdot j}) - (\overline{X}_{i\cdot} - \overline{X}_{\cdot\cdot})]$$

and sum each term in the inner summation, as grouped here, to get zero.

9.4-5. A psychology student was interested in testing how food consumption by rats would be affected by a particular drug. She used two levels of one attribute, namely, drug and placebo, and four levels of a second attribute, namely, male (M), castrated (C), female (F), and ovariectomized (O). For each cell, she observed five rats. The amount of food consumed in grams per 24 hours is listed in the following table:

	M	C	F	O
Drug	22.56	16.54	18.58	18.20
	25.02	24.64	15.44	14.56
	23.66	24.62	16.12	15.54
	17.22	19.06	16.88	16.82
	22.58	20.12	17.58	14.56
Placebo	25.64	22.50	17.82	19.74
	28.84	24.48	15.76	17.48
	26.00	25.52	12.96	16.46
	26.02	24.76	15.00	16.44
	23.24	20.62	19.54	15.70

(a) Use the 5% significance level and test H_{AB}: $\gamma_{ij} = 0$, $i = 1, 2, j = 1, 2, 3, 4$.

(b) Use the 5% significance level and test H_A: $\alpha_1 = \alpha_2 = 0$.

(c) Use the 5% significance level and test H_B: $\beta_1 = \beta_2 = \beta_3 = \beta_4 = 0$.

(d) How could you modify this model so that there are three attributes of classification, each with two levels?

9.4-6. With $a = 3$ and $b = 5$, find μ, α_i, β_j, and γ_{ij} if μ_{ij}, $i = 1, 2, 3$ and $j = 1, 2, 3, 4, 5$, are given by

7	8	8	9	15
11	4	12	13	11
9	6	10	11	13

Note the difference between the layout here and that in Exercise 9.4-2. Does the interaction help explain the difference?

9.4-7. In order to test whether four brands of gasoline give equal performance in terms of mileage, each of three cars was driven with each of the four brands of gasoline. Then each of the $(3)(4) = 12$ possible combinations was repeated four times. The number of miles per gallon for each of the four repetitions in each cell is recorded in the following table:

Car	Brand of Gasoline															
	1				2				3				4			
1	31.0	24.9	26.3	30.0	25.8	29.4	27.8	27.3								
	26.2	28.8	25.2	31.6	24.5	24.8	28.2	30.4								
2	30.6	29.5	25.5	26.8	26.6	23.7	28.1	27.1								
	30.8	28.9	27.4	29.4	28.2	26.1	31.5	29.1								
3	24.2	23.1	27.4	28.1	25.2	26.7	26.3	26.4								
	26.8	27.4	26.4	26.9	27.7	28.1	27.9	28.8								

Test the hypotheses H_{AB}: no interaction, H_A: no row effect, and H_B: no column effect, each at the 5% significance level.

9.4-8. There is another way of looking at Exercise 9.3-6, namely, as a two-factor analysis-of-variance problem with the levels of gender being female and male, the levels of age being less than 50 and at least 50, and the measurement for each subject being their cholesterol level. The data would then be set up as follows:

	Age	
Gender	<50	≥50
	221	262
	213	193
	202	224
Female	183	201
	185	161
	197	178
	162	265
	271	192
	192	253
	189	248
Male	209	278
	227	232
	236	267
	142	289

(a) Test H_{AB}: $\gamma_{ij} = 0$, $i = 1, 2$; $j = 1, 2$ (no interaction).

(b) Test H_A: $\alpha_1 = \alpha_2 = 0$ (no row effect).

(c) Test H_B: $\beta_1 = \beta_2 = 0$ (no column effect).

Use a 5% significance level for each test.

9.4-9. Ledolter and Hogg (see References) report that volunteers who had a smoking history classified as heavy, moderate, and nonsmoker were accepted until nine men were in each category. Three men in each category were randomly assigned to each of the following three stress tests: bicycle ergometer, treadmill, and step tests. The time until maximum oxygen uptake was recorded in minutes as follows:

	Test		
Smoking History	Bicycle	Treadmill	Step Test
Nonsmoker	12.8, 13.5, 11.2	16.2, 18.1, 17.8	22.6, 19.3, 18.9
Moderate	10.9, 11.1, 9.8	15.5, 13.8, 16.2	20.1, 21.0, 15.9
Heavy	8.7, 9.2, 7.5	14.7, 13.2, 8.1	16.2, 16.1, 17.8

(a) Analyze the results of this experiment. Obtain the ANOVA table and test for main effects and interactions.

(b) Use box plots to compare the data graphically.

9.5* GENERAL FACTORIAL AND 2^K FACTORIAL DESIGNS

In Section 9.4, we studied two-factor experiments in which the A factor is performed at a levels and the B factor has b levels. Without replications, we need ab-level combinations, and with c replications with each of these combinations, we need a total of abc experiments.

Let us now consider a situation with three factors—say, A, B, and C, with a, b, and c levels, respectively. Here there are a total of abc-level combinations, and if, at each of these combinations, we have d replications, there is a need for $abcd$ experiments. Once these experiments are run, in some random order, and the data collected, there are computer programs available to calculate the entries in the ANOVA table, as in Table 9.5-1.

The main effects (A, B, and C) and the two-factor interactions (AB, AC, and BC) have the same interpretations as in the two-factor ANOVA. The three-factor interaction represents that part of the model for the means μ_{ijh}, $i = 1, 2, \ldots, a$; $j = 1, 2, \ldots, b$; $h = 1, 2, \ldots, c$, that cannot be explained by a model including only the main effects and two-factor interactions. In particular, if, for each fixed h, the "plane" created by μ_{ijh} is "parallel" to the "plane" created by every other fixed h, then the three-factor interaction is equal to zero. Usually, higher-order interactions tend to be small.

Table 9.5-1 ANOVA table

Source	SS	d.f.	MS	F
A	SS(A)	$a-1$	MS(A)	MS(A)/MS(E)
B	SS(B)	$b-1$	MS(B)	MS(B)/MS(E)
C	SS(C)	$c-1$	MS(C)	MS(C)/MS(E)
AB	SS(AB)	$(a-1)(b-1)$	MS(AB)	MS(AB)/MS(E)
AC	SS(AC)	$(a-1)(c-1)$	MS(AC)	MS(AC)/MS(E)
BC	SS(BC)	$(b-1)(c-1)$	MS(BC)	MS(BC)/MS(E)
ABC	SS(ABC)	$(a-1)(b-1)(c-1)$	MS(ABC)	MS(ABC)/MS(E)
Error	SS(E)	$abc(d-1)$	MS(E)	
Total	SS(TO)	$abcd-1$		

In the testing sequence, we test the three-factor interaction first by checking to see whether or not

$$\text{MS(ABC)}/\text{MS(E)} \geq F_\alpha[(a-1)(b-1)(c-1), abc(d-1)].$$

If this inequality holds, the ABC interaction is significant at the α level. We would then not continue testing the two-factor interactions and the main effects with those F values, but analyze the data otherwise. For example, for each fixed h, we could look at a two-factor ANOVA for factors A and B. Of course, if the inequality does not hold, we next check the two-factor interactions with the appropriate F values. If these are not significant, we check the main effects, A, B, and C.

Factorial analyses with three or more factors require many experiments, particularly if each factor has several levels. Often, in the health, social, and physical sciences, experimenters want to consider several factors (maybe as many as 10, 20, or even hundreds), and they cannot afford to run that many experiments. This is particularly true with preliminary or screening investigations, in which they want to detect the factors that seem most important. In these cases, they often consider factorial experiments such that each of k factors is run at just two levels, frequently without replication. We consider only this situation, although the reader should recognize that it has many variations. In particular, there are methods for investigating only *fractions of these* 2^k *designs*. The reader interested in more information should refer to a good book on the design of experiments, such as that by Box, Hunter, and Hunter (see References). Many statisticians in industry believe that these statistical methods are the most useful in improving product and process designs. Hence, this is clearly an extremely important topic, as many industries are greatly concerned about the quality of their products.

In factorial experiments in which each of the k factors is considered at only two levels, those levels are selected at some reasonable low and high values. That is, with the help of someone in the field, the typical range of each factor is considered. For instance, if we are considering baking temperatures in the range from 300° to 375°, a representative low is selected—say, 320°—and a representative high is selected—say, 355°. There is no formula for these selections, and someone familiar with the experiment would help make them. Often, it happens that only two different types of a material (e.g., fabric) are considered and one is called low and the other high.

Thus, we select a low and high for each factor and code them as -1 and $+1$ or, more simply, $-$ and $+$, respectively. We give three 2^k designs, for $k = 2, 3$, and 4, in standard order in Tables 9.5-2, 9.5-3, and 9.5-4, respectively. From these three tables, we can easily note what is meant by standard order. The A column starts with a minus sign and then the sign alternates. The B column begins with two minus signs and then the signs alternate in blocks of two. The C column has 4 minus signs and then 4 plus signs, and so on. The D column starts with 8 minus signs and then 8 plus signs. It is easy to extend this idea to 2^k designs, where $k \geq 5$. To illustrate, under the E column in a 2^5 design, we have 16 minus signs followed by 16 plus signs, which together account for the 32 experiments.

To be absolutely certain what these runs mean, consider run number 12 in Table 9.5-4: A is set at its high level, B at its high, C at its low, and D at its high level. The value X_{12} is the random observation resulting from this one combination of these four settings. It must be emphasized that the runs are not necessarily performed in the order $1, 2, 3, \ldots, 2^k$; in fact, they should be performed in a random

Table 9.5-2 2^2 Design

	2^2 Design		
Run	A	B	Observation
1	$-$	$-$	X_1
2	$+$	$-$	X_2
3	$-$	$+$	X_3
4	$+$	$+$	X_4

Table 9.5-3 2^3 Design

	2^3 Design			
Run	A	B	C	Observation
1	$-$	$-$	$-$	X_1
2	$+$	$-$	$-$	X_2
3	$-$	$+$	$-$	X_3
4	$+$	$+$	$-$	X_4
5	$-$	$-$	$+$	X_5
6	$+$	$-$	$+$	X_6
7	$-$	$+$	$+$	X_7
8	$+$	$+$	$+$	X_8

Table 9.5-4 2^4 Design

Run	\multicolumn{4}{c}{2^4 Design}	Observation

Run	A	B	C	D	Observation
1	−	−	−	−	X_1
2	+	−	−	−	X_2
3	−	+	−	−	X_3
4	+	+	−	−	X_4
5	−	−	+	−	X_5
6	+	−	+	−	X_6
7	−	+	+	−	X_7
8	+	+	+	−	X_8
9	−	−	−	+	X_9
10	+	−	−	+	X_{10}
11	−	+	−	+	X_{11}
12	+	+	−	+	X_{12}
13	−	−	+	+	X_{13}
14	+	−	+	+	X_{14}
15	−	+	+	+	X_{15}
16	+	+	+	+	X_{16}

order if at all possible. That is, in a 2^3 design, we might perform the experiment in the order 3, 2, 8, 6, 5, 1, 4, 7 if this, in fact, was a random selection of a permutation of the first eight positive integers.

Once all 2^k experiments have been run, it is possible to consider the total sum of squares

$$\sum_{i=1}^{2^k} (X_i - \overline{X})^2$$

and decompose it very easily into $2^k - 1$ parts, which represent the respective measurements (estimators) of the k main effects, $\binom{k}{2}$ two-factor interactions, $\binom{k}{3}$ three-factor interactions, and so on, until we have the one k-factor interaction. We illustrate this decomposition with the 2^3 design in Table 9.5-5. Note that column AB is found by formally multiplying the elements of column A by the corresponding ones in B. Likewise, AC is found by multiplying the elements of column A by the corresponding ones in column C, and so on, until column ABC is the product of the corresponding elements of columns A, B, and C. Next, we construct

Table 9.5-5 2^3 Design decomposition

Run	A	B	C	AB	AC	BC	ABC	Observation
1	−	−	−	+	+	+	−	X_1
2	+	−	−	−	−	+	+	X_2
3	−	+	−	−	+	−	+	X_3
4	+	+	−	+	−	−	−	X_4
5	−	−	+	+	−	−	+	X_5
6	+	−	+	−	+	−	−	X_6
7	−	+	+	−	−	+	−	X_7
8	+	+	+	+	+	+	+	X_8

seven linear forms, using these seven columns of signs with the corresponding observations. The resulting measures (estimates) of the main effects (A, B, C), the two-factor interactions (AB, AC, BC), and the three-factor interaction (ABC) are then found by dividing the linear forms by $2^k = 2^3 = 8$. (Some statisticians divide by $2^{k-1} = 2^{3-1} = 4$.) These are denoted by

$$[A] = (-X_1 + X_2 - X_3 + X_4 - X_5 + X_6 - X_7 + X_8)/8,$$

$$[B] = (-X_1 - X_2 + X_3 + X_4 - X_5 - X_6 + X_7 + X_8)/8,$$

$$[C] = (-X_1 - X_2 - X_3 - X_4 + X_5 + X_6 + X_7 + X_8)/8,$$

$$[AB] = (+X_1 - X_2 - X_3 + X_4 + X_5 - X_6 - X_7 + X_8)/8,$$

$$[AC] = (+X_1 - X_2 + X_3 - X_4 - X_5 + X_6 - X_7 + X_8)/8,$$

$$[BC] = (+X_1 + X_2 - X_3 - X_4 - X_5 - X_6 + X_7 + X_8)/8,$$

$$[ABC] = (-X_1 + X_2 + X_3 - X_4 + X_5 - X_6 - X_7 + X_8)/8.$$

With assumptions of normality, mutual independence, and common variance σ^2, under the overall null hypothesis of the equality of all the means, each of these measures has a normal distribution with mean zero and variance $\sigma^2/8$ (in general, $\sigma^2/2^k$). This implies that the square of each measure divided by $\sigma^2/8$ is $\chi^2(1)$. Moreover, it can be shown (see Exercise 9.5-2) that

$$\sum_{i=1}^{8} \left(X_i - \overline{X}\right)^2 = 8\left([A]^2 + [B]^2 + [C]^2 + [AB]^2 + [AC]^2 + [BC]^2 + [ABC]^2\right).$$

So, by Theorem 9.3-1, the terms on the right-hand side, divided by σ^2, are mutually independent random variables, each being $\chi^2(1)$. While it requires a little more theory, it follows that the linear forms [A], [B], [C], [AB], [AC], [BC], and [ABC] are mutually independent $N(0, \sigma^2/8)$ random variables.

Since we have assumed that we have not run any replications, how can we obtain an estimate of σ^2 to see if any of the main effects or interactions are significant? To help us, we fall back on the use of a q–q plot because, under the overall null hypothesis, those seven measures are mutually independently, normally distributed variables with the same mean and variance. Thus, a q–q plot of the normal percentiles against the corresponding ordered values of the measures should be about on a straight line if, in fact, the null hypothesis is true. If one of these points is "out of line," we might believe that the overall null hypothesis is not true and that the effect associated with the factor represented with that point is significant. It is possible that two or three points might be out of line; then all corresponding effects (main or interaction) should be investigated. Clearly, this is not a formal test, but it has been extremely successful in practice.

As an illustration, we use the data from an experiment designed to evaluate the effects of laundering on a certain fire-retardant treatment for fabrics. These data, somewhat modified, were taken from *Experimental Statistics, National Bureau of Standards Handbook 91*, by Mary G. Natrella (Washington, DC: U.S. Government Printing Office, 1963). Factor A is the type of fabric (sateen or monk's cloth), factor B corresponds to two different fire-retardant treatments, and factor C describes the laundering conditions (no laundering, after one laundering). The observations are

Table 9.5-6 Seven measures ordered			
Identity of Effect	Ordered Effect	Percentile	Percentile from $N(0,1)$
[A]	−8.06	12.5	−1.15
[AB]	−2.19	25.0	−0.67
[AC]	−0.31	37.5	−0.32
[ABC]	0.31	50.0	0.00
[C]	0.56	62.5	0.32
[BC]	0.81	75.0	0.67
[B]	1.56	87.5	1.15

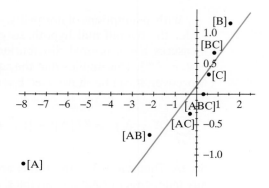

Figure 9.5-1 A q–q plot of normal percentiles versus estimated effects

inches burned, measured on a standard-size fabric after a flame test. They are as follows, in standard order:

$$x_1 = 41.0, \qquad x_2 = 30.5, \qquad x_3 = 47.5, \qquad x_4 = 27.0,$$
$$x_5 = 39.5, \qquad x_6 = 26.5, \qquad x_7 = 48.0, \qquad x_8 = 27.5.$$

Thus, the measures of the effects are

$$[A] = (-41.0 + 30.5 - 47.5 + 27.0 - 39.5 + 26.5 - 48.0 + 27.5)/8 = -8.06,$$

$$[B] = (-41.0 - 30.5 + 47.5 + 27.0 - 39.5 - 26.5 + 48.0 + 27.5)/8 = 1.56,$$

$$[C] = (-41.0 - 30.5 - 47.5 - 27.0 + 39.5 + 26.5 + 48.0 + 27.5)/8 = 0.56,$$

$$[AB] = (+41.0 - 30.5 - 47.5 + 27.0 + 39.5 - 26.5 - 48.0 + 27.5)/8 = -2.19,$$

$$[AC] = (+41.0 - 30.5 + 47.5 - 27.0 - 39.5 + 26.5 - 48.0 + 27.5)/8 = -0.31,$$

$$[BC] = (+41.0 + 30.5 - 47.5 - 27.0 - 39.5 - 26.5 + 48.0 + 27.5)/8 = 0.81,$$

$$[ABC] = (-41.0 + 30.5 + 47.5 - 27.0 + 39.5 - 26.5 - 48.0 + 27.5)/8 = 0.31.$$

In Table 9.5-6, we order these seven measures, determine their percentiles, and find the corresponding percentiles of the standard normal distribution.

The q–q plot is given in Figure 9.5-1. Each point has been identified with its effect. A straight line fits six of those points reasonably well, but the point associated with $[A] = -8.06$ is far from this straight line. Hence, the main effect of factor A (the type of fabric) seems to be significant. It is interesting to note that the laundering factor, C, does not seem to be a significant factor.

Exercises

9.5-1. Write out a 2^2 design, displaying the A, B, and AB columns for the four runs.

(a) If X_1, X_2, X_3, and X_4 are the four observations for the respective runs in standard order, write out the three linear forms, $[A]$, $[B]$, and $[AB]$, that measure the two main effects and the interaction. These linear forms should include the divisor $2^2 = 4$.

(b) Show that $\sum_{i=1}^{4} (X_i - \overline{X})^2 = 4([A]^2 + [B]^2 + [AB]^2)$.

(c) Under the null hypothesis that all the means are equal and with the usual assumptions (normality, mutual independence, and common variance), what can you say about the distributions of the expressions in (b) after each is divided by σ^2?

9.5-2. Show that, in a 2^3 design,

$$\sum_{i=1}^{8} (X_i - \overline{X})^2$$
$$= 8 \left([A]^2 + [B]^2 + [C]^2 + [AB]^2 + [AC]^2 + [BC]^2 + [ABC]^2 \right).$$

HINT: Since both the right and the left members of this equation are symmetric in the variables $X_1, X_2, \ldots, X_8$, it is necessary to show only that the corresponding coefficients of $X_1 X_i$, $i = 1, 2, \ldots, 8$, are the same in each member of the equation. Of course, recall that $\overline{X} = (X_1 + X_2 + \cdots + X_8)/8$.

9.5-3. Show that the unbiased estimator of the variance σ^2 from a sample of size $n = 2$ is one half of the square of the difference of the two observations. Thus, show that, if a 2^k design is replicated, say, with X_{i1} and X_{i2}, $i = 1, 2, \ldots, 2^k$, then the estimate of the common σ^2 is

$$\frac{1}{2^{k+1}} \sum_{i=1}^{2^k} (X_{i1} - X_{i2})^2 = \text{MS(E)}.$$

Under the usual assumptions, this equation implies that each of $2^k[A]^2/\text{MS(E)}$, $2^k[B]^2/\text{MS(E)}$, $2^k[AB]^2/\text{MS(E)}$, and so on has an $F(1, 2^k)$ distribution under the null hypothesis. This approach, of course, would provide tests for the significance of the various effects, including interactions.

9.5-4. Ledolter and Hogg (see References) note that percent yields from a certain chemical reaction for changing temperature (factor A), reaction time (factor B), and concentration (factor C) are $x_1 = 79.7$, $x_2 = 74.3$, $x_3 = 76.7$, $x_4 = 70.0$, $x_5 = 84.0$, $x_6 = 81.3$, $x_7 = 87.3$, and $x_8 = 73.7$, in standard order with a 2^3 design.

(a) Estimate the main effects, the three two-factor interactions, and the three-factor interaction.

(b) Construct an appropriate q–q plot to see if any of these effects seem to be significantly larger than the others.

9.5-5. Box, Hunter, and Hunter (see References) studied the effects of catalyst charge (10 pounds = -1, 20 pounds = $+1$), temperature (220 °C = -1, 240 °C = $+1$), pressure (50 psi = -1, 80 psi = $+1$), and concentration (10% = -1, 12% = $+1$) on percent conversion (X) of a certain chemical. The results of a 2^4 design, in standard order, are

$$x_1 = 71, \quad x_2 = 61, \quad x_3 = 90, \quad x_4 = 82, \quad x_5 = 68, \quad x_6 = 61,$$

$$x_7 = 87, \quad x_8 = 80, \quad x_9 = 61, \quad x_{10} = 50, x_{11} = 89,$$

$$x_{12} = 83, x_{13} = 59, x_{14} = 51, x_{15} = 85, x_{16} = 78.$$

(a) Estimate the main effects and the two-, three-, and four-factor interactions.

(b) Construct an appropriate q–q plot and assess the significance of the various effects.

9.6* TESTS CONCERNING REGRESSION AND CORRELATION

In Section 6.5, we considered the estimation of the parameters of a very simple regression curve, namely, a straight line. We can use confidence intervals for the parameters to test hypotheses about them. For example, with the same model as that in Section 6.5, we could test the hypothesis H_0: $\beta = \beta_0$ by using a t random variable that was used for a confidence interval with β replaced by β_0, namely,

$$T_1 = \frac{\widehat{\beta} - \beta_0}{\sqrt{\dfrac{n\widehat{\sigma^2}}{(n-2)\sum_{i=1}^{n}(x_i - \overline{x})^2}}}.$$

The null hypothesis, along with three possible alternative hypotheses, is given in Table 9.6-1; these tests are equivalent to stating that we reject H_0 if β_0 is not in certain confidence intervals. For example, the first test is equivalent to rejecting H_0 if β_0 is not in the one-sided confidence interval with lower bound

$$\widehat{\beta} - t_\alpha(n-2)\sqrt{\frac{n\widehat{\sigma^2}}{(n-2)\sum_{i=1}^{n}(x_i - \overline{x})^2}}.$$

Often we let $\beta_0 = 0$ and test the hypothesis H_0: $\beta = 0$. That is, we test the null hypothesis that the slope is equal to zero.

Table 9.6-1 Tests about the slope of the regression line		
H_0	H_1	Critical Region
$\beta = \beta_0$	$\beta > \beta_0$	$t_1 \geq t_\alpha(n-2)$
$\beta = \beta_0$	$\beta < \beta_0$	$t_1 \leq -t_\alpha(n-2)$
$\beta = \beta_0$	$\beta \neq \beta_0$	$|t_1| \geq t_{\alpha/2}(n-2)$

Example 9.6-1

Let x equal a student's preliminary test score in a psychology course and y equal the same student's score on the final examination. With $n = 10$ students, we shall test H_0: $\beta = 0$ against H_1: $\beta \neq 0$. At the 0.01 significance level, the critical region is $|t_1| \geq t_{0.005}(8) = 3.355$. Using the data in Example 6.5-1, we find that the observed value of T_1 is

$$t_1 = \frac{0.742 - 0}{\sqrt{10(21.7709)/8(756.1)}} = \frac{0.742}{0.1897} = 3.911.$$

Thus, we reject H_0 and conclude that a student's score on the final examination is related to his or her preliminary test score. ∎

We consider tests about the correlation coefficient ρ of a bivariate normal distribution. Let X and Y have a bivariate normal distribution. We know that if the correlation coefficient ρ is zero, then X and Y are independent random variables. Furthermore, the value of ρ gives a measure of the linear relationship between X and Y. We now give methods for using the sample correlation coefficient to test the hypothesis H_0: $\rho = 0$ and also to form a confidence interval for ρ.

Let $(X_1, Y_1), (X_2, Y_2), \ldots, (X_n, Y_n)$ denote a random sample from a bivariate normal distribution with parameters μ_X, μ_Y, σ_X^2, σ_Y^2, and ρ. That is, the n pairs of (X, Y) are independent, and each pair has the same bivariate normal distribution. The **sample correlation coefficient** is

$$R = \frac{[1/(n-1)] \sum_{i=1}^{n} (X_i - \overline{X})(Y_i - \overline{Y})}{\sqrt{[1/(n-1)] \sum_{i=1}^{n} (X_i - \overline{X})^2} \sqrt{[1/(n-1)] \sum_{i=1}^{n} (Y_i - \overline{Y})^2}} = \frac{S_{XY}}{S_X S_Y}.$$

We note that

$$R \frac{S_Y}{S_X} = \frac{S_{XY}}{S_X^2} = \frac{[1/(n-1)] \sum_{i=1}^{n} (X_i - \overline{X})(Y_i - \overline{Y})}{[1/(n-1)] \sum_{i=1}^{n} (X_i - \overline{X})^2}$$

is exactly the solution that we obtained for $\widehat{\beta}$ in Section 6.5 when the X-values were fixed at $X_1 = x_1, X_2 = x_2, \ldots, X_n = x_n$. Let us consider these values fixed temporarily so that we are considering conditional distributions, given $X_1 = x_1, \ldots, X_n = x_n$. Moreover, if H_0: $\rho = 0$ is true, then $Y_1, Y_2, \ldots, Y_n$ are independent of $X_1, X_2, \ldots, X_n$ and $\beta = \rho \sigma_Y / \sigma_X = 0$. Under these conditions, the conditional distribution of

$$\widehat{\beta} = \frac{\sum_{i=1}^{n} (X_i - \overline{X})(Y_i - \overline{Y})}{\sum_{i=1}^{n} (X_i - \overline{X})^2},$$

given that $X_1 = x_1, X_2 = x_2, \ldots, X_n = x_n$, is $N[0, \sigma_Y^2/(n-1)s_x^2]$ when $s_x^2 > 0$. Moreover, recall from Section 6.5 that the conditional distribution of

$$\frac{\sum_{i=1}^{n} [Y_i - \overline{Y} - (S_{XY}/S_X^2)(X_i - \overline{X})]^2}{\sigma_Y^2} = \frac{(n-1)S_Y^2(1 - R^2)}{\sigma_Y^2},$$

given that $X_1 = x_1, \ldots, X_n = x_n$, is $\chi^2(n-2)$ and is independent of $\widehat{\beta}$. (See Exercise 9.6-6.) Thus, when $\rho = 0$, the conditional distribution of

$$T = \frac{(RS_Y/S_X)/(\sigma_Y/\sqrt{n-1}\,S_X)}{\sqrt{[(n-1)S_Y^2(1 - R^2)/\sigma_Y^2][1/(n-2)]}} = \frac{R\sqrt{n-2}}{\sqrt{1 - R^2}}$$

is t with $n - 2$ degrees of freedom. However, since the conditional distribution of T, given that $X_1 = x_1, \ldots, X_n = x_n$, does not depend on $x_1, x_2, \ldots, x_n$, the unconditional distribution of T must be t with $n-2$ degrees of freedom, and T and $(X_1, X_2, \ldots, X_n)$ are independent when $\rho = 0$.

REMARK It is interesting to note that in the discussion about the distribution of T, the assumption that (X, Y) has a bivariate normal distribution can be relaxed. Specifically, if X and Y are independent and Y has a normal distribution, then T has a t distribution regardless of the distribution of X. Obviously, the roles of X and Y can be reversed in all of this development. In particular, if X and Y are independent, then T and $Y_1, Y_2, \ldots, Y_n$ are also independent. ∎

Now T can be used to test H_0: $\rho = 0$. If the alternative hypothesis is H_1: $\rho > 0$, we would use the critical region defined by the observed $T \geq t_\alpha(n-2)$, since large T implies large R. Obvious modifications would be made for the alternative hypotheses H_1: $\rho < 0$ and H_1: $\rho \neq 0$, the latter leading to a two-sided test.

Using the pdf $h(t)$ of T, we can find the distribution function and pdf of R when $-1 < r < 1$, provided that $\rho = 0$:

$$G(r) = P(R \leq r) = P\left(T \leq \frac{r\sqrt{n-2}}{\sqrt{1-r^2}}\right)$$

$$= \int_{-\infty}^{r\sqrt{n-2}/\sqrt{1-r^2}} h(t)\, dt$$

$$= \int_{-\infty}^{r\sqrt{n-2}/\sqrt{1-r^2}} \frac{\Gamma[(n-1)/2]}{\Gamma(1/2)\,\Gamma[(n-2)/2]} \frac{1}{\sqrt{n-2}} \left(1 + \frac{t^2}{n-2}\right)^{-(n-1)/2} dt.$$

The derivative of $G(r)$, with respect to r, is (see Appendix D.4)

$$g(r) = h\left(\frac{r\sqrt{n-2}}{\sqrt{1-r^2}}\right) \frac{d(r\sqrt{n-2}/\sqrt{1-r^2})}{dr},$$

which equals

$$g(r) = \frac{\Gamma[(n-1)/2]}{\Gamma(1/2)\,\Gamma[(n-2)/2]} (1-r^2)^{(n-4)/2}, \qquad -1 < r < 1.$$

Thus, to test the hypothesis H_0: $\rho = 0$ against the alternative hypothesis H_1: $\rho \neq 0$ at a significance level α, select either a constant $r_{\alpha/2}(n-2)$ or a constant $t_{\alpha/2}(n-2)$ so that

$$\alpha = P(|R| \geq r_{\alpha/2}(n-2); H_0) = P(|T| \geq t_{\alpha/2}(n-2); H_0),$$

depending on the availability of R or T tables.

It is interesting to graph the pdf of R. Note in particular that if $n = 4$, $g(r) = 1/2$, $-1 < r < 1$, and if $n = 6$, $g(r) = (3/4)(1 - r^2)$, $-1 < r < 1$. The graphs of the pdf of R when $n = 8$ and when $n = 14$ are given in Figure 9.6-1. Recall that this is the pdf of R when $\rho = 0$. As n increases, R is more likely to equal values close to 0.

Table IX in Appendix B lists selected values of the distribution function of R when $\rho = 0$. For example, if $n = 8$, then the number of degrees of freedom is 6 and $P(R \leq 0.7887) = 0.99$. Also, if $\alpha = 0.10$, then $r_{\alpha/2}(6) = r_{0.05}(6) = 0.6215$. [See Figure 9.6-1(a).]

It is also possible to obtain an approximate test of size α by using the fact that

$$W = \frac{1}{2} \ln \frac{1+R}{1-R}$$

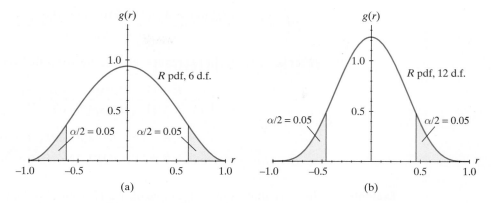

Figure 9.6-1 R pdfs when $n = 8$ and $n = 14$

has an approximate normal distribution with mean $(1/2) \ln[(1 + \rho)/(1 - \rho)]$ and variance $1/(n - 3)$. We accept this statement without proof. (See Exercise 9.6-8.) Thus, a test of H_0: $\rho = \rho_0$ can be based on the statistic

$$Z = \frac{\dfrac{1}{2} \ln \dfrac{1+R}{1-R} - \dfrac{1}{2} \ln \dfrac{1+\rho_0}{1-\rho_0}}{\sqrt{\dfrac{1}{n-3}}},$$

which has a distribution that is approximately $N(0, 1)$ under H_0. Notice that this approximate size-α test can be used to test a null hypothesis specifying a nonzero population correlation coefficient, whereas the exact size-α test may be used only in conjunction with the null hypothesis H_0: $\rho = 0$. Also, notice that the sample size must be at least $n = 4$ for the approximate test, but $n = 3$ is sufficient for the exact test.

Example 9.6-2

We would like to test the hypothesis H_0: $\rho = 0$ against H_1: $\rho \neq 0$ at an $\alpha = 0.05$ significance level. A random sample of size 18 from a bivariate normal distribution yielded a sample correlation coefficient of $r = 0.35$. From Table XI in Appendix B, since $0.35 < 0.4683$, H_0 is accepted (not rejected) at an $\alpha = 0.05$ significance level. Using the t distribution, we would reject H_0 if $|t| \geq 2.120 = t_{0.025}(16)$. Since

$$t = \frac{0.35\sqrt{16}}{\sqrt{1 - (0.35)^2}} = 1.495,$$

H_0 is not rejected. If we had used the normal approximation for Z, H_0 would be rejected if $|z| \geq 1.96$. Because

$$z = \frac{(1/2) \ln[(1 + 0.35)/(1 - 0.35)] - 0}{\sqrt{1/(18 - 3)}} = 1.415,$$

H_0 is not rejected. ∎

To develop an approximate $100(1 - \alpha)\%$ confidence interval for ρ, we use the normal approximation for the distribution of Z. Thus, we select a constant $c = z_{\alpha/2}$ from Table V in Appendix B so that

$$P\left(-c \le \frac{(1/2)\ln[(1+R)/(1-R)] - (1/2)\ln[(1+\rho)/(1-\rho)]}{\sqrt{1/(n-3)}} \le c\right) \approx 1 - \alpha.$$

After several algebraic manipulations, this formula becomes

$$P\left(\frac{1+R-(1-R)\exp(2c/\sqrt{n-3})}{1+R+(1-R)\exp(2c/\sqrt{n-3})} \le \rho \le \right.$$

$$\left.\frac{1+R-(1-R)\exp(-2c/\sqrt{n-3})}{1+R+(1-R)\exp(-2c/\sqrt{n-3})}\right) \approx 1 - \alpha.$$

Example 9.6-3   Suppose that a random sample of size 12 from a bivariate normal distribution yielded a correlation coefficient of $r = 0.6$. An approximate 95% confidence interval for ρ would be

$$\left[\frac{1+0.6-(1-0.6)\exp\left(\frac{2(1.96)}{3}\right)}{1+0.6+(1-0.6)\exp\left(\frac{2(1.96)}{3}\right)}, \frac{1+0.6-(1-0.6)\exp\left(\frac{-2(1.96)}{3}\right)}{1+0.6+(1-0.6)\exp\left(\frac{-2(1.96)}{3}\right)}\right]$$

$$= [0.040, 0.873].$$

If the sample size had been $n = 39$ and $r = 0.6$, the approximate 95% confidence interval would have been $[0.351, 0.770]$.   ∎

Exercises

(In some of the exercises that follow, we must make assumptions of normal distributions with the usual notation.)

9.6-1. For the data given in Exercise 6.5-3, use a t test to test H_0: $\beta = 0$ against H_1: $\beta > 0$ at the $\alpha = 0.025$ significance level.

9.6-2. For the data given in Exercise 6.5-4, use a t test to test H_0: $\beta = 0$ against H_1: $\beta > 0$ at the $\alpha = 0.025$ significance level.

9.6-3. A random sample of size $n = 10$ from a bivariate normal distribution yielded a sample correlation coefficient of $r = 0.732$. Would the hypothesis $H_0 : \rho = 0$ be rejected in favor of $H_1 : \rho \ne 0$ at an $\alpha = 0.05$ significance level?

9.6-4. In bowling, it is often possible to score well in the first game and then bowl poorly in the second game, or vice versa. The following five pairs of numbers give the scores of the first and second games bowled by the same person on five consecutive Monday evenings:

Game 1:	167	195	186	192	205
Game 2:	188	174	177	200	198

Assume a bivariate normal distribution, and use these scores to test the hypothesis $H_0 : \rho = 0$ against $H_1 : \rho \ne 0$ at $\alpha = 0.01$.

9.6-5. A random sample of size 28 from a bivariate normal distribution yielded a sample correlation coefficient of $r = 0.65$. Find an approximate 90% confidence interval for ρ.

9.6-6. By squaring the binomial expression $[(Y_i - \overline{Y}) - (S_{xY}/s_x^2)(x_i - \overline{x})]$, show that

$$\sum_{i=1}^{n}[(Y_i - \overline{Y}) - (S_{xY}/s_x^2)(x_i - \overline{x})]^2$$

$$= \sum_{i=1}^{n}(Y_i - \overline{Y})^2 - 2\left(\frac{S_{xY}}{s_x^2}\right)\sum_{i=1}^{n}(x_i - \overline{x})(Y_i - \overline{Y})$$

$$+ \frac{S_{xY}^2}{s_x^4}\sum_{i=1}^{n}(x_i - \overline{x})^2$$

equals $(n-1)S_Y^2(1-R^2)$, where $X_1 = x_1, X_2 = x_2, \ldots,$ $X_n = x_n$. HINT: Replace $S_{xY} = \sum_{i=1}^{n}(x_i - \overline{x})(Y_i - \overline{Y})/(n-1)$ by Rs_xS_Y.

9.6-7. To help determine whether gallinules selected their mate on the basis of weight, 14 pairs of gallinules were captured and weighed. Test the null hypothesis H_0: $\rho = 0$ against a two-sided alternative at an $\alpha = 0.01$ significance level. Given that the male and female weights for the $n = 14$ pairs of birds yielded a sample correlation coefficient of $r = -0.252$, would H_0 be rejected?

9.6-8. In sampling from a bivariate normal distribution, it is true that the sample correlation coefficient R has an approximate normal distribution $N[\rho, (1 - \rho^2)^2/n]$ if the sample size n is large. Since, for large n, R is close to ρ, use two terms of the Taylor's expansion of $u(R)$ about ρ and determine that function $u(R)$ such that it has a variance which is (essentially) free of ρ. (The solution of this exercise explains why the transformation $(1/2)\ln[(1+R)/(1-R)]$ was suggested.)

9.6-9. Show that when $\rho = 0$,

(a) The points of inflection for the graph of the pdf of R are at $r = \pm 1/\sqrt{n-5}$ for $n \geq 7$.

(b) $E(R) = 0$.

(c) $\text{Var}(R) = 1/(n-1), n \geq 3$. HINT: Note that $E(R^2) = E[1 - (1 - R^2)]$.

9.6-10. In a college health fitness program, let X equal the weight in kilograms of a female freshman at the beginning of the program and let Y equal her change in weight during the semester. We shall use the following data for $n = 16$ observations of (x, y) to test the null hypothesis H_0: $\rho = 0$ against a two-sided alternative hypothesis:

(61.4, −3.2)	(62.9, 1.4)	(58.7, 1.3)	(49.3, 0.6)
(71.3, 0.2)	(81.5, −2.2)	(60.8, 0.9)	(50.2, 0.2)
(60.3, 2.0)	(54.6, 0.3)	(51.1, 3.7)	(53.3, 0.2)
(81.0, −0.5)	(67.6, −0.8)	(71.4, −0.1)	(72.1, −0.1)

(a) What is the conclusion if $\alpha = 0.10$?

(b) What is the conclusion if $\alpha = 0.05$?

9.6-11. Let X and Y have a bivariate normal distribution with correlation coefficient ρ. To test H_0: $\rho = 0$ against H_1: $\rho \neq 0$, a random sample of n pairs of observations is selected. Suppose that the sample correlation coefficient is $r = 0.68$. Using a significance level of $\alpha = 0.05$, find the smallest value of the sample size n so that H_0 is rejected.

9.6-12. In Exercise 6.5-5, data are given for horsepower, the time it takes a car to go from 0 to 60, and the weight in pounds of a car, for 14 cars. Those data are repeated here:

Horsepower	0–60	Weight	Horsepower	0–60	Weight
230	8.1	3516	282	6.2	3627
225	7.8	3690	300	6.4	3892
375	4.7	2976	220	7.7	3377
322	6.6	4215	250	7.0	3625
190	8.4	3761	315	5.3	3230
150	8.4	2940	200	6.2	2657
178	7.2	2818	300	5.5	3518

(a) Let ρ be the correlation coefficient of horsepower and weight. Test H_0: $\rho = 0$ against H_1: $\rho \neq 0$.

(b) Let ρ be the correlation coefficient of horsepower and "0–60." Test H_0: $\rho = 0$ against H_1: $\rho < 0$.

(c) Let ρ be the correlation coefficient of weight and "0–60." Test H_0: $\rho = 0$ against H_1: $\rho \neq 0$.

9.7* STATISTICAL QUALITY CONTROL

Statistical methods can be used in many scientific fields, such as medical research, engineering, chemistry, and psychology. Often, it is necessary to compare two ways of doing something—say, the old way and a possible new way. We collect data on each way, quite possibly in a laboratory situation, and try to decide whether the new way is actually better than the old. Needless to say, it would be terrible to change to the new way at great expense, only to find out that it is really not any better than the old. That is, suppose the lab results indicate, by some statistical method, that the new is seemingly better than the old. Can we actually extrapolate those outcomes in the lab to the situations in the real world? Clearly, statisticians cannot make these decisions, but they should be made by some professional who knows both statistics and the specialty in question very well. The statistical analysis might provide helpful guidelines, but we still need the expert to make the final decision.

However, even before investigating possible changes in any process, it is extremely important to determine exactly what the process in question is doing at the present time. Often, people in charge of an organization do not understand the capabilities of many of its processes. Simply measuring what is going on frequently leads to improvements. In many cases, measurement is easy, such as determining the diameter of a bolt, but sometimes it is extremely difficult, as in evaluating good teaching or many other service activities. But if at all possible, we encourage those involved to begin to "listen" to their processes; that is, they should measure what is going on in their organization. These measurements alone often are the beginning of desirable improvements. While most of our remarks in this chapter concern measurements made in manufacturing, service industries frequently find them just as useful.

At one time, some manufacturing plants would make parts to be used in the construction of some piece of equipment. Say a particular line in the plant, making a certain part, might produce several hundreds of them each day. These items would then be sent on to an inspection cage, where they would be checked for goodness, often several days or even weeks later. Occasionally, the inspectors would discover many defectives among the items made, say, two weeks ago. There was little that could be done at that point except scrap or rework the defective parts, both expensive outcomes.

In the 1920s, W. A. Shewhart, who was working for AT&T Bell Labs, recognized that this was an undesirable situation and suggested that, with some suitable frequency, a sample of these parts should be taken as they were being made. If the sample indicated that the items were satisfactory, the manufacturing process would continue. But if the sampled parts were not satisfactory, corrections should be made then so that things became satisfactory. This idea led to what are commonly called *Shewhart control charts*—the basis of what was called *statistical quality control* in those early days; today it is often referred to as *statistical process control*.

Shewhart control charts consist of calculated values of a statistic, say, $\bar{x}$, plotted in sequence. That is, in making products, every so often (each hour, each day, or each week, depending upon how many items are being produced) a sample of size n of them is taken, and they are measured, resulting in the observations $x_1, x_2, \ldots, x_n$. The average $\bar{x}$ and the standard deviation s are computed. This is done k times, and the k values of $\bar{x}$ and s are averaged, resulting in $\bar{\bar{x}}$ and $\bar{s}$, respectively; usually, k is equal to some number between 10 and 30.

The central limit theorem states that if the true mean μ and standard deviation σ of the process were known, then almost all of the $\bar{x}$-values would plot between $\mu - 3\sigma/\sqrt{n}$ and $\mu + 3\sigma/\sqrt{n}$, unless the system has actually changed. However, suppose we know neither μ nor σ, and thus μ is estimated by $\bar{\bar{x}}$ and $3\sigma/\sqrt{n}$ by $A_3\bar{s}$, where $\bar{\bar{x}}$ and $\bar{s}$ are the respective means of the k observations of $\bar{x}$ and s, and where A_3 is a factor depending upon n that can be found in books on statistical quality control. A few values of A_3 (and some other constants that will be used later) are given in Table 9.7-1 for typical values of n.

The estimates of $\mu \pm 3\sigma/\sqrt{n}$ are called the *upper control limit* (UCL), $\bar{\bar{x}} + A_3\bar{s}$, and the *lower control limit* (LCL), $\bar{\bar{x}} - A_3\bar{s}$, and $\bar{\bar{x}}$ provides the estimate of the centerline. A typical plot is given in Figure 9.7-1. Here, in the 13th sampling period, $\bar{x}$ is outside the control limits, indicating that the process has changed and that some investigation and action are needed to correct this change, which seems like an upward shift in the process.

Note that there is a control chart for the s values, too. From sampling distribution theory, values of B_3 and B_4 have been determined and are given in Table 9.7-1, so we know that almost all the s-values should be between $B_3\bar{s}$ and $B_4\bar{s}$ if there is no

n	A_3	B_3	B_4	A_2	D_3	D_4
4	1.63	0	2.27	0.73	0	2.28
5	1.43	0	2.09	0.58	0	2.11
6	1.29	0.03	1.97	0.48	0	2.00
8	1.10	0.185	1.815	0.37	0.14	1.86
10	0.98	0.28	1.72	0.31	0.22	1.78
20	0.68	0.51	1.49	0.18	0.41	1.59

Table 9.7-1 Some constants used with control charts

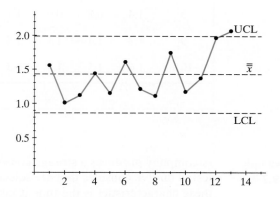

Figure 9.7-1 Typical control chart

change in the underlying distribution. So again, if an individual s-value is outside these control limits, some action should be taken, as it seems as if there has been a change in the variation of the underlying distribution.

Often, when these charts are first constructed after $k = 10$ to 30 sampling periods, many points fall outside the control limits. A team consisting of workers, the manager of the process, the supervisor, an engineer, and even a statistician should try to find the reasons that this has occurred, and the situation should be corrected. After this is done and the points plot within the control limits, the process is "in statistical control." However, being in statistical control is not a guarantee of satisfaction with the products. Since $A_3\bar{s}$ is an estimate of $3\sigma/\sqrt{n}$, it follows that $\sqrt{n}A_3\bar{s}$ is an estimate of 3σ, and with an underlying distribution close to a normal one, almost all items would be between $\bar{\bar{x}} \pm \sqrt{n}A_3\bar{s}$. If these limits are too wide, then corrections must be made again.

If the variation is under control (i.e., if $\bar{x}$ and s are within their control limits), we say that the variations seen in $\bar{x}$ and s are due to common causes. If products made under such a system with these existing common causes are satisfactory, then production continues. If either $\bar{x}$ or s, however, is outside the control limits, that is an indication that some special causes are at work, and they must be corrected. That is, a team should investigate the problem and some action should be taken.

Table 9.7-2 Console opening times							
Group	x_1	x_2	x_3	x_4	x_5	$\bar{x}$	s	R
1	1.2	1.8	1.7	1.3	1.4	1.480	0.259	0.60
2	1.5	1.2	1.0	1.0	1.8	1.300	0.346	0.80
3	0.9	1.6	1.0	1.0	1.0	1.100	0.283	0.70
4	1.3	0.9	0.9	1.2	1.0	1.060	0.182	0.40
5	0.7	0.8	0.9	0.6	0.8	0.760	0.114	0.30
6	1.2	0.9	1.1	1.0	1.0	1.040	0.104	0.30
7	1.1	0.9	1.1	1.0	1.4	1.100	0.187	0.50
8	1.4	0.9	0.9	1.1	1.0	1.060	0.207	0.50
9	1.3	1.4	1.1	1.5	1.6	1.380	0.192	0.50
10	1.6	1.5	1.4	1.3	1.5	1.460	0.114	0.30
						$\bar{\bar{x}} = 1.174$	$\bar{s} = 0.200$	$\bar{R} = 0.49$

Example 9.7-1

A company produces a storage console. Twice a day, nine critical characteristics are tested on five consoles that are selected randomly from the production line. One of these characteristics is the time it takes the lower storage component door to open completely. Table 9.7-2 lists the opening times in seconds for the consoles that were tested during one week. Also included in the table are the sample means, the sample standard deviations, and the ranges.

The upper control limit (UCL) and the lower control limit (LCL) for $\bar{x}$ are found using A_3 in Table 9.7-1 with $n = 5$ as follows:

$$\text{UCL} = \bar{\bar{x}} + A_3\bar{s} = 1.174 + 1.43(0.20) = 1.460$$

and

$$\text{LCL} = \bar{\bar{x}} - A_3\bar{s} = 1.174 - 1.43(0.20) = 0.888.$$

These control limits and the sample means are plotted on the $\bar{x}$ chart in Figure 9.7-2. There should be some concern about the fifth sampling period; thus, there should be an investigation to determine why that particular $\bar{x}$ is below the LCL.

The UCL and LCL for s are found using B_3 and B_4 in Table 9.7-1 with $n = 5$ as follows:

$$\text{UCL} = B_4\bar{s} = 2.09(0.200) = 0.418$$

and

$$\text{LCL} = B_3\bar{s} = 0(0.200) = 0.$$

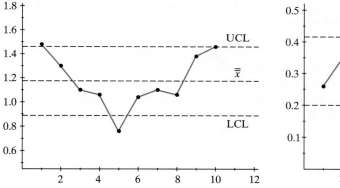
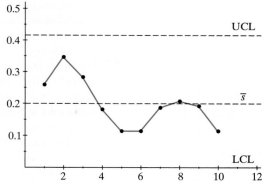

Figure 9.7-2 The $\bar{x}$ chart and s chart

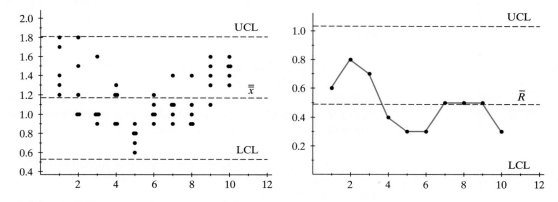

Figure 9.7-3 Plot of 50 console opening times and R chart

These control limits and the sample standard deviations are plotted on the s chart in Figure 9.7-2.

Almost all of the observations should lie between $\bar{\bar{x}} \pm \sqrt{n}\, A_3 \bar{s}$; namely,

$$1.174 + \sqrt{5}\,(1.43)(0.20) = 1.814$$

and

$$1.174 - \sqrt{5}\,(1.43)(0.20) = 0.534.$$

This situation is illustrated in Figure 9.7-3, in which all 50 observations do fall within these control limits. ■

In most books on statistical quality control, there is an alternative way of constructing the limits on an $\bar{x}$ chart. For each sample, we compute the range, R, which is the absolute value of the difference of the extremes of the sample. This computation is much easier than that for calculating s. After k samples are taken, we compute the average of these R-values, obtaining $\bar{R}$ as well as $\bar{\bar{x}}$. The statistic $A_2 \bar{R}$ serves as an estimate of $3\sigma/\sqrt{n}$, where A_2 is found in Table 9.7-1. Thus, the estimates of $\mu \pm 3\sigma/\sqrt{n}$, namely, $\bar{\bar{x}} \pm A_2 \bar{R}$, can be used as the UCL and LCL of an $\bar{x}$ chart.

In addition, $\sqrt{n}A_2\overline{R}$ is an estimate of 3σ; so, with an underlying distribution that is close to a normal one, we find that almost all observations are within the limits $\overline{\overline{x}} \pm \sqrt{n}A_2\overline{R}$.

Moreover, an R chart can be constructed with centerline $\overline{R}$ and control limits equal to $D_3\overline{R}$ and $D_4\overline{R}$, where D_3 and D_4 are given in Table 9.7-1 and were determined so that almost all R-values should be between the control limits if there is no change in the underlying distribution. Thus, a value of R falling outside those limits would indicate a change in the spread of the underlying distribution, and some corrective action should be considered.

The use of R, rather than s, is illustrated in the next example.

Example 9.7-2

Using the data in Example 9.7-1, we compute UCL and LCL for an $\overline{x}$ chart. We use $\overline{\overline{x}} \pm A_2\overline{R}$ as follows:

$$\text{UCL} = \overline{\overline{x}} + A_2\overline{R} = 1.174 + 0.58(0.49) = 1.458$$

and

$$\text{LCL} = \overline{\overline{x}} - A_2\overline{R} = 1.174 - 0.58(0.49) = 0.890.$$

Note that these values are very close to the limits that we found for the $\overline{x}$ chart in Figure 9.7-2 using $\overline{\overline{x}} \pm A_3\overline{s}$. In addition, almost all of the observations should lie within the limits $\overline{\overline{x}} + \sqrt{n}A_2\overline{R}$, which are

$$\text{UCL} = 1.174 + \sqrt{5}\,(0.58)(0.49) = 1.809$$

and

$$\text{LCL} = 1.174 - \sqrt{5}\,(0.58)(0.49) = 0.539.$$

Note that these are almost the same as the limits found in Example 9.7-1 and plotted in Figure 9.7-3.

An R chart can be constructed with centerline $\overline{R} = 0.49$ and control limits given by

$$\text{UCL} = D_4\overline{R} = 2.11(0.49) = 1.034$$

and

$$\text{LCL} = D_3\overline{R} = 0(0.49) = 0.$$

Figure 9.7-3 illustrates this control chart for the range, and we see that its pattern is similar to that of the s chart in Figure 9.7-2. ■

There are two other Shewhart control charts: the p and c charts. The central limit theorem, which provided a justification for the three-sigma limits in the $\overline{x}$ chart, also justifies the control limits in the p chart. Suppose the number of defectives among n items that are selected randomly—say, D—has a binomial distribution $b(n, p)$. Then the limits $p \pm 3\sqrt{p(1-p)/n}$ should include almost all of the D/n-values.

However, p must be approximated by observing k values of D—say, $D_1, D_2, \ldots, D_k$—and computing what is called $\bar{p}$ in the statistical quality control literature, namely,

$$\bar{p} = \frac{D_1 + D_2 + \cdots + D_k}{kn}.$$

Thus, the LCL and UCL for the fraction defective, D/n, are respectively given by

$$\text{LCL} = \bar{p} - 3\sqrt{\bar{p}(1-\bar{p})/n}$$

and

$$\text{UCL} = \bar{p} + 3\sqrt{\bar{p}(1-\bar{p})/n}.$$

If the process is in control, almost all D/n-values are between the LCL and UCL. Still, this may not be satisfactory and improvements might be needed to decrease $\bar{p}$. If it is satisfactory, however, let the process continue under these common causes of variation until a point, D/n, outside the control limits would indicate that some special cause has changed the variation. [Incidentally, if D/n is below the LCL, this might very well indicate that some type of change for the better has been made, and we want to find out why. In general, outlying statistics can often suggest that good (as well as bad) breakthroughs have been made.]

The next example gives the results of a simple experiment that you can easily duplicate.

Example 9.7-3

Let D_i equal the number of yellow candies in a 1.69-ounce bag. Because the number of pieces of candy varies slightly from bag to bag, we shall use an average value for n when we construct the control limits. Table 9.7-3 lists, for 20 packages, the number of pieces of candy in the package, the number of yellow ones, and the proportion of yellow ones.

For these data,

$$\sum_{i=1}^{20} n_i = 1124 \qquad \text{and} \qquad \sum_{i=1}^{20} D_i = 219.$$

It follows that

$$\bar{p} = \frac{219}{1124} = 0.195 \qquad \text{and} \qquad \bar{n} = \frac{1124}{20} \approx 56.$$

Thus, the LCL and UCL are respectively given by

$$\text{LCL} = \bar{p} - 3\sqrt{\bar{p}(1-\bar{p})/56} = 0.195 - 3\sqrt{0.195(0.805)/56} = 0.036$$

and

$$\text{UCL} = \bar{p} + 3\sqrt{\bar{p}(1-\bar{p})/56} = 0.195 + 3\sqrt{0.195(0.805)/56} = 0.354.$$

The control chart for p is depicted in Figure 9.7-4. (For your information the "true" value for p is 0.20.) ■

Consider the following explanation of the c chart: Suppose the number of flaws, say, C, on some product has a Poisson distribution with parameter λ. If λ is

Table 9.7-3 Data on yellow candies

Package	n_i	D_i	D_i/n_i	Package	n_i	D_i	D_i/n_i
1	56	8	0.14	11	57	10	0.18
2	55	13	0.24	12	59	8	0.14
3	58	12	0.21	13	54	10	0.19
4	56	13	0.23	14	55	11	0.20
5	57	14	0.25	15	56	12	0.21
6	54	5	0.09	16	57	11	0.19
7	56	14	0.25	17	54	6	0.11
8	57	15	0.26	18	58	7	0.12
9	54	11	0.20	19	58	12	0.21
10	55	13	0.24	20	58	14	0.24

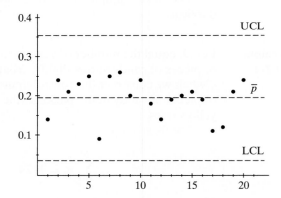

Figure 9.7-4 The p chart

sufficiently large, as in Example 5.7-5, we consider approximating the discrete Poisson distribution with the continuous $N(\lambda, \lambda)$ distribution. Thus, the interval from $\lambda - 3\sqrt{\lambda}$ to $\lambda + 3\sqrt{\lambda}$ contains virtually all of the C-values. Since λ is unknown, however, it must be approximated by $\bar{c}$, the average of the k values, $c_1, c_2, \ldots, c_k$. Hence the two control limits for C are computed as

$$\text{LCL} = \bar{c} - 3\sqrt{\bar{c}} \qquad \text{and} \qquad \text{UCL} = \bar{c} + 3\sqrt{\bar{c}}.$$

The remarks made about the $\bar{x}$ and $\bar{p}$ charts apply to the c chart as well, but we must remember that each c-value is the number of flaws on one manufactured item, not an average $\bar{x}$ or a fraction defective D/n.

Exercises

9.7-1. It is important to control the viscosity of liquid dishwasher soap so that it flows out of the container but does not run out too rapidly. Thus, samples are taken randomly throughout the day and the viscosity is measured. Use the following 20 sets of 5 observations for this exercise:

Observations					$\bar{x}$	s	R
158	147	158	159	169	158.20	7.79	22
151	166	151	143	169	156.00	11.05	26
153	174	151	164	185	165.40	14.33	34
168	140	180	176	154	163.60	16.52	40
160	187	145	164	158	162.80	15.29	42
169	153	149	144	157	154.40	9.48	25
156	183	157	140	162	159.60	15.47	43
158	160	180	154	160	162.40	10.14	26
164	168	154	158	164	161.60	5.55	14
159	153	170	158	170	162.00	7.65	17
150	161	169	166	154	160.00	7.97	19
157	138	155	134	165	149.80	13.22	31
161	172	156	145	153	157.40	10.01	27
143	152	152	156	163	153.20	7.26	20
179	157	135	172	143	157.20	18.63	44
154	165	145	152	145	152.20	8.23	20
171	189	144	154	147	161.00	18.83	45
187	147	159	167	151	162.20	15.85	40
153	168	148	188	152	161.80	16.50	40
165	155	140	157	176	158.60	13.28	36

(a) Calculate the values of $\bar{\bar{x}}$, $\bar{s}$, and $\bar{R}$.

(b) Use the values of A_3 and $\bar{s}$ to construct an $\bar{x}$ chart.

(c) Construct an s chart.

(d) Use the values of A_2 and $\bar{R}$ to construct an $\bar{x}$ chart.

(e) Construct an R chart.

(f) Do the charts indicate that viscosity is in statistical control?

9.7-2. It is necessary to control the percentage of solids in a product, so samples are taken randomly throughout the day and the percentage of solids is measured. Use the following 20 sets of 5 observations for this exercise:

Observations					$\bar{x}$	s	R
69.8	71.3	65.6	66.3	70.1	68.62	2.51	5.7
71.9	69.6	71.9	71.1	71.7	71.24	0.97	2.3
71.9	69.8	66.8	68.3	64.4	68.24	2.86	7.5
64.2	65.1	63.7	66.2	61.9	64.22	1.61	4.3
66.1	62.9	66.9	67.3	63.3	65.30	2.06	4.4
63.4	67.2	67.4	65.5	66.2	65.94	1.61	4.0
67.5	67.3	66.9	66.5	65.5	66.74	0.79	2.0
63.9	64.6	62.3	66.2	67.2	64.84	1.92	4.9
66.0	69.8	69.7	71.0	69.8	69.26	1.90	5.0
66.0	70.3	65.5	67.0	66.8	67.12	1.88	4.8
67.6	68.6	66.5	66.2	70.4	67.86	1.71	4.2
68.1	64.3	65.2	68.0	65.1	66.14	1.78	3.8
64.5	66.6	65.2	69.3	62.0	65.52	2.69	7.3
67.1	68.3	64.0	64.9	68.2	66.50	1.96	4.3
67.1	63.8	71.4	67.5	63.7	66.70	3.17	7.7
60.7	63.5	62.9	67.0	69.6	64.74	3.53	8.9
71.0	68.6	68.1	67.4	71.7	69.36	1.88	4.3
69.5	61.5	63.7	66.3	68.6	65.92	3.34	8.0
66.7	75.2	79.0	75.3	79.2	75.08	5.07	12.5
77.3	67.2	69.3	67.9	65.6	69.46	4.58	11.7

(a) Calculate the values of $\bar{\bar{x}}$, $\bar{s}$, and $\bar{R}$.

(b) Use the values of A_3 and $\bar{s}$ to construct an $\bar{x}$ chart.

(c) Construct an s chart.

(d) Use the values of A_2 and $\bar{R}$ to construct an $\bar{x}$ chart.

(e) Construct an R chart.

(f) Do the charts indicate that the percentage of solids in this product is in statistical control?

9.7-3. It is important to control the net weight of a packaged item; thus, items are selected randomly throughout the day from the production line and their weights are recorded. Use the following 20 sets of 5 weights (in grams) for this exercise (note that a weight recorded here is the actual weight minus 330):

Observations					$\bar{x}$	s	R
7.97	8.10	7.73	8.26	7.30	7.872	0.3740	0.96
8.11	7.26	7.99	7.88	8.88	8.024	0.5800	1.62
7.60	8.23	8.07	8.51	8.05	8.092	0.3309	0.91
8.44	4.35	4.33	4.48	3.89	5.098	1.8815	4.55
5.11	4.05	5.62	4.13	5.01	4.784	0.6750	1.57
4.79	5.25	5.19	5.23	3.97	4.886	0.5458	1.28
4.47	4.58	5.35	5.86	5.61	5.174	0.6205	1.39
5.82	4.51	5.38	5.01	5.54	5.252	0.5077	1.31
5.06	4.98	4.13	4.58	4.35	4.620	0.3993	0.93
4.74	3.77	5.05	4.03	4.29	4.376	0.5199	1.28
4.05	3.71	4.73	3.51	4.76	4.152	0.5748	1.25
3.94	5.72	5.07	5.09	4.61	4.886	0.6599	1.78
4.63	3.79	4.69	5.13	4.66	4.580	0.4867	1.34
4.30	4.07	4.39	4.63	4.47	4.372	0.2079	0.56
4.05	4.14	4.01	3.95	4.05	4.040	0.0693	0.19
4.20	4.50	5.32	4.42	5.24	4.736	0.5094	1.12
4.54	5.23	4.32	4.66	3.86	4.522	0.4999	1.37
5.02	4.10	5.08	4.94	5.18	4.864	0.4360	1.08
4.80	4.73	4.82	4.69	4.27	4.662	0.2253	0.55
4.55	4.76	4.45	4.85	4.02	4.526	0.3249	0.83

(a) Calculate the values of $\bar{\bar{x}}, \bar{s},$ and $\bar{R}$.

(b) Use the values of A_3 and $\bar{s}$ to construct an $\bar{x}$ chart.

(c) Construct an s chart.

(d) Use the values of A_2 and $\bar{R}$ to construct an $\bar{x}$ chart.

(e) Construct an R chart.

(f) Do the charts indicate that these fill weights are in statistical control?

9.7-4. A company has been producing bolts that are about $\bar{p} = 0.02$ defective, and this is satisfactory. To monitor the quality of the process, 100 bolts are selected at random each hour and the number of defective bolts counted. With $\bar{p} = 0.02$, compute the UCL and LCL of the $\bar{p}$ chart. Then suppose that, over the next 24 hours, the following numbers of defective bolts are observed:

4 1 1 0 5 2 1 3 4 3 1 0 0 4 1 1 6 2 0 0 2 8 7 5

Would any action have been required during this time?

9.7-5. To give some indication of how the values in Table 9.7-1 are calculated, values of A_3 are found in this exercise. Let $X_1, X_2, \ldots, X_n$ be a random sample of size n from the normal distribution $N(\mu, \sigma^2)$. Let S^2 equal the sample variance of this random sample.

(a) Use the fact that $Y = (n-1)S^2/\sigma^2$ has a distribution that is $\chi^2(n-1)$ to show that $E[S^2] = \sigma^2$.

(b) Using the $\chi^2(n-1)$ pdf, find the value of $E(\sqrt{Y})$.

(c) Show that

$$E\left[\frac{\sqrt{n-1}\,\Gamma\left(\dfrac{n-1}{2}\right)}{\sqrt{2}\,\Gamma\left(\dfrac{n}{2}\right)} S \right] = \sigma.$$

(d) Verify that

$$\frac{3}{\sqrt{n}} \left[\frac{\sqrt{n-1}\,\Gamma\left(\dfrac{n-1}{2}\right)}{\sqrt{2}\,\Gamma\left(\dfrac{n}{2}\right)} \right] = A_3,$$

found in Table 9.7-1 for $n = 5$ and $n = 6$. Thus, $A_3\,\bar{s}$ approximates $3\sigma/\sqrt{n}$.

9.7-6. 250-yard pieces of wool are inspected in a mill. In the last 15 observations, the following numbers of flaws were found:

4 3 0 1 2 5 1 2 0 3 1 4 2 3 1

(a) Compute the control limits of the c chart and draw this control chart.

(b) Is the process in statistical control?

9.7-7. In the past, $n = 50$ fuses are tested each hour and $\bar{p} = 0.03$ have been found defective. Calculate the UCL and LCL. After a production error, say the true p shifts to $p = 0.05$.

(a) What is the probability that the next observation exceeds the UCL?

(b) What is the probability that at least one of the next five observations exceeds the UCL? HINT: Assume independence and compute the probability that none of the next five observations exceeds the UCL.

9.7-8. Snee (see References) has measured the thickness of the "ears" of paint cans. (The "ear" of a paint can is the tab that secures the lid of the the the can.) At periodic intervals, samples of five paint cans are taken from a hopper that collects the production from two machines, and the thickness of each ear is measured. The results (in inches $\times$ 1000) of 30 such samples are as follows:

Observations					$\bar{x}$	s	R
29	36	39	34	34	34.4	3.64692	10
29	29	28	32	31	29.8	1.64317	4
34	34	39	38	37	36.4	2.30217	5
35	37	33	38	41	36.8	3.03315	8
30	29	31	38	29	31.4	3.78153	9
34	31	37	39	36	35.4	3.04959	8
30	35	33	40	36	34.8	3.70135	10
28	28	31	34	30	30.2	2.48998	6
32	36	38	38	35	35.8	2.48998	6
35	30	37	35	31	33.6	2.96648	7
35	30	35	38	35	34.6	2.88097	8
38	34	35	35	31	34.6	2.50998	7
34	35	33	30	34	33.2	1.92354	5
40	35	34	33	35	35.4	2.70185	7
34	35	38	35	30	34.4	2.88097	8
35	30	35	29	37	33.2	3.49285	8
40	31	38	35	31	35.0	4.06202	9
35	36	30	33	32	33.2	2.38747	6
35	34	35	30	36	34.0	2.34521	6
35	35	31	38	36	35.0	2.54951	7
32	36	36	32	36	34.4	2.19089	4
36	37	32	34	34	34.6	1.94936	5
29	34	33	37	35	33.6	2.96648	8
36	36	35	37	37	36.2	0.83666	2
36	30	35	33	31	33.0	2.54951	6
35	30	29	38	35	33.4	3.78153	9

Observations					$\bar{x}$	s	R
35	36	30	34	36	34.2	2.48998	6
35	30	36	29	35	33.0	3.24037	7
38	36	35	31	31	34.2	3.11448	7
30	34	40	28	30	32.4	4.77493	12

(a) Calculate the values of $\bar{\bar{x}}$, $\bar{s}$, and $\bar{R}$.

(b) Use the values of A_3 and $\bar{s}$ to construct an $\bar{x}$ chart.

(c) Construct an s chart.

(d) Use the values of A_2 and $\bar{R}$ to construct an $\bar{x}$ chart.

(e) Construct an R chart.

(f) Do the charts indicate that these fill weights are in statistical control?

9.7-9. Ledolter and Hogg (see References) report that, in the production of stainless steel pipes, the number of defects per 100 feet should be controlled. From 15 randomly selected pipes of length 100 feet, the following data on the number of defects were observed:

$$6\ \ 10\ \ 8\ \ 1\ \ 7\ \ 9\ \ 7\ \ 4\ \ 5\ \ 10\ \ 3\ \ 4\ \ 9\ \ 8\ \ 5$$

(a) Compute the control limits of the c chart and draw this control chart.

(b) Is the process in statistical control?

9.7-10. Suppose we find that the number of blemishes in 80-foot tin strips averages about $\bar{c} = 1.9$. Calculate the control limits. Say the process has gone out of control and this average has increased to 3.4.

(a) What is the probability that the next observation will exceed the UCL?

(b) What is the probability that at least 3 of the next 10 observations will exceed the UCL?

HISTORICAL COMMENTS Chi-square tests were the invention of Karl Pearson, except that he had it wrong in the case in which parameters are estimated. When R. A. Fisher was a brash young man, he told his senior, Pearson, that he should reduce the number of degrees of freedom of the chi-square distribution by 1 for every parameter that was estimated. Pearson never believed this (of course, Fisher was correct), and, as editor of the very prestigious journal *Biometrika*, Pearson blocked Fisher in his later professional life from publishing in that journal. Fisher was disappointed, and the two men battled during their lifetimes; however, later Fisher saw this conflict to be to his advantage, as it made him consider applied journals in which to publish, and thus he became a better, more well-rounded scientist.

Another important item in this chapter is the analysis of variance (ANOVA). This is just the beginning of what is called the design of experiments, developed by R. A. Fisher. In our simple cases in this section, he shows how to test for the best levels of factors in the one-factor and two-factor cases. We study a few important generalizations in Section 9.5. The analysis of designed experiments was a huge contribution by Fisher.

Quality improvement made a substantial change in manufacturing beginning in the 1920s, with Walter A. Shewhart's control charts. In fairness, it should be noted that the British started a similar program about the same time. Statistical quality control, as described in Section 9.7, really had a huge influence during World War II, with many universities giving short courses in the subject. These courses continued after the war, but the development of the importance of total quality improvement lagged behind. W. Edwards Deming complained that the Japanese used his quality ideas beginning in the 1950s, but the Americans did not adopt them until 1980. That year NBC televised a program entitled *If Japan Can, Why Can't We?*, and Deming was the "star" of that broadcast. He related that the next day his phone "started ringing off the hook." Various companies requested that he spend one day with them to get them started on the right path. According to Deming, they all wanted "instant pudding," and he noted that he had asked the Japanese to give him five years to make the improvements he pioneered. Actually, using his philosophy, many of these companies did achieve substantial results in quality sooner than that. However, it was after the NBC program that Deming started his famous four-day courses, and he taught his last one in December of 1993, about 10 days before his death at the age of 93.

Many of these quality efforts in the 1970s and 1980s used the name "Total Quality Management" or, later, "Continuous Process Improvements." However, it was Motorola's Six Sigma program, which started in the late 1980s and has continued for over 20 years since then, that has had the biggest impact. In addition to Motorola, GE, Allied, and a large number of companies have used this system. In our opinion, Six Sigma is the leading development in the quality improvement effort.

EPILOGUE

Clearly, there is much more to applied and theoretical statistics than can be studied in one book. As mentioned in the Prologue, there is a huge demand for statistical scientists in many fields to make sense out of the increasing volumes of data. Statistics is needed to turn quality data into useful information upon which decisions can be made. The striking thing about any data set is that there is variation; all the data points simply do not lie on the pattern. It is the statistician's job to find that pattern and describe the variation about it. Done properly, this clearly helps the decision maker significantly.

One observation about variation should be noted before any major adjustments or decisions are made. Frequently, persons in charge jump at conclusions too quickly; that is, major decisions are often made after too few observations. For illustration, we know that if X_1, X_2, X_3 are independent and identically distributed continuous-type observations, then

$$P(X_1 < X_2 < X_3) = \frac{1}{3!} = \frac{1}{6}.$$

Yet if this occurred and these observations were taken on sales, say, and plotted in time sequence, the fact that these three points were "going up" might suggest to management that "the company is on a roll." In some cases, only two increasing points might cause this reaction. If there is no change in the system, two or three increasing points have respective probabilities of 1/2 and 1/6, and those probabilities do not warrant that sort of reaction. If, on the other hand, we observe four or five such points, with respective probabilities of 1/24 and 1/120, then an appropriate reaction is in order.

An interesting question to ask is why statisticians treat $1/20 = 0.05$ as the value at which the probability of an event is considered small and often suggests some type of action. Possibly it was because Ronald A. Fisher suggested that 1 out of 20 seemed small enough. Obviously the value of 0.05 is not written in stone, but statisticians seem to look for differences of two or three standard deviations as a guide for action. Since many estimators have approximate normal distributions, such differences do have small probabilities.

For illustration, suppose a candidate believes that he or she has at least 50% of the votes. Yet in a poll of $n = 400$, only 160 favor that candidate; thus, the standardized value

$$\frac{160 - 400(1/2)}{\sqrt{400(1/2)(1/2)}} = -4$$

suggests that the candidate does not, in fact, have 50% of the votes. Depending upon the financial situation, the candidate must change the approach and/or work harder, or possibly even consider dropping from the race. However, we note that this simple statistical analysis can be a guide in the decision process.

Another example is provided by W. A. Shewhart's efforts at Bell Telephone Laboratories in the 1920s in his quality improvement through statistical methods. He conceived of the idea of sampling occasionally during the manufacturing process rather than waiting until the items arrived at an inspection center at the end of the line. To monitor the need for possible adjustments, certain statistics, like $\overline{X}$ and S, were taken. If one of these had an *unusual* value, corrective action was taken before too many more defective items were produced. If T is one of those statistics, an unusual value of T was one that was outside the *control limits* $\mu_T \pm 3\sigma_T$, where the mean μ_T and the standard deviation σ_T of T would often be estimated from past data coming from a satisfactory (*in control*) process.

While the three sigma limits have proved to be an effective guide, many companies are now considering six sigma limits. The Six Sigma program was first started by Motorola, and it has its statistical underpinnings in the following. There are specifications in manufacturing items that are usually set by the engineers—or possibly by customers. These are usually given in terms of a target value and upper and lower specification limits, USL and LSL, respectively. The mean μ of the values resulting from the process is hopefully close to the target value. The "specs" are six standard deviations, 6σ, away from the target, where σ is the standard deviation associated with the process. However, the mean μ is often dynamic, and these Six Sigma companies try to keep it at least 4.5σ from the closest spec. If the values of the items are distributed normally and nearest spec is 4.5σ from μ, there are only 3.4 defectives per million. This is the goal of the Six Sigma companies, and they use many of Deming's ideas in their attempts to achieve this very worthwhile goal.

While an understanding of quality improvement ideas and basic statistical methods, like those in Section 9.7, is extremely important to the Six Sigma programs, possibly the major factor is the attitude of the CEOs and other important administrators. There were many total quality management programs in the 1980s, but they were not as successful as Six Sigma, for now each CEO is demanding, "Show me the money!" That is, companies hire a Six Sigma expert to come in for four one-week periods, about one month apart, for a fee of about $15,000 per person. If they have 20 participants, this cost is $300,000 plus the four weeks of time "lost" to the work process by each trainee. That is a great deal of money. However, each of these individuals (sometimes a pair) has a project associated with some process that has not been very efficient. They work on these projects using statistical methods during the "off months" and report to the Six Sigma expert during the training weeks to get advice. Often these projects, if successful, will save millions of dollars for a company, and the expenses are well worth the benefits. As a matter of fact, if a participant's project saves the company at least $1 million, he or she earns a "Six Sigma Black Belt." So the CEO does see the money saved, and thus the bottom line looks extremely good to him or her.

While the reader has now studied enough statistics to appreciate the importance of understanding variation, there are many more useful statistical techniques to be studied if the reader is so inclined. For example, there are courses in regression and time series in which we learn how to predict future observations. Or a study of design of experiments can help an investigator select the most efficient levels of the various factors. After all, if we have 10 factors and we run each at only two levels, we have created $2^{10} = 1024$ runs. Can we perform only a fraction of these runs without losing too much information? Additional study of multivariate analysis can lead to interesting problems in classification. Say a doctor takes several measurements on a patient and then classifies the patent's disease as one of many possible diseases. There are errors of misclassification, and statisticians can help reduce the probabilities of those errors. Doctors—and statisticians—can make mistakes, and second or third opinions should be asked for if there is some doubt.

As mentioned in the Prologue, the computer has opened the door to a wide variety of new statistical techniques; and researchers, computer scientists, and statisticians are working together to reduce huge amounts of data into nuggets of quality information on which important decisions can be made. Statistics is an exciting field that finds many useful applications in the social, health, and physical sciences. The authors have found statistics to be a great profession; we hope a few of you find it that way too. In any case, we hope that statistical thinking will make you more aware of the need for understanding variation, which can have a great influence on your daily life.

REFERENCES

Aspin, A. A., "Tables for Use in Comparisons Whose Accuracy Involves Two Variances, Separately Estimated," *Biometrika*, **36** (1949), pp. 290–296.

Agresti, A., and B. A. Coull, "Approximate Is Better than 'Exact' for Interval Estimation of Binomial Proportions," *Amer. Statist.*, **52**, 2(1998), pp. 119–126.

Barnett, A., "How Numbers Can Trick You," *Technology Review*, 1994.

Basu, D., "On Statistics Independent of a Complete Sufficient Statistic," *Sankhya*, **15** (1955), pp. 377–380.

Bernstein, P. L., *Against the Gods: The Remarkable Story of Risk*. New York: John Wiley & Sons, Inc., 1996.

Berry, D. A. *Statistics: A Bayesian Perspective*. Belmont, CA: Duxbury Press, an Imprint of Wadsworth Publishing Co., 1996.

Box, G. E. P., and M. E. Muller, "A Note on the Generation of Random Normal Deviates," *Ann. Math. Statist.*, **29** (1958), pp. 610–611.

Box, G. E. P., J. S. Hunter, and W. G. Hunter, *Statistics for Experimenters: Design, Innovation, and Discovery*, 2nd ed. New York: John Wiley & Sons, Inc., 2005.

Hogg, R. V., "Testing the Equality of Means of Rectangular Populations," *Ann. Math. Statist.*, **24** (1953), p. 691.

Hogg, R. V., and A. T. Craig, "On the Decomposition of Certain Chi-Square Variables," *Ann. Math. Statist.*, **29** (1958), pp. 608–610.

Hogg, R. V., and A. T. Craig, *Introduction to Mathematical Statistics*, 3rd ed. New York: Macmillan Publishing Co., Inc., 1970.

Hogg, R. V., J. W. McKean, and A. T. Craig, *Introduction to Mathematical Statistics*, 7th ed. Upper Saddle River, NJ: Prentice Hall, 2013.

Johnson, V. E., and J. H. Albert, *Ordinal Data Modeling*. New York: Springer-Verlag, 1999.

Karian, Z. A., and E. A. Tanis, *Probability & Statistics: Explorations with MAPLE*, 2nd ed. Upper Saddle River, NJ: Prentice Hall, 1999.

Keating, J. P., and D. W. Scott, "Ask Dr. STATS," *Stats, The Magazine for Students of Statistics*, **25**, Spring (1999), pp. 16–22.

Latter, O. H. "The Cuckoo's Egg," *Biometrika*, **1** (1901), pp. 164–176.

Ledolter, J., and R. V. Hogg, *Applied Statistics for Engineers and Scientists*, 3rd ed. Upper Saddle River, NJ: Prentice Hall, 2010.

Lurie, D., and H. O. Hartley, "Machine-Generation of Order Statistics for Monte Carlo Computations," *Amer. Statist.*, February (1972), pp. 26–27.

Montgomery, D. C., *Design and Analysis of Experiments*, 2nd ed. New York: John Wiley & Sons, Inc., 1984.

Natrella, M. G., *Experimental Statistics, National Bureau of Standards Handbook 91*. Washington, DC: U.S. Government Printing Office, 1963.

Nicol, S. J. "Who's Picking Up the Pieces?" *Primus*, **IV**, 2, June (1994), pp. 182–184.

Pearson, K., "On the Criterion That a Given System of Deviations from the Probable in the Case of a Correlated System of Variables Is Such That It Can Be Reasonably Supposed to Have Arisen from Random Sampling," *London, Edinburgh, Dublin Phil. Mag. J. Sci.*, Series 5, **50** (1900), pp. 157–175.

Putz, J., "The Golden Section and the Piano Sonatas of Mozart," *Mathematics Magazine*, **68**, 4, October (1995), pp. 275–282.

Quain, J. R., "Going Mainstream," *PC Magazine*, February (1994).

Rafter, J. A., M. L. Abell, and J. P. Braselton, *Statistics with Maple*. Amsterdam and Boston: Academic Press, an imprint of Elsevier Science (USA), 2003.

Raspe, R. E., *The Surprising Adventures of Baron Munchausen*. IndyPublish.com, 2001.

Snee, R. D., "Graphical Analysis of Process Variation Studies," *J. Qual. Technol.*, **15**, April (1983), pp. 76–88.

Snee, R. D., L. B. Hare, and J. R. Trout, *Experiments in Industry*. Milwaukee: American Society of Quality Control, 1985.

Stigler, S. M., *The History of Statistics: The Measurement of Uncertainty Before 1900*. Cambridge, MA: Harvard University Press, 1986.

Tanis, E. A., "Maple Integrated into the Instruction of Probability and Statistics," *Proceedings of the Statistical Computing Section* (1998), American Statistical Association, pp. 19–24.

Tanis, E. A., and R. V. Hogg, *A Brief Course in Mathematical Statistics*. Upper Saddle River, NJ: Prentice Hall, 2008.

Tate, R. F., and G. W. Klett, "Optimum Confidence Intervals for the Variance of a Normal Distribution," *J. Am. Statist. Assoc.*, **54** (1959), pp. 674–682.

Tsutakawa, R. K., G. L. Shoop, and C. J. Marienfeld, "Empirical Bayes estimation of cancer mortality rates," *Stat. Med.*, **4** (1985), pp. 201–212.

Tukey, J. W., *Exploratory Data Analysis*. Reading, MA: Addison-Wesley Publishing Company, 1977.

Velleman, P. F., and D. C. Hoaglin, *Applications, Basics, and Computing of Exploratory Data Analysis*. Boston: Duxbury Press, 1981.

Wilcoxon, F., "Individual Comparisons by Ranking Methods," *Biometrics Bull.*, **1** (1945), pp. 80–83.

Zerger, M., "Mean Meets Variance," *Primus*, **IV**, 2, June (1994), pp. 106–108.

TABLES

B

Table I Binomial Coefficients

$$\binom{n}{r} = \frac{n!}{r!(n-r)!} = \binom{n}{n-r}$$

n	$\binom{n}{0}$	$\binom{n}{1}$	$\binom{n}{2}$	$\binom{n}{3}$	$\binom{n}{4}$	$\binom{n}{5}$	$\binom{n}{6}$	$\binom{n}{7}$	$\binom{n}{8}$	$\binom{n}{9}$	$\binom{n}{10}$	$\binom{n}{11}$	$\binom{n}{12}$	$\binom{n}{13}$
0	1													
1	1	1												
2	1	2	1											
3	1	3	3	1										
4	1	4	6	4	1									
5	1	5	10	10	5	1								
6	1	6	15	20	15	6	1							
7	1	7	21	35	35	21	7	1						
8	1	8	28	56	70	56	28	8	1					
9	1	9	36	84	126	126	84	36	9	1				
10	1	10	45	120	210	252	210	120	45	10	1			
11	1	11	55	165	330	462	462	330	165	55	11	1		
12	1	12	66	220	495	792	924	792	495	220	66	12	1	
13	1	13	78	286	715	1,287	1,716	1,716	1,287	715	286	78	13	1
14	1	14	91	364	1,001	2,002	3,003	3,432	3,003	2,002	1,001	364	91	14
15	1	15	105	455	1,365	3,003	5,005	6,435	6,435	5,005	3,003	1,365	455	105
16	1	16	120	560	1,820	4,368	8,008	11,440	12,870	11,440	8,008	4,368	1,820	560
17	1	17	136	680	2,380	6,188	12,376	19,448	24,310	24,310	19,448	12,376	6,188	2,380
18	1	18	153	816	3,060	8,568	18,564	31,824	43,758	48,620	43,758	31,824	18,564	8,568
19	1	19	171	969	3,876	11,628	27,132	50,388	75,582	92,378	92,378	75,582	50,388	27,132
20	1	20	190	1,140	4,845	15,504	38,760	77,520	125,970	167,960	184,756	167,960	125,970	77,520
21	1	21	210	1,330	5,985	20,349	54,264	116,280	203,490	293,930	352,716	352,716	293,930	203,490
22	1	22	231	1,540	7,315	26,334	74,613	170,544	319,770	497,420	646,646	705,432	646,646	497,420
23	1	23	253	1,771	8,855	33,649	100,947	245,157	490,314	817,190	1,144,066	1,352,078	1,352,078	1,144,066
24	1	24	276	2,024	10,626	42,504	134,596	346,104	735,471	1,307,504	1,961,256	2,496,144	2,704,156	2,496,144
25	1	25	300	2,300	12,650	53,130	177,100	480,700	1,081,575	2,042,975	3,268,760	4,457,400	5,200,300	5,200,300
26	1	26	325	2,600	14,950	65,780	230,230	657,800	1,562,275	3,124,550	5,311,735	7,726,160	9,657,700	10,400,600

For $r > 13$ you may use the identity $\binom{n}{r} = \binom{n}{n-r}$.

Table II The Binomial Distribution

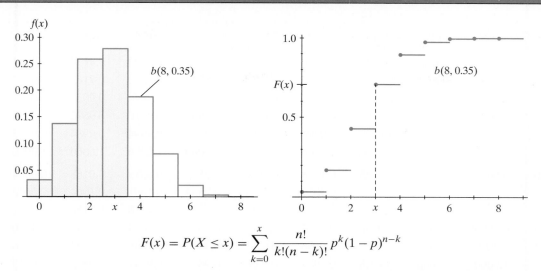

$$F(x) = P(X \le x) = \sum_{k=0}^{x} \frac{n!}{k!(n-k)!} p^k (1-p)^{n-k}$$

							p				
n	x	0.05	0.10	0.15	0.20	0.25	0.30	0.35	0.40	0.45	0.50
2	0	0.9025	0.8100	0.7225	0.6400	0.5625	0.4900	0.4225	0.3600	0.3025	0.2500
	1	0.9975	0.9900	0.9775	0.9600	0.9375	0.9100	0.8775	0.8400	0.7975	0.7500
	2	1.0000	1.0000	1.0000	1.0000	1.0000	1.0000	1.0000	1.0000	1.0000	1.0000
3	0	0.8574	0.7290	0.6141	0.5120	0.4219	0.3430	0.2746	0.2160	0.1664	0.1250
	1	0.9928	0.9720	0.9392	0.8960	0.8438	0.7840	0.7182	0.6480	0.5748	0.5000
	2	0.9999	0.9990	0.9966	0.9920	0.9844	0.9730	0.9571	0.9360	0.9089	0.8750
	3	1.0000	1.0000	1.0000	1.0000	1.0000	1.0000	1.0000	1.0000	1.0000	1.0000
4	0	0.8145	0.6561	0.5220	0.4096	0.3164	0.2401	0.1785	0.1296	0.0915	0.0625
	1	0.9860	0.9477	0.8905	0.8192	0.7383	0.6517	0.5630	0.4752	0.3910	0.3125
	2	0.9995	0.9963	0.9880	0.9728	0.9492	0.9163	0.8735	0.8208	0.7585	0.6875
	3	1.0000	0.9999	0.9995	0.9984	0.9961	0.9919	0.9850	0.9744	0.9590	0.9375
	4	1.0000	1.0000	1.0000	1.0000	1.0000	1.0000	1.0000	1.0000	1.0000	1.0000
5	0	0.7738	0.5905	0.4437	0.3277	0.2373	0.1681	0.1160	0.0778	0.0503	0.0312
	1	0.9774	0.9185	0.8352	0.7373	0.6328	0.5282	0.4284	0.3370	0.2562	0.1875
	2	0.9988	0.9914	0.9734	0.9421	0.8965	0.8369	0.7648	0.6826	0.5931	0.5000
	3	1.0000	0.9995	0.9978	0.9933	0.9844	0.9692	0.9460	0.9130	0.8688	0.8125
	4	1.0000	1.0000	0.9999	0.9997	0.9990	0.9976	0.9947	0.9898	0.9815	0.9688
	5	1.0000	1.0000	1.0000	1.0000	1.0000	1.0000	1.0000	1.0000	1.0000	1.0000
6	0	0.7351	0.5314	0.3771	0.2621	0.1780	0.1176	0.0754	0.0467	0.0277	0.0156
	1	0.9672	0.8857	0.7765	0.6553	0.5339	0.4202	0.3191	0.2333	0.1636	0.1094
	2	0.9978	0.9842	0.9527	0.9011	0.8306	0.7443	0.6471	0.5443	0.4415	0.3438
	3	0.9999	0.9987	0.9941	0.9830	0.9624	0.9295	0.8826	0.8208	0.7447	0.6562
	4	1.0000	0.9999	0.9996	0.9984	0.9954	0.9891	0.9777	0.9590	0.9308	0.8906
	5	1.0000	1.0000	1.0000	0.9999	0.9998	0.9993	0.9982	0.9959	0.9917	0.9844
	6	1.0000	1.0000	1.0000	1.0000	1.0000	1.0000	1.0000	1.0000	1.0000	1.0000
7	0	0.6983	0.4783	0.3206	0.2097	0.1335	0.0824	0.0490	0.0280	0.0152	0.0078
	1	0.9556	0.8503	0.7166	0.5767	0.4449	0.3294	0.2338	0.1586	0.1024	0.0625

Table II *continued*

n	x	0.05	0.10	0.15	0.20	0.25	0.30	0.35	0.40	0.45	0.50
						p					
	2	0.9962	0.9743	0.9262	0.8520	0.7564	0.6471	0.5323	0.4199	0.3164	0.2266
	3	0.9998	0.9973	0.9879	0.9667	0.9294	0.8740	0.8002	0.7102	0.6083	0.5000
	4	1.0000	0.9998	0.9988	0.9953	0.9871	0.9712	0.9444	0.9037	0.8471	0.7734
	5	1.0000	1.0000	0.9999	0.9996	0.9987	0.9962	0.9910	0.9812	0.9643	0.9375
	6	1.0000	1.0000	1.0000	1.0000	0.9999	0.9998	0.9994	0.9984	0.9963	0.9922
	7	1.0000	1.0000	1.0000	1.0000	1.0000	1.0000	1.0000	1.0000	1.0000	1.0000
8	0	0.6634	0.4305	0.2725	0.1678	0.1001	0.0576	0.0319	0.0168	0.0084	0.0039
	1	0.9428	0.8131	0.6572	0.5033	0.3671	0.2553	0.1691	0.1064	0.0632	0.0352
	2	0.9942	0.9619	0.8948	0.7969	0.6785	0.5518	0.4278	0.3154	0.2201	0.1445
	3	0.9996	0.9950	0.9786	0.9437	0.8862	0.8059	0.7064	0.5941	0.4770	0.3633
	4	1.0000	0.9996	0.9971	0.9896	0.9727	0.9420	0.8939	0.8263	0.7396	0.6367
	5	1.0000	1.0000	0.9998	0.9988	0.9958	0.9887	0.9747	0.9502	0.9115	0.8555
	6	1.0000	1.0000	1.0000	0.9999	0.9996	0.9987	0.9964	0.9915	0.9819	0.9648
	7	1.0000	1.0000	1.0000	1.0000	1.0000	0.9999	0.9998	0.9993	0.9983	0.9961
	8	1.0000	1.0000	1.0000	1.0000	1.0000	1.0000	1.0000	1.0000	1.0000	1.0000
9	0	0.6302	0.3874	0.2316	0.1342	0.0751	0.0404	0.0207	0.0101	0.0046	0.0020
	1	0.9288	0.7748	0.5995	0.4362	0.3003	0.1960	0.1211	0.0705	0.0385	0.0195
	2	0.9916	0.9470	0.8591	0.7382	0.6007	0.4628	0.3373	0.2318	0.1495	0.0898
	3	0.9994	0.9917	0.9661	0.9144	0.8343	0.7297	0.6089	0.4826	0.3614	0.2539
	4	1.0000	0.9991	0.9944	0.9804	0.9511	0.9012	0.8283	0.7334	0.6214	0.5000
	5	1.0000	0.9999	0.9994	0.9969	0.9900	0.9747	0.9464	0.9006	0.8342	0.7461
	6	1.0000	1.0000	1.0000	0.9997	0.9987	0.9957	0.9888	0.9750	0.9502	0.9102
	7	1.0000	1.0000	1.0000	1.0000	0.9999	0.9996	0.9986	0.9962	0.9909	0.9805
	8	1.0000	1.0000	1.0000	1.0000	1.0000	1.0000	0.9999	0.9997	0.9992	0.9980
	9	1.0000	1.0000	1.0000	1.0000	1.0000	1.0000	1.0000	1.0000	1.0000	1.0000
10	0	0.5987	0.3487	0.1969	0.1074	0.0563	0.0282	0.0135	0.0060	0.0025	0.0010
	1	0.9139	0.7361	0.5443	0.3758	0.2440	0.1493	0.0860	0.0464	0.0233	0.0107
	2	0.9885	0.9298	0.8202	0.6778	0.5256	0.3828	0.2616	0.1673	0.0996	0.0547
	3	0.9990	0.9872	0.9500	0.8791	0.7759	0.6496	0.5138	0.3823	0.2660	0.1719
	4	0.9999	0.9984	0.9901	0.9672	0.9219	0.8497	0.7515	0.6331	0.5044	0.3770
	5	1.0000	0.9999	0.9986	0.9936	0.9803	0.9527	0.9051	0.8338	0.7384	0.6230
	6	1.0000	1.0000	0.9999	0.9991	0.9965	0.9894	0.9740	0.9452	0.8980	0.8281
	7	1.0000	1.0000	1.0000	0.9999	0.9996	0.9984	0.9952	0.9877	0.9726	0.9453
	8	1.0000	1.0000	1.0000	1.0000	1.0000	0.9999	0.9995	0.9983	0.9955	0.9893
	9	1.0000	1.0000	1.0000	1.0000	1.0000	1.0000	1.0000	0.9999	0.9997	0.9990
	10	1.0000	1.0000	1.0000	1.0000	1.0000	1.0000	1.0000	1.0000	1.0000	1.0000
11	0	0.5688	0.3138	0.1673	0.0859	0.0422	0.0198	0.0088	0.0036	0.0014	0.0005
	1	0.8981	0.6974	0.4922	0.3221	0.1971	0.1130	0.0606	0.0302	0.0139	0.0059
	2	0.9848	0.9104	0.7788	0.6174	0.4552	0.3127	0.2001	0.1189	0.0652	0.0327
	3	0.9984	0.9815	0.9306	0.8389	0.7133	0.5696	0.4256	0.2963	0.1911	0.1133
	4	0.9999	0.9972	0.9841	0.9496	0.8854	0.7897	0.6683	0.5328	0.3971	0.2744
	5	1.0000	0.9997	0.9973	0.9883	0.9657	0.9218	0.8513	0.7535	0.6331	0.5000
	6	1.0000	1.0000	0.9997	0.9980	0.9924	0.9784	0.9499	0.9006	0.8262	0.7256

Table II *continued*

n	x	p									
		0.05	0.10	0.15	0.20	0.25	0.30	0.35	0.40	0.45	0.50
	7	1.0000	1.0000	1.0000	0.9998	0.9988	0.9957	0.9878	0.9707	0.9390	0.8867
	8	1.0000	1.0000	1.0000	1.0000	0.9999	0.9994	0.9980	0.9941	0.9852	0.9673
	9	1.0000	1.0000	1.0000	1.0000	1.0000	1.0000	0.9998	0.9993	0.9978	0.9941
	10	1.0000	1.0000	1.0000	1.0000	1.0000	1.0000	1.0000	1.0000	0.9998	0.9995
	11	1.0000	1.0000	1.0000	1.0000	1.0000	1.0000	1.0000	1.0000	1.0000	1.0000
12	0	0.5404	0.2824	0.1422	0.0687	0.0317	0.0138	0.0057	0.0022	0.0008	0.0002
	1	0.8816	0.6590	0.4435	0.2749	0.1584	0.0850	0.0424	0.0196	0.0083	0.0032
	2	0.9804	0.8891	0.7358	0.5583	0.3907	0.2528	0.1513	0.0834	0.0421	0.0193
	3	0.9978	0.9744	0.9078	0.7946	0.6488	0.4925	0.3467	0.2253	0.1345	0.0730
	4	0.9998	0.9957	0.9761	0.9274	0.8424	0.7237	0.5833	0.4382	0.3044	0.1938
	5	1.0000	0.9995	0.9954	0.9806	0.9456	0.8822	0.7873	0.6652	0.5269	0.3872
	6	1.0000	0.9999	0.9993	0.9961	0.9857	0.9614	0.9154	0.8418	0.7393	0.6128
	7	1.0000	1.0000	0.9999	0.9994	0.9972	0.9905	0.9745	0.9427	0.8883	0.8062
	8	1.0000	1.0000	1.0000	0.9999	0.9996	0.9983	0.9944	0.9847	0.9644	0.9270
	9	1.0000	1.0000	1.0000	1.0000	1.0000	0.9998	0.9992	0.9972	0.9921	0.9807
	10	1.0000	1.0000	1.0000	1.0000	1.0000	1.0000	0.9999	0.9997	0.9989	0.9968
	11	1.0000	1.0000	1.0000	1.0000	1.0000	1.0000	1.0000	1.0000	0.9999	0.9998
	12	1.0000	1.0000	1.0000	1.0000	1.0000	1.0000	1.0000	1.0000	1.0000	1.0000
13	0	0.5133	0.2542	0.1209	0.0550	0.0238	0.0097	0.0037	0.0013	0.0004	0.0001
	1	0.8646	0.6213	0.3983	0.2336	0.1267	0.0637	0.0296	0.0126	0.0049	0.0017
	2	0.9755	0.8661	0.6920	0.5017	0.3326	0.2025	0.1132	0.0579	0.0269	0.0112
	3	0.9969	0.9658	0.8820	0.7473	0.5843	0.4206	0.2783	0.1686	0.0929	0.0461
	4	0.9997	0.9935	0.9658	0.9009	0.7940	0.6543	0.5005	0.3530	0.2279	0.1334
	5	1.0000	0.9991	0.9924	0.9700	0.9198	0.8346	0.7159	0.5744	0.4268	0.2905
	6	1.0000	0.9999	0.9987	0.9930	0.9757	0.9376	0.8705	0.7712	0.6437	0.5000
	7	1.0000	1.0000	0.9998	0.9988	0.9944	0.9818	0.9538	0.9023	0.8212	0.7095
	8	1.0000	1.0000	1.0000	0.9998	0.9990	0.9960	0.9874	0.9679	0.9302	0.8666
	9	1.0000	1.0000	1.0000	1.0000	0.9999	0.9993	0.9975	0.9922	0.9797	0.9539
	10	1.0000	1.0000	1.0000	1.0000	1.0000	0.9999	0.9997	0.9987	0.9959	0.9888
	11	1.0000	1.0000	1.0000	1.0000	1.0000	1.0000	1.0000	0.9999	0.9995	0.9983
	12	1.0000	1.0000	1.0000	1.0000	1.0000	1.0000	1.0000	1.0000	1.0000	0.9999
	13	1.0000	1.0000	1.0000	1.0000	1.0000	1.0000	1.0000	1.0000	1.0000	1.0000
14	0	0.4877	0.2288	0.1028	0.0440	0.0178	0.0068	0.0024	0.0008	0.0002	0.0001
	1	0.8470	0.5846	0.3567	0.1979	0.1010	0.0475	0.0205	0.0081	0.0029	0.0009
	2	0.9699	0.8416	0.6479	0.4481	0.2811	0.1608	0.0839	0.0398	0.0170	0.0065
	3	0.9958	0.9559	0.8535	0.6982	0.5213	0.3552	0.2205	0.1243	0.0632	0.0287
	4	0.9996	0.9908	0.9533	0.8702	0.7415	0.5842	0.4227	0.2793	0.1672	0.0898
	5	1.0000	0.9985	0.9885	0.9561	0.8883	0.7805	0.6405	0.4859	0.3373	0.2120
	6	1.0000	0.9998	0.9978	0.9884	0.9617	0.9067	0.8164	0.6925	0.5461	0.3953
	7	1.0000	1.0000	0.9997	0.9976	0.9897	0.9685	0.9247	0.8499	0.7414	0.6047
	8	1.0000	1.0000	1.0000	0.9996	0.9978	0.9917	0.9757	0.9417	0.8811	0.7880
	9	1.0000	1.0000	1.0000	1.0000	0.9997	0.9983	0.9940	0.9825	0.9574	0.9102
	10	1.0000	1.0000	1.0000	1.0000	1.0000	0.9998	0.9989	0.9961	0.9886	0.9713

Table II *continued*

n	x	0.05	0.10	0.15	0.20	0.25	0.30	0.35	0.40	0.45	0.50
							p				
	11	1.0000	1.0000	1.0000	1.0000	1.0000	1.0000	0.9999	0.9994	0.9978	0.9935
	12	1.0000	1.0000	1.0000	1.0000	1.0000	1.0000	1.0000	0.9999	0.9997	0.9991
	13	1.0000	1.0000	1.0000	1.0000	1.0000	1.0000	1.0000	1.0000	1.0000	0.9999
	14	1.0000	1.0000	1.0000	1.0000	1.0000	1.0000	1.0000	1.0000	1.0000	1.0000
15	0	0.4633	0.2059	0.0874	0.0352	0.0134	0.0047	0.0016	0.0005	0.0001	0.0000
	1	0.8290	0.5490	0.3186	0.1671	0.0802	0.0353	0.0142	0.0052	0.0017	0.0005
	2	0.9638	0.8159	0.6042	0.3980	0.2361	0.1268	0.0617	0.0271	0.0107	0.0037
	3	0.9945	0.9444	0.8227	0.6482	0.4613	0.2969	0.1727	0.0905	0.0424	0.0176
	4	0.9994	0.9873	0.9383	0.8358	0.6865	0.5155	0.3519	0.2173	0.1204	0.0592
	5	0.9999	0.9978	0.9832	0.9389	0.8516	0.7216	0.5643	0.4032	0.2608	0.1509
	6	1.0000	0.9997	0.9964	0.9819	0.9434	0.8689	0.7548	0.6098	0.4522	0.3036
	7	1.0000	1.0000	0.9994	0.9958	0.9827	0.9500	0.8868	0.7869	0.6535	0.5000
	8	1.0000	1.0000	0.9999	0.9992	0.9958	0.9848	0.9578	0.9050	0.8182	0.6964
	9	1.0000	1.0000	1.0000	0.9999	0.9992	0.9963	0.9876	0.9662	0.9231	0.8491
	10	1.0000	1.0000	1.0000	1.0000	0.9999	0.9993	0.9972	0.9907	0.9745	0.9408
	11	1.0000	1.0000	1.0000	1.0000	1.0000	0.9999	0.9995	0.9981	0.9937	0.9824
	12	1.0000	1.0000	1.0000	1.0000	1.0000	1.0000	0.9999	0.9987	0.9989	0.9963
	13	1.0000	1.0000	1.0000	1.0000	1.0000	1.0000	1.0000	1.0000	0.9999	0.9995
	14	1.0000	1.0000	1.0000	1.0000	1.0000	1.0000	1.0000	1.0000	1.0000	1.0000
	15	1.0000	1.0000	1.0000	1.0000	1.0000	1.0000	1.0000	1.0000	1.0000	1.0000
16	0	0.4401	0.1853	0.0743	0.0281	0.0100	0.0033	0.0010	0.0003	0.0001	0.0000
	1	0.8108	0.5147	0.2839	0.1407	0.0635	0.0261	0.0098	0.0033	0.0010	0.0003
	2	0.9571	0.7892	0.5614	0.3518	0.1971	0.0994	0.0451	0.0183	0.0066	0.0021
	3	0.9930	0.9316	0.7899	0.5981	0.4050	0.2459	0.1339	0,0651	0.0281	0.0106
	4	0.9991	0.9830	0.9209	0.7982	0.6302	0.4499	0.2892	0.1666	0.0853	0.0384
	5	0.9999	0.9967	0.9765	0.9183	0.8103	0.6598	0.4900	0.3288	0.1976	0.1051
	6	1.0000	0.9995	0.9944	0.9733	0.9204	0.8247	0.6881	0.5272	0.3660	0.2272
	7	1.0000	0.9999	0.9989	0.9930	0.9729	0.9256	0.8406	0.7161	0.5629	0.4018
	8	1.0000	1.0000	0.9998	0.9985	0.9925	0.9743	0.9329	0.8577	0.7441	0.5982
	9	1.0000	1.0000	1.0000	0.9998	0.9984	0.9929	0.9771	0.9417	0.8759	0.7728
	10	1.0000	1.0000	1.0000	1.0000	0.9997	0.9984	0.9938	0.9809	0.9514	0.8949
	11	1.0000	1.0000	1.0000	1.0000	1.0000	0.9997	0.9987	0.9951	0.9851	0.9616
	12	1.0000	1.0000	1.0000	1.0000	1.0000	1.0000	0.9998	0.9991	0.9965	0.9894
	13	1.0000	1.0000	1.0000	1.0000	1.0000	1.0000	1.0000	0.9999	0.9994	0.9979
	14	1.0000	1.0000	1.0000	1.0000	1.0000	1.0000	1.0000	1.0000	0.9999	0.9997
	15	1.0000	1.0000	1.0000	1.0000	1.0000	1.0000	1.0000	1.0000	1.0000	1.0000
	16	1.0000	1.0000	1.0000	1.0000	1.0000	1.0000	1.0000	1.0000	1.0000	1.0000
20	0	0.3585	0.1216	0.0388	0.0115	0.0032	0.0008	0.0002	0.0000	0.0000	0.0000
	1	0.7358	0.3917	0.1756	0.0692	0.0243	0.0076	0.0021	0.0005	0.0001	0.0000
	2	0.9245	0.6769	0.4049	0.2061	0.0913	0.0355	0.0121	0.0036	0.0009	0.0002
	3	0.9841	0.8670	0.6477	0.4114	0.2252	0.1071	0.0444	0.0160	0.0049	0.0013
	4	0.9974	0.9568	0.8298	0.6296	0.4148	0.2375	0.1182	0.0510	0.0189	0.0059

Table II *continued*

							p				
n	x	0.05	0.10	0.15	0.20	0.25	0.30	0.35	0.40	0.45	0.50
	5	0.9997	0.9887	0.9327	0.8042	0.6172	0.4164	0.2454	0.1256	0.0553	0.0207
	6	1.0000	0.9976	0.9781	0.9133	0.7858	0.6080	0.4166	0.2500	0.1299	0.0577
	7	1.0000	0.9996	0.9941	0.9679	0.8982	0.7723	0.6010	0.4159	0.2520	0.1316
	8	1.0000	0.9999	0.9987	0.9900	0.9591	0.8867	0.7624	0.5956	0.4143	0.2517
	9	1.0000	1.0000	0.9998	0.9974	0.9861	0.9520	0.8782	0.7553	0.5914	0.4119
	10	1.0000	1.0000	1.0000	0.9994	0.9961	0.9829	0.9468	0.8725	0.7507	0.5881
	11	1.0000	1.0000	1.0000	0.9999	0.9991	0.9949	0.9804	0.9435	0.8692	0.7483
	12	1.0000	1.0000	1.0000	1.0000	0.9998	0.9987	0.9940	0.9790	0.9420	0.8684
	13	1.0000	1.0000	1.0000	1.0000	1.0000	0.9997	0.9985	0.9935	0.9786	0.9423
	14	1.0000	1.0000	1.0000	1.0000	1.0000	1.0000	0.9997	0.9984	0.9936	0.9793
	15	1.0000	1.0000	1.0000	1.0000	1.0000	1.0000	1.0000	0.9997	0.9985	0.9941
	16	1.0000	1.0000	1.0000	1.0000	1.0000	1.0000	1.0000	1.0000	0.9997	0.9987
	17	1.0000	1.0000	1.0000	1.0000	1.0000	1.0000	1.0000	1.0000	1.0000	0.9998
	18	1.0000	1.0000	1.0000	1.0000	1.0000	1.0000	1.0000	1.0000	1.0000	1.0000
	19	1.0000	1.0000	1.0000	1.0000	1.0000	1.0000	1.0000	1.0000	1.0000	1.0000
	20	1.0000	1.0000	1.0000	1.0000	1.0000	1.0000	1.0000	1.0000	1.0000	1.0000
25	0	0.2774	0.0718	0.0172	0.0038	0.0008	0.0001	0.0000	0.0000	0.0000	0.0000
	1	0.6424	0.2712	0.0931	0.0274	0.0070	0.0016	0.0003	0.0001	0.0000	0.0000
	2	0.8729	0.5371	0.2537	0.0982	0.0321	0.0090	0.0021	0.0004	0.0001	0.0000
	3	0.9659	0.7636	0.4711	0.2340	0.0962	0.0332	0.0097	0.0024	0.0005	0.0001
	4	0.9928	0.9020	0.6821	0.4207	0.2137	0.0905	0.0320	0.0095	0.0023	0.0005
	5	0.9988	0.9666	0.8385	0.6167	0.3783	0.1935	0.0826	0.0294	0.0086	0.0020
	6	0.9998	0.9905	0.9305	0.7800	0.5611	0.3407	0.1734	0.0736	0.0258	0.0073
	7	1.0000	0.9977	0.9745	0.8909	0.7265	0.5118	0.3061	0.1536	0.0639	0.0216
	8	1.0000	0.9995	0.9920	0.9532	0.8506	0.6769	0.4668	0.2735	0.1340	0.0539
	9	1.0000	0.9999	0.9979	0.9827	0.9287	0.8106	0.6303	0.4246	0.2424	0.1148
	10	1.0000	1.0000	0.9995	0.9944	0.9703	0.9022	0.7712	0.5858	0.3843	0.2122
	11	1.0000	1.0000	0.9999	0.9985	0.9893	0.9558	0.8746	0.7323	0.5426	0.3450
	12	1.0000	1.0000	1.0000	0.9996	0.9966	0.9825	0.9396	0.8462	0.6937	0.5000
	13	1.0000	1.0000	1.0000	0.9999	0.9991	0.9940	0.9745	0.9222	0.8173	0.6550
	14	1.0000	1.0000	1,0000	1.0000	0.9998	0.9982	0.9907	0.9656	0.9040	0.7878
	15	1.0000	1.0000	1.0000	1.0000	1.0000	0.9995	0.9971	0.9868	0.9560	0.8852
	16	1.0000	1.0000	1.0000	1.0000	1.0000	0.9999	0.9992	0.9957	0.9826	0.9461
	17	1.0000	1.0000	1.0000	1.0000	1.0000	1.0000	0.9998	0.9988	0.9942	0.9784
	18	1.0000	1.0000	1.0000	1.0000	1.0000	1.0000	1.0000	0.9997	0.9984	0.9927
	19	1.0000	1.0000	1.0000	1.0000	1.0000	1.0000	1.0000	0.9999	0.9996	0.9980
	20	1.0000	1.0000	1.0000	1.0000	1.0000	1.0000	1.0000	1.0000	0.9999	0.9995
	21	1.0000	1.0000	1.0000	1.0000	1.0000	1.0000	1.0000	1.0000	1.0000	0.9999
	22	1.0000	1.0000	1.0000	1.0000	1.0000	1.0000	1.0000	1.0000	1.0000	1.0000
	23	1.0000	1.0000	1.0000	1.0000	1.0000	1.0000	1.0000	1.0000	1.0000	1.0000
	24	1.0000	1.0000	1.0000	1.0000	1.0000	1.0000	1.0000	1.0000	1.0000	1.0000
	25	1.0000	1.0000	1.0000	1.0000	1.0000	1.0000	1.0000	1.0000	1.0000	1.0000

Table III The Poisson Distribution

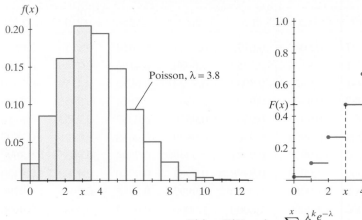

$$F(x) = P(X \le x) = \sum_{k=0}^{x} \frac{\lambda^k e^{-\lambda}}{k!}$$

	$\lambda = E(X)$									
x	0.1	0.2	0.3	0.4	0.5	0.6	0.7	0.8	0.9	1.0
0	0.905	0.819	0.741	0.670	0.607	0.549	0.497	0.449	0.407	0.368
1	0.995	0.982	0.963	0.938	0.910	0.878	0.844	0.809	0.772	0.736
2	1.000	0.999	0.996	0.992	0.986	0.977	0.966	0.953	0.937	0.920
3	1.000	1.000	1.000	0.999	0.998	0.997	0.994	0.991	0.987	0.981
4	1.000	1.000	1.000	1.000	1.000	1.000	0.999	0.999	0.998	0.996
5	1.000	1.000	1.000	1.000	1.000	1.000	1.000	1.000	1.000	0.999
6	1.000	1.000	1.000	1.000	1.000	1.000	1.000	1.000	1.000	1.000

x	1.1	1.2	1.3	1.4	1.5	1.6	1.7	1.8	1.9	2.0
0	0.333	0.301	0.273	0.247	0.223	0.202	0.183	0.165	0.150	0.135
1	0.699	0.663	0.627	0.592	0.558	0.525	0.493	0.463	0.434	0.406
2	0.900	0.879	0.857	0.833	0.809	0.783	0.757	0.731	0.704	0.677
3	0.974	0.966	0.957	0.946	0.934	0.921	0.907	0.891	0.875	0.857
4	0.995	0.992	0.989	0.986	0.981	0.976	0.970	0.964	0.956	0.947
5	0.999	0.998	0.998	0.997	0.996	0.994	0.992	0.990	0.987	0.983
6	1.000	1.000	1.000	0.999	0.999	0.999	0.998	0.997	0.997	0.995
7	1.000	1.000	1.000	1.000	1.000	1.000	1.000	0.999	0.999	0.999
8	1.000	1.000	1.000	1.000	1.000	1.000	1.000	1.000	1.000	1.000

x	2.2	2.4	2.6	2.8	3.0	3.2	3.4	3.6	3.8	4.0
0	0.111	0.091	0.074	0.061	0.050	0.041	0.033	0.027	0.022	0.018
1	0.355	0.308	0.267	0.231	0.199	0.171	0.147	0.126	0.107	0.092
2	0.623	0.570	0.518	0.469	0.423	0.380	0.340	0.303	0.269	0.238
3	0.819	0.779	0.736	0.692	0.647	0.603	0.558	0.515	0.473	0.433
4	0.928	0.904	0.877	0.848	0.815	0.781	0.744	0.706	0.668	0.629
5	0.975	0.964	0.951	0.935	0.916	0.895	0.871	0.844	0.816	0.785
6	0.993	0.988	0.983	0.976	0.966	0.955	0.942	0.927	0.909	0.889
7	0.998	0.997	0.995	0.992	0.988	0.983	0.977	0.969	0.960	0.949
8	1.000	0.999	0.999	0.998	0.996	0.994	0.992	0.988	0.984	0.979
9	1.000	1.000	1.000	0.999	0.999	0.998	0.997	0.996	0.994	0.992
10	1.000	1.000	1.000	1.000	1.000	1.000	0.999	0.999	0.998	0.997
11	1.000	1.000	1.000	1.000	1.000	1.000	1.000	1.000	0.999	0.999
12	1.000	1.000	1.000	1.000	1.000	1.000	1.000	1.000	1.000	1.000

Table III *continued*

x	4.2	4.4	4.6	4.8	5.0	5.2	5.4	5.6	5.8	6.0
0	0.015	0.012	0.010	0.008	0.007	0.006	0.005	0.004	0.003	0.002
1	0.078	0.066	0.056	0.048	0.040	0.034	0.029	0.024	0.021	0.017
2	0.210	0.185	0.163	0.143	0.125	0.109	0.095	0.082	0.072	0.062
3	0.395	0.359	0.326	0.294	0.265	0.238	0.213	0.191	0.170	0.151
4	0.590	0.551	0.513	0.476	0.440	0.406	0.373	0.342	0.313	0.285
5	0.753	0.720	0.686	0.651	0.616	0.581	0.546	0.512	0.478	0.446
6	0.867	0.844	0.818	0.791	0.762	0.732	0.702	0.670	0.638	0.606
7	0.936	0.921	0.905	0.887	0.867	0.845	0.822	0.797	0.771	0.744
8	0.972	0.964	0.955	0.944	0.932	0.918	0.903	0.886	0.867	0.847
9	0.989	0.985	0.980	0.975	0.968	0.960	0.951	0.941	0.929	0.916
10	0.996	0.994	0.992	0.990	0.986	0.982	0.977	0.972	0.965	0.957
11	0.999	0.998	0.997	0.996	0.995	0.993	0.990	0.988	0.984	0.980
12	1.000	0.999	0.999	0.999	0.998	0.997	0.996	0.995	0.993	0.991
13	1.000	1.000	1.000	1.000	0.999	0.999	0.999	0.998	0.997	0.996
14	1.000	1.000	1.000	1.000	1.000	1.000	0.999	0.999	0.999	0.999
15	1.000	1.000	1.000	1.000	1.000	1.000	1.000	1.000	1.000	0.999
16	1.000	1.000	1.000	1.000	1.000	1.000	1.000	1.000	1.000	1.000

x	6.5	7.0	7.5	8.0	8.5	9.0	9.5	10.0	10.5	11.0
0	0.002	0.001	0.001	0.000	0.000	0.000	0.000	0.000	0.000	0.000
1	0.011	0.007	0.005	0.003	0.002	0.001	0.001	0.000	0.000	0.000
2	0.043	0.030	0.020	0.014	0.009	0.006	0.004	0.003	0.002	0.001
3	0.112	0.082	0.059	0.042	0.030	0.021	0.015	0.010	0.007	0.005
4	0.224	0.173	0.132	0.100	0.074	0.055	0.040	0.029	0.021	0.015
5	0.369	0.301	0.241	0.191	0.150	0.116	0.089	0.067	0.050	0.038
6	0.527	0.450	0.378	0.313	0.256	0.207	0.165	0.130	0.102	0.079
7	0.673	0.599	0.525	0.453	0.386	0.324	0.269	0.220	0.179	0.143
8	0.792	0.729	0.662	0.593	0.523	0.456	0.392	0.333	0.279	0.232
9	0.877	0.830	0.776	0.717	0.653	0.587	0.522	0.458	0.397	0.341
10	0.933	0.901	0.862	0.816	0.763	0.706	0.645	0.583	0.521	0.460
11	0.966	0.947	0.921	0.888	0.849	0.803	0.752	0.697	0.639	0.579
12	0.984	0.973	0.957	0.936	0.909	0.876	0.836	0.792	0.742	0.689
13	0.993	0.987	0.978	0.966	0.949	0.926	0.898	0.864	0.825	0.781
14	0.997	0.994	0.990	0.983	0.973	0.959	0.940	0.917	0.888	0.854
15	0.999	0.998	0.995	0.992	0.986	0.978	0.967	0.951	0.932	0.907
16	1.000	0.999	0.998	0.996	0.993	0.989	0.982	0.973	0.960	0.944
17	1.000	1.000	0.999	0.998	0.997	0.995	0.991	0.986	0.978	0.968
18	1.000	1.000	1.000	0.999	0.999	0.998	0.096	0.993	0.988	0.982
19	1.000	1.000	1.000	1.000	0.999	0.999	0.998	0.997	0.994	0.991
20	1.000	1.000	1.000	1.000	1.000	1.000	0.999	0.998	0.997	0.995
21	1.000	1.000	1.000	1.000	1.000	1.000	1.000	0.999	0.999	0.998
22	1.000	1.000	1.000	1.000	1.000	1.000	1.000	1.000	0.999	0.999
23	1.000	1.000	1.000	1.000	1.000	1.000	1.000	1.000	1.000	1.000

Table III *continued*

x	11.5	12.0	12.5	13.0	13.5	14.0	14.5	15.0	15.5	16.0
0	0.000	0.000	0.000	0.000	0.000	0.000	0.000	0.000	0.000	0.000
1	0.000	0.000	0.000	0.000	0.000	0.000	0.000	0.000	0.000	0.000
2	0.001	0.001	0.000	0.000	0.000	0.000	0.000	0.000	0.000	0.000
3	0.003	0.002	0.002	0.001	0.001	0.000	0.000	0.000	0.000	0.000
4	0.011	0.008	0.005	0.004	0.003	0.002	0.001	0.001	0.001	0.000
5	0.028	0.020	0.015	0.011	0.008	0.006	0.004	0.003	0.002	0.001
6	0.060	0.046	0.035	0.026	0.019	0.014	0.010	0.008	0.006	0.004
7	0.114	0.090	0.070	0.054	0.041	0.032	0.024	0.018	0.013	0.010
8	0.191	0.155	0.125	0.100	0.079	0.062	0.048	0.037	0.029	0.022
9	0.289	0.242	0.201	0.166	0.135	0.109	0.088	0.070	0.055	0.043
10	0.402	0.347	0.297	0.252	0.211	0.176	0.145	0.118	0.096	0.077
11	0.520	0.462	0.406	0.353	0.304	0.260	0.220	0.185	0.154	0.127
12	0.633	0.576	0.519	0.463	0.409	0.358	0.311	0.268	0.228	0.193
13	0.733	0.682	0.629	0.573	0.518	0.464	0.413	0.363	0.317	0.275
14	0.815	0.772	0.725	0.675	0.623	0.570	0.518	0.466	0.415	0.368
15	0.878	0.844	0.806	0.764	0.718	0.669	0.619	0.568	0.517	0.467
16	0.924	0.899	0.869	0.835	0.798	0.756	0.711	0.664	0.615	0.566
17	0.954	0.937	0.916	0.890	0.861	0.827	0.790	0.749	0.705	0.659
18	0.974	0.963	0.948	0.930	0.908	0.883	0.853	0.819	0.782	0.742
19	0.986	0.979	0.969	0.957	0.942	0.923	0.901	0.875	0.846	0.812
20	0.992	0.988	0.983	0.975	0.965	0.952	0.936	0.917	0.894	0.868
21	0.996	0.994	0.991	0.986	0.980	0.971	0.960	0.947	0.930	0.911
22	0.999	0.997	0.995	0.992	0.989	0.983	0.976	0.967	0.956	0.942
23	0.999	0.999	0.998	0.996	0.994	0.991	0.986	0.981	0.973	0.963
24	1.000	0.999	0.999	0.998	0.997	0.995	0.992	0.989	0.984	0.978
25	1.000	1.000	0.999	0.999	0.998	0.997	0.996	0.994	0.991	0.987
26	1.000	1.000	1.000	1.000	0.999	0.999	0.998	0.997	0.995	0.993
27	1.000	1.000	1.000	1.000	1.000	0.999	0.999	0.998	0.997	0.996
28	1.000	1.000	1.000	1.000	1.000	1.000	0.999	0.999	0.999	0.998
29	1.000	1.000	1.000	1.000	1.000	1.000	1.000	1.000	0.999	0.999
30	1.000	1.000	1.000	1.000	1.000	1.000	1.000	1.000	1.000	0.999
31	1.000	1.000	1.000	1.000	1.000	1.000	1.000	1.000	1.000	1.000
32	1.000	1.000	1.000	1.000	1.000	1.000	1.000	1.000	1.000	1.000
33	1.000	1.000	1.000	1.000	1.000	1.000	1.000	1.000	1.000	1.000
34	1.000	1.000	1.000	1.000	1.000	1.000	1.000	1.000	1.000	1.000
35	1.000	1.000	1.000	1.000	1.000	1.000	1.000	1.000	1.000	1.000

Table IV The Chi-Square Distribution

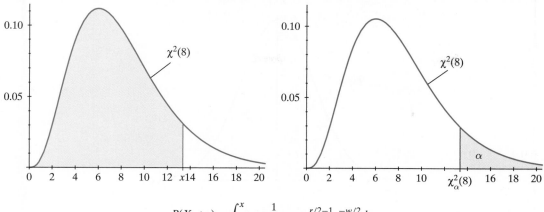

$$P(X \le x) = \int_0^x \frac{1}{\Gamma(r/2)2^{r/2}} w^{r/2-1} e^{-w/2} dw$$

	$P(X \le x)$							
	0.010	0.025	0.050	0.100	0.900	0.950	0.975	0.990
r	$\chi^2_{0.99}(r)$	$\chi^2_{0.975}(r)$	$\chi^2_{0.95}(r)$	$\chi^2_{0.90}(r)$	$\chi^2_{0.10}(r)$	$\chi^2_{0.05}(r)$	$\chi^2_{0.025}(r)$	$\chi^2_{0.01}(r)$
1	0.000	0.001	0.004	0.016	2.706	3.841	5.024	6.635
2	0.020	0.051	0.103	0.211	4.605	5.991	7.378	9.210
3	0.115	0.216	0.352	0.584	6.251	7.815	9.348	11.34
4	0.297	0.484	0.711	1.064	7.779	9.488	11.14	13.28
5	0.554	0.831	1.145	1.610	9.236	11.07	12.83	15.09
6	0.872	1.237	1.635	2.204	10.64	12.59	14.45	16.81
7	1.239	1.690	2.167	2.833	12.02	14.07	16.01	18.48
8	1.646	2.180	2.733	3.490	13.36	15.51	17.54	20.09
9	2.088	2.700	3.325	4.168	14.68	16.92	19.02	21.67
10	2.558	3.247	3.940	4.865	15.99	18.31	20.48	23.21
11	3.053	3.816	4.575	5.578	17.28	19.68	21.92	24.72
12	3.571	4.404	5.226	6.304	18.55	21.03	23.34	26.22
13	4.107	5.009	5.892	7.042	19.81	22.36	24.74	27.69
14	4.660	5.629	6.571	7.790	21.06	23.68	26.12	29.14
15	5.229	6.262	7.261	8.547	22.31	25.00	27.49	30.58
16	5.812	6.908	7.962	9.312	23.54	26.30	28.84	32.00
17	6.408	7.564	8.672	10.08	24.77	27.59	30.19	33.41
18	7.015	8.231	9.390	10.86	25.99	28.87	31.53	34.80
19	7.633	8.907	10.12	11.65	27.20	30.14	32.85	36.19
20	8.260	9.591	10.85	12.44	28.41	31.41	34.17	37.57
21	8.897	10.28	11.59	13.24	29.62	32.67	35.48	38.93
22	9.542	10.98	12.34	14.04	30.81	33.92	36.78	40.29
23	10.20	11.69	13.09	14.85	32.01	35.17	38.08	41.64
24	10.86	12.40	13.85	15.66	33.20	36.42	39.36	42.98
25	11.52	13.12	14.61	16.47	34.38	37.65	40.65	44.31
26	12.20	13.84	15.38	17.29	35.56	38.88	41.92	45.64
27	12.88	14.57	16.15	18.11	36.74	40.11	43.19	46.96
28	13.56	15.31	16.93	18.94	37.92	41.34	44.46	48.28
29	14.26	16.05	17.71	19.77	39.09	42.56	45.72	49.59
30	14.95	16.79	18.49	20.60	40.26	43.77	46.98	50.89
40	22.16	24.43	26.51	29.05	51.80	55.76	59.34	63.69
50	29.71	32.36	34.76	37.69	63.17	67.50	71.42	76.15
60	37.48	40.48	43.19	46.46	74.40	79.08	83.30	88.38
70	45.44	48.76	51.74	55.33	85.53	90.53	95.02	100.4
80	53.34	57.15	60.39	64.28	96.58	101.9	106.6	112.3

This table is abridged and adapted from Table III in *Biometrika Tables for Statisticians*, edited by E.S.Pearson and H.O.Hartley.

Table Va The Standard Normal Distribution Function

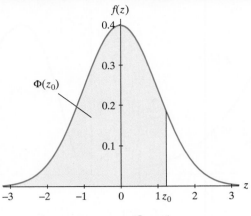

$$P(Z \le z) = \Phi(z) = \int_{-\infty}^{z} \frac{1}{\sqrt{2\pi}} e^{-w^2/2}\, dw$$

$$\Phi(-z) = 1 - \Phi(z)$$

z	0.00	0.01	0.02	0.03	0.04	0.05	0.06	0.07	0.08	0.09
0.0	0.5000	0.5040	0.5080	0.5120	0.5160	0.5199	0.5239	0.5279	0.5319	0.5359
0.1	0.5398	0.5438	0.5478	0.5517	0.5557	0.5596	0.5636	0.5675	0.5714	0.5753
0.2	0.5793	0.5832	0.5871	0.5910	0.5948	0.5987	0.6026	0.6064	0.6103	0.6141
0.3	0.6179	0.6217	0.6255	0.6293	0.6331	0.6368	0.6406	0.6443	0.6480	0.6517
0.4	0.6554	0.6591	0.6628	0.6664	0.6700	0.6736	0.6772	0.6808	0.6844	0.6879
0.5	0.6915	0.6950	0.6985	0.7019	0.7054	0.7088	0.7123	0.7157	0.7190	0.7224
0.6	0.7257	0.7291	0.7324	0.7357	0.7389	0.7422	0.7454	0.7486	0.7517	0.7549
0.7	0.7580	0.7611	0.7642	0.7673	0.7703	0.7734	0.7764	0.7794	0.7823	0.7852
0.8	0.7881	0.7910	0.7939	0.7967	0.7995	0.8023	0.8051	0.8078	0.8106	0.8133
0.9	0.8159	0.8186	0.8212	0.8238	0.8264	0.8289	0.8315	0.8340	0.8365	0.8389
1.0	0.8413	0.8438	0.8461	0.8485	0.8508	0.8531	0.8554	0.8577	0.8599	0.8621
1.1	0.8643	0.8665	0.8686	0.8708	0.8729	0.8749	0.8770	0.8790	0.8810	0.8830
1.2	0.8849	0.8869	0.8888	0.8907	0.8925	0.8944	0.8962	0.8980	0.8997	0.9015
1.3	0.9032	0.9049	0.9066	0.9082	0.9099	0.9115	0.9131	0.9147	0.9162	0.9177
1.4	0.9192	0.9207	0.9222	0.9236	0.9251	0.9265	0.9279	0.9292	0.9306	0.9319
1.5	0.9332	0.9345	0.9357	0.9370	0.9382	0.9394	0.9406	0.9418	0.9429	0.9441
1.6	0.9452	0.9463	0.9474	0.9484	0.9495	0.9505	0.9515	0.9525	0.9535	0.9545
1.7	0.9554	0.9564	0.9573	0.9582	0.9591	0.9599	0.9608	0.9616	0.9625	0.9633
1.8	0.9641	0.9649	0.9656	0.9664	0.9671	0.9678	0.9686	0.9693	0.9699	0.9706
1.9	0.9713	0.9719	0.9726	0.9732	0.9738	0.9744	0.9750	0.9756	0.9761	0.9767
2.0	0.9772	0.9778	0.9783	0.9788	0.9793	0.9798	0.9803	0.9808	0.9812	0.9817
2.1	0.9821	0.9826	0.9830	0.9834	0.9838	0.9842	0.9846	0.9850	0.9854	0.9857
2.2	0.9861	0.9864	0.9868	0.9871	0.9875	0.9878	0.9881	0.9884	0.9887	0.9890
2.3	0.9893	0.9896	0.9898	0.9901	0.9904	0.9906	0.9909	0.9911	0.9913	0.9916
2.4	0.9918	0.9920	0.9922	0.9925	0.9927	0.9929	0.9931	0.9932	0.9934	0.9936
2.5	0.9938	0.9940	0.9941	0.9943	0.9945	0.9946	0.9948	0.9949	0.9951	0.9952
2.6	0.9953	0.9955	0.9956	0.9957	0.9959	0.9960	0.9961	0.9962	0.9963	0.9964
2.7	0.9965	0.9966	0.9967	0.9968	0.9969	0.9970	0.9971	0.9972	0.9973	0.9974
2.8	0.9974	0.9975	0.9976	0.9977	0.9977	0.9978	0.9979	0.9979	0.9980	0.9981
2.9	0.9981	0.9982	0.9982	0.9983	0.9984	0.9984	0.9985	0.9985	0.9986	0.9986
3.0	0.9987	0.9987	0.9987	0.9988	0.9988	0.9989	0.9989	0.9989	0.9990	0.9990

α	0.400	0.300	0.200	0.100	0.050	0.025	0.020	0.010	0.005	0.001
z_α	0.253	0.524	0.842	1.282	1.645	1.960	2.054	2.326	2.576	3.090
$z_{\alpha/2}$	0.842	1.036	1.282	1.645	1.960	2.240	2.326	2.576	2.807	3.291

Table Vb The Standard Normal Right-Tail Probabilities

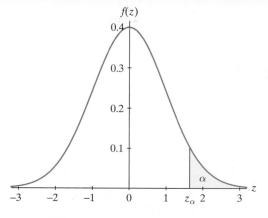

$$P(Z > z_\alpha) = \alpha$$
$$P(Z > z) = 1 - \Phi(z) = \Phi(-z)$$

z_α	0.00	0.01	0.02	0.03	0.04	0.05	0.06	0.07	0.08	0.09
0.0	0.5000	0.4960	0.4920	0.4880	0.4840	0.4801	0.4761	0.4721	0.4681	0.4641
0.1	0.4602	0.4562	0.4522	0.4483	0.4443	0.4404	0.4364	0.4325	0.4286	0.4247
0.2	0.4207	0.4168	0.4129	0.4090	0.4052	0.4013	0.3974	0.3936	0.3897	0.3859
0.3	0.3821	0.3783	0.3745	0.3707	0.3669	0.3632	0.3594	0.3557	0.3520	0.3483
0.4	0.3446	0.3409	0.3372	0.3336	0.3300	0.3264	0.3228	0.3192	0.3156	0.3121
0.5	0.3085	0.3050	0.3015	0.2981	0.2946	0.2912	0.2877	0.2843	0.2810	0.2776
0.6	0.2743	0.2709	0.2676	0.2643	0.2611	0.2578	0.2546	0.2514	0.2483	0.2451
0.7	0.2420	0.2389	0.2358	0.2327	0.2296	0.2266	0.2236	0.2206	0.2177	0.2148
0.8	0.2119	0.2090	0.2061	0.2033	0.2005	0.1977	0.1949	0.1922	0.1894	0.1867
0.9	0.1841	0.1814	0.1788	0.1762	0.1736	0.1711	0.1685	0.1660	0.1635	0.1611
1.0	0.1587	0.1562	0.1539	0.1515	0.1492	0.1469	0.1446	0.1423	0.1401	0.1379
1.1	0.1357	0.1335	0.1314	0.1292	0.1271	0.1251	0.1230	0.1210	0.1190	0.1170
1.2	0.1151	0.1131	0.1112	0.1093	0.1075	0.1056	0.1038	0.1020	0.1003	0.0985
1.3	0.0968	0.0951	0.0934	0.0918	0.0901	0.0885	0.0869	0.0853	0.0838	0.0823
1.4	0.0808	0.0793	0.0778	0.0764	0.0749	0.0735	0.0721	0.0708	0.0694	0.0681
1.5	0.0668	0.0655	0.0643	0.0630	0.0618	0.0606	0.0594	0.0582	0.0571	0.0559
1.6	0.0548	0.0537	0.0526	0.0516	0.0505	0.0495	0.0485	0.0475	0.0465	0.0455
1.7	0.0446	0.0436	0.0427	0.0418	0.0409	0.0401	0.0392	0.0384	0.0375	0.0367
1.8	0.0359	0.0351	0.0344	0.0336	0.0329	0.0322	0.0314	0.0307	0.0301	0.0294
1.9	0.0287	0.0281	0.0274	0.0268	0.0262	0.0256	0.0250	0.0244	0.0239	0.0233
2.0	0.0228	0.0222	0.0217	0.0212	0.0207	0.0202	0.0197	0.0192	0.0188	0.0183
2.1	0.0179	0.0174	0.0170	0.0166	0.0162	0.0158	0.0154	0.0150	0.0146	0.0143
2.2	0.0139	0.0136	0.0132	0.0129	0.0125	0.0122	0.0119	0.0116	0.0113	0.0110
2.3	0.0107	0.0104	0.0102	0.0099	0.0096	0.0094	0.0091	0.0089	0.0087	0.0084
2.4	0.0082	0.0080	0.0078	0.0075	0.0073	0.0071	0.0069	0.0068	0.0066	0.0064
2.5	0.0062	0.0060	0.0059	0.0057	0.0055	0.0054	0.0052	0.0051	0.0049	0.0048
2.6	0.0047	0.0045	0.0044	0.0043	0.0041	0.0040	0.0039	0.0038	0.0037	0.0036
2.7	0.0035	0.0034	0.0033	0.0032	0.0031	0.0030	0.0029	0.0028	0.0027	0.0026
2.8	0.0026	0.0025	0.0024	0.0023	0.0023	0.0022	0.0021	0.0021	0.0020	0.0019
2.9	0.0019	0.0018	0.0018	0.0017	0.0016	0.0016	0.0015	0.0015	0.0014	0.0014
3.0	0.0013	0.0013	0.0013	0.0012	0.0012	0.0011	0.0011	0.0011	0.0010	0.0010
3.1	0.0010	0.0009	0.0009	0.0009	0.0008	0.0008	0.0008	0.0008	0.0007	0.0007
3.2	0.0007	0.0007	0.0006	0.0006	0.0006	0.0006	0.0006	0.0005	0.0005	0.0005
3.3	0.0005	0.0005	0.0005	0.0004	0.0004	0.0004	0.0004	0.0004	0.0004	0.0003
3.4	0.0003	0.0003	0.0003	0.0003	0.0003	0.0003	0.0003	0.0003	0.0003	0.0002

Table VI The *t* Distribution

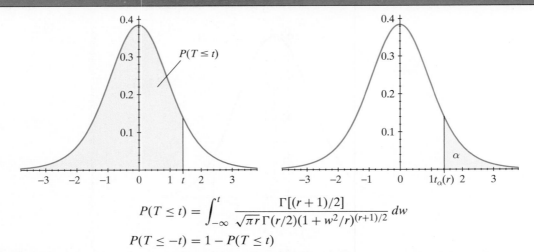

$$P(T \leq t) = \int_{-\infty}^{t} \frac{\Gamma[(r+1)/2]}{\sqrt{\pi r}\, \Gamma(r/2)(1 + w^2/r)^{(r+1)/2}}\, dw$$

$$P(T \leq -t) = 1 - P(T \leq t)$$

			$P(T \leq t)$				
	0.60	0.75	0.90	0.95	0.975	0.99	0.995
r	$t_{0.40}(r)$	$t_{0.25}(r)$	$t_{0.10}(r)$	$t_{0.05}(r)$	$t_{0.025}(r)$	$t_{0.01}(r)$	$t_{0.005}(r)$
1	0.325	1.000	3.078	6.314	12.706	31.821	63.657
2	0.289	0.816	1.886	2.920	4.303	6.965	9.925
3	0.277	0.765	1.638	2.353	3.182	4.541	5.841
4	0.271	0.741	1.533	2.132	2.776	3.747	4.604
5	0.267	0.727	1.476	2.015	2.571	3.365	4.032
6	0.265	0.718	1.440	1.943	2.447	3.143	3.707
7	0.263	0.711	1.415	1.895	2.365	2.998	3.499
8	0.262	0.706	1.397	1.860	2.306	2.896	3.355
9	0.261	0.703	1.383	1.833	2.262	2.821	3.250
10	0.260	0.700	1.372	1.812	2.228	2.764	3.169
11	0.260	0.697	1.363	1.796	2.201	2.718	3.106
12	0.259	0.695	1.356	1.782	2.179	2.681	3.055
13	0.259	0.694	1.350	1.771	2.160	2.650	3.012
14	0.258	0.692	1.345	1.761	2.145	2.624	2.997
15	0.258	0.691	1.341	1.753	2.131	2.602	2.947
16	0.258	0.690	1.337	1.746	2.120	2.583	2.921
17	0.257	0.689	1.333	1.740	2.110	2.567	2.898
18	0.257	0.688	1.330	1.734	2.101	2.552	2.878
19	0.257	0.688	1.328	1.729	2.093	2.539	2.861
20	0.257	0.687	1.325	1.725	2.086	2.528	2.845
21	0.257	0.686	1.323	1.721	2.080	2.518	2.831
22	0.256	0.686	1.321	1.717	2.074	2.508	2.819
23	0.256	0.685	1.319	1.714	2.069	2.500	2.807
24	0.256	0.685	1.318	1.711	2.064	2.492	2.797
25	0.256	0.684	1.316	1.708	2.060	2.485	2.787
26	0.256	0.684	1.315	1.706	2.056	2.479	2.779
27	0.256	0.684	1.314	1.703	2.052	2.473	2.771
28	0.256	0.683	1.313	1.701	2.048	2.467	2.763
29	0.256	0.683	1.311	1.699	2.045	2.462	2.756
30	0.256	0.683	1.310	1.697	2.042	2.457	2.750
∞	0.253	0.674	1.282	1.645	1.960	2.326	2.576

This table is taken from Table III of Fisher and Yates: *Statistical Tables for Biological, Agricultrual, and Medical Research*, published by Longman Group Ltd., London (previously published by Oliver and Boyd, Edinburgh).

Table VII The F Distribution

$$P(F \leq f) = \int_0^f \frac{\Gamma[(r_1 + r_2)/2](r_1/r_2)^{r_1/2} w^{r_1/2-1}}{\Gamma(r_1/2)\Gamma(r_2/2)(1 + r_1 w/r_2)^{(r_1+r_2)/2}} \, dw$$

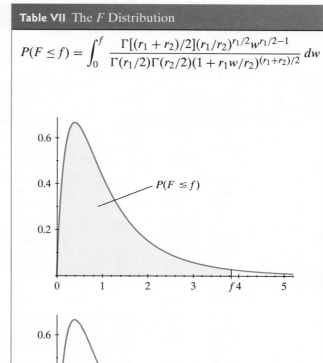

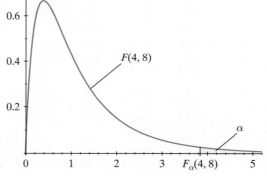

Table VII *continued*

$$P(F \le f) = \int_0^f \frac{\Gamma[(r_1 + r_2)/2](r_1/r_2)^{r_1/2} w^{r_1/2-1}}{\Gamma(r_1/2)\Gamma(r_2/2)(1 + r_1 w/r_2)^{(r_1+r_2)/2}}\, dw$$

α	$P(F \le f)$	Den. d.f. r_2	Numerator Degrees of Freedom, r_1									
			1	2	3	4	5	6	7	8	9	10
0.05	0.95	1	161.4	199.5	215.7	224.6	230.2	234.0	236.8	238.9	240.5	241.9
0.025	0.975		647.79	799.50	864.16	899.58	921.85	937.11	948.22	956.66	963.28	968.63
0.01	0.99		4052	4999.5	5403	5625	5764	5859	5928	5981	6022	6056
0.05	0.95	2	18.51	19.00	19.16	19.25	19.30	19.33	19.35	19.37	19.38	19.40
0.025	0.975		38.51	39.00	39.17	39.25	39.30	39.33	39.36	39.37	39.39	39.40
0.01	0.99		98.50	99.00	99.17	99.25	99.30	99.33	99.36	99.37	99.39	99.40
0.05	0.95	3	10.13	9.55	9.28	9.12	9.01	8.94	8.89	8.85	8.81	8.79
0.025	0.975		17.44	16.04	15.44	15.10	14.88	14.73	14.62	14.54	14.47	14.42
0.01	0.99		34.12	30.82	29.46	28.71	28.24	27.91	27.67	27.49	27.35	27.23
0.05	0.95	4	7.71	6.94	6.59	6.39	6.26	6.16	6.09	6.04	6.00	5.96
0.025	0.975		12.22	10.65	9.98	9.60	9.36	9.20	9.07	8.98	8.90	8.84
0.01	0.99		21.20	18.00	16.69	15.98	15.52	15.21	14.98	14.80	14.66	14.55
0.05	0.95	5	6.61	5.79	5.41	5.19	5.05	4.95	4.88	4.82	4.77	4.74
0.025	0.975		10.01	8.43	7.76	7.39	7.15	6.98	6.85	6.76	6.68	6.62
0.01	0.99		16.26	13.27	12.06	11.39	10.97	10.67	10.46	10.29	10.16	10.05
0.05	0.95	6	5.99	5.14	4.76	4.53	4.39	4.28	4.21	4.15	4.10	4.06
0.025	0.975		8.81	7.26	6.60	6.23	5.99	5.82	5.70	5.60	5.52	5.46
0.01	0.99		13.75	10.92	9.78	9.15	8.75	8.47	8.26	8.10	7.98	7.87
0.05	0.95	7	5.59	4.74	4.35	4.12	3.97	3.87	3.79	3.73	3.68	3.64
0.025	0.975		8.07	6.54	5.89	5.52	5.29	5.12	4.99	4.90	4.82	4.76
0.01	0.99		12.25	9.55	8.45	7.85	7.46	7.19	6.99	6.84	6.72	6.62
0.05	0.95	8	5.32	4.46	4.07	3.84	3.69	3.58	3.50	3.44	3.39	3.35
0.025	0.975		7.57	6.06	5.42	5.05	4.82	4.65	4.53	4.43	4.36	4.30
0.01	0.99		11.26	8.65	7.59	7.01	6.63	6.37	6.18	6.03	5.91	5.81
0.05	0.95	9	5.12	4.26	3.86	3.63	3.48	3.37	3.29	3.23	3.18	3.14
0.025	0.975		7.21	5.71	5.08	4.72	4.48	4.32	4.20	4.10	4.03	3.96
0.01	0.99		10.56	8.02	6.99	6.42	6.06	5.80	5.61	5.47	5.35	5.26
0.05	0.95	10	4.96	4.10	3.71	3.48	3.33	3.22	3.14	3.07	3.02	2.98
0.025	0.975		6.94	5.46	4.83	4.47	4.24	4.07	3.95	3.85	3.78	3.72
0.01	0.99		10.04	7.56	6.55	5.99	5.64	5.39	5.20	5.06	4.94	4.85

Table VII *continued*

$$P(F \leq f) = \int_0^f \frac{\Gamma[(r_1 + r_2)/2](r_1/r_2)^{r_1/2} w^{r_1/2 - 1}}{\Gamma(r_1/2)\Gamma(r_2/2)(1 + r_1 w/r_2)^{(r_1 + r_2)/2}} \, dw$$

α	$P(F \leq f)$	Den. d.f. r_2	Numerator Degrees of Freedom, r_1									
			1	2	3	4	5	6	7	8	9	10
0.05	0.95	12	4.75	3.89	3.49	3.26	3.11	3.00	2.91	2.85	2.80	2.75
0.025	0.975		6.55	5.10	4.47	4.12	3.89	3.73	3.61	3.51	3.44	3.37
0.01	0.99		9.33	6.93	5.95	5.41	5.06	4.82	4.64	4.50	4.39	4.30
0.05	0.95	15	4.54	3.68	3.29	3.06	2.90	2.79	2.71	2.64	2.59	2.54
0.025	0.975		6.20	4.77	4.15	3.80	3.58	3.41	3.29	3.20	3.12	3.06
0.01	0.99		8.68	6.36	5.42	4.89	4.56	4.32	4.14	4.00	3.89	3.80
0.05	0.95	20	4.35	3.49	3.10	2.87	2.71	2.60	2.51	2.45	2.39	2.35
0.025	0.975		5.87	4.46	3.86	3.51	3.29	3.13	3.01	2.91	2.84	2.77
0.01	0.99		8.10	5.85	4.94	4.43	4.10	3.87	3.70	3.56	3.46	3.37
0.05	0.95	24	4.26	3.40	3.01	2.78	2.62	2.51	2.42	2.36	2.30	2.25
0.025	0.975		5.72	4.32	3.72	3.38	3.15	2.99	2.87	2.78	2.70	2.64
0.01	0.99		7.82	5.61	4.72	4.22	3.90	3.67	3.50	3.36	3.26	3.17
0.05	0.95	30	4.17	3.32	2.92	2.69	2.53	2.42	2.33	2.27	2.21	2.16
0.025	0.975		5.57	4.18	3.59	3.25	3.03	2.87	2.75	2.65	2.57	2.51
0.01	0.99		7.56	5.39	4.51	4.02	3.70	3.47	3.30	3.17	3.07	2.98
0.05	0.95	40	4.08	3.23	2.84	2.61	2.45	2.34	2.25	2.18	2.12	2.08
0.025	0.975		5.42	4.05	3.46	3.13	2.90	2.74	2.62	2.53	2.45	2.39
0.01	0.99		7.31	5.18	4.31	3.83	3.51	3.29	3.12	2.99	2.89	2.80
0.05	0.95	60	4.00	3.15	2.76	2.53	2.37	2.25	2.17	2.10	2.04	1.99
0.025	0.975		5.29	3.93	3.34	3.01	2.79	2.63	2.51	2.41	2.33	2.27
0.01	0.99		7.08	4.98	4.13	3.65	3.34	3.12	2.95	2.82	2.72	2.63
0.05	0.95	120	3.92	3.07	2.68	2.45	2.29	2.17	2.09	2.02	1.96	1.91
0.025	0.975		5.15	3.80	3.23	2.89	2.67	2.52	2.39	2.30	2.22	2.16
0.01	0.99		6.85	4.79	3.95	3.48	3.17	2.96	2.79	2.66	2.56	2.47
0.05	0.95	∞	3.84	3.00	2.60	2.37	2.21	2.10	2.01	1.94	1.88	1.83
0.025	0.975		5.02	3.69	3.12	2.79	2.57	2.41	2.29	2.19	2.11	2.05
0.01	0.99		6.63	4.61	3.78	3.32	3.02	2.80	2.64	2.51	2.41	2.32

Table VII *continued*

$$P(F \leq f) = \int_0^f \frac{\Gamma[(r_1 + r_2)/2](r_1/r_2)^{r_1/2} w^{r_1/2-1}}{\Gamma(r_1/2)\Gamma(r_2/2)(1 + r_1 w/r_2)^{(r_1+r_2)/2}} \, dw$$

		Den. d.f.	Numerator Degrees of Freedom, r_1								
α	$P(F \leq f)$	r_2	12	15	20	24	30	40	60	120	∞
0.05	0.95	1	243.9	245.9	248.0	249.1	250.1	251.1	252.2	253.3	254.3
0.025	0.975		976.71	984.87	993.10	997.25	1001.4	1005.6	1009.8	1014.0	1018.3
0.01	0.99		6106	6157	6209	6235	6261	6287	6313	6339	6366
0.05	0.95	2	19.41	19.43	19.45	19.45	19.46	19.47	19.48	19.49	19.50
0.025	0.975		39.42	39.43	39.45	39.46	39.47	39.47	39.48	39.49	39.50
0.01	0.99		99.42	99.43	99.45	99.46	99.47	99.47	99.48	99.49	99.50
0.05	0.95	3	8.74	8.70	8.66	8.64	8.62	8.59	8.57	8.55	8.53
0.025	0.975		14.34	14.25	14.17	14.12	14.08	14.04	13.99	13.95	13.90
0.01	0.99		27.05	26.87	26.69	26.60	26.50	26.41	26.32	26.22	26.13
0.05	0.95	4	5.91	5.86	5.80	5.77	5.75	5.72	5.69	5.66	5.63
0.025	0.975		8.75	8.66	8.56	8.51	8.46	8.41	8.36	8.31	8.26
0.01	0.99		14.37	14.20	14.02	13.93	13.84	13.75	13.65	13.56	13.46
0.05	0.95	5	4.68	4.62	4.56	4.53	4.50	4.46	4.43	4.40	4.36
0.025	0.975		6.52	6.43	6.33	6.28	6.23	6.18	6.12	6.07	6.02
0.01	0.99		9.89	9.72	9.55	9.47	9.38	9.29	9.20	9.11	9.02
0.05	0.95	6	4.00	3.94	3.87	3.84	3.81	3.77	3.74	3.70	3.67
0.025	0.975		5.37	5.27	5.17	5.12	5.07	5.01	4.96	4.90	4.85
0.01	0.99		7.72	7.56	7.40	7.31	7.23	7.14	7.06	6.97	6.88
0.05	0.95	7	3.57	3.51	3.41	3.41	3.38	3.34	3.30	3.27	3.23
0.025	0.975		4.67	4.57	4.47	4.42	4.36	4.31	4.25	4.20	4.14
0.01	0.99		6.47	6.31	6.16	6.07	5.99	5.91	5.82	5.74	5.65
0.05	0.95	8	3.28	3.22	3.15	3.12	3.08	3.04	3.01	2.97	2.93
0.025	0.975		4.20	4.10	4.00	3.95	3.89	3.84	3.78	3.73	3.67
0.01	0.99		5.67	5.52	5.36	5.28	5.20	5.12	5.03	4.95	4.86
0.05	0.95	9	3.07	3.01	2.94	2.90	2.86	2.83	2.79	2.75	2.71
0.025	0.975		3.87	3.77	3.67	3.61	3.56	3.51	3.45	3.39	3.33
0.01	0.99		5.11	4.96	4.81	4.73	4.65	4.57	4.48	4.40	4.31

Table VII *continued*

$$P(F \leq f) = \int_0^f \frac{\Gamma[(r_1 + r_2)/2](r_1/r_2)^{r_1/2} w^{r_1/2-1}}{\Gamma(r_1/2)\Gamma(r_2/2)(1 + r_1 w/r_2)^{(r_1+r_2)/2}}\, dw$$

α	$P(F \leq f)$	Den. d.f. r_2	12	15	20	24	30	40	60	120	∞
						Numerator Degrees of Freedom, r_1					
0.05	0.95	10	2.91	2.85	2.77	2.74	2.70	2.66	2.62	2.58	2.54
0.025	0.975		3.62	3.52	3.42	3.37	3.31	3.26	3.20	3.14	3.08
0.01	0.99		4.71	4.56	4.41	4.33	4.25	4.17	4.08	4.00	3.91
0.05	0.95	12	2.69	2.62	2.54	2.51	2.47	2.43	2.38	2.34	2.30
0.025	0.975		3.28	3.18	3.07	3.02	2.96	2.91	2.85	2.79	2.72
0.01	0.99		4.16	4.01	3.86	3.78	3.70	3.62	3.54	3.45	3.36
0.05	0.95	15	2.48	2.40	2.33	2.29	2.25	2.20	2.16	2.11	2.07
0.025	0.975		2.96	2.86	2.76	2.70	2.64	2.59	2.52	2.46	2.40
0.01	0.99		3.67	3.52	3.37	3.29	3.21	3.13	3.05	2.96	2.87
0.05	0.95	20	2.28	2.20	2.12	2.08	2.04	1.99	1.95	1.90	1.84
0.025	0.975		2.68	2.57	2.46	2.41	2.35	2.29	2.22	2.16	2.09
0.01	0.99		3.23	3.09	2.94	2.86	2.78	2.69	2.61	2.52	2.42
0.05	0.95	24	2.18	2.11	2.03	1.98	1.94	1.89	1.84	1.79	1.73
0.025	0.975		2.54	2.44	2.33	2.27	2.21	2.15	2.08	2.01	1.94
0.01	0.99		3.03	2.89	2.74	2.66	2.58	2.49	2.40	2.31	2.21
0.05	0.95	30	2.09	2.01	1.93	1.89	1.84	1.79	1.74	1.68	1.62
0.025	0.975		2.41	2.31	2.20	2.14	2.07	2.01	1.94	1.87	1.79
0.01	0.99		2.84	2.70	2.55	2.47	2.39	2.30	2.21	2.11	2.01
0.05	0.95	40	2.00	1.92	1.84	1.79	1.74	1.69	1.64	1.58	1.51
0.025	0.975		2.29	2.18	2.07	2.01	1.94	1.88	1.80	1.72	1.64
0.01	0.99		2.66	2.52	2.37	2.29	2.20	2.11	2.02	1.92	1.80
0.05	0.95	60	1.92	1.84	1.75	1.70	1.65	1.59	1.53	1.47	1.39
0.025	0.975		2.17	2.06	1.94	1.88	1.82	1.74	1.67	1.58	1.48
0.01	0.99		2.50	2.35	2.20	2.12	2.03	1.94	1.84	1.73	1.60
0.05	0.95	120	1.83	1.75	1.66	1.61	1.55	1.50	1.43	1.35	1.25
0.025	0.975		2.05	1.95	1.82	1.76	1.69	1.61	1.53	1.43	1.31
0.01	0.99		2.34	2.19	2.03	1.95	1.86	1.76	1.66	1.53	1.38
0.05	0.95	∞	1.75	1.67	1.57	1.52	1.46	1.39	1.32	1.22	1.00
0.025	0.975		1.94	1.83	1.71	1.64	1.57	1.48	1.39	1.27	1.00
0.01	0.99		2.18	2.04	1.88	1.79	1.70	1.59	1.47	1.32	1.00

Table VIII Random Numbers on the Interval $(0, 1)$

3407	1440	6960	8675	5649	5793	1514
5044	9859	4658	7779	7986	0520	6697
0045	4999	4930	7408	7551	3124	0527
7536	1448	7843	4801	3147	3071	4749
7653	4231	1233	4409	0609	6448	2900
6157	1144	4779	0951	3757	9562	2354
6593	8668	4871	0946	3155	3941	9662
3187	7434	0315	4418	1569	1101	0043
4780	1071	6814	2733	7968	8541	1003
9414	6170	2581	1398	2429	4763	9192
1948	2360	7244	9682	5418	0596	4971
1843	0914	9705	7861	6861	7865	7293
4944	8903	0460	0188	0530	7790	9118
3882	3195	8287	3298	9532	9066	8225
6596	9009	2055	4081	4842	7852	5915
4793	2503	2906	6807	2028	1075	7175
2112	0232	5334	1443	7306	6418	9639
0743	1083	8071	9779	5973	1141	4393
8856	5352	3384	8891	9189	1680	3192
8027	4975	2346	5786	0693	5615	2047
3134	1688	4071	3766	0570	2142	3492
0633	9002	1305	2256	5956	9256	8979
8771	6069	1598	4275	6017	5946	8189
2672	1304	2186	8279	2430	4896	3698
3136	1916	8886	8617	9312	5070	2720
6490	7491	6562	5355	3794	3555	7510
8628	0501	4618	3364	6709	1289	0543
9270	0504	5018	7013	4423	2147	4089
5723	3807	4997	4699	2231	3193	8130
6228	8874	7271	2621	5746	6333	0345
7645	3379	8376	3030	0351	8290	3640
6842	5836	6203	6171	2698	4086	5469
6126	7792	9337	7773	7286	4236	1788
4956	0215	3468	8038	6144	9753	3131
1327	4736	6229	8965	7215	6458	3937
9188	1516	5279	5433	2254	5768	8718
0271	9627	9442	9217	4656	7603	8826
2127	1847	1331	5122	8332	8195	3322
2102	9201	2911	7318	7670	6079	2676
1706	6011	5280	5552	5180	4630	4747
7501	7635	2301	0889	6955	8113	4364
5705	1900	7144	8707	9065	8163	9846
3234	2599	3295	9160	8441	0085	9317
5641	4935	7971	8917	1978	5649	5799
2127	1868	3664	9376	1984	6315	8396

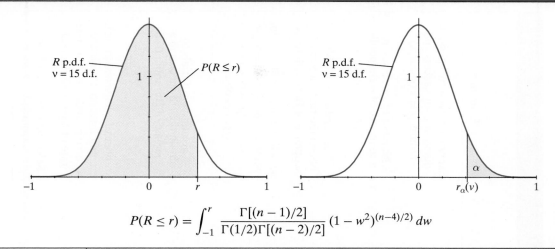

$$P(R \le r) = \int_{-1}^{r} \frac{\Gamma[(n-1)/2]}{\Gamma(1/2)\Gamma[(n-2)/2]} (1 - w^2)^{(n-4)/2} \, dw$$

$\nu = n - 2$ degrees of freedom	$P(R \le r)$			
	0.95	0.975	0.99	0.995
	$r_{0.05}(\nu)$	$r_{0.025}(\nu)$	$r_{0.01}(\nu)$	$r_{0.005}(\nu)$
1	0.9877	0.9969	0.9995	0.9999
2	0.9000	0.9500	0.9800	0.9900
3	0.8053	0.8783	0.9343	0.9587
4	0.7292	0.8113	0.8822	0.9172
5	0.6694	0.7544	0.8329	0.8745
6	0.6215	0.7067	0.7887	0.8343
7	0.5822	0.6664	0.7497	0.7977
8	0.5493	0.6319	0.7154	0.7646
9	0.5214	0.6020	0.6850	0.7348
10	0.4972	0.5759	0.6581	0.7079
11	0.4761	0.5529	0.6338	0.6835
12	0.4575	0.5323	0.6120	0.6613
13	0.4408	0.5139	0.5922	0.6411
14	0.4258	0.4973	0.5742	0.6226
15	0.4123	0.4821	0.5577	0.6054
16	0.4000	0.4683	0.5425	0.5897
17	0.3887	0.4555	0.5285	0.5750
18	0.3783	0.4437	0.5154	0.5614
19	0.3687	0.4328	0.5033	0.5487
20	0.3597	0.4226	0.4920	0.5367
25	0.3232	0.3808	0.4450	0.4869
30	0.2959	0.3494	0.4092	0.4487
35	0.2746	0.3246	0.3809	0.4182
40	0.2572	0.3044	0.3578	0.3931
45	0.2428	0.2875	0.3383	0.3721
50	0.2306	0.2732	0.3218	0.3541
60	0.2108	0.2500	0.2948	0.3248
70	0.1954	0.2318	0.2736	0.3017
80	0.1829	0.2172	0.2565	0.2829
90	0.1725	0.2049	0.2422	0.2673
100	0.1638	0.1946	0.2300	0.2540

Table X Discrete Distributions

Probability Distribution and Parameter Values	Probability Mass Function	Moment-Generating Function	Mean $E(X)$	Variance $\text{Var}(X)$	Examples
Bernoulli $0 < p < 1$ $q = 1 - p$	$p^x q^{1-x}$, $x = 0,1$	$q + pe^t$, $-\infty < t < \infty$	p	pq	Experiment with two possible outcomes, say success and failure, $p = P(\text{success})$
Binomial $n = 1,2,3,\ldots$ $0 < p < 1$	$\binom{n}{x} p^x q^{n-x}$, $x = 0,1,\ldots,n$	$(q + pe^t)^n$, $-\infty < t < \infty$	np	npq	Number of successes in a sequence of n Bernoulli trials, $p = P(\text{success})$
Geometric $0 < p < 1$ $q = 1 - p$	$q^{x-1} p$, $x = 1,2,\ldots$	$\dfrac{pe^t}{1 - qe^t}$ $t < -\ln(1-p)$	$\dfrac{1}{p}$	$\dfrac{q}{p^2}$	The number of trials to obtain the first success in a sequence of Bernoulli trials
Hypergeometric $x \le n, x \le N_1$ $n - x \le N_2$ $N = N_1 + N_2$ $N_1 > 0,\ N_2 > 0$	$\dfrac{\binom{N_1}{x}\binom{N_2}{n-x}}{\binom{N}{n}}$		$n\left(\dfrac{N_1}{N}\right)$	$n\left(\dfrac{N_1}{N}\right)\left(\dfrac{N_2}{N}\right)\left(\dfrac{N-n}{N-1}\right)$	Selecting n objects at random without replacement from a set composed of two types of objects
Negative Binomial $r = 1,2,3,\ldots$ $0 < p < 1$	$\binom{x-1}{r-1} p^r q^{x-r}$, $x = r, r+1,\ldots$	$\dfrac{(pe^t)^r}{(1 - qe^t)^r}$, $t < -\ln(1-p)$	$\dfrac{r}{p}$	$\dfrac{rq}{p^2}$	The number of trials to obtain the rth success in a sequence of Bernoulli trials
Poisson $\lambda > 0$	$\dfrac{\lambda^x e^{-\lambda}}{x!}$, $x = 0,1,\ldots$	$e^{\lambda(e^t - 1)}$ $-\infty < t < \infty$	λ	λ	Number of events occurring in a unit interval, events are occurring randomly at a mean rate of λ per unit interval
Uniform $m > 0$	$\dfrac{1}{m}$, $x = 1,2,\ldots,m$		$\dfrac{m+1}{2}$	$\dfrac{m^2 - 1}{12}$	Select an integer randomly from $1,2,\ldots,m$

Table XI Continuous Distributions

Probability Distribution and Parameter Values	Probability Density Function	Moment-Generating Function	Mean $E(X)$	Variance $\mathrm{Var}(X)$	Examples
Beta $\alpha > 0$ $\beta > 0$	$\dfrac{\Gamma(\alpha+\beta)}{\Gamma(\alpha)\Gamma(\beta)}x^{\alpha-1}(1-x)^{\beta-1},$ $0 < x < 1$		$\dfrac{\alpha}{\alpha+\beta}$	$\dfrac{\alpha\beta}{(\alpha+\beta+1)(\alpha+\beta)^2}$	$X = X_1/(X_1 + X_2)$, where X_1 and X_2 have independent gamma distributions with same θ
Chi-square $r = 1, 2, \ldots$	$\dfrac{x^{r/2-1}e^{-x/2}}{\Gamma(r/2)2^{r/2}},$ $0 < x < \infty$	$\dfrac{1}{(1-2t)^{r/2}},\ t < \dfrac{1}{2}$	r	$2r$	Gamma distribution, $\theta = 2$, $\alpha = r/2$; sum of squares of r independent $N(0,1)$ random variables
Exponential $\theta > 0$	$\dfrac{1}{\theta}e^{-x/\theta},\ 0 \le x < \infty$	$\dfrac{1}{1-\theta t},\ t < \dfrac{1}{\theta}$	θ	θ^2	Waiting time to first arrival when observing a Poisson process with a mean rate of arrivals equal to $\lambda = 1/\theta$
Gamma $\alpha > 0$ $\theta > 0$	$\dfrac{x^{\alpha-1}e^{-x/\theta}}{\Gamma(\alpha)\theta^\alpha},$ $0 < x < \infty$	$\dfrac{1}{(1-\theta t)^\alpha},\ t < \dfrac{1}{\theta}$	$\alpha\theta$	$\alpha\theta^2$	Waiting time to αth arrival when observing a Poisson process with a mean rate of arrivals equal to $\lambda = 1/\theta$
Normal $-\infty < \mu < \infty$ $\sigma > 0$	$\dfrac{e^{-(x-\mu)^2/2\sigma^2}}{\sigma\sqrt{2\pi}},$ $-\infty < x < \infty$	$e^{\mu t + \sigma^2 t^2/2}$ $-\infty < t < \infty$	μ	σ^2	Errors in measurements; heights of children; breaking strengths
Uniform $-\infty < a < b < \infty$	$\dfrac{1}{b-a},\ a \le x \le b$	$\dfrac{e^{tb}-e^{ta}}{t(b-a)},\ t \ne 0$ $1,\qquad t = 0$	$\dfrac{a+b}{2}$	$\dfrac{(b-a)^2}{12}$	Select a point at random from the interval $[a, b]$

Table XII Tests and Confidence Intervals

Distribution	θ: The parameter of interest	W: The variable used to test $H_0: \theta = \theta_0$	Two-sided $1 - \alpha$ Confidence Interval for θ	Comments
$N(\mu, \sigma^2)$ or n large σ^2 known	μ	$\dfrac{\bar{X} - \theta_0}{\sigma/\sqrt{n}}$	$\bar{x} \pm z_{\alpha/2} \dfrac{\sigma}{\sqrt{n}}$	W is $N(0,1)$; $P(W \geq z_{\alpha/2}) = \alpha/2$
$N(\mu, \sigma^2)$ σ^2 unknown	μ	$\dfrac{\bar{X} - \theta_0}{S/\sqrt{n}}$	$\bar{x} \pm t_{\alpha/2}(n-1) \dfrac{s}{\sqrt{n}}$	W has a t distribution with $n-1$ degrees of freedom; $P[W \geq t_{\alpha/2}(n-1)] = \alpha/2$
Any distribution with known variance, σ^2	μ	$\dfrac{\bar{X} - \theta_0}{\sigma/\sqrt{n}}$	$\bar{x} \pm z_{\alpha/2} \dfrac{\sigma}{\sqrt{n}}$	W has an approximate $N(0,1)$ distribution for n sufficiently large
$N(\mu_X, \sigma_X^2)$ $N(\mu_Y, \sigma_Y^2)$ σ_X^2, σ_Y^2 known	$\mu_X - \mu_Y$	$\dfrac{\bar{X} - \bar{Y} - \theta_0}{\sqrt{\dfrac{\sigma_X^2}{n} + \dfrac{\sigma_Y^2}{m}}}$	$\bar{x} - \bar{y} \pm z_{\alpha/2} \sqrt{\dfrac{\sigma_X^2}{n} + \dfrac{\sigma_Y^2}{m}}$	W is $N(0,1)$
$N(\mu_X, \sigma_X^2)$ $N(\mu_Y, \sigma_Y^2)$ σ_X^2, σ_Y^2 unknown	$\mu_X - \mu_Y$	$\dfrac{\bar{X} - \bar{Y} - \theta_0}{\sqrt{\dfrac{S_X^2}{n} + \dfrac{S_Y^2}{m}}}$	$\bar{x} - \bar{y} \pm z_{\alpha/2} \sqrt{\dfrac{s_x^2}{n} + \dfrac{s_y^2}{m}}$	W is approximately $N(0,1)$ if sample sizes are large
$N(\mu_X, \sigma_X^2)$ $N(\mu_Y, \sigma_Y^2)$ $\sigma_X^2 = \sigma_Y^2$, unknown	$\mu_X - \mu_Y$	$\dfrac{\bar{X} - \bar{Y} - \theta_0}{\sqrt{\dfrac{(n-1)S_X^2 + (m-1)S_Y^2}{n+m-2}\left(\dfrac{1}{n} + \dfrac{1}{m}\right)}}$	$\bar{x} - \bar{y} \pm t_{\alpha/2}(n+m-2) s_p \sqrt{\dfrac{1}{n} + \dfrac{1}{m}}$ $s_p = \sqrt{\dfrac{(n-1)s_x^2 + (m-1)s_y^2}{n+m-2}}$	W has a t distribution with $r = n+m-2$ degrees of freedom
$D = X - Y$ is $N(\mu_X - \mu_Y, \sigma_D^2)$ X and Y dependent	$\mu_X - \mu_Y$	$\dfrac{\bar{D} - \theta_0}{S_D/\sqrt{n}}$	$\bar{d} \pm t_{\alpha/2}(n-1) \dfrac{s_d}{\sqrt{n}}$	W has a t distribution with $n-1$ degrees of freedom

Table XII *continued*

Distribution	θ: The parameter of interest	W: The variable used to test $H_0: \theta = \theta_0$	Two-sided $1 - \alpha$ Confidence Interval for θ	Comments
$N(\mu,\sigma^2)$ μ unknown	σ^2	$\dfrac{(n-1)S^2}{\theta_0}$	$\dfrac{(n-1)s^2}{\chi^2_{\alpha/2}(n-1)}, \dfrac{(n-1)s^2}{\chi^2_{1-\alpha/2}(n-1)}$	W is $\chi^2(n-1)$, $P[W \le \chi^2_{1-\alpha/2}(n-1)] = \alpha/2$, $P[W \ge \chi^2_{\alpha/2}(n-1)] = \alpha/2$
$N(\mu,\sigma^2)$ μ unknown	σ	$\dfrac{(n-1)S^2}{\theta_0^2}$	$\sqrt{\dfrac{(n-1)s^2}{\chi^2_{\alpha/2}(n-1)}}, \sqrt{\dfrac{(n-1)s^2}{\chi^2_{1-\alpha/2}(n-1)}}$	W is $\chi^2(n-1)$. $P[W \le \chi^2_{1-\alpha/2}(n-1)] = \alpha/2$, $P[W \ge \chi^2_{\alpha/2}(n-1)] = \alpha/2$
$N(\mu_X,\sigma_X^2)$ $N(\mu_Y,\sigma_Y^2)$ μ_X, μ_Y unknown	$\dfrac{\sigma_X^2}{\sigma_Y^2}$	$\dfrac{S_Y^2}{S_X^2}\theta_0$	$\dfrac{s_x^2/s_y^2}{F_{\alpha/2}(n-1,m-1)}, F_{\alpha/2}(m-1,n-1)\dfrac{s_x^2}{s_y^2}$	W has an F distribution with $m-1$ and $n-1$ degrees of freedom
$b(n,p)$	p	$\dfrac{\dfrac{Y}{n} - \theta_0}{\sqrt{\left(\dfrac{Y}{n}\right)\left(1-\dfrac{Y}{n}\right)/n}}$	$\dfrac{y}{n} \pm z_{\alpha/2}\sqrt{\left(\dfrac{y}{n}\right)\left(1-\dfrac{y}{n}\right)/n}$	W is approximately $N(0,1)$ for n sufficiently large
$b(n,p)$	p		$\tilde{p} \pm z_{\alpha/2}\sqrt{\tilde{p}(1-\tilde{p})/(n+4)}$ $\tilde{p} = (y+2)/(n+4)$	W is approximately $N(0,1)$ for n sufficiently large
$b(n_1,p_1)$ $b(n_2,p_2)$	$p_1 - p_2$	$\dfrac{\dfrac{Y_1}{n_1} - \dfrac{Y_2}{n_2} - \theta_0}{\sqrt{\left(\dfrac{Y_1+Y_2}{n_1+n_2}\right)\left(1-\dfrac{Y_1+Y_2}{n_1+n_2}\right)\left(\dfrac{1}{n_1}+\dfrac{1}{n_2}\right)}}$	$\dfrac{y_1}{n_1} - \dfrac{y_2}{n_2} \pm$ $z_{\alpha/2}\sqrt{\dfrac{y_1}{n_1}\left(1-\dfrac{y_1}{n_1}\right)/n_1 + \dfrac{y_2}{n_2}\left(1-\dfrac{y_2}{n_2}\right)/n_2}$	W is approximately $N(0,1)$ when n_1 and n_2 are sufficiently large

Answers to Odd-Numbered Exercises

Chapter 1

1.1-1 0.70.

1.1-3 (a) 12/52; (b) 2/52; (c) 16/52; (d) 1; (e) 0.

1.1-5 (a) 1/6; (b) 5/6; (c) 1.

1.1-7 0.63.

1.1-9 (a) $3(1/3) - 3(1/3)^2 + (1/3)^3$;

(b) $P(A_1 \cup A_2 \cup A_3) = 1 - [1 - 3(1/3) + 3(1/3)^2 - (1/3)^3] = 1 - (1 - 1/3)^3$.

1.1-11 (a) $S = \{00, 0, 1, 2, 3, \ldots, 36\}$;

(b) $P(A) = 2/38$;

(c) $P(B) = 4/38$;

(d) $P(D) = 18/38$.

1.1-13 2/3.

1.2-1 4096.

1.2-3 (a) 6,760,000; (b) 17,576,000.

1.2-5 (a) 24; (b) 256.

1.2-7 (a) 0.0024; (b) 0.0012; (c) 0.0006; (d) 0.0004.

1.2-9 (a) 2; (b) 8; (c) 20; (d) 40.

1.2-11 (a) 362,880; (b) 84; (c) 512.

1.2-13 (a) 0.00539; (b) 0.00882; (c) 0.00539; (d) Yes.

1.2-17 (a) 0.00024; (b) 0.00144; (c) 0.02113; (d) 0.04754;

(e) 0.42257.

1.3-1 (a) 5000/1,000,000; (b) 78,515/1,000,000;

(c) 73,630/995,000; (d) 4,885/78,515.

1.3-3 (a) 5/12; (b) 9/12; (c) 5/7; (d) 3/5; (e) Left.

1.3-5 (a) $S = \{(R, R), (R, W), (W, R), (W, W)\}$; (b) 1/3.

1.3-7 4/15.

1.3-9 (f) $1 - 1/e$.

1.3-11 (a) 365^r; (b) $_{365}P_r$; (c) $1 - {}_{365}P_r/365^r$; (d) 23.

1.3-13 (b) 8/36; (c) 5/11; (e) $8/36 + 2[(5/36)(5/11) + (4/36)(4/10) + (3/36)(3/9)] = 0.49293$.

1.3-15 11.

1.4-1 (a) 0.14; (b) 0.76; (c) 0.86.

1.4-3 (a) 1/6; (b) 1/12; (c) 1/4; (d) 1/4; (e) 1/2.

1.4-5 Yes; $0.9 = 0.8 + 0.5 - (0.8)(0.5)$.

1.4-7 (a) 0.9; (b) 0.39.

1.4-9 (a) 0.36; (b) 0.49; (c) 0.01.

1.4-11 (a) No, unless $P(A) = 0$ or $P(B) = 0$;

(b) Only if $P(A) = 0$ or $P(B) = 1$.

1.4-13 $(2/3)^3(1/3)^2$; $(2/3)^3(1/3)^2$.

1.4-15 (a) $1/16, 1/8, 5/32, 5/32$;

(b) $14/323, 35/323, 105/646, 60/323$;

(c) Neither model is very good.

1.4-17 (a) $1 - (11/12)^{12}$; (b) $1 - (11/12)^{11}$.

1.4-19 (b) $1 - 1/e$.

1.5-1 (a) 21/32; (b) 16/21.

1.5-3 15.1%.

1.5-5 $60/95 = 0.632$.

1.5-7 0.8182.

1.5-9 (a) $495/30,480 = 0.016$; (b) $29,985/30,480 = 0.984$.

1.5-11 1/4.

1.5-13 0.4.

Chapter 2

2.1-3 (a) 10; (b) 1/55; (c) 3; (d) 1/30; (e) $n(n+1)/2$; (f) 1.

2.1-5 (b)

x	Frequency	Relative Frequency	$f(x)$
1	38	0.38	0.40
2	27	0.27	0.30
3	21	0.21	0.20
4	14	0.14	0.10

2.1-7 (a) $f(x) = \dfrac{13 - 2x}{36}$, $x = 1, 2, 3, 4, 5, 6$;

(b) $g(0) = \dfrac{6}{36}$, $g(y) = \dfrac{12 - 2y}{36}$, $y = 1, 2, 3, 4, 5$.

2.1-11 0.5132.

2.1-13 (a) 0.073; (b) 0.33; (c) 0.8462.

2.1-15 (c)

x	Frequency	Relative Frequency	$f(x)$
0	13	0.325	0.2532
1	16	0.400	0.4220
2	9	0.225	0.2509
3	2	0.050	0.0660
4	0	0.000	0.0076
5	0	0.000	0.0003

2.1-17 78.

2.2-1 (a) 3; (b) 7; (c) 4/3; (d) 7/3; (e) $(2n+1)/3$; (f) $E(X) = +\infty$, so does not exist.

2.2-3 \$360.

2.2-5 (a) $h(z) = (4 - z^{1/3})/6$, $z = 1, 8, 27$; (b) 23/3 of a dollar; (c) 7/3 of a dollar.

2.2-7 \$31,000.

2.2-9 (a) $-\$1/19$; (b) $-\$1/37$.

2.2-11 $-\$0.01414$.

2.3-1 (a) 15; 50; (b) 5; 0; (c) 5/3; 5/9.

2.3-3 (a) 16; (b) 6; (c) 16.

2.3-5 $\mu = 7$.

2.3-7 $m = 7$.

2.3-9 \$1809.80.

2.3-11 $\mu = 2, \sigma^2 = 4/5$,

$$f(x) = \begin{cases} 2/5, & x = 1, \\ 1/5, & x = 2, \\ 2/5, & x = 3. \end{cases}$$

2.3-13 $(4/5)^3(1/5)$.

2.3-15 (a) 0.4604; (b) 0.5580; (c) 0.0184.

2.3-17 (a) $f(x) = (x-1)/2^x$, $x = 2, 3, \ldots$; (c) $\mu = 4, \sigma^2 = 4$; (d) (i) 1/2, (ii) 5/16, (iii) 1/4.

2.3-19 (a) $\mu = 1, \sigma^2 = 1$; (b) 19/30.

2.4-1 $f(x) = (7/18)^x(11/18)^{1-x}$, $x = 0, 1$; $\mu = 7/18$; $\sigma^2 = 77/324$.

2.4-3 (a) $(1/5)^2(4/5)^4 = 0.0164$;
(b) $\dfrac{6!}{2!4!}(1/5)^2(4/5)^4 = 0.2458$.

2.4-5 (a) 0.4207; (b) 0.5793; (c) 0.1633; (d) $\mu = 5, \sigma^2 = 4$, $\sigma = 2$.

2.4-7 (a) $b(2000, \pi/4)$; (b) 1570.796, 337.096, 18.360; (c) π; (f) $V_n = \pi^{n/2}/\Gamma(n/2 + 1)$ is the volume of a ball of radius 1 in n-space.

2.4-9 (a) $b(20, 0.80)$; (b) $\mu = 16, \sigma^2 = 3.2, \sigma = 1.789$; (c) (i) 0.1746, (ii) 0.6296, (iii) 0.3704.

2.4-11 0.1268.

2.4-13 (a) 0.6513; (b) 0.7941.

2.4-15 0.178.

2.4-17 (a) 0.0778; (b) 0.3456; (c) 0.9898.

2.4-19 (a) $b(1, 2/3)$; (b) $b(12, 0.75)$.

2.5-1 (a) $0.9^{12} = 0.2824$; (b) 0.0236.

2.5-3 (a) $\mu = 10/0.60, \sigma^2 = 4/0.36, \sigma = 2/0.60$; (b) 0.1240.

2.5-7 $M(t) = e^{5t}$, $-\infty < t < \infty$, $f(5) = 1$.

2.5-9 25/3.

2.6-1 (a) 0.448; (b) 0.990; (c) 0.029.

2.6-3 0.540.

2.6-5 0.558.

2.6-7 0.947.

2.6-9 (a) 2.681; (b) $n = 6$.

2.6-11 (a) 0.564 using binomial, 0.560 using Poisson approximation.
(b) \$598.56 using binomial, \$613.90 using Poisson approximation.

2.6-13 21/16.

Chapter 3

3.1-3 (a) $f(x) = 1/10$, $0 < x < 10$; (b) 0.2; (c) 0.6; (d) $\mu = 5$; (e) $\sigma^2 = 25/3$.

3.1-5 (a) $G(w) = (w - a)/(b - a)$, $a \le w \le b$; (b) $U(a, b)$.

3.1-7 (a) (i) 3; (ii) $F(x) = x^4$, $0 \le x \le 1$; (iv) $\mu = 4/5, \sigma^2 = 2/75$;
(b) (i) 3/16; (ii) $F(x) = (1/8)x^{3/2}$, $0 \le x \le 4$; (iv) $\mu = 12/5, \sigma^2 = 192/175$;
(c) (i) 1/4; (ii) $F(x) = x^{1/4}$, $0 \le x \le 1$; (iv) $\mu = 1/5, \sigma^2 = 16/225$.

3.1-9 (b)
$$F(x) = \begin{cases} 0, & x < 0, \\ x(2 - x), & 0 \le x < 1, \\ 1, & 1 \le x. \end{cases}$$
(c) (i) 3/4, (ii) 1/2, (iii) 0, (iv) 1/16.

3.1-11 (a) $d = 3$; (b) $E(Y) = 0.75$; (c) $E(Y^2) = 0.60$.

3.1-13 $f(x) = \dfrac{e^{-x}}{(1 + e^{-x})^2} = \dfrac{e^{-x}}{(1 + e^{-x})^2}\dfrac{e^{2x}}{e^{2x}} = \dfrac{e^x}{(e^x + 1)^2} = f(-x)$.

3.1-15 (a) $1/e$; (b) $1/e^{19/8}$.

3.1-17 \$740.74.

3.1-19 (a) $\mu = \$28,571.43, \sigma = \$15,971.91$; (b) 0.6554.

3.1-21 (a) $\int_{-\infty}^{\infty} \sum_{i=1}^{k} c_i f_i(x)\, dx = \sum_{i=1}^{k} c_i \int_{-\infty}^{\infty} f_i(x)\, dx = \sum_{i=1}^{k} c_i = 1$;
(b) $\mu = \sum_{i=1}^{k} c_i \mu_i$, $\sigma^2 = \sum_{i=1}^{k} c_i(\sigma_i^2 + \mu_i^2) - \mu^2$.

3.2-1 **(a)** $f(x) = (1/3)e^{-x/3}$, $0 < x < \infty$; $\mu = 3$; $\sigma^2 = 9$;
(b) $f(x) = 3e^{-3x}$, $0 < x < \infty$; $\mu = 1/3$; $\sigma^2 = 1/9$.

3.2-3 $P(X > x + y \mid X > x) = \dfrac{P(X > x + y)}{P(X > x)} = \dfrac{e^{-(x+y)/\theta}}{e^{-x/\theta}} = P(X > y)$.

3.2-5 **(a)** $F(x) = 1 - e^{-(x-\delta)/\theta}$, $\delta \le x < \infty$;
(b) $\theta + \delta$; θ^2.

3.2-9 $f(x) = \dfrac{1}{\Gamma(20)7^{20}} x^{19} e^{-x/7}$, $0 \le x < \infty$; $\mu = 140$; $\sigma^2 = 980$.

3.2-11 **(a)** 0.025; **(b)** 0.05; **(c)** 0.94; **(d)** 8.672; **(e)** 30.19.

3.2-13 **(a)** 0.80; **(b)** $a = 11.69, b = 38.08$; **(c)** $\mu = 23$, $\sigma^2 = 46$; **(d)** 35.17, 13.09.

3.2-15 **(a)** $r - 2$; **(b)** $x = r - 2 \pm \sqrt{2r - 4}$, $r \ge 4$.

3.2-17 0.9444.

3.2-19 1.96, or 1,960 units per day, yields an expected profit of $3,304.96.

3.2-21 $e^{-1/2}$.

3.2-23 $M = 83.38$.

3.3-1 **(a)** 0.4192; **(b)** 0.4772; **(c)** 0.3382; **(d)** 0.0044;
(e) 0.9500; **(f)** 0.6826; **(g)** 0.9544; **(h)** 0.9974.

3.3-3 **(a)** 1.96; **(b)** 1.96; **(c)** 1.645; **(d)** 1.645.

3.3-5 **(a)** 0.3849; **(b)** 0.5403; **(c)** 0.0603; **(d)** 0.0013;
(e) 0.6826; **(f)** 0.9544; **(g)** 0.9974; **(h)** 0.9869.

3.3-7 **(a)** 0.6326; **(b)** 50.

3.3-9 **(a)** Gamma ($\alpha = 1/2$, $\theta = 8$); **(b)** Gamma ($\alpha = 1/2$, $\theta = 2\sigma^2$).

3.3-11 **(a)** 0.0401; **(b)** 0.8159.

3.3-13 0.1437.

3.3-15 **(a)** $\sigma = 0.043$; **(b)** $\mu = 12.116$.

3.3-17 The three respective distributions are exponential with $\theta = 4$, $\chi^2(4)$, and $N(4,1)$. Each has a mean of 4, so the slopes of the mgfs equal 4 at $t = 0$.

3.4-1 $e^{-(5/10)^2} = e^{-1/4} = 0.7788$.

3.4-3 Weibull with parameters α and $\beta/3^{1/\alpha}$.

3.4-5 **(a)** 0.5; **(b)** 0; **(c)** 0.25; **(d)** 0.75; **(e)** 0.625; **(f)** 0.75.

3.4-7 **(b)** $\mu = 31/24, \sigma^2 = 167/567$; **(c)** 15/64; 1/4; 0; 11/16.

3.4-9 **(a)**
$$F(x) = \begin{cases} 0, & x < 0, \\ x/2, & 0 \le x < 1, \\ 1/2, & 1 \le x < 2, \\ 4/6, & 2 \le x < 4, \\ 5/6, & 4 \le x < 6, \\ 1, & 6 \le x. \end{cases}$$
(b) $2.25.

3.4-11 $3 + 5e^{-3/5} = 5.744$.

3.4-13 $\mu = \$226.21$, $\sigma = \$1,486.92$.

3.4-15 $\mu = \$345.54$, $\sigma = \$780.97$.

3.4-17 $g(y) = \dfrac{c}{3y^{4/3}}$ for $e^{0.12} < y < e^{0.24}$; $c = 26.54414$.

3.4-19 0.4219.

3.4-21 **(a)** $e^{-(125/216)}$; **(b)** $120 * \Gamma(4/3) = 107.1575$.

Chapter 4

4.1-1 **(a)** 1/33; **(b)** 1/24; **(c)** 1/18; **(d)** 6.

4.1-3 **(a)** $f_X(x) = (2x + 5)/16$, $x = 1, 2$;
(b) $f_Y(y) = (2y + 3)/32$, $y = 1, 2, 3, 4$;
(c) 3/32; **(d)** 9/32; **(e)** 3/16; **(f)** 1/4; **(g)** Dependent;
(h) $\mu_X = 25/16$; $\mu_Y = 45/16$; $\sigma_X^2 = 63/256$; $\sigma_Y^2 = 295/256$.

4.1-5 **(b)** $f(x, y) = 1/16$, $x = 1, 2, 3, 4$; $y = x + 1, x + 2, x + 3, x + 4$;
(c) $f_X(x) = 1/4$, $x = 1, 2, 3, 4$;
(d) $f_Y(y) = (4 - |y - 5|)/16$, $y = 2, 3, 4, 5, 6, 7, 8$;
(e) Dependent because the space is not rectangular.

4.1-7 **(b)** $b(6, 1/2)$, $b(6, 1/2)$.

4.1-9 **(a)** $f(x, y) = \dfrac{15!}{x! \, y! \, (15 - x - y)!} \left(\dfrac{6}{10}\right)^x \left(\dfrac{3}{10}\right)^y \left(\dfrac{1}{10}\right)^{15-x-y}$, $0 \le x + y \le 15$;
(b) No, because the space is not rectangular;
(c) 0.0735;
(d) X is $b(15, 0.6)$;
(e) 0.9095.

4.2-1 $\mu_X = 25/16$; $\mu_Y = 45/16$; $\sigma_X^2 = 63/256$; $\sigma_Y^2 = 295/256$;
Cov$(X, Y) = -5/256$; $\rho = -\sqrt{2,065}/1,239 = -0.0367$.

4.2-3 **(a)** $\mu_X = 5/2$; $\mu_Y = 5$; $\sigma_X^2 = 5/4$; $\sigma_Y^2 = 5/2$; Cov$(X, Y) = 5/4$; $\rho = \sqrt{2}/2$;
(b) $y = x + 5/2$.

4.2-5 $a = \mu_Y - \mu_X b$, $b = \text{Cov}(X, Y)/\sigma_X^2$.

4.2-7 **(a)** No; **(b)** Cov$(X, Y) = 0$, $\rho = 0$.

4.2-9 **(a)** $c = 1/154$;
(c) $f_X(0) = 6/77$, $f_X(1) = 21/77$, $f_X(2) = 30/77$, $f_X(3) = 20/77$;
$f_Y(0) = 30/77$, $f_Y(1) = 32/77$, $f_Y(2) = 15/77$;
(d) No;
(e) $\mu_X = 141/77, \sigma_X^2 = 4836/5929$;
(f) $\mu_Y = 62/77, \sigma_Y^2 = 3240/5929$;
(g) Cov$(X, Y) = 1422/5929$;
(h) $\rho = 79\sqrt{12090}/24180$;
(i) $y = 215/806 + (237/806)x$.

4.3-1 (d) 9/14, 7/18, 5/9; (e) 20/7, 55/49.

4.3-3 (a) $f(x, y) = \dfrac{50!}{x!\,y!\,(50 - x - y)!}(0.02)^x(0.90)^y(0.08)^{50-x-y}$, $0 \le x + y \le 50$;

(b) Y is $b(50, 0.90)$;

(c) $b(47, 0.90/0.98)$;

(d) 2115/49; (e) $\rho = -3/7$.

4.3-5 (a) $E(Y \mid x) = 2(2/3) - (2/3)x$, $x = 1, 2$; (b) Yes.

4.3-7 $E(Y \mid x) = x + 5/2$, $x = 1, 2, 3, 4$; yes.

4.3-9 (a) $f_X(x) = 1/8$, $x = 0, 1, \ldots, 7$;

(b) $h(y \mid x) = 1/3$, $y = x, x + 1, x + 2$, for $x = 0, 1, \ldots, 7$;

(c) $E(Y \mid x) = x + 1$, $x = 0, 1, \ldots, 7$;

(d) $\sigma_Y^2 = 2/3$;

(e)
$$f_Y(y) = \begin{cases} 1/24, & y = 0, 9, \\ 2/24, & y = 1, 8, \\ 3/24, & y = 2, 3, 4, 5, 6, 7. \end{cases}$$

4.3-11 $f_X(x) = 1/5$ and $h(y \mid x) = 1/[5(x + 1)]$, for $x = 0, 1, 2, 3, 4$, and $y = 0 \ldots x$; $P(X + Y > 4) = 13/50$.

4.4-1 (a) $f_X(x) = x/2$, $0 \le x \le 2$; $f_Y(y) = 3y^2/8$, $0 \le y \le 2$;

(b) Yes, because $f_X(x)f_Y(y) = f(x, y)$;

(c) $\mu_X = 4/3$; $\mu_Y = 3/2$; $\sigma_X^2 = 2/9$; $\sigma_Y^2 = 3/20$;

(d) 3/5.

4.4-3 $f_X(x) = 2e^{-2x}$, $0 < x < \infty$; $f_Y(y) = 2e^{-y}(1 - e^{-y})$, $0 < y < \infty$; no.

4.4-5 (a) $c = 1/8$; (b) $c = 2$; (c) $c = 6$; (d) $c = 6/5$.

4.4-7 (b) 1/3.

4.4-9 11/30.

4.4-11 (a) $c = 8$; (b) 29/93.

4.4-13 (a) $f_X(x) = 4x(1 - x^2)$, $0 \le x \le 1$; $f_Y(y) = 4y^3$, $0 \le y \le 1$;

(b) $\mu_X = 8/15$; $\mu_Y = 4/5$; $\sigma_X^2 = 11/225$; $\sigma_Y^2 = 2/75$; $\text{Cov}(X, Y) = 4/225$; $\rho = 2\sqrt{66}/33$;

(c) $y = 20/33 + (4/11)x$.

4.4-15 $E(Y \mid x) = x$ and $E(X) = 0.700$; thus, $700.

4.4-17 (b) $f_X(x) = 1/10$, $0 \le x \le 10$;

(c) $h(y \mid x) = 1/4$, $10 - x \le y \le 14 - x$ for $0 \le x \le 10$;

(d) $E(Y \mid x) = 12 - x$.

4.4-19 (a) $f(x, y) = 1/(2x^2)$, $0 < x < 2$, $0 < y < x^2$;

(b) $f_Y(y) = (2 - \sqrt{y})/(4\sqrt{y})$, $0 < y < 4$;

(c) $E(X \mid y) = [2\sqrt{y}\ln(2/\sqrt{y})]/[2 - \sqrt{y}]$;

(d) $E(Y \mid x) = x^2/2$.

4.5-1 (a) 0.6006; (b) 0.7888; (c) 0.8185; (d) 0.9371.

4.5-3 (a) 0.2347; (b) 0.221.

4.5-5 (a) $N(86.4, 40.96)$; (b) 0.4192.

4.5-7 (a) 0.8248;

(b) $E(Y \mid x) = 457.1735 - 0.2655x$;

(c) $\text{Var}(Y \mid x) = 645.9375$;

(d) 0.8079.

4.5-9 $a(x) = x - 11$, $b(x) = x + 5$.

4.5-11 (a) 0.3015; (b) $\mu_{Y \mid x} = 3.3822 - 0.876x$;

(c) $\sigma_{Y \mid x}^2 = 0.216$; (d) 0.8009.

4.5-13 (a) 0.3721; (b) 0.1084.

Chapter 5

5.1-1 $g(y) = 2/9(y - 1)$, $1 < y < 4$.

5.1-3 $g(y) = (1/8)y^5 e^{-y^2/2}$, $0 < y < \infty$.

5.1-5 Exponential distribution with mean 2.

5.1-7 (a) $F(x) = (\ln x - \ln c - 0.03)/0.04$ and $f(x) = 1/(0.04x)$, $ce^{0.03} \le x \le ce^{0.07}$, where $c = 50{,}000$;

(b) The interest for each of n equal parts is R/n. The amount at the end of the year is $50{,}000(1 + R/n)^n$; the limit as $n \to \infty$ is $50{,}000\,e^R$.

5.1-9 (a) $G(y) = P(Y \le y) = P(X \le \ln y) = 1 - e^{-y}$, $0 < y < \infty$;

(b) $G(y) = 1 - \exp[-e^{(\ln y - \theta_1)/\theta_2}]$, $0 < y < \infty$; $g(y) = \exp[-e^{(\ln y - \theta_1)/\theta_2}][e^{(\ln y - \theta_1)/\theta_2}][1/\theta_2 y]$, $0 < y < \infty$;

(c) A Weibull distribution with $G(y) = 1 - e^{-(y/\beta)^\alpha}$, $0 < y < \infty$; $g(y) = (\alpha y^{\alpha-1}/\beta^\alpha)e^{-(y/\beta)^\alpha}$, $0 < y < \infty$, where α is the shape parameter and β is the scale parameter;

(d) $\exp(-e^{-2}) = 0.873$.

5.1-11 (a) $\dfrac{1}{2} - \dfrac{\arctan 1}{\pi} = 0.25$;

(b) $\dfrac{1}{2} - \dfrac{\arctan 5}{\pi} = 0.0628$;

(c) $\dfrac{1}{2} - \dfrac{\arctan 10}{\pi} = 0.0317$.

5.1-13 (b) (i) $\exp(\mu + \sigma^2/2)$, (ii) $\exp(2\mu + 2\sigma^2)$, (iii) $\exp(2\mu + 2\sigma^2) - \exp(2\mu + \sigma^2)$.

5.1-15 (a) $g(y) = \dfrac{1}{\sqrt{2\pi y}}\exp(-y/2)$, $0 < y < \infty$;

(b) $g(y) = \dfrac{3}{2}\sqrt{y}$, $0 < y < 1$.

5.2-1 $g(y_1, y_2) = (1/4)e^{-y_2/2}$, $0 < y_1 < y_2 < \infty$; $g_1(y_1) = (1/2)e^{-y_1/2}$, $0 < y_1 < \infty$; $g_2(y_2) = (y_2/4)e^{-y_2/2}$, $0 < y_2 < \infty$; no.

5.2-3 $\mu = \dfrac{r_2}{r_2 - 2}, r_2 > 2;$

$\sigma^2 = \dfrac{2r_2^2(r_1 + r_2 - 2)}{r_1(r_2 - 2)^2(r_2 - 4)}, r_2 > 4.$

5.2-5 **(a)** 4.00; **(b)** $1/3 = 0.3333;$ **(c)** 0.90.

5.2-9 840.

5.2-11 **(a)** 0.1792; **(b)** 0.1792.

5.2-13 **(a)**

$$G(y_1, y_2) = \int_0^{y_1} \int_u^{y_2} 2(1/1000^2) \exp[-(u + v)/1000]\, dv\, du$$

$$= 2\exp[-(y_1 + y_2)/1000] - \exp[-y_1/500]$$

$$- 2\exp[-y_2/1000] + 1, \quad 0 < y_1 < y_2 < \infty;$$

(b) $2e^{-6/5} - e^{-12/5} \approx 0.5117.$

5.3-1 **(a)** 0.0182; **(b)** 0.0337.

5.3-3 **(a)** 36/125; **(b)** 2/7.

5.3-5

$$g(y) = \begin{cases} 1/36, & y = 2, \\ 4/36, & y = 3, \\ 10/36, & y = 4, \\ 12/36, & y = 5, \\ 9/36, & y = 6; \end{cases}$$

$\mu = 14/3, \sigma^2 = 10/9.$

5.3-7 2/5.

5.3-9 **(a)** 729/4096; **(b)** $\mu = 3/2; \sigma^2 = 3/40.$

5.3-11 **(a)** 0.0035; **(b)** 8; **(c)** $\mu_Y = 6, \sigma_Y^2 = 4.$

5.3-13 $1 - e^{-3/100} \approx 0.03.$

5.3-15 0.0384.

5.3-17 $21,816.

5.3-19 5.

5.3-21 **(c)** Using *Maple*, we obtain $\mu = 13,315,424/3,011,805 = 4.4211.$

(d) $E(Y) = 5.377$ with 16 coins, $E(Y) = 6.355$ with 32 coins.

5.4-1 **(a)**

$$g(y) = \begin{cases} 1/64, & y = 3, 12, \\ 3/64, & y = 4, 11, \\ 6/64, & y = 5, 10, \\ 10/64, & y = 6, 9, \\ 12/64, & y = 7, 8. \end{cases}$$

5.4-3 **(a)** $M(t) = e^{7(e^t - 1)};$ **(b)** Poisson, $\lambda = 7;$ **(c)** 0.800.

5.4-5 0.925.

5.4-7 **(a)** $M(t) = 1/(1 - 5t)^{21}, t < 1/5;$

(b) gamma distribution, $\alpha = 21, \theta = 5.$

5.4-11 **(a)** $g(w) = 1/12, w = 0, 1, 2, \ldots, 11;$ **(b)** $h(w) = 1/36,$
$w = 0, 1, 2, \ldots, 35.$

5.4-13 **(a)**

$$h_1(w_1) = \begin{cases} 1/36, & w_1 = 0, \\ 4/36, & w_1 = 1, \\ 10/36, & w_1 = 2, \\ 12/36, & w_1 = 3, \\ 9/36, & w_1 = 4; \end{cases}$$

(b) $h_2(w) = h_1(w);$

(c)

$$h(w) = \begin{cases} 1/1296, & w = 0, \\ 8/1296, & w = 1, \\ 36/1296, & w = 2, \\ 104/1296, & w = 3, \\ 214/1296, & w = 4, \\ 312/1296, & w = 5, \\ 324/1296, & w = 6, \\ 216/1296, & w = 7, \\ 81/1296, & w = 8; \end{cases}$$

(d) With denominators equal to $6^8 = 1,679,616,$ the respective numerators of $0, 1, \ldots, 16$ are 1, 16, 136, 784, 3,388, 11,536, 31,864, 72,592, 137,638, 217,776, 286,776, 311,472, 274,428, 190,512, 99,144, 34,992, 6,561;

(e) They are becoming more symmetrical as the value of n increases.

5.4-15 **(b)** $\mu_Y = 25/3, \sigma_Y^2 = 130/9;$

(c)

$$P(Y = y) = \begin{cases} 96/1024, & y = 4, \\ 144/1024, & y = 5, \\ 150/1024, & y = 6, \\ 135/1024, & y = 7. \end{cases}$$

5.4-17 $Y - X + 25$ is $b(50, 1/2);$

$$P(Y - X \geq 2) = \sum_{k=27}^{50} \binom{50}{k}\left(\frac{1}{2}\right)^{50} = 0.3359.$$

5.4-19 $1 - 17/2e^3 = 0.5768.$

5.4-21 0.4207.

5.5-1 **(a)** 0.8664; **(b)** 0.7800.

5.5-3 **(a)** 46.58, 2.56; **(b)** 0.8447.

5.5-5 **(b)** 0.05466; 0.3102.

5.5-7 0.9830.

5.5-9 **(a)** 0.3085; **(b)** 0.2267.

5.5-11 0.8413 > 0.7734, select X.

5.5-13 (a) $t(2)$; (c) $\mu_V = 0$; (d) $\sigma_V = 1$;

(e) In part (b), numerator and denominator are not independent.

5.5-15 (a) 2.567; (b) -1.740; (c) 0.90.

5.6-1 0.4772.

5.6-3 0.8185.

5.6-5 (a) $\chi^2(18)$; (b) 0.0756, 0.9974.

5.6-7 0.6247.

5.6-9 $P(1.7 \le Y \le 3.2) = 0.6749$; the normal approximation is 0.6796.

5.6-11 \$444,338.13.

5.6-13 0.9522.

5.6-15 (a) $\int_0^{25} \frac{1}{\Gamma(13)2^{13}} y^{13-1} e^{-y/2}\, dy = 0.4810$;

(b) 0.4448 using normal approximation.

5.7-1 (a) 0.2878, 0.2881; (b) 0.4428, 0.4435; (c) 0.1550, 0.1554.

5.7-3 0.9759 using normal approximation.

5.7-5 0.3085.

5.7-7 0.6247 using normal approximation, 0.6148 using Poisson.

5.7-9 (a) 0.5548; (b) 0.3823.

5.7-11 0.6813 using normal approximation, 0.6788 using binomial.

5.7-13 (a) 0.3802; (b) 0.7571.

5.7-15 0.4734 using normal approximation; 0.4749 using Poisson approximation with $\lambda = 50$; 0.4769 using $b(5000, 0.01)$.

5.7-17 0.6455 using normal approximation, 0.6449 using Poisson.

5.8-1 (a) 0.84; (b) 0.082.

5.8-3 $k = 1.464$; 8/15.

5.8-5 (a) 0.25; (b) 0.85; (c) 0.925.

5.8-7 (a) $E(W) = 0$; the variance does not exist.

5.9-1 (a) 0.9984; (b) 0.998.

5.9-3 $M(t) = \left[1 - \dfrac{2t\sigma^2}{n-1} \right]^{-(n-1)/2} \to e^{\sigma^2 t}$.

Chapter 6

6.1-1 (a) $\bar{x} = 1.1$; (b) $s^2 = 0.035$; (c) $s = 0.1871$.

6.1-3 (a) $\bar{x} = 16.706$, $s = 1.852$;

(b) Frequencies: [1, 1, 2, 14, 18, 16, 23, 10, 7, 2, 1, 1].

6.1-5 (a) $\bar{x} = 112.12$; $s = 231.3576$;

(d) Half of the observations are less than 48.

6.1-7 (b) $\bar{x} = 7.275$, $s = 1.967$; (c) 47, 61.

6.1-9 (a) With class boundaries $90.5, 108.5, 126.5, \ldots, 306.5$:
Frequencies: [8, 11, 4, 1, 0, 0, 0, 1, 2, 12, 12, 3];

(b) $\bar{x} = 201$;

(c) With class boundaries $47.5, 52.5, \ldots, 107.5$:
Frequencies: [4, 4, 9, 4, 4, 0, 3, 9, 7, 5, 4, 1];

(d) $\bar{x} = 76.35$.

6.1-11 (a) 0.261, 0.196;

(b) 0.327, 0.315;

(c) 0.270, 0.273;

(d) Example of Simpson's paradox.

6.2-1 (a)

Stems	Leaves	Frequency
11	9	1
12	3	1
13	6 7	2
14	1 1 4 4 4 6 7 8 8 8 8 9 9 9	14
15	0 0 1 3 4 5 5 6 6 6 6 7 8 8 9 9 9	18
16	1 1 1 1 1 3 3 5 5 6 6 6 7 7 8 9	16
17	0 1 1 1 1 1 2 2 2 3 4 4 4 6 6 7 7 8 8 8 8 8 8	23
18	0 0 0 1 1 4 5 8 9 9	10
19	0 1 3 3 4 7 8	7
20	2 8	2
21	5	1
22	1	1

(Multiply numbers by 10^{-1}.)

(b) 11.9, 15.5, 16.65, 17.8, 22.1;

(c) There are three suspected outliers: 11.9, 21.5, 22.1.

6.2-3 (a) Frequencies for males: [1, 1, 3, 4, 20, 23, 16, 10, 3, 0, 1],
Frequencies for females: [5, 14, 32, 36, 13];

(c) Five-number summary for males: 1.4, 3.5, 4.0, 4.525, 6.5;
Five-number summary for females: 0.5, 1.325, 1.7, 2.0, 2.7.

6.2-5 (b) Five-number summary: 5, 35/2, 48, 173/2, 1,815;

(d) Inner fence at 190, outer fence at 293.5;

(e) The median.

6.2-7 (a)

Stems	Leaves	Frequency
127	8	1
128	8	1
129	5 8 9	3
130	8	1
131	2 3 4 4 5 5 7	7
132	2 7 7 8	4
133	7 9	2
134	8	1

(Multiply numbers by 10^{-1}.)

(b) 131.3, 7.0, 2.575, 131.45, 131.47, 3.034;

(c) Five-number summary: 127.8, 130.125, 131.45, 132.70, 134.8.

6.2-9 (a)

Stems	Leaves	Frequency
$30f$	5	1
$30s$		0
$30\bullet$	8 8	2
$31*$	0 0 1	3
$31t$	2 3 3	3
$31f$	4 4 4	3
$31s$	6 6 7 7 7	5
$31\bullet$	8 8 8 9 9	5
$32*$	0 0 0 0 0 1	6
$32t$	2 2 2 3 3	5
$32f$	4 4 4 4 5 5	6
$32s$	6 7 7	3
$32\bullet$	8 8 9 9 9	5
$33*$	0 1	2
$33t$		0
$33f$	5	1

(b) Five-number summary: 305, 315.5, 320, 325, 335.

6.3-1 (b) $\tilde{m} = 146, \tilde{\pi}_{0.80} = 270$;
(c) $\tilde{q}_1 = 95, \tilde{q}_3 = 225$.

6.3-3 (a) $g_3(y) = 10(1 - e^{-y/3})^2 e^{-y}, \ 0 < y < \infty$;
(b) $5(1 - e^{-5/3})^4 e^{-5/3} + (1 - e^{-5/3})^5 = 0.7599$;
(c) $e^{-5/3} = 0.1889$.

6.3-5 (a) 0.2553; **(b)** 0.7483.

6.3-7 (a) $g_1(y) = 19(e^{-y/\theta})^{18} \dfrac{1}{\theta} e^{-y/\theta}, \ 0 < y < \infty$;
(b) 1/20.

6.3-9 (a) $g_r(y) = \dfrac{n!}{(r-1)!(n-r)!}(1 - e^{-y})^{r-1}(e^{-y})^{n-r}e^{-y}, \ 0 < y < \infty$;
(b) A beta pdf with $\alpha = n - r + 1, \beta = r$.

6.3-11 (a) $g(y_1, y_n) = \dfrac{n!}{(n-2)!}(y_n - y_1)^{(n-2)}, \ 0 < y_1 < y_n < 1$;
(b) $h(w_1, w_2) = n(n-1)w_2^{n-1}(1 - w_1)^{n-2}, \ 0 < w_1 < 1, \ 0 < w_2 < 1$;
$h_1(w_1) = (n-1)(1 - w_1)^{n-2}, \ 0 < w_1 < 1$;
$h_2(w_2) = nw_2^{n-1}, \ 0 < w_2 < 1$;
(c) Yes.

6.3-13 Both could be normal because of the linearity of the q–q plots.

6.4-3 (b) $\bar{x} = 89/40 = 2.225$.

6.4-5 (a) $\hat{\theta} = \overline{X}/2$; **(b)** $\hat{\theta} = \overline{X}/3$; **(c)** $\hat{\theta}$ equals the sample median.

6.4-7 (c) (i) $\hat{\theta} = 0.5493, \ \tilde{\theta} = 0.5975$, **(ii)** $\hat{\theta} = 2.2101, \ \tilde{\theta} = 2.4004$, **(iii)** $\hat{\theta} = 0.9588, \ \tilde{\theta} = 0.8646$.

6.4-9 (c) $\bar{x} = 3.48$.

6.4-13 (a) $\tilde{\theta} = \overline{X}$; **(b)** Yes; **(c)** 7.382; **(d)** 7.485.

6.4-15 $\bar{x} = \alpha\theta, \ v = \alpha\theta^2$ so $\tilde{\theta} = v/\bar{x} = 0.0658, \ \tilde{\alpha} = \bar{x}^2/v = 102.4991$.

6.4-17 (b) $\tilde{\theta} = 2\overline{X}$; **(c)** 0.74646.

6.4-19 $\hat{\mu} = \dfrac{\sum_{i=1}^{n} y_i/x_i^2}{\sum_{i=1}^{n} 1/x_i^2}; \ \widehat{\gamma^2} = \dfrac{1}{n}\sum_{i=1}^{n}\dfrac{(y_i - \hat{\mu})^2}{x_i^2}$.

6.5-3 (a) $\hat{y} = 86.8 + (842/829)(x - 74.5)$;
(c) $\widehat{\sigma^2} = 17.9998$.

6.5-5 (a) $\hat{y} = 10.6 - 0.015x$;
(c) $\hat{y} = 5.47 + 0.0004x$;
(e) Horsepower.

6.5-7 (a) $\hat{y} = 0.819x + 2.575$;
(c) $\hat{\alpha} = 10.083; \ \hat{\beta} = 0.819; \ \widehat{\sigma^2} = 3.294$.

6.5-9 (a) $\hat{y} = 46.59 + 1.085x$.

6.6-1 (b) 2; **(c)** 0.20.

6.6-3 (a) $2\theta^2/n$; **(b)** $N(\theta, 2\theta^2/n)$; **(c)** $\chi^2(n)$.

6.7-1 (a) $\sum_{i=1}^{n} X_i^2$; **(b)** $\widehat{\sigma^2} = \left(\dfrac{1}{n}\right)\sum_{i=1}^{n} X_i^2$; **(c)** Yes.

6.7-7 (a) $f(x;p) = \exp\{x \ln(1 - p) + \ln[p/(1-p)]\}$;
$K(x) = x; \ \sum_{i=1}^{n} X_i$ is sufficient;
(b) $\overline{X}$.

6.8-1 (a) $k(\theta|y) \propto \theta^{2+y} e^{-\theta}(n + 1/2)$.
Thus, the posterior pdf of θ is gamma with parameters $3 + y$ and $1/(n + 1/2)$;
(b) $w(y) = E(\theta|y) = (3 + y)/(n + 1/2)$;
(c) $w(y) = \left(\dfrac{y}{n}\right)\left(\dfrac{n}{n + 1/2}\right) + 6\left(\dfrac{1/2}{n + 1/2}\right)$.

6.8-3 (a) $E[\{w(Y) - \theta\}^2] = \{E[w(Y) - \theta]\}^2 + \text{Var}[w(Y)]$
$= (74\theta^2 - 114\theta + 45)/500$;
(b) $\theta = 0.569$ to $\theta = 0.872$.

6.8-5 The posterior median (or posterior mean), because the posterior pdf is symmetric.

6.8-7 $d = 2/n$.

6.9-5 (c) $2yz$ is $\chi^2(2n)$; $\left(\dfrac{\chi_{1-\alpha/2}^2(2n)}{2y}, \dfrac{\chi_{\alpha/2}^2(2n)}{2y}\right)$ is the interval for z, and the interval for θ is $\left(\dfrac{2y}{\chi_{\alpha/2}^2(2n)}, \dfrac{2y}{\chi_{1-\alpha/2}^2(2n)}\right)$.

Chapter 7

7.1-1 [71.35, 76.25].

7.1-3 (a) $\bar{x} = 15.757$; **(b)** $s = 1.792$; **(c)** [14.441, 17.073].

7.1-5 [48.467, 72.266] or [48.076, 72.657].

7.1-7 $[19.47, 22.33]$.

7.1-9 $[21.373, \infty)$.

7.1-11 $[22.74, 25.48]$.

7.1-13 (a) 29.49, 3.41; **(b)** $[0, 31.259]$;

(c) Yes, because of the linearity of the q–q plot of the data and the corresponding normal quantiles.

7.1-15 (a) $\bar{x} = 25.475$, $s = 2.4935$; **(b)** $[24.059, \infty)$.

7.1-17 25.

7.2-1 $[-59.725, -43.275]$.

7.2-3 $[-5.845, 0.845]$.

7.2-5 $(-\infty, -1.828]$.

7.2-7 (a) Yes, since a 95% confidence lower bound for μ_X is 25.511;

(b) $[11.5, 13.7]$;

(c) Do not change since a 95% confidence lower bound for μ_Y is 12.238.

7.2-9 (a) $[-0.556, 1.450]$;

(b) $[0.367, 1.863]$;

(c) No for men because the confidence interval contains 0; Yes for women because the confidence interval does not contain 0.

7.2-11 $[157.227, \infty)$.

7.2-13 $[-5.599, -1.373]$ assuming equal variances, otherwise, $[-5.577, -1.394]$.

7.3-1 (a) 0.0374; **(b)** $[0.0227, 0.0521]$; **(c)** $[0.0252, 0.0550]$; **(d)** $[0.0251, 0.0554]$; **(e)** $[0, 0.0497]$.

7.3-3 (a) 0.5061; **(b)** $[0.4608, 0.5513]$ or $[0.4609, 0.5511]$ or $[0.4610, 0.5510]$; **(c)** Not necessarily because there is a danger of self-selection bias.

7.3-5 (a) 0.1800; **(b)** $[0.0735, 0.2865]$; **(c)** $[0.0977, 0.3080]$; **(d)** $[0.0963, 0.3111]$.

7.3-7 $[0.207, 0.253]$.

7.3-9 (a) 0.2115; **(b)** $[0.1554, 0.2676]$.

7.3-11 $[0.011, 0.089]$.

7.4-1 117.

7.4-3 (a) 1083; **(b)** $[6.047, 6.049]$; **(c)** \$58,800; **(d)** 0.0145.

7.4-5 (a) 257; **(b)** Yes.

7.4-7 (a) 1068; **(b)** 2401; **(c)** 752.

7.4-9 2305.

7.4-11 (a) 38; **(b)** $[0.621, 0.845]$.

7.4-13 235.

7.4-15 144.

7.5-1 (a) 0.7812; **(b)** 0.7844; **(c)** 0.4528.

7.5-3 (a) $(6.31, 7.40)$; **(b)** $(6.58, 7.22)$, 0.8204.

7.5-5 $(15.40, 17.05)$.

7.5-7 (a)

Stems	Leaves	Frequency
101	. 7	1
102	0 0 0	3
103		0
104		0
105	8 9	2
106	1 3 3 6 6 7 7 8 8	9
107	3 7 9	3
108	8	1
109	1 3 9	3
110	0 2 2	3

(b) $\tilde{\pi}_{0.25} = 106.0$, $\tilde{m} = 106.7$, $\tilde{\pi}_{0.75} = 108.95$;

(c) (i) $(102.0, 106.6)$, 89.66%; **(ii)** $(106.3, 107.7)$, 89.22%; **(iii)** $(107.3, 110.0)$, 89.66%;

(d) $[106.3, 107.7]$, 89.22%; $[105.87, 107.63]$, 90%.

7.5-9 (a) $\tilde{\pi}_{0.50} = \tilde{m} = 0.92$;

(b) $(y_{41}, y_{60}) = (0.91, 0.93)$; 0.9426 using normal approximation, 0.9431 using binomial;

(c) $\tilde{\pi}_{0.25} = 0.89$;

(d) $(y_{17}, y_{34}) = (0.88, 0.90)$; 0.9504 using normal approximation, 0.9513 using binomial;

(e) $\tilde{\pi}_{0.75} = 0.97$;

(f) $(y_{67}, y_{84}) = (0.95, 0.98)$; 0.9504 using normal approximation, 0.9513 using binomial.

7.5-11 $y_4 = 5.08 < \pi_{0.25} < y_{15} = 5.27$, $y_{14} = 5.27 < \pi_{0.5} < y_{26} = 5.31$, $y_{24} = 5.30 < \pi_{0.75} < y_{35} = 5.35$. The corresponding exact confidence coefficients are 0.9503, 0.9467, and 0.9602, respectively.

7.6-1 (a) Normal with mean α_1 and variance
$$\sigma^2 \left(\frac{1}{n} + \frac{\bar{x}^2}{\sum_{i=1}^n (x_i - \bar{x})^2} \right);$$

(b) $\hat{\alpha}_1 \pm h t_{\gamma/2}(n-2)$, where
$$h = \hat{\sigma} \sqrt{\frac{n}{n-2}} \sqrt{\frac{1}{n} + \frac{\bar{x}^2}{\sum_{i=1}^n (x_i - \bar{x})^2}}.$$

7.6-3 (a) $[75.283, 85.113]$, $[83.838, 90.777]$, $[89.107, 99.728]$; **(b)** $[68.206, 92.190]$, $[75.833, 98.783]$, $[82.258, 106.577]$.

7.6-5 (a) $[4.897, 8.444]$, $[9.464, 12.068]$, $[12.718, 17.004]$; **(b)** $[1.899, 11.442]$, $[6.149, 15.383]$, $[9.940, 19.782]$.

7.6-7 (a) $[19.669, 26.856]$, $[22.122, 27.441]$, $[24.048, 28.551]$, $[25.191, 30.445]$, $[25.791, 32.882]$; **(b)** $[15.530, 30.996]$, $[17.306, 32.256]$, $[18.915, 33.684]$, $[20.351, 35.285]$, $[21.618, 37.055]$.

7.6-9 $\widehat{y} = 1.1037 + 2.0327x - 0.2974x^2 + 0.6204x^3$.

7.6-11 (a) $r = 0.143$;

    (b) $\widehat{y} = 37.68 + 0.83x$;

    (d) No;

    (e) $\widehat{y} = 12.845 + 22.566x - 3.218x^2$;

    (f) Yes.

7.6-17 $[83.341, 90.259]$, $[0.478, 1.553]$, $[10.265, 82.578]$.

7.6-19 $[29.987, 31.285]$, $[0.923, 1.527]$, $[0.428, 3.018]$.

Chapter 8

8.1-1 (a) $1.4 < 1.645$, do not reject H_0;

    (b) $1.4 > 1.282$, reject H_0.

    (c) p-value $= 0.0808$.

8.1-3 (a) $z = (\bar{x} - 170)/2, z \geq 1.645$;

    (b) $1.260 < 1.645$, do not reject H_0;

    (c) 0.1038.

8.1-5 (a) $t = (\bar{x} - 3,315)/(s/\sqrt{30}) \leq -1.699$;

    (b) $-1.414 > -1.699$, do not reject H_0;

    (c) $0.05 < p$-value < 0.10 or p-value ≈ 0.08.

8.1-7 (a) $t = (\bar{x} - 47)/(s/\sqrt{20}) \leq -1.729$;

    (b) $-1.789 < -1.729$, reject H_0;

    (c) $0.025 < p$-value < 0.05, p-value ≈ 0.045.

8.1-9 (a) $-4.60, p$-value < 0.0001;

    (b) Clearly, reject H_0;

    (c) $[0, 14.573]$.

8.1-11 $1.477 < 1.833$, do not reject H_0.

8.1-13 (a) $t \leq -1.729$;

    (b) $t = -1.994 < -1.729$, so we reject H_0;

    (c) $t = -1.994 > -2.539$, so we would fail to reject H_0;

    (d) $0.025 < p$-value < 0.05. In fact, p-value $= 0.0304$.

8.1-15 (a) $\chi^2 \geq 28.87$ or $s^2 \geq 48.117$;

    (b) $\beta \approx 0.10$.

8.2-1 (a) (a) $t \leq -1.734$; **(b)** $t = -2.221 < -1.734$, reject H_0.

8.2-3 (a) $|t| = 0.374 < 2.086$, do not reject H_0 at $\alpha = 0.05$.

8.2-5 (a) $t < -1.706$; **(b)** $-1.714 < -1.706$, reject H_0;

    (c) $0.025 < p$-value < 0.05.

8.2-7 (a) $t < -2.552$; **(b)** $t = -3.638 < -2.552$, reject H_0.

8.2-9 (a) $t = -1.67$, $0.05 < p$-value < 0.10, p-value $= 0.054$, fail to reject H_0.

8.2-11 (a) $z = 2.245 > 1.645$, reject H_0; **(b)** p-value $= 0.0124$.

8.2-13 (a) $t = 3.440, p$-value < 0.005, reject H_0.

8.2-15 (c) $F_{0.01}(30, 30) = 2.39$.

8.2-17 About $n = 54$ and $c = 5.6$.

8.3-1 (a) 0.3032 using $b(100, 0.08)$, 0.313 using Poisson approximation, 0.2902 using normal approximation and continuity correction;

    (b) 0.1064 using $b(100, 0.04)$, 0.111 using Poisson approximation, 0.1010 using normal approximation and continuity correction.

8.3-3 (a) $\alpha = 0.1056$; **(b)** $\beta = 0.3524$.

8.3-5 (a) $z = 2.269 > 1.645$, reject H_0;

    (b) $z = 2.269 < 2.326$, do not reject H_0;

    (c) p-value $= 0.0116$.

8.3-7 (a) $z = \dfrac{y/n - 0.40}{\sqrt{(0.40)(0.60)/n}} \geq 1.645$;

    (b) $z = 2.215 > 1.645$, reject H_0.

8.3-9 (a) $H_0: p = 0.037, H_1: p > 0.037$;

    (b) $z \geq 2.326$;

    (c) $z = 2.722 > 2.326$, reject H_0.

8.3-11 (a) $|z| = \dfrac{|\widehat{p}_1 - \widehat{p}_2|}{\sqrt{\widehat{p}(1-\widehat{p})(1/n_1 + 1/n_2)}} \geq 1.960$;

    (b) $1.726 < 1.960$, do not reject H_0.

8.3-13 $[0.007, 0.070]$; yes, because $z = 2.346 > 1.96$.

8.3-15 (a) P(at least one match) $= 1 - P$(no matches) $= 1 - \dfrac{52}{52} \cdots \dfrac{47}{52} = 0.259$.

8.4-1 (a) $-55 < -47.08$, reject H_0; **(b)** 0.0296;

    (c) $9 < 10$, do not reject H_0; **(d)** p-value $= 0.1334$.

8.4-3 (a) $y = 17, p$-value $= 0.0539$;

    (b) $w = 171, p$-value $= 0.0111$;

    (c) $t = 2.608, p$-value $= 0.0077$.

8.4-5 $w = 54, z = 1.533, p$-value $= 0.0626$ for a one-sided alternative, do not reject H_0.

8.4-7 (a) $z \geq 1.282$;

    (b) $w = 66$, reject H_0;

    (c) p-value $= 0.0207$, making a unit correction for continuity.

8.4-9 (a) $w = 145 > 130.9$ or $z = 3.024 > 1.96$, reject H_0; p-value ≈ 0.0025.

8.4-11 (a) $C = \{w: w \leq 79$ or $w \geq 131\}, \alpha \approx 0.0539, w = 95$, do not reject H_0.

8.4-13 (a) $C = \{w: w \leq 79$ or $w \geq 131\}, \alpha \approx 0.0539, w = 107.5, p$-value $= 0.8798$, do not reject H_0.

8.4-15 $C = \{w: w \leq 184$ or $w \geq 280\}, \alpha \approx 0.0489, w = 241$, do not reject H_0.

8.4-17 (a) $w = 71$, reject H_0, p-value $= 0.0057$;

    (b) $w = 101$, do not reject H_0, p-value $= 0.7913$;

    (c) $w = 108$, do not reject H_0, p-value $= 0.8501$.

8.4-19 (b) $w = 223.5 < 224.25$, reject H_0;

    (c) p-value ≈ 0.01;

    (d) Reject H_0;

    (e) The p-values are approximately equal.

8.5-1 (a) $K(\mu) = \Phi\left(\dfrac{22.5 - \mu}{3/2}\right)$; $\alpha = 0.0478$;

    (b) $\bar{x} = 24.1225 > 22.5$, do not reject H_0;

    (c) 0.2793.

8.5-3 **(a)** $K(\mu) = \Phi\left(\dfrac{510.77 - \mu}{15}\right)$;

(b) $\alpha = 0.10$;

(c) 0.5000;

(d) (i) 0.0655, **(ii)** 0.0150.

8.5-5 $n = 25, c = 1.6$.

8.5-7 $n = 40, c = 678.38$.

8.5-9 **(a)** $K(p) = \displaystyle\sum_{y=14}^{25} \binom{25}{y} p^y (1-p)^{25-y}$, $0.40 \le p \le 1.0$;

(b) $\alpha = 0.0778$;

(c) $0.1827, 0.3450, 0.7323, 0.9558, 0.9985, 1.0000$;

(e) Yes;

(f) 0.0344.

8.5-11 With $n = 130, c = 8.5, \alpha \approx 0.055, \beta \approx 0.094$.

8.6-1 **(a)** $\dfrac{L(80)}{L(76)} = \exp\left[\dfrac{6}{128}\displaystyle\sum_{i=1}^{n} x_i - \dfrac{624n}{128}\right] \le k$ or

$\bar{x} \le c$;

(b) $n = 43, c = 78$.

8.6-3 **(a)** $\dfrac{L(3)}{L(5)} \le k$ if and only if $\displaystyle\sum_{i=1}^{n} x_i \ge (-15/2)[\ln(k) -$

$\ln(5/3)^n] = c$;

(b) $\bar{x} \ge 4.15$;

(c) $\bar{x} \ge 4.15$;

(d) Yes.

8.6-5 **(a)** $\dfrac{L(50)}{L(\mu_1)} \le k$ if and only if $\bar{x} \le \dfrac{(-72)\ln(k)}{2n(\mu_1 - 50)} +$

$\dfrac{50 + \mu_1}{2} = c$.

8.6-7 **(a)** $\dfrac{L(0.5)}{L(\mu)} \le k$ if and only if $\displaystyle\sum_{i=1}^{n} x_i \ge$

$\dfrac{\ln(k) + n(0.05 - \mu)}{\ln(0.5/\mu)} = c$;

(b) $\displaystyle\sum_{i=1}^{10} x_i \ge 9$.

8.6-9 $K(\theta) = P(Y \le 1) = (1 - \theta)^5 + 5\theta(1 - \theta)^4 = (1 - \theta)^4(1 + 4\theta)$, $0 < \theta \le 1/2$.

8.7-1 **(a)** $|-1.80| > 1.645$, reject H_0;

(b) $|-1.80| < 1.96$, do not reject H_0;

(c) p-value $= 0.0718$.

8.7-3 **(a)** $\bar{x} \ge 230 + 10z_\alpha/\sqrt{n}$ or $\dfrac{\bar{x} - 230}{10/\sqrt{n}} \ge z_\alpha$;

(b) Yes; **(c)** $1.04 < 1.282$, do not reject H_0; **(d)** p-value $= 0.1492$.

8.7-5 **(a)** $|2.10| < 2.306$, do not reject H_0; $0.05 < p$-value < 0.10.

8.7-7 $2.20 > 1.282$, reject H_0; p-value $= 0.0139$.

8.7-9 **(a)** When $\mu_X = \mu_Y = \mu$ and $\sigma_X^2 = \sigma_Y^2 = \sigma^2$,

$$\widehat{\mu} = \dfrac{\sum_{i=1}^{n} x_i + \sum_{i=1}^{m} y_i}{n + m},$$

$$\widehat{\sigma^2} = \dfrac{\sum_{i=1}^{n}(x_i - \widehat{\mu})^2 + \sum_{i=1}^{m}(y_i - \widehat{\mu})^2}{n + m}.$$

When $\mu_X \ne \mu_Y$ and $\sigma_X^2 = \sigma_Y^2 = \sigma^2$,

$\widehat{\mu}_X = \bar{x}$, $\widehat{\mu}_Y = \bar{y}$,

$$\widehat{\sigma^2} = \dfrac{\sum_{i=1}^{n}(x_i - \bar{x})^2 + \sum_{i=1}^{m}(y_i - \bar{y})^2}{n + m}.$$

$$\lambda = \dfrac{1}{\{1 + (\bar{x} - \bar{y})^2/[\sum_{i=1}^{n}(x_i - \bar{x})^2 + \sum_{i=1}^{m}(y_i - \bar{y})^2]\}^{(n+m)/2}},$$

which is a function of a t random variable with $n + m - 2$ degrees of freedom,

$$t = c\dfrac{\bar{x} - \bar{y}}{\sqrt{\sum_{i=1}^{n}(x_i - \bar{x})^2 + \sum_{i=1}^{m}(y_i - \bar{y})^2}};$$

(b) When H_0 is true, $\widehat{\mu}_X = \bar{x}$, $\widehat{\mu}_Y = \bar{y}$,

$$\widehat{\sigma^2} = \dfrac{\sum_{i=1}^{n}(x_i - \bar{x})^2 + \sum_{i=1}^{m}(y_i - \widehat{y})^2}{n + m}.$$

When H_1 is true, $\widehat{\mu}_X = \bar{x}$, $\widehat{\mu}_Y = \bar{y}$,

$$\widehat{\sigma_X^2} = \dfrac{1}{n}\sum_{i=1}^{n}(x_i - \bar{x})^2, \quad \widehat{\sigma_Y^2} = \dfrac{1}{m}\sum_{i=1}^{m}(y_i - \bar{y})^2.$$

$$\lambda = \dfrac{(n + m)^{(n+m)/2}}{n^{n/2}m^{m/2}}\dfrac{\left[\sum_{i=1}^{m}(y_i - \bar{y})^2/\sum_{i=1}^{n}(x_i - \bar{x})^2\right]^{m/2}}{\left[1 + \sum_{i=1}^{m}(y_i - \bar{y})^2/\sum_{i=1}^{n}(x_i - \bar{x})^2\right]^{(n+m)/2}}.$$

This is a function of an F random variable with $m - 1$ and $n - 1$ degrees of freedom,

$$F = \dfrac{\sum_{i=1}^{m}(y_i - \bar{y})^2/(m - 1)}{\sum_{i=1}^{n}(x_i - \bar{x})^2/(n - 1)}.$$

8.7-11 **(a)** $\widehat{\beta} = \dfrac{\sum_{i=1}^{n} x_i y_i}{\sum_{i=1}^{n} x_i^2}$; $\widehat{\sigma^2} = \dfrac{1}{n}\sum_{i=1}^{n}(y_i - \widehat{\beta} x_i)^2$;

$$\lambda = \left[\dfrac{1}{1 + \widehat{\beta}^2 \sum_{i=1}^{n} x_i^2/\sum_{i=1}^{n}(y_i - \widehat{\beta} x_i)^2}\right]^{n/2}.$$

(b) λ is a function $T = c\dfrac{\widehat{\beta}\sqrt{\sum_{i=1}^{n} x_i^2}}{\sqrt{\sum_{i=1}^{n}(y_i - \widehat{\beta} x_i)^2}}$,

a t random variable with $n - 1$ degrees of freedom.

Chapter 9

9.1-1 $1.63 < 7.815$, do not reject if $\alpha = 0.05$; p-value ≈ 0.10.

9.1-3 $7.60 < 16.92$, do not reject H_0.

9.1-5 **(a)** $q_3 \ge 7.815$;

(b) $q_3 = 17 > 7.815$, reject H_0.

9.1-7 Grouping last two classes: $2.75 < 9.210$, not grouping: $3.46 < 11.34$; in either case, do not reject H_0.

9.1-9 Using 10 sets of equal probability, $4.44 < 14.07 = \chi^2_{0.05}(7)$, so do not reject H_0.

9.1-11 $\bar{x} = 320.10, s^2 = 45.56$; using class boundaries $303.5, 307.5, \ldots, 335.5$, $q = 3.21 < 11.07 = \chi^2_{0.05}(5)$, do not reject.

9.2-1 $3.23 < 11.07$, do not reject H_0.

9.2-3 $2.40 < 5.991$, do not reject H_0.

9.2-5 $10.0891 > \chi^2_{0.05}(1) = 3.841$, reject the null hypothesis.

9.2-7 $2.09728 < \chi^2_{0.05}(2) = 7.815$, do not reject the null hypothesis.

9.2-9 $4.149 > \chi^2_{0.05}(1) = 3.841$, reject the null hypothesis; $0.025 < p\text{-value} < 0.05$; $p\text{-value} \approx 0.042$.

9.2-11 $23.78 > 21.03$, reject hypothesis of independence.

9.2-13 **(a)** $39.591 > 9.488$, reject hypothesis of independence;
(b) $7.117 > 5.991$, reject hypothesis of independence;
(c) $11.399 > 9.488$, reject hypothesis of independence.
(d) $\approx 0, 0.0285, 0.0224$.

9.3-1 $7.875 > 4.26$, reject H_0.

9.3-3 $13.773 > 4.07$, reject H_0.

9.3-5 **(a)**

Source	SS	DF	MS	F	p-value
Treatment	31.112	2	15.556	22.33	0.000
Error	29.261	42	0.697		
Total	60.372	44			

(b) The respective means are 23.114, 22.556, and 21.120, with the eggs of the shortest lengths in the nests of the smallest bird.

9.3-7 $14.757 > 2.87$, reject H_0.

9.3-9 **(a)** $F \geq 4.07$;
(b)

Source	SS	DF	MS	F	p-value
Treatment	3214.9	3	1071.6	4.1059	0.0489
Error	2088.0	8	261.0		
Total	5302.9	11			

$4.1059 > 4.07$, reject H_0;
(c) $4.1059 < 5.42$, do not reject H_0;
(d) $0.025 < p\text{-value} < 0.05$, $p\text{-value} \approx 0.05$.

9.3-11 $10.224 > 4.26$, reject H_0.

9.3-13 **(a)** $F \geq 5.61$;

(b)

Source	SS	DF	MS	F	p-value
Treatment	1.6092	2	0.8046	6.3372	0.0062
Error	3.0470	24	0.1270		
Total	4.6562	26			

$6.3372 > 5.61$, reject H_0.

9.3-15 **(a)** $F = 12.47$, there seems to be a difference in feed supplements;
(b) Yes, supplement B looks best and supplement C the poorest.

9.4-1 $18.00 > 5.14$, reject H_A.

9.4-3 **(a)** $7.624 > 4.46$, reject H_A;
(b) $15.539 > 3.84$, reject H_B.

9.4-5 **(a)** $1.723 < 2.90$, accept H_{AB};
(b) $5.533 > 4.15$, reject H_A;
(c) $28.645 > 2.90$, reject H_B.

9.4-7 **(a)** $1.727 < 2.36$, do not reject H_{AB};
(b) $2.238 < 3.26$, do not reject H_A;
(c) $2.063 < 2.87$, do not reject H_B.

9.4-9 **(a)**

Source	SS	DF	MS	F	p-value
Smoking History	84.899	2	42.449	12.90	0.000
Test	298.072	2	149.036	45.28	0.000
Interaction	2.815	4	0.704	0.21	0.927
Error	59.247	18	3.291		
Total	445.032	26			

9.5-1

2^2 Design				
Run	A	B	AB	Observations
1	−	−	+	X_1
2	+	−	−	X_2
3	−	+	−	X_3
4	+	+	+	X_4

(a) $[A] = (-X_1 + X_2 - X_3 + X_4)/4$,
$[B] = (-X_1 - X_2 + X_3 + X_4)/4$,
$[AB] = (X_1 - X_2 - X_3 + X_4)/4$;
(b) It is sufficient to compare the coefficients on both sides of the equations of X_1^2, X_1X_2, X_1X_3, and X_1X_4, which are $3/4, -1/2, -1/2$, and $-1/2$, respectively;
(c) Each is $\chi^2(1)$.

9.5-3 [A] is $N(0, \sigma^2/2)$, so $E[(X_2 - X_1)^2/4] = \sigma^2/2$ or $E[(X_2 - X_1)^2/2] = \sigma^2$.

9.5-5 **(a)** [A] $= -4$, [B] $= 12$, [C] $= -1.125$, [D] $= -2.75$, [AB] $= 0.5$, [AC] $= 0.375$, [AD] $= 0$, [BC] $= -0.625$, [BD] $= 2.25$, [CD] $= -0.125$, [ABC] $= -0.375$, [ABD] $= 0.25$, [ACD] $= -0.125$, [BCD] $= -0.375$, [ABCD] $= -0.125$;

(b) There is clearly a temperature (B) effect. There is also a catalyst charge (A) effect and probably a concentration (D) and a temperature–concentration (BD) effect.

9.6-1 $4.359 > 2.306$, reject H_0.

9.6-3 $2.47 > 1.96$, reject H_0.

9.6-5 $[0.419, 0.802]$.

9.6-7 $|r| = 0.252 < 0.6613$, do not reject H_0.

9.6-11 $n = 9$.

9.7-1 **(a)** $\bar{\bar{x}} = 158.97$, $\bar{s} = 12.1525$, $\bar{R} = 30.55$; **(f)** Yes.

9.7-3 **(a)** $\bar{\bar{x}} = 5.176 + 330 = 335.176$, $\bar{s} = 0.5214$, $\bar{R} = 1.294$; **(f)** No.

9.7-5 **(b)** $E(\sqrt{Y}) = \dfrac{\sqrt{2}\,\Gamma\left(\dfrac{n}{2}\right)}{\Gamma\left(\dfrac{n-1}{2}\right)}$;

(c) $S = \dfrac{\sigma\sqrt{Y}}{\sqrt{n-1}}$ so $E(S) = \dfrac{\sqrt{2}\,\Gamma\left(\dfrac{n}{2}\right)}{\sqrt{n-1}\,\Gamma\left(\dfrac{n-1}{2}\right)}\,\sigma$.

9.7-7 $\text{LCL} = 0$, $\text{UCL} = 0.1024$; **(a)** 0.0378; **(b)** 0.1752.

9.7-9 **(a)** $\text{LCL} = 0$, $\text{UCL} = 13.99$; **(b)** Yes.

REVIEW OF SELECTED MATHEMATICAL TECHNIQUES

Appendix

D

D.1 Algebra of Sets

D.2 Mathematical Tools for the Hypergeometric Distribution

D.3 Limits

D.4 Infinite Series

D.5 Integration

D.6 Multivariate Calculus

D.1 ALGEBRA OF SETS

The totality of objects under consideration is called the **universal set** and is denoted by S. Each object in S is called an **element** of S. If a set A is a collection of elements that are also in S, then A is said to be a **subset** of S. In applications of probability, S usually denotes the **sample space**. An **event** A is a collection of possible outcomes of an experiment and is a subset of S. We say that event A *has occurred* if the outcome of the experiment is an element of A. The set or event A may be described by listing all of its elements or by defining the properties that its elements must satisfy.

Example D.1-1

A four-sided die, called a tetrahedron, has four faces that are equilateral triangles. These faces are numbered 1, 2, 3, 4. When the tetrahedron is rolled, the outcome of the experiment is the number of the face that is down. If the tetrahedron is rolled twice and we keep track of the first roll and the second roll, then the sample space is that displayed in Figure D.1-1.

Let A be the event that the second roll is a 1 or a 2. That is,

$$A = \{(x,y): \ y = 1 \text{ or } y = 2\}.$$

Let

$$B = \{(x,y): \ x + y = 6\} = \{(2,4),(3,3),(4,2)\},$$

and let

$$C = \{(x,y): \ x + y \geq 7\} = \{(4,3),(3,4),(4,4)\}.$$

Events A, B, and C are shown in Figure D.1-1. ■

When a is an element of A, we write $a \in A$. When a is not an element of A, we write $a \notin A$. So, in Example D.1-1, we have $(3,1) \in A$ and $(1,3) \notin A$. If every

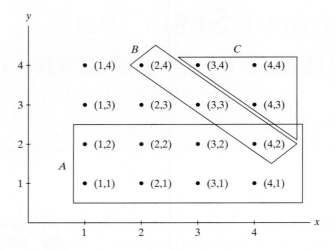

Figure D.1-1 Sample space for two rolls of a four-sided die

element of a set A is also an element of a set B, then A is a **subset** of B and we write $A \subset B$. In probability, if event B occurs whenever event A occurs, then $A \subset B$. The two sets A and B are equal (i.e., $A = B$) if $A \subset B$ and $B \subset A$. Note that it is always true that $A \subset A$ and $A \subset S$, where S is the universal set. We denote the subset that contains no elements by $\emptyset$. This set is called the **null**, or **empty**, set. For all sets A, $\emptyset \subset A$.

The set of elements of either A or B or possibly both A and B is called the **union** of A and B and is denoted $A \cup B$. The set of elements of both A and B is called the **intersection** of A and B and is denoted $A \cap B$. The **complement** of a set A is the set of elements of the universal set S that are not in the set A and is denoted A'. In probability, if A and B are two events, the event that at least one of the two events has occurred is denoted by $A \cup B$, and the event that both events have occurred is denoted by $A \cap B$. The event that A has not occurred is denoted by A', and the event that A has not occurred but B has occurred is denoted by $A' \cap B$. If $A \cap B = \emptyset$, we say that A and B are **mutually exclusive**. In Example D.1-1, $B \cup C = \{(x,y): x + y \geq 6\}$, $A \cap B = \{(4,2)\}$, and $A \cap C = \emptyset$. Note that A and C are mutually exclusive. Also, $C' = \{(x,y): x + y \leq 6\}$.

The operations of union and intersection may be extended to more than two sets. Let $A_1, A_2, \ldots, A_n$ be a finite collection of sets. Then the **union**

$$A_1 \cup A_2 \cup \cdots \cup A_n = \bigcup_{k=1}^{n} A_k$$

is the set of elements that belong to at least one A_k, $k = 1, 2, \ldots, n$. The **intersection**

$$A_1 \cap A_2 \cap \cdots \cap A_n = \bigcap_{k=1}^{n} A_k$$

is the set of all elements that belong to every A_k, $k = 1, 2, \ldots, n$. Similarly, let $A_1, A_2, \ldots, A_n, \ldots$ be a countable collection of sets. Then x belongs to the **union**

$$A_1 \cup A_2 \cup A_3 \cup \cdots = \bigcup_{k=1}^{\infty} A_k$$

if x belongs to at least one A_k, $k = 1, 2, 3, \ldots$. Also, x belongs to the **intersection**

$$A_1 \cap A_2 \cap A_3 \cap \cdots = \bigcap_{k=1}^{\infty} A_k$$

if x belongs to every A_k, $k = 1, 2, 3, \ldots$.

Example D.1-2

Let

$$A_k = \left\{ x \colon \frac{10}{k+1} \leq x \leq 10 \right\}, \qquad k = 1, 2, 3, \ldots.$$

Then

$$\bigcup_{k=1}^{8} A_k = \left\{ x \colon \frac{10}{9} \leq x \leq 10 \right\};$$

$$\bigcup_{k=1}^{\infty} A_k = \{ x \colon 0 < x \leq 10 \}.$$

Note that the number zero is not in this latter union, since it is not in at least one of the sets $A_1, A_2, A_3, \ldots$. Also,

$$\bigcap_{k=1}^{8} A_k = \{ x \colon 5 \leq x \leq 10 \} = A_1$$

and

$$\bigcap_{k=1}^{\infty} A_k = \{ x \colon 5 \leq x \leq 10 \} = A_1,$$

since $A_1 \subset A_k$, $k = 1, 2, 3, \ldots$. ∎

A convenient way to illustrate operations on sets is with a **Venn** diagram. In Figure D.1-2, the universal set S is represented by the rectangle and its interior, and the subsets of S are represented by the points enclosed by the ellipses, as well as by the complement of the union of those subsets. The sets under consideration are the shaded regions.

Set operations satisfy several properties. For example, if A, B, and C are subsets of S, we have the following laws:

Commutative Laws: $A \cup B = B \cup A$

$A \cap B = B \cap A$

Associative Laws: $(A \cup B) \cup C = A \cup (B \cup C)$

$(A \cap B) \cap C = A \cap (B \cap C)$

Distributive Laws $A \cap (B \cup C) = (A \cap B) \cup (A \cap C)$

$A \cup (B \cap C) = (A \cup B) \cap (A \cup C)$

De Morgan's Laws: $(A \cup B)' = A' \cap B'$

$(A \cap B)' = A' \cup B'$.

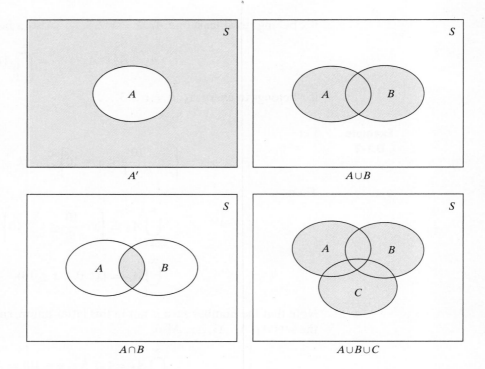

Figure D.1-2 Algebra of sets

A Venn diagram will be used to justify the first of De Morgan's laws. In Figure D.1-3(a), $A \cup B$ is represented by horizontal lines, and thus $(A \cup B)'$ is the region represented by vertical lines. In Figure D.1-3(b), A' is indicated with horizontal lines and B' is indicated with vertical lines. An element belongs to $A' \cap B'$ if it belongs to both A' and B'. Thus, the crosshatched region represents $A' \cap B'$. Clearly, this crosshatched region is the same as that shaded with vertical lines in Figure D.1-3(a).

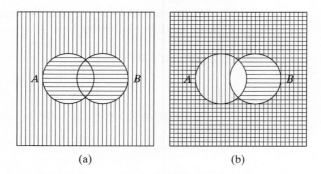

Figure D.1-3 Venn diagrams illustrating De Morgan's laws

D.2 MATHEMATICAL TOOLS FOR THE HYPERGEOMETRIC DISTRIBUTION

Let X have a hypergeometric distribution. That is, the pmf of X is

$$f(x) = \frac{\binom{N_1}{x}\binom{N_2}{n-x}}{\binom{N_1+N_2}{n}}$$

$$= \frac{\binom{N_1}{x}\binom{N_2}{n-x}}{\binom{N}{n}}, \qquad x \le n,\ x \le N_1,\ n - x \le N_2.$$

To show that $\sum_{x=0}^{n} f(x) = 1$ and to find the mean and variance of X, we use the following theorem.

Theorem D.2-1

$$\binom{N}{n} = \sum_{x=0}^{n} \binom{N_1}{x}\binom{N_2}{n-x},$$

where $N = N_1 + N_2$ and it is understood that $\binom{k}{j} = 0$ if $j > k$.

Proof Because $N = N_1 + N_2$, we have the identity

$$(1+y)^N \equiv (1+y)^{N_1}(1+y)^{N_2}. \qquad \text{(D.2-1)}$$

We will expand each of these binomials, and since the polynomials on each side are identically equal, the coefficients of y^n on each side of Equation D.2-1 must be equal. Using the binomial expansion, we find that the expansion of the left side of Equation D.2-1 is

$$(1+y)^N = \sum_{k=0}^{N} \binom{N}{k} y^k$$

$$= \binom{N}{0} + \binom{N}{1}y + \cdots + \binom{N}{n}y^n + \cdots + \binom{N}{N}y^N.$$

The right side of Equation D.2-1 becomes

$$(1+y)^{N_1}(1+y)^{N_2} = \left[\binom{N_1}{0} + \binom{N_1}{1}y + \cdots + \binom{N_1}{n}y^n + \cdots + \binom{N_1}{N_1}y^{N_1}\right]$$

$$\times \left[\binom{N_2}{0} + \binom{N_2}{1}y + \cdots + \binom{N_2}{n}y^n + \cdots + \binom{N_2}{N_2}y^{N_2}\right].$$

The coefficient of y^n in this product is

$$\binom{N_1}{0}\binom{N_2}{n} + \binom{N_1}{1}\binom{N_2}{n-1} + \cdots + \binom{N_1}{n}\binom{N_2}{0} = \sum_{x=0}^{n}\binom{N_1}{x}\binom{N_2}{n-x},$$

and this sum must be equal to $\binom{N}{n}$, the coefficient of y^n on the left side of Equation D.2-1. $\qquad\square$

Using Theorem D.2-1, we find that if X has a hypergeometric distribution with pmf $f(x)$, then

$$\sum_{x=0}^{n} f(x) = \sum_{x=0}^{n} \frac{\binom{N_1}{x}\binom{N_2}{n-x}}{\binom{N}{n}} = 1.$$

To find the mean and variance of a hypergeometric random variable, it is useful to note that, with $n > 0$,

$$\binom{N}{n} = \frac{N!}{n!\,(N-n)!} = \frac{N}{n} \cdot \frac{(N-1)!}{(n-1)!\,(N-n)!} = \frac{N}{n}\binom{N-1}{n-1}.$$

The mean of a hypergeometric random variable X is

$$\mu = \sum_{x=0}^{n} xf(x)$$

$$= \frac{\displaystyle\sum_{x=1}^{n} x \cdot \frac{N_1!}{x!\,(N_1-x)!} \cdot \frac{N_2!}{(n-x)!\,(N_2-n+x)!}}{\binom{N}{n}}$$

$$= \frac{N_1 \displaystyle\sum_{x=1}^{n} \frac{(N_1-1)!}{(x-1)!\,(N_1-x)!} \cdot \frac{N_2!}{(n-x)!\,(N_2-n+x)!}}{\binom{N}{n}}.$$

If we now make the change of variables $k = x - 1$ in the summation and replace

$$\binom{N}{n} \qquad \text{with} \qquad \binom{N}{n}\binom{N-1}{n-1}$$

in the denominator, the previous equation becomes

$$\mu = \frac{N_1}{\binom{N}{n}} \frac{\displaystyle\sum_{k=0}^{n-1} \frac{(N_1 - 1)!}{k!\,(N_1 - 1 - k)!} \cdot \frac{N_2!}{(n-k-1)!\,(N_2 - n + k + 1)!}}{\binom{N-1}{n-1}}$$

$$= n\left(\frac{N_1}{N}\right) \frac{\displaystyle\sum_{k=0}^{n-1} \binom{N_1 - 1}{k}\binom{N_2}{n-1-k}}{\binom{N-1}{n-1}} = n\left(\frac{N_1}{N}\right),$$

because, from Theorem D.2-1, the summation in the expression for μ is equal to $\binom{N-1}{n-1}$.

Note that

$$\mathrm{Var}(X) = \sigma^2 = E[(X - \mu)^2]$$

$$= E[X^2] - \mu^2$$

$$= E[X(X - 1)] + E(X) - \mu^2.$$

So, to find the variance of X, we first find $E[X(X - 1)]$:

$$E[X(X - 1)] = \sum_{x=0}^{n} x(x - 1)f(x)$$

$$= \frac{\displaystyle\sum_{x=2}^{n} x(x - 1) \frac{N_1!}{x!\,(N_1 - x)!} \cdot \frac{N_2!}{(n-x)!\,(N_2 - n + x)!}}{\binom{N}{n}}$$

$$= N_1(N_1 - 1) \frac{\displaystyle\sum_{x=2}^{n} \frac{(N_1 - 2)!}{(x-2)!\,(N_1 - x)!} \cdot \frac{N_2!}{(n-x)!\,(N_2 - n + x)!}}{\binom{N}{n}}.$$

In the summation, let $k = x - 2$, and in the denominator, note that

$$\binom{N}{n} = \frac{N!}{n!(N-n)!} = \frac{N(N-1)}{n(n-1)}\binom{N-2}{n-2}.$$

Thus, from Theorem D.2-1,

$$E[X(X-1)] = \frac{N_1(N_1-1)}{\frac{N(N-1)}{n(n-1)}} \sum_{k=0}^{n-2} \frac{\binom{N_1-2}{k}\binom{N_2}{n-2-k}}{\binom{N-2}{n-2}}$$

$$= \frac{N_1(N_1-1)(n)(n-1)}{N(N-1)}.$$

Hence, the variance of a hypergeometric random variable is, after some algebraic manipulations,

$$\sigma^2 = \frac{N_1(N_1-1)(n)(n-1)}{N(N-1)} + \frac{nN_1}{N} - \left(\frac{nN_1}{N}\right)^2$$

$$= n\left(\frac{N_1}{N}\right)\left(\frac{N_2}{N}\right)\left(\frac{N-n}{N-1}\right).$$

D.3 LIMITS

We refer the reader to the many fine books on calculus for the definition of a limit and the other concepts used in that subject. Here we simply remind you of some of the techniques we find most useful in probability and statistics.

Early in a calculus course, the existence of the following limit, denoted by the letter e, is discussed:

$$e = \lim_{t\to 0}(1+t)^{1/t} = \lim_{n\to\infty}\left(1+\frac{1}{n}\right)^n.$$

Of course, e is an irrational number, which, to six significant figures, equals 2.71828.

Often, it is rather easy to see the value of certain limits. For example, with $-1 < r < 1$, the sum of the geometric progression allows us to write

$$\lim_{n\to\infty}(1+r+r^2+\cdots+r^{n-1}) = \lim_{n\to\infty}\left(\frac{1-r^n}{1-r}\right) = \frac{1}{1-r}.$$

That is, the limit of the ratio $(1-r^n)/(1-r)$ is not difficult to find because $\lim_{n\to\infty} r^n = 0$ when $-1 < r < 1$.

However, it is not that easy to determine the limit of every ratio; for example, consider

$$\lim_{b\to\infty}(be^{-b}) = \lim_{b\to\infty}\left(\frac{b}{e^b}\right).$$

Since both the numerator and the denominator of the latter ratio are unbounded, we can use **L'Hôpital's rule**, taking the limit of the ratio of the derivative of the numerator to the derivative of the denominator. We then have

$$\lim_{b\to\infty}\left(\frac{b}{e^b}\right) = \lim_{b\to\infty}\left(\frac{1}{e^b}\right) = 0.$$

This result can be used in the evaluation of the integral

$$\int_0^\infty xe^{-x}\,dx = \lim_{b\to\infty} \int_0^b xe^{-x}\,dx$$

$$= \lim_{b\to\infty} \left[-xe^{-x} - e^{-x}\right]_0^b$$

$$= \lim_{b\to\infty} \left[1 - be^{-b} - e^{-b}\right] = 1.$$

Note that

$$\frac{d}{dx}\left[-xe^{-x} - e^{-x}\right] = xe^{-x} - e^{-x} + e^{-x} = xe^{-x};$$

that is, $-xe^{-x} - e^{-x}$ is the antiderivative of xe^{-x}.

Another limit of importance is

$$\lim_{n\to\infty}\left(1 + \frac{b}{n}\right)^n = \lim_{n\to\infty} e^{n\ln(1+b/n)},$$

where b is a constant.

Since the exponential function is continuous, the limit can be taken to the exponent. That is,

$$\lim_{n\to\infty} \exp[n\ln(1 + b/n)] = \exp[\lim_{n\to\infty} n\ln(1 + b/n)].$$

By L'Hôpital's rule, the limit in the exponent is equal to

$$\lim_{n\to\infty} \frac{\ln(1 + b/n)}{1/n} = \lim_{n\to\infty} \frac{\dfrac{-b/n^2}{1 + b/n}}{-1/n^2} = \lim_{n\to\infty} \frac{b}{1 + b/n} = b.$$

Since this limit is equal to b, the original limit is

$$\lim_{n\to\infty}\left(1 + \frac{b}{n}\right)^n = e^b.$$

Applications of this limit in probability occur with $b = -1$, yielding

$$\lim_{n\to\infty}\left(1 - \frac{1}{n}\right)^n = e^{-1}.$$

D.4 INFINITE SERIES

A function $f(x)$ possessing derivatives of all orders at $x = b$ can be expanded in the following **Taylor series**:

$$f(x) = f(b) + \frac{f'(b)}{1!}(x - b) + \frac{f''(b)}{2!}(x - b)^2 + \frac{f'''(b)}{3!}(x - b)^3 + \cdots.$$

If $b = 0$, we obtain the special case that is often called the **Maclaurin series**;

$$f(x) = f(0) + \frac{f'(0)}{1!}x + \frac{f''(0)}{2!}x^2 + \frac{f'''(0)}{3!}x^3 + \cdots.$$

For example, if $f(x) = e^x$, so that all derivatives of $f(x) = e^x$ are $f^{(r)}(x) = e^x$, then $f^{(r)}(0) = 1$, for $r = 1, 2, 3, \ldots$. Thus, the Maclaurin series expansion of $f(x) = e^x$ is

$$e^x = 1 + \frac{x}{1!} + \frac{x^2}{2!} + \frac{x^3}{3!} + \frac{x^4}{4!} + \cdots .$$

The **ratio test**,

$$\lim_{n \to \infty} \left| \frac{x^n/n!}{x^{n-1}/(n-1)!} \right| = \lim_{n \to \infty} \left| \frac{x}{n} \right| = 0,$$

shows that the Maclaurin series expansion of e^x converges for all real values of x.
Note, for examples, that

$$e = 1 + \frac{1}{1!} + \frac{1}{2!} + \frac{1}{3!} + \cdots$$

and

$$e^{-1} = 1 - \frac{1}{1!} + \frac{1}{2!} - \frac{1}{3!} + \cdots + \frac{(-1)^n}{n!} + \cdots .$$

As another example, consider

$$h(w) = (1 - w)^{-r},$$

where r is a positive integer. Here

$$h'(w) = r(1 - w)^{-(r+1)},$$

$$h''(w) = (r)(r + 1)(1 - w)^{-(r+2)},$$

$$h'''(w) = (r)(r + 1)(r + 2)(1 - w)^{-(r+3)},$$

$$\vdots .$$

In general, $h^{(k)}(0) = (r)(r + 1) \cdots (r + k - 1) = (r + k - 1)!/(r - 1)!$. Thus,

$$(1 - w)^{-r} = 1 + \frac{(r + 1 - 1)!}{(r - 1)! \, 1!} w + \frac{(r + 2 - 1)!}{(r - 1)! \, 2!} w^2 + \cdots + \frac{(r + k - 1)!}{(r - 1)! \, k!} w^k + \cdots$$

$$= \sum_{k=0}^{\infty} \binom{r + k - 1}{r - 1} w^k.$$

This is often called the negative binomial series. Using the ratio test, we obtain

$$\lim_{n \to \infty} \left| \frac{w^n (r + n - 1)!/[(r - 1)! \, n!]}{w^{n-1}(r + n - 2)!/[(r - 1)! \, (n - 1)!]} \right| = \lim_{n \to \infty} \left| \frac{w(r + n - 1)}{n} \right| = |w|.$$

Thus, the series converges when $|w| < 1$, or $-1 < w < 1$.

A negative binomial random variable receives its name from this negative binomial series. Before showing that relationship, we note that, for $-1 < w < 1$,

$$h(w) = \sum_{k=0}^{\infty} \binom{r+k-1}{r-1} w^k = (1-w)^{-r},$$

$$h'(w) = \sum_{k=1}^{\infty} \binom{r+k-1}{r-1} k w^{k-1} = r(1-w)^{-r-1},$$

$$h''(w) = \sum_{k=2}^{\infty} \binom{r+k-1}{r-1} k(k-1) w^{k-2} = r(r+1)(1-w)^{-r-2}.$$

The pmf of a negative binomial random variable X is

$$g(x) = \binom{x-1}{r-1} p^r q^{x-r}, \qquad x = r, r+1, r+2, \ldots.$$

In the series expansion for $h(w) = (1-w)^{-r}$, let $x = k + r$. Then

$$\sum_{x=r}^{\infty} \binom{x-1}{r-1} w^{x-r} = (1-w)^{-r}.$$

Letting $w = q$ in this equation, we see that

$$\sum_{x=r}^{\infty} g(x) = \sum_{x=r}^{\infty} \binom{x-1}{r-1} p^r q^{x-r} = p^r (1-q)^{-r} = 1.$$

That is, $g(x)$ does satisfy the properties of a pmf.

To find the mean of X, we first find

$$E(X - r) = \sum_{x=r}^{\infty} (x-r) \binom{x-1}{r-1} p^r q^{x-r} = \sum_{x=r+1}^{\infty} (x-r) \binom{x-1}{r-1} p^r q^{x-r}.$$

Letting $k = x - r$ in this latter summation and using the expansion of $h'(w)$ gives us

$$E(X - r) = \sum_{k=1}^{\infty} (k) \binom{r+k-1}{r-1} p^r q^k$$

$$= p^r q \sum_{k=1}^{\infty} (k) \binom{r+k-1}{r-1} k q^{k-1}$$

$$= p^r q r (1-q)^{-r-1} = r\left(\frac{q}{p}\right).$$

Thus,

$$E(X) = r + r\left(\frac{q}{p}\right)$$

$$= r\left(1 + \frac{q}{p}\right) = r\left(\frac{1}{p}\right).$$

Similarly, using $h''(w)$, we can show that

$$E[(X-r)(X-r-1)] = \left(\frac{q^2}{p^2}\right)(r)(r+1).$$

Hence,

$$\text{Var}(X) = \text{Var}(X-r) = \left(\frac{q^2}{p^2}\right)(r)(r+1) + r\left(\frac{q}{p}\right) - r^2\left(\frac{q^2}{p^2}\right) = r\left(\frac{q}{p^2}\right).$$

A special case of the negative binomial series occurs when $r = 1$, whereupon we obtain the well-known geometric series

$$(1-w)^{-1} = 1 + w + w^2 + w^3 + \cdots,$$

provided that $-1 < w < 1$.

The geometric series gives its name to the geometric probability distribution. Perhaps you recall the geometric series

$$g(r) = \sum_{k=0}^{\infty} ar^k = \frac{a}{1-r}, \tag{D.4-1}$$

for $-1 < r < 1$. To find the mean and the variance of a geometric random variable X, simply let $r = 1$ in the respective formulas for the mean and the variance of a negative binomial random variable. However, if you want to find the mean and variance directly, you can use

$$g'(r) = \sum_{k=1}^{\infty} akr^{k-1} = \frac{a}{(1-r)^2} \tag{D.4-2}$$

and

$$g''(r) = \sum_{k=2}^{\infty} ak(k-1)r^{k-2} = \frac{2a}{(1-r)^3} \tag{D.4-3}$$

to find $E(X)$ and $E[X(X-1)]$, respectively.

In applications associated with the geometric random variable, it is also useful to recall that the nth partial sum of a geometric series is

$$s_n = \sum_{k=0}^{n-1} ar^k = \frac{a(1-r^n)}{1-r}.$$

A bonus in this section is a logarithmic series that produces a useful tool in daily life. Consider

$$f(x) = \ln(1+x),$$

$$f'(x) = (1+x)^{-1},$$

$$f''(x) = (-1)(1+x)^{-2},$$

$$f'''(x) = (-1)(-2)(1+x)^{-3},$$

$$\vdots$$

Thus, $f^{(r)}(0) = (-1)^{r-1}(r-1)!$ and

$$\ln(1+x) = \frac{0!}{1!}x - \frac{1!}{2!}x^2 + \frac{2!}{3!}x^3 - \frac{3!}{4!}x^4 + \cdots$$

$$= x - \frac{x^2}{2} + \frac{x^3}{3} - \frac{x^4}{4} + \cdots ,$$

which converges for $-1 < x < 1$.

Now consider the following question: "How long does it take money to double in value if the interest rate is i?" Assuming that compounding is on an annual basis and that you begin with $1, after one year you have $(1 + i)$, and after two years the number of dollars you have is

$$(1 + i) + i(1 + i) = (1 + i)^2.$$

Continuing this process, we find that the equation that we have to solve is

$$(1 + i)^n = 2,$$

the solution of which is

$$n = \frac{\ln 2}{\ln(1 + i)}.$$

To approximate the value of n, recall that $\ln 2 \approx 0.693$ and use the series expansion of $f(x) = \ln(1 + x)$ to obtain

$$n \approx \frac{0.693}{i - \frac{i^2}{2} + \frac{i^3}{3} - \cdots}.$$

Due to the alternating series in the denominator, the denominator is a little less than i. Frequently, brokers increase the numerator a little (say, to 0.72) and simply divide by i, obtaining the "well-known Rule of 72," namely,

$$n \approx \frac{72}{100i}. \tag{D.4-4}$$

For example, if $i = 0.08$, then $n \approx 72/8 = 9$ provides an excellent approximation. (The answer is about 9.006.) Many people find that the Rule of 72 is extremely useful in dealing with money matters.

D.5 INTEGRATION

Say $F'(t) = f(t)$, $a \le t \le b$. Then

$$\int_a^b f(t)\,dt = F(b) - F(a).$$

Thus, if $u(x)$ is such that $u'(x)$ exists and $a \le u(x)$, then

$$\int_a^{u(x)} f(t)\,dt = F[u(x)] - F(a).$$

Taking derivatives of this latter equation, we obtain

$$\frac{d}{dx}\left[\int_a^{u(x)} f(t)\, dt\right] = F'[u(x)]u'(x) = f[u(x)]u'(x).$$

For example, with $0 < v$,

$$\frac{d}{dv}\left[2\int_0^{\sqrt{v}} \frac{1}{\sqrt{2\pi}} e^{-z^2/2}\, dz\right] = \left(\frac{2}{\sqrt{2\pi}} e^{-v/2}\right)\frac{1}{2\sqrt{v}} = \frac{v^{(1/2)-1} e^{-v/2}}{\sqrt{\pi}\, 2^{1/2}}.$$

This formula is needed in proving that if Z is $N(0,1)$, then Z^2 is $\chi^2(1)$.

The preceding example could be worked by first changing variables in the integral—that is, first using the fact that

$$\int_a^b f(x)\, dx = \int_{u(a)}^{u(b)} f[w(y)]\, w'(y)\, dy,$$

where the monotonically increasing (decreasing) function $x = w(y)$ has derivative $w'(y)$ and inverse function $y = u(x)$. In that example, $a = 0$, $b = \sqrt{v}$, $z = \sqrt{t}$, $z' = 1/2\sqrt{t}$, and $t = z^2$, so that

$$2\int_0^{\sqrt{v}} \frac{1}{\sqrt{2\pi}} e^{-z^2/2}\, dz = 2\int_0^v \frac{1}{\sqrt{2\pi}} e^{-t/2}\left(\frac{1}{2\sqrt{t}}\right) dt.$$

The derivative of the latter, by one form of the fundamental theorem of calculus, is

$$2\frac{1}{\sqrt{2\pi}} e^{-v/2}\left(\frac{1}{2\sqrt{v}}\right) = \frac{v^{(1/2)-1} e^{-v/2}}{\sqrt{\pi}\, 2^{1/2}}.$$

Integration by parts is frequently needed. It is based upon the derivative of the product of two functions of x—say, $u(x)$ and $v(x)$. The derivative is

$$\frac{d}{dx}[u(x)v(x)] = u(x)v'(x) + v(x)u'(x).$$

Thus,

$$[u(x)v(x)]_a^b = \int_a^b u(x)v'(x)\, dx + \int_a^b v(x)u'(x)\, dx$$

or, equivalently,

$$\int_a^b u(x)v'(x)\, dx = [u(x)v(x)]_a^b - \int_a^b v(x)u'(x)\, dx.$$

For example, letting $u(x) = x$ and $v'(x) = e^{-x}$, we obtain

$$\int_0^b xe^{-x}\, dx = \left[-xe^{-x}\right]_0^b - \int_0^b (1)(-e^{-x})\, dx$$

$$= -be^{-b} + \left[-e^{-x}\right]_0^b = -be^{-b} - e^{-b} + 1,$$

because $u'(x) = 1$ and $v(x) = -e^{-x}$.

With some thought about the product rule of differentiation, we see that it is not always necessary to assign $u(x)$ and $v'(x)$, however. For example, an integral such as

$$\int_0^b x^3 e^{-x}\, dx$$

would require integration by parts three times, the first of which would assign $u(x) = x^3$ and $v'(x) = e^{-x}$. But note that

$$\frac{d}{dx}(-x^3 e^{-x}) = x^3 e^{-x} - 3x^2 e^{-x}.$$

That is, $-x^3 e^{-x}$ is "almost" the antiderivative of $x^3 e^{-x}$—except for the undesirable term $-3x^2 e^{-x}$. Clearly,

$$\frac{d}{dx}(-x^3 e^{-x} - 3x^2 e^{-x}) = x^3 e^{-x} - 3x^2 e^{-x} + 3x^2 e^{-x} - 6x e^{-x} = x^3 e^{-x} - 6x e^{-x}.$$

So we eliminated that undesirable term $-3x^2 e^{-x}$, but got another one, namely, $-6x e^{-x}$. However,

$$\frac{d}{dx}(-x^3 e^{-x} - 3x^2 e^{-x} - 6x e^{-x}) = x^3 e^{-x} - 6e^{-x},$$

and finally,

$$\frac{d}{dx}(-x^3 e^{-x} - 3x^2 e^{-x} - 6x e^{-x} - 6e^{-x}) = x^3 e^{-x}.$$

That is,

$$-x^3 e^{-x} - 3x^2 e^{-x} - 6x e^{-x} - 6e^{-x}$$

is the antiderivative of $x^3 e^{-x}$ and can be written down without ever assigning u and v.

As practice in this technique, consider

$$\int_0^{\pi/2} x^2 \cos x \, dx = \left[x^2 \sin x + 2x \cos x - 2\sin x \right]_0^{\pi/2}.$$

Now, $x^2 \sin x$ is our first guess because we obtain $x^2 \cos x$ when we differentiate the $\sin x$ factor. But we get the undesirable term $2x \sin x$. That is why we add $2x \cos x$, as the derivative of $\cos x$ is $-\sin x$ and $-2x \sin x$ eliminates $2x \sin x$. But the second term of the derivative of $2x \cos x$ is $2 \cos x$, which we get rid of by taking the derivative of the next term, $-2 \sin x$.

Possibly the best advice is to take the derivative of the right-hand member, here

$$x^2 \sin x + 2x \cos x - 2\sin x,$$

and note how the terms cancel, leaving only $x^2 \cos x$. Then practice on integrals such as

$$\int x^4 e^{-x} \, dx, \quad \int x^3 \sin x \, dx, \quad \int x^5 e^x \, dx.$$

D.6 MULTIVARIATE CALCULUS

We really only make some suggestions about functions of two variables, say,

$$z = f(x, y).$$

But these remarks can be extended to more than two variables. The two first *partial derivatives* with respect to x and y, denoted, respectively, by $\dfrac{\partial z}{\partial x}$ and $\dfrac{\partial z}{\partial y}$, can be found in the usual manner of differentiating by treating the "other" variable as a constant. For instance,

$$\frac{\partial(x^2y + \sin x)}{\partial x} = 2xy + \cos x$$

and

$$\frac{\partial(e^{xy^2})}{\partial y} = (e^{xy^2})(2xy).$$

The second partial derivatives are simply first partial derivatives of the first partial derivatives. If $z = e^{xy^2}$, then

$$\frac{\partial}{\partial x}\left(\frac{\partial z}{\partial y}\right) = \frac{\partial}{\partial x}(2xye^{xy^2}) = 2xye^{xy^2}(y^2) + 2ye^{xy^2}.$$

For notation, we use

$$\frac{\partial}{\partial x}\left(\frac{\partial z}{\partial x}\right) = \frac{\partial^2 z}{\partial x^2}, \qquad \frac{\partial}{\partial x}\left(\frac{\partial z}{\partial y}\right) = \frac{\partial^2 z}{\partial x \partial y},$$

$$\frac{\partial}{\partial y}\left(\frac{\partial z}{\partial x}\right) = \frac{\partial^2 z}{\partial y \partial x}, \qquad \frac{\partial}{\partial y}\left(\frac{\partial z}{\partial y}\right) = \frac{\partial^2 z}{\partial y^2},$$

In general,

$$\frac{\partial^2 z}{\partial x \partial y} = \frac{\partial^2 z}{\partial y \partial x},$$

provided that the partial derivatives involved are continuous functions.

As you might guess, at a relative maximum or minimum of $z = f(x, y)$, we have

$$\frac{\partial z}{\partial x} = 0 \qquad \text{and} \qquad \frac{\partial z}{\partial y} = 0,$$

provided that the derivatives exist. To assure us that we have a maximum or minimum, we need

$$\left(\frac{\partial^2 z}{\partial x \partial y}\right)^2 - \left(\frac{\partial^2 z}{\partial x^2}\right)\left(\frac{\partial^2 z}{\partial y^2}\right) < 0.$$

Moreover, we have a relative minimum if $\frac{\partial^2 z}{\partial x^2} > 0$ and a relative maximum if $\frac{\partial^2 z}{\partial x^2} < 0$.

A major problem in statistics, called least squares, is to find a and b such that the point (a, b) minimizes

$$K(a, b) = \sum_{i=1}^{n} (y_i - a - bx_i)^2.$$

Thus, the solution of the two equations

$$\frac{\partial K}{\partial a} = \sum_{i=1}^{n} 2(y_i - a - bx_i)(-1) = 0$$

and

$$\frac{\partial K}{\partial b} = \sum_{i=1}^{n} 2(y_i - a - bx_i)(-x_i) = 0$$

could give us a point (a, b) that minimizes $K(a, b)$. Taking second partial derivatives, we obtain

$$\frac{\partial^2 K}{\partial a^2} = \sum_{i=1}^{n} 2(-1)(-1) = 2n > 0,$$

$$\frac{\partial^2 K}{\partial b^2} = \sum_{i=1}^{n} 2(-x_i)(-x_i) = 2\sum_{i=1}^{n} x_i^2 > 0,$$

and

$$\frac{\partial^2 K}{\partial a \partial b} = \sum_{i=1}^{n} 2(-1)(-x_i) = 2\sum_{i=1}^{n} x_i,$$

and note that

$$\left(2\sum_{i=1}^{n} x_i\right)^2 - (2n)\left(2\sum_{i=1}^{n} x_i^2\right) < 0$$

because $(\sum_{i=1}^{n} x_i)^2 < n \sum_{i=1}^{n} x_i^2$ provided that not all the x_i are equal. Noting that $\frac{\partial^2 z}{\partial x^2} > 0$, we see that the solution of the two equations, $\frac{\partial K}{\partial a} = 0$ and $\frac{\partial K}{\partial b} = 0$, provides the only minimizing solution.

The **double integral**

$$\iint\limits_{A} f(x, y)\, dx\, dy$$

can usually be evaluated by an iteration—that is, by evaluating two successive single integrals. For example, say $A = \{(x, y) : 0 \leq x \leq 1, 0 \leq y \leq x\}$, as given in Figure D.6-1.

Then

$$\iint\limits_{A} (x + x^3 y^2)\, dx\, dy = \int_0^1 \left[\int_0^x (x + x^3 y^2)\, dy\right] dx$$

$$= \int_0^1 \left[xy + \frac{x^3 y^3}{3}\right]_0^x dx$$

$$= \int_0^1 \left(x^2 + \frac{x^6}{3}\right) dx = \left[\frac{x^3}{3} + \frac{x^7}{3 \cdot 7}\right]_0^1$$

$$= \frac{1}{3} + \frac{1}{21} = \frac{8}{21}.$$

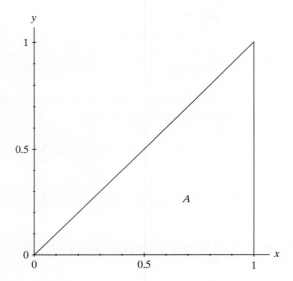

Figure D.6-1 $A = \{(x, y) : 0 \leq x \leq 1, 0 \leq y \leq x\}$

When placing the limits on the iterated integral, note that, for each fixed x between 0 and 1, y is restricted to the interval from 0 to x. Also, in the inner integral on y, x is treated as a constant.

In evaluating this double integral, we could have restricted y to the interval from 0 to 1. Then x would be between y and 1. That is, we would have evaluated the iterated integral

$$\int_0^1 \left[\int_y^1 (x + x^3 y^2)\, dx \right] dy = \int_0^1 \left[\frac{x^2}{2} + \frac{x^4 y^2}{4} \right]_y^1 dy$$

$$= \int_0^1 \left[\frac{1}{2} + \frac{y^2}{4} - \frac{y^2}{2} - \frac{y^6}{4} \right] dy$$

$$= \left[\frac{y}{2} - \frac{y^3}{3 \cdot 4} - \frac{y^7}{7 \cdot 4} \right]_0^1$$

$$= \frac{1}{2} - \frac{1}{12} - \frac{1}{28} = \frac{8}{21}.$$

Finally, we can *change variables* in a double integral

$$\iint_A f(x, y)\, dx\, dy.$$

If $f(x, y)$ is a joint pdf of random variables X and Y of the continuous type, then the double integral represents $P[(X, Y) \in A]$. Consider only one-to-one transformations—say, $z = u_1(x, y)$ and $w = u_2(x, y)$—with inverse transformation given by $x = v_1(z, w)$ and $y = v_2(z, w)$. The determinant of order 2,

$$J = \begin{vmatrix} \dfrac{\partial x}{\partial z} & \dfrac{\partial x}{\partial w} \\[2ex] \dfrac{\partial y}{\partial z} & \dfrac{\partial y}{\partial w} \end{vmatrix},$$

is called the **Jacobian** of the inverse transformation. Moreover, say the region A maps onto the region B in (z, w) space. Since we are usually dealing with probabilities in this book, we fixed the *sign* of the integral so that it is positive (by using the absolute value of the Jacobian). Then it follows that

$$\iint\limits_{A} f(x, y)\, dx\, dy = \iint\limits_{B} f[v_1(z, w), v_2(z, w)]\, |J|\, dz\, dw.$$

To illustrate, let

$$f(x, y) = \frac{1}{2\pi} e^{-(x^2 + y^2)/2}, \qquad -\infty < x < \infty, \quad -\infty < y < \infty,$$

which is the joint pdf of two independent normal variables, each with mean 0 and variance 1. Say $A = \{(x, y): 0 \le x^2 + y^2 \le 1\}$, and consider

$$P(A) = \iint\limits_{A} f(x, y)\, dx\, dy.$$

This integration is impossible to deal with directly in the x, y variables. However, consider the inverse transformation to polar coordinates, namely,

$$x = r\cos\theta, \qquad y = r\sin\theta,$$

with Jacobian

$$J = \begin{vmatrix} \cos\theta & -r\sin\theta \\ \sin\theta & r\cos\theta \end{vmatrix} = r(\cos^2\theta + \sin^2\theta) = r.$$

Since A maps onto $B = \{(r, \theta): 0 \le r \le 1, 0 \le \theta < 2\pi\}$, we have

$$P(A) = \int_0^{2\pi} \left(\int_0^1 \frac{1}{2\pi} e^{-r^2/2}\, r\, dr \right) d\theta$$

$$= \int_0^{2\pi} \left[-\frac{1}{2\pi} e^{-r^2/2} \right]_0^1 d\theta$$

$$= \int_0^{2\pi} \frac{1}{2\pi} (1 - e^{-1/2})\, d\theta$$

$$= \frac{1}{2\pi} (1 - e^{-1/2})\, 2\pi = 1 - e^{-1/2}.$$

INDEX